ENGINEERING MATHEMATICS

VOLUME 1

ENGINEERING MATHEMATICS

VOLUME 1

(Revised Second Edition)

Dr. H.S.G. Rao, Ph.D.

Professor and HOD, Deptt. of Mathematics
and Research Centre in
Computational Mathematics
Dr. Ambedkar Institute of Technology
Aided by Govt. of Karnataka, Bangalore

NEW AGE INTERNATIONAL (P) LIMITED, PUBLISHERS

New Delhi • Bangalore • Chennai • Cochin • Guwahati • Hyderabad
Jalandhar • Kolkata • Lucknow • Mumbai • Ranchi

Published by New Age International (P) Ltd., Publishers
First Edition : 2002
Second Edition : 2006

ISBN : 81-224-1888-0

Rs. 295.00

C-06-05-711

Printed in India at Ajit Printer, New Delhi.
Typeset at Goswami Printer, Delhi

PUBLISHING FOR ONE WORLD

NEW AGE INTERNATIONAL (P) LIMITED, PUBLISHERS
(formerly Wiley Eastern Limited)
4835/24, Ansari Road, Daryaganj, New Delhi - 110002
Visit us at **www.newagepublishers.com**

Dedicated to

my

Parents

(Late) N. Shama Rao
S.P. Padmavathamma

Foreword

Mathematics is an important subject for all technical students. Presenting it in a book form is challenging. The author Dr. H.S. Govinda Rao, with rich experience in teaching, has presented the subject clearly in a simple and lucid manner. This will help the student to understand and grasp the subject. He has presented several illustrations and worked examples. I wish the best for the users of the book and the author.

Dr. V. Seenappa
Principal
Dr. Ambedkar Institute of Technology
Bangalore

Preface to the Second Edition

It gives me immense pleasure in presenting the second edition of the book in a short period. The book has been thoroughly revised and enlarged according to the latest syllabus prescribed by all Indian Universities.

Suggestions for the improvement off the book will be thankfully acknowledged.

Dr. H.S. Govinda Rao

Preface to the First Edition

This book has been prepared with a view to help the students preparing for B.E. classes of all Indian Universities and various Competitive Examinations and deals with all chapters exhaustively, plenty of solved VTU examination questions are included in every topic of all the parts (A, B, C and D) and a large number of problems of various standards (easy, difficult, engineering oriented) in the exercises (with answers) are included.

I cannot find adequate words to express my gratitude to Dr. V. Seenappa, Principal Dr. Ambedkar Institute of Technology for his support and encouragement and am ever grateful to him, and also to Mr. K.R. Krishnamurthy (Administrative Officer, Dr. Ambedkar Institute of Technology, Bangalore) for his kind support.

I express my sincere thanks to New Age International, Publishers, New Delhi and Managing Director for providing me with an opportunity to publish my first work on the subject.

My sincere thanks to Mr. V.R. Babu, Mr. Srinivas and Mr. Mahesh of New Age International, Publishers, Bangalore for their efforts and co-operation in bringing out this book.

The book would serve as an excellent text for undergraduate (B.E.) Classes of all Indian Universties,.

Suggestions, critical evaluations for improvement of this book will be highly appreciated and thankfully acknowledged.

Dr. H.S. Govinda Rao

Contents

PART-D

PART-A

1

Analytical Geometry in Three Dimensions

1.1 INTRODUCTION

In a plane, the position of a point is determined by two numbers x and y, obtained with reference to two straight lines in the plane intersecting at right angles. The position of a point (location) in space can be determined in terms of its perpendicular distances (known as rectangular cartesian coordinates) from three mutually perpendicular planes (known as coordinate planes). The lines of intersection of these three coordinate planes are known as coordinate axes and their point of intersection, 'the origin'.

The three dimensional analytic geometry is the study of the geometrical objects (regular shapes or irregular shapes) in the three-dimensional Euclidean space by three cartesian coordinates: The x-coordinate, the y-coordinate and the z-coordinate. Thus, the idea gathered will be useful in the various branches of Engineering.

Definition 1 (Analytic Geometry). The geometry in which the position is represented analytically (by coordinates) and algebraic methods of reasoning are used for the most part is known as analytic geometry.

Definition 2 (Solid Geometry). The branch of geometry which studies figures in space (three dimensional space) whose plane sections are the figures studied in plane elementary geometry, such as, cubes, spheres, polyhedrons and angles between planes is called solid geometry.

1.2 CARTESIAN COORDINATES IN SPACE

In the figure 1.1, we have

$$x = PL, \quad y = PM, \quad z = PN$$

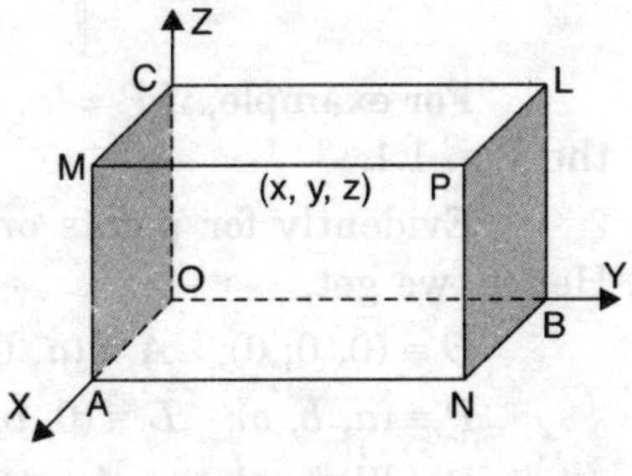

The point P whose X-co-ordinate is = x, Y-co-ordinate is = y, and Z-co-ordinate = z, is written $P \equiv (x, y, z)$.

Fig. 1.1

The co-ordinate planes are the YOZ plane (or the X plane), the ZOX plane (or the Y plane) and the XOY plane (or the Z plane) the planes ZOX and XOY intersect along OX or the x-axis; and so on.

Thus, the co-ordinate axes are:

$OX \equiv$ the x-axis, OY $\equiv$ y-axis, $OZ \equiv z$-axis.

A plane divides space into two parts, and so, the three co-ordinate planes divide space into $2.2.2 = 2^3 = 8$ parts, called octants.

If P ≡ (x, y, z), then we have

$$\begin{Bmatrix} x = LP = CM = OA = BN \\ y = MP = CL = OB = AN \\ z = NP = AM = OC = BL \end{Bmatrix}$$

Equations of co-ordinate planes and axes.

For any point on the *YOZ* plane, the perpendicular distance from the *YOZ* plane is evidently equal to zero; that is, the *x*-co-ordinate is zero. Thus, $x = 0$ is the characteristic property of all points on the *YOZ* plane. Hence, $x = 0$ is the equation of the *YOZ* plane. Similarly $y = 0$ is the equation of *ZOX*; and $z = 0$ is the equation of the *XOY* plane.

The *x*-axis, or *OX* is the interesection of the planes *ZOX* and *XOY* ; *i.e.*, of $y = 0$, and $z = 0$. The co-ordinates of all points on the *x*-axis, satisfy the relations $y = 0$, $z = 0$. Hence, we get,

$$\begin{Bmatrix} \text{Equations of the } x\text{-axis is} : y = 0 = z \\ \text{Equations of the } y\text{-axis is} : z = 0 = x \\ \text{Equations of the } z\text{-axis is} : x = 0 = y \end{Bmatrix}$$

For any point on the *x*-axis. only the *x* co-ordinate is present, since $y = 0$ and the *z* co-ordinate is also zero. Hence we can write

$$\begin{Bmatrix} \text{any point on the } x\text{-axis} \equiv (x_1, 0, 0) \\ \text{any point on the } y\text{-axis} \equiv (0, y_1, 0) \\ \text{any point on the } z\text{-axis} \equiv (0, 0, z_1) \end{Bmatrix}$$

All points on a plane parallel to the *YOZ* plane are at the same distance from that plane. Hence, their *x* co-ordinates must be the same. So,

$$\begin{Bmatrix} x = k, \text{ is a plane parallel to } YOZ \\ y = k, \text{ is a plane parallel to } ZOX \\ z = k, \text{ is a plane parallel to } XOY \end{Bmatrix}$$

Similarly $y = m$, $z = n$; represents all the points whose *y* and *z* co-ordinate remain unchanged ; and only the *x*-co-ordinate changes. So, it gives a straight line parallel to the *x*-axis.

$$\begin{Bmatrix} \text{Thus } y = m, z = n; \text{ is a straight line} \parallel x \text{ axis,} \\ z = n,\ x = l; \text{ is a straight line} \parallel y \text{ axis,} \\ x = l,\ \ y = m; \text{ is straight line} \parallel z \text{ axis.} \end{Bmatrix}$$

For example, if $P = (a, b, c)$, we can write down the coordinates of the gereral points in the Fig. 1.1.

Evidently for points on the *x*-plane, $x = 0$; and for points on the *x*-axis, $y = 0 = z$. Hence, we get,

$O \equiv (0, 0; 0)$; $A \equiv (a, 0, 0)$; $B \equiv (0, b, 0)$; $C \equiv (0, 0, c)$

$P \equiv (a, b, c)$; $L \equiv (0, b, c)$; $M \equiv (a, 0, c)$; $N \equiv (a, b, 0)$

(1) *Write down the equations of the planes parallel to the co-ordinate planes, and passing through the point* $P \equiv (2, 3, -5)$.

Any plane parallel to the *YOZ* plane, is $x = k$; here $P \equiv (2, 3, -5)$ lies on the plane. $\therefore 2 = k$. Hence, the plane through *P*, parallel to *YOZ* is, $x = 2$. Similarly, plane through *P* parallel to *ZOX* is, $y = 3$; and parallel to *XOY* is, $z = -5$.

We can write down the equations of straight lines through $P \equiv (2, 3, -5)$, parallel to the co-ordinate axes respectively are :

$$\left\{\begin{array}{l}(i)\ \text{parallel to } x \text{ axis} : y = 3,\quad z = -5 \\ (ii)\ \text{parallel to } y \text{ axis} : z = -5, x = 2 \\ (iiii)\ \text{parallel to } z \text{ axis} ; x = 2,\quad y = 3\end{array}\right\}$$

1. Distance Formula

If $P = (x_1, y_1, z_1)$, $Q = (x_2, y_2, z_2)$ be any two points in space then the distance between and B is given by

$$PQ = \sqrt{(x_2 - x_1)^2 + (y_2 - y_1)^2 + (z_2 - z_1)^2}$$

2. Division Formula

The coordinates of a point dividing the line joining

$P(x_1, y_1, z_1)$, $Q(x_2, y_2, z_2)$ in the ratio $l : m$ is given by

$$R = \left(\frac{lx_2 + mx_1}{l+m}, \frac{ly_2 + my_1}{l+m}, \frac{lz_2 + mz_1}{l+m}\right) \quad \text{(For internal division)}$$

$$R^1 = \left(\frac{lx_2 - mx_1}{l-m}, \frac{ly_2 - my_1}{l-m}, \frac{lz_2 - mz_1}{l-m}\right) \quad \text{(For external division)}$$

(*a*) Coordinates of any point on the line segment PQ.

Let $l : m = k : l$ or $l/m = k$ in the internal division formula.

putting $l = k\,m$ we have

$$R = \left(\frac{kmx_2 + mx_1}{km + m}, \frac{kmy_2 + my_1}{km + m}, \frac{kmz_2 + mz_1}{km + m}\right)$$

i.e.,

$$R = \left(\frac{kx_2 + x_1}{k+1}, \frac{ky_2 + y_1}{k+1}, \frac{kz_2 + z_1}{k+1}\right)$$

WORKED EXAMPLES

1. Using the distance formula show that the points $P(1, 2, 3)$, $Q\,(3, 7, 7)$ and $R\,(5, 12, 11)$ are collinear.

Solution:

$$PQ^2 = (x_2 - x_1)^2 + (y_2 - y_1)^2 + (z_2 - z_1)^2$$
$$= (3 - 1)^2 + (7 - 2)^2 + (7 - 3)^2$$
$$= 45$$

$\therefore \quad PQ = 3\sqrt{5}$

$QR^2 = (5 - 3)^2 + (12 - 7)^2 + (11 - 7)^2 = 45$

$\therefore \quad QR = 3\sqrt{5}$

$RP^2 = (5 - 1)^2 + (12 - 2)^2 + (11 - 3)^2 = 180.$

$RP = 6\sqrt{5}.$

Clearly, $PR = PQ + QR$

So, P, Q, R are collinear.

2. Find the coordinates of the point which divides the line joining (2, –3, 1) and (3, 4, –5) in the ratio 1 : 3.

Solution: Using the section formula,

(3, 4, – 5) (2, – 3, 1)
1 3

A 1 : 3 B
(2, – 3, 1) P (3, 4, – 5)

Let $P(x, y, z)$ be the required point,

then $$x = \frac{1 \cdot 3 + 3 \cdot 2}{1+3} = \frac{9}{4}$$

$$y = \frac{1 \cdot 4 + 3(-3)}{1+3} = \frac{-5}{4} \quad \text{and} \quad z = \frac{1(-5)+3(1)}{1+3} = -\frac{1}{2}$$

Hence $$P = \left(\frac{9}{4}, \frac{-5}{4}, \frac{-1}{2}\right).$$

3. Find the points of trisection of PQ where $P = (3, -6, 12)$, $Q = (9, 3, 6)$.

Solution: Clearly, there are two points of intersection, R and S

where $PR : RQ = 1 : 2$

and $PS : SQ = 2 : 1$

$$\therefore \quad R = \left[\frac{(1)(9)+(2)(3)}{1+2}, \frac{(1)(3)+2(-6)}{1+2}, \frac{1(6)+2(12)}{1+2}\right] = (5, -3, 10)$$

$$S = \left[\frac{2(9)+1(3)}{2+1}, \frac{2(3)+1(-6)}{2+1}, \frac{2(6)+1(12)}{2+1}\right] = (7, 0, 8).$$

Similar Problems

1. Find the coordinates of the point which divide internally and externally, the line joining the point $(a + b, a - b)$ to the point $(a - b, a + b)$ in the ratio $a : b$.
2. Find the ratio in which the *yz*-plane divides the line joining the points A (–3, 4, 5) and B(2, –1, –2) Also find its coordinates
3. Show that the points A(3, 4, 7), B = (5, 6, 2), C = (1, 2, 1), D = (–1, 0, 6) are the vertices of the parallelogram.

1.3 DIRECTION COSINES AND DIRECTION RATIOS

Definition 1. If α, β, γ are the direction angles made by a line with the positive direction of x, y, z axes respectively, then $l = \cos\alpha$, $m = \cos\beta$, and $n = \cos\gamma$ are called direction cosines of the line.

Direction cosines are not independent, when two of them are given, the third can be found, except for sign, by use of the pythogorean relation,

$$\cos^2\alpha + \cos^2\beta + \cos^2\gamma = 1.$$

Note: Direction cosines, (D.C's) here referred to as actual D.C's, denoted by $(l : m : n)$

Definition 2. Direction ratios (direction numbers), of a line in space: (D.R's)

Any three numbers, not all zero, proportional to the direction cosines of the line, are called direction ratios or direction numbers of the line.

If a line passes through the points (x_1, y_1, z_1) and (x_2, y_2, z_2), its direction numbers are proportional to (or, D.r's) are: $(x_2 - x_1) : (y_2 - y_1) : (z_2 - z_1)$, and its direction cosines are

$$\frac{x_2 - x_1}{D}, \frac{y_2 - y_1}{D}, \frac{z_2 - z_1}{D}$$

when $D = \sqrt{(x_2 - x_1)^2 + (y_2 - y_1)^2 + (z_2 - z_1)^2}$, the distance between the points.

A Simple and Useful Result

Let $OP = r$, if $(l : m : n)$ are the actual D.C's of OP, then, from the figure:

$$x = r\cos\alpha = lr$$
$$y = r\cos\beta = mr$$
$$z = r\cos\gamma = nr$$

Then the coordinates of P are (lr, mr, nr)

i.e. $P = (lr, mr, nr)$ [VTU, Aug., 1999] [1]

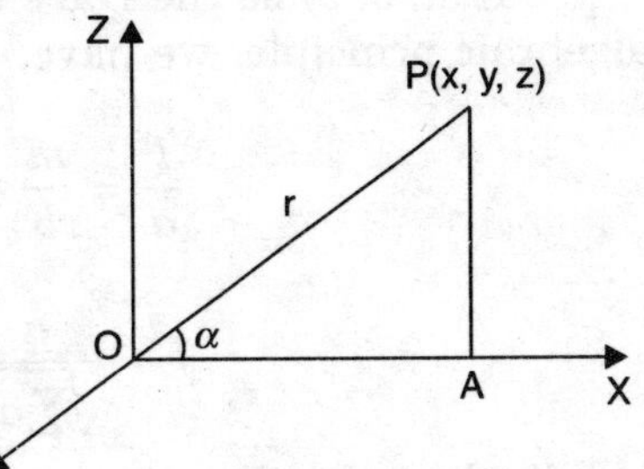

Fig. 1.2

If $P = (x, y, z)$

Then the actual D.C's of OP are:

$$\left(\frac{x}{r}, \frac{y}{r}, \frac{z}{r}\right) = (l\,;\, m;\, n)$$

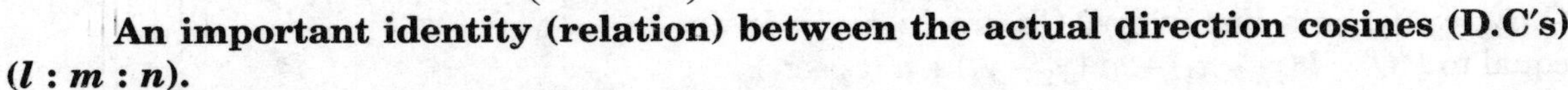

An important identity (relation) between the actual direction cosines (D.C's) $(l : m : n)$.

Let $OP = r = 1$ unit

Let AB be the given line whose D.C's are $(l : m : n)$. Through the origin $O = (0, 0, 0)$, draw a line OP parallel to AB. Hence, D.C'.s of OP are also $(l : m : n)$

(*If* the lines are parallel, then their D.C's are equal).

Now cut off $OP = 1$ unit. Then coordinates of $P = (l, m, n)$, (using [1], $r = 1$).

By distance formula:

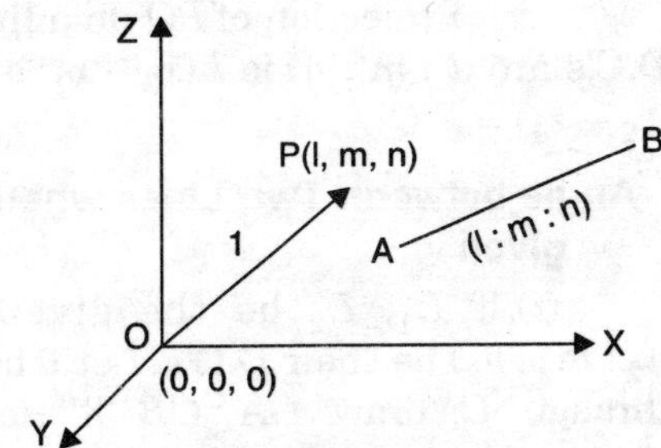

Fig. 1.3

$$OP = \sqrt{(l - o)^2 + (m - o)^2 + (n - o)^2}$$

or $$1 = \sqrt{l^2 + m^2 + n^2}$$

Squaring both sides, we get

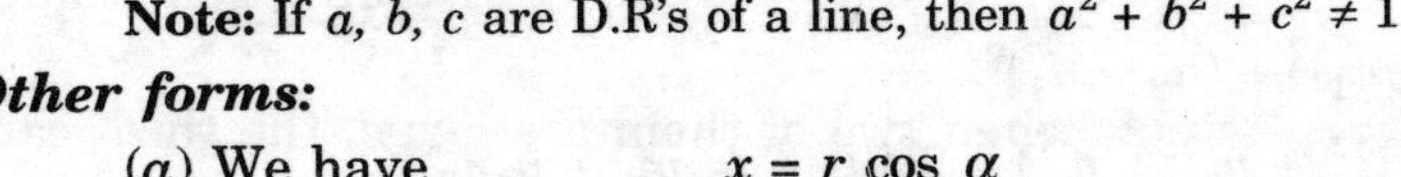

$$\boxed{l^2 + m^2 + n^2 = 1} \qquad [2]$$

Note: If a, b, c are D.R's of a line, then $a^2 + b^2 + c^2 \neq 1$

Other forms:

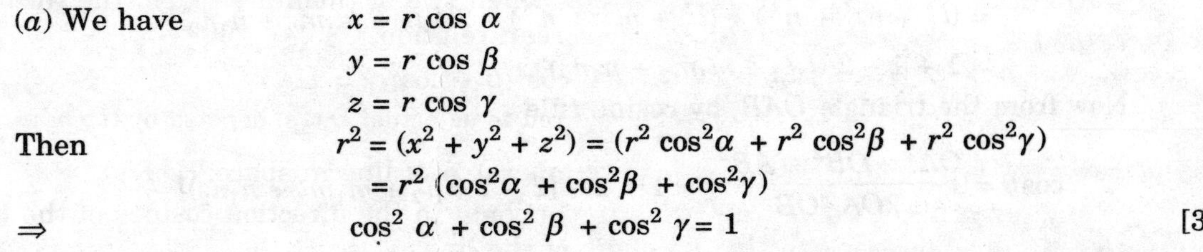

(*a*) We have $x = r\cos\alpha$

$$y = r\cos\beta$$
$$z = r\cos\gamma$$

Then $r^2 = (x^2 + y^2 + z^2) = (r^2\cos^2\alpha + r^2\cos^2\beta + r^2\cos^2\gamma)$

$$= r^2(\cos^2\alpha + \cos^2\beta + \cos^2\gamma)$$

$\Rightarrow$ $$\cos^2\alpha + \cos^2\beta + \cos^2\gamma = 1 \qquad [3]$$

(*b*) The identity [3] can also be written as

$$(1 - \sin^2 \alpha) + (1 - \sin^2 \beta) + (1 - \sin^2 \gamma) = 1$$

or
$$3 - (\sin^2 \alpha + \sin^2 \beta + \sin^2 \gamma) = 1$$

$\Rightarrow$
$$\sin^2 \alpha + \sin^2 \beta + \sin^2 \gamma = 2 \qquad \text{[VTU, Mar., 2000] [4]}$$

Rule to find the actual direction cosines of a line whose proportional D.C'.s (*i.e.* direction ratios) are given:

If (a, b, c) be the D.R's of a line, and if $(l : m : n)$ be actual D.C's of the line, then by an algebraic principle, we have,

$$\frac{l}{a} = \frac{m}{b} = \frac{n}{c} = \frac{\sqrt{l^2 + n^2 + m^2}}{\sqrt{a^2 + b^2 + c^2}} = \frac{1}{\sqrt{\sum a^2}} \qquad (\because l^2 + m^2 + n^2 = 1)$$

$$l = \frac{a}{\sqrt{\sum a^2}}, m = \frac{b}{\sqrt{\sum a^2}}, n = \frac{c}{\sqrt{\sum a^2}}$$

i.e. dividing the direction ratios a, b, c each by their Square root of sum of Squares *viz.* $\sqrt{\sum a^2}$, to get the actual D.C's

Projection of a Line

The join of two points $p(x_1, y_1, z_1)$ and $Q\ (x_2, y_2, z_2)$ on a line with D.C's are $l : m : n$ is equal to $P'Q' = l\,(x_2 - x_1) + m\,(y_2 - y_1) + n\,(z_2 - z_1)$.

$\therefore$ Projection of PQ on a line is $P'Q'$ whose D.C's are $(l : m : n)$ is $l\,(x_2 - x_1) + m\,(y_2 - y_1) + n\,(z_2 - z_1) = P'Q'$

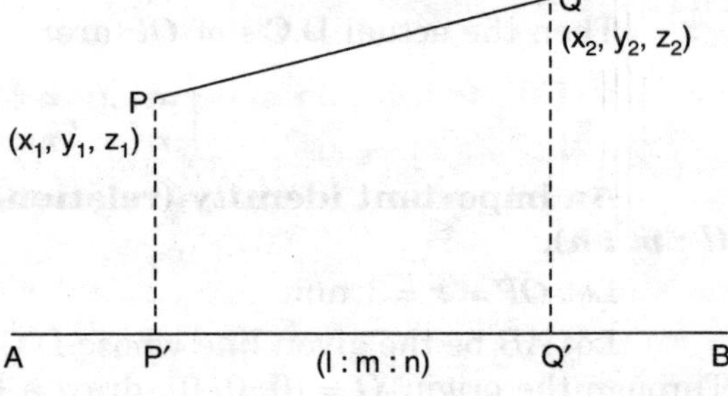

Fig. 1.4

Angle between Two Lines when Their D.Cs are given [VTU, Mar., 1999]

(*i*) If L_1, L_2 be the given lines and $(l_1 : m_1 : n_1)$, $(l_2 : m_2 : n_2)$ be their D.C's. Let θ be the angle between them, through O, draw OA, OB ||[el] to L_1 and L_2 respectively, then

$$A\hat{O}B = \theta.$$

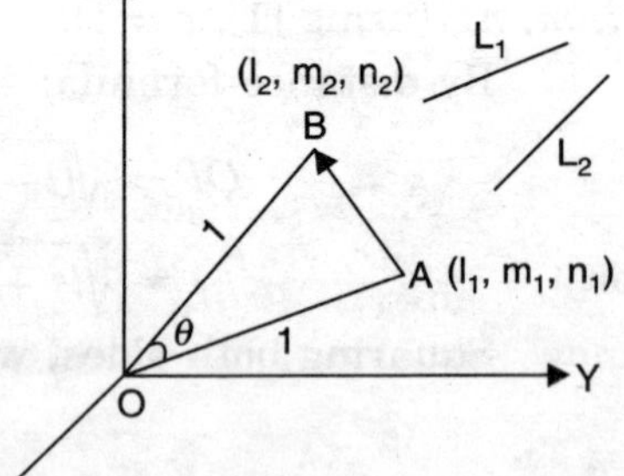

Fig. 1.5

To find an expression for the angle between the lines whose direction cosines are $(l_1 : m_1 : n_1)$ and $(l_2 : m_2 : n_2)$

Let $OA = OB = 1$ unit.

$\therefore$ Coordinates of A and B are (l_1, m_1, n_1) and (l_2, m_2, n_2).

$$AB^2 = (l_2 - l_1)^2 + (m_2 - m_1)^2 + (n_2 - n_1)^2.$$

$$= (l_1^2 + m_1^2 + n_1^2) + (l_2^2 + m_2^2 + n_2^2) - 2(l_1 l_2 + m_1 m_2 + n_1 n_2)$$

$$= 1 + 1 - 2\,(l_1 l_2 + m_1 m_2 + n_1 n_2)$$

Now from the triangle OAB, by cosine rule

$$\cos\theta = \frac{OA^2 + OB^2 - AB^2}{2OA \cdot OB} = 1 + 1 - [2 - 2\,(l_1 l_2 + m_1 m_2 + n_1 n_2)]$$

$\Rightarrow \qquad \cos\theta = (l_1 l_2 + m_1 m_2 + n_1 n_2)$ [1]

Hence $\qquad \theta = \cos^{-1}(l_1 l_2 + m_1 m_2 + n_1 n_2)$

(*ii*) Expression for $\sin\theta$ and $\tan\theta$:

$$\sin\theta = \sqrt{(l_1 m_2 - l_2 m_1)^2 + (m_2 n_1 - m_1 n_2)^2 + (n_2 l_1 - n_1 l_2)^2}$$ [VTU, Mar., 1999]

Proof:

$$\begin{aligned}\sin^2\theta &= 1 - \cos^2\theta \\ &= 1 - [l_1 l_2 + m_1 m_2 + n_1 n_2]^2 \qquad \text{(using [1])} \\ &= (l_1^2 + m_1^2 + n_1^2)(l_2^2 + m_2^2 + n_2^2) - (l_1 l_2 + m_1 m_2 + n_1 n_2)^2 \\ &= (l_1 m_2 - m_1 l_2)^2 + (m_1 n_2 - n_1 m_2)^2 + (n_1 l_2 - l_1 n_2)^2\end{aligned}$$

$$(\because\ l_1^2 + m_1^2 + n_1^2 = 1 = l_2^2 + m_2^2 + n_2^2)$$

$$\therefore \qquad \sin\theta = \sqrt{\Sigma(l_1 m_2 - m_1 l_2)^2} \qquad [2]$$

and $$\tan\theta = \frac{\sin\theta}{\cos\theta} = \frac{\sqrt{\Sigma(l_1 m_2 - m_1 l_2)^2}}{l_1 l_2 + m_1 m_2 + n_1 n_2}$$

(*iii*) **Condition of perpendicularity**

If the two lines are $\perp^r$, then, $\theta = 90°$

$\therefore \qquad \cos\theta = \cos 90° = 0$

$\Rightarrow \qquad l_1 l_2 + m_1 m_2 + n_1 n_2 = 0$

(*iv*) **Condition of parallelism.**

If the two lines are parallel, then, $\theta = 0$, so that $\sin\theta = 0$

$\Rightarrow \qquad (l_1 m_2 - m_1 l_2)^2 + (m_1 n_2 - n_1 m_2)^2 + (n_1 l_2 - n_2 l_1)^2 = 0$

$$\Rightarrow \left.\begin{aligned} l_1 m_2 - m_1 l_2 = 0, &\quad \text{or} \quad \frac{l_1}{l_2} = \frac{m_1}{m_2} \\ m_1 n_2 - n_1 m_2 = 0, &\quad \text{or} \quad \frac{m_1}{m_2} = \frac{n_1}{n_2} \\ n_1 l_2 - l_1 n_2 = 0, &\quad \text{or} \quad \frac{n_1}{n_2} = \frac{l_1}{l_2} \end{aligned}\right\} \qquad [3]$$

Note: (*i*) If (a_1, b_1, c_1) and (a_2, b_2, c_2) are sets of number, then we have an identity known as Lagrange's identity given by:

$$\begin{aligned}&(a_1^2 + b_1^2 + c_1^2)(a_2^2 + b_2^2 + c_2^2) - (a_1 a_2 + b_1 b_2 + c_1 c_2)^2 \\ &= (a_1 b_2 - a_2 b_1)^2 + (b_1 c_2 - c_1 b_2)^2 + (c_1 a_2 - a_1 c_2)^2\end{aligned}$$

which is same as $\sqrt{\Sigma(a_1 b_2 - a_2 b_1)^2}$ which is similar to the expression for sine in equation [2] above.

(*ii*) The result for sine given in [2] above can be remembered as: write down the D.C's in two rows as:

$$\left\{\begin{matrix} m_1 & n_1 & l_1 & m_1 \\ & \times & \times & \times \\ m_2 & n_2 & l_2 & m_2 \end{matrix}\right\}$$ to get $m_1 n_2 - n_1 m_2$ etc. Square and add.

$$[1] \quad \Rightarrow \quad \frac{l_1}{l_2} = \frac{m_1}{m_2} = \frac{n_1}{n_2} = \frac{\sqrt{l_1^2 + m_1^2 + n_1^2}}{\sqrt{l_2^2 + m_2^2 + n_2^2}} = \frac{1}{1}$$

$$\Rightarrow \quad l_1 = l_2,\ m_1 = m_2,\ n_1 = n_2$$

(*a*) Find the angle between the two lines whose D.R's are $(a_1 : b_1 : c_1)$ and $(a_2 : b_2 : c_2)$.

[VTU, Aug., 1999]

Let $(l_1 : m_1 : n_1)$ be the actual D.C's of line whose D.R's are $(a_1 : b_1 : c_1)$

$$\therefore \quad l_1 = \frac{a_1}{\sqrt{\Sigma a_1^2}},\ m_1 = \frac{b_1}{\sqrt{\Sigma a_1^2}},\ n_1 = \frac{c_1}{\sqrt{\Sigma a_1^2}}$$

Let $(l_2 : m_2 : n_2)$ be the actual D.C's of the line whose D.R's are $(a_2 : b_2 : c_2)$,

$$\therefore \quad l_2 = \frac{a_2}{\sqrt{\Sigma a_2^2}},\ m_2 = \frac{b_2}{\sqrt{\Sigma a_2^2}},\ n_2 = \frac{c_2}{\sqrt{\Sigma a_2^2}}$$

Thus we have, from results [1], that

$$\cos\theta = \frac{a_1a_2 + b_1b_2 + c_1c_2}{\sqrt{a_1^2 + b_1^2 + c_1^2}\ \sqrt{a_2^2 + b_2^2 + c_2^2}}$$

If the lines are $\perp^r$, $\quad \theta = 90°$

$$\therefore \quad \cos 90° = 0 = \frac{a_1a_2 + b_1b_2 + c_1c_2}{\sqrt{\Sigma a_1^2}\ \sqrt{\Sigma a_2^2}}$$

$$\therefore \quad a_1a_2 + b_1b_2 + c_1c_2 = 0$$

If the lines are $||^{el}$,

$$\therefore \quad l_1 = l_2,\ m_1 = m_2,\ n_1 = n_2$$

or

$$\frac{a_1}{\sqrt{\Sigma a_1^2}} = \frac{a_1}{\sqrt{\Sigma a_2^2}}$$

$$\therefore \quad \frac{a_1}{a_2} = \frac{\sqrt{\Sigma a_1^2}}{\sqrt{\Sigma a_2^2}};\ |||^{ly},\ \frac{b_1}{b_2} = \frac{\sqrt{\Sigma a_1^2}}{\sqrt{\Sigma a_2^2}}$$

and

$$\frac{c_1}{c_2} = \frac{\sqrt{\Sigma a_1^2}}{\sqrt{\Sigma a_2^2}};$$

$$\therefore \quad \boxed{\frac{a_1}{a_2} = \frac{b_1}{b_2} = \frac{c_1}{c_2}}$$

Note:

D.C's of the *x*-axis = (1:0:0)
D.C's of the *y*-axis = (0:1:0)
D.C's of the *z*-axis = (0:0:1)

WORKED EXAMPLES

1. Find the coordinates of the foot of the $\perp^r$ from $A = (1, 8, 4)$ to the line joining $B\,(0, -11, 4)$ and $C\,(2, -3, 1)$ (V.T.U.J/A, 2004)

Solution: Let D be the foot of the perpendicular from A on the line joining BC.

Let the point D divide BC in the ratio $\lambda : 1$

$$\therefore \quad P = \left(\frac{2\lambda}{\lambda+1}, \frac{-3\lambda-1}{\lambda+1}, \frac{\lambda+4}{\lambda+1}\right)$$

We find that the lines AD and BC are $\perp^r$.

Now, D.R's of BC are:

$$(0-2) : (-11+3) : (4-1)$$

i.e. $\quad -2, -8, 3$

or, $\quad 2, 8, -3 \qquad [1]$

and D.R's of AD are

$$\left(\frac{2\lambda}{\lambda+1}-1\right) : \left(\frac{-3\lambda-11}{\lambda+1}-8\right) : \left(\frac{\lambda+4}{\lambda+1}-4\right)$$

$$\Rightarrow \quad \frac{\lambda-1}{\lambda+1}, \frac{-11\lambda-19}{\lambda+1}, \frac{-3\lambda}{\lambda+1}$$

$$\Rightarrow \quad \lambda-1, -11\lambda-19, -3\lambda \qquad [2]$$

Since the lines are $\perp^r$, from [1] and [2], we get

$$2(\lambda-1) + 8(-11\lambda-19) - 3(-3\lambda) = 0$$

$$\therefore \quad \Rightarrow \quad \lambda = -2$$

Hence the coordinates of D are $(4, +5, -2)$.

2. Find the actual direction cosines of the line joining the two points $P = (2, 3, -5)$ and $Q = (5, 7, 7)$

Solution: The D.R's are equal to:

$$(5-2) : (7-3) : (7-(-5)) = (3 : 4 : 12)$$

The actual D.C's are equal to

$$\frac{3}{\sqrt{3^2+4^2+12^2}}, \frac{4}{\sqrt{3^2+4^2+12^2}}, \frac{12}{\sqrt{3^2+4^2+12^2}} = \left(\frac{3}{13}, \frac{4}{13}, \frac{12}{13}\right).$$

3. Find the D.C's of a st. line equally inclined to the three coordinate axes.

Solution: It is given that

$$\alpha = \beta = \gamma$$

$$\therefore \quad \cos\alpha = \cos\beta = \cos\gamma$$

We know that $\cos^2\alpha + \cos^2\beta + \cos^2\gamma = 1$

i.e. $\quad 3\cos^2\alpha = 1;$

$$\therefore \qquad \cos\alpha = \pm \frac{1}{\sqrt{3}}$$

$$\therefore \qquad \cos\alpha = \cos\beta = \cos\gamma = \pm \frac{1}{\sqrt{3}}$$

Hence, the D.C's are

$$\left(\pm\frac{1}{\sqrt{3}} \pm \frac{1}{\sqrt{3}} \pm \frac{1}{\sqrt{3}}\right) = (1:1:1)$$

4. Find the D.C's of the line perpendicular to the lines whose D.C's are proportional to (1, –1, 2) and (2, 1, –1)

Solution: If (l; m; n) be the actual D.C's of the line, then since the line is $\perp^r$ to the other two lines, we have

$$l \cdot 1 - 1 \cdot m + 2 \cdot n = 0$$

and

$$2l + 1 \cdot m - 1 \cdot n = 0$$

Solving, by Cross multiplications,

we get $\dfrac{l}{1-2}, \dfrac{-m}{-4-1}, \dfrac{n}{1+2}$

$$\Rightarrow \qquad \frac{l}{-1} = \frac{m}{5} = \frac{n}{3} = \frac{1}{\sqrt{35}} = \frac{1}{\sqrt{(-1)^2 + 5^2 + 3^2}}$$

$$\Rightarrow \quad \text{D.C's are } \left[-\frac{1}{\sqrt{35}}, \frac{5}{\sqrt{35}}, \frac{3}{\sqrt{35}}\right].$$

5. Find the angle P of the triangle $P\hat{Q}R$, where P = (2, 3, 5) Q = (3, 5, 2) and R = (5, 2, 3).

Solution: ***PQR*** **forms a triangle. The angle** ***P*** **of the triangle is the angle between the lines** ***PQ*** **and** ***PR*****.**

Now, D.R's of PQ are: $(3-2):(5-3):(2-5) = (1:2:-3)$.

D.R's of PR are: $(5-2):(2-3):(3-5) = (3:-1:-2)$

$$\therefore \qquad \cos P = \pm \frac{a_1 a_2 + b_1 b_2 + c_1 c_2}{\sqrt{a_1^2 + b_1^2 + c_1^2}\ \sqrt{a_2^2 + b_2^2 + c_2^2}}$$

$$= \frac{1 \cdot 3 + 2(-1) + (-3)(-2)}{\sqrt{[(1+4+9)(9+1+4)]}} = \frac{3-2+6}{\sqrt{14}\ \sqrt{14}} = \frac{7}{14} = \frac{1}{2}$$

$$\therefore \qquad P = 60°.$$

6. Find the angle between the two lines whose direction cosines are given by the equations.

$$l + m + n = 0;\ l^2 + m^2 - n^2 = 0.$$

we have

$$l + m + n = 0 \qquad [1]$$

$$l^2 + m^2 - n^2 = 0 \qquad [2]$$

From [1], $l = -m - n$.

Then from [2], we get

$$[-m-n]^2 + m^2 - n^2 = 0$$

$$\Rightarrow \qquad 2m^2 + 2mn = 0 \qquad \Rightarrow \qquad 2m(m+n) = 0$$

This gives $m = 0$ or $m = -n$.

From [1], if $m = 0$, then $l = -n$

and if $m = -n$ then $l = 0$.

Hence, DR's of the line

are $(-n, 0, n)$ and $(0, -n, n)$

or $(-1, 0, 1)$ and $(0, -1, 1)$

$\therefore$ The actual D.C's are

$$\left(-\frac{1}{\sqrt{2}}, 0, \frac{1}{\sqrt{2}}\right) \quad \text{and} \quad \left(0, -\frac{1}{\sqrt{2}}, \frac{1}{\sqrt{2}}\right)$$

Thus, θ is given by

$$\cos\theta = \left(-\frac{1}{\sqrt{2}}\right)(0) + (0)\left(-\frac{1}{\sqrt{2}}\right) + \left(\frac{1}{\sqrt{2}}\right)\left(\frac{1}{\sqrt{2}}\right) = \frac{1}{2} \text{ or } \theta = 60°.$$

7. Find whether the lines whose direction cosines are given by

$$2l - m + 2n = 0$$

and $lm + mn + nl = 0$ are at right angles.

Solution:

Given: $m = 2l + 2n$. and therefore the second equation becomes:

$$l(2l + 2n) + n(2l + 2n) + nl = 0 \quad \Rightarrow \quad 2l^2 + 5ln + 2n^2 = 0$$

$$\Rightarrow \quad (2l + n)(l + 2n) = 0 \quad \Rightarrow \quad l = -\frac{n}{2} \text{ or } l = -2n$$

$$\Rightarrow \quad \frac{l}{1} = \frac{n}{-2} \quad \text{or} \quad \frac{l}{2} = \frac{n}{-1} \quad \Rightarrow \quad m = (2)(1) + 2(-2) = -2$$

or $m = 2(2) + 2(-1) = 2$

The DR's of the lines are

$$(l_1 : m_1 : n_1) = (1, -2, -2)$$

and $(l_2 : m_2 : n_2) = (2, -1, 2)$

Hence $l_1 l_2 + m_1 m_2 + n_1 n_2 = 0$

$\Rightarrow$ $1 \cdot 2 + (-2)(-1) + (-2)(2) = 0.$

$\Rightarrow$ Lines are perpendicular.

8. Show that the lines whose D.C.s satisfy the relations $l + m + 4n = 0$ and $mn + nl + lm = 0$ are parallel. [V.T.U. F/M 2005]

Solution: $l + m + 4n = 0$ ...[1]

From [1], $mn + nl + lm = 0$ ...[2]

From [1], $l = -(m + 4n)$, substituting in [2]

$$mn - n(m + 4n) - (m + 4n)m = 0$$

i.e., $mn - mn - 4n^2 - m^2 - 4mn = 0$

$\Rightarrow$ $m^2 + 4mn + 4n^2 = 0$

or $(m + 2n)^2 = 0$ or $m = -2n$

i.e., $m_1 = -2n_1$ and $m_2 = -2n_2$

or $$\frac{m_1}{n_1} = \frac{m_2}{n_2} = -2 \quad \text{or} \quad \frac{m_1}{m_2} = \frac{n_1}{n_2} = -2$$

Also, we have from (1),

$$l_1 + m_1 + 4n_1 = 0 \quad \text{and} \quad l_2 + m_2 + 4n_2 = 0$$

$$\therefore \quad \frac{l_1}{l_2} = \frac{-(m_1 + 4n_1)}{-(m_2 + 4n_2)}.$$

But $m_1 = -2n_1$ and $m_2 = -2n_2$

Hence $$\frac{l_1}{l_2} = \frac{-(-2n_1 + 4n_1)}{-(-2n_2 + 4n_2)} = \frac{n_1}{n_2} \quad \Rightarrow \quad \frac{m_1}{m_2} = \frac{n_1}{n_2} = \frac{l_1}{l_2}$$

or $$\frac{m_1}{n_1} = \frac{m_2}{n_2}, \frac{n_1}{l_1} = \frac{n_2}{l_2}, \frac{l_1}{m_1} = \frac{l_2}{m_2} \quad \Rightarrow \quad \text{the lines are parallel.}$$

9. Show that the straight lines whose D.R's are given by the equations

$l + m + n = 0$ and $2mn + 3nl - 5lm = 0$ are at right angles.

we have $$l + m + n = 0$$

$$\Rightarrow \quad l = -(m + n) \qquad [1]$$

Also, $$2mn + 3nl - 5lm = 0$$

$$\Rightarrow \quad 2mn - 3mn - 3n^2 + 5m^2 + 5mn = 0.$$

$$\Rightarrow \quad 5m^2 + 4mn - 3n^2 = 0;$$

$$\therefore \quad m = \frac{-4n \pm \sqrt{16n^2 + 60n^2}}{10} = \frac{-4 \pm 2\sqrt{19}}{10}.n = \frac{-2 \pm \sqrt{19}}{5} n$$

Let $$m_1 = \frac{-2 + \sqrt{19}}{5} n_1 \quad \text{and} \quad m_2 = \frac{-2 - \sqrt{19}}{5} n_2$$

and from [1], $$l_1 = -\frac{-3 + \sqrt{19}}{5} n_1 \quad \text{and} \quad l_2 = -\frac{-3 - \sqrt{19}}{5} n_2$$

If 'θ' be the angle between two lines, then

$$\cos\theta = l_1 l_2 + m_1 m_2 + n_1 n_2$$

$$= \frac{9 - 19}{25} n_1 n_2 + \frac{4 - 19}{25} n_1 n_2 + n_1 n_2$$

$$= \frac{n_1 n_2}{25} (-10 - 15 + 25) = 0.$$

$$\Rightarrow \quad \theta = \frac{\pi}{2}$$

10. Two lines are connected by the relations $l - 5m + 3n = 0$, and $7l^2 + 5m^2 - 3n^2 = 0$. Find D.C's ($l : m : n$).

Solution:

Given: $$l - 5m + 3n = 0 \qquad [1]$$

$$7l^2 + 5m^2 - 3n^2 = 0 \qquad [2]$$

From [1], $$l = 5m - 3n$$

Substitute the value of l in [2], we get

$$7(5m-3n)^2+5m^2-3n^2=0$$

$$\Rightarrow \quad 180m^2-210mn+60n^2=0$$

$$\Rightarrow \quad 6m^2-7mn+2n^2=0$$

$$\Rightarrow \quad (3m-2n)(2m-n)=0$$

$$\Rightarrow \quad \text{Either } 3m-2n=0 \text{ or } 2m-n=0$$

when $3m-2n=0$

$$\Rightarrow \quad \frac{m}{2}=\frac{n}{3}=\lambda \text{ (say)}$$

$$\therefore \quad m=2\lambda,\ n=3\lambda$$

$\therefore$ From [1],

$$l-10\lambda+9\lambda=0 \quad \Rightarrow \quad l=\lambda \quad \Rightarrow \quad \frac{l}{1}=\lambda$$

$$\therefore \quad \frac{l}{1}=\frac{m}{2}=\frac{n}{3}$$

$\Rightarrow$ D.R's of one line are: (1 : 2 : 3)

Actual D.C's are: $\left(\frac{1}{\sqrt{14}},\frac{2}{\sqrt{14}},\frac{3}{\sqrt{14}}\right)$

When $2m-n=0$

$$\Rightarrow \quad \frac{m}{1}=\frac{n}{2}=\lambda \text{ (say)}$$

$$\therefore \quad m=\lambda$$

$$n=2\lambda$$

$\therefore$ From [1], $l-5\lambda+6\lambda=0$

$$\Rightarrow \quad l=-\lambda \quad \Rightarrow \quad \frac{l}{-1}=\lambda$$

$$\therefore \quad \frac{l}{-1}=\frac{m}{1}=\frac{n}{2}$$

Hence D.R's of 2nd line are: (–1 : 1 : 2)

$\therefore$ The D.C's of 2nd line are:

$$\left(\frac{-1}{\sqrt{6}}:\frac{1}{\sqrt{6}}:\frac{2}{\sqrt{6}}\right).$$

11. Show that the straight lines whose D.C's are given by the equations:

$$ul+vm+wn=0,\ al^2+bm^2+cn^2=0$$

are parallel if

$$\frac{u^2}{a}+\frac{v^2}{b}+\frac{w^2}{c}=0.$$

and $\perp^r$ if $u^2(b+c)+v^2(c+a)+w^2(a+b)=0$.

Solution: D.C's of the lines are given by

$$ul + vm + wn = 0 \quad (i)$$
$$al^2 + bm^2 + cn^2 = 0 \quad (ii)$$

Eliminating 'n' from (i) and (ii),

by putting $n = -\left(\dfrac{ul + vm}{w}\right)$ from (i) in equation (ii), we have

$$al^2 + bm^2 + c\left(\frac{ul + vm}{w}\right)^2 = 0$$

$$\Rightarrow \quad aw^2l^2 + bw^2m^2 + c(u^2l^2 + v^2m^2 + 2uvlm) = 0$$

or $$l^2(aw^2 + cu^2) + 2uvclm + m^2(bw^2 + cv^2) = 0$$

Now, dividing both sides by m^2, we have

$$\frac{l^2}{m^2}(aw^2 + cu^2) + 2uvc\frac{l}{m} + (bw^2 + cv^2) = 0 \quad [3]$$

which is quadratic in (l/m).

If the lines are parallel, then their direction cosines are equal. Hence the roots of [3] must be equal $\therefore$ the lines will be parallel

if $$4c^2u^2v^2 = 4(aw^2 + cu^2)(bw^2 + cv^2)$$

if $$c^2u^2v^2 - (abw^4 + acw^2v^2 + bcu^2w^2 + c^2u^2v^2) = 0$$

if $$-w^2(abw^2 + bcu^2 + cav^2) = 0$$

if $$abw^2 + bcu^2 + acv^2 = 0$$

if $$\frac{u^2}{a} + \frac{v^2}{b} + \frac{w^2}{c} = 0$$

Again, if $(l_1 : m_1 : n_1)$ and $(l_2 : m_2 : n_2)$ are

The D.C's of two lines, then $\dfrac{l_1}{m_1}, \dfrac{l_2}{m_2}$ are the roots of [3].

$\therefore$ From [3],

Product of roots $= \dfrac{l_1l_2}{m_1m_2} = \dfrac{bw^2 + cv^2}{aw^2 + cu^2}$

$$\because \quad \alpha\beta = \frac{c}{a}$$

or $$\frac{l_1l_2}{bw^2 + cv^2} = \frac{m_1m_2}{aw^2 + cu^2} = \frac{n_1n_2}{av^2 + bu^2} = k \text{ (say)}$$

(By symmetry, changing a, b, c and u, v, w in cyclic order).

If the lines are perpendicular, then, $l_1l_2 + m_1m_2 + n_1n_2 = 0$

if $$k(bw^2 + cv^2) + k(aw^2 + cu^2) + k(av^2 + bu^2) = 0$$

or $$bw^2 + cv^2 + aw^2 + cu^2 + av^{2s} + bu^2 = 0$$

(cancelling k)

if $$(b + c)u^2 + (c + a)v^2 + (a + b)w^2 = 0$$

12. Show that the straight lines whose direction cosines are given by $al + bm + cn = 0$ and $fmn + gnl + hlm = 0$ are perpendicular if (*i*) $\frac{f}{a} + \frac{g}{b} + \frac{h}{c} = 0$ and parallel if $\sqrt{af} \pm \sqrt{bg} \pm \sqrt{ch} = 0.$

Solution: We have

$$al + bm + cn = 0;$$

i.e.

$$n = -\frac{al + bm}{c}$$

and

$$fmn + gnl + hlm = 0.$$

$$\Rightarrow \quad -fm \cdot \frac{al+bm}{c} - gl \cdot \frac{al+bm}{c} + hlm = 0$$

$$\Rightarrow \quad agl^2 + lm\,(af + bg - ch) + bfm^2 = 0$$

$$\Rightarrow \quad ag \cdot \frac{l^2}{m^2} + (af + bg - ch)\left(\frac{l}{m}\right) + bf = 0 \qquad [1]$$

Let $\frac{l_1}{m_1}$ and $\frac{l_2}{m_2}$ be the roots of this equation.

$\therefore$ product of roots $= \left(\frac{l_1}{m_1}\right)\left(\frac{l_2}{m_2}\right) = \frac{bf}{ag}$,

or $\frac{l_1 l_2}{bf} = \frac{m_1 m_2}{ag}$ or $\frac{l_1 l_2}{\frac{f}{a}} = \frac{m_1 m_2}{\frac{g}{b}}$

|||ly $\frac{l_1 l_2}{\frac{f}{a}} = \frac{n_1 n_2}{\frac{h}{c}}$ or $\frac{l_1 l_2}{f/a} = \frac{m_1 m_2}{g/b} = \frac{n_1 n_2}{h/c}$

the lines are perpendicular if (**i**) $l_1 l_2 + m_1 m_2 + n_1 n_2 = 0$

i.e. $\frac{f}{a} + \frac{g}{b} + \frac{h}{c} = 0$

Again, the lines are parallel

$$\text{if } \frac{l_1}{l_2} = \frac{m_1}{m_2} = \frac{n_1}{n_2}.$$

taking first two,

$$\frac{l_1}{l_2} = \frac{m_1}{m_2}.$$

Hence, Eqn. (1) has equal roots

$$\therefore \quad (af + bg - ch)^2 = 4ag \cdot bf$$

or

$$af + bg - ch = \pm 2\sqrt{abgf}$$

or $af + bg \mp 2\sqrt{abgf} = ch$ or $(\sqrt{af} \mp \sqrt{bg})^2 = (\sqrt{ch})^2$

$\therefore$ $\sqrt{af} \mp \sqrt{bg} = \pm \sqrt{ch}$; or $\sqrt{af} \mp \sqrt{bg} \mp \sqrt{ch} = 0$

13. Find the angle between two diagonals of a cube.

[VTU, Aug., 1999 ; VTU J/A, 2003 ; VTU, J/F, 2003, 2004]

Solution: Take *OP* and *MM′* be any two diagonals of the cube, where *M*(*a*, 0, 0), *P*(*a*, *a*, *a*) and M′ (0, *a*, *a*). The DC's of *OP* and *MM′* are

$$\frac{1}{\sqrt{3}}, \frac{1}{\sqrt{3}}, \frac{1}{\sqrt{3}} \text{ and } -\frac{1}{\sqrt{3}}, \frac{1}{\sqrt{3}}, \frac{1}{\sqrt{3}}$$

If θ is the angle between these diagonals *OP* and *MM′* then

$$\cos\theta = \frac{1}{\sqrt{3}}\left(-\frac{1}{\sqrt{3}}\right) + \frac{1}{\sqrt{3}} \cdot \frac{1}{\sqrt{3}} + \frac{1}{\sqrt{3}} \cdot \frac{1}{\sqrt{3}} = \frac{1}{3}.$$

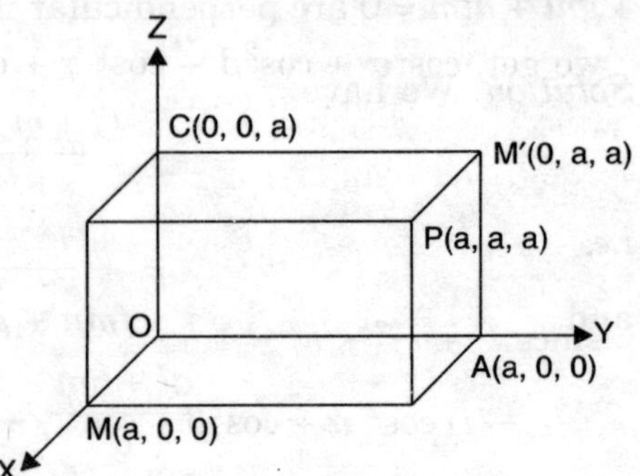

Fig. 1.9

14. If a straight line makes angle α, β, γ and δ with the four diagonals of a cube, then show that

(*i*) $\cos^2\alpha + \cos^2\beta + \cos^2\gamma + \cos^2\delta = \frac{4}{3}$. [VTU, Aug., 2001]

(*ii*) $\sin^2\alpha + \sin^2\beta + \sin^2\gamma + \sin^2\delta = \frac{8}{3}$ [VTU, Mar., 2001]

Solution: Let the origin be one of the corner of the cube and three edges *OA, OB* and *OC* be along the coordinate axes. Let $OA = OB = OC = a$. Then

$$O = (0, 0, 0);\ P = (a, a, a)$$
$$A = (a, 0, 0);\ L = (0, a, a)$$
$$B = (0, a, 0);\ M = (a, 0, a)$$
$$C = (0, 0, a);\ N = (a, a, 0)$$

Let α, β, γ, δ be the angle between the diagonals, *OP, AL, BM* and *CN* respectively with the given line having D.C's ($l : m : n$), then

$$\cos\alpha = \frac{al + am + an}{\sqrt{a^2 + a^2 + a^2}} = \frac{l + m + n}{\sqrt{3}}.$$

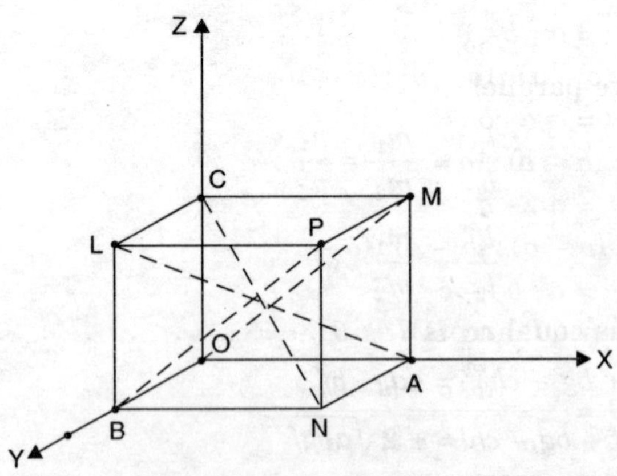

Fig. 1.10

$$|||^{ly},\ \cos\beta = \frac{-l + m + n}{\sqrt{3}},\ \cos\gamma = \frac{l + m - n}{\sqrt{3}}$$

$$\cos\delta = \frac{l-m+n}{\sqrt{3}}, \text{ squaring and adding,}$$

we get, $\cos^2\alpha + \cos^2\beta + \cos^2\gamma + \cos^2\delta$

$$= \frac{(l+m+n)^2 (-l+m+n)^2 + (l+m-n)^2 + (l-m+n)^2}{3}$$

$$= \frac{4(l^2+m^2+n^2)}{3} = \frac{4}{3}.$$

since $l^2 + m^2 + n^2 = 1$.

(*i*) $\cos^2\alpha + \cos^2\beta + \cos^2\gamma + \cos^2\delta = \frac{4}{3}$.

Also, $(1-\sin^2\alpha) + (1-\sin^2\beta) + [1-\sin^2\gamma) + [1-\sin^2\delta) = \frac{4}{3}$

$$4 - (\sin^2\alpha + \sin^2\beta + \sin^2\gamma + \sin^2\delta) = \frac{4}{3}.$$

$\Rightarrow$ $(\sin^2\alpha + \sin^2\beta + \sin^2\gamma + \sin^2\delta) = \frac{8}{3}$, which proves (*ii*).

15. If the edges of a rectangular parallelopiped are of lengths *a, b, c*. Show that the angles between four diagonals will be given by $\cos^{-1}\left(\frac{\pm a^2 \pm b^2 \pm c^2}{a^2+b^2+c^2}\right)$

Solution: The diagonals are *OP, AL, BM* and *CN* respectively, then

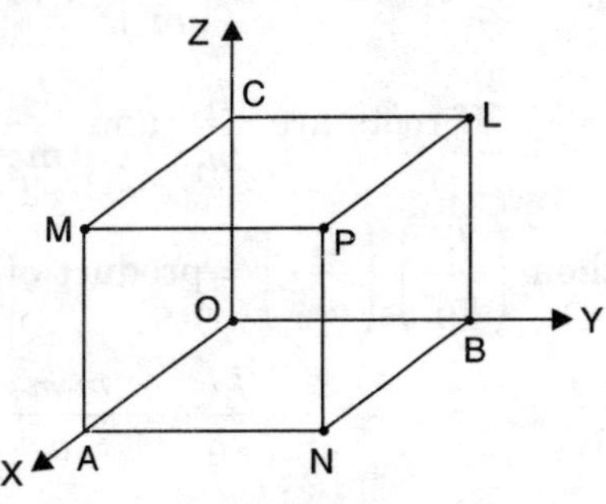

Fig. 1.11

$O\,(0, 0, 0);\ P\,(a, b, c)$
$A\,(a, 0, 0);\ L\,(o, b, c)$
$B\,(o, b, o);\ M\,(a, o, c)$
$C\,(0, 0, c);\ N\,(a, b, o)$

D.R's of *OP* are: $(a-o) : (b-o) : (c-o)$
$= a : b : c$

D.R's of *AL* are: $(o-a) : (b-o) : (c-o)$
$= -a : b : c$

D.R's of *BM* are: $(a-o) : (o-b) : (c-o)$
$= a : -b : c$

D.R's of *CN* are: $(a-o) : (b-o) : (o-c)$
$= a : b : -c$

The angle between *OP* and *BM* is θ,

where $$\cos\theta = \frac{a \cdot a + b(-b) + c(c)}{\sqrt{a^2+b^2+c^2}\,\sqrt{a^2+b^2+c^2}} = \frac{a^2-b^2+c^2}{(a^2+b^2+c^2)}$$

Similarly, by taking other pair of diagonals *AL* and *CN*, we can prove the other results, namely, $\cos\theta = \frac{a^2+b^2-c^2}{(a^2+b^2+c^2)}$ and $\cos\theta = \frac{-a^2+b^2+c^2}{(a^2+b^2+c^2)}$.

All the results are consolidated in the single statement as:

$$\cos\theta = \frac{\pm a^2 \pm b^2 \pm c^2}{(a^2+b^2+c^2)}$$

$$\theta = \cos^{-1}\left\{\frac{\pm a^2 \pm b^2 \pm c^2}{a^2+b^2+c^2}\right\}.$$

16. Prove that the lines whose D.C's are given by the relations

$$al + bm + cn = 0 \quad \text{and} \quad mn + nl + lm = 0 \text{ are}$$

(*i*) Perpendicular if $\frac{1}{a}+\frac{1}{b}+\frac{1}{c} = 0$

and (*ii*) Parallel if $\sqrt{a}+\sqrt{b}+\sqrt{c} = 0$

Solution:

Let $(l_1 : m_1 : n_1)$ and $(l_2 : m_2 : n_2)$ be the D.C's of the two lines.

Eliminating '*n*' between the given equations, we get

$$m\left[\frac{-(al+bm)}{c}\right] + l\left[\frac{-(al+bm)}{c}\right] + lm = 0$$

or $$-alm - bm^2 - al^2 - blm + clm = 0 \quad \div m^2, \text{ we have}$$

$$a\left(\frac{l}{m}\right)^2 + (a+b-c)\left(\frac{l}{m}\right) + b = 0 \qquad [1]$$

If roots are $\frac{l_1}{m_1}$ and $\frac{l_2}{m_2}$,

then $\left(\frac{l_1}{m_1}\right)\cdot\left(\frac{l_2}{m_2}\right)$ = product of the roots $= \frac{b}{a}$

$$\Rightarrow \frac{l_1 l_2}{b} = \frac{m_1 m_2}{a} \Rightarrow \frac{l_1 l_2}{\left(\frac{1}{a}\right)} = \frac{m_1 m_2}{\left(\frac{1}{b}\right)} = \frac{n_1 n_2}{\left(\frac{1}{c}\right)} \quad \text{(By symmetry)}$$

If the lines are perpendicular, then

$$l_1 l_2 + m_1 m_2 + n_1 n_2 = 0$$

$\Rightarrow \frac{1}{a}+\frac{1}{b}+\frac{1}{c} = 0$. If the lines are parallel, then their D.C's must be same *i.e.*, the roots of eqn (*i*) must be equal.

$$\therefore \quad (a+b+c)^2 = 4ab, \qquad (\because \ b^2 = 4ac)$$

or $$(a+b) - c = \pm 2\sqrt{a}\sqrt{b}, \text{ (using equation [1])}$$

or $$(a + b \mp 2\sqrt{ab}) = c \quad \text{or} \quad (\sqrt{a} \pm \sqrt{b})^2 = (\sqrt{c})^2$$

or $$\sqrt{a} \mp \sqrt{b} = \pm\sqrt{c} \quad \text{or} \quad \sqrt{a}+\sqrt{b}+\sqrt{c} = 0.$$

17. Show that the lines whose D.C's are given by the equations

$$l + m + n = 0 \quad \text{and} \quad al^2 + bm^2 + cn^2 = 0$$

are (*i*) Perpendicular if $a + b + c = 0$

(*ii*) Parallel if $\frac{1}{a} + \frac{1}{b} + \frac{1}{c} = 0$

Solution: We have, $l + m + n = 0$ (*i*)

$al^2 + bm^2 + cn^2 = 0$ (*ii*)

Eliminating n from (*i*) and (*ii*), we get

$$al^2 + bm^2 + c(-l - m)^2 = 0$$

i.e. $$(a + c)\, l^2 + 2clm + (b + c)\, m^2 = 0$$

i.e. $$(a + c)\left(\frac{l}{m}\right)^2 + 2c\left(\frac{l}{m}\right) + (b + c) = 0 \quad \text{(iii)}$$

which is quadratic in l/m. Let l_1/m_1 and l_2/m_2 be its roots.

Then $$\frac{l_1 l_2}{m_1 m_2} = \frac{b + c}{a + c} \quad \text{(iv)}$$

|||ly by eliminating m from (*i*) and (*ii*) we can prove that

$$\frac{l_1 l_2}{n_1 n_2} = \frac{b + c}{a + b} \quad \text{(v)}$$

From (*iv*) and (*v*), we get

$$\frac{l_1 l_2}{b + c} = \frac{m_1 m_2}{a + c} = \frac{n_1 n_2}{a + b} = \lambda \quad \text{(say)}$$

(*i*) The lines whose D.C's are

$(l_1 : m_1 : n_1)$ and $[(l_2 : m_2 : n_2)$ are $\perp^r$ if $l_1 l_2 + m_1 m_2 + n_1 n_2 = 0$.

i.e. $$\lambda(b + c) + \lambda(a + c) + \lambda(a + b) = 0$$

or $$a + b + c = 0.$$

(*ii*) The lines are parallel if

$$\frac{l_1}{l_2} = \frac{m_1}{m_2} = \frac{n_1}{n_2}$$

This will hold good if the roots of (*iii*) are equal

$\Rightarrow$ discriminant $= 0$ $\Rightarrow$ $4c^2 - 4(a + c)(b + c) = 0$

i.e. $$ab + bc + ca = 0$$

($\div\, abc$), we get $\frac{1}{a} + \frac{1}{b} + \frac{1}{c} = 0$

is the condition for the two lines to be parallel.

18. If P (4, 2, 3) and Q (1, –3, 4), then find the projection of PQ on a line which makes 30°, 120° and 90° with coordinate axes.

The actual D.C's of the line are:

cos 30°, cos 120°, cos 90° the D.C's of the line are $\frac{\sqrt{3}}{2}, \frac{-1}{2}$, 0.

The projection of PQ on the line whose D.C's are $\frac{\sqrt{3}}{2}, \frac{-1}{2}, 0.$

$$= l(x_2 - x_1) + m(y_2 - y_1) + n(z_2 - z_1)$$

$$= \frac{\sqrt{3}}{2}(1-4) + \left(\frac{-1}{2}\right)(-3-2) + 0(4-3)$$

$$= \frac{-3\sqrt{3}}{2} + \frac{5}{2} + 0 = \frac{1}{2}(5 - 3\sqrt{3})$$

19. If the actual D.C's of two lines are connected by $l - 3m + n = 0$, $l^2 - 3m^2 - 2n^2 = 0$. Find the D.C's.

Solution: **Given:**

$$l - 3m + n = 0$$

$$l^2 - 3m^2 - 2n^2 = 0$$

$\div\, n$ and n^2, and put $\frac{l}{n} = p$, $\frac{m}{n} = q$

we get $\frac{l}{n} - 3 \cdot \frac{m}{n} + 1 = 0$

$$\frac{l^2}{n^2} - 3 \cdot \frac{m^2}{n^2} - 2 = 0$$

$\Rightarrow$ $P - 3q + 1 = 0$ and $p^2 - 3q^2 - 2 = 0$

From the first, $p = 3q - 1$, Substituting in the second,

we get $(3q-1)^2 - 3q^2 - 2 = 0$ or $6q^2 - 6q - 1 = 0$

Solving, $q = \frac{6 \pm \sqrt{36+24}}{2 \cdot 6} - 1 = \frac{3 \pm \sqrt{15}}{6}$

But $p = 3q - 1$

$\therefore$ $p = 3\left(\frac{3 \pm \sqrt{15}}{6}\right) - 1 = \frac{9 \pm 3\sqrt{15} - 6}{6} = \frac{3 \pm 3\sqrt{15}}{6}$

$\Rightarrow$ $\frac{l}{n} = \frac{3 \pm 3\sqrt{15}}{6}$, $\frac{m}{n} = \frac{3 \pm \sqrt{15}}{6}$

$(l : m : n)$ are proportional to $3 \pm 3\sqrt{15}$, $3 \pm \sqrt{15}$, 6

But $\sqrt{(3 \pm 3\sqrt{15})^2 + (3 \pm \sqrt{15})^2 + 6^2} = \sqrt{204 \pm 24\sqrt{15}}$

$\therefore$ Actual D.C's are

$$\frac{3 \pm 3\sqrt{15}}{\sqrt{204 \pm 24\sqrt{15}}}, \frac{3 \pm \sqrt{15}}{\sqrt{204 \pm 24\sqrt{15}}}, \frac{6}{\sqrt{204 \pm 24\sqrt{15}}}$$

20. Find the area of ΔPQR where $P = (1, 0, 0)$, $Q = (0, 2, 0)$, $R = (0, 0, 3)$ using $PQ \cdot PR \sin Q\hat{P}R$.

D.R's of $PQ = (0-1) : (2-0) ; (0-0) = -1 : 2 : 0$

D.R's of $PR = (0-1) : (0-0) : (3-0) = -1 : 0 : 3$

$$\therefore \quad \cos\theta = \frac{(-1)(-1)+(2)(0)+(0)(3)}{\sqrt{1+4+0}\,\sqrt{1+0+9}} = \frac{1+0+0}{\sqrt{5}\,\sqrt{10}} = \frac{1}{\sqrt{50}}$$

Now, $\sin\theta = \sqrt{1-\cos^2\theta} = \sqrt{1-\frac{1}{50}} = \frac{7}{\sqrt{50}}$.

Also $PQ = \sqrt{(0-1)^2+(2-0)^2+(0-0)^2} = \sqrt{1+4} = \sqrt{5}$

$PR = \sqrt{(0-1)^2+(0-0)^2+(3-0)^2} = \sqrt{1+9} = \sqrt{10}$

Area of $\Delta PQR = \frac{1}{2}\, PQ \cdot PR \cdot \sin Q\hat{P}R$

$$= \frac{1}{2}\cdot(\sqrt{5})(\sqrt{10})\frac{7}{\sqrt{50}} = \frac{7}{2}\cdot\frac{\sqrt{50}}{\sqrt{50}} = \frac{7}{2} \text{ Sq. units}$$

21. Find the direction ratios of the bisectors of the angle between the lines whose direction cosines are $(l_1 : m_1 : n_1)$ and $(l_2 : m_2 : n_2)$.

Solution: Let L_1, L_2 be the lines with D.C's $(l_1 : m_1 : n_1)$ and $(l_2 : m_2 : n_2)$ respectively.

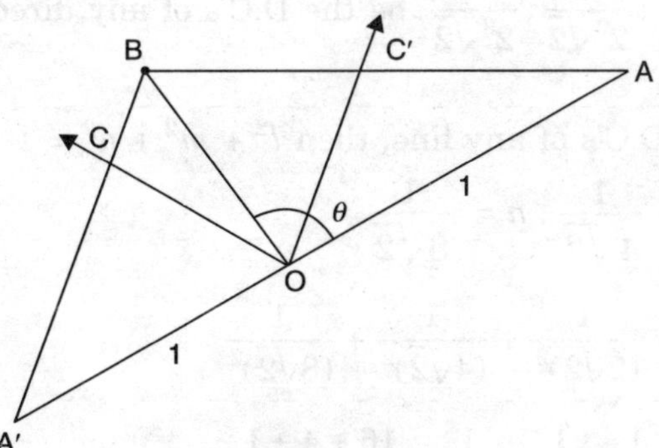

Fig. 1.12

Let OA, OB be lines thro' the origin such that OA is in direction $O + L_1$ and OB is in direction of L_2.

If θ is the angle between L_1 and L_2 then $A\hat{O}B = \theta$

Let $OA = OB = 1$ unit.

Then $A = (l_1, m_1, n_1)$, $B = (l_2, m_2, n_2)$

Let A be the point on AO produced such that $OA' = 1$

then $A' = (-l_1, -m_1, -n_1)$

Now, if C and C' are the mid points of AB and $A'B$ then OC and OC' are along the internal bisector and external bisector of the angle $A\hat{O}B = \theta$.

Now, $C = \left[\frac{l_1+l_2}{2}, \frac{m_1+m_2}{2}, \frac{n_1+n_2}{2}\right]$

$C^1 = \left[\frac{l_2-l_1}{2}, \frac{m_2-m_1}{2}, \frac{n_2-n_1}{2}\right]$

$\therefore$ D.R's of the internal bisector is the D.R's of the line joining O and C.

$(l_1 + l_2,\ m_1 + m_2,\ n_1 + n_2)$

|||ly DR's of external bisector are

$$(l_2 - l_1, m_2 - m_1, n_2 - n_1)$$

22. It $A = (1, 3, 5)$, $B = (6, 4, 3)$, $C = (2, -1, 4)$ and $D = (0, 1, 5)$ find the projection of AB on CD.

Solution: The D.R's of CD are $(-2, 2, 1)$

[Using the formula $(x_2 - x_1) : (y_2 - y_1) : (z_2 - z_1)$]

Now, the D.Cs of CD are

$$\left[\frac{-2}{\sqrt{4+4+1}}, \frac{2}{\sqrt{4+4+1}}, \frac{1}{\sqrt{4+4+1}}\right] = \left(\frac{-2}{3}, \frac{2}{3}, \frac{1}{3}\right)$$

$\therefore$ the projection of AB on CD

$$= -\frac{2}{3}(6-1) + \frac{2}{3}(4-3) + \frac{1}{3}(3-5) = -\frac{10}{3} + \frac{2}{3} - \frac{2}{3} = -\frac{10}{3}.$$

Projection: $l(x_2 - x_1) + m[y_2 - y_1) + n[z_2 - z_1)$.

23. Can the numbers $\frac{1}{2\sqrt{2}}, \frac{1}{2^2\sqrt{2}}, \frac{1}{2^3\sqrt{2}}$ be the D.C's of any directed line? Give reasons for your answer.

Solution: If $(l : m : n)$ are D.C's of any line, then $l^2 + m^2 + n^2 = 1$

Here $l = \frac{1}{2\sqrt{2}}$, $m = \frac{1}{4\sqrt{2}}$, $n = \frac{1}{8\sqrt{2}}$

$$\text{Since } l^2 + m^2 + n^2 = \frac{1}{(2\sqrt{2})^2} + \frac{1}{(4\sqrt{2})^2} + \frac{1}{(8\sqrt{2})^2}$$

$$= \frac{1}{8} + \frac{1}{32} + \frac{1}{128} = \frac{16+4+1}{128} \neq 1.$$

So, the given numbers are not the D.C's of the directed line.

24. Find the angle between the lines whose direction cosines are proportional to 1, 2, 1 and 2, –3, 6.

Solution: Have

$$a_1 = 1, b_1 = 2, c_1 = 1$$

$$a_2 = 2, b_2 = -3, c_2 = 6$$

If 'θ' be the required angle

$$\therefore \quad \cos\theta = \frac{a_1a_2 + b_1b_2 + c_1c_2}{\sqrt{a_1^2 + b_1^2 + c_1^2}\sqrt{a_2^2 + b_2^2 + c_2^2}}$$

$$= \frac{1\times 2 + 2(-3) + 1\times 6}{\sqrt{1^2 + 2^2 + 1^2}\sqrt{2^2 + (-3)^2 + 6^2}} = \frac{2-6+6}{\sqrt{6}\sqrt{49}} = \frac{2}{(\sqrt{6})7} = \frac{2}{7(\sqrt{6})}$$

$$\therefore \quad \theta = \cos^{-1}\left\{\frac{2}{7\sqrt{2}}\right\}$$

25. A line makes an angle of 45° with OX and 60° with OY. What angle does it make with OZ?

Solution: If the line makes an angle γ with the positive z-axis, Then since the line makes angles 45°, 60° and γ with the axes:,

D.C's are : $(\cos 45° : \cos 60° : \cos \gamma)$

or $$\left(\frac{1}{\sqrt{2}} : \frac{1}{2} : \cos \gamma\right)$$

But we know $l^2 + m^2 + n^2 = 1 \quad \Rightarrow \quad \frac{1}{2} + \frac{1}{4} + \cos^2 \gamma = 1$

$$\cos^2\gamma = 1 - \frac{3}{4} = \frac{1}{4}$$

$$\therefore \quad \cos^2 \gamma = \pm \frac{1}{2}$$

$$\therefore \quad \gamma = 60°; \quad \text{or} \quad 120°,$$

26. What are the D.C's of the line equally included to the axes? How many such lines are there?

Solution:

Since $\alpha = \beta = \gamma$; (Given)

$\Rightarrow \quad \cos \alpha = \cos \beta = \cos \gamma \quad$ or $\quad l = m = n$

Since $l^2 + m^2 + n^2 = 1 \quad \therefore \quad l^2 + l^2 + l^2 = 1$

$3l^2 = 1 \quad \therefore \quad l = \pm \frac{1}{\sqrt{3}}$

$\therefore$ D.C's are $\left(\frac{\pm 1}{\sqrt{3}} : \frac{\pm 1}{\sqrt{3}} : \frac{\pm 1}{\sqrt{3}}\right)$

Now consider

$$\begin{bmatrix} + & + & + \\ + & - & + \\ + & + & - \end{bmatrix}$$

−	+	+	
+	−	−	(or − + +)
−	−	+	(or + + −)
−	+	−	(or + − +)
−	−	−	(or + + +)

$\because (l : m : n), (-l : -m : -n)$ are same. So there can be 4 different lines which make equal angle with the axis.

27. If a straight line makes an angle of $\frac{\pi}{4}$ with each of x-axis, y-axis, then what angle does it make with the z-axis?

Solution: Let γ be the angle it makes with z-axis, since

$$\cos^2 \alpha + \cos^2 \beta + \cos^2 \gamma = 1$$

$$\therefore \quad \cos^2\frac{\pi}{4} + \cos^2\frac{\pi}{4} + \cos^2 \gamma = 1$$

$$\Rightarrow \quad \frac{1}{2}+\frac{1}{2}+\cos^2\gamma = 1$$

$$\cos^2\gamma = 0$$

$$\cos\gamma = 0 = \cos \pi/2$$

$$\gamma = \pi/2.$$

28. $(l_1 : m_1 : n_1)$, $(l_2 : m_2 : n_2)$ are the D.C's of two mutually $\perp^r$ lines, show that the D.C's of the line $\perp$ to both of them are $(m_1n_2 - n_1m_2)$, $(n_1l_2 - n_2l_1)$, $(l_1m_2 - m_1l_2)$.

Solution: If $(l : m : n)$ be the D.C's of the line $\perp^r$ two given lines, then

$$ll_1 + mm_1 + nn_1 = 0$$

$$ll_2 + mm_2 + nn_2 = 0$$

$$\therefore \quad \frac{l}{m_1n_2 - n_1m_2} = \frac{m}{n_1l_2 - l_1n_2} = \frac{n}{l_1m_2 - m_1l_2}$$

$$= \frac{\sqrt{\Sigma l^2}}{\sqrt{\Sigma(m_1n_1 - n_1m_2)^2}} \quad \text{(By the principle of algebra)}$$

$$= \frac{1}{\sin\theta} = \frac{1}{\sin 90^\circ} = 1$$

where $\theta = 90°$ is the angle between the given lines

$$\therefore \quad l = n_2m_1 - n_1m_2,$$

$$m = n_1l_2 - l_1n_2$$

$$n = l_1m_2 - m_1l_2$$

i.e. the D.C's are $(m_1n_1 - n_1m_2)$, $(n_1l_2 - n_2l_1)$, $(l_1m_2 - m_1l_2)$.

29. Show that the pair of lines whose D.C's satisfy the equations,

$l + m + n = 0$, $2l^2 + 2m^2 - n^2 = 0$ are parallel.

Solution: Let $(l_1 : m_1 : n_1)$, $(l_2 : m_2 : n_2)$ be D.C's of the given lines then these satisfy the given equations

$$l + m + n = 0 \qquad [1]$$

$$2l^2 + 2m^2 - n^2 = 0 \qquad [2]$$

Eliminating 'n' from these eqn, we get $2l^2 + 2m^2 - (l + m)^2 = 0$

$$2l^2 + 2m^2 - (l + m)^2 = 0$$

or $\quad (l - m)^2 = 0 \quad$ or $\quad l = m$

Thus, we have $\quad l_1 = m_1,\ l_2 = m_2$

so that $\dfrac{l_1}{l_2} = \dfrac{m_1}{m_2} = 1 \qquad [3]$

Now using [1] and [3], we get

$$\frac{n_1}{n_2} = \frac{-(l_1+m_1)}{-(l_2+m_2)} = 1$$

Expressions [3] and [4], give

$$l_1 = l_2,\ m_1 = m_2,\ n_1 = n_2.$$

$\Rightarrow$ lines are parallel.

30. The coordinates of the angular points *A, B, C, D* of a tetrahedron are (–2, 1, 3) (3, –1, 2), (2, 4, –1) and (1, 2, 3) respectively. Find the angle between the edges *AC* and *BD*.

Solution: The D.R's of *AC* are (4 : 3 : – 4).

Actual D.C's are $\left(\frac{4}{\sqrt{41}} : \frac{3}{\sqrt{41}} : \frac{-4}{\sqrt{41}}\right)$

and the Actual D.C's of *BD* are $\left(\frac{-2}{\sqrt{14}} : \frac{3}{\sqrt{14}} : \frac{1}{\sqrt{14}}\right)$

If $\hat{\theta}$ is the angle between *AC* and *BD*, then

$$\cos\theta = \frac{-8+9-4}{\sqrt{14\times 41}} = \frac{-3}{\sqrt{574}}; \quad \Rightarrow \quad \theta = \cos^{-1}\left[\frac{-3}{\sqrt{574}}\right]$$

31. Find the angle between the straight line which intersection of the planes $2x + 2y - z + 15 = 0 = 4y + z + 29$ and the line

$$\frac{x+4}{4} = \frac{y-3}{-3} = \frac{z+1}{1}.$$

[VTU J/A, 2003]

Solution. Let *l, m, n* be the D.C's of the line of intersection of the planes $2x + 2y - z + 15 = 0$.

$4y + z + 29 = 0$. Then $\frac{l}{3} = \frac{m}{-1} = \frac{n}{4}$

D.C's of the second line are 4, – 3, 1. If θ is the angle between these two lines, then

$$\cos\theta = \frac{3(4)+(-1)(-3)+(4)(1)}{\sqrt{9+1+16}\,\sqrt{16+9+1}} = \frac{19}{26} \quad \text{or} \quad \theta = \cos^{-1}\left(\frac{19}{26}\right).$$

EXERCISES

1. Find the angle between the lines whose D.R's are (2 : 3 : 4) and (1 : – 2 : 1) $\left(Ans: \theta = \frac{\pi}{2}\right)$

2. *A, B, C* are the points (1, 4, 2), (– 2, 1, 2), (2, – 3, 4). Find the angles of the triangle *ABC*.

   $\left(Ans: 90°, \cos^{-1}\frac{1}{\sqrt{3}}, \cos^{-1}\left(\frac{\sqrt{2}}{\sqrt{3}}\right)\right)$

3. Find the D.C's of the line which is $\perp^r$ to the lines with D.C's proportional to (1 : –2 : –2) and (O : 2 : 1) $\left(Ans: \text{Actual D.C's are}\left(\frac{2}{3}, \frac{-1}{3}, \frac{2}{3}\right)\right)$

4. If $(l_1 : m_1 : n_1)$, $(l_2 : m_2 : n_2)$ and $(l_3 : m_3 : n_3)$ are the D.C's of 3 mutually $\perp^r$ lines, then show that the line whose D.C's are $\frac{l_1+l_2+l_3}{3}, \frac{m_1+m_2+m_3}{3}$, and $\frac{n_1+n_2+n_3}{3}$ makes equal angles with them.

5. If $(l_1 : m_1 : n_1)$, $(l_2 : m_2 : n_2)$ be two directions inclined as an angle θ, show that actual D.C's of the direction bisecting them are

   $$\frac{1}{2}(l_1 + l_2)\sec\theta/2; \ \frac{1}{2}(m_1 + m_2)\sec\theta/2; \ \frac{1}{2}(n_1 + n_2)\sec\theta/2.$$

6. If P (1, 2, 3), Q (4, 5, 7), R (– 4, 3, – 6) and S (2, 9, 2).
S.T. PQ and RS are parallel.

7. If A (4, 7, –2), B (2, 0, 3), C(– 5, – 8, – 9), D (– 9, – 4, – 5). S.T. AB is $\perp^r$ to CD.

8. Find the D.C's of a st. line equally inclined to the coordinate axes.

9. If P(2, 3, 5), Q (3, 5, 2), R(5, 2, 3) are the vertices of a triangle find the angle P.

10. A straight line is inclined with the axes of y and z at angles of 45° θ and 60°. Find the inclination of the line to the x-axis.

1.4 PLANES

A plane is a surface such that a straight line joining any two of its points lies entirely in the surface.

Equations of Planes

An equation of first degree: $Ax + By + Cz + D = 0$ (1) in x, y, z always represents a plane.

This equation is called the General equation of a plane.

1. *Any plane through* (x_1, y_1, z_1)
(one point form)

Let $$Ax + By + Cz + D = 0 \quad [1]$$

Let it pass through P (x_1, y_1, z_1).

Then, we have $$Ax_1 + By_1 + Cz_1 + D = 0 \quad [2]$$

Subtracting [2] from [1], we get

$$A(x - x_1) + B(y - y_1) + C(z - z_1) + D - D = 0$$

or $$A(x - x_1) + B(y - y_1) + C(z - z_1) = 0$$

is the equation of the plane through (x_1, y_1, z_1).

2. *Normal Form* [VTU, Aug., 1999]

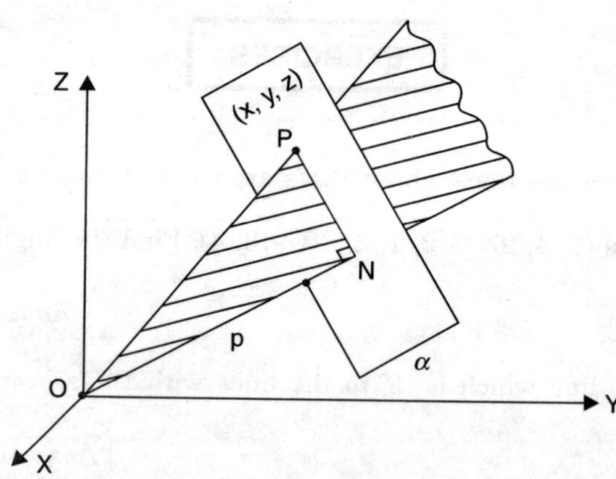

Fig. 1.13

Let α be a plane.

Let ON be a normal to plane

Let $ON = p$ and (l, m, n) be D.C's of the normal to the plane.

Then, the equation of the plane is

$$lx + my + nz = p \qquad \text{or} \qquad x\cos\alpha + y\cos\beta + z\cos\gamma = p$$

where (l, m, n) are D.C's of the normal to the plane, p, is the length of the $\perp^r$ from the origin to the plane and p is always positive.

3. ***Intercept Form*** [VTU, Aug., 2001]

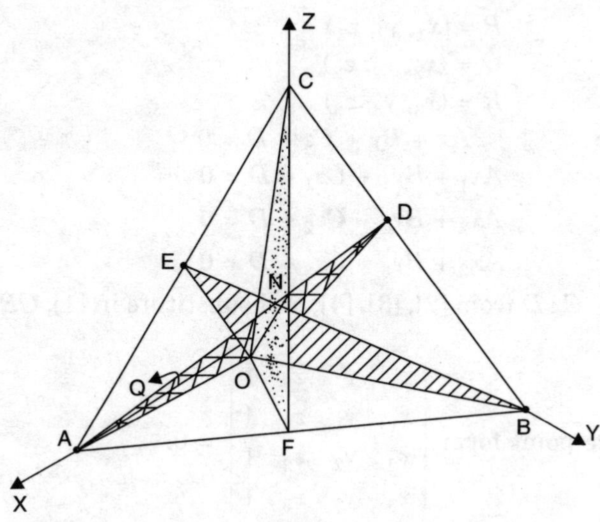

Fig. 1.14

We have the equation of plane in normal form:

$$x \cos \alpha + y \cos \beta + z \cos \gamma = 0 \qquad [1]$$

We find from the figure,

ONA, *ONB* and on *ONC* are right angled Δ^s

$$\left.\begin{aligned} ON &= p; \\ OA &= f \\ ONA &= \alpha; \end{aligned}\right\} \quad \begin{aligned} &\Rightarrow \quad \cos \alpha = \left(\frac{p}{f}\right); \\ &\Rightarrow \quad f = p \sec \alpha \end{aligned}$$

$|||^{ly}$, we get the relations

$$\cos \alpha = (p/f), \cos\beta = (p/g), \cos \gamma = (p/h) \qquad [2]$$

$$\Rightarrow \qquad f = p \sec\alpha, g = p \sec \beta, h = p \sec\gamma$$

Substituting [2] in [1], we get

$$x\frac{p}{f} + y\frac{p}{g} + z\frac{p}{h} = p \; ; \quad \frac{x}{f} + \frac{y}{g} + \frac{z}{h} = 1$$

4. ***Reduction of the general equation***

$Ax + By + Cz + D = 0$ to the normal form

Let $Ax + By + Cz + D = 0$ be the eqn. to a given plane.

or, $\qquad Ax + By + Cz = -D, \left(\div \pm \sqrt{\sum A^2}\right)$, we get

$$\pm \frac{A}{\sqrt{\sum A^2}} \cdot x \pm \frac{B}{\sqrt{\sum A^2}} \cdot y \pm \frac{C}{\sqrt{\sum A^2}} \cdot z = \pm \frac{D}{\sqrt{\sum A^2}}$$

Since in the normal form *P* is always positive

$\therefore \qquad \pm \dfrac{D}{\sqrt{\sum A^2}}$ should be positive.

5. ***Plane through three given points (three point form)*** [V.T.U. F/M. 2005]

Let $P = (x_1, y_1, z_1)$

$Q = (x_2, y_2, z_2)$

$R = (x_3, y_3, z_3)$

Let the plane be $Ax + By + Cz + D = 0$ [1]

P, lies on it $\therefore$ $Ax_1 + By_1 + Cz_1 + D = 0$ [2]

Q lies on it $\therefore$ $Ax_2 + By_2 + Cz_2 + D = 0$ [3]

R lies on it $\therefore$ $Ax_3 + By_3 + Cz_3 + D = 0$ [4]

Solving for $A : B : C : D$ from [2], [3], [4]; and substitute in [1], *OR*, eliminate $A : B : C : D$ from [1] to [4],

We get, the three point form $\begin{vmatrix} x & y & z & 1 \\ x_1 & y_1 & z_1 & 1 \\ x_2 & y_2 & z_2 & 1 \\ x_3 & y_3 & z_3 & 1 \end{vmatrix} = 0.$

6. *Deduction of equation of the plane in the intercept form, from the 3 point form.* [V.T.U. F/M 2005]

If we choose in the equation of the plane in three point form,

$(x_1 = f, y_1 = 0, z_1 = 0);$

$(x_2 = 0, y_2 = g, z_2 = 0)$

$(x_3 = 0, y_3 = 0, z_3 = h).$

Then we get the equation of the plane passing through the points $(f, 0, 0)$; $(0, g, h)$ and $(0, 0, h)$, *i.e.* a plane making intercepts f, g and h, on the three axes. The equation of the plane then becomes:

$$\begin{vmatrix} x & y & z & 1 \\ f & 0 & 0 & 1 \\ 0 & g & 0 & 1 \\ 0 & 0 & h & 1 \end{vmatrix} = 0; \qquad \text{or,} \qquad x\begin{vmatrix} 0 & 0 & 1 \\ g & 0 & 1 \\ 0 & h & 1 \end{vmatrix} - f\begin{vmatrix} y & z & 1 \\ g & 0 & 1 \\ 0 & h & 1 \end{vmatrix} = 0$$

or, $$x\begin{vmatrix} g & 0 \\ 0 & h \end{vmatrix} - fy\begin{vmatrix} 0 & 1 \\ h & 1 \end{vmatrix} + fg\begin{vmatrix} z & 1 \\ h & 1 \end{vmatrix} + 0 = 0$$

$\therefore$ $xgh - fy\,(-h) + fg\,(z - h) = 0$

$(\div\, fgh)$

$$\frac{x}{f} + \frac{y}{g} + \frac{z}{h} + (-1) = 0 \Rightarrow \frac{x}{f} + \frac{y}{g} + \frac{z}{h} = 1$$

7. *Condition for four points to be coplanar*

Let $A = (x_1, y_1, z_1)$, $B = (x_2, y_2, z_2)$

$C = (x_3, y_3, z_3)$ and $D = (x_4, y_4, z_4)$

be any four points. Then the equation of the plane passing through three points A, B, C is

$$\begin{vmatrix} x - x_1 & y - y_1 & z - z_1 \\ x_2 - x_1 & y_2 - y_1 & z_2 - z_1 \\ x_3 - x_1 & y_3 - y_1 & z_3 - z_1 \end{vmatrix} = 0 \quad [1]$$

Since $D = (x_4, y_4, z_4)$ also lies on [1]

$$\therefore \quad \begin{vmatrix} x_4 - x_1 & y_4 - y_1 & z_4 - z_1 \\ x_2 - x_1 & y_2 - y_1 & z_2 - z_1 \\ x_3 - x_1 & y_3 - y_1 & z_3 - z_1 \end{vmatrix} = 0 = \begin{vmatrix} x_2 - x_1 & y_2 - y_1 & z_2 - z_1 \\ x_3 - x_1 & y_3 - y_1 & z_3 - z_1 \\ x_4 - x_1 & y_4 - y_1 & z_4 - z_1 \end{vmatrix}$$

is the condition for A, B, C, D to be coplanar.

WORKED EXAMPLES

31. Find the intercepts made on the coordinate axes by the plane $x + 2y - 2z = 9$. Find the length of the normal from the origin to the plane $x + 2y - 2z = 9$ and also the D.R's of the normal.

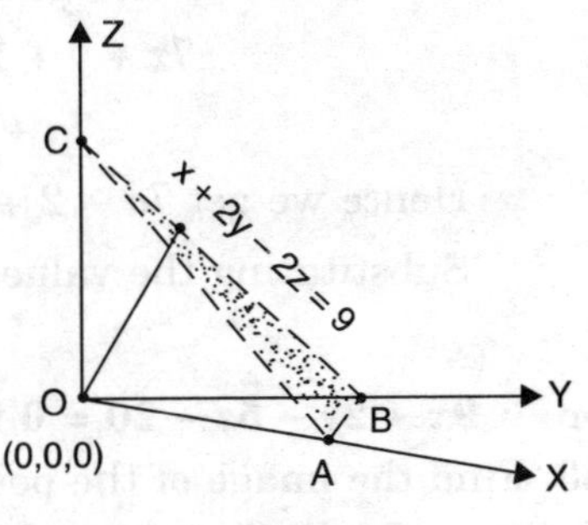

Fig. 1.13

Solution: To find the intercepts on the axes, we write the equation in the form:

$$\frac{x}{f} + \frac{y}{g} + \frac{z}{h} = 1 \qquad [1]$$

Given plane: $\quad x + 2y - 2z = 9$

i.e.,
$$\frac{x}{9} + \frac{2y}{9} - \frac{2z}{9} = 1$$

$$\frac{x}{9} + \frac{y}{\left(\frac{9}{2}\right)} - \frac{z}{\left(-\frac{9}{2}\right)} = 1$$

$\Rightarrow$ the intercepts are $9, \frac{9}{2}, \frac{-9}{2}$, the given plane is

$$x + 2y - 2z = 9$$

DR's of normal to plane $(1 : 2 : -2)$.

$\therefore$ The actual D.C's

$$\frac{1}{\sqrt{1^2 + 2^2 (-2)^2}}, \frac{2}{\sqrt{1^2 + 2^2 + (-2)^2}} \frac{-2}{\sqrt{1^2 + 2^2 + (-2)^2}}$$

$$\Rightarrow \quad \frac{1}{3}, \frac{2}{3}, \frac{-2}{3}.$$

We now write the equations of the given plane in the normal form

$$lx + my + nz = p$$

$\therefore$ The eqn. of the plane:

$$x + 2y - 2z = 9$$

can be written as:

$$\frac{x}{3} + \frac{2y}{3} - \frac{2z}{3} = \frac{9}{3};$$

i.e.,
$$\frac{1}{3}x + \frac{2}{3}y - \frac{2z}{3} = 3.$$

Hence the length of the normal from the origin to the plane is 3.

32. Find the equation of the plane having y intercept 10, z interecept-4 and perpendicular to the plane $7x + y + 13z - 17 = 0$. [V.T.U. F/M. 2005]

Solution: Given y-intercept 10 ; z-intercept–4

$\Rightarrow$ Plane passes through (0, 10, 0) and (0, 0, – 4).

$\therefore$ $$a(x-0) + b(y-10) + c(z-0) = 0 \quad [1]$$

Hence, $$a \cdot 0 + b(-10) + c(-4) = 0$$

i.e., $$5b + 2c = 0 \Rightarrow \frac{b}{c} = \frac{-2}{5} \text{ or } b : c = -2 : 5$$

Further the plane given by (1) is perpendicular to the plane

$$7x + y + 13z - 17 = 0$$

$\therefore$ $7a + b + 13c = 0$. But $b : c = -2 : 5$

Hence we get $7a - 2 + 65 = 0$ or $7a + 63 = 0$ $\therefore$ $a = -9$

Substituting the values of a, b, c in (1), we get

$$-9x - 2(y - 10) + 5z = 0$$

or **$9x + 2y - 5z - 20 = 0$** is the required equation.

33. Find the image of the point (1, 2, 3) in the plane: $x + y + z = 9$

Solution: Let P (1, 2, 3) and Q (x_1, y_1, z_1) be the image of the point P in the given plane $x + y + z = 9$.

Let R be the point on the plane where the line PQ cuts the plane. Then R is the mid point of PQ.

$\therefore$ $$R = \left(\frac{x_1+1}{2}, \frac{y_1+2}{2}, \frac{z_1+3}{2}\right)$$

(mid point formula)

This point lies on the plane

$$x + y + z = 9$$

$$\Rightarrow \quad \frac{x_1+1}{2} + \frac{y_1+2}{2} + \frac{z_1+3}{2} = 9$$

$$\Rightarrow \quad x_1 + y_1 + z_1 = 12 \quad [1]$$

Now D.R's of PQ are

$$[(x_1 - 1) : (y_1 - 2) : (z_1 - 3)]$$

Now PQ is $\perp^r$ to the plane and D.R's of Normal to the plane are (1 : 1 : 1)

Thus, we have

$$\frac{x_1-1}{1} = \frac{y_1-2}{1} = \frac{z_1-3}{1} = k$$

$$\Rightarrow \quad \left.\begin{aligned} x_1 &= k+1, \\ y_1 &= k+2, \\ z_1 &= k+3 \end{aligned}\right\} \quad (2)$$

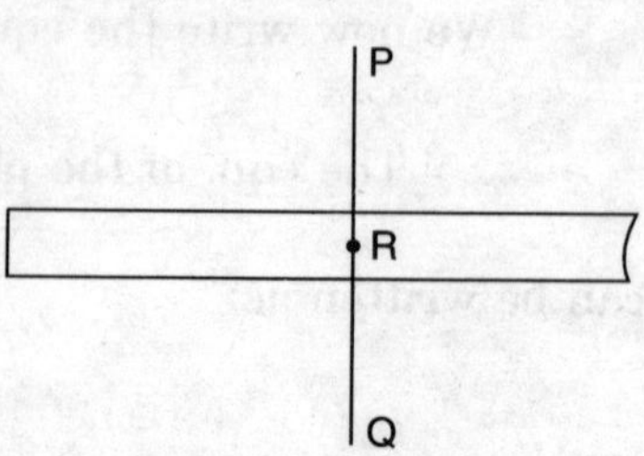

Fig. 1.16(a)

$\therefore$ (1) $\Rightarrow k + 1 + k + 2 + k + 3 = 12$

$\Rightarrow$ $3k = 6$

$k = 2$

Now by putting $k = 2$ in [2], we get the image point

$$Q = (3, 4, 5)$$

33(*a*). Find the image of the point (1, – 1, 2) in the plane

$$2x + 2y + z = 11$$ [V.T.U. F/M, 2005]

Solution: Let P(1, – 1, 2) and Q (x_1, y_1, z_1) be the image of the point P in the given plane

$$2x + 2y + z = 11$$

Let R be the point on the plane where the line PQ cuts the plane. Then R is the mid point of PQ.

P(1,–1,2)

R

Q(x_1,y_1,z_1)

$$\therefore \quad R = \left(\frac{x_1+1}{2}, \frac{y_1-1}{2} = \frac{z_1+2}{2}\right)$$

lies in the plane : $2x + 2y + z = 11.$

$$\Rightarrow \quad 2 \cdot \frac{x_1+1}{2} + 2 \cdot \frac{y_1-1}{2} + \frac{z_1+2}{2} = 11.$$

$$\Rightarrow \quad 2x_1 + 2y_1 + z_1 = 20 \qquad (1)$$

Now, DR's of PQ are $[(x_1 - 1) : (y_1 + 1) : (z_1 - 2)]$.

and PQ is $\perp^r$ to the plane. DR's of normal to the plane are (2 : 2 : 1)

$$\therefore \quad \frac{x_1-1}{2} = \frac{y_1+1}{2} = \frac{z_1-2}{1} = k \text{ (say)},$$

$$\Rightarrow \quad (x_1 = 2k + 1, y_1 = 2k - 1, z_1 = k + 1) \qquad (2)$$

Equation (1) become $\Rightarrow$ $4k + 2 + 4k - 2 + k + 2 = 20$

$\Rightarrow$ $9k = 18$ or $k = 2$

Substituting $k = 2$ in (2), we get

$x_1 = 5, y_1 = 3, z_1 = 4.$ therefore, the required image $= Q = (5, 3, 4).$

33 (*b*). Find the equation of the plane passing through the points (9, 0, 6), (2, 1, 1) and perpendicular to the plane $2x + 6y + 6z = 9$. [V.T.U. J/A, 2003]

Solution: Equation of plane be

$$A(x - x_1) + B(y - y_1) + C(z - z_1) = 0$$

Equation passes through (9, 0, 6) so

$$A(x - 9) + B(y - 0) + c(z - 6) = 0$$

Equating passes through (2, 1, 1) so

$$-7A + B - 5C = 0 \qquad (1)$$

Since equation is $\perp^r$ to the plane $2x + 6y + 6z = 9$.

We have $\quad 2A + 6B + 6C = 0 \qquad (2)$

Solving (1) and (2)

$$\frac{A}{-36} = \frac{B}{-32} = \frac{C}{44} \quad \text{or} \quad \frac{a}{9} = \frac{b}{8} = \frac{c}{-11}$$

The required equation of plane is

$$9(x - 9) + 8y - 11(z - 6) = 0 \quad \text{or} \quad 9x + 8y - 11z = 15.$$

34. Find the D.C's of the normal to, and the length of the perpendicular from the origin on, the plane whose equation is $Ax + By + Cz + D = 0$.

Solution:

Let $Ax + By + Cz + D = 0$ be the equation of the plane [1]

Equation in Normal form is:

$$x \cos\alpha + y \cos\beta + z \cos\gamma = p \quad [2]$$

If [1] is put in the normal form, [1] and [2] represent the same plane, and so the two equations are equivalent. Hence the coefficients are proportional

$$\therefore \quad \frac{\cos\alpha}{A} = \frac{\cos\beta}{B} = \frac{\cos\gamma}{C} = \frac{-p}{D} \quad [3]$$

[1] $\therefore \cos\alpha : \cos\beta : \cos\gamma = A : B : C$

i.e D.C's of the normal to the plane

$$= (\cos\alpha : \cos\beta : \cos\gamma)$$

$$= A : B : C$$

$$= (x\text{-coeff} : y\text{-coeff} : z\text{-coeff}) \quad [4]$$

[2] Also, $$\frac{-p}{D} = \frac{\cos\alpha}{A} = \frac{\cos\beta}{B} = \frac{\cos\gamma}{C}$$

$$= \frac{\sqrt{(\cos^2\alpha + \cos^2\beta + \cos^2\gamma)}}{\sqrt{A^2 + B^2 + C^2}} = \frac{+1}{\sqrt{A^2 + B^2 + C^2}}$$

$$\therefore \quad \frac{-p}{D} = \frac{1}{\sqrt{A^2 + B^2 + C^2}} \quad \therefore \quad p = \frac{-D}{\sqrt{A^2 + B^2 + C^2}}$$

Note:

(*a*) D.C's of Normal:

$$= \text{coeff } x : \text{coeff } y : \text{coeff } z$$

(*b*) The position is specified by $p = \dfrac{-D}{\sqrt{A^2 + B^2 + C^2}}$.

(*c*) $\perp^r$ from the origin $= \dfrac{-D}{\sqrt{A^2 + B^2 + C^2}}$. The $\perp^r$ in magnitude

$$= \frac{|-D|}{\sqrt{A^2 + B^2 + C^2}} = \frac{D}{\sqrt{A^2 + B^2 + C^2}}$$

35. Find the length of the perpendicular from the point (x_1, y_1, z_1) on the plane.

$$(ax + by + cz + d) = 0$$

We know that:

$$p = \left[\frac{|-D|}{\sqrt{A^2 + B^2 + C^2}}\right] \quad [1]$$

where p = length of the $\perp^r$ from the origin O (0, 0, 0).

Hence, we require the $\perp^r$ from $P(x_1, y_1, z_1)$ to the plane

$$(ax + by + cz + d) = 0 \quad [2]$$

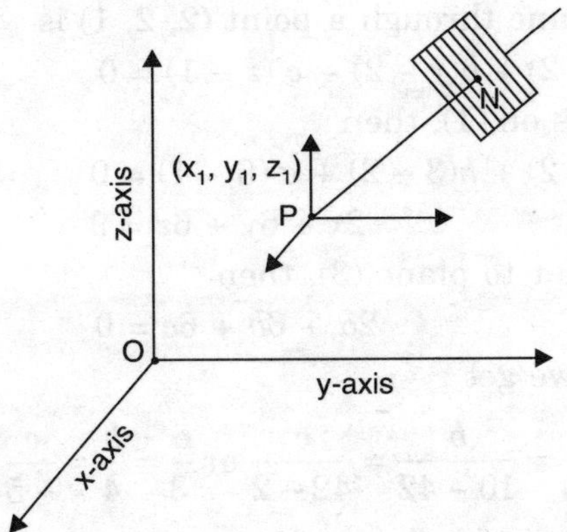

Fig. 1.16(b)

Shifting the origin to the point

(x_1, y_1, z_1), *i.e.* put
$$\left.\begin{aligned} x &= (X + x_1) \\ y &= (Y + y_1) \\ z &= (Z + z_1) \end{aligned}\right\} \quad [3]$$

We get $a(X + x_1) + b(Y + y_1) + c(Z + z_1) = d = 0$

$$\Rightarrow \quad aX + bY + cZ + (ax_1 + by_1 + cz_1 + d) = 0$$

$$\Rightarrow \quad aX + bY + cZ + k = 0$$

$$k = (ax_1 + by_1 + cz_1 + d) \quad [4]$$

The perpendicular distance of p from the plane [2] is the same as the perpendicular from the new origin to the plane [4].

Hence apply Eqn. [1], we get

$$p = \frac{-k}{\sqrt{a^2 + b^2 + c^2}} = \frac{-(ax_1 + by_1 + cz_1 + d)}{\sqrt{a^2 + b^2 + c^2}}$$

Note: The perpendicular distance is, in magnitude equal to the numerical value

$$\frac{|ax_1 + by_1 + cz_1 + d|}{\sqrt{a^2 + b^2 + c^2}}.$$

36. Find the D.C's of the normal to the plane

$$3x + 2y + 6z + 14 = 0.$$

Also obtain the perpendicular distance of the origin from this plane.

Solution: D.C's of the Normal

$$= \text{coeff } x : \text{coeff } y : \text{coeff } z$$

$$= 3 : 2 : 6$$

$$p = \frac{|-D|}{\sqrt{A^2 + B^2 + C^2}} = \frac{|-14|}{\sqrt{3^2 + 2^2 + 6^2}} = \frac{|-14|}{7} = 2.$$

37. Find the equation of the plane through the points (2, 2, 1) and (9, 3, 6) and perpendicular to the plane $2x + 6y + 6z = 9$. [V.T.U. J/F, 2004]

Solution: The equation of a plane through a point (2, 2, 1) is

$$a\,(x-2)+b(y-2)+c\,(z-1)=0 \quad ...(1)$$

A point (9, 3, 6) also lies on (1), then

$$a\,(9-2)+b(3-2)+c\,(6-1)=0 \quad \Rightarrow \quad 7a+b+5c=0 \quad ...(2)$$

$$2x+6y+6z=9 \quad ...(3)$$

Plane (1) is perpendicular to plane (3), then

$$2a+6b+6c=0 \quad ...(4)$$

On solving (2) and (4), we get

$$\frac{a}{6-30}=\frac{b}{10-42}=\frac{c}{42-2} \text{ or } \frac{a}{3}=\frac{b}{4}=\frac{c}{-5}$$

On substituting the value of a, b, c in (1) we get

$$3\,(x-2)+4\,(y-2)-5\,(z-1)=0 \quad \text{or} \quad 3x+4y-5z=9.$$

38. Prove that the equation of the plane through the points (1, – 2, 4) and (3,– 4, 5) and perpendicular and (3, – 4, 5) and perpendicular to yz-plane or parallel to x-axis is $y + 2z = 6$.

Solution: Any plane through (1, – 2, 4) is

$$A(x-1)+B(y+2)+C(z-4)=0 \quad [1]$$

where $A : B : C$ are D.R's of Normal. Since [1] passes through (3, – 4, 5)

$$\therefore \quad A(3-1)+B(-4+2)+C(5-4)=0$$

or

$$A\cdot 2+B\cdot(-2)+C\cdot 1=0 \quad [2]$$

Again, Since [1] is parallel to x-axis (DC's 1 : 0 : 0)

$\therefore$ Normal to [1] is perpendicular to x–axis

$$\therefore \quad A\cdot 1+B\cdot 0+C\cdot 0=0 \quad [3]$$

Eliminating ($A : B : C$) from [1], [2] and [3], determinantally,

we get

$$\begin{vmatrix} x-1 & y+2 & z-4 \\ 2 & -2 & 1 \\ 1 & 0 & 0 \end{vmatrix}=0$$

$$\Rightarrow \quad 1(y+2)+2(z-4)=0$$

$$\Rightarrow \quad y+2z=6 \text{ Hence proved.}$$

39. Find the equation of the plane through the point $P(2, 3, -1)$ and perpendicular to the line OP, where O is the origin.

Solution: $O = (0, 0, 0)$

$P = (2, 3, -1)$

$\therefore$ DR's of $OP = (2-0) : (3-0) : (-1-0) = (2 : 3 : -1)$

Any plane perpendicular OP

$$= 2x+3y+(-1)\,z=k$$

P lies on it.

$$\therefore \quad 2\cdot 2+3\cdot(3)+(-1)(-1)=k=14$$

plane: $(2x+3y-z)=14$

40. Find the plane through the point (α, β, γ) and parallel parallel to

$$Ax+By+Cz+D=0$$

Solution: Any plane parallel to $Ax+By+Cz+D=0$

is $Ax+By+Cz+k=0$

(α, β, γ) lies on it).

$\therefore \quad A\alpha + B\beta + C\gamma + k = 0$

Subtracting we get

$$A(x - \alpha) + B(y - \beta) + C(z - \gamma) = 0$$

41. Find the plane through the point (3, 2, 5) and parallel to the plane

$$2x + 4y + 3z - 7 = 0.$$

Solution: Any plane parallel to $2x + 4y + 3z - 7 = 0$ has the same normal.

Hence D.R's of Normal = (2 : 4 : 3)

Any plane parallel to the above

$$= 2x + 4y + 3z + k = 0$$

(3, 2, 5) lies on it

$\therefore \quad 2 \cdot 3 + 4 \cdot 2 + 3 \cdot 5 + k = 0$

$\therefore \quad k = -29$

$\therefore$ the plane is : $2x + 4y + 3z - 29 = 0$

42. Find the equation of the plane which passes through the point (1, – 2, 1) and is $\perp^r$ to each of the planes

$$3x + y + z - 2 = 0 \text{ and}$$
$$x - 2y + z + 4 = 0.$$

Solution: Let $\quad Ax + By + Cz + D = 0 \quad$ [1]

be the equation of the plane. It passes through (1, – 2, 1). So, we get

$$A - 2B + C + D = 0 \quad [2]$$

Since [1] is perpendicular to the planes

$$3x + y + z - 2 = 0 \quad \text{and} \quad x - 2y + z + 4 = 0$$

we have $\quad 3A + B + C = 0 \quad$ [3]

$$A - 2B + C = 0 \quad [4]$$

$\therefore \quad \dfrac{A}{3} = \dfrac{B}{-2} = \dfrac{C}{-7} = k$ (say)

$\therefore \quad A = 3k; B = -2k; C = -7k.$

Substituting in [2], we get

$$D = 0$$

$\therefore$ Equation of the required plane is $3x - 2y - 7z = 0$.

43. Find the distance between the parallel planes

$$2x + 3y - z + 4 = 0 \quad \text{and} \quad 4x + 6y - 2z - 5 = 0$$

Solution: The given planes are

$$2x + 3y - z + 4 = 0 \quad [1]$$
$$4x + 6y - 2z - 5 = 0 \quad [2]$$

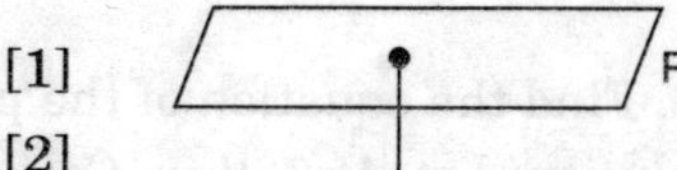

Fig. 1.16(c)

We shall find any one point on one plane say [1]

To this end put $x = 0$, $y = 0$ is [1], we get $z = 4$

$\therefore$ one point on the plane [1] is (0, 0, 4)

Distance between the parallel planes = perpendicular distance of (0, 0, 4) from plane [2]

$$= \left| \frac{4(0) + 6(0) - 2(4) - 5}{\sqrt{16 + 36 + 4}} \right| = \frac{13}{\sqrt{56}}.$$

44. If the foot of the perpendicular from the origin to the plane is (2, – 1, 2). Find the equation of the plane.

Solution: Let $Ax + By + Cz + D = 0$ [1]

be the equation of the required plane. Since D.R's of the normal to [1] is $(A : B : C)$, also OP is a normal to that plane

$\therefore$ the D.R's of normal to the plane are $(2 : -1 : 2)$

The plane becomes

$$2x - y + 2z + D = 0$$

(2, – 1, 2) lies on it $\therefore$ $4 + 1 + 4 + D = 0$

$\Rightarrow$ $D = -9$

$\therefore$ $2x - y + 2z - 9 = 0$

is the required plane.

45. Prove that the four points (– 6, 3, 2), (3, – 2, 4) (5, 7, 3) and (– 13, 17, – 1) lie in one plane. Find the equation of the plane containing them.

Solution: Let $A(-6, 3, 2)$

$B(3, -2, 4)$

$C(5, 7, 3)$

and $D(-13, 17, -1)$

The equation to the plane containing the points A, B and C is

$$\begin{vmatrix} x+6 & y-3 & z-2 \\ 3+6 & -2-3 & 4-2 \\ 5+6 & 7-3 & 3-2 \end{vmatrix} = 0 \quad \Rightarrow \quad \begin{vmatrix} x+6 & y-3 & z-2 \\ 9 & -5 & 2 \\ 11 & 4 & 1 \end{vmatrix} = 0$$

$\Rightarrow$ $(x + 6)(-5 - 8) - (4 - 3)(9 - 22) + (z - 2)(36 + 55) = 0$

$\Rightarrow$ $-13(x + 6) + 13(4 - 3) + 91(z - 2) = 0$

$\Rightarrow$ $-13x + 13y + 91z - 299 = 0$

$\Rightarrow$ $x - y - 7z + 23 = 0$

Put $x = -13$, $y = 17$, $z = -1$ in the *LHS* of this equation,

we get $-13 - 17 + 7 + 23 = -30 + 30$

$= 0 = \text{RHS}$

$\therefore$ $D(-13, 17, -1)$ lies on this plane

Thus, the 4 points lie on the same plane whose eqn is

$$x - y - 7z + 23 = 0$$

46. Find the equation of the plane through (2, 3, – 4) and (1, – 1, 3) and parallel to the x-axis

Solution: Let $Ax + By + Cz + D = 0$ be the plane [1]

If crosses through (2, 3, – 4) and (1, – 1, 3)

$\therefore$ $2A + 3B - 4C + D = 0$ [2]

and $A - B + 3C + D = 0$ [3]

From (2) – (3), $A + 4B - 7C = 0$ [4]

. P(2, 3, – 4)
P : plane
. Q(1, – 1, 3)

Since [1] is parallel to the x-axis whose D.C's are $(1:0:0)$

$\therefore \quad A \cdot 1 + B \cdot 0 + C \cdot 0 = 0 \quad \Rightarrow \quad A = 0$

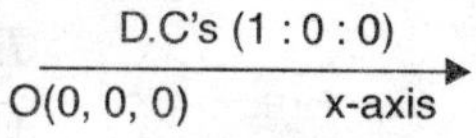

D.C's (1 : 0 : 0)

O(0, 0, 0) x-axis

Fig. 1.16(d)

Putting this value in [4],

we get, $4B = 7C$

$$B = \frac{7}{4}C$$

From [2], $2 \cdot 0 + \frac{21}{4}C - 4C + D = 0$

$$\Rightarrow \quad D = -\frac{5}{4}C$$

Putting this values in [1], we get

$$\frac{7}{4}Cy + Cz - \frac{5}{4}C = 0$$

$\Rightarrow \quad 7y + 4z - 5 = 0$ is the required plane

47. Find the equation of the plane through the points (1, – 2, 4) and (3, – 4, 5) and perpendicular to yz-plane

Solution: The equation of the given yz-plane is $x = 0$. [1]

Let the equation of the plane $\perp^r$ to [1] be

$$Ax + By + Cz + D = 0 \quad [2]$$

Then $A \cdot 1 + B \cdot 0 + C \cdot 0 = 0$

i.e. $A = 0$ [3]

$(\because aa_1 + bb_1 + cc_1 = 0)$ [3]

By data, [2] passes thro'

(1, –2, 4) and (3, – 4, 5)

$\therefore \quad A - 2B + 4C + D = 0$

i.e. $-2B + 4C + D = 0$ [4]

Using (3) $3A - 4B + 5C + D = 0$

i.e. $-4B + 5C + D = 0$ [5]

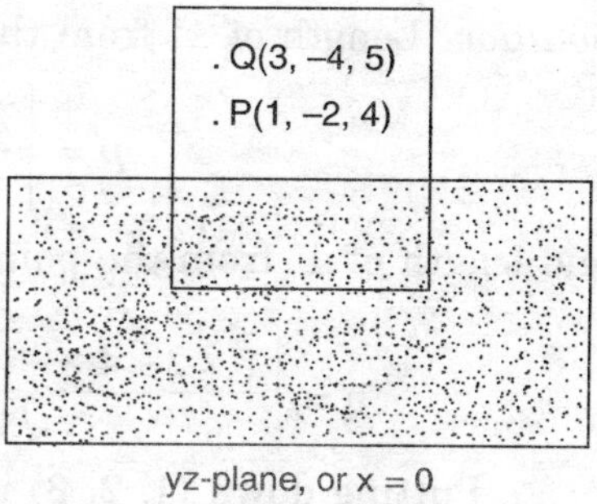

PLANE

. Q(3, –4, 5)

. P(1, –2, 4)

yz-plane, or x = 0

Fig. 1.16(e)

Using [3]

From [4] and [5], we get by Cross-multiplication

$$\frac{B}{4-5} = \frac{C}{-4+2} = \frac{D}{-10+10}$$

$$\Rightarrow \quad \frac{B}{-1} = \frac{C}{-2} = \frac{D}{6} = k \text{ (say)}$$

$\therefore \quad B = -k, C = -2k, D = +6k$

putting these values is [2],

$$-ky - 2kz + 6k = 0$$

i.e. $y + 2z - 6 = 0$, is the desired eqn of the plane.

48. A plane meets the coordinate axes at A, B, C such that the centroid of the triangle ΔABC is the point (α, β, γ). Show that the equation of the plane is $\frac{x}{\alpha} + \frac{y}{\beta} + \frac{z}{\gamma} = 3$.

Solution: Let A $(a, 0, 0)$
B $(0, b, 0)$ then the
C $(0, 0, c)$

equation of the plane is

$$\frac{x}{a} + \frac{y}{b} + \frac{z}{c} = 1 \qquad [1]$$

Since the centroid of the triangle is $\left(\frac{a}{3}, \frac{b}{3}, \frac{c}{3}\right)$ but it is given to be (α, β, γ)

We have $\frac{a}{3} = \alpha, \frac{b}{3} = \beta, \frac{c}{3} = \gamma.$

$\therefore$ $a = 3\alpha, b = 3\beta, c = 3\gamma$

and [1] becomes $\frac{x}{3\alpha} + \frac{y}{3\beta} + \frac{z}{3\gamma} = 1$

i.e. $\frac{x}{\alpha} + \frac{y}{\beta} + \frac{z}{\gamma} = 3$

49. Find the length of the perpendicular from the points (1, 2, 3) and (2, – 1, 2) from the plane $3x - 4y + 5z = 12$.

Are these points on the same side of the plane?

Solution: Length of $\perp^r$ from the point (1, 2, 3) to the plane:

$$p = \left|\frac{3 - 4\times 2 + 5\times 3 - 12}{\sqrt{9+16+25}}\right| = \frac{1-21}{\sqrt{50}} = \frac{\sqrt{2}}{5}$$

and length of $\perp^r$ from the point (2, –1, 2) to the plane:

$$p = \left|\frac{3\times 2 - 4\times -1 + 5\times 2 - 12}{\sqrt{50}}\right| = \frac{|8|}{\sqrt{50}} = \frac{4}{5}\sqrt{2}. \qquad \left(\because \quad \frac{8}{\sqrt{50}} = \frac{2\cdot 2\cdot 2}{\sqrt{2}\cdot(5)} = \frac{4\sqrt{2}}{5}\right)$$

Putting down (1, 2, 3) in the eqn of plane we find –ve value and putting (2, –1, 2) in the eqn of the plane we find +ve value. Thus, they are on opposite sides of the plane.

50. Find the distance between the parallel planes

$$2x + 3y - z + 4 = 0$$

and $$4x + 6y - 2z - 5 = 0$$

Solution: The given planes are:

$$2x + 3y - z + 4 = 0 \qquad [1]$$

$$4x + 6y - 2z - 5 = 0 \qquad [2]$$

We shall find any one point on one plane say [1]. To find the same we put $x = 0$, $y = 0$ in [1], we get $z = 4$

$\therefore$ (0, 0, 4) is a point on plane [1]

$\therefore$ Distance between the parallel plane

$= \perp^r$ distance of (0, 0, 4) from the plane [2]

$$= \frac{|4(0) + 6(0) - 2(4) - 5|}{\sqrt{16+36+4}} = \frac{13}{\sqrt{56}}$$

51. Show that the two points $P = (1, -1, 3)$ and $Q = (3, 3, 3)$ are equally distant from the plane $(5x + 2y - 7z + 9) = 0$ and are on opposite sides of the plane.

Solution: **Given plane:** $(5x + 2y - 7z + 9) = 0$ [1]

$P = (1, -1, 3)$

$Q = (3, 3, 3)$

Let PM and QN be the perpendiculars from P and Q on the plane [1].

$$PM = \frac{-(A\alpha + B\beta + C\gamma + D)}{\sqrt{A^2 + B^2 + C^2}} = \frac{-[5\cdot1 + 2(-1) + (-7)(3) + 9]}{\sqrt{5^2 + 2^2 + (-7)^2}} = \frac{9}{\sqrt{78}}$$

$$QN = \frac{-[5\cdot3 + 2\cdot4 + (-7)(3) + 9]}{\sqrt{5^2 + 2^2 + (-7)^2}} = \frac{-9}{\sqrt{78}}$$

Now, PM and QN are equal in magnitude, hence, the two points are equally distant from the plane.

Also PM and QN have opposite signs, and hence the two points lie on opposite sides of the plane.

EXERCISES

1. Find the equation of the plane
   (*i*) thro' the point $(1, -1, 2)$ and parallel to the xy-plane (*Ans.* $z - 2 = 0$)
   (*ii*) thro' the point $(2, -1, 6)$, $(1, -2, 4)$ and perpendicular to the plane $x - 2y - 2z + 9 = 0$
   (*Ans.* $2x + 4y - 3z + 18 = 0$)
2. Find the image of the line $\frac{x-1}{2} = \frac{y-2}{1} = \frac{z-3}{4}$ in the plane $2x + y + z = -2$.
   $\left(\textit{Ans. } \frac{x+5}{-4} = \frac{y+1}{-2} = \frac{z-0}{1}\right)$ (each $= k$)
3. Find the distance between the parallel planes $2x - 2y + 2 + 3 = 0$ and $4x - 4y + 2z - 5 = 0$.
   $\left(\textit{Ans. } \text{perpendicular to distance} = \frac{11}{6}\right)$
4. Show that the form points
   $(0, -1, 0)$, $(2, 1, -1)$, $(1, 1, 1)$
   and $(3, 3, 0)$ are coplanar. Find the equation of the plane thro' them. (*Ans.* $4x - 3y + 2z = 3$)
5. Find the distances of the points $(2, 3, -5)$, $(3, 4, 7)$ from the plane $x + 2y - 2z = 9$, are the points on the same side of the plane? (*Ans.* $3, -4$) Since the signs are opposite, the two points are on opposite sides of the plane.
6. If the plane $ax + by + cz + d = 0$ is parallel to the line with D.R's $(l : m : n)$. Show that
   $al + bm + cn = 0$.
7. Find the equation of the plane thro' the points $(2, 2, 1)$ $(1, -2, 3)$ and parallel to x–axis
   (*Ans.* $y + 2z - 4 = 0$)

Any Plane Through the Line of Intersection two Planes: (Coaxal Planes)

If $P_1 = 0$ and $P_2 = 0$ are any two given planes then the equation to any plane passing through the intersection of the planes

$P_1 = 0$ and $P_2 = 0$ is of the form:

$$P_1 + k(P_2) = 0$$

$\Rightarrow$ one plane + k (other plane) = 0.

If $P_1: A_1x + B_1y + C_1z + D_1 = 0$

and $P_2: A_2x + B_2y + C_2z + D_2 = 0$ are the equations of two planes, then

$$A_1x + B_1y + C_1z + D_1 + k(A_2x + B_2y + C_2z + D_2) = 0 \quad (1)$$

is the plane passes thro', the intersection of the given planes.

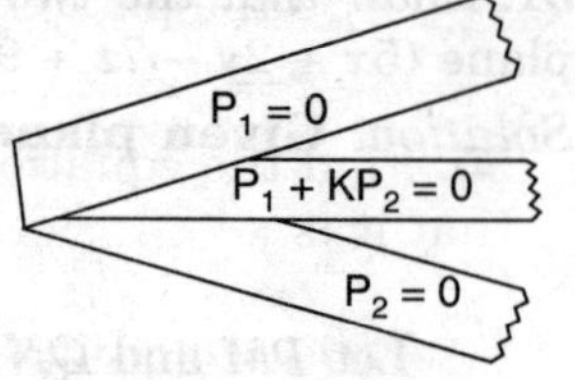

Fig. 1.16(f)

WORKED EXAMPLES

52. Find the equation of the plane passing through the line of intersection of

$$P_1: 3x + 4y - 2z + 6 = 0 \quad [1]$$

$$P_2: x + 2y + z - 2 = 0 \quad [2]$$

and passing thro' the origin.

Solution: Any plane through the point of intersection of [1] and [2] is of the form

$$(P_1 + kP_2 = 0)$$

i.e. $(3x + 4y - 2z + 6) + K(x + 2y + z - 2) = 0$

(0, 0, 0) lies on it.

$\therefore$ $(0 + 0 + 0 + 6) + k(0 + 0 + 0 - 2) = 0$

$\therefore$ $6 - 2k = 0$

$\therefore$ $K = 3.$

Hence the plane is:

$$(3x + 4y - 2z + 6) + 3(x + 2y + z - 2) = 0$$

$\Rightarrow$ $6x + 10y + z = 0$

53. Find the plane passing through the line of intersection of $3x + 4y - 2z + 6 = 0$ and $x + 2y + z - 2 = 0$ and passing through (3, – 4, 2).

Solution: Any plane through the intersection of [1] and [2] is of the form

$$P_1 + kP_2 = 0$$

$$(3x + 4y - 2z + 6) + k(x + 2y + z - 2) = 0$$

(3, – 4, 2) lies on it

$\therefore$ $9 - 16 - 4 + 6 + k(3 - 8 + 2 - 2) = 0$

so $k = -1.$

$\therefore$ The plane: $(2x + 2y - 3z + 8) = 0.$

54. Find the plane through the same line given Eqn (2) and parallel to the z-axis.

$$(3x + 4y - 2z + 6) + k(x + 2y + z - 2) = 0$$

$$x(3 + k) + y(4 + 2k) + z(-2 + k) + (6 - 2k) = 0$$

This is parallel to Z-axis

$\therefore$ coeff of z is zero

or $(-2 + k) = 0$

So the plane: $x(3 + 2) + y(4 + 4) + z(-2 + 2) + (6 - 4) = 0$

$\Rightarrow$ $5x + 8y + 2 = 0$

55. Find the equation to the plane containing the line $2x - 5y + 2z = 6$, $2x + 3y - z = 5$ such that it is

(*i*) parallel to the line $\dfrac{x}{1} = \dfrac{y}{-6} = \dfrac{z}{7}$.

(*ii*) perpendicular to the plane $x + y - z = 7$.

Solution: Any plane through the intersection of the given planes is

$$2x - 5y + 2z - 6 + k(2x + 3y - z - 5) = 0$$

$\Rightarrow$ $(2 + 2k)x + (3k - 5)y + (2 - k)z - 6 - 5k = 0$ [1]

[1] is parallel to the line

$$\frac{x}{1} = \frac{y}{-6} = \frac{z}{7}$$

$\Rightarrow$ $1(2 + 2k) - 6(3x - 5) + 7(2 - k) = 0$

$\Rightarrow$ $k = 2$

Putting down for k in [1], we get $6x + y - 16 = 0$.

Again [1] is perpendicular to plane $x + y - z = 7$

$\Rightarrow$ $1(2 + 2k) + 1(3k - 5) - (2 - k) = 0$

$\Rightarrow$ $k = 1.$

Putting down for k in eqn [1], we get

$$4x - 2y + z - 11 = 0$$

56. Find the equation of the plane thro' the line of intersection of the planes

$$3x + 2y + 4z + 6 = 0$$

and $(2x - y + z + 2) = 0,$

so as to be parallel to the y-axis.

Solution: $\left.\begin{aligned} P_1 &= (3x + 2y + 4z + 6) = 0 \\ P_2 &= (2x - y + z + 2) = 0 \end{aligned}\right\}$ [1]

Any plane thro' the line of intersection of the two planes above, is of the form $P_1 + kP_2 = 0$.

i.e. $(3x + 2y + 4z + 6) + k(2x - y + z + 2) = 0$ [2]

$\Rightarrow$ $x(3 + 2k) + y(2 - k) + z(4 + k) + (6 + 2k) = 0$ [3]

which is parallel to the y-axis (Given)

Hence coeff of y should be zero

$\therefore$ *i.e.* $(2 - k) = 0$ $\Rightarrow$ $k = 2$ [4]

Substitute in [3], we get the equation of the required plane:

$$x(3 + 2 \cdot 2) + y(2 - 2) + z(4 + 2) + (6 + 2 \cdot 2) = 0$$

$\Rightarrow$ $(7x + 6z + 10) = 0$

57. Find the plane thro' the line of intersection of

$$(3x + 2y + 4z - 5) = 0,$$

and $(2x - y + 2z + 3) = 0,$

so as to be perpendicular to the latter plane.

Solution: $\left.\begin{array}{l} P_1 = (3x + 2y + 4z - 5) = 0 \\ P_2 = (2x - y + 2z + 3) = 0 \end{array}\right\}$ [1]

Any plane thro' the line of intersection of the two planes above, is of the form

$$(P_1 + kP_2) = 0$$

$\Rightarrow$ $(3x + 2y + 4z - 5) + k\,(2x - y + 2z + 3) = 0$ [2]

$x\,(3 + 2k) + y(2 - k) + z\,(4 + 2k) + (-5 + 3k) = 0$ [3]

This is required to be perpendicular to the latter plane of the two $P_1 = 0$ and $P_2 = 0$, namely $P_2 = 0$

$\Rightarrow$ $2x - y + 2z + 3 = 0$ [4]

$\therefore$ $(3 + 2k)\,(2) + (2 - k)\,(-1) + (4 + 2k)\,(2)$ [5]

$\Rightarrow$ $k(4 + 1 + 4) + (6 - 2 + 8) = 0$

$\Rightarrow$ $k = -4/3$ [6]

Substituting for k, in [2], we get

$$(3x + 2y + 4z - 5) - (4/3)\,(2x - y + 2z + 3) = 0$$

$\Rightarrow$ $x + 10y + 4z - 27 = 0$

58. Find, the equation of the plane passing thro' the intersection of the planes $x + 3y - z = 4$ and $2x + 2y + 2z = 1$ and perpendicular to plane.

$$x + y - 4z = 0$$

Solution: The eqn of plane:

$$x + 3y - z - 4 + k\,(2x + 2y + 2z - 1) = 0$$

$\Rightarrow$ $(1 + 2k)\,x + (3 + 2k)y + (-1 + 2k)\,z - 4 = 0$

Since it is perpendicular to $x + y - 4z = 0$,

we have

$$(1 + 2k)\,1 + (3 + 2k)\,1 + (-1 + 2k)\,(-4) = 0$$

$\Rightarrow$ $k = 2$

$\therefore$ The required plane is

$$x + 3y - z - 4 + 2\,(2x + 2y + 2z - 1) = 0$$

$\Rightarrow$ $5x + 7y + 3z - 6 = 0.$

EXERCISES

1. Obtain the equation of the line of intersection of the planes

$$4x + 4y - 5z = 12 \text{ and } 8x + 12y - 13z = 32$$

in the symmetric form. $\left(Ans: \dfrac{x-1}{2} = \dfrac{y-2}{3} = \dfrac{z}{4}\right)$

2. Find the equation of the line passing thro, the point (1, 2, 3) and parallel to the line of intersection of the planes

$$x - y + 2z = 5 \text{ and } 3x + y + z = 6.$$

3. Show that the straight lines

$$3x + 2y + z - 5 = 0 = x + y - 2z - 3$$

and $$8x - 4y - 4z = 0 = 7x + 10y - 8z$$

are at right angles.

4. Find the equation of the plane thro' the intersection of the planes

$$3x - y + 2z - 4 = 0$$

and $$x + y + z - 2 = 0$$

and passing thro' (2, 2, 1). (*Ans:* $7x - 5y + 4z - 8 = 0$)

5. Find the equation of the plane passing thro' the intersection of the planes

$$x + 3y - z = 4$$

and $$2x + 2y + 2z = 1$$

and perpendicular to the plane $x + y - 4z = 0$ (*Ans:* $5x + 7y + 3z - 6 = 0$)

1.5 STRAIGHT LINES

A straight line may be regarded as the common line of intersection of the planes

$$A_1x + B_1y + C_1z + D_1 = 0$$

and $$A_2x + B_2y + C_2Z + D_2 = 0.$$

Equations of Straight Lines

1. *Symmetrical form of equation of a line*

To find the equation of the straight line passing through the point (α, β, γ) and having direction cosines $(l : m : n)$

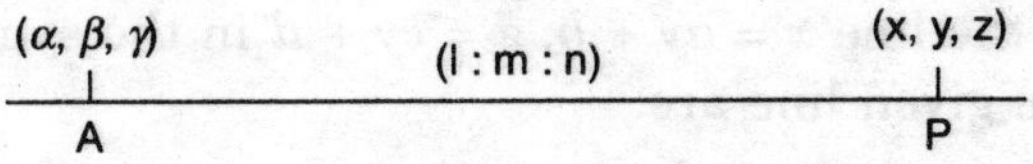

(α, β, γ) (l : m : n) (x, y, z)

A P

Fig. 1.16(g)

Let $A = (\alpha, \beta, \gamma)$ be the given point. Consider any point $P = (x, y, z)$ on the straight line.

Then D.R's $AP = (x - \alpha) : (y - \beta) : (z - \gamma)$

But it is given that the D.C's are $(l : m : n)$.

Evidently these two are equivalent.

$$\Rightarrow \quad (x - \alpha) : (y - \beta) : (z - \gamma) = l : m : n$$

$$\Rightarrow \quad \frac{x - \alpha}{l} = \frac{y - \beta}{m} = \frac{z - \gamma}{n}.$$

The above equation is called the 'Symmetric Form' of the equations to a line.

2. *Parametric form*

We know that $$\frac{x - \alpha}{l} = \frac{y - \beta}{m} = \frac{z - \gamma}{n} \qquad [1]$$

Put each ratio equal to r,

$$\therefore \quad \frac{x - \alpha}{l} = \frac{y - \beta}{m} = \frac{z - \gamma}{n} = r \qquad [2]$$

i.e. $$(x - \alpha) = lr, (y - \beta) = mr, z - \gamma = nr$$

or, $$[x = (\alpha + lr),$$
$$y = (\beta + mr)$$
$$z = (\gamma + nr)] \qquad [3]$$

Any point on the straight line can be written in the form

$$[(\alpha + lr, \beta + mr, \gamma + nr)].$$

For different values of r, we get different points on the line.

3. *Two point form of a straight line:* $\dfrac{x-x_1}{x_2-x_1}=\dfrac{y-y_1}{y_2-y_1}=\dfrac{z-z_1}{z_2-z_1}$.

If a line passes through the points (x_1, y_1, z_1) and (x_2, y_2, z_2), its direction ratios are given by

$$[(x_2-x_1):(y_2-y_1):(z_2-z_1)]$$

Hence if we choose the points A and B and D.R's as a, b, c of the line then, the 'Symmetric form' of the equation of the line:

$$\frac{x-x_1}{a}=\frac{y-y_1}{b}=\frac{z-z_1}{c}\text{; becomes }\frac{x-x_1}{x_2-x_1}=\frac{y-y_1}{y_2-y_1}=\frac{z-z_1}{z_2-z_1}.$$

(a : b : c)

A(x₁, y₁, z₁) P(x₂, y₂, z₂)

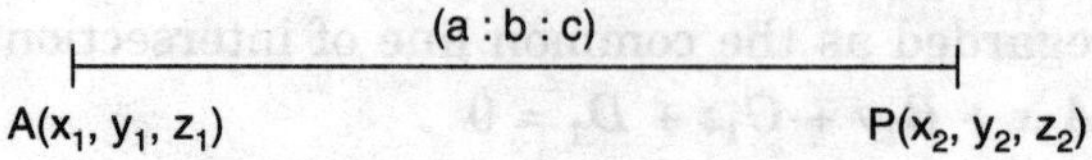

Fig. 1.16(h)

Note: In particular, equation of a line passing through the origin and having the direction cosines $(l: m : n)$ are given by

$$\frac{x-0}{l}=\frac{y-0}{m}=\frac{z-0}{n}\quad\text{or}\quad\frac{x}{l}=\frac{y}{m}=\frac{z}{n}.$$

59. Find the equations of the line $x = ay + b$, $z = cy + d$ in the symmetrical form.

Solution: Equations of the given line are:

$$x = ay + b \qquad [1]$$

$$z = cy + d \qquad [2]$$

From [1], we get

$$ay = x - b$$

$$\therefore \quad \frac{y}{1}=\frac{x-b}{a} \qquad [3]$$

From [2], we get

$$cy = z - d$$

$$\therefore \quad \frac{y}{1}=\frac{z-d}{c} \qquad [4]$$

$\therefore$ From [3] and [4], we get

$$\frac{x-b}{a}=\frac{y}{1}=\frac{z-d}{c}$$

is the required symmetrical form of the given equations.

60. Find the equations of the lines $x + 5y - z - 7 = 0$;

$$2x - 5y + 3z + 1 = 0$$

in symmetrical form.

Solution: **Given:**

$$x + 5y - z - 7 = 0 \qquad [1]$$

$$2x - 5y + 3z + 1 = 0 \qquad [2]$$

If $(a : b : c)$ be the DR's of the required line.

Since the line is perpendicular to the normal of [1] and [2] we have

$$a + 5b - c = 0 \qquad [3]$$

$$2a - 5b + 3c = 0 \qquad [4]$$

$$\therefore \quad \frac{a}{10} = \frac{b}{-5} = \frac{c}{-15}$$

$\therefore$ D.R's are $(2 : -1 : -3)$

Let us find one point on the line. This can be done by taking $z = 0$ i.e. a the point $(x, y, 0)$ where (x, y).

Satisfy: $x + 5y = 7$ [5]

$2x - 5y = -1$ [6]

Solving [5] and [6], we get $x = 2, y = 1$

$\therefore$ A point on the line is (2, 1, 0).

$\therefore$ Equations of the line are

$$\frac{x-2}{2} = \frac{y-1}{-1} = \frac{z}{-3}$$

61. Put the equations

$$4x + 4y - 5z = 12,$$
$$8x + 12y - 13z = 32$$

of a straight line in symmetrical form

Solution: If $(l : m : n)$ be the direction cosines of the line, then

$$4l + 4m - 5n = 0$$
$$8l + 12m - 13n = 0$$

$$\therefore \quad \frac{l}{-52+60} = \frac{m}{-40+52} = \frac{n}{48-32}$$

i.e. $$\frac{l}{8} = \frac{m}{12} = \frac{n}{16}$$

$$\Rightarrow \quad \frac{l}{2} = \frac{m}{3} = \frac{n}{4}.$$

Thus D.R's of the line are proportional to $(2 : 3 : 4)$.

Next, put $z = 0$ in the given

eqns. $4x + 4y - 12 = 0; \quad x + y - 3 = 0$

$8x + 12y - 32 = 0; \quad 2x + 3y - 8 = 0$

$$\therefore \quad \frac{x}{-8+9} = \frac{y}{-6+8} = \frac{1}{3-2}$$

$$\Rightarrow \quad x = 1, y = 2.$$

Hence, required eqns of the line in symmetrical form are $\frac{x-1}{2} = \frac{y-2}{3} = \frac{z}{4}$

62. Find the equation of the line joining the points (3, 0, 2) and (1, –2, 3).

The eqn. of the line thro' the points (x_1, y_1, z_1) and (x_2, y_2, z_2) is given by

$$\frac{x-x_1}{x_1-x_2} = \frac{y-y_1}{y_1-y_2} = \frac{z-z_1}{z_1-z_2}$$

Hence the required eqn is $\frac{x-3}{3-1} = \frac{y-0}{0+2} = \frac{z-2}{z-3}$

i.e. $$\frac{x-3}{2} = \frac{y}{2} = \frac{z-2}{-1}.$$

63. Find the line thro' (1, 2, 3) and $\perp^r$ to both the lines

$$\frac{x-3}{2}=\frac{y-4}{3}=\frac{z-6}{4} \quad \text{and} \quad \frac{x}{4}=\frac{y}{5}=\frac{z}{7} \qquad \text{point: } (1, 2, 3)$$

Let the D.C's of the required line $(l : m : n)$

It is perpendicular to both the lines

$$\left.\begin{aligned} 2l + 3m + 4n &= 0 \\ 4l + 5m + 7n &= 0 \end{aligned}\right\}$$

$\therefore$ $$(l : m : n) = 1 : 2 : 2$$

$\therefore$ Straight lines

$\therefore$ $$\frac{x-1}{1}=\frac{y-2}{2}=\frac{z-3}{-2}$$

EXERCISES

1. While down the equations of the line thro' the point (2, – 3, – 7) and having D.R.'s (3 : – 4 : 5).

$$\left(Ans: \frac{x-2}{3}=\frac{y+3}{-4}=\frac{z+7}{5}\right)$$

2. Find the equations of a line through (1, 4, – 2) and parallel to the planes

$$6x + 2y + 2z + 3 = 0$$

and $$x + 2y - 6z + 4 = 0. \qquad \left(Ans: \frac{x-1}{-8}=\frac{y-4}{19}=\frac{z+2}{5}\right)$$

3. Obtain the equations of the line of intersection of the planes

$$x + y - 2z = 8 \quad \text{and} \quad 3x - y + 4z = 12$$

in the symmetric form. $\left(Ans: \frac{x-5}{1}=\frac{y-3}{-5}=\frac{z}{-2}\right)$

4. Write the equations

$$x + 2y + 4z = 0 = 2x + 4y - z$$

of a line in the symmetric form $\left(Ans: \frac{x}{2}=\frac{y}{-1}=\frac{z}{0}\right)$

5. Prove that the points (3, 2, 4), (4, 5, 2) and (5, 8, 0) are collinear.

Find the equations of the line thro' them. $\left(Ans: \frac{x-3}{1}=\frac{y-2}{3}=\frac{z-4}{2}\right)$

1.6 ANGLE BETWEEN PLANES/STRAIGHT LINES

The angle between two planes, is equal to the angle between their normals.

Let the equations to the two planes be:

$$Ax + By + Cz + D = 0 \qquad [1]$$

and $$A'x + B'y + C'z + D' = 0 \qquad [2]$$

The D.R's of the normal to the first plane is

$(A : B : C)$ and the D.R's of the normal to the 2nd plane is $(A' : B' : C')$. Hence, if θ is the angle between the two planes, then it is the same as the angle between $(A : B : C)$ and $(A' : B' : C')$

$\therefore$ $$\cos\theta = \frac{(AA' + BB' + CC'}{\sqrt{(\Sigma A^2)}\sqrt{(\Sigma B^2)}}$$

(A) The Plane and the Straight Line

The line $\frac{x-\alpha}{l}=\frac{y-\beta}{m}=\frac{z-\gamma}{n}$ will lie in the plane

$$Ax + By + Cz + D = 0$$

if $$A\alpha + B\beta + C\gamma + D = 0$$

and $$Al + Bm + Cn = 0$$

$$\therefore \quad A(x-\alpha) + B(y-\beta) + C(z-\gamma) = 0$$

is the plane through the line if $Al + Bm + Cn = 0$

(B) Plane Through the Lines

We have, $$\frac{x-\alpha}{l_1}=\frac{y-\beta}{m_1}=\frac{z-\gamma}{n_1}$$

and $$\frac{x-\alpha^1}{l_2}=\frac{y-\beta^1}{m_2}=\frac{z-\gamma^1}{n_2}$$

is $$\begin{vmatrix} x-\alpha & y-\beta & z-\gamma \\ l_1 & m_1 & n_1 \\ l_2 & m_2 & n_2 \end{vmatrix} = 0$$

(C) Angle Between the Line and the Plane

To find the expression for the angle between:

line: $$\frac{x-\alpha}{l}=\frac{y-\beta}{m}=\frac{z-\gamma}{n} \quad [1]$$

and plane: $Ax + By + Cz + D = 0$ [2]

The actual D.C's of the normal plane $Ax + By + Cz + D = 0$ are proportional to A, B, C.

The actual D.C's of the line are proportional to l, m, n

The angle between the normal to the plane and the line is given by

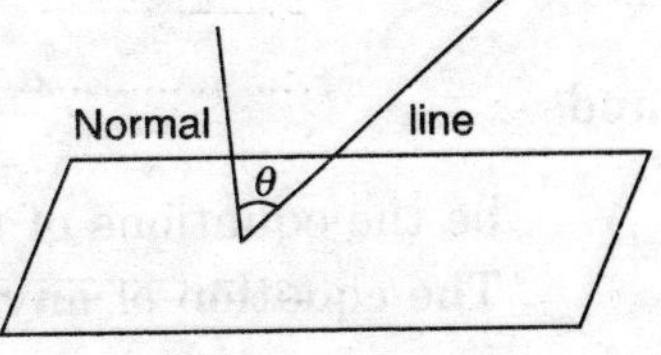

Fig. 1.17

$$\cos(90° - \theta) = \frac{Al + Bm + Cn}{\sqrt{A^2+B^2+C^2}\sqrt{l^2+m^2+n^2}} \quad [3]$$

as the angle between the line and the plane is $(90° - \theta)$

$$\therefore \quad \sin\theta = \frac{Al + Bm + Cn}{\sqrt{A^2+B^2+C^2}\sqrt{l^2+m^2+n^2}}$$

$$\Rightarrow \quad \theta = \sin^{-1}\left\{\frac{Al + Bm + Cn}{\sqrt{(\Sigma A^2)}\sqrt{(\Sigma l^2)}}\right\}$$

1.7 COPLANAR LINES (TWO INTERSECTING STRAIGHT LINES)

Two straight lines may be coplanar (or intersect) *i.e.* they lie in the same plane.

Condition for the two lines to intersect (or to be coplanar)

Let $$\frac{x-\alpha}{l}=\frac{y-\beta}{m}=\frac{z-\gamma}{n} \qquad [1]$$

and $$\frac{x-\alpha'}{l'}=\frac{y-\beta'}{m'}=\frac{z-\gamma'}{n'} \qquad [2]$$

Since the lines [1] and [2] intersect, if some point of [1] lies on [2].

Any point on [1] can be

written: $$(\alpha+lr,\ \beta+mr,\ \gamma+nr) \qquad [3]$$

Substitute and test in [2],

$$\frac{\alpha+lr-\alpha'}{l'}=\frac{\beta+mr-\beta'}{m'}=\frac{\gamma+nr-\gamma'}{n'}=k \qquad [4]$$

$$\Rightarrow \quad \alpha+lr-\alpha'=l'k$$

$$\Rightarrow \quad \left.\begin{aligned}(\alpha-\alpha')+lr-l'k&=0\\(\beta-\beta')+mr-m'k&=0\\(\gamma-\gamma')+nr-n'k&=0\end{aligned}\right\} \qquad [5]$$

|||ly, we get

Eliminating r and k, we get

$$\begin{vmatrix}\alpha-\alpha' & l & l'\\ \beta-\beta' & m & m'\\ \gamma-\gamma' & n & n'\end{vmatrix}=0 \quad \text{or} \quad \begin{vmatrix}\alpha-\alpha' & \beta-\beta' & \gamma-\gamma'\\ l & m & n\\ l' & m' & n'\end{vmatrix}=0 \qquad [6]$$

Note: to get the point of intersection, substitute the common value of r satisfying [4], in [3].

To find the plane containing them

Let $$\frac{x-\alpha}{l}=\frac{y-\beta}{m}=\frac{z-\gamma}{n} \qquad [1]$$

and $$\frac{a-\alpha'}{l'}=\frac{y-\beta'}{m'}=\frac{z-\gamma'}{n'} \qquad [2]$$

be the equations of two lines.

The equation of any plane containing the line [1]

is $$A(x-\alpha)+B(y-\beta)+C(z-\gamma)=0 \qquad [3]$$

where $$Al+Bm+Cn=0 \qquad [4]$$

The plane [3] contains the lines [2] if it is parallel to the line [2] and a point of [2] namely $(\alpha', \beta', \gamma')$ lies on the plane [3].

$$\therefore \quad Al'+bm'+cn'=0 \qquad [5]$$

and $$A(\alpha'-\alpha)+B(\beta'-\beta)+C(\gamma'-\gamma)=0 \qquad [6]$$

Eliminating

$A : B : C$ from [4], [5], [6], we get

$$\begin{vmatrix}\alpha'-\alpha & \beta'-\beta & \gamma'-\gamma\\ l & m & n\\ l' & m' & n'\end{vmatrix}=0 \qquad [7]$$

Which is the condition for the coplanarity of the lines [1] and [2].

Also Eliminate $A : B : C$ from [3], [4], [5], to get

$$\begin{vmatrix} x-\alpha & y-\beta & z-\gamma \\ l & m & n \\ l' & m' & n' \end{vmatrix} = 0 \text{ is the reqd plane containing the lines [1] and [2].}$$

WORKED EXAMPLES

64. Find the angle between the planes $2x - y + z = 6$ and $x + y + 2z = 3$.

Solution: Here $A = 2, B = -1, C = 1, D = -6$

$$A' = 1, B' = 1, C' = 2, D' = -3$$

If θ be the angle between the planes, then

$$\text{Cos } \theta = \frac{2(1) + (-1)(1) + 1(2)}{\sqrt{2^2 + (-1)^2 + 1^2}\sqrt{1^2 + 1^2 + 2^2}} = \frac{3}{\sqrt{6}\sqrt{6}} = \frac{1}{2}$$

$\therefore$ $\theta = 60°$

65. Find the equation of the plane through $(0, 1, -2)$ parallel to the plane $2x - 3y + 4z = 0$.

Solution: Any plane parallel to the given plane is

$$2x - 3y + 4z + \lambda = 0 \qquad [1]$$

Since [1] passes through $(0, 1, -2)$

$\therefore$ $2(0) - 3(1) + 4(-2) + \lambda = 0$

$\therefore$ $\lambda = 11$

required plane is $2x - 3y + 4z + 11 = 0$

66. Find the equation of the plane through $(-1, 1, 1)$ and $(1, -1, 1)$ perpendicular to the plane

$$x + 2y + 2z = 5.$$

Solution: Let the equation of the plane be $Ax + By + Cz + D = 0$ [1]

Since [1] is $\perp^r$ to the plane

$$x + 2y + 2z = 5,$$

$$A + 2B + 2C = 0 \qquad [2]$$

Since (1) passes through $(-1, 1, 1)$ and $(1, -1, 1)$,

$$-A + B + C + D = 0 \qquad [3]$$

$$A - B + C + D = 0 \qquad [4]$$

[3] – [4] gives $-2A + 2B = 0 \Rightarrow A = B$

From [2], we get $3B = -2C$

$\therefore$ [1] becomes $B = \dfrac{-2C}{3}$

$\therefore$ $A = \dfrac{-2C}{3}$

$\therefore$ $D = \dfrac{-2C}{3} + \dfrac{2C}{3} - C = -C$

∴ [1] becomes

$$-\frac{2C}{3}x-\frac{2C}{3}y+Cz-C=0$$

i.e. $$2x+2y-3z+3=0$$

67. Find the angle between the line $\frac{x+1}{-3}=\frac{y-3}{2}=\frac{z+2}{1}$ and the plane

$$x+2y+3z+4=0$$

Solution: Here $A=1, B=2, C=3, D=4$

∴ $l=-3, m=2, n=1$

$\alpha=-1, \beta=3, \gamma=2$

$$\theta=\sin^{-1}\left\{\frac{Al+\beta m+Cn}{\sqrt{A^2+B^2+C^2}\sqrt{l^2+m^2+n^2}}\right\}$$

$$=\sin^{-1}\left\{\frac{1(-3)+2(2)+(3)(1)}{\sqrt{1^2+2^2+3^2}\sqrt{(-3)^2+2^2+1^2}}\right\}$$

$$=\sin^{-1}\left[\frac{4}{\sqrt{14}\sqrt{14}}\right]=\sin^{-1}\left(\frac{2}{7}\right)$$

68. Show that the lines

$$\frac{1}{2}(x+3)=\frac{1}{3}(y+5)=-\frac{1}{3}(z-7)$$

and $$\frac{1}{4}(n+1)=\frac{1}{5}(y+1)=-(z+1)$$

are coplanar. Find the equation of the plane containing them.

Solution: The given lines are:

$$\frac{x+3}{2}=\frac{y+5}{3}=\frac{z-7}{-3} \qquad [1]$$

and $$\frac{x+1}{4}=\frac{y+1}{5}=\frac{z+1}{-1} \qquad [2]$$

The equation of the plane which contain the lines [1] and is parallel to [2] is

$$\begin{vmatrix} x+3 & y+5 & z-7 \\ 2 & 3 & -3 \\ 4 & 5 & -1 \end{vmatrix}=0$$

$$=(x+3)(-3+15)-(y+5)(-2+12)+(z-7)(10-12)=0$$

$$12x+36-10y-50-2z+14=0$$

$$6x-5y-z=0 \qquad [3]$$

The plane [3] passes thro' (–1, –1, –1) a point on the line [2], if

$-6+5+1=0$ or, $0=0$ which is true.

Hence the two lines are coplanar and the equations of the plane containing the given two lines is $6x-5y-z=0$.

69. Find the equation of the plane which contains the line $\frac{x-1}{2}=\frac{y+1}{-1}=\frac{z-3}{4}$ and is $\perp^r$ to the plane $x + 2y + z = 12$.

Solution: **Given:**
$$\frac{x-1}{2}=\frac{y+1}{-1}=\frac{z-3}{4} \quad [1]$$

Any plane through the given line is of the form
$$A(x-1)+B(y+1)+C(z-3)=0$$
where
$$2A - B + 4C = 0$$

Again, by data [1] must be $\perp^r$ to the given plane
$$x + 2y + z = 12$$
$$\therefore \quad A \cdot 1 + B \cdot 2 + C \cdot 1 = 0$$
(using $a_1a_2 + b_1b_2 + c_1c_2 = 0$)

or,
$$A + 2B + C = 0 \quad [3]$$

Solving [2] and [3] by the method of cross multiplication, we have
$$\frac{A}{-1-8}=\frac{B}{4-2}=\frac{C}{4+1} \quad [1]$$
$$\Rightarrow \quad \frac{A}{-9}=\frac{B}{2}=\frac{C}{5} \quad [2]$$

Putting these values in [1], we get
$$-9(x-1)+2(y+1)+3(2-3)=0$$
$$\Rightarrow \quad 9x - 2y - 5z + 4 = 0$$

This is the eqn of the required plane.

70. Find the equation of the plane through the points (1, 0, –1), (3, 2, 2) and parallel to the line $x-1=\frac{1-y}{2}=\frac{z-2}{3}$.

Solution: Equation of the plane through (1, 0, –1) is
$$A(x-1)+B(y-0)+C(z+1)=0 \quad [1]$$

Since this passes through (3, 2, 2),

we have
$$2A + 2B + 3C = 0 \quad [2]$$

The plane [1] is parallel to the line $\frac{x-1}{1}=\frac{y-1}{-2}=\frac{z-2}{+3}$. Normal to this plane is $\perp^r$ to this line.

Hence
$$A - 2B + 3C = 0 \quad [3]$$

Solving [2] and [3], we get
$$\frac{A}{4}=\frac{B}{-1}=\frac{C}{-2}$$

From [1] required eqn of the plane is,
$$4(x-1)-1(y-0)-2(z+1)=0$$
$$\Rightarrow \quad 4x - y - 2z - 6 = 0$$

71. Prove that the lines $\frac{x}{1}=\frac{y-2}{2}=\frac{z-3}{3}$ and $\frac{x-2}{2}=\frac{y-6}{3}=\frac{z-3}{4}$ are coplanar and find the equation of the plane in which they lie and the point of intersection.

Solution: Condition of coplanarity is

$$\begin{vmatrix} \alpha'-\alpha & \beta'-\beta & \gamma'-\gamma \\ l & m & n \\ l' & m' & n' \end{vmatrix} = 0$$

By data, the lines are

$$\frac{x-0}{1}=\frac{y-2}{2}=\frac{z-3}{3} \quad \text{and} \quad \frac{x-2}{2}=\frac{y-6}{3}=\frac{z-3}{4}$$

$$\therefore \quad (\alpha, \beta, \gamma) = (0, 2, 3)$$

$$(\alpha', \beta', \gamma') = (2, 6, 3)$$

$$(l, m, n) = (1, 2, 3)$$

$$(l', m', n') = (2, 3, 4)$$

Substituting, we get

$$\begin{vmatrix} -2 & -4 & 0 \\ 1 & 2 & 3 \\ 2 & 3 & 4 \end{vmatrix}$$

$$= -2(8-9) - (-4)(4-6) \neq 0$$

$$= -2(-1) + 4(-2) = 2 - 8$$

$$= -6 \neq 0$$

$\Rightarrow$ Condition of coplanarity is not satisfied.

$\therefore$ the lines are not coplanar and hence not intersecting also.

72. Find the coordinates of a point of intersection of the line

$$\frac{x+2}{3}=\frac{y-1}{-2}=\frac{z-4}{1}$$

with the plane $\quad x + 5y + 4z - 7 = 0$

Solution: Let $\quad \frac{x+2}{3}=\frac{y-1}{-2}=\frac{z-y}{1}=r$

$\Rightarrow \quad (3r-2, -2r+1, r+4) = P$

a parametric point on the line.

then '*P*' lies in the given plane $x + 5y + 4z - 7 = 0$

$\Rightarrow \quad (3r-2) + 5(-2r+1) + 4(r+9) - 4 = 0$

$\Rightarrow \quad -3r + 12 = 0$

$\Rightarrow \quad r = 4$

$\therefore$ $P = (10, -7, 8)$ are the coordinates of a pt. of intersection.

73. Find the condition that the three straight lines given below are coplanar: $\frac{x}{\alpha}=\frac{y}{\beta}=\frac{z}{\gamma}$;

$$\frac{x}{a\alpha}=\frac{y}{b\beta}=\frac{z}{c\gamma};\ \frac{x}{l}=\frac{y}{m}=\frac{z}{n}.$$

Solution: Clearly, origin (0, 0, 0) is a point on all the 3 straight lines. If the 3 are to be coplanar, they have a common normal.

Let its D.C's be $(A : B : C)$

$$\therefore \quad A\alpha + B\beta + C\gamma = 0$$

$$Aa\alpha + Bb\beta + Cc\gamma = 0$$

$$Al + Bm + Cn = 0$$

or
$$\begin{vmatrix} \alpha & \beta & \gamma \\ a\alpha & b\beta & c\gamma \\ l & m & n \end{vmatrix} = 0$$

$$\Rightarrow \quad a\alpha(\beta n - \gamma m) - b\beta(\alpha n - \gamma l) + C\gamma(\alpha m - \beta l) = 0.$$

74. Show that the two straight lines

$$\frac{x-a}{a'} = \frac{y-b}{b'} = \frac{z-c}{c'}\ ;\ \frac{x-a'}{a} = \frac{y-b'}{b} = \frac{z-c'}{c}$$

are coplanar. Find the point of intersection and the plane containing them.

Solution: We know that two straight lines intersect if we can find some point that lies on both the given lines.

Consider the point $P = [(a + a'), (b + b'), (c + c')]$ [1]

Examine it with reference to the two lines,

$$\frac{x-a}{a'} = \frac{y-b}{b'} = \frac{z-c}{c'} \quad [2]$$

$$\frac{x-a'}{a} = \frac{y-b'}{b} = \frac{z-c'}{c} \quad [3]$$

$$\frac{(a+a'-a)}{a'} = \frac{b+b'-b}{b'} = \frac{c+c'-c}{c'} = 1$$

and
$$\frac{a+a'-a'}{a} = \frac{b+b'-b'}{b} = \frac{c+c'-c'}{c} = 1$$

Hence, P lies on the lines [2] and [3]. Thus, the two lines intersect, and the point of intersection is

$$[(a + a'), (b + b'), (c + c')]$$

plane containing the lines

is:
$$\begin{vmatrix} x-a & y-b & z-c \\ a' & b' & c' \\ a & b & c \end{vmatrix} = 0$$

$$\xrightarrow{\text{Apply } R_1 + R_2} \begin{vmatrix} x & y & z \\ a' & b' & c' \\ a & b & c \end{vmatrix} = 0$$

$$\Rightarrow \quad x(bc' - b'c) + y(ca' - c'a) + z(ab' - a'b) = 0.$$

75. Find the angle between the line $\dfrac{x+1}{-3} = \dfrac{y-3}{2} = \dfrac{z+2}{1}$ and the plane $x + 2y + 3z + 4 = 0$.

Solution: $A = 1, B = 2, C = 3, D = 4, l = -3, m = 2, n = 1, \alpha = -1, \beta = 3, \gamma = -2$

$$\therefore \quad \theta = \sin^{-1}\left\{\frac{Al + Bm + Cn}{\sqrt{\Sigma A^2}\sqrt{\Sigma l^2}}\right\} = \sin^{-1}\left\{\frac{1(-3) + 2(2) + (3)(1)}{\sqrt{14}\sqrt{14}}\right\}$$

$$= \sin^{-1}\left\{\frac{4}{\sqrt{14}\sqrt{14}}\right\} = \sin^{-1}\left\{\frac{2}{7}\right\}.$$

76. Show that the lines $\frac{x+1}{1} = \frac{y+1}{2} = \frac{z+1}{3}$ and $x + 2y + 3z - 8 = 0 = 2x + 3y + 4z - 11$ intersect. Find their point of intersection. [V.T.U. F/M 2005]

Solution: Let $\frac{x+1}{1} = \frac{y+1}{2} = \frac{z+1}{3} = k$ (say), any point on the plane is therefore is $x = k - 1$, $y = 2k - 1$, $z = 3k - 1$ which lie on the line of intersection of the two planes :

$$\left\{\begin{array}{l}(k-1) + 2(2k-1) + 3(3k-1) - 8 = 0 \text{ and} \\ 2(k-1) + 3(2k-1) + 4(3k-1) - 11 = 0 \\ 14k - 14 = 0 \text{ and } 20k - 20 = 0\end{array}\right\}$$

Thus $k = 1$ and $k = 1$.

Since the value of k is equal we conclude that the lines intersect and the point of intersection from (1) is given by $x = 0, y = 1, z = 2$

$\therefore$ (0, 1, 2) is the point of intersection.

77. Show that the two straight lines

$$(3x + 2y - 3z + 2) = 0 = (x + 2y - 2z + 1)$$

and

$$(9x - 2y - 3z - 34) = 0 = (2x - z - 9)$$

are coplanar and find the plane containing them.

Solution: D.C's of the lines:

Line [1]: $\left\{\begin{array}{l}3l + 2m - 3n = 0 \\ l + 2m - 2n = 0\end{array}\right\}$

$\Rightarrow \quad l : m : n = 2 : 3 : 4$

Line [2]: $\left\{\begin{array}{l}9l - 2m - 3n = 0 \\ 2l + 0m - n = 0\end{array}\right\}$

$\Rightarrow \quad (l : m : n) = 2 : 3 : 4$

The D.C's are same. So the 2 lines are parallel. Hence, they are evidently coplanar.

The plane containing them:

Any plane thro' to line [2] is of the trun:

$$7x - 10y + 4z + 1 = 0$$

PART-A

EXERCISES

1. Find the angle between the line $\frac{x+1}{2}=\frac{y}{3}=\frac{z-3}{6}$ and the plane

$$3x + 2y + z = 7$$

$\left[Ans: \alpha = \sin^{-1}\left(\frac{18}{7\sqrt{14}}\right)\right]$

2. Find the angle between planes $2x - y + z = 6$ and $x + y + 2z = 7$

(*Ans:* $\alpha = 90°$, plane perpendicular)

3. Find the angle between the lines $\frac{x-7}{2}=\frac{y+3}{-1}=\frac{z-4}{1}$ and $6x + 4y - 5z - 4 = 0 = x - 5y + 2z - 1z$

$\left[Ans: \theta = \cos^{-1}\left(\frac{1}{2}\right) = 60°\right]$

4. Prove that the lines $\frac{x-4}{1}=\frac{y+3}{-4}=\frac{z+1}{7}$ and $\frac{x-1}{2}=\frac{y+1}{-3}=\frac{z+10}{8}$ intersect, and find the coordinates of their point of intersection.

5. Show that the lines $\frac{x+4}{3}=\frac{y+6}{5}=\frac{z-1}{-2}$ and $3x - 2y + z + 5 = 0 = 2x + 3y + 4z - 4$ are coplanar. Find their point of intersection and the plane in which they lie.

1.8 SHORTEST DISTANCE BETWEEN SKEW LINES

(*Non-intersecting, non-parallel lines in space*)

Two lines are skew if and only if they do not lie in a common plane.

To show that the length of the line intercepted between two lines which is perpendicular to both is the shortest distance between them.

Abbre: (shortest distance is usually written as S.D.)

Let
$$\frac{x-\alpha}{l}=\frac{y-\beta}{m}=\frac{z-\gamma}{n} \quad [1]$$

and
$$\frac{x-\alpha'}{l'}=\frac{y-\beta'}{m'}=\frac{z-\gamma'}{n'} \quad [2]$$

Lines lying in two different planes are said to be skew. The shortest distance line is the line which is perpendicular to both [1] and [2]. It is also equal to the perpendicular distance from any point on [1], say $(\alpha, \beta, \gamma) = A$, upon the line through [2] parallel to [1]. (see figure)

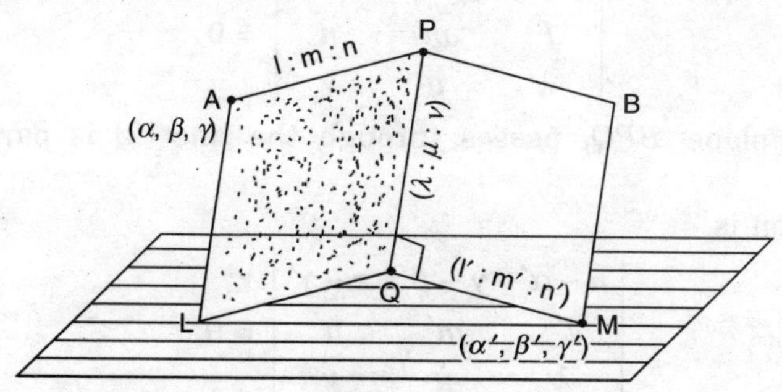

Fig. 1.18

Let $(\lambda : \mu : \nu)$ be the D.R's of the line of shortest distance (S.D.). This is perpendicular to [1] and [2]

$$(\lambda : \mu : \nu) \perp (l : m : n)$$

$$(\lambda : \mu : \nu) \perp (l^1 : m^1 : n^1)$$

$$\therefore \quad \left.\begin{aligned} \lambda l + \mu m + \nu n &= 0 \\ \lambda l' + \mu m' + \nu n' &= 0 \end{aligned}\right\} \quad [3]$$

$$\therefore \quad \frac{\lambda}{(mn' - m'n)} = \frac{\mu}{(nl' - n'l)} = \frac{\nu}{(lm' - l'm)} \quad [4]$$

Putting each equal to $\left(\frac{1}{k}\right)$, we get

$$\left.\begin{aligned} mn' - m'n &= \lambda k \\ nl' - n'l &= \mu k \\ lm' - l'm &= k\nu \end{aligned}\right\} \quad [5]$$

Again, the plane through the line [2], parallel to [1] is given by

$$\begin{vmatrix} x - \alpha' & y - \beta' & z - \gamma' \\ l' & m' & n' \\ l & m & n \end{vmatrix} = 0 \quad [6]$$

$$[x - \alpha'] [m'n - mn'] - [y - \beta'] [l'n - ln'] + [z - \gamma'] [l'm - lm'] = 0$$

$$[x - \alpha'] [mn' - m'n] + [y - \beta'] [l'n - ln'] + [z - \gamma'] [l'm - lm'] = 0$$

i.e. $$[x - \alpha'] \lambda k + [y - \beta'] \mu k + [z - \gamma'] \nu k = 0$$

$$\Rightarrow \quad \lambda[x - \alpha'] + \mu[y - \beta'] + \gamma[z - \gamma'] = 0 \quad [7]$$

Shortest distance (S.D.)

$$= PQ = \frac{\lambda[\alpha - \alpha'] + \mu[\beta - \beta'] + \nu[\gamma - \gamma']}{\sqrt{[\lambda^2 + \mu^2 + \nu^2]}} \quad [8]$$

To find the equations of the S.D. line

The S.D. line PQ may be regarded as the common line of intersection of the planes APQ and BPQ.

The plane APQ passes through AP (i.e. [1]) and is parallel to PQ [i.e. to the direction $(\lambda : \mu : \nu)$]

Hence its equation is given by

$$\begin{vmatrix} x - \alpha & y - \beta & z - \gamma \\ l & m & n \\ \lambda & \mu & \nu \end{vmatrix} = 0 \quad [9]$$

Similarly, the plane BPQ, passes through the line [2] is parallel to the direction $[\lambda : \mu : \nu]$

∴ Its equation is

$$\begin{vmatrix} n - \alpha' & y - \beta' & z - \gamma' \\ l' & m' & n' \\ \lambda & \mu & \nu \end{vmatrix} = 0 \quad [10]$$

WORKED EXAMPLES

78. Find the magnitude and equations to the line of shortest distance between the lines

$$\frac{x-6}{3}=\frac{y-7}{-1}=\frac{z-4}{1} \text{ and } \frac{x}{-3}=\frac{y+9}{2}=\frac{z-2}{4}.$$ [V.T.U. J/A 2003]

Solution. Let l, m, n be the DC's of the line of shortest distance between the two lines.

Since S.D. is perpendicular to both the lines, we have

$$3l - m + n = 0$$
$$-3l + 2m + 4n = 0$$

Solving,
$$\frac{l}{2}=\frac{m}{5}=\frac{n}{-1}=\frac{1}{\sqrt{2^2+5^2+1^2}}=\frac{1}{\sqrt{30}}$$

Length of S.D.
$$=(6-0)\frac{2}{\sqrt{30}}+(7+9)-\frac{5}{\sqrt{30}}+(4-2)\left(-\frac{1}{\sqrt{30}}\right)=3\sqrt{30}$$

Equation of the line of shortest distance is

$$\begin{vmatrix} x-6 & y-7 & z-4 \\ 3 & 1 & 1 \\ 2 & 5 & -1 \end{vmatrix} = 4x-5y+17z+79=0$$

and
$$\begin{vmatrix} x & y+9 & z-2 \\ -3 & 2 & 4 \\ 2 & 5 & -1 \end{vmatrix} = 22x-5y+19z-7=0.$$

79. Find the shortest distance between the lines.

$$\frac{x-3}{3}=\frac{y-2}{4}=\left(\frac{z-1}{-2}\right) \quad [1]$$
$$P_1=(3x+5y+4z+7)=0 \quad [2]$$
$$P_2=x-2y+8z+2=0$$

Any plane through [2] is:

$$P_1+kP_2=0$$
$$\Rightarrow \quad x(3+k)+y(5-2k)+2(4+8k)+(7+2k)=0 \quad [3]$$

[3] is parallel to [1], if normal is perpendicular to [1]

i.e. $(3+k)\,3+(5-2k)\,4+(4+8k)(-2)=0$

$\Rightarrow$ $k=1$ [4]

Plane: $x(3+1)+y(5-2)+z(4+8)+(7+2)=0$

$\Rightarrow$ $4x+3y+12z+9=0$ [5]

S.D. = perpendicular from any point of [1] *i.e.*

of (3, 2, 1) on [5]

$$=\frac{4\cdot 3+3\cdot 2+12\cdot 1+9}{\sqrt{4^2+3^2+12^2}}=\frac{39}{13}=3$$

80. Find the length and equation of line of shortest distance between the lines

$$\left.\begin{array}{l}3x + 2y + 4z - 8 = 0\\ x - 4y + z + 2 = 0\end{array}\right\} \qquad [1]$$

and $$\frac{x-5}{3} = \frac{y-4}{-2} = \frac{z-1}{+2} \qquad [2]$$

Solution: Here, the length as well as the equation is required. So, we reduce the eqn [1] also to the symmetrical form, first: Line [1]

$$3x + 2y + 4z - 8 = 0 = x - 4y + z + 2 \qquad [1]$$

D.C's: $3l + 2m + 4n = 0$

$$l - 4m + ln = 0$$

$$l : m : n = 18 : 1 : -14 \qquad [2]$$

To find point

Put $z = 0$:

$$3x + 2y - 8 = 0$$

$$x - 4y + 2 = 0$$

Solve $x = 2, y = 1, \therefore$ point [2, 1, 0]. [3]

St. line [1]: $$\frac{x-2}{18} = \frac{y-1}{1} = \frac{z-0}{-14} \qquad [4]$$

St. line (2): $$\frac{x-5}{3} = \frac{y-4}{-2} = \frac{z-1}{2} \qquad [5]$$

Length of S.D.

Let D.C's be $(\lambda : \mu : v)$

$\therefore$ $$18\lambda + \mu - 14\,v = 0$$

$$3\lambda - 2\mu + 2v = 0$$

$$\lambda : \mu : v = -\,26 : -\,78 : -\,39 = 2 : 6 : 3 \qquad [6]$$

$$\text{S.D} = \frac{\lambda[\alpha - \alpha'] + \mu[\beta - \beta'] + [v - v']}{\sqrt{\lambda^2 + \mu^2 + v^2}}$$

$$= \frac{2(2-5) + 6(1-4) + 3(0-1)}{\sqrt{2^2 + 6^2 + 3^2}}$$

$$= 21/7 = 3 \text{ units}$$

Eqn: $$\begin{vmatrix} x-\alpha & y-\beta & z-\gamma \\ l & m & n \\ \lambda & \mu & v \end{vmatrix} = 0 = \begin{vmatrix} (x-\alpha') & (y-\beta') & (z-\gamma') \\ l' & m' & n' \\ \lambda & \mu & v \end{vmatrix}$$

$$\begin{vmatrix} x-2 & y-1 & z \\ 18 & 1 & -14 \\ 2 & 6 & 3 \end{vmatrix} = 0 = \begin{vmatrix} x-5 & y-4 & z-1 \\ 3 & -2 & 2 \\ 2 & 6 & 3 \end{vmatrix}$$

$\Rightarrow$ $$87x - 82y + 106z - 92 = 0 = 11x + 5y - 22z - 18$$

81. Find the shortest distance between the axis of z and the line

$$x + 2y + z + 1 = 0 = 2x + 2y - z + 9$$

Solution: The z-axis passes thro' the origin (0, 0, 0); and has D.C's (0 : 0 : 1).

Hence, it can be written

$$\frac{x-0}{0} = \frac{y-0}{0} = \frac{z-0}{1} \quad [1]$$

The second st. line is $\left.\begin{aligned} P_1 &= x + 2y + z + 1 = 0 \\ P_2 &= 2x + 2y - z + 9 = 0 \end{aligned}\right\}$ [2]

The shortest distance between the two lines [1] and [2] is the length of the $\perp^r$ from any point on [1] say (0, 0, 0) on the plane thro' – [2], parallel to [1]

Any plane thro' [2] is of the form $P_1 + kP_2 = 0$

$$(x + 2y + z + 1) + k\,(2x + 2y - z + 9) = 0$$

$$\Rightarrow \quad x\,(1 + 2k) + y(2 + 2k) + z(1 - k) + (1 + 9k) = 0 \quad [3]$$

This is to be parallel to the line [1]; hence the normal to [3] is $\perp^r$ to [1].

$\Rightarrow$ $(1 + 2k) : (2 + 2k) : (1 - k)$

is $\perp^r$ to $(0 : 0 : 1)$

$\Rightarrow$ $(1 + 2k)\,0 + (2 + 2k)\,0 + (1 - k)\,(1) = 0$

$\Rightarrow$ $k = 1$ [4]

Hence, the plane is

[from [3] and [4]]

$$x\,(1 + 2 \cdot 1) + y(2 + 2 \cdot 1) + 2(1 - 1) + (1 + 9 \cdot 1) = 0$$

$$\Rightarrow \quad 3x + 4y + 10 = 0 \quad [5]$$

The S.D. = perp. from (0, 0, 0) on [5]

$$= \frac{D}{\sqrt{A^2 + B^2 + C^2}} = \frac{-10}{\sqrt{3^2 + 4^2 + 0^2}} = \left|\frac{-10}{5}\right| = 2 \text{ units}$$

82. Find the coordinates of the point of intersection of the line of S.D. with the lines $\frac{x+3}{2} = \frac{y-6}{3} = \frac{z-3}{-2}$ and $\frac{x}{2} = \frac{y-6}{2} = \frac{z}{-1}$ *hence find its length.* [V.T.U. F/M 2005]

Solution: Suppose $\frac{x+3}{2} = \frac{y-6}{3} = \frac{z-3}{-2} = p$ (say)

and $\frac{x}{2} = \frac{y-6}{2} = \frac{z}{-1} = q$ (say)

$\therefore$ $x = 2p - 3,\ y = 3p + 6,$

$z = -\,2p + 3$ ...(1)

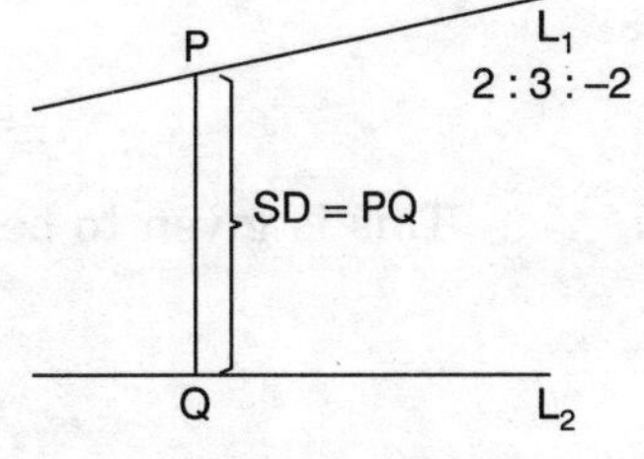

Fig. 1.19

and any point on L_1, L_2 and $x = 2q$, $y = 2q + 6$, $z = -\,q$ ...(2)

If P and Q are on L_1, L_2 respectively, then,

$\therefore$ DR's of PQ = $2q - 2p + 3,\ 2q - 3p,\ -\,q + 2p - 3$

PQ being the line of S.D. is perpendicular to L_1 whose D.R.s are (2 : 3 : – 2) and also to L_2 whose d.r.s are (2 : 2 : – 1)

$$\begin{aligned} &\therefore \\ &\text{and} \end{aligned} \quad \left\{\begin{aligned} 2\,(2q - 2p + 3) + 3\,(2q - 3p) - 2\,(-\,q + 2p - 3) = 0 \\ 2\,(2q - 2p + 3) + 2\,(2q - 3p) - 1\,(-\,q + 2p - 3) = 0 \end{aligned}\right\}$$

or $$12q - 17p + 12 = 0 \text{ and } 9q - 12p + 9 = 0$$

on solving, we get $p = 0$ and $q = -1$

Substituting the value of p in (1) we get $P = (-3, 6, 3)$ and substituting the value of q in (2) we get $Q = (-2, 4, 1)$.

$\therefore$ S.D. = Distance between P and Q (nearest to each other)

$$= \sqrt{(-2+3)^2 + (4-6)^2 + (1-3)^2} = \sqrt{9} = 3 \text{ units.}$$

83. If $2d$ is the shortest distance between the lines

$$y/b + z/c = 1,\ x = 0 \text{ and}$$
$$x/a - z/c = 1,\ y = 0$$

P.T. $$\frac{1}{a^2} + \frac{1}{b^2} + \frac{1}{c^2} = \frac{1}{d^2}$$

Solution: The symmetric form of the eqns of the lines are:

$$\left.\begin{aligned} \frac{x}{0} &= \frac{y-b}{a} = \frac{z}{-c} \\ \text{and} \quad \frac{x-a}{a} &= \frac{y}{0} = \frac{z}{c} \end{aligned}\right\} \quad [1]$$

$\therefore$ if $(l : m : n)$ be the D.C's of the shortest distance, then

$$l \cdot 0 + m \cdot b + n(-c) = 0$$

and $$l \cdot a + m \cdot 0 + n \cdot c = 0$$

$$\therefore \quad \frac{l}{bc} = \frac{-m}{ac} = \frac{n}{-ab}$$

We find from [1], that $B(0, b, 0)$, $A(a, 0, 0)$ lie on the two lines.

Hence S.D. between the two lines

$$= \text{projection of } BA \text{ on the line } (l : m : n)$$

$$= \frac{(a-0)\,bc + (0-b)(-ac) + (0-0)(-ab)}{\sqrt{(b^2c^2 + c^2a^2 + a^2b^2)}}$$

$$= 2abc \Big/ \sqrt{b^2c^2 + c^2a^2 + a^2b^2}$$

This is given to be $= 2d$.

$$\therefore \quad \frac{1}{d^2} = \frac{b^2c^2 + c^2a^2 + a^2b^2}{a^2b^2c^2} = \frac{1}{a^2} + \frac{1}{b^2} + \frac{1}{c^2}$$

$$\Rightarrow \quad \frac{1}{a^2} + \frac{1}{b^2} + \frac{1}{c^2} = \frac{1}{d^2}$$

EXERCISES

1. Find the S.D. between the lines

$$\frac{x-5}{3} = \frac{y-7}{-16} = \frac{z-3}{7} \text{ and } \frac{x-9}{3} = \frac{y-13}{8} = \frac{z-15}{-5}$$ *(Ans: 14)*

2. Find the shortest distance between the lines

$$2x - 2y + 3z - 12 = 0 = 2x + 2y + z$$

$$2x - z = 0 = 5x - 2y + 9 \qquad \textit{(Ans: 6.)}$$

3. Show that the S.D. between any two opposite edges of the tetrahedron formed by the planes $y + z = 0;\ z + x = 0;\ x + y = 0;\ x + y + z = a$ is $\dfrac{2a}{\sqrt{6}}$.

4. Find the length and equations of the line of S.D. between the lines $\dfrac{x-8}{3} = \dfrac{y+9}{-16} = \dfrac{z-10}{7}$ and $\dfrac{x-15}{3} = \dfrac{y-29}{8} = \dfrac{z-5}{-5}$. (*Ans:* 14; $117x + 4y - 41z - 490 = 0 = 9x - 4y - z - 14$)

5. Find the length and equations of S.D. between the two lines $\dfrac{x-1}{2} = \dfrac{y-3}{4} = z + 2$ and $3x - y - 2z + 4 = 0 = 2x + y + z + 1$ $\left(\textit{Ans.}\ \dfrac{8}{\sqrt{14}};\ x - y + 2z + 6 = 0 = 19x\right)$

1.9 RIGHT CIRCULAR CONE AND RIGHT CIRCULAR CYLINDER

(A) Right Circular Cone

A right circular cone is the surface generated by a straight line revolving about another line which is fixed, the two lines intersecting at a constant angle θ. The fixed line is called the 'axis' of the cone; the point of intersection is the 'vertex' of the cone; the angle between the two lines is 'semi-vertical angle' of the cone. The straight line which revolves and generates the cone is called the 'generator' of the cone.

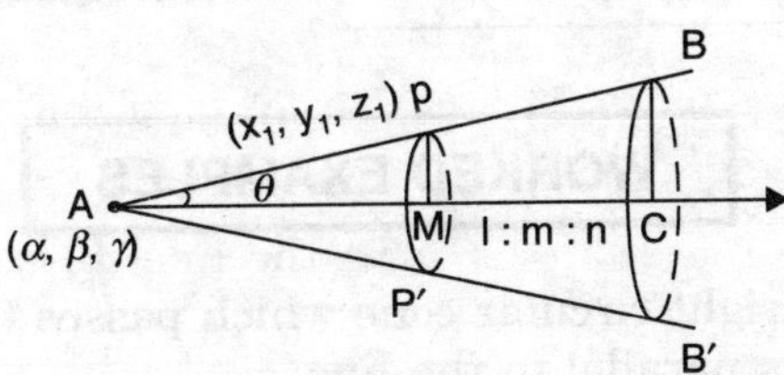

Fig. 1.19

To find equations of a right circular cone

Let $A\ (\alpha, \beta, \gamma)$ be the vertex of the cone, $\dfrac{x-\alpha}{l} = \dfrac{y-\beta}{m} = \dfrac{z-\gamma}{n}$ [1]

be the axis, the axis has the D.C's $(l : m : n)$ of the cone.

Let $p(x_1, y_1, z_1)$ be any point on the cone; and AP makes an angle 'θ' with the axis [1].

D.R's of $AP = (x_1 - \alpha) : (y_1 - \beta) : (z_1 - \gamma)$

D.R's of Axis $= (l : m : n)$

The two are inclined at an angle θ,

where $$\cos\theta = \frac{(x_1 - \alpha)l + (y_1 - \beta)m + (z_1 - \gamma)n}{\sqrt{(l^2 + m^2 + n^2)[(x_1 - \alpha)^2 + (y_1 - \beta)^2 + (z_1 - \gamma)^2]}}$$

$$\therefore \quad \cos^2\theta = \frac{[(x_1 - \alpha)l + (y_1 - \beta)m + (z_1 - \gamma)n]}{(l^2 + m^2 + n^2)[(x_1 - \alpha)^2 + (y_1 - \beta)^2 + (z_1 - \gamma)^2]}$$

or dropping the suffixes, we get

$$[l(x-\alpha)+m(y-\beta)+n[z-\gamma)]^2=\cos^2\theta(p^2+m^2+n^2)[(x-\alpha)^2+(y-\beta)^2+(l-\gamma)^2]$$

To find the equation of the right circular cone, whose vertex is at the origin, and axis along a line of (z–axis), D.R's $(l:m:n)$, and whose semivertical angle is equal to α

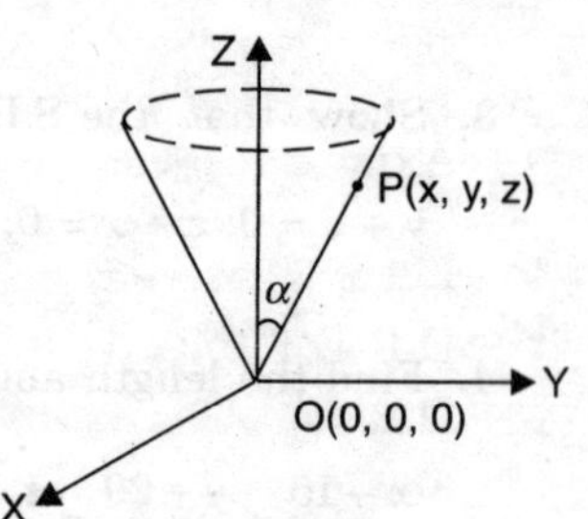

Fig. 1.20

Vertex = (0, 0, 0)

$P = (x, y, z)$, any point on RCC

$\therefore \quad Z\hat{O}P = \alpha.$

D.R's of OP are:

$[(x-0):(y-0):(z-0)] = (x:y:z)$

D.C's of the axis (z-axis) of the cone are $(0:0:1)$

$$\therefore \quad \cos\alpha = \frac{x(0)+y(0)+z(1)}{\sqrt{x^2+y^2+z^2}} \qquad \left(\because \cos\theta = \frac{a_1a_2+b_1b_2+c_1c_2}{\sqrt{\Sigma a_1^2}\sqrt{\Sigma a_2^2}}\right)$$

or
$$\cos\alpha = \frac{z}{\sqrt{x^2+y^2+z^2}} \Rightarrow (x^2+y^2+z^2)\cos^2\alpha = z^2$$

$$\therefore \quad x^2+y^2+z^2 = \frac{z^2}{\cos^2\alpha} = z^2\sec^2\alpha$$

$$\therefore \quad x^2+y^2 = z^2\sec^2\alpha - z^2 = z^2(\sec^2\alpha - 1)$$

or $\boxed{x^2+y^2 = z^2\tan^2\alpha}$ is the equation of the cone.

WORKED EXAMPLES

84. Find the equation of the right circular cone which passes through the point (2, 1, 3), has its vertex at (1, 1, 2) and axis parallel to the line. [V.T.U. J/A 2003]

$$\frac{x-2}{2} = \frac{y-1}{-4} = \frac{z+2}{3}$$

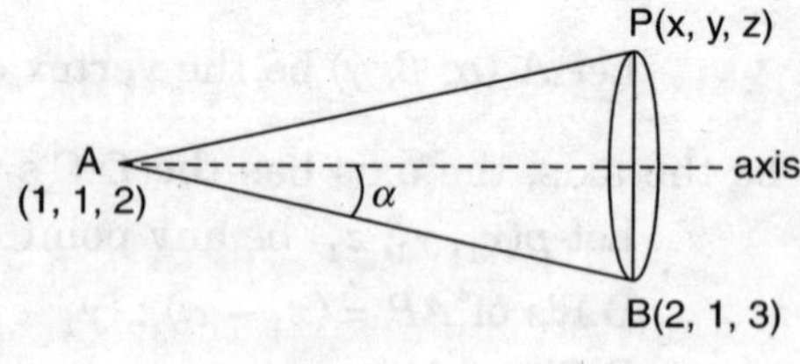

Solution: DR's of axis are 2, – 4, 3. DR's of the generator AB are 1, 0, 1. If α is the semi-vertical angle between the axis and AB, then

$$\text{coa } \alpha = \frac{2.1+(-4).0+3.1}{\sqrt{1+0+1}\sqrt{4+16+9}}$$

$$= \frac{5}{\sqrt{2}\sqrt{29}} \qquad (1)$$

If $P(x, y, z)$ is any point on the cone, then DR's of AP are $x-1$, $y-1$, $z-2$. So

$$\cos\alpha = \frac{(x-1)2+(y-1)(-4)+(z-2)3}{\sqrt{(x-1)^2+(y-1)^2+(z-2)^2}\sqrt{29}} = \frac{5}{\sqrt{2}\sqrt{29}} \text{ from (1)}$$

Squaring, the required equation of cone is

$$2[2(x-1)-4(y-1)+3(z-2)]^2 = 25[(x-1)^2+(y-1)^2+(z-2)^2].$$

85. Find the equation of a right circular cone whose vertex is at the point (1, 2, 3) and whose axis is the straight line $\frac{x-1}{2} = \frac{y-2}{3} = \frac{z-3}{4}$, given that its semivertical angle is 30°.

Vertex = (1, 2, 3);

Axis: $\frac{x-1}{2} = \frac{y-2}{3} = \frac{z-3}{4}$ [1]

Let $P = (x, y, z)$ be any point on the cone. Then

D.R's of $PA = (x-1) : (y-2) : (z-3)$ [2]

D.R's of Axis = 2 : 3 : 4 [3]

For any point $P(x, y, z)$ on the cone. PA makes an angle 30°, with the axis of the cone.

$$\therefore \quad \frac{\sqrt{3}}{2} = \cos 30°$$

$$= \frac{(x-1)2 + (y-2)3 + (z-3)4}{\sqrt{(2^2+3^2+4^2\,[(x-1)^2+(y-2)^2+(z-3)^2}}$$

$$\Rightarrow \quad \frac{3}{4} = \frac{(2x+3y+4z-20)^2}{29\,[(x-1)^2+(y-2)^2+(z-3)^2]}$$

$$87\,[(x-1)^2 + (y-2)^2 + (z-3)^2]$$
$$= 4\,(2x + 3y + 4z - 20)^2$$

86. Find the equation of the right circular cone of semi vertical angle 30°, whose vertex is at the point (1, 2, 3) and whose axis is parallel to the line $(x = y = z)$.

We know that the equation of the cone with vertex (α, β, γ) axis with D.C's $(l : m : n)$ and semivertical angle θ,

$$[l\,(x-\alpha) + m\,(y-\beta) + n\,(z-\gamma)]^2$$
$$= \cos^2\theta\,(l^2+m^2+n^2)\,[(x-\alpha)^2 + (y-\beta)^2 + [z-\gamma)^2] \quad [1]$$

Here vertex $(\alpha, \beta, \gamma) = (1, 2, 3).$

$\therefore \quad (\alpha = 1, \beta = 2, \gamma = 3)$

Axis has D.C's = (1 : 1 : 1)

$\theta = 30°$

$$\therefore \quad [1(x-1) + 1\,(y-2) + 1\,(z-3)]^2$$
$$= (\cos^2 30°)\,(1^2+1^2+1^2)\,[(x-1)^2 + (y-2)^2 + (z-3)^2]$$

i.e. $$(x+y+z-6)^2 = \left(\frac{3}{4}\right) \cdot 3\,[(x-1)^2 + (y-2)^2 + (z-3)^2]$$

$$4\,(x+y+z-6)^2 = 9[(x-1)^2 + (y-2)^2 + (z-3)^2]$$

87. Find the equation of the right circular cone whose vertex is (3, 1, 2) axis has D.C's 1 : 2 : 3 and the semi-vertical angle is $\cos^{-1}\left(\frac{1}{\sqrt{7}}\right)$.

Solution: V = Vertex = (3, 1, 2).

Let $P(x, y, z)$ be any general point on the surface of the cone. The D.R's of VP are

$$(x-3) : (y-1) : (z-2).$$

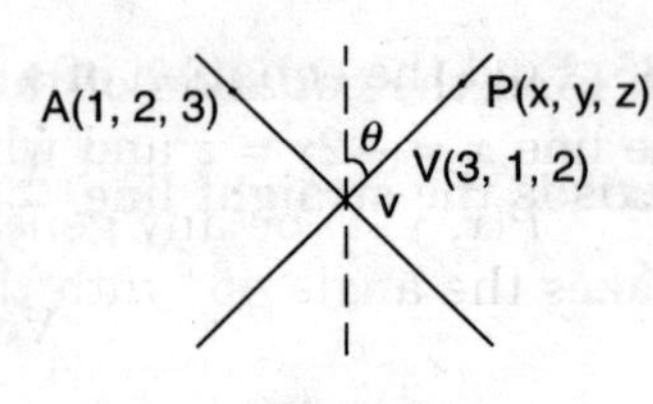

Fig. 1.21

D.R's of axis are given to be

$$(1 : 2 : 3)$$

Angle between them = semi-vertical angle θ

$$= \cos^{-1} \frac{1}{\sqrt{7}}.$$

$$\cos \theta = \frac{1(x-3) + 2(y-1) + 3(z-2)}{\sqrt{1^2 + 2^2 + 3^3} \sqrt{(x-3)^2 + (y-1)^2 + (z-2)^2}}$$

But $\quad \theta = \cos^{-1} \frac{1}{\sqrt{7}};$

$\therefore \quad \cos \theta = \frac{1}{\sqrt{7}}$

Hence $\frac{1}{\sqrt{7}} = \frac{x + 2y + 3z - 11}{\sqrt{14} \cdot \sqrt{(x-3)^2 + (y-1)^2 + (z-1)^2}}$

Squaring, cancelling 7 and cross-multiplying,

$$2\,[(x-3)^2 + (y-1)^2 + (z-2)^2] = 17y + 20z + 2 = (x + 2y + 3z - 11)^2$$

is the required eqn of cone.

88. Find the equation of the right circular cone with origin as vertex, and the axis of coordinates as three generators.

Solution: Given that the axis coordinates are 3 generators, the axis of the cone has to make equal angles with the coordinate axes.

$\therefore$ we can take the D.R's of axis as 1 : 1 : 1. It makes angles $\cos^{-1} \frac{1}{\sqrt{3}}$ with *x, y, z* axes.

$\therefore$ Semivertical angle $\theta = \cos^{-1} \frac{1}{\sqrt{3}}$

i.e. $\quad \cos\theta = \frac{1}{\sqrt{3}}$

Let $P(x, y, z)$ be any general point on the cone.

D.R's of the line joining P to the vertex $V(0, 0, 0)$ are

$(x-0) : (y-0) : (z-0) = x : y : z$ D.R's of axis are $(1 : 1 : 1)$

$\therefore \quad \cos \theta = \frac{(1)(x) + (1)(y) + (1)(z)}{\sqrt{1^2 + 1^2 + 1^2} \sqrt{x^2 + y^2 + z^2}}$

But $\quad \cos\theta = \frac{1}{\sqrt{3}}$

$$\frac{1}{\sqrt{3}} = \frac{x + y + z}{\sqrt{3} \cdot \sqrt{x^2 + y^2 + z^2}}$$

Simplifying $\quad x^2 + y^2 + z^2 = (x + y + z)^2$

$\Rightarrow \quad 2xy + 2yz + 2zx = 0$

$\Rightarrow \quad xy + yz + zx = 0$ is the eqn of the cone.

89. **Find the equation of the right circular cone whose vertex is at the origin, whose axis is the line $x = -2y = z$ and which has semivertical angle 45° $V = (0, 0, 0)$ is the vertex.**

$P(x, y, z)$ be any general point on the surface of the cone. Then VP is generator and if makes the angle 45° with the axis.

$$\text{D.R's of } VP = (x - 0) : (y - 0) : (z - 0) = x : y : z$$

Eqns. of axis are $\quad x = -2y = z$

or $\quad \dfrac{x}{-2} = \dfrac{y}{1} = \dfrac{z}{-2}$ (on dividing by -2)

$\therefore$ D.R's it axis are $-2 : 1 : -2$

Angle between them = 45°

$$\therefore \quad \cos 45° = \frac{-(2x - y + 2z)}{\sqrt{9} \cdot \sqrt{x^2 + y^2 + z^2}}$$

$\Rightarrow \quad 9(x^2 + y^2 + 2^2) = 2(2x - y + 2z)^2$ is the equation of the cone.

90. **Find the equation of the right circular cone of semi-vertical angle $\dfrac{\pi}{4}$, and vertex (0, 0, 1) given that the axis of the cone is parallel to the x-axis. Also, show that, the section of the above cone by the plane $z = 0$ is a rectangular Hyperbola.**

Solution: Axis is parallel to x-axis.

$$\therefore \quad \text{D.C's} = (1 : 0 : 0)$$

In the eqn of the general cone we put $(\alpha : \beta : \gamma) = (0 : 0 : 1)$

$$\left[\theta = \frac{\pi}{4}, \cos\theta = \frac{1}{\sqrt{2}}\right]$$

and $\quad (l : m : n) = (1 : 0 : 0)$

$$[\Sigma l(x - \alpha)]^2 = \cos^2\theta\ \Sigma(l^2)\ [\Sigma(x - \alpha)^2] \qquad [1]$$

i.e. $\quad [1(x - 0) + 0 + 0]^2 = \left(\dfrac{1}{\sqrt{2}}\right)(1^2 + 0^2 + 0^2)\ [(x - 0)^2 + (y - 0)^2 + (z - 0)^2]$

$$\Rightarrow \quad 2x^2 = x^2 + y^2 + (z - 1)^2 \text{ i.e. } x^2 - y^2 - (z - 1)^2 = 0 \qquad [2]$$

To find the section by XOY plane, put $z = 0$ in [2], we get

$$x^2 - y^2 - (0 - 1)^2 = 0$$

i.e. $\quad x^2 - y^2 = 1$

(a rectangular Hyperbola)

91. **Find the cone got by rotating the straight line ($z = 0$, $y = 2x$) about the x–axis.**

Let us put the given straight lines the Std. symmetrical form.

Solution: **Given:** ($z = 0$, $y = 2x$); $\dfrac{x - 0}{1} = \dfrac{y - 0}{2} = \dfrac{z - 0}{0}$.

$\Rightarrow$ (0, 0, 0) is a point on the line as well as on the x-axis (axis of rotation).

Hence the vertex is at the origin.

Axis of the cone = x–axis;

has D.C's = (1 : 0 : 0)

Semi-vertical angle between
(1 : 2 : 0) and (1 : 0 : 0)

$$\cos\theta = \frac{1 \cdot 1 + 2 \cdot 0 + 0 \cdot 0}{\sqrt{1^2 + 0^2 + 0^2}\ \sqrt{1^2 + 2^2 + 5^2}} = \frac{1}{\sqrt{5}}$$

$$\therefore \qquad \cos^2\theta = (\Sigma l^2)\,[\Sigma(x - \alpha)^2] = [\Sigma l\,(x - \alpha)]^2$$

$$(1/5)\,(1^2 + 0^2 + 0^2)\,[x^2 + y^2 + z^2] = [1\,(x - 0) + 0(y - 0) + 0\,(z - 0)]^2$$

$$\therefore \qquad (x^2 + y^2 + z^2) = 5(x^2)$$

$$4x^2 - y^2 - z^2 = 0$$

92. The axis of right circular cone, vertex 0, makes equal angles with the coordinate axes, and the cone passes through the line drawn through 0, with D.C's proportional to (3 : 4 : 5). Find the eqn of the cone

D.C's of axis = (1 : 1 : 1)
Vertex = (0, 0, 0).

Semivertical angle is the same as that between (3 : 4 : 5) and (1 : 1 : 1).

$$\therefore \qquad \cos\theta = \frac{3 \cdot 1 + 4 \cdot 1 + 5 \cdot 1}{\sqrt{1^2 + 1^2 + 1^2}\ \sqrt{(3^2 + 4^2 + 5^2)}} = \frac{12}{\sqrt{150}} = \frac{2\sqrt{6}}{5}$$

Eqn of cone is

$$\cos^2\theta\,(l^2 + m^2 + n^2)\,[\Sigma(x - \alpha)^2] = [\Sigma l(x - \alpha)]^2$$

$$(24/25)\,[1 + 1 + 1]\,[x^2 + y^2 + z^2] = [1\,(x - 0) + 1\,(y - 0) + 1[z - 0)]^2$$

$$72\,(x^2 + y^2 + z^2) = 25\,(x + y + z)^2$$

93. Find the equation of the right circular cone with vertex (2, – 3, – 4), semivertical angle 30° and whose axis is equally inclined to the coordinate axes. [V.T.U. F/M 2005]

Solution: It is given that the axis of the cone is equally inclined with the coordinate axes we have basically $\alpha = \beta = \gamma \ \Rightarrow\ \cos\alpha\,(l) = \cos\beta\,(m) = \cos\gamma\,(n)$

Since $l^2 + m^2 + n^2 = 1$ we get $3l^2 = 1$ or $l = 1/\sqrt{3}$

$$\Rightarrow \qquad l = m = n = \frac{1}{\sqrt{3}}$$ Hence, DR's of the axis of cone are 1, 1, 1.

By question, $V = (2, -3, -4)$ is the vertex of the cone and let $P\,(x, y, z)$ be any point on the cone.

$\therefore$ DR's of $VP = (x - 2 : y + 3 : z + 4)$.

Also D.R.'s of the axis = 1, 1, 1.

Since 30° is the angle between these two lines we have,

$$\cos 30° = \frac{1(x - 2) + 1(y + 3) + 1(z + 4)}{\sqrt{1 + 1 + 1}\ \sqrt{(x - 2)^2 + (y + 3)^2 + (z + 4)^2}}$$

i.e.,

$$\frac{\sqrt{3}}{2} = \frac{x + y + z + 5}{\sqrt{3}\ \sqrt{(x - 2)^2 + (y + 3)^2 + (z + 4)^2}}$$

or

$$\boxed{9\,[(x - 2)^2 + (y + 3)^2 + (z + 4)^2] = 4\,(x + y + z + 5)^2}$$

94. If α is the semi-vertical angle of the right circular cone which passes thro' the lines OX, OY, $x = y = z$, show that

$$\cos \alpha = (9 - 4\sqrt{3})^{-1/2}$$

Solution: If $(l : m : n)$ are the D.C's of the axis of the cone, since the axis makes the same angle α with each of the lines OX (D.C's 1, 0, 0), OY (D.C's 0, 1, 0) and $x = y = z$ (D.C's proportional to 1 : 1 : 1)

$$\therefore \quad \cos \alpha = l(1) + m(0) + n(0) = l(0) + m(1) + n(0)$$

$$= \frac{l(1) + m(1) + n(1)}{\sqrt{1+1+1}}$$

$$\Rightarrow \quad l = \cos \alpha, \; m = \cos \alpha$$

and $\quad l + m + n = \sqrt{3} \cos\alpha \quad$ or $\quad n = \sqrt{3} \cos\alpha - m - l$

$$= \sqrt{3} \cos \alpha - 2 \cos\alpha = (\sqrt{3} - 2) \cos \alpha$$

But $l^2 + m^2 + n^2 = 1$.

$$\therefore \quad \cos^2\alpha + \cos^2\alpha + (\sqrt{3} - 2)^2 \cos^2\alpha = 1$$

or $\quad [(1 + 1 + (\sqrt{3} - 2)^2] \cos^2\alpha = 1 \quad$ or $\quad (2 + 3 + 4 - 4\sqrt{3}) \cos^2\alpha = 1$

or $\quad (9 - 4\sqrt{3}) \cos^2\alpha = 1 \quad$ or $\quad \cos^2\alpha = \dfrac{1}{9 - 4\sqrt{3}}$

$$\therefore \quad \cos\alpha = \frac{1}{(9 - 4\sqrt{3})^{1/2}} = (9 - 4\sqrt{3})^{-1/2}$$

Hence the result.

EXERCISES

Find the equation of the right circular cone for which:

(*i*) Vertex = (0, 0, 0), axis: $\dfrac{x}{4} = \dfrac{y}{2} = \dfrac{z}{3}$; semivertical angle = 30°.

(*Ans:* $4(4x + 2y + 3z)^2 = 87 (x^2 + y^2 + z^2)$)

(*ii*) Vertex = (0, 0, 0), axis: $2x = 3y = 4z$; semivertical angle = 45°.

(*Ans:* $2(6x + 4y + 3z)^2 = 61(x^2 + y^2 + z^2)$)

(*iii*) Vertex = (2, 3, 5), axis makes equal angles with the coordinate axis and the semivertical angle is $\cos^{-1} \dfrac{\sqrt{2}}{3}$.

(*Ans:* $3(x + y + z - 10)^2 = 2[(x - 2)^2 + (y - 3)^2 + (z - 5)^2]$)

(B) RIGHT CIRCULAR CYLINDER

Definition: A right circular cylinder is the surface generated by a line (generator) revolves about a fixed line (axis), parallel to it, at a constant distance from it.

Thus any plane perpendicular to the axis of a right circular cylinder, cuts the cylinder in a circle.

Equations of a Right Circular Cylinder (R.C.C.)

In order to find the equation of R.C.C. we must know the equations of the axis, a point on the axis, D.R's of the axis and base radius r of the cylinder.

Let $A(\alpha, \beta, \gamma)$ be a point on the axis of the cylinder whose equation is

$$= \frac{x-\alpha}{l} = \frac{y-\beta}{m} = \frac{z-\gamma}{n} \quad [1]$$

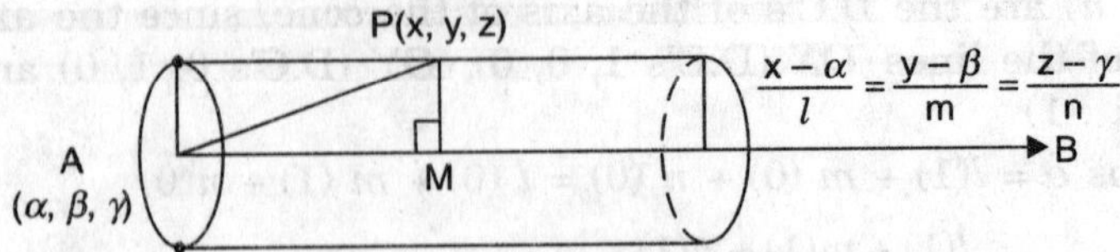

Fig. 1.22

Let $P(x, y, z)$ be any point on the cylinder.

Draw $PM\perp^r$ on the axis AB.

$$\frac{x-\alpha}{l} = \frac{y-\beta}{m} = \frac{z-\gamma}{n}$$

$A(\alpha, \beta, \gamma)$ is a point on the line [1]

In ΔAPM, $\hat{M}$ is st. angle

$$AP^2 = AM^2 + MP^2$$

But
$$AP = \sqrt{(x-\alpha)^2 + (y-\beta)^2 + (z-\gamma)^2}$$

$$AM = \text{projection of } AP \text{ on } AB$$

$$= \frac{(x-\alpha)l + (y-\beta)m + (z-\gamma)n}{\sqrt{l^2+m^2+n^2}}$$

and
$$MP = r$$

$\therefore$ From [2],

$$\left[\sqrt{(x-\alpha)^2 + (y-\beta)^2 + (z-\gamma)^2}\right]^2 = \left[\frac{(x-\alpha)l + (y-\beta)m + (z-\gamma)n}{l^2+m^2+n^2}\right]^2 + r^2$$

$$\Rightarrow \quad (x-\alpha)^2 + (y-\beta)^2 + (z-\gamma)^2 = \frac{[(x-\alpha)l + (y-\beta)m + (z-\gamma)n]^2}{l^2+m^2+n^2} + r^2$$

which is the required equation of the right-circular cylinder.

WORKED EXAMPLES

95. Find the equation of the right circular cylinder of radius 2 whose axis is

$$\frac{x-1}{2} = \frac{y}{3} = \frac{z-3}{1}$$

Solution: Axis:
$$\frac{x-1}{2} = \frac{y}{3} = \frac{z-3}{1}$$

passes thro: A (1, 0, 3)

D.R's: (2 : 3 : 1)

If $P(x, y, z)$ be any point on the surface of the cylinder and PM be drawn perpendicular to the axis.

$$AP = \sqrt{(x-1)^2 + (y-0)^2 + (z-3)^2}$$

$\therefore$ AM = projection of line joining A (1, 0, 3) to P (x, y, z) on the axis

D.R's are (2 : 3 : 1)

$$\therefore \quad AM = \frac{l(x_2 - x_1) + m(y_2 - y_1) + n(z_2 - z_1)}{\sqrt{l^2 + m^2 + n^2}}$$

$$= \frac{2(x-1) + 3(y-0) + 1(z-3)}{\sqrt{2^2 + 3^2 + 1^2}} = \frac{2x + 3y + z - 5}{\sqrt{14}}$$

Also MP = radius of the cylinder = 2.

By pythogorean theorem $\quad AP^2 = AM^2 + MP^2$

$$\Rightarrow \quad (x-1)^2 + (y-0)^2 + (z-3)^2 = \left(\frac{2x + 3y + z - 5}{\sqrt{14}}\right)^2 + 2^2$$

$$14\,[x^2 + y^2 + z^2 - 2x - 6z + 10] = (2x + 3y + z - 5)^2 + 56$$

$$\Rightarrow \quad 10x^2 + 5y^2 + 13z^2 - 6yz - 4zx - 12xy - 8x + 30y - 74z + 59 = 0$$

96. Find the equation of the right circular cylinder whose axis passes through the origin and generator is the line $\frac{x-3}{2} = \frac{y}{3} = \frac{z-4}{6}$. [V.T.U. J/A 2003]

Solution: The generator passes through the point $A(3, 0, 4)$ and has D.R's. 2, 3, 6 or D.C.'s $\frac{2}{7}, \frac{3}{7}, \frac{6}{7}$. The radius of the cylinder $r = AO$ is the perpendicular distance from $O(0, 0, 0)$ to the line AP. Thus $r^2 = (x_1 - \alpha)^2 + (y_1 - \beta)^2 + (z_1 - \gamma)^2 - [l(x_1 - \alpha) + m(y - \beta) + n(z - \gamma)]^2$. Here $\alpha = 0, \beta = 0, \gamma = 0, x_1 = 3, y_1 = 0,$

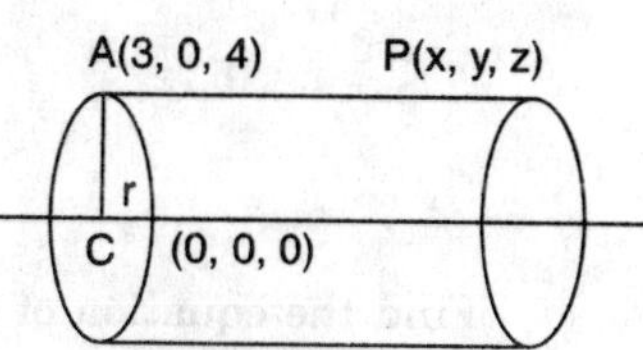

$z_1 = 4,\ l = \frac{2}{7},\ m = \frac{3}{7},\ n = \frac{6}{7}$. So $r^2 = 3^2 + 0^2 + 4^2 - \left\{\frac{6}{7} + 0 + \frac{24}{7}\right\}^2$ or $r^2 = 25 - \left(\frac{30}{7}\right)^2$.

Equation of cylinder of radius r is

$$(x - \alpha)^2 + (y - \beta)^2 + (z - \gamma^2)$$

$$= \frac{l(x - \alpha) + m(y - \beta) + n(z - \gamma)^2}{l^2 + m^2 + n^2} + r^2$$

$$x^2 + y^2 + z^2 = \frac{\left(\frac{2}{7}x + \frac{3}{7}y + \frac{6}{7}z\right)^2}{1} + 25 - \left(\frac{30}{7}\right)^2$$

$$45x^2 + 40y^2 + 13z^2 - 12xy - 36yz - 24xz = 325.$$

97. Find the radius of cross section of the Right Circular cylinder passing through the point (5, 4, – 1) with the axis along the line

$$\frac{x-1}{2} = \frac{y-0}{9} = \frac{z}{5}.$$

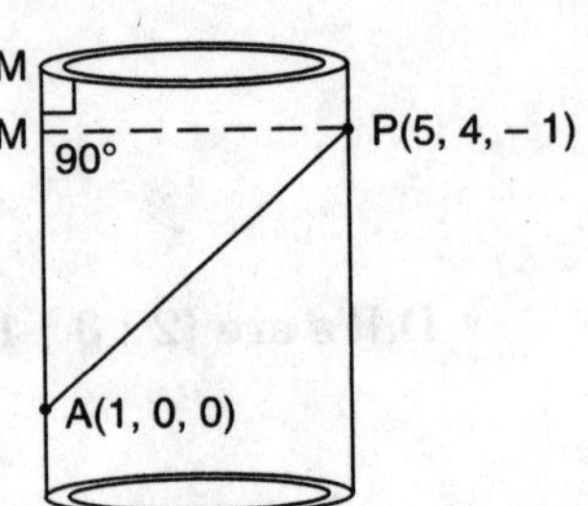

Fig. 1.23

Solution: The axis is given by

$$\frac{x-1}{2}=\frac{y-0}{9}=\frac{z-0}{5}$$

$\therefore$ A(1, 0, 0) is a point on the axis & D.R's are (2 : 9 : 5)

Now, $AP = \sqrt{(5-1)^2+(4-0)^2+(1-0)^2} = \sqrt{33}$

AM = projection of the line joining (1, 0, 0) to P(5, 4, – 1) on the axis whose D.R's (2 : 9 : 5)

$$= \frac{l(x_2-x_1)+m(y_2-y_1)+n(z_2-z_1)}{\sqrt{l^2+m^2+n^2}}$$

$$= \frac{2(5-1)+9(4-0)+5(-1-0)}{\sqrt{2^2+9^2+5^2}} = \frac{39}{\sqrt{110}}$$

By pythagoras, theorem:

$$PM^2 = AP^2 - AM^2$$

$$= \left(\sqrt{33}\right)^2 - \left(\frac{39}{\sqrt{110}}\right)^2 = \frac{2109}{110}$$

$\therefore$ $PM = \sqrt{\dfrac{2109}{110}}$ is the reqd. radius

EXERCISES

Find the equation of the right circular cylinder for which :

(*i*) radius is 2; axis passes thro (1, 2, 3) has D.R's (2 : – 3 : 6)

(*ii*) radius is 2, axis passes thro (1, –3, 2) and was D.R's (2 : –1 : 5).

[*Ans:* (*i*) $49\,[(x-1)^2+(y-2)^2+(z-3)^2]-(2x-3y+5z-14)^2=196$

(*ii*) $30\,[(x-1)^2+(y+3)^2+[z-2)^2]-(2x-y+5z-15)^2=120$

(*iii*) Find the equations of the right circular cylinder whose axis passes thro' the origin, and which has $\dfrac{x-3}{2}=\dfrac{y}{3}=\dfrac{z-4}{6}$ as a generator. [*Ans:* $49\,(x^2+y^2+z^2)-(2x+3y+6z)^2=325$]

(*iv*) Find the *R.C.C* for which

(*a*) $r = 1,\ \dfrac{x-1}{2}=\dfrac{y}{3}=\dfrac{z-3}{1}$ (*b*) $r = 2\ \dfrac{x-1}{2}=\dfrac{y-2}{1}=\dfrac{z-3}{2}$.

[*Ans:* (*a*) $10x^2+5y^2+13z^2-6yz-4zx-12xy-8x+30y-74z+101=0$

(*b*) $5x^2+8y^2-4xy-4yz-8zx+22x-16y-14z-10=0$]

PART-B

2

Differential Calculus

Differential calculus is essentially concerned with the rate of change of a dependent variables w.r.t. an independent variable.

Determination of nth derivatives of standard functions, Leibnitz's theorem.

If $y = f(x)$, then the result of differentiating the function n times (*i.e.* the nth derivative) is written as:

$$D^n y = \left(\frac{d^n y}{dx^n}\right) = f^{(n)}(x) = y_n$$

2.1 THE nth DERIVATIVES OF SOME STANDARD FUNCTIONS

1. To find the nth derivative of x^m.

Let $\quad y = x^m$, (m is a positive integer)

then $\quad Dy = y_1 = mx^{m-1}$

$$D[Dy] = D^2 y = Dy_1 = D\,(mx^{m-1}) = m\,(m-1)x^{m-2}$$

$$D^3 y = D[D^2 y] = D\,[Dy_1] = D\,[m\,(m-1)\,x^{m-2}]$$

$$= m\,(m-1)\,(m-2)\,x^{m-3}$$

$$\vdots \qquad \vdots \qquad \vdots$$

$$\therefore \quad D^n y = y_n = [m\,(m-1)\,(m-2)\ldots \text{upto } n \text{ factors}]\cdot x^{m-n}$$

$$= m\,(m-1)\,(m-2)\ldots(m-n+1)\,x^{m-n}, \text{ where } n < m$$

Hence $D^n\,(x^m) = m\,(m-1)\ldots(m-n+1)\,x^{m-n}$, $(n < m)$

Multiply above and below $(m-n)!$

We get, $\quad D^n y = \left[\dfrac{m!}{(m-n)!}\right]\cdot x^{m-n}$

Cor. If $\quad m = n$ (m is positive integer)

then $\quad y_m = m\,(m-1)\ldots(m-m+1)\,x^{m-m}$

$$= m\,(m-1)\,(m-2)\ldots 1$$

$$= m!$$

As $y_m = m!$ is constant

$\therefore$ y_{m+1} which is the derivative of y_m will be zero and so on.

Thus, $$D^n(x^m) = \left[\frac{m!}{(m-n)!}\right].x^{m-n}, \text{ if } (n < m)$$
$$= n!, \text{ if } (n = m)$$
$$= 0 \text{ if } (n > m)$$
(m is a positive Integer)

2.(a) To find the nth derivative of $(ax + b)^m$, where $m > n$,

Let $y = (ax + b)^m$

$$y_1 = m(ax + b)^{m-1} \cdot a$$
$$y_2 = m(m-1)(ax + b)^{m-2} \cdot a^2$$
$$y_3 = m(m-1)(m-2)(ax + b)^{m-3} \cdot a^3$$
::: ::: :::
$$y_n = m(m-1)(m-2)\ldots(m-n+1)(ax+b)^{m-n} \cdot a^n$$

Now, $m(m-1)\ldots(m-n+1)\cdot(m-n)(m-n-1)\ldots 3.2.1$

$$= \frac{m(m-1)\ldots(m-n+1)\cdot(m-n)(m-n-1)\ldots 3.2.1}{(m-n)\cdot(m-n-1)\ldots 3.2.1} = \frac{m!}{(m-n)!}$$

$$\therefore \quad y_n = \frac{m!}{(m-n)!} \cdot a^n \cdot (ax+b)^{m-n} \qquad (m > n) \quad [1]$$

Where 'm' is a positive integer greater than n.

Hence $$y_n = \frac{d^n}{dx^n}[(ax+b)^m] = \frac{m!}{(m-n)!} a^n \cdot (ax+b)^{m-n} \qquad (m > n)$$

Note: If we put $a = 1$, $b = 0$ and $m = \frac{p}{q}$ in (p, q are integers $q \neq 0$) $(ax + b)^m$, this will become $x^{p/q}$ · we find $D^n[x^{p/q}]$ as:

$$y_1 = Dy = \frac{p}{q} x^{p/q-1} = \frac{p}{q} x^{\frac{p-q}{q}}$$

$$y_2 = \frac{p}{q}\left(\frac{p}{q} - 1\right) x^{\frac{p-q}{q}-1}$$

$$\Rightarrow \quad y_2 = \frac{p(p-1)}{q^2} x^{\frac{p-2q}{q}} \quad \text{and so on}$$

$$D^n[x^{p/q}] = \frac{p(p-q)\ldots[p-(n-1)q]}{q^n} \cdot x^{\frac{(p-nq)}{q}}$$

2.(b) To find the nth derivative of $(ax + b)^m$ when $(m = n)$

Let $y = (ax + b)^m$

then, $$y_1 = m(ax+b)^{m-1} \cdot a$$
$$y_2 = m(m-1)(ax+b)^{m-2} \cdot a^2$$
$$y_3 = m(m-1)(m-2)(ax+b)^{m-3} \cdot a^3$$
::: ::: :::
$$y_n = [m(m-1)(m-2)\ldots \text{upto } n \text{ factors}] \cdot [ax+b]^{m-n} \cdot a^n$$

If $m = n$, then,

$$= m(m-1)(m-2)\ldots(m-m+1)\cdot(ax+b)^{m-m}\cdot a^m$$

$$= m(m-1)\ldots 1\cdot a^m$$

$= m!\cdot a^m$ = constant and therefore $(ax+b)^m$ can at the most have mth d.c. and not $(m+1)$th etc. . .

i.e. $y_{m+1} = y_{m+2} = \ldots = 0$ (each)

3. To find the nth differential coefficient of $\dfrac{1}{(ax+b)}$, $\left(x \neq -\dfrac{b}{a}\right)$

Let $\quad y = \dfrac{1}{(ax+b)} = (ax+b)^{-1}$

Then $\quad y_1 = -1\cdot(ax+b)^{-2}\cdot a$

$$y_2 = (-1)^2\cdot 1\cdot 2\cdot(ax+b)^{-3}\cdot a^2$$

$$= (-1)^2\cdot 2!\cdot(ax+b)^{-3}\cdot a^2$$

$$y_3 = (-1)^3\cdot 3\cdot 2!\cdot(ax+b)^{-4}\cdot a^3$$

$$= (-1)^3\cdot 3!\cdot(ax+b)^{-4}\cdot a^3$$

$$\vdots \qquad \vdots \qquad \vdots$$

$$y_n = (-1)^n\cdot n!\cdot(ax+b)^{-(n+1)}\cdot a^n$$

$$y_n = \frac{(-1)^n\cdot n!\cdot a^n}{(ax+b)^{n+1}}$$

4. To find the nth derivative of $\dfrac{1}{(ax+b)^m}$ $\left(x \neq -\dfrac{b}{a}\right)$

Let $\quad y = \dfrac{1}{(ax+b)^m} = (ax+b)^{-m}$

$$y_1 = -ma(ax+b)^{-(m+1)}$$

$$y_2 = (-m)\{-(m+1)\}\,a^2\cdot(ax+b)^{-(m+2)}$$

$$= (-1)^2\cdot m\cdot(m+1)\cdot a^2(ax+b)^{-(m+2)}$$

$$y_3 = (-1)^3\cdot m\cdot(m+1)\cdot(m+2)\,a^3(ax+b)^{-(m+3)}$$

$$\vdots \qquad \vdots \qquad \vdots$$

$$y_n = \frac{(-1)^n\cdot m(m+1)(m+2)\ldots(m+n-1)a^n}{(ax+b)^{m+n}}$$

Now, $m(m+1)(m+2)\ldots(m+n-1)$

$$= \frac{1\cdot 2\cdot 3\ldots(m-1)\cdot m\cdot(m+1)(m+2)\ldots(m+n-1)}{1\cdot 2\cdot 3\ldots(m-1)} = \frac{(m+n-1)!}{(m-1)!}$$

Hence, $\quad D^n y = y_n = \dfrac{d^n}{dx^n}\left[\dfrac{1}{(ax+b)^m}\right] = (-1)^n\cdot\dfrac{(m+n-1)!}{(m-1)!}\cdot\dfrac{a^n}{(ax+b)^{m+n}}$

Ex. 1. $(a^2-x^2)^{-1}$, find y_n

Let $\quad y = (a^2-x^2)^{-1} = \dfrac{1}{(a^2-x^2)} = \dfrac{1}{(a+x)(a-x)}$

$$= \frac{1}{(a+x)[a-(-a)]} + \frac{1}{(a+a)(a-x)}$$

$$= \frac{1}{2a}\left[\frac{1}{(a+x)} + \frac{1}{(a-x)}\right]$$

$$\therefore \quad y_n = \frac{1}{2a}\left[\frac{(-1)^n . n!}{(a+x)^{n+1}} + \frac{(-1)^n . n!(-1)^n}{(a-x)^{n+1}}\right]$$

$$= \frac{n!}{2a}\left[\frac{(-1)^n}{(a+x)^{n+1}} + \frac{1}{(a-x)^{n+1}}\right] \qquad (\because (-1)^{2n} = 1)$$

Ex 2. $(2x + 3)^7$ find y_4, & y_7.

Let $y = (2x + 3)^7$

then $y_4 = \frac{7!}{(7-4)!} \cdot 2^4 (2x+3)^{7-4}$ using nth derivative in result [2]

$$= \frac{7!}{3!} \cdot 2^4 (2x+3)^3 \text{ and}$$

$$y_7 = (7! \cdot 2^7)$$

and $y_8 = y_9 = 9_{10}$ $\therefore = 0$ (each)

5. To find the *n*th derivative of (i) sin (*ax* + *b*) (ii) cos (*ax* + *b*).

(*i*) Let $y = \sin(ax + b)$

$$y_1 = a\cos(ax+b) \qquad \left(\because \sin\left(\frac{\pi}{2}+\theta\right) = \cos\theta\right)$$

$$= a\sin\left(ax + b + \frac{\pi}{2}\right)$$

$$y_2 = a^2\cos\left(ax + b + \frac{1}{2}\pi\right)$$

$$= a^2\sin\left(ax + b + \frac{1}{2}\pi + \frac{1}{2}\pi\right) = a^2\sin\left(ax + b + 2\cdot\frac{1}{2}\pi\right) \qquad (i)$$

$$y_3 = a^3\cos\left(ax + b + 2\cdot\frac{1}{2}\pi\right) \qquad (ii)$$

(Note that after differentiating (*i*), we get (*ii*), further, we express (*ii*) as:

$$= a^3\sin\left(ax + b + 2\cdot\frac{1}{2}\pi + \frac{1}{2}\pi\right)$$

$$= a^3\sin\left(ax + b + 3\cdot\frac{1}{2}\pi\right)$$

proceeding as above, we get,

$$y_n = a^n\sin\left(ax + b + n\cdot\frac{1}{2}\pi\right)$$

Hence, $D^n[\sin(ax+b)] = a^n \cdot \sin\left(ax + b + \frac{n\pi}{2}\right)$

Note: 1 Putting $b = 0$, we get

$$y = \sin(ax) \text{ and hence}$$

$$y_n = a^n \sin\left(ax + n\cdot\frac{1}{2}\pi\right)$$

Note: 2 If $a = 1$, $b = 0$, we get

$$D^n[\sin x] = \sin\left(x + \frac{n\pi}{2}\right)$$

(*ii*) Let $\quad$ $\boldsymbol{y = \cos(ax + b)}$

$$y_1 = -a\sin(ax+b)$$

$$= a\cos\left(ax+b+\frac{\pi}{2}\right) \qquad \left(\because \cos\left[\left(\frac{\pi}{2}\right)+\theta\right] = -\sin\theta\right)$$

$$y_2 = -a^2\sin\left(ax+b+\frac{\pi}{2}\right) = a^2\cos\left(ax+b+\frac{\pi}{2}+\frac{\pi}{2}\right)$$

$$= a^2\cos\left(ax+b+2\,.\frac{\pi}{2}\right)$$

$$\therefore \quad y_3 = -a^3\sin\left(ax+b+2\,.\frac{\pi}{2}\right) = a^3\cos\left(ax+b+2\,.\frac{\pi}{2}+\frac{\pi}{2}\right)$$

$$= a^3\cos\left(ax+b+3\cdot\frac{\pi}{2}\right)$$

$$\vdots \qquad \vdots \qquad \vdots$$

$$y_n = a^n\cos\left(ax+b+n\cdot\frac{\pi}{2}\right)$$

Note: 1. If $b = 0$, then $y = \cos(ax)$

$$\therefore \quad y_n = a^n\cos\left(ax + n\cdot\frac{\pi}{2}\right)$$

Note: 2. If $b = 0$, $a = 1$

then $\quad y = \cos x$

$$\therefore \quad y_n = \cos\left(x + n\cdot\frac{\pi}{2}\right).$$

6. (*i*) **To find the *n*th derivative of e^{ax}.**

Let $\quad y = e^{ax}$

$$y_1 = ae^{ax}$$

$$y_2 = a^2\cdot e^{ax}$$

$$y_3 = a^3\cdot e^{ax}$$

$$\vdots \qquad \vdots$$

$$y_n = a^n\cdot e^{ax}$$

Note: If $y = e^x$, $a = 1$

$\therefore$ $\quad y_n = e^x$.

(*ii*) **To find the *n*th derivative of a^{mx}.**

Let $\quad y = a^{mx}$

$$y_1 = m \cdot a^{mx} \cdot (\log a)$$
$$y_2 = m^2 \cdot a^{mx} (\log a)^2$$
$$y_3 = m^3 \cdot a^{mx} \cdot (\log a)^3$$
$$\vdots \qquad \vdots \qquad \vdots$$
$$y_n = m^n \cdot a^{mx} \cdot (\log a)^n$$

Note: 1. If $y = a^x$, where $m = 1$

$\therefore \quad y_n = a^x (\log a)^n$

Note: 2. $A^B = e^{B\log A}$, so we can also write $y = a^{mx} = e^{mx \log a} = e^{x\,(m \log a)} = e^{x\,(A)}$

where $A = m \log a$ and proceed as in (*i*).

7. To find the *n*th derivative of log (*ax* + *b*).

Let $\quad y = \log (ax + b)$

$$y_1 = \frac{1}{(ax + b)} \cdot a = a\,(ax + b)^{-1}$$
$$y_2 = (-1)\, a^2\, (ax + b)^{-2}$$
$$y_3 = (-1)\,(-2)\, a^3\, (ax + b)^{-3} = (-1)^2 \cdot 2!\, a^3\, (ax + b)^{-3}$$
$$y_4 = (-1)^3 \cdot 3 \cdot 2! \cdot a^4\, (ax + b)^{-4} = (-1)^3 \cdot 3!\, a^4\, (ax + b)^{-4}$$
$$y_n = (-1)^{n-1} \cdot (n - 1)!\, a^n\, (ax + b)^{-n}$$

Hence
$$y_n = \frac{(-1)^{n-1} \cdot (n-1)! \cdot a^n}{(ax + b)^n}$$

8. To find the *n*th derivative of:

(*i*) e^{ax} . sin (*bx* + *c*), (VTU, Mar., 2000)

(*ii*) e^{ax} . cos (*bx* + *c*), (VTU, Mar., 1999)

(*i*) Let $\quad y = e^{ax} \sin (bx + c)$

$$y_1 = e^{ax} \cdot \cos (bx + c) \cdot b + \sin (bx + c) \cdot e^{ax} \cdot a$$
$$= e^{ax} [a \sin (bx + c) + b \cos (bx + c)]$$

put $\quad a = r \cos \theta;\ b = r \sin \theta,$

so that $\quad a^2 + b^2 = r^2 (\cos^2 \theta + \sin^2 \theta) = r^2$

$\therefore \quad r = \sqrt{a^2 + b^2}$

and $\quad \tan \theta = \dfrac{b}{a},$

$\therefore \quad \theta = \tan^{-1} \dfrac{b}{a}$

$\therefore$
$$y_1 = e^{ax} [r \cos \theta \sin (bx + c) + r \sin \theta \cos (bx + c)]$$
$$= r \cdot e^{ax} [\sin (bx + c) \cos \theta + \cos (bx + c) \sin \theta]$$
$$= r \cdot e^{ax} \sin (bx + c + \theta) \qquad (1)\ (\because S_{A+B} = S_A C_B + C_A S_B)$$

Thus, the result of differentiation of the given function is equivalent to multiplying it by a constant r and adding an angle $\theta = \tan^{-1} \left(\dfrac{b}{a}\right)$.

Repeating the same procedure for the differentiation of y_1, we have

$$y_2 = r^2 \cdot e^{ax} \sin (bx + c + 2\theta)$$

similarly,

$$y_3 = r^3 \cdot e^{ax} \sin (bx + c + 3\theta)$$

$$\vdots \quad \vdots \quad \vdots$$

$$y_n = r^n e^{ax} \sin (bx + c + n\theta)$$

Hence,

$$y_n = (a^2 + b^2)^{n/2} \cdot e^{ax} \cdot \sin \left[bx + c + n \tan^{-1} \frac{b}{a} \right]$$

(*ii*) Let

$$y = e^{ax} \cos (bx + c)$$

$$y_1 = Dy = e^{ax} [- \sin (bx + c)] \cdot b + \cos (bx + c) \cdot e^{ax} \cdot a]$$

$$= e^{ax} [a \cos (bx + c) - b \sin (bx + c)]$$

put

$$a = r \cos \theta;$$

$$b = r \sin \theta \quad \Rightarrow \quad a^2 + b^2 = r^2$$

$$\therefore \quad r = \sqrt{a^2 + b^2}$$

and

$$\tan \theta = \frac{b}{a}$$

$$\therefore \quad \theta = \tan^{-1} \left(\frac{b}{a} \right)$$

$$\therefore \quad y_1 = e^{ax} [r \cos \theta \cos (bx + c) - r \sin \theta \sin (bx + c)]$$

$$= r \cdot e^{ax} \cos (bx + c) \cos \theta - \sin (bx + c) \sin \theta]$$

$$= r \cdot e^{ax} \cos (bx + c + \theta) \qquad (\because C_{A+B} = C_A C_B - S_A S_B)$$

Thus, the result of differentiation of the given function is equivalent to multiplying it by a constant r and adding an angle $\theta = \tan^{-1} \left(\frac{b}{a} \right)$.

Repeating the same procedure for the differentiation of y_1, we have

$$y_2 = r^2 \cdot e^{ax} \cdot \cos (bx + c + 2\theta)$$

|||ly,

$$y_3 = r^3 \cdot e^{ax} \cdot \cos (bx + c + 3\theta)$$

$$\vdots \quad \vdots \quad \vdots$$

$$y_n = r^n \cdot e^{ax} \cos (bx + c + n\theta)$$

Hence,

$$y_n = D^n y = (a^2 + b^2)^{n/2} \cdot e^{ax} \cdot \cos (bx + c + n \tan^{-1} b/a).$$

A. Examples (1-10)

[Relating to forms and by spliting into partial fractions]:

$$(i)\ D^n \left[\frac{1}{ax + b} \right] = \left[\frac{(-1)^n \cdot n! \cdot a^n}{(ax + b)^{n+1}} \right] \qquad (ii)\ D^n \left[\frac{1}{x} \right] = \frac{(-1)^n \cdot n!}{x^{n+1}}$$

Find the nth derivatives of following functions.

1. $\dfrac{1}{(3 - 2x)}$

Solution: Let $y = \dfrac{1}{(3 - 2x)}$, $y_1 = - \dfrac{1 \cdot (-2)}{(3 - 2x)^2}$, etc.

$$\therefore \quad y_n = \frac{(-1)^1 \cdot n! \cdot (-2)^n}{(3-2x)^{n+1}} = \frac{n! \cdot 2^n}{(3-2x)^{n+1}} \qquad (\because (-1)^{2n} = 1)$$

2. $\dfrac{1}{1-5x+6x^2}$

Solution: Let $y = \dfrac{1}{1-5x+6x^2} = \dfrac{1}{6x^2-5x+1} = \dfrac{1}{(3x-1)(2x-1)}$

$$= \frac{1}{(3x-1)\left(2\cdot\frac{1}{3}-1\right)} + \frac{1}{\left(3\cdot\frac{1}{2}-1\right)(2x-1)}$$

(By the application of the Rule of Everywhere but not there).

$$= \frac{1}{(3x-1)\left(-\frac{1}{3}\right)} + \frac{1}{\frac{1}{2}(2x-1)} = \frac{-3}{3x-1} + \frac{2}{2x-1}$$

$$\therefore \quad y_n = -3 \cdot \frac{(-1)^n \cdot n!\, 3^n}{(3x-1)^{n+1}} + 2 \cdot \frac{(-1)^n \cdot n! \cdot 2^n}{(2x-1)^{n+1}}$$

$$= (-1)^n \cdot n! \left[\frac{-3^{n+1}}{(3x-1)^{n+1}} + \frac{2^{n+1}}{(2x-1)^{n+1}}\right]$$

$$= (-1)^n \cdot n! \left[\frac{2^{n+1}}{(2x-1)^{n+1}} - \frac{3^{n+1}}{(3x-1)^{n+1}}\right]$$

3. $\dfrac{(2x+3)}{(x^2+3x+2)}$

Solution: Let $y = \dfrac{(2x+3)}{(x^2+3x+2)} = \dfrac{2x+3}{(x+1)(x+2)}$

$$= \frac{A}{(x+1)} + \frac{B}{(x+2)}$$

$\therefore \quad 2x + 3 = A(x+2) + B(x+1)$

put $\quad x = -1;\ (-2+3) = A(-1+2) + 0$

$\therefore \quad A = 1$

$\quad x = -2;\ (-4+3) = 0 + B(-2+1)$

$\therefore \quad B = 1$

$$\therefore \quad y = \left[\frac{1}{(x+1)}\right] + \left[\frac{1}{(x+2)}\right]$$

$$\therefore \quad D^n y = \left[\frac{(-1)^n \cdot n!}{(x+1)^{n+1}}\right] + \left[\frac{(-1)^n \cdot n!}{(x+2)^{n+1}}\right]$$

$$= (-1)^n\, n! \left\{\left[\frac{1}{(x+1)^{n+1}}\right] + \left[\frac{1}{(x+2)^{n+1}}\right]\right\}.$$

4. $\dfrac{1}{(x^2+3x+2)}$.

Solution: Let $y = \dfrac{1}{(x^2+3x+2)} = \dfrac{1}{(x+1)(x+2)} = \dfrac{A}{(x+1)} + \dfrac{B}{(x+2)}$

$\therefore \quad 1 = A(x+2) + B(x+1)$

put $\quad x = -1;\ 1 = A(-1+2) + 0;\ A = 1$

$= -2;\ 1 = 0 + B(-2+1)\ B = -1$

$\therefore \quad y = \left[\dfrac{1}{(x+1)}\right] - \left[\dfrac{1}{(x+2)}\right]$

$\therefore \quad D^n y = [(-1)^n \cdot n!/(x+1)^{n+1}] - [(-1)^n \cdot n!/(x+2)^{n+1}]$

$$= (-1)^n \cdot n! \left[\left\{\frac{1}{(x+1)^{n+1}}\right\} - \left\{\frac{1}{(x+2)^{n+1}}\right\}\right]$$

5. $\dfrac{x^2+1}{(x-1)(x-2)(x-3)}$

Solution: Let $y = \dfrac{x^2+1}{(x-1)(x-2)(x-3)} = \dfrac{A}{(x-1)} + \dfrac{B}{(x-2)} + \dfrac{C}{(x-3)}$

Now by the rule of "Everywhere but not there" or method of supression where A by the method of supression is

$$\frac{1^2+1}{(1-3)(1-2)} = \frac{2}{2} = 1.$$

B by the method of supression is

$$\frac{2^2+1}{(2-1)(2-3)} = \frac{5}{-1} = -5.$$

C by the method of supression is

$$\frac{3^2+1}{(3-1)(3-2)} = \frac{10}{1} = 5.$$

Thus, $\quad y_n = D^n\left[\dfrac{x^2+1}{(x-1)(x-2)(x-3)}\right] = D^n\left[\dfrac{1}{(x-1)} - \dfrac{5}{(x-2)} + \dfrac{5}{(x-3)}\right]$

$$= (-1)^n \cdot n!\left[\frac{1}{(x-1)^{n+1}} - \frac{5}{(x-2)^{n+1}} + \frac{5}{(x-3)^{n+1}}\right]$$

6. $\dfrac{x^2}{2x^2+7x+6} = \dfrac{x^2}{(x+2)(2x+3)}$ **(B.U. Jan., 1993, 1996)**

Solution: Since, degree of $\quad N^r = 2$

and degree of $\quad D^r = 2$

$\therefore$ By actual division, we get

$$2x^2 + 7x + 6 \,\big)\, x^2 \,\big(\, \frac{1}{2}$$

$$\begin{array}{l} -\left(x^2 + \frac{7}{2}x + 3\right) \\ \hline -\frac{7}{2}(x+6) \end{array}$$

$$\therefore \quad \frac{x^2}{2x^2 + 7x + 6} = \frac{1}{2} - \frac{1}{2}\left(\frac{7x+6}{(x+2)(2x+3)}\right) \quad [1]$$

Splitting $\frac{7x+6}{(x+2)(2x+3)}$ into P.F.

we get, $$\frac{7x+6}{(x+2)(2x+3)} = \left(\frac{A}{(x+2)} + \frac{B}{(2x+3)}\right)$$

$$\Rightarrow \quad 7x + 6 = A(2x+3) + B(x+2)$$

put $x = -2$, we get

$$-8 = A(-4+3) + 0 \quad \therefore \quad A = 8$$

put $$x = -\frac{3}{2}$$

$$\frac{-21}{2} + 6 = A(-3+3) + B\left(-\frac{3}{2} + 2\right)$$

$$\frac{-9}{2} = \frac{B}{2}; \quad \therefore B = -9$$

$$\therefore \quad D^n\left[\frac{x^2}{(x+2)(2x+3)}\right] = D^n\left[\frac{1}{2} - 4\cdot\frac{1}{x+2} + \frac{9}{2}\cdot\frac{1}{(2x+3)}\right] \quad \text{(using [1])}$$

$$= 0 - 4\,\frac{(-1)^n \cdot n!}{(x+2)^{n+1}} + \frac{9}{2}\,\frac{(-1)^n\, n!\, 2^n}{(2x+3)^{n+1}}$$

$$= (-1)^n \cdot n!\left[\frac{9\cdot 2^{n-1}}{(2x+3)^{n+1}} - \frac{4}{(x+2)^{n+1}}\right]$$

7. $\dfrac{x^4}{(x-1)(x-2)}$

Solution: degree of $N^r = 4$; degree of $D^r = 2$

Let $$\frac{x^4}{(x-1)(x-2)} = \frac{x^4}{x^2 - 3x + 2}$$

By Actual Division Method

$$\begin{array}{r} x^2 - 3x + 2 \,\big)\; x^4 \;\big(x^2 + 3x + 7 \\ -(x^4 - 3x^3 + 2x^2) \\ \hline 3x^3 - 2x^2 \\ -(3x^3 - 9x^2 + 6x) \\ \hline 7x^2 - 6x \\ -(7x^2 - 21x + 14) \\ \hline 15x - 14 \end{array}$$

$$\therefore \quad \frac{x^4}{(x-1)(x-2)} = x^2 + 3x + 7 + \frac{15x-14}{(x-1)(x-2)} \qquad [1]$$

Now, $$\frac{15x-14}{(x-1)(x-2)} = \frac{15(1)-14}{(x-1)(1-2)} + \frac{15(2)-14}{(2-1)(x-2)}$$

(By the rule of everywhere but not there)

$$= \frac{-1}{(x-1)} + \frac{16}{(x-2)}$$

$\therefore$ [1] becomes

$$D^n\left[\frac{x^4}{(x-1)(x-2)}\right] = D^n\left[x^2+3x+7-\frac{1}{x-1}+\frac{16}{x-2}\right]$$

$$= -\frac{(-1)^n \cdot n!}{(x-1)^{n+1}} + 16 \cdot \frac{(-1)^n \cdot n!}{(x-2)^{n+1}}$$

$$= (-1)^n \cdot n!\left[\frac{16}{(x-2)^{n+1}} - \frac{1}{(x-1)^{n+1}}\right]$$

8. $\dfrac{x^3}{(x^2-1)}$.

Here: degree of $N^r = 3$

degree of $D^r = 2$

$\therefore$ By actual division method, we have

$$x^2-1\overline{)\;x^3\;}(x$$
$$\underline{-(x^3-x)}$$
$$x$$

Now, Let $$y = \frac{x^3}{x^2-1} = x + \frac{x}{x^2-1} \qquad [1]$$

and $$\frac{x}{x^2-1} = \frac{x}{(x-1)(x+1)} = \frac{A}{(x-1)} + \frac{B}{(x+1)}$$

$\Rightarrow$ $x = A(x+1) + B(x-1)$

put $x = -1$, $-1 = A(0) + B(-2)$ $\therefore B = \dfrac{1}{2}$

put $x = 1$, we get

$1 = A(2) + B(0)$

$\therefore$ $A = \dfrac{1}{2}$

Using in [1] the values of A and B

we get $$D^n y = D^n\left[\frac{x^3}{(x^2-1)}\right] = D^n\left[x + \frac{1}{2}\left(\frac{1}{x-1}+\frac{1}{x+1}\right)\right] \qquad (n>1)$$

$$= \frac{(-1)^n \cdot n!}{2}\left[\frac{1}{(x-1)^{n+1}} + \frac{1}{(x+1)^{n+1}}\right]$$

PART-B

Note: $[D^n y]_{x=0} = (y_n)_0 = \frac{(-1)^n \cdot n!}{2}\left[\frac{1}{(-1)^{n+1}} + \frac{1}{(1)^{n+1}}\right]$

9. Prove that $\frac{d^{2n}}{dx^{2n}}(x^2 - 1)^n = (2n)!$

Solution: Expanding by binomial theorem, the function $y = (x^2 - 1)^n$, we find:

$$(x^2 - 1)^n = (x^2)^n + {}^nC_1 (x^2)^{n-1}(-1) + \ldots + {}^nC_{n-1}(x^2)(-1)^{n-1} + {}^nC_n(-1)^n$$
$$= x^{2n} - n \cdot x^{2n-2} + {}^nC_2\, x^{2n-4} - \ldots {}^nC_{n-1}\, x^2 (-1)^{n-1} + (-1)^n$$

Differentiate both sides of above expression $2n$ times (and not n times), we get

$$D^{2n}(x^2 - 1)^n = D^{2n}(x^{2n}) - n \cdot D^{2n}(x^{2n-2}) + \ldots + D^{2n}(-1)^n$$
$$= (2n)! - 0 + 0 \ldots + 0 + \ldots$$

(By the concept of differentiation)

i.e., $D^n(x^m) = 0$, if $(n > m)$

(if n is greater than m)

$D^n(x^m) = D^n(x^n) = n!$, if $(m = n)$

Hence $\frac{d^{2n}}{dx^{2n}}(x^2 - 1)^n = (2n)!$

10. $\frac{1}{(4x^2 - 1)}$.

Let $y = \frac{1}{(4x^2 - 1)} = \frac{1}{(2x+1)(2x-1)} = \frac{A}{(2x+1)} + \frac{B}{(2x-1)}$

$\therefore$ $1 = A(2x - 1) + B(2x + 1)$

put $2x = 1;\ 1 = 0 + B(1 + 1);\ B = \frac{1}{2}$

$2x = -1;\ 1 = A(-1 - 1) + 0;\ A = -\frac{1}{2}$

$$\therefore \quad y = \left(-\frac{1}{2}\right)\left[\frac{1}{(2x+1)}\right] + \left(\frac{1}{2}\right)\left[\frac{1}{(2x-1)}\right]$$

$$\therefore \quad D^n y = \left\{\left(-\frac{1}{2}\right)(-1)^n \cdot n!\frac{2^n}{(2x+1)^{n+1}}\right\} + \left\{\left(\frac{1}{2}\right)(-1)^n\, n!\frac{2^n}{(2x-1)^{n+1}}\right\}$$
$$= (-1)^n \cdot n!\, 2^{n-1}\left[\left\{\frac{1}{(2x-1)^{n+1}}\right\} - \left\{\frac{1}{(2x+1)^{n+1}}\right\}\right]$$

B. Examples (11-20)

(Related to the forms:

(*i*) $$D^n\left[\frac{1}{(ax+b)^m}\right] = \left[\frac{(-1)^n (m+n-1)!}{(m-1)!} \cdot \frac{a^n}{(ax+b)^{m+n}}\right]$$

(*ii*) $$D^n\left[\frac{1}{x^2}\right] = \left[(-1)^n \frac{(n+1)!}{x^{n+2}}\right]$$

$$D^n\left[\frac{1}{x^3}\right] = \left[(-1)^n \frac{(n+2)!}{2!\, x^{n+3}}\right]$$

11. $\dfrac{1}{x(x+1)^2}$

[Since the denominator has a pair of repeated factors, and so in the partial fractions, we have one fraction for each of the powers of the repeated factors and noting $N^r = 1 =$ constant.]

Soluion: Let $y = \dfrac{1}{x(x+1)^2} = \dfrac{A}{x} + \dfrac{B}{(x+1)} + \dfrac{C}{(x+1)^2}$ [1]

$\therefore$ $1 = A(x+1)^2 + Bx(x+1) + C(x)$

put $x = 0, 1 = A \cdot 1 + 0 + 0, \therefore A = 1$

$x = -1; 1 = 0 + 0 + C(-1) \therefore C = -1$

$x = 1; 1 = A(1+1)^2 + B(1+1) + C(-1)$

$1 = 4 + 2B - 1$

$\therefore$ $B = -1$

(using [1]), we get

$$y = \left[\frac{1}{x}\right] - \left[\frac{1}{(x+1)}\right] - \left[\frac{1}{(x+1)^2}\right]$$

$$\therefore \quad D^n y = \frac{(-1)^n \cdot n!}{x^{n+1}} - \frac{(-1)^n \cdot n!}{(x+1)^{n+1}} - \frac{(-1)^n \cdot (n+1)!}{(x+1)^{n+2}}$$

$$= (-1)^n \cdot n! \left[\frac{1}{x^{n+1}} - \frac{1}{(x+1)^{n+1}} - \frac{(n+1)}{(x+1)^{n+2}}\right]$$

12. $\dfrac{1}{(x+3)(4-x)^2}$

D^r = repeated factor.

Let $y = \dfrac{1}{(x+3)(4-x)^2} = \dfrac{1}{(x+3)(x-4)^2}$

$= \dfrac{A}{(x+3)} + \dfrac{B}{(x-4)} + \dfrac{C}{(x-4)^2}$ [1]

$\Rightarrow$ $1 = A(x-4)^2 + B(x+3)(x-4) + C(x+3)$

put $x = 4; 1 = 0 + 0 + C(4+3)$:

$\therefore$ $C = \left(\dfrac{1}{7}\right)$

$x = -3; 1 = A(-3-4)^2 + 0 + 0$

$\therefore$ $A = \left(\dfrac{1}{49}\right)$

$x = 0; 1 = A(16) + B(-12) + C(3)$

$= \left(\dfrac{16}{49}\right) - 12B + \left(\dfrac{3}{7}\right); B = -\dfrac{1}{49}$

$\therefore$ By using [1], we have

$$D^n y = (-1)^n \cdot n! \left[\frac{1}{49} \frac{1}{(x+3)^{n+1}} - \frac{1}{49} \frac{1}{(x-9)^{n+1}} + \frac{(n+1)}{7(x-4)^{n+2}} \right]$$

13. (a) $\dfrac{x^2}{(x-1)^2 (x+2)}$.

Let $x - 1 = z$, *i.e.* $x = z + 1$

$$\therefore \quad y = \frac{(z+1)^2}{z^2(3+z)} = \frac{1}{z^2} \cdot \frac{1+2z+z^2}{3+z}$$

By Actual Division

$$= \frac{1}{z^2}\left[\frac{1}{3} + \frac{5z}{9} + \frac{4}{9} \cdot \frac{z^2}{3+z}\right]$$

$$= \frac{1}{3z^2} + \frac{5}{9z} + \frac{4}{9} \cdot \frac{1}{3+z} = \frac{1}{3(x-1)^2} + \frac{5}{9(x-1)} + \frac{4}{9(x+2)}$$

Hence, $$y_n = \frac{(n+1)!(-1)^n}{3(x-1)^{n+2}} + \frac{5n!(-1)^n}{9(x-1)^{n+1}} + \frac{4n!(-1)^n}{9(x+2)^{n+1}}$$

13. (b) $\dfrac{1}{(x-1)^3 (x-2)}$.

Method (I): Let $y = \dfrac{1}{(x-1)^3 (x-2)}$ and

$$\frac{1}{(x-1)^3 (x-2)} = \frac{A}{(x-1)} + \frac{B}{(x-1)^2} + \frac{C}{(x-1)^3} + \frac{D}{(x-2)}$$

or $1 = A(x-1)^2(x-2) + B(x-1)(x-2) + C(x-2) + D(x-1)^3$ [1]

$1 = A(x^2 - 2x + 1)(x-2) + B(x^2 - 3x + 2) + C(x-2) + D(x^3 - 3x^2 + 3x - 1)$ [2]

Putting $x = 1$ is [1], we get

$1 = -C$ or $C = -1$

Putting $x = 2$ in [1], we get $1 = D$

Equating the coefficients of x^3 on both sides in [2],

we get $0 = A + D = A + 1$

or $A = -1$

putting $x = 0$ in [2], we get

$1 = -2A + 2B - 2C - D = 2 + 2B + 2 - 1$

or $2B = -2$ or $B = -1$

$$\therefore \quad y = \frac{-1}{(x-1)} - \frac{1}{(x-1)^2} - \frac{1}{(x-1)^3} + \frac{1}{(x-2)}$$

Differentiating 'n' times w.r.t x, we get

$$y_n = -\frac{(-1)^n\, n!}{(x-1)^{n+1}} - \frac{(-1)^n\,(n+1)!}{1!(x-1)^{n+2}} - \frac{(-1)^n\,(n+2)!}{2!(x-1)^{n+3}} + \frac{(-1)^n\, n!}{(x-2)^{n+1}}.$$

Method (II): We split the given fraction into partial fraction by putting

$x - 1 = z$; $\therefore$ $x = z + 1$;

$\therefore$ $x - 2 = z - 1$

Let $$y = \frac{1}{(x-1)^3\,(x-2)} = \frac{1}{z^3\,(z-1)} = -\frac{1}{z^3\,(1-z)} \qquad [1]$$

$$= -\frac{1}{z^3}\left[1 + z + z^2 + \frac{z^3}{(1-z)}\right]$$

Now, by actual division method for $\frac{1}{(1-z)}$ in Eqn [1],

we have:

$$\begin{array}{r|l} 1-z\,) & 1 \quad (1+z+z^2 \\ & -(1-z) \\ \hline & z \\ & -(z-z^2) \\ \hline & z^2 \\ & -(z^2-z^3) \\ \hline & z^3 \end{array}$$

$\therefore$ $$\frac{1}{(1-z)} = 1 + z + z^2 + \frac{z^3}{(1-z)}$$

$\therefore$ [1] becomes

$$y = -\frac{1}{z^3} - \frac{1}{z^2} - \frac{1}{z} - \frac{1}{1-z}$$

$$= -\frac{1}{(x-1)^3} - \frac{1}{(x-1)^2} - \frac{1}{(x-1)} + \frac{1}{(x-2)}$$

$$y_n = -(-1)^n \cdot \frac{(n+2)!}{2!} \cdot \frac{1}{(x-1)^{n+3}} - (-1)^n \cdot \frac{(n+1)!}{1!} \cdot \frac{1}{(x-1)^{n+2}}$$

$$- \frac{(-1)^n \cdot n!}{(x-1)^{n+1}} + \frac{(-1)^n \cdot n!}{(x-2)^{n+1}}$$

$$= (-1)^{n+1} \cdot n! \left[\frac{(n+2)(n+1)}{2\cdot(x-1)^{n+3}} + \frac{(n+1)}{(x-1)^{n+2}} + \frac{1}{(x-1)^{n+1}} - \frac{1}{(x-2)^{n+1}}\right]$$

14. $\frac{1}{(bx+a)^2}$.

Let $$y = \frac{1}{(bx+a)^2} = (bx+a)^{-2}$$

$\therefore$ $$y_1 = Dy = (-2)\,(bx+a)^{-3} \cdot b = \frac{(-1)\cdot 2!\cdot b}{(bx+a)^3}$$

$$y_2 = (-2)\,(-3)\,(bx+a)^{-4} \cdot b^2 = \frac{(-1)^2 \cdot 3! \cdot b^2}{(bx+a)^4}$$

PART-B

$$y_3 = \frac{(-1)^3 \cdot 4! \cdot b^3}{(bx+a)^5}$$

$$\vdots \qquad \vdots \qquad \vdots$$

$$y_n = \frac{(-1)^n \cdot (n+1)! \cdot b^n}{(bx+a)^{n+2}}.$$

15. $\dfrac{1}{(3-2x)^3}$.

Let
$$y = \frac{1}{(3-2x)^3} = (3-2x)^{-3}$$

$\therefore$
$$y_1 = (-3)(3-2x)^{-4} \cdot (-2)$$
$$= \frac{3!}{(3-2x)^4}$$
$$y_2 = (-3)(-4)(3-2x)^{-5}(-2)^2$$
$$= \frac{4! \cdot 2}{(3-2x)^5}$$
$$y_3 = (-3)(-4)(-5)(3-2x)^{-6}(-2)^3$$
$$= \frac{5! \cdot 2^2}{(3-2x)^6} \text{ etc.}$$

$$\vdots \qquad \vdots \qquad \vdots$$

$$y_n = \frac{(n+2)! \cdot 2^{n-1}}{(3-2x)^{n+3}}.$$

16. $\sqrt{x+a}$

Let
$$y = (x+a)^{1/2}$$

$\therefore$
$$y_1 = \frac{1}{2}(x+a)^{-1/2}$$
$$y_2 = \frac{1}{2}\left(-\frac{1}{2}\right)(x+a)^{-3/2}$$
$$= (-1) \cdot \frac{1}{2^2}(x+a)^{\frac{1}{2}-2}$$
$$y_3 = \frac{1}{2}\left(-\frac{1}{2}\right)\left(-\frac{3}{2}\right)(x+a)^{-5/2}$$
$$= (-1)^2 \cdot \frac{1 \cdot 3}{2^3}(x+a)^{\frac{1}{2}-3}$$
$$y_4 = \frac{1}{2}\left(-\frac{1}{2}\right)\left(-\frac{3}{2}\right)\left(-\frac{5}{2}\right)(x+a)^{-7/2}$$

$$= (-1)^3 \cdot \frac{1 \cdot 3 \cdot 5}{2^4} \cdot (x+a)^{\frac{1}{2}-4}$$

$$\vdots \quad \vdots \quad \vdots$$

$$y_n = (-1)^{n-1} \cdot \frac{1 \cdot 3 \cdot 5 \ldots \text{upto } (n-1) \text{ factors}}{2^n} (x+a)^{\frac{1}{2}-n}$$

$$= (-1)^{n-1} \cdot \frac{1 \cdot 3 \cdot 5 \ldots (2n-3)}{2^n} (x+a)^{-\left(\frac{2n-1}{2}\right)}$$

17. $\dfrac{4x}{(x-1)^2 (x+1)}$. (B.U. Aug., 1994, 1996)

Let $\dfrac{4x}{(x-1)^2 (x+1)} = \dfrac{A}{(x-1)} + \dfrac{B}{(x-1)^2} + \dfrac{C}{(x+1)}$ [1]

$\Rightarrow \quad 4x = A(x-1)(x+1) + B(x+1) + C(x-1)^2$

put $x = 1$

$\therefore \quad 4 = B(1+1)$

$2B = 4 \quad \therefore \quad B = 2$

put $x = -1 \quad \therefore \quad -4 = C(-1-1)^2$

or $4C = -4 \quad \therefore \quad C = -1$

put $x = 0$,

$\therefore \quad 0 = A(-1)(1) + B(1) + C(-1)^2 \quad \Rightarrow \quad -A + B + C = 0$

$\therefore \quad -A + 2 - 1 = 0 \quad \Rightarrow \quad A = 1$

Hence $D^n \left[\dfrac{4x}{(x-1)^2 (x+1)}\right]$

$$= D^n \left[\frac{1}{(x-1)}\right] + 2D^n \left[\frac{1}{(x-1)^2}\right] - D^n \left[\frac{1}{(x+2)}\right] \quad \text{(using [1])}$$

$$= \left[\frac{(-1)^n \, n!}{(x-1)^{n+1}}\right] + 2\left[\frac{(-1)^n \, (n+1)!}{(x-1)^{n+2}}\right] - \left[\frac{(-1)^n \, n!}{(x+1)^{n+1}}\right]$$

18. $\dfrac{1}{6x^2 + 11x + 3}$.

Let $y = \dfrac{1}{6x^2 + 9x + 2x + 3} = \dfrac{1}{3x(2x+3) + 1(2x+3)}$

$$= \frac{1}{(3x+1)(2x+3)} = \frac{1}{7}\left[\frac{3}{3x+1} - \frac{2}{2x+3}\right] \quad \text{(by partial fractions)}$$

Hence, $y_n = D^n y$

$$= \frac{3}{7} \frac{(-1)^n \cdot \lfloor n \cdot 3^n}{(3x+1)^{n+1}} - \frac{2}{7} \frac{(-1)^n \cdot \lfloor n \cdot 2^n}{(2x+3)^{n+1}}$$

$$= \frac{(-1)^n}{7} \lfloor n \left[\frac{3^{n+1}}{(3x+1)^{n+1}} - \frac{2^{n+1}}{(2x+3)^{n+1}}\right]$$

PART-B

19. $\dfrac{1}{(x^2-a^2)(x^2-b^2)}$

Let
$$y = \frac{1}{(x^2-a^2)(x^2-b^2)}$$
$$= \frac{1}{(x^2-a^2)(a^2-b^2)} + \frac{1}{(b^2-a^2)(x^2-b^2)}$$
$$= \frac{1}{(a^2-b^2)}\left[\frac{1}{(x^2-a^2)} - \frac{1}{(x^2-b^2)}\right]$$
$$= \frac{1}{(a^2-b^2)}\left[\frac{1}{(x+a)(x-a)} - \frac{1}{(x+b)(x-b)}\right]$$
$$= \frac{1}{(a^2-b^2)}\left[\frac{1}{2a}\left(\frac{1}{(x-a)} - \frac{1}{(x+a)}\right) - \frac{1}{2b}\left(\frac{1}{(x-b)} - \frac{1}{(x+b)}\right)\right]$$

$\therefore$
$$y_n = D^n y$$
$$= \frac{1}{(a^2-b^2)}\left[\frac{1}{2a}\left\{\frac{(-)^n\, n!}{(x-a)^{n+1}} - \frac{(-)^n\, n!}{(x+a)^{n+1}}\right\} - \frac{1}{2b}\left\{\frac{(-)^n\, n!}{(x-b)^{n+1}} - \frac{(-1)^n\, n!}{(x+b)^{n+1}}\right\}\right]$$
$$= \frac{(-1)^n\, n!}{a^2-b^2}\left[\frac{1}{2a}\left\{\frac{1}{(x-a)^{n+1}} - \frac{1}{(x+a)^{n+1}}\right\} - \frac{1}{2b}\left\{\frac{1}{(x-b)^{n+1}} - \frac{1}{(x+b)^{n+1}}\right\}\right]$$

20. $\dfrac{1}{(a-bx)^4}$.

Let
$$y = \frac{1}{(a-bx)^4}$$

$\therefore$
$$y_n = D^n y$$
$$= (-1)^n \cdot \frac{(4+n-1)!}{(4-1)!} \cdot \frac{(-b)^n}{(a-bx)^{n+4}}$$
$$= (-1)^{2n} \cdot \frac{(n+3)!}{3!} \cdot \frac{b^n}{(a-bx)^{n+4}} = \frac{(n+3)!}{3!} \cdot \frac{b^n}{(a-bx)^{n+4}}$$

21. Find the nth derivative of
$$y = \frac{x}{(x-1)(x-2)(x-3)}$$

Solution: Use of Rule of "Everywhere but not there", to split it into P.F.

Let $y = \dfrac{x}{(x-1)(x-2)(x-3)} = \dfrac{1}{(x-1)(-1)(-2)} + \dfrac{2}{(1)(x-2)(-1)} + \dfrac{3}{(2)(1)(x-3)}$

$$\therefore \quad y = \frac{1}{2}\frac{1}{(x-1)} - \frac{2}{(x-2)} + \frac{3}{2}\frac{1}{(x-3)}$$

$$\therefore \quad y_n = \frac{1}{2}\cdot\frac{(-1)^n\cdot n!}{(x-1)^{n+1}} - 2\cdot\frac{(-1)^n\cdot n!}{(x-2)^{n+1}} + \frac{3}{2}\cdot\frac{(-1)^n\cdot n!}{(x-3)^{n+1}}$$

C. Examples (22-34)

Problems related to the forms:

(*i*) $D^n[\sin(ax+b)] = a^n \sin\left(ax+b+\frac{n\pi}{2}\right)$

(*ii*) $D^n[\cos(ax+b)] = a^n \cos\left(ax+b+\frac{n\pi}{2}\right)$

(*iii*) $D^n[e^{ax}\sin(bx+c)] = (a^2+b^2)^{n/2}\,.\,e^{ax}\sin(bx+c+n\alpha)$

(*iv*) $D^n[e^{ax}\cos(bx+c)] = (a^2+b^2)^{n/2}\,.\,e^{ax}\cos(bx+c+n\alpha)$

where $\alpha = \tan^{-1}(b/a)$

22. (*i*) $\sin^2 x$ (*ii*) $\sin^3 x$ (*iii*) $\sin^4 x$

(*i*) Let $y = \sin^2 x = \frac{1}{2}(1-\cos 2x)$

$$\therefore \quad y_n = D^n y = \frac{1}{2}\left[0 - 2^n\cos\left(2x+\frac{n\pi}{2}\right)\right] = -\,2^{n-1}\cos\left(2x+\frac{n\pi}{2}\right).$$

(*ii*) We know that $\sin(3x) = 3\sin x - 4\sin^3 x$ [1]

Now,

Let $y = \sin^3 x$

$$= \frac{1}{4}[3\sin x - \sin 3x]; \qquad \text{(using [1])}$$

$$\therefore \quad y_n = \frac{3}{4}D^n(\sin x) - \frac{1}{4}D^n(\sin 3x)$$

$$= \frac{3}{4}\sin\left(x+\frac{n\pi}{2}\right) - \frac{1}{4}\cdot 3^n\cdot\sin\left(3x+\frac{n\pi}{2}\right)$$

(*iii*) Let $y = \sin^4 x = (\sin^2 x)^2$

$$= \left[\frac{1-\cos 2x}{2}\right]^2 = \frac{1}{4}[1 - 2\cos 2x + \cos^2 2x]$$

$$= \frac{1}{4}\left[1 - 2\cos 2x + \frac{1+\cos 4x}{2}\right]$$

$$= \frac{1}{8}[2 - 4\cos 2x + 1 + \cos 4x]$$

$$= \frac{3}{8} - \frac{1}{2}\cos 2x + \frac{1}{8}\cos 4x$$

$$\therefore \quad y_n = 0 - \frac{1}{2}\cdot 2^n\cos\left[2x + n\cdot\frac{\pi}{2}\right] + \frac{1}{8}\cdot 4^n\cdot\cos\left[4x+n\cdot\frac{\pi}{2}\right]$$

$$= -2^{n-1}\cos\left[2x+n\cdot\frac{\pi}{2}\right] + \frac{1}{2^3}\cdot 2^{2n}\cos\left[4x+n\cdot\frac{\pi}{2}\right]$$

PART-B

23. (*i*) $\cos^2 x$ (*ii*) $\cos^3 x$ (*iii*) $\cos^4 x$

(*i*) Let $y = \cos^2 x = \dfrac{1+\cos 2x}{2} = \dfrac{1}{2} + \dfrac{1}{2}\cos 2x$

$$\therefore \quad y_n = D^n y = 0 + \frac{1}{2}\cdot 2^n \cos\left(2x + n\cdot\frac{\pi}{2}\right) = 2^{n-1}\cos\left(2x + n\cdot\frac{\pi}{2}\right).$$

(*ii*) Let $y = \cos^3 x$

we know that

$$\cos 3x = 4\cos^3 x - 3\cos x$$

$$\Rightarrow \quad 4\cos^3 x = \cos 3x + 3\cos x$$

$$\Rightarrow \quad \cos^3 x = \frac{1}{4}(\cos 3x + 3\cos x)$$

$$\Rightarrow \quad y = \frac{1}{4}\cos 3x + \frac{3}{4}\cos x$$

$$\therefore \quad y_n = D^n y = \frac{1}{4}\cdot 3^n\cdot\cos\left(3x + n\cdot\frac{\pi}{2}\right) + \frac{3}{4}\cos\left(x + n\cdot\frac{\pi}{2}\right)$$

(*iii*) Let $y = \cos^4 x = (\cos^2 x)^2$

$$= \frac{1}{4}(1+\cos 2x)^2$$

$$= \frac{1}{4}[1 + 2\cos 2x + \cos^2(2x)]$$

$$= \frac{1}{4}\left[1 + 2\cos 2x + \frac{1}{2}(1+\cos 4x)\right]$$

$$= \frac{1}{4} + \frac{1}{2}\cos 2x + \frac{1}{8} + \frac{1}{8}\cos 4x$$

$$\therefore \quad y_n = D^n y = \frac{1}{2}\cdot 2^n\cos\left(2x + \frac{n\pi}{2}\right) + \frac{1}{8}4^n\cos\left(4x + \frac{n\pi}{2}\right)$$

$$y_n = 2^{n-1}\cos\left(2x + \frac{n\pi}{2}\right) + 2^{2n-3}\cos\left(4x + \frac{n\pi}{2}\right).$$

24. (*a*) $\sin(x)\sin(3x)$, (*b*) $\sin(x)\sin(2x)$, (*c*) $\sin(x)\sin(2x)\sin(3x)$

(*a*) Let $y = \sin x \sin 3x$

$$= \frac{1}{2}(2\sin 3x \sin x)$$

We know that:

$$2S_A S_B = C_{A-B} - C_{A+B}$$

$$= \frac{1}{2}[\cos 2x - \cos 4x]$$

$$\therefore \quad y_n = D^n y = \frac{1}{2}\left[2^n\cos\left(2x + n\cdot\frac{\pi}{2}\right) - 4^n\cos\left(4x + n\cdot\frac{\pi}{2}\right)\right]$$

(*b*) Let $y = \sin x \sin 2x$

$= \frac{1}{2}(2 \sin x \sin 2x)$

$= \frac{1}{2}[2 \sin 2x \sin x]$

We know that

$$2\, S_A S_B = C_{A-B} - C_{A+B}$$

$$y = \frac{1}{2}[\cos(x) - \cos(3x)]$$

$$\therefore \quad y_n = D^n y = \frac{1}{2}\left[\cos\left(x + n\cdot\frac{\pi}{2}\right) - 3^n \cdot \cos\left(3x + n\cdot\frac{\pi}{2}\right)\right]$$

(*c*) Let $y = \sin x \sin 2x \sin 3x$

$= \sin x \sin 3x \sin 2x$

$= \frac{1}{2}\sin x(\cos x - \cos 5x)$ $\quad (\because 2S_A S_B = C_{A-B} - C_{A+B})$

$\therefore \quad y = \frac{1}{2}\sin x \cos x - \frac{1}{2}\cos 5x \sin x$

$= \frac{1}{4}\sin 2x - \frac{1}{4}(\sin 6x - \sin 4x),$

$(\therefore 2S_A C_A = S_{2A})$ and $(2\, C_A S_B = S_{A+B} - S_{A-B})$

$$\therefore \quad y = \frac{1}{4}(\sin 2x - \sin 6x + \sin 4x)$$

Hence $y_n = D^n y = \frac{1}{4}\left[2^n \sin\left(2x + \frac{n\pi}{2}\right) - 6^n \sin\left(6x + \frac{n\pi}{2}\right) + 4^n \sin\left(4x + \frac{n\pi}{2}\right)\right].$

25. (*a*) $\cos x \cos 2x$ (*b*) $\cos x \cdot \cos 2x \cdot \cos 3x$

(*a*) Let $y = \cos x \cos 2x$

$= \frac{1}{2}(2 \cos 2x \cos x)$

We know that

$$2\, C_A C_B = C_{A+B} + C_{A-B}$$

$= \frac{1}{2}[\cos 3x + \cos x]$

$$\therefore \quad y_n = D^n y = \frac{1}{2}\left[3^n \cos\left(3x + n\cdot\frac{\pi}{2}\right) + \cos\left(x + n\cdot\frac{\pi}{2}\right)\right]$$

(*b*) Let $y = \cos x \cos 2x \cos 3x$

$= \frac{1}{2}(2 \cos 3x \cos x) \cdot \cos 2x$

$= \frac{1}{2}(\cos 4x + \cos 2x)\cos 2x$

$$= \frac{1}{4} [2 \cos 4x \cos 2x + 2 \cos^2 2x]$$

$$= \frac{1}{4} [\cos 6x + \cos 2x + 1 + \cos 4x]$$

$$= \frac{1}{4} [1 + \cos 2x + \cos 4x + \cos 6x]$$

$$\therefore \quad y_n = \frac{1}{4} \left[2^n \cos\left(2x + \frac{n\pi}{2}\right) + 4^n \cos\left(4x + \frac{n\pi}{2}\right) + 6^n \cos\left(6x + \frac{n\pi}{2}\right)\right]$$

26. (*a*) $\sin x \cos 2x$ (*b*) $\cos 2x \sin 3x$

(*a*) Let $y = \sin x \cos 2x$

$$= \frac{1}{2} [2 \cos 2x \sin x]$$

We know that

$$2\, C_A S_B = S_{(A+B)} - S_{(A-B)}$$

$$= \frac{1}{2} [\sin 3x - \sin x]$$

$$\therefore \quad y_n = \frac{1}{2}\left[3^n \cdot \sin\left(3x + \frac{n\pi}{2}\right) - \sin\left(x + \frac{n\pi}{2}\right)\right]$$

(*b*) Let $y = \cos 2x \sin 3x$

$$= \frac{1}{2} (2 \sin 3x \cos 2x)$$

We know that

$$2 S_A C_B = S_{(A+B)} + S_{(A-B)}$$

$$= \frac{1}{2} [\sin 5x + \sin x]$$

$$\therefore \quad y_n = \frac{1}{2}\left[5^n \sin\left(5x + \frac{n\pi}{2}\right) + \sin\left(x + \frac{n\pi}{2}\right)\right]$$

27. (*a*) $\sin^2 x \cos^3 x$
(*b*) $\cos^2 x \sin^3 x$ (B.U. Jan., 1993)

(*a*) Let $y = \sin^2 x \cos^3 x$ [1]

We know that

$$y = \sin^2 x \cos^3 x$$

$$= \frac{1}{8} (1 - \cos 2x) (3 \cos x + \cos 3x)$$

$$= \frac{1}{8} [3 \cos x + \cos 3x - 3 \cos 2x \cos x - \cos 3x \cos 2x]$$

$$\because \quad \sin^2 x = \frac{1}{2} (1 - \cos 2x)$$

and $$\cos^3 x = \frac{3}{4} \cos x + \frac{1}{4} \cos 3x$$

But $2C_AC_B = C_{(A+B)} + C_{(A-B)}$

$$= \frac{1}{8}\left\{3\cos x + \cos 3x - \frac{3}{2}(\cos 3x + \cos x) - \frac{1}{2}(\cos 5x + \cos x)\right\}$$

$$= \frac{1}{8}\cos x - \frac{1}{16}\cos 3x - \frac{1}{16}\cos 5x$$

$\therefore \quad y_n = D^n y$

$$= \frac{1}{8}D^n(\cos x) - \frac{1}{16}D^n(\cos 3x) - \frac{1}{16}D^n(\cos 5x)$$

$$= \frac{1}{8}\cdot\cos\left(x + \frac{n\pi}{2}\right) - \frac{1}{16}3^n\cdot\cos\left(3x + \frac{n\pi}{2}\right) - \frac{1}{16}\cdot 5^n\cos\left(5x + \frac{n\pi}{2}\right)$$

(*b*) Let $\quad y = \cos^2 x \sin^3 x = \sin^3 x \cos^2 x$

$$= \frac{1}{4}\cdot(4\sin^2 x\cos^2 x)\sin x$$

$$= \frac{1}{4}\cdot\sin^2(2x)\sin x$$

$$= \frac{1}{4}\cdot\frac{1-\cos 4x}{2}\cdot\sin x$$

$$= \frac{1}{8}(\sin x - \cos 4x\sin x)$$

$$= \frac{1}{16}(2\sin x - 2\cos 4x\sin x)$$

$$= \frac{1}{16}[2\sin x - (\sin 5x - \sin 3x)]$$

$$= \frac{1}{16}[2\sin x + \sin 3x - \sin 5x]$$

$$\therefore \quad y_n = \frac{1}{16}\left[2\sin\left(x + n\frac{\pi}{2}\right) + 3^n\sin\left(3x + \frac{n\pi}{2}\right) - 5^n\sin\left(5x + \frac{n\pi}{2}\right)\right]$$

28. (*a*) $\sin x\,.\,\cos^2 x$ (*b*) $\sin x\cdot\cos^3 x$

(*a*) Let $\quad y = \sin\cdot\cos^2 x$

$$= \sin x\cdot\left[\left(\frac{1}{2}\right)(1+\cos 2x)\right]$$

$$= \left(\frac{1}{2}\right)\sin x + \frac{1}{2}\sin x\cos 2x$$

$$= \left(\frac{1}{2}\right)\sin x + \left(\frac{1}{4}\right)(2\sin x\cos 2x)$$

$$= \left(\frac{1}{2}\right)\sin x + \left(\frac{1}{4}\right)(\sin 3x - \sin x)$$

$$= \left(\frac{1}{2}\right)\sin x + \left(\frac{1}{4}\right)\sin 3x$$

$\therefore \quad y_n = D^n y = \left(\frac{1}{4}\right) D^n (\sin x) + \left(\frac{1}{4}\right) D^n . (\sin 3x)$

$$= \frac{1}{4}\left[\sin\left(x + \frac{n\pi}{2}\right) + 3^n \sin\left(3x + \frac{n\pi}{2}\right)\right]$$

(*b*) Let $\quad y = \sin x \cdot \cos^3 x$

since, $\quad \cos 3x = (4\cos^3 x - 3\cos x)$

$\therefore \quad \cos^3 x = (\cos 3x + 3\cos x)\left(\frac{1}{4}\right)$

Hence, $\quad y = \sin x \cdot \cos^3 x = \sin x\,(\cos 3x + 3\cos x)\left(\frac{1}{4}\right)$

$$= \left(\frac{1}{4}\right)[\sin x \cdot \cos 3x + 3\sin x \cdot \cos x]$$

$$= \left(\frac{1}{8}\right)[2\sin x\cos 3x + 2\cdot 3\cdot \sin x\cdot \cos x]$$

$$= \left(\frac{1}{8}\right)[(\sin 4x - \sin 2x) + 3\sin 2x]$$

$$= \left(\frac{1}{8}\right)[\sin 4x + 2\sin 2x]$$

$\therefore \quad y_n = D^n y$

$$= \left(\frac{1}{8}\right)[D^n (\sin 4x) + 2\cdot D^n (\sin 2x)]$$

$$= \left(\frac{1}{8}\right)\left[4^n \sin\left(4x + \frac{n\pi}{2}\right) + 2\cdot 2^n \sin\left(2x + \frac{n\pi}{2}\right)\right]$$

29. (*a*) If $\quad y = \sin(mx) + \cos(mx)$, show that

$$y_r = D^r y = m^r\,[1 + (-1)^r \sin 2m\,x]^{1/2}$$

(*b*) Find the *n*th derivative of $e^{x\cos\alpha}\cos(x\sin\alpha)$

(*c*) $a^x \cos x$

(*a*) Let $\quad y = \sin(mx) + \cos(mx)$

$\therefore \quad y_r = D^r y = m^r\left[\sin\left(mx + \frac{r\pi}{2}\right) + \cos\left(mx + \frac{r\pi}{2}\right)\right]$

$$= m^r\,[(\sin\theta + \cos\theta)^2]^{1/2}$$

where $\theta = \left(mx + \frac{r\pi}{2}\right)$.

$$= m^r\,[1 + 2\sin\theta\cos\theta]^{1/2}$$

$$= m^r\,[1 + \sin 2\theta]^{1/2}$$

$$= m^r\,[1 + \sin(2\,mx + r\pi)]^{1/2}$$

$$= m^r\,[1 + (-1)^r \sin 2\,mx]$$

Note: $\sin(\pi + \theta) = -\sin\theta = (-1)^1 \sin\theta$

$\sin(2\pi + \theta) = \sin\theta = (-1)^2 \sin\theta$

$\sin(3\pi + \theta) = -\sin\theta = (-1)^3 \sin\theta$

$\sin(4\pi + \theta) = \sin\theta = (-1)^4 \sin\theta$

::: ::: :::

$\sin(n\pi + \theta) = (-1)^n \cdot \sin\theta$

(*b*) Let $y = e^{x\cos\alpha} \cdot \cos(x \sin\alpha) = e^{(\cos\alpha)x} \cdot \cos[(\sin\alpha)\cdot x]$

Compare with $e^{ax}\cos(bx + c)$,

here, $a = \cos\alpha,\ b = \sin\alpha,\ c = 0$

$\therefore \quad y_n = D^n y$

$$= (\cos^2\alpha + \sin^2\alpha)^{n/2} \cdot e^{(\cos\alpha)\,x} \times \cos\left[(\sin\alpha)\cdot x + n\tan^{-1}\left(\frac{\sin\alpha}{\cos\alpha}\right)\right]$$

$$= e^{x\cos\alpha} \cdot \cos[x\sin\alpha + n\tan^{-1}(\tan\alpha)]$$

$$= e^{x\cos\alpha} \cdot \cos[x\sin\alpha + n\alpha]$$

(*c*) Let $y = a^x \cos x = e^{\log a^x} \cdot \cos x = e^{x\log a} \cdot \cos x \qquad (\because\ e^{\log f(x)} = f(x))$

Compare with

$$D^n[e^{ax}\cos x] = [a^2 + 1^2]^{n/2} \cdot e^{ax}\cos\left(x + n\cdot\tan^{-1}\frac{1}{\log a}\right) \qquad [1]$$

$\therefore \quad y_n = D^n y$

$$= [(\log a)^2 + 1^2]^{n/2}\, e^{x\log a}\cos\left(x + n\tan^{-1}\frac{1}{\log a}\right)$$

$$= [1 + (\log a)^2]^{n/2} \cdot a^x \cos\left(x + n\tan^{-1}\frac{1}{\log a}\right).$$

30. $e^{ax}\cos bx$

Let $y = e^{ax}\cos bx$

$\therefore \quad y_1 = Dy = ae^{ax}\cos bx + e^{ax}(-b\sin bx) = e^{ax}[a\cos bx - b\sin bx];$

put $(a = r\cos\alpha,\ b = r\sin\alpha)$

$$= e^{ax}[r\cos\alpha\cos bx - r\sin\alpha\sin bx]$$

$$= re^{ax}[\cos bx\cos\alpha - \sin bx\sin\alpha]$$

$$= re^{ax}\cos(bx + \alpha) \qquad (\because\ C_A C_B - S_A S_B = C_{A+B})$$

$\therefore \quad D^2 y = D(Dy)$

$$= D[re^{ax}\cos(bx + \alpha)] = r\cdot D[e^{ax}\cos(bx + \alpha)]$$

$$= r\cdot[re^{ax}\cos(bx + \alpha + \alpha)] = r^2 e^{ax}\cos(bx + 2\alpha)$$

$\therefore \quad D^n y = r^n e^{ax}\cos(bx + n\alpha);$

where $(a = r\cos\alpha,\ b = r\sin\alpha)$

and $r = \left(\sqrt{a^2 + b^2}\right)$

$$\alpha = \tan^{-1}\left(\frac{b}{a}\right)$$

31. (*a*) $\cos h\,(3x)\,.\,\sin(4x)$ (*b*) $e^{2x}\cos^3 x$ (VTU. Aug., 1999)

(*a*) Let $y = \cos h\,(3x)\sin(4x)$

$$= \left[\frac{(e^{3x} + e^{-3x})}{2}\right](\sin 4x)$$

$$= \left(\frac{1}{2}\right) e^{3x}\sin 4x + (1/2)\, e^{-3x}\sin 4x \qquad [1]$$

$\therefore$ $y_n = D^n y$

$$= (1/2)\, D^n\,(e^{3x}\sin 4x) + \left(\frac{1}{2}\right) D^n\,(e^{-3x}\sin 4x)$$

$(3 = r\cos\alpha;\ 4 = r\sin\alpha)$

$\therefore$ $\left[r = \sqrt{3^2 + 4^2} = 5;\ \tan\alpha = \left(\frac{4}{3}\right)\right].$ [2]

$$D^n y = \left(\frac{1}{2}\right) 5^n e^{3x}\sin(4x + n\alpha) + \left(\frac{1}{2}\right) 5^n e^{-3x}\sin[4x + n(\pi - \alpha)]$$

$$= \left(\frac{1}{2}\right)(5)^n\,[e^{3x}\sin(4x + n\alpha) + e^{-3x}\sin(4x + n(\pi - \alpha)]$$

(*b*) Let $y = e^{2x}\cos^3 x = e^{2x}\left\{\frac{\cos 3x + 3\cos x}{4}\right\}$

We have, if $y = e^{ax}\cos(bx + c)$

then $y_n = D^n y = r^n e^{ax}\cos(bx + c + n\alpha)$

Where $r = \sqrt{a^2 + b^2}$, $\alpha = \tan^{-1}\left(\frac{b}{a}\right)$

$\therefore$ If $y = \frac{1}{4}\,\{e^{2x}\cos 3x + 3e^{2x}\cos x\}$

then $y_n = D^n y$

$$= \frac{1}{4}\left\{(13)^{n/2}\, e^{2x}\cos\left(3x + n\tan^{-1}\left(\frac{3}{2}\right)\right)\right\} + 3\cdot 5^{n/2}\cdot e^{2x}\cos\left(x + \tan^{-1}\frac{1}{2}\right)\bigg\}$$

32. (*a*) $e^x\sin^2 x$, (*b*) $e^x\,(\cos^2 x\,.\,\sin x)$

Since, $\sin^2 x = \dfrac{1 - \cos 2x}{2}$

Let $y = \dfrac{e^x - e^x\cos 2x}{2}$

$\therefore$ $y_n = D^n y = \frac{1}{2}\,\{e^x - r^n e^x\cos(2x + n\alpha)\}$

Where $r = \sqrt{(1^2 + 2^2)} = \sqrt{5}$ and $\alpha = \tan^{-1}\left(\frac{2}{1}\right)$

$$y_n = \frac{1}{2}\, e^x\,\{1 - 5^{n/2}\cos(2x + n\tan^{-1}(2)\}$$

(*b*) Let $$y = e^x \cos^2 x \cdot \sin x$$ [1]

$\therefore$ $$4y = 4e^x \cos^2 x \sin x = 2\, e^x \sin x\,(2\cos^2 x)$$
$$= 2e^x \sin x\,(1 + \cos 2x) = 2e^x \sin x + 2e^x \sin x \cos 2x$$
$$= 2e^x \sin x + e^x\,[2 \sin x \cos 2x]$$
$$= 2e^x \sin x + e^x\,[\sin 3x - \sin x] \qquad (\because 2\, S_A C_B = S_{A+B} - S_{A-B})$$
$$= e^x \sin x + e^x \sin 3x$$ [2]

$\therefore$ $$4D^n y = D^n\,(e^x \sin x) + D^n\,(e^x \sin 3x)$$ [3]
$$= (\sqrt{2}\,)^n\, e^x \sin\left[x + n\left(\frac{\pi}{4}\right)\right] + \left(\sqrt{10}\right)^n e^x \sin\,(3x + n\alpha)\Big]$$

where $\alpha = \tan^{-1}$ (3)

$\therefore$ $$D^n y = \left(\frac{1}{4}\right)\left[\left(\sqrt{2}\right) e^x \sin\left(x + n\left(\frac{\pi}{4}\right) + \left(\sqrt{10}\right)^n e^x \sin\,(3x + n\alpha)\right]$$

33. (*a*) $\sin h\ 2x \cos 4x$, (*b*) $e^{-x} \cos x$

(*a*) Let $$y = \sin h\ 2x \cos 4x$$
$$= \left[\frac{e^{2x} - e^{-2x}}{2}\right] \cos 4x$$
$$= \frac{1}{2}\,[e^{2x} \cos 4x - e^{-2x} \cos 4x]$$

$\therefore$ $$D^n\,[\sin hx \cos 4x] = \frac{1}{2}\,[D^n\,(e^{2x} \cos 4x) - D^n\,(e^{-2x} \cos 4x)]$$
$$= \frac{1}{2}\,[r^n e^{2x} \cos\,(4x + n\alpha) - R^n e^{-2x} \cos\,(4x + n\theta)]$$

where $$r = \sqrt{20}\,;\ \alpha = \tan^{-1} \frac{4}{2} = \tan^{-1}(2)$$

and $$R = \sqrt{20}\,;\ \theta = \tan^{-1} \frac{4}{2} = \tan^{-1}(-2)$$

$\therefore$ $$D^n\,[\sin h\ 2x \cos 4x] = \frac{20^{n/2}}{2}\,[e^{2x} \cos\,(4x + n \tan^{-1}(2)) - e^{-2x} \cos\,(4x - n\tan^{-1}(2))]$$

(*b*) Let $$y = e^{-x} \cos x$$

we know, $$D^n(e^{ax} \cos bx) = r^n e^{ax} \cos\,(bx + n\alpha);$$
$$a = r \cos \alpha,\ b = r \sin \alpha,$$

Here $$r \cos \alpha = a = -1;\ r \sin \alpha = b = 1$$

$\therefore$ $$r = \sqrt{a^2 + b^2} = \sqrt{1+1} = \sqrt{2}$$
$$\cos \alpha = \left(\frac{9}{r}\right) = -\left(\frac{1}{\sqrt{2}}\right);$$
$$\sin \alpha = \left(+\frac{1}{\sqrt{2}}\right)$$
$$\cos \alpha = -\text{ve},\ \sin\alpha = +\text{ve}$$

$\therefore$ α is in the 2nd quadrant.

PART-B

Hence $\alpha = \left[\pi - \left(\frac{\pi}{4}\right)\right] = \left(\frac{3\pi}{4}\right)$

$$D^n (e^{-x} \cos x) = \left(\sqrt{2}\right)^n e^{-x} \cos\left(x + n\frac{3\pi}{4}\right)$$

34. $e^{ax} \cos^2 x \sin x$

Let $y = e^{ax} \cos^2 x \sin x$ [1]

we know that

$$\cos^2 x \sin x = \frac{1}{2}(1 + \cos 2x) \sin x = \frac{1}{2}\sin x + \frac{1}{4} \cdot 2 \sin x \cos 2x$$

$$= \frac{1}{2}\sin x + \frac{1}{4}(\sin 3x - \sin x) = \frac{1}{4}\sin x + \frac{1}{4}\sin 3x$$

$$\therefore \quad y_n = D^n y = \frac{1}{4} D^n (e^{ax} \sin x) + \frac{1}{4} D^n (e^{ax} \sin 3x)$$

$$= \frac{1}{4} e^{ax}\left[(a^2+1)^{n/2} \sin\left(x + \tan^{-1}\frac{1}{a}\right) + (a^2+9)^{n/2} \sin (3x + n\tan^{-1}\left(\frac{3}{a}\right)\right]$$

D. Examples (35-46)

Problems related to the forms:

(i) $D^n [\log (ax + b)] = \dfrac{(-1)^{n-1} (n-1)! \cdot a^n}{(ax+b)^n}$

(ii) $D^n [e^{ax}] = a^n . e^{ax}$

35. $x^5 + \log_{10} (3x^2 + 5x - 2)$ (VTU Aug., 2001)

Let $y = x^5 + \log_{10} (3x^2 + 5x - 2) = x^5 + \dfrac{\log_e (3x^2+5x-2)}{\log_e 10}$

$$= x^5 + \log_{10} e\log_e (3x-1)(x+2)$$

$$= x^5 + \log_{10} e \{\log_e (3x-1) + \log_e (x+2)\}$$

$$\therefore \quad y_n = D^n y = D^n (x^5) + \log_{10} e \{D^n \log_e (3x-1) + D^n \log_e (x+2)\}$$

$$= 0 + \log_{10} e \left\{\frac{(-1)^{n-1}(n-1)!\, 3^n}{(3x-1)^n} + \frac{(-1)^{n-1}(n-1)!}{(x+2)^n}\right\} \quad \text{if } n > 5$$

$$= (-1)^{n-1}(n-1)! \log_{10} e \left\{\frac{3^n}{(3x-1)^n} + \frac{1}{(x+2)^n}\right\}, \quad \text{if } n > 5$$

36. $\log [(ax + b)(c - dx)]$.

Let $y = \log [(ax+b)(c-dx)] = \log (ax+b) + \log (c-dx)$

$$\therefore \quad y_n = \frac{(-1)^{n-1} \cdot (n-1)!\, a^n}{(ax+b)^n} + \frac{(-1)^{n-1} \cdot (n-1)!\, (-d)^n}{(c-dx)^n}$$

$$= (-1)^{n-1} \cdot (n-1)! \left\{\frac{a^n}{(ax+b)^n} + \frac{(-1)^n d^n}{(c-dx)^n}\right\}$$

37. $\log \sqrt{\dfrac{2x+1}{x-2}}$.

Let $$y = \log \sqrt{\frac{2x+1}{x-2}} = \frac{1}{2} \log \frac{2x+1}{x-2} = \frac{1}{2} [\log (2x+1) - \log (x-2)]$$

$$\therefore \quad y_n = \frac{1}{2}\left[\frac{(-1)^{n-1} \cdot (n-1)! \cdot 2^n}{(2x+1)^n} - \frac{(-1)^{n-1} (n-1)!}{(x-2)^n}\right]$$

$$= \frac{1}{2} \cdot (-1)^{n-1} \cdot (n-1)! \left[\frac{2^n}{(2x+1)^n} - \frac{1}{(x-2)^n}\right]$$

38. $\log (2x)$.

Let $y = \log 2x = \log 2 + \log x$

We know that

$$D^n (\log x) = \left[\frac{(-1)^{n-1} (n-1)!}{x^n}\right]$$

$$\therefore \quad D^n y = D^n \log 2 + D^n (\log x) = 0 + [(-1)^{n-1} (n-1)!\, x^n]$$

$$= \left[\frac{(-1)^{n-1} (n-1)!}{x^n}\right]$$

39. $\tan h^{-1} x$.

We know that $\tan h^{-1} x = \dfrac{1}{2} \log \left[\dfrac{(1+x)}{(1-x)}\right]$ and $D^n [\log (ax+b)]$

$$= \frac{(-1)^{n-1} (n-1)!\, a^n}{(ax+b)^n}$$

Let $$y = \tan h^{-1} (x) = \frac{1}{2} \log \left[\frac{(1+x)}{(1-x)}\right]$$

$$= \frac{1}{2} [\log (1+x) - \log (1-x)]$$

$$\therefore \quad D^n y = (1/2) [D^n \log (1+x) - D^n \log (1-x)]$$

$$= \left(\frac{1}{2}\right)\left[\frac{(-1)^{n-1} (n-1)!}{(1+x)^n} - \frac{(-1)^{n-1} (n-1)! (-1)^n}{(1-x)^n}\right]$$

$$(-1)^n (-1)^{n-1} = (-1)^{2n-1} = (-1)^{2n} (-1) = (+1)(-1) = -1$$

$$\therefore \quad D^n y = \left[\frac{(-1)^{n-1} (n-1)!}{(1+x)^n} - \frac{(-1)^{2n-1} (n-1)!}{(1-x)^n}\right]$$

$$= \left(\frac{1}{2}\right)\left[\frac{(-1)^{n-1} (n-1)!}{(1+x)^n} - \frac{(-1)(n-1)!}{(1-x)^n}\right]$$

$$= \frac{(n-1)!}{2}\left[\frac{1}{(1-x)^n} + \frac{(-1)^{n-1}}{(1+x)^n}\right]$$

PART-B

40. $\log [(4x-1)\, e^{\sin 2x}]$ (B.U. Mar., 1995)

Let $y = \log [(4x-1)\, e^{\sin 2x}] = \log (4x-1) + \sin 2x$

But we know that

$$(i)\ D^n [\log (ax+b)] = \frac{(-1)^{n-1}(n-1)!\, a^n}{(ax+b)^n}$$

$$(ii)\ D^n [\sin (ax+b)] = a^n \sin \left(ax + b + \frac{n\pi}{2} \right)$$

$$\therefore \quad y_n = D^n [\log (4x-1) + \sin 2x] = \frac{(-1)^{n-1}(n-1)!\, 4^n}{(4x-1)^n} + 2^n \sin \left(2x + \frac{n\pi}{2} \right)$$

41. Find the nth derivative of

$$\{x^4 + \log_{10}(3x^2 + 5x - 2)\}$$

Solution: Let $y = x^4 + \log_{10}(3x^2 + 5x - 2) = u + v$ where $u = x^4$ and $v = \log_{10}(3x^2 + 5x - 2)$.

Differentiating once w.r.t. x,

$$\frac{dv}{dx} = \frac{1}{3x^2 + 5x - 2} \log_{10} e = \log_{10} e \cdot \frac{1}{(x - x_1)(x - x_2)}$$

where x_1, x_2 are the roots of $3x^2 + 5x - 2 = 0$

i.e.
$$x_{1,2} = \frac{-5 \pm \sqrt{25 + 24}}{2} = \frac{-5 \pm \sqrt{49}}{2}$$

$$\frac{dv}{dx} = \log_{10} e \cdot \frac{1}{x_2 - x_1} \cdot \left[\frac{1}{x - x_2} - \frac{1}{x - x_1} \right]$$

Differentiating $(n-1)$ times

$$\frac{d^n v}{dx^n} = \log_{10} e \cdot \frac{1}{(x_2 - x_1)} \times \left[(-1)^{n-1}(n-1)!(1)^n \left\{ \frac{1}{(x - x_2)^n} - \frac{1}{(x - x_1)^n} \right\} \right]$$

nth derivative of $u = x^4$ is zero for $n > 4$. Thus

$$\frac{d^n v}{dx^n} = \frac{1}{x_2 - x_1} \cdot \log_{10} e \times \left[(-1)^{n-1}(n-1)! \left\{ \frac{1}{(x - x_2)^n} - \frac{1}{(x - x_1)^n} \right\} \right].$$

42. Find the nth derivative of

(*a*) $(3)^{2x}$ (*b*) $(4)^x$ (*c*) $(100)^x$

(*a*) Let $y = 3^{2x}$

We know that $A^B = e^{B \log A}$

$\therefore$ $y_n = 2^n \cdot 3^{2x} \cdot (\log 3)^n$

(*b*) Let $y = 4^x$

$y_n = 4^x \cdot (\log 4)^n$

(*c*) $(100)^x$

Let $y = a^x$;

then $\log y = x \log a$

$\therefore \qquad \left(\frac{1}{y}\right)\left(\frac{dy}{dx}\right) = (1 \cdot \log a);$

Hence $\qquad \left(\frac{dy}{dx}\right) = y \log a$

i.e. $\qquad D(a^x) = a^x \cdot \log a$

$\therefore \qquad D^n(a^x) = a^x (\log a)^n$

Here $\qquad y = 100^x$

$\therefore \qquad D^n y = (100)^x (\log 100)^n$:

44. (*a*) $(e^{2x})^3$, (*b*) e^{2x+3}, (*c*) 3^{2x+3}.

(*a*) We know that

$$D^n(e^{ax}) = a^n \cdot e^{ax}$$

Here, $\qquad y = (e^{2x})^3 = e^{6x}$

$\therefore \qquad D^n y = D^n e^{6x} = 6^n \cdot e^{6x}$

(*b*) Let $\qquad y = e^{2x+3}$

$y_n = 2^n \cdot e^{2x+3}$

(*c*) Let $\qquad y = 3^{2x+3}$

$\therefore \qquad y_n = 2^n \cdot (\log 3)^n \cdot 3^{2x+3}$

45. (*a*) e^{ax+b} (*b*) a^{lx+m}.

(*a*) Let $\qquad y = e^{ax+b}$

$\therefore \qquad y_n = a^n \cdot e^{(ax+b)}$

(*b*) Let $\qquad y = a^{lx+m}$

$\therefore \qquad y_n = l^n (\log a)^n \cdot a^{lx+m}$

46. $e^{2x} + e^{-2x}$

Let $\qquad y = e^{2x} + e^{-2x}$

(compare with $\qquad y = e^{ax}; a = 2, -2)$

$\therefore \qquad y_n = 2^n \cdot e^{2x} + (-2)^n \cdot e^{-2x} = 2^n \cdot e^{2x} + (-1)^n \cdot 2^n e^{-2x} = 2^n [e^{2x} + (-1)^n \cdot e^{-2x}]$.

E. Examples (47-55)

Use of De' Moivre's theorem.

[Related to the application of De' Moivre's theorem:

$(\cos \theta + i \sin \theta)^n = \cos n\theta + i \sin n\theta$

n is a Positive Integer

in the simplification parts of functions to find n^{th} derivatives]

47. $\dfrac{1}{x^2 + a^2}$.

Here $\qquad N^r = 1 = \text{constant}$

$D^r = (x^2 + a^2) = (x^2 - i^2 a^2)$

$\therefore$ Let $\qquad y = \dfrac{1}{x^2 + a^2} = \dfrac{1}{x^2 - i^2 a^2} = \dfrac{1}{(x - ia)(x + ia)}$

Now, $$\frac{1}{(x-ia)(x+ia)} = \frac{A}{(x-ia)} + \frac{B}{(x+ia)}$$

$$\frac{1}{(x+ia)(x-ia)} = \frac{A}{(x-ia)} + \frac{B}{(x+ia)} \quad [1]$$

$\Rightarrow$ $1 = A(x+ia) + B(x-ia)$

Put $x = ia$;

$\therefore$ $1 = A(ia+ia) + B(0)$

$\therefore$ $A = \dfrac{1}{2ia}$

put $x = -ia$ $1 = A(0) + B(-ia-ia)$

$\Rightarrow$ $B = -\dfrac{1}{2ia}$

$$\therefore \quad y = \frac{1}{2ia}\left[\frac{1}{(x-ai)} - \frac{1}{(x+ai)}\right] \quad \text{(using [1])}$$

and $$y_n = D^n y = \frac{1}{2ia}\left[\frac{(-1)^n \cdot n!}{(x-ai)^{n+1}} - \frac{(-1)^n \cdot n!}{(x+ai)^{n+1}}\right] \quad [2]$$

To eliminate i, we substitute

$$x = r\cos\theta;\ a = r\sin\theta; \quad \therefore\ r = \sqrt{x^2+a^2}\ \ \tan\theta = \frac{a}{x} \quad [3]$$

$$\therefore \quad y_n = \frac{(-1)^n \cdot n!}{2ai}\left[r^{-(n+1)}(\cos\theta - i\sin\theta)^{-(n+1)} - r^{-(n+1)}(\cos\theta + i\sin\theta)^{-(n+1)}\right]$$

$$= \frac{(-1)^n \cdot n!}{2ai \cdot r^{n+1}}\left[\{\cos(n+1)\theta + i\sin(n+1)\theta\}\right.$$
$$\left. - \{\cos(n+1)\theta - i\sin(n+1)\theta\}\right]$$

$$= \frac{(-1)^n \cdot n!}{2ai \cdot r^{n+1}} \cdot 2i\sin(n+1)\theta$$

$$= \frac{(-1)^n \cdot n!}{a\left(\dfrac{a}{\sin\theta}\right)^{n+1}} \cdot \sin(n+1)\theta \quad \left(\because\ r = \frac{a}{\sin\theta}\right)$$

$$= \frac{(-1)^n \cdot n!}{a^{n+2}} \cdot \sin^{n+1}\theta \sin(n+1)\theta$$

where $\theta = \tan^{-1}\left(\dfrac{a}{x}\right)$.

48. $\dfrac{x}{x^2+a^2}$.

N^r = variable x

$D^r = x^2 + a^2 = x^2 - i^2a^2 = (x-ia)(x+ia)$

$\therefore$ Let $\quad y = \dfrac{x}{x^2 + a^2} = \dfrac{x}{(x + ai)(x - ai)}$

$$= \frac{1}{2}\left[\frac{1}{(x - ai)} + \frac{1}{(x + ai)}\right] \quad \text{(By Partial fractions)}$$

$$\therefore \quad y_n = D^n\left[\frac{x}{x^2 + a^2}\right] = \frac{1}{2}\left[\frac{(-1)^n\, n!}{(x - ai)^{n+1}} + \frac{(-1)^n \cdot n!}{(x + ai)^{n+1}}\right]$$

$$= \frac{1}{2}\ (-1)^n\ n!\ [(x - ai)^{-(n+1)} + (x + ia)^{-(n+1)}] \qquad [1]$$

To eliminate i, we substitute

$$x = r \cos\theta,\ a = r \sin\theta,$$

then $\quad r^2 = x^2 + a^2,$

$$\tan\theta = \frac{a}{x} \qquad [2]$$

and $\quad (x - ai)^{-(n+1)} = (r \cos\theta - ir \sin\theta)^{-(n+1)}$

$$= \frac{1}{r^{n+1}}\ [\cos (n + 1)\ \theta + i \sin (n + 1)\ \theta] \qquad [3]$$

(using De' Moivre's theorem)

Similarly $\quad (x + ai)^{-(n+1)} = \dfrac{1}{r^{n+1}}\ [\cos (n + 1)\ \theta - i \sin (n + 1)\ \theta] \qquad [4]$

putting [3] and [4] into equation (1), we get

$$y_n = D^n\left[\frac{x}{x^2 + a^2}\right] = \frac{(-1)^n\ n!}{r^{n+1}}\ \cos (n + 1)\ \theta$$

$$= \frac{(-1)^n \cdot n!}{(x^2 + a^2)^{(n+1)/2}}\ \cos\left[(n + 1)\left(\tan^{-1}\left(\frac{a}{x}\right)\right)\right]$$

49. $\tan^{-1}(x)$

Let $\quad y = \tan^{-1} x$, then

$$y_1 = \frac{1}{1 + x^2} = \frac{1}{(x + i)(x - i)} = \frac{1}{2i}\left[\frac{1}{(x - i)} - \frac{1}{(x + i)}\right] \qquad [1]$$

Taking $(n - 1)$th derivative on both sides, we get

$$y_n = D^{n-1}\ y_1 = \frac{1}{2i}\left\{D^{n-1}\left[\frac{1}{(x - i)}\right] - D^{n-1}\left[\frac{1}{(x + i)}\right]\right\}$$

$$= \frac{1}{2i}\left[\frac{(-1)^{n-1}\ (n - 1)!}{(x - i)^n} - \frac{(-1)^{n-1}\ (n - 1)!}{(x + i)^n}\right]$$

i.e. $\quad D^n y = \dfrac{(-1)^{n-1}\ (n - 1)!}{2i}\left[\dfrac{1}{(x - i)^n} - \dfrac{1}{(x + i)^n}\right] \qquad [2]$

To eliminate i, we substitute

$$x = r \cos\theta;\ 1 = r \sin\theta$$

$$\tan\theta = \frac{1}{x};\ \theta = \tan^{-1}\left(\frac{1}{x}\right)$$

$$\therefore\ y_n = \frac{1}{2i}\ (-1)^{n-1}\cdot(n-1)!\ [r^{-n}(\cos\theta - i\sin\theta)^{-n} - r^{-n}(\cos\theta + i\sin\theta)^{-n}]$$

$$= \frac{1}{2ir^n}\ (-1)^{n-1}\cdot(n-1)!\qquad [\cos n\theta + i\sin n\theta - \cos n\theta + i\sin n\theta]$$

$$= \frac{1}{2ir^n}\ (-1)^{n-1}\cdot(n-1)!\ (n-1)!\ 2i\sin n\theta$$

$$= \frac{(-1)^{n-1}(n-1)!}{r^n}\cdot\sin(n\theta)\qquad \left(\because\ r = \frac{1}{\sin\theta}\right)$$

$$= (-1)^n\cdot(n-1)!\ \sin^n\theta\ \sin n\theta,\ \text{where } \theta = \tan^{-1}\left(\frac{1}{x}\right)$$

50. $\tan^{-1}\dfrac{1+x}{1-x}$.

Let $$y = \tan^{-1}\frac{1+x}{1-x}$$

$$y_1 = \frac{1}{1+\left(\frac{1+x}{1-x}\right)^2}\cdot\frac{1(1-x)-(1+x)(-1)}{(1-x)^2}$$

$$= \frac{2}{(1-x)^2+(1+x)^2} = \frac{2}{2x^2+2} = \frac{1}{x^2+1};$$

$$\therefore\quad y_1 = \frac{1}{x^2+1} = \frac{1}{x^2-i^2} = \frac{1}{(x-i)(x+i)}$$

$$= \frac{1}{2i}\left(\frac{1}{(x-i)} - \frac{1}{(x+i)}\right)$$

$$y_n = \frac{1}{2i}\left[\frac{(-1)^{n-1}\lfloor n-1}{(x-i)^n} - \frac{(-1)^n\lfloor n-1}{(x+i)^n}\right] = \frac{(-1)^{n-1}\lfloor n-1}{2i}\left[\frac{1}{(x-i)^n} - \frac{1}{(x+i)^n}\right]$$

$$= \frac{(-1)^{n-1}\lfloor n-1}{2i}\ [(x-i)^{-n} - (x+i)^{-n}]$$

put $$x = r\cos\theta;\ 1 = r\sin\theta$$

$$\therefore\quad 1 + x^2 = r^2(\sin^2\theta + \cos^2\theta) = r^2$$

$$\therefore\quad r = (1+x^2)^{1/2}$$

$$\tan\theta = \frac{1}{x};\ \theta = \tan^{-1}\frac{1}{x}.$$

Hence, $$y_n = \frac{(-1)^{n-1}\lfloor n-1}{2i}\ [(r\cos\theta - ir\sin\theta)^{-n} - (r\cos\theta + ir\sin\theta)^{-n}]$$

$$= \frac{(-1)^{n-1}\lfloor n-1}{2i}\cdot[r^{-n}(\cos\theta - i\sin\theta)^{-n} - r^{-n}(\cos\theta + i\sin\theta)^{-n}]$$

$$= \frac{(-1)^{n-1} \cdot \lfloor n-1}{2i} \cdot r^{-n} [(\cos\theta - i\sin\theta)^{-n} - (\cos\theta + i\sin\theta)^{-n}]$$

$$= \frac{(-1)^{n-1} \cdot \lfloor n-1}{2ir^n} \cdot [\cos h\,\theta + i\sin n\theta - (\cos n\theta - i\sin\theta)]$$

(From the theorem of De Moivre)

$$= \frac{(-1)^{n-1} \cdot \lfloor n-1}{2ir^n} \cdot 2i\sin n\theta = \frac{(-1)^{n-1} \cdot \lfloor n-1}{r^n} \sin n\,\theta$$

$$= \frac{(-1)^{n-1} \lfloor n-1}{\left(\frac{1}{\sin\theta}\right)^n} \cdot \sin n\theta, \qquad (\because\ 1 = r\sin\theta)$$

$$= (-1)^{n-1} \cdot \lfloor n-1 \ \sin^n\theta \cdot \sin n\theta,$$

where $\theta = \tan^{-} \frac{1}{x}$

51. $\log(x^2 + a^2)$

Let $y = \log(x^2 + a^2)$

$$\therefore \quad y_1 = \frac{1}{(x^2 + a^2)} \cdot 2x = \left[\frac{1}{(x - ai)} + \frac{1}{(x + ai)}\right]$$

To find y_n, we should differentiate y_1, $(n - 1)$ times

$$y_n = D^n y = D^{n-1}(y_1) = (-1)^{n-1}(n-1)! \cdot \left[\frac{1}{(x - ai)^n} + \frac{1}{(x + ia)^n}\right]$$

put $\left.\begin{aligned} x &= r\cos\theta \\ a &= r\sin\theta \end{aligned}\right\}$ & Simplify

$$\therefore \quad y_n = \frac{2(-1)^{n-1}(n-1)!}{a^n} \sin^n\theta \cos n\theta$$

where $\theta = \tan^{-1}\left(\frac{a}{x}\right)$.

52. $\frac{(x+1)^2}{(x^2+1)}$.

Let $y = \frac{(x+1)^2}{(x^2+1)}$ [1]

By actual division in an elegant form; we have

$$x^2 + 1 \,\Big)\, x^2 + 2x + 1 \,\Big(\, 1$$
$$\frac{-(x^2+1)}{2x}$$

$$\therefore \quad y = \frac{x^2 + 1 + 2x}{x^2 + 1} = 1 + \frac{2x}{x^2 + 1}$$

Now let us find nth derivative of y

Also $$\frac{2x}{x^2+1} = \left[\frac{1}{(x-i)} + \frac{1}{(x+i)}\right]$$

$$\therefore \quad y_n = 1 + \left[\frac{1}{(x-i)} + \frac{1}{(x+i)}\right]$$

$$y_n = D^n y = (-1)^n \cdot n!\,[\sin^{n+1}\theta \cos(n+1)\theta]$$

where $$\theta = \tan^{-1}\left(\frac{1}{x}\right).$$

53. (*a*) $\tan^{-1}\dfrac{2x}{1-x^2}$, (*b*) $\sin^{-1}\dfrac{2x}{1+x^2}$, (*c*) $\cos^{-1}\dfrac{(1-x^2)}{(1+x^2)}$, (*d*) $\tan^{-1}\left(\dfrac{\sqrt{1+x^2}-1}{x}\right)$.

Solution: In each case put ($x = \tan\theta$)

(*a*) Let $$y = \tan^{-1}\frac{2x}{1-x^2} \qquad \text{(put } x = \tan\theta)$$

$$= \tan^{-1}\left(\frac{2\tan\theta}{1-\tan^2\theta}\right) = \tan^{-1}(\tan 2\theta) = 2\theta = 2\tan^{-1}(x)$$

$$\therefore \quad y_1 = \frac{2}{1+x^2} = \frac{2}{(x+i)(x-i)} = \frac{1}{i}\cdot\frac{(x+i)-(x-i)}{(x+i)(x-i)} = \frac{1}{i}\left[\frac{1}{(x-i)} - \frac{1}{(x+i)}\right]$$

Diff. $(n-1)$ times w.r.t. x we get

$$y_n = \frac{1}{i}\left[\frac{(-1)^{n-1}(n-1)!}{(x-i)^n} - \frac{(-1)^{n-1}(n-1)!}{(x+i)^n}\right]$$

$$= \frac{(-1)^{n-1}(n-1)!}{i}\,[(x-i)^{-n} - (x+i)^{-n}]$$

putting $$x = r\cos\theta,\ 1 = r\sin\theta$$

$$\therefore \quad \tan\theta = \frac{1}{x}$$

$$\therefore \quad y_n = \frac{(-1)^{n-1}(n-1)!}{i}\,[r^{-n}(\cos\theta - i\sin\theta)^{-n} - r^{-n}(\cos\theta + i\sin\theta)^{-n}]$$

$$= \frac{(-1)^{n-1}(n-1)!}{ir^n}\,[(\cos n\theta + i\sin n\theta) - (\cos n\theta - i\sin n\theta)]$$

$$= \frac{(-1)^{n-1}(n-1)!}{ir^n}\cdot 2i\sin n\theta = \frac{2(-1)^{n-1}(n-1)!}{\left(\frac{1}{\sin\theta}\right)^n}.\sin n\theta$$

$$= 2(-1)^{n-1}(n-1)!\,\sin^n\theta\,\sin n\theta,$$

where $$\theta = \tan^{-1}\left(\frac{1}{x}\right)$$

(*b*) Let $$y = \sin^{-1}\left(\frac{2x}{1+x^2}\right)$$

put $\qquad x = \tan\theta$

$$y = \sin^{-1}\left(\frac{2\tan\theta}{1+\tan^2\theta}\right) = \sin^{-1}(\sin 2\theta) = 2\theta = 2\tan^{-1} x$$

Now, proceed as in (*a*)

(*c*) Let $\quad y = \cos^{-1}\left(\frac{1-x^2}{1+x^2}\right) \qquad (x = \tan\theta)$

$$= \cos^{-1}(\cos 2\theta) = 2\theta = 2\tan^{-1}(x) \qquad \left(\because \frac{1-\tan^2\theta}{1+\tan^2\theta} = \cos 2\theta\right)$$

Now, proceed as in part (*a*)

(*d*) Let $\quad y = \tan^{-1}\frac{\sqrt{1+x^2}-1}{x} = \tan^{-1}\left[\frac{\sqrt{1+\tan^2\theta}-1}{\tan\theta}\right] \qquad (x = \tan\theta)$

$$= \tan^{-1}\left(\frac{\sec\theta-1}{\tan\theta}\right) = \tan^{-1}\left(\frac{\frac{1}{\cos\theta}-1}{\frac{\sin\theta}{\cos\theta}}\right)$$

$$= \tan^{-1}\left(\frac{1-\cos\theta}{\sin\theta}\right) = \tan^{-1}\left(\frac{2\sin^2\left(\frac{\theta}{2}\right)}{2\sin\frac{\theta}{2}\cos\frac{\theta}{2}}\right)$$

$$= \tan^{-1}\left(\tan\frac{\theta}{2}\right) = \frac{\theta}{2} = \frac{1}{2}\tan^{-1}(x)$$

$$\therefore \qquad y_1 = \frac{1}{2}\cdot\frac{1}{1+x^2}$$

Further, proceed as in part (a)

Ans: $\frac{1}{2}(-1)^{n-1}(n-1)!\sin^n\theta\sin n\theta,\ \theta = \cot^{-1}(x) = \tan^{-1}\frac{1}{x}$

54. $\cos^6 x$

Let $\qquad z = \cos x + i\sin x$

$$\frac{1}{z} = (\cos x + i\sin x)^{-1} = (\cos x - i\sin x)$$

$$\therefore \qquad \left(z+\frac{1}{z}\right) = 2\cos x$$

Hence $(2\cos x)^6 = \left(z+\frac{1}{2}\right)^6 = z^6 + 6\cdot z^4 + 15.z^2 + 20 + 15\cdot\frac{1}{z^2} + 6\cdot\frac{1}{z^4} + \frac{1}{z^6}$

$$= \left(z^6+\frac{1}{z^6}\right) + 6\left(z^4+\frac{1}{z^4}\right) + 15\left(z^2+\frac{1}{z^2}\right) + 20$$

$$= (2\cos 6x) + 6(2\cos 4x) + 15(2\cos 2x) + 20$$

PART-B

$$\therefore \quad y = \cos^6 x = \frac{1}{2^5}\ [\cos 6x + 6\cos 4x + 15\cos 2x + 10].$$

$$\therefore \quad y_n = \frac{1}{2^5}\left[6^n \cos\left(6x + \frac{n\pi}{2}\right) + 6\cdot 4^n \cos\left(4x + \frac{n\pi}{2}\right) + 15\cdot 2^n \cdot \cos\left(2x + \frac{n\pi}{2}\right)\right]$$

55. $\cos^5 x$

Solution: Put $z = \cos x + i \sin x$;

$$\frac{1}{z} = \cos x - i \sin x$$

$$\therefore \quad z + \frac{1}{z} = 2\cos x$$

$$\Rightarrow \quad \cos x = \frac{1}{2}\left(z + \frac{1}{z}\right)$$

Also $\quad z^n + \dfrac{1}{z^n} = 2\cos nx$

Hence

$$y = \cos^5 x = \left[\frac{1}{2}\left(z + \frac{1}{z}\right)\right]^5$$

$$= \frac{1}{2^5}\left[z^5 + 5\cdot z^4 \cdot \frac{1}{z} + 10\cdot z^3 \cdot \frac{1}{z^2} + 10\cdot z^2\left(\frac{1}{z^3}\right) + 5\cdot z\cdot\left(\frac{1}{z^4}\right) + \frac{1}{z^5}\right]$$

$$= \frac{1}{2^5}\left[z^5 + 5\cdot z^3 + 10\cdot z + 10\cdot\left(\frac{1}{z}\right) + 5\cdot\left(\frac{1}{z^3}\right) + \left(\frac{1}{z^5}\right)\right]$$

$$= \frac{1}{2^5}\left[\left(z^5 + \frac{1}{z^5}\right) + 5\left(z^3 + \frac{1}{z^3}\right) + 10\left(z + \frac{1}{z}\right)\right]$$

$$= \frac{1}{2^5}\ [2\cos 5x + 5\cdot 2\cos 3x + 10\cdot 2\cdot \cos x]$$

$$= \frac{1}{2^4}\cdot \cos 5x + \frac{5}{2^4}\cdot \cos 3x + \frac{10}{2^4}\cdot \cos x$$

$$\therefore \quad y_n = \left(\frac{1}{2^4}\right)\cdot 5^n \cos\left(5x + \frac{n\pi}{2}\right) + \left(\frac{5}{2^4}\right)\cdot 3^n \cdot \cos\left(3x + \frac{n\pi}{2}\right) + \left(\frac{10}{24}\right)\cdot \cos\left(x + \frac{n\pi}{2}\right)$$

EXERCISES

Find the nth derivatives of the following functions:

(1) $\cos^2 x$

(2) $\sin^3 x$

(3) $\cot^{-1} x$

(4) $\sin 3x \cos x$

(5) $\log(4 - x^2)$

(6) $e^{ax} \cos^2 bx$

(7) If $y = e^{x/\sqrt{2}} \cos\left(\frac{x}{\sqrt{2}}\right)$ P.T.

$$y_3 = e^{x/\sqrt{2}} \cos\left[\frac{x}{\sqrt{2}} + \frac{3\pi}{4}\right]$$

(8) $2 \sin (3x) \sin (4x)$

(9) $\sin x . \cos^2 x$

(10) $\sin x . \cos^3 x$

(11) $e^{-x} \cos x$

(12) $e^{2x} \sin^2 x$

(13) $e^{4x} \sin 4x \sin x$

(14) $\frac{1}{x^2 + 3x + 2}$

(15) $\frac{1}{(x-2)^3 (x-3)}$

(16) $\frac{1}{(x+a)(x+b)}$

(17) $\frac{x^2}{(x-1)^2 (x+2)}$

(18) $\frac{1}{x(x+1)^2}$

(19) $\frac{ax+b}{(cx+d)}$

(20) $\frac{x^3}{(x-a)(x-b)(x-c)}$

(21) $e^{5x} \sin^3 ax$

(22) $e^x \sin^4 x$

(23) $\frac{x}{(x-1)^2 (x+2)}$

(24) $\frac{1}{(x^2+a^2)(x^2+b^2)}$

(25) $\sin h\ 2x . \sin 4x.$

(26) $e^x \cos^4 x$

(27) $\frac{x^2 - 4x + 1}{x^3 + 2x^2 - x - 2}$

(28) $\frac{x^4}{(x-1)(x-2)}$

(29) $\log \sqrt{\frac{x-1}{2x+3}}$

(30) $\log\left(\frac{a-x}{a+x}\right)$

(31) $\sin^5 x . \cos^3 x$

(32) $e^{ax} . \cos^3 bx$ $\left(Hint : \cos^3 bx = \left[\frac{\cos 3bx + 3 \cos bx}{4}\right]\right)$

2.2 LEIBNITZ'S THEOREM FOR *n*th DERIVATIVE OF THE PRODUCT OF TWO FUNCTIONS

Statement: If V and U are two functions of x possessing derivatives of any desired order, and if $y = UV$, then

$$D^n y = D^n (UV)$$
$$= U_n V + {}^nC_1 U_{n-1} V_1 + {}^nC_2 U_{n-2} V_2 + \ldots + {}^nC_n U V_n$$
$$= VnU + {}^nC_1 V_{n-1} U_1 + {}^nC_2 V_{n-2} U_2 + \ldots + {}^nC_n V U_n$$

where $U_n = \frac{d^n}{dx^n} (U),\ V_n = \frac{d^n}{dx^n} (V).$

(The proof of this theorem is beyond the scope of the syllabus)

2.3 EXAMPLES (56-70)

(Problems based on the determination of the *n*th derivative with the help of Leibnitz's theorem.)

PART-B

56. Find the nth derivative of $x^2 . \cos x$.

Let $y = x^2 \cos x,\ V = x^2,\ v_1 = 2x,$

$V_2 = 2,\ V_3 = 0$

$$\therefore \quad y_n = D^n y = D^n (x^2 \cos x)$$

$$= \cos\left(x + \frac{n\pi}{2}\right) \cdot x^2 + {}^nC_1 \cos\left(x + (n-1)\frac{\pi}{2}\right) \cdot 2x$$

$$+ {}^nC_2 \cos\left(x + (n-2)\frac{\pi}{2}\right) \cdot 2 \qquad \text{(From the theorem of Leibnitz's)}$$

$$= x^2 \cdot \cos\left(x + \frac{n\pi}{2}\right) + 2nx \cos\left[x + (n-1)\frac{\pi}{2}\right]$$

$$+ n(n-1) \cos\left(x + (n-2)\frac{\pi}{2}\right)$$

57. Find the nth derivative of $x^2 e^{ax}$.

Let $V = x^2,\ V_1 = 2x,\ V_2 = 2,\ V_3 = 0.$

By Leibnitz's theorem

$$D^n (x^2 e^{ax}) = x^2 \cdot D^n (e^{ax}) + {}^nC_1 \cdot 2x \cdot D^{n-1} (e^{ax}) + {}^nC_2 \cdot 2 \cdot D^{n-2} (e^{ax})$$

(other terms being zero)

$$= x^2 (e^{ax} \cdot a^n) + n \cdot 2x \cdot (e^{ax} \cdot a^{n-1}) + \frac{n(n-1)}{1 \cdot 2} \cdot 2 \cdot (e^{ax} a^{n-2})$$

$$= [a^2x^2 + 2nax + n(n-1)]\, a^{n-2} \cdot e^{ax}$$

58. Find the nth derivative of $y = x^2 \sin x$

Let $V = x^2, \quad U = \sin x$

$V_1 = 2x \quad U_n = \sin\left(x + \frac{n\pi}{2}\right)$

$V_2 = 2 \quad U_{n-1} = \sin\left(x + \frac{(n-2)\pi}{2}\right)$

$V_3 = 0 \quad U_{n-2} = \sin\left(x + \frac{(n-2)\pi}{2}\right)$

$$\therefore \quad y_n = D^n (x^2 \sin x)$$

$$= \sin\left(x + \frac{n\pi}{2}\right) \cdot x^2 + n \sin\left[x + \frac{(n-1)\pi}{2}\right] 2x$$

$$+ \frac{n(n-1)}{1 \cdot 2} \cdot \sin\left[x + \frac{(n-2)\pi}{2}\right] \cdot 2 \qquad \text{(other terms being zero)}$$

$$= x^2 \sin\left(x + \frac{n\pi}{2}\right) + 2nx \sin\left[x + \frac{(n-1)\pi}{2}\right] + n(n-1) \sin\left[x + \frac{(n-1)\pi}{2}\right]$$

59. $x^3 e^{4x}$.

Let $V = x^3,\ U = e^{4x}$

$V_1 = 3x^2,\ V_2 = 6x,\ V_3 = 6,\ V_4 = 0,\ V_n = 4^n e^{4x}$

Then by Leibnitz's theorem, we have $D^n\,[x^3e^{4x}]$

$$= 4^n \cdot e^{4x} \cdot x^3 + n \cdot 4^{n-1} \cdot e^{4x} \cdot (3x^2) + \left\{\frac{n(n-1)}{1\cdot 2}\right\} (6x) \cdot 4^{n-2}\, e^{4x}$$

$$+ \left\{\frac{n\,(n-1)\,(n-2)}{1\cdot 2\cdot 3}\right\} \cdot (6)\; 4^{n-3} \cdot e^{4x}$$

(other terms being zero)

$$= 4^{n-3}\, e^{4x}\, [64x^3 + 48nx^2 + 12n\,(n-1)\,x + n\,(n-1)\,(n-2)]$$

60. Find the nth derivatives of x^2e^{3x} sin$x$

Let $V = x^2$ $\quad U = e^{3x} \sin x$

$V_1 = 2x$ $\quad U_n = r^n e^{3x} \sin(x + n\alpha)$

$V_2 = 2$ $\quad r = \sqrt{9+1} = \sqrt{10}$

$V_3 = 0$ $\tan\alpha = \dfrac{1}{3}$

$$D^n\,[e^{3x} \cdot \sin x \cdot x^2]$$

$$= (10)^{\frac{n}{2}} \cdot e^{3x} \sin(x + n\alpha) \cdot x^2 + {}^nC_1 \cdot (10)^{\frac{n-1}{2}} \cdot e^{3x} \cdot \sin[x + (n-1)\,\alpha] \cdot 2x$$

$$+ {}^nC_2 \cdot (10)^{\frac{n-2}{2}} \cdot e^{3x} \cdot \sin[x + (n-2)\,\alpha] \cdot (2)$$

(other terms being zero).

61. $(x^2 + 3x)\, e^{-2x}$

$$D_n\,(UV) = UV_n + n \cdot UVn_{-1} + \left[\frac{n\,(n-1)}{1\cdot 2}\right] U_2V_{n-2} + \ldots$$

Here $U = (x^2 + 3x)$

$U_1 = (2x + 3)$

$U_2 = 2,\; U_3 = 0$

$V = e^{-2x}$

$V_n = (-2)^n \cdot e^{-2x};$

$V_{n-1} = (-2)^{n-1}\, e^{-2x};$

$\therefore$ $D^n\,(UV) = D^n\,[(x^2 + 3x)\, e^{-2x}]$

$$= (x^2 + 3x)\,(-2)^n\, e^{-2x} + n \cdot (2x + 3)\,(-2)^{n-1} \cdot e^{-2x}$$

$$+ \left[\frac{n\,(n-1)}{1\cdot 2}\right] \cdot 2 \cdot (-2)^{n-2} \cdot e^{-2x} + 0 + 0 + \ldots$$

$$= (-2)^{n-2} \cdot e^{-2x}\, [(-2)^2\,(x^2 + 3x) + n \cdot (2x + 3)\,(-2) + n\,(n-1)]$$

$$= (-2)^{n-2}\, e^{-2x}\, [4x^2 + 12x - 4nx - 6n + n^2 - n]$$

$$= (-2)^{n-2} \cdot e^{-2x}\, [4x^2 + x\,(12 - 4x) + (n^2 - 7n)]$$

62. $e^{ax}\, [a^2x^2 - 2nax + n(n + 1)]$

$$D^n\,[UV] = \left\{UV_n + n \cdot U_1V_{n-1} + \left[\frac{n\,(n-1)}{1\cdot 2}\right] U_2V_{n-2} + \ldots\right\}$$

PART-B

$$U = a^2x^2 - 2nax + n\,[n+1]$$
$$U_1 = (2a^2x - 2na)$$
$$U_2 = 2a^2$$
$$V = e^{ax}$$
$$V_n = a^n e^{ax}$$
$$V_{n-1} = a^{n-1}\,e^{ax}$$
$$V_{n-2} = a^{n-2}\cdot e^{ax}\ \ldots$$

$$\therefore\quad D^n\,[a^2x^2 - 2nax + n\,(n+1)]\,e^{ax}$$
$$= [a^2x^2 - 2nax + n\,(n+1)]\,a^n e^{ax} + n\cdot[2a^2x - 2na]\,a^{n-1}$$
$$\cdot\, e^{ax} + \left[\frac{n\,(n-1)}{1\cdot 2}\right]\cdot[2a^2]\cdot a^{n-2}\,e^{ax} + 0 + 0 + \ldots$$
$$= a^{n-2}\,e^{ax}\begin{cases} a^2\,(a^2x^2 - 2nax + n^2 + n \\ + a\,(2na^2x - 2n^2a) + n\,(n-1)\,a^2 \end{cases}$$
$$= a^{n-2}\,e^{ax}\begin{bmatrix} a^4x^2 - 2na^3x + n^2a^2 + na^2 \\ + 2na^3x - 2n^2a^2 + n^2a^2 - na^2 \end{bmatrix}$$
$$= a^{n-2}\,e^{ax}\,a^4x^2$$
$$= a^{n+2}\,x^2e^{ax}$$

63. $(ax + 2b)\sin 3x$

$$D^n\,(UV) = \left[UV_n + n\cdot U_1V_{n-1} + \left[\frac{n\,(n-1)}{1\cdot 2}\right]U_2V_{n-2} + \ldots\right.$$

Here
$$\begin{cases} U = (ax + 2b); \\ U_1 = a \\ U_2 = 0 \end{cases}$$
$$\begin{cases} V = \sin 3x \\ V_n = 3^n \sin\left[3x + n\left(\frac{\pi}{2}\right)\right] \end{cases}$$

$$\therefore\quad D^n\,[UV] = (ax + 2b)\cdot 3^n \sin\left[3x + n\left(\frac{\pi}{2}\right)\right] + n\cdot a\cdot 3^{n-1}\sin\left[ax + (n-1)\left(\frac{\pi}{2}\right)\right] + 0$$
$$= 3^{n-1}\left\{3\,(ax + b)\sin\left[3x + n\left(\frac{\pi}{2}\right)\right] + na\sin\left[3x + (n-1)\left(\frac{\pi}{2}\right)\right]\right\}$$

64. $x\log x$

We know that

$$D^n\,[UV] = \left\{UV_n + n\cdot U_1V_{n-1} + \left[\frac{n\,(n-1)}{1\cdot 2}\right]U_2V_{n-2} + \ldots\right\}$$

Also, $$D^n\,[\log x] = \frac{(-1)^{n-1}\,(n-1)!}{x^n}$$

$$D^n[x \cdot \log x] = x\,(D^n \log x) + n \cdot 1 \cdot D^{n-1}(\log x)$$

$$= x\left[\frac{(-1)^{n-1}\cdot(n-1)!}{x^n}\right] + \left[\frac{(-1)^{n-2}\cdot(n-2)!}{x^{n-1}}\right]$$

$$= \left[\frac{(-1)^{n-1}\,(n-1)!}{x^{n-1}}\right] + \left[\frac{(-1)^{n-2}\,n\,(n-2)!}{x^{n-1}}\right]$$

$$= \left[(-1)^{n-1}\,\frac{(n-1)\,(n-2)!}{x^{n-1}}\right] + \left[(-1)^n\,\frac{n\,(n-2)!}{x^{n-1}}\right]$$

$$= \left[-(-1)^n\,\frac{(n-1)\,(n-2)!}{x^{n-1}}\right] + \left[\frac{(-1)^n\,n\,(n-2)!}{x^{n-1}}\right]$$

$$= \left[\frac{(-1)^n\,(n-2)!}{x^{n-1}}\right][\,n-(n-1)]$$

$$= \left[\frac{(-1)^n\,(n-2)!}{x^{n-1}}\right]$$

65. $x^2 \,.\, \log x$

Let $\quad U = \log x \qquad V = x^2$

$$U_n = \left[\frac{(-1)^{n-1}\,(n-1)!}{x^n}\right] \qquad V_1 = 2x$$

$$U_{n-1} = \left[\frac{(-1)^{n-2}\,(n-2)!}{x^{n-1}}\right] \qquad V_2 = 2$$

$$U_{n-2} = \left[\frac{(-1)^{n-3}\,(n-3)!}{x^{n-2}}\right] \qquad V_3 = 0$$

$\therefore$ **By Leibnitz's theorem**

$$D^n[UV] = \left[\frac{(-1)^{n-1}\cdot(n-1)!}{x^n}\right]\cdot x^2 + n\cdot\left[\frac{(-1)^{n-2}\cdot(n-2)!}{x^{n-1}}\right]\cdot 2x$$

$$+\left[\frac{n(n-1)}{1\cdot 2}\right](2)\cdot\left[\frac{(-1)^{n-3}\cdot(n-3)!}{x^{n-2}}\right]$$

$$= \left[\frac{(-1)^{n-1}\cdot n!}{nx^{n-2}}\right] + 2\cdot\left[\frac{(-1)^{n-2}\cdot n!}{(n-1)x^{n-2}}\right] + \frac{(-1)^{n-3}\cdot n!}{(n-2)\,x^{n-2}}$$

$$= \frac{(-1)^{n-1}\cdot n!}{x^{n-2}}\left[\frac{1}{n} - \frac{2}{n-1} + \frac{1}{n-2}\right]$$

66. $x^3 \,.\, \log x$

$$D^n[UV] = \left\{UV_n + n\cdot U_1V_{n-1} + \left[\frac{n(n-1)}{1\cdot 2}\right]U_2V_{n-2} + \ldots\right\} \qquad [1]$$

$$\left.\begin{aligned} U &= x^3 \\ U_1 &= 3x^2 \\ U_2 &= 6x \\ U_3 &= 6 \\ U_4 &= 0 \end{aligned}\right\} \qquad [2]$$

$$D^n(\log x) = \frac{(-1)^{n-1}(n-1)!}{x^n}$$

$$D^{n-1}(\log x) = \frac{(-1)^{n-2}\cdot(n-2)!}{x^{n-1}} \qquad [3]$$

Also $\left.\begin{aligned} (-)^{n-4} &= (-1)^{n-2} = (-1)^n; \\ (-1)^{n-3} &= (-1)^{n-1} \end{aligned}\right\}$ [4]

$$\therefore \quad D^n(x^3\cdot\log x) = \frac{x^3(-1)^{n-1}\cdot(n-1)!}{x^n} + \frac{n\cdot(3x^2)(-1)^{n-2}\cdot(n-2)!}{x^{n-1}}$$

$$+\left[\frac{n(n-1)}{1\cdot 2}\right]\frac{(6x)\cdot(-1)^{n-3}(n-3)!}{x^{n-2}}$$

$$+\left[\frac{n(n-1)(n-2)}{1\cdot 2\cdot 3}\right]\frac{(6)(-1)^{n-4}(n-4)!}{x^{n-3}} + 0$$

$$= -\frac{(-1)^n(n-1)!}{x^{n-3}} + \frac{(-1)^n\cdot 3n(n-2)!}{x^{n-3}}$$

$$-\frac{n(n-1)(n-3)!(-1)^n\cdot 3}{x^{n-3}} + \frac{n(n-1)(n-2)(-1)^n(n-4)!}{x^{n-3}}$$

$$= \frac{(-1)^n(n-4)!}{x^{(n-3)}}\begin{cases} -(n-1)(n-2)(n-3) \\ +3n(n-2)(n-3) \\ -3n(n-1)(n-3) + n(n-1)(n-2) \end{cases}$$

$$= \left\{\left[\frac{1}{x^{n-3}}\right]\cdot(-1)^n\cdot(n-4)!\,6\right\}$$

67. $x^4 . \log x$

Let $\quad y = x^4\cdot\log x$. Then

$$Dy = x^4\cdot\frac{1}{x} + 4x^3\cdot\log x = x^3(1+4\log x)$$

$$D^2y = x^3\left(\frac{4}{x}\right) + (1+4\log x)\cdot 3x^2$$

$$= x^2(4) + 3x^2(1+4\log x) = x^2[4+3(1+4\log x)]$$

$$= x^2[4+3+12\log x] = x^2[7+12\log x]$$

$$D^3y = x^2\left[\frac{12}{x}\right] + [7+12\log x](2x)$$

$$= x[12+14+24\log x] = x[26+24\log x]$$

$$D^4y = x\left[\frac{24}{x}\right] + 1\cdot[26 + 24\log x] = 50 + 24\log x$$

Taking $(n - 4)$th derivative on both sides, we get

$$D^ny = D^{n-4}(50 + 24\cdot D^{n-4}(\log x)$$

$$= 24\cdot\frac{(n-5)!\,(-1)^{n-5}}{x^{n-4}} = \left[\frac{(-1)^{n-1}}{x^{n-4}}\right](n-5)!\cdot 24$$

68. If $y = x^{n-1}\log x$, prove that

$$xy_1 = (n-1)y + x^{n-1}$$

By differentiating this result $(n - 1)$ times

Show that $y_n = \dfrac{(n-1)!}{x}$

Solution: $y = x^{n-1}\log x$

$\Rightarrow$ $xy = x^n\log x$ (Diff. once)

$$xy_1 + y\cdot 1 = nx^{n-1}\log x + x^n\cdot\frac{1}{x}$$

$$xy_1 + y = ny + x^{n-1} \qquad (\because x^{n-1}\log x = y)$$

$$xy_1 = (n-1)\,y + x^{n-1}$$

Diff. $(n - 1)$ times, using Leibnitz's theorem

$$xy_n + (n-1)\cdot 1\cdot y_{n-1} = (n-1)\,y_{n-1} + (n-1)!$$

Cancelling off $(n-1)\cdot y_{n-1}$ on both sides,

$$xy_n = (n-1)! \quad\text{or}\quad y_n = \frac{(n-1)!}{x}$$

69. If $y = x^n\log x$, prove that

$$y_{n+1} = \frac{n!}{x}$$

Solution: $y = x^n\log x$ [1]

$$\therefore\quad y_1 = x^n\cdot\frac{1}{x} + nx^{n-1}\cdot\log x$$

or $xy_1 = x^n + nx^n\log x$ or $xy_1 = x^n + ny$

Diff. 'n' times by Leibnitz's theorem

$$y_{n+1}\,x + n\cdot y_n = n! + ny_n$$

or $xy_{n+1} = n!$ or $y_{n+1} = \dfrac{n!}{x}$.

70. If $y_n = D^n\,[x^n\,.\log x]$ prove that

$$y_n = (n-1)! + ny_{n-1}$$

Hence deduce that

$$y_n = n!\left\{\log x + 1 + \frac{1}{2} + \frac{1}{3} + \ldots + \frac{1}{n}\right\}$$ [VTU Aug., 1999]

Solution: We have $y_n = D^n\{x^n \cdot \log x\} = D^{n-1}\{D(x^n \cdot \log x\}$

$$= D^{n-1}\{nx^{n-1} \cdot \log x + x^{n-1}\} = nD^{n-1}(x^{n-1}\log x) + D^{n-1}(x^{n-1})$$

$$y_n = ny_{n-1} + (n-1)! \qquad [1]$$

Deduction:

Substituting $n = 1, 2, 3, \ldots$ in (1) we get

$$y_1 = y_0 + 0! = \log x + 1$$

$$y_2 = 2y_1 + 1! = 2(\log x + 1) + 1 = 2!\left\{\log x + 1 + \frac{1}{2}\right\}$$

$$y_3 = 3y_2 + 2! = 3 \cdot 2!\left\{\log x + 1 + \frac{1}{2}\right\} + 2 = 3!\left\{\log x + 1 + \frac{1}{2} + \frac{1}{3}\right\}$$

$$\vdots \qquad \vdots \qquad \vdots$$

Proceeding in this way, we find that

$$y_n = n!\left\{\log x + 1 + \frac{1}{2} + \frac{1}{3} + \ldots + \frac{1}{n}\right\}$$

Examples (71-84)

71. If $\cos^{-1}\left(\frac{y}{b}\right) = \log\left(\frac{x}{n}\right)^p$ prove that:

$$x^2 y_{n+2} + (2n+1)\,xy_{n+1} + (n^2 + p^2)\,y_n = 0$$

[VTU Aug., 2001
VTU Aug., 2002]

Solution: **Given:** $\cos^{-1}\left(\frac{y}{b}\right) = p(\log x - \log n)$

Diff. w.r.t. 'x', we get

$$\frac{-1}{\sqrt{1-\left(\frac{y}{b}\right)^2}} \cdot \frac{y_1}{b} = p\left(\frac{1}{x} - 0\right) \qquad \left(\because \frac{d}{dx}\cos^{-1}(x) = \frac{-1}{\sqrt{1-x^2}}\right)$$

i.e.

$$\frac{-y_1}{\sqrt{b^2 - y^2}} = \frac{p}{x} \quad \text{or} \quad xy_1 = -p\sqrt{b^2 - y^2}$$

Squaring we get

$$x^2 y_1^2 = p^2(b^2 - y^2)$$

Diff. again w.r.t x we get

$$x^2\, 2y_1 y_2 + y_1^2\,(2x) = p^2(-2yy_1)$$

i.e.

$$x^2 y_2 + xy_1 + p^2 y = 0 \qquad [1]$$

Applying Leibnitz's Theorem term by term to differentiate n times, we get

$$\left[\left\{x^2 y_{n+2} + n(2x)\,y_{n+1} + \frac{n(n-1)}{1 \cdot 2}(2)\,y_n\right\} + \{xy_{n+1} + n(1)y_n\} + p^2 y_n = 0\right.$$

i.e.,

$$x^2 y_{n+2} + (2n+1)\,xy_{n+1} + (n^2 + p^2)\,y_n = 0$$

Hence proved.

72. If $y = (x^2 - 1)^n$ show that

$$(1 - x^2)\, y_{n+2} - 2xy_{n+1} + n(n + 1)y_n = 0$$ [VTU, Aug., 2000]

Solution: **Given:** $y = (x^2 - 1)^n$

Now $\log y = n \log (x^2 - 1)$

Differentiate w.r.t. x we get

$$\frac{y_1}{y} = n \cdot \frac{2x}{x^2 - 1}$$

i.e., $(x^2 - 1)y_1 - 2nxy = 0$ [1]

Applying Leibnitz's theorem term by term to differentiate '$(n + 1)$' times we get

$$(x^2 - 1)\, y_{n+2} + (n + 1) \cdot 2x \cdot y_{n+1} + (n + 1)\, n \cdot 2 \cdot y_n - 2n\,\{x \cdot y_{n+1} + (n + 1) \cdot 1 \cdot y_n\} = 0$$

i.e. $(x^2 - 1)y_{n+2} + 2xy_{n+1} - n(n + 1)\, y_n = 0$

or $(1 - x^2)\, y_{n+2} - 2x\, y_{n+1} + n(n + 1)y_n = 0$

73. Prove that

$$\frac{d^n}{dx^n}\left(\frac{\log x}{x}\right) = (-1)^n \cdot \frac{n!}{x^{n+1}}\left[\log x - 1 - \frac{1}{2} - \frac{1}{3} - \ldots - \frac{1}{n}\right]$$ [VTU, March, 2001]

By using Leibnitz's theorem, we get,

$$D^n\left[\frac{\log x}{x}\right] = D^n\left[\log x \cdot \frac{1}{x}\right]$$

$$= \log x \cdot D^n\left[\frac{1}{x}\right] + n \cdot D\,(\log x) \cdot D^{n-1}\left(\frac{1}{x}\right) + \ldots + D^n\,(\log x) \cdot \frac{1}{x}$$

$$= \log x \cdot \frac{(-1)^n \cdot n!}{x^{n+1}} + n \cdot \frac{1}{x} \cdot \frac{(-1)^{n-1}\,(n-1)!}{x^n}$$

$$+ \frac{n(n-1)}{1 \cdot 2}\left(-\frac{1}{x^2}\right)\frac{(-1)^{n-2}\,(n-2)!}{x^{n-1}} + \ldots. + \frac{(-1)^{n-1}\,(n-1)!}{x^n} \cdot \frac{1}{x}$$

$$= \frac{(-1)^n \cdot n!}{x^{n+1}} \cdot \log x - 1\,\frac{(-1)^n \cdot n!}{x^{n+1}} - \frac{(-1)^n \cdot n!}{x^{n+1}} \cdot \frac{1}{2} \ldots \frac{(-1)^n \cdot n!}{x^{n+1}} \cdot \frac{1}{n}$$

$$= \frac{(-1)^n \cdot n!}{x^{n+1}}\left[\log x - 1 - \frac{1}{2} - \ldots \frac{1}{n}\right]$$

Hence proved.

74. If $y = a \cos (\log x) + b \sin (\log x)$ prove that

$$x^2 y_{n+2} + (2n + 1)\, xy_{n+1} + (n^2 + 1)\, y_n = 0$$ [VTU, Adv. Maths-I, Aug., 2002]

Let $y = a \cos (\log x) + b \sin (\log x)$ [1]

(Differentiate once)

$$\therefore \quad y_1 = a\,[-\sin (\log x)]\left(\frac{1}{x}\right) + b\,[\cos (\log x)]\left(\frac{1}{x}\right)$$

(Multiply by x)

$$\therefore \quad xy_1 = -\,a \sin (\log x) + b \cos (\log x)$$

(Differentiate again)

$$(xy_2 + 1 \cdot y_1) = -a\,[\cos(\log x)]\left(\frac{1}{x}\right) + b\,[-\sin(\log x)]\left(\frac{1}{x}\right)$$

(Multiply by x)

$$(x^2y_2 + xy_1) = -a\cos(\log x) - b\sin(\log x)$$

$$= -[a\cos(\log x) + b\sin(\log x)] = -y$$

$$\therefore \quad x^2y + xy_1 + y = 0 \qquad [2]$$

To get the required result, we differentiate [2], n times

$$D^n(x^2y_2) + D^n(xy_1) + D^n(y) = 0.$$

Each term of this is obtained by applying Leibnitz's theorem. Thus, we get

$$\left.\begin{aligned} D^n(x^2y_2) &= x^2y_{n+2} + n\cdot 2x\cdot y_{n+1} + n(n-1)\,y_n \\ D^n(xy_1) &= xy_{n+1} + n\cdot 1\cdot y_n \\ D^n(y) &= y_n \end{aligned}\right] = 0$$

Adding, we get

$$x^2y_{n+2} + xy_{n+1}(2n+1) + y_n[n(n-1) + n + 1] = 0$$

i.e.
$$x^2y_{n+2} + x(2n+1)\,y_{n+1} + (n^2+1)\,y_n = 0 \qquad [3]$$

75. If $y = \left(x - \sqrt{x^2-1}\right)^m$ show that

$$(x^2-1)\,y_{n+2} + (2n+1)\,xy_{n+1} + (n^2-m^2)y_n = 0 \qquad \text{[B.U. Jan., 1993]}$$

We have

$$y = \left(x - \sqrt{x^2-1}\right)^m \qquad [1]$$

$$\therefore \quad y_1 = Dy = m\left(x - \sqrt{x^2-1}\right)^m \cdot \left[1 - \frac{2x}{2\sqrt{x^2-1}}\right]$$

$$= m\left(x - \sqrt{x^2-1}\right)^{m-1} \cdot \left[\frac{\sqrt{x^2-1} - x}{\sqrt{x^2-1}}\right]$$

$$= -\frac{m}{\sqrt{x^2-1}}\left[x - \sqrt{x^2-1}\right]^{m-1} \cdot \left[x - \sqrt{x^2-1}\right]$$

$$y_1 = -\frac{my}{\sqrt{x^2-1}} \qquad \left[\because \left(x - \sqrt{x^2-1}\right)^m = y\right]$$

$$\sqrt{x^2-1}\cdot y_1 = -my$$

Squaring both sides we get

$$(x^2-1)y_1^2 - m^2y^2 = 0$$

Diff. both sides w.r.t. x, we get

$$(x^2-1)\,2y_1y_2 + 2xy_1^2 - m^2 2yy_1 = 0$$

$$(x^2-1)y_2 + xy_1 - m^2y = 0 \qquad [2]$$

Differentiating '*n*' times by using Leibnitz's theorem, we get

$$(x^2-1)\,y_{n+2}+n\cdot 2x\cdot y_{n+1}+\frac{n\,(n-1)}{1\cdot 2}\cdot 2\cdot y_n+xy_{n+1}+n\cdot 1\cdot y_n-m^2y_n=0$$

$$(x^2-1)\,y_{n+2}+(2n+1)\,xy_{n+1}+[n\,(x-1)+n-m^2]\,y_n=0$$

$$(x^2-1)\,y_{n+2}+(2n+1)\,xy_{n+1}+[n^2-m^2]y_n=0$$

76. If $y=\tan^{-1}x$ show that

$$(1+x^2)\,y_{n+2}+2\,(n+1)\,xy_{n+1}+n(n+1)y_n=0 \qquad \text{[B.U. July, 1993]}$$

Let $y=\tan^{-1}(x)$

$\therefore$ $y_1=\dfrac{1}{1+x^2}$ or $(1+x^2)y_1=1$

Differentiating w.r.t. x, we get

$$(1+x^2)\,y_2+2xy_1=0$$

Diff. w.r.t. x, using Leibnitz's theorem, we get

$$(1+x^2)\,y_{n+2}+n2xy_{n+1}+\frac{n\,(n-1)}{2}\,2y_n+2xy_{n+1}+n2y_n=0$$

$$(1+x^2)y_{n+2}+2(n+1)\,xy_{n+1}+[n(n-1)+2n]\,y_n=0$$

$$(1+x^2)\,y_{n+2}+2(n+1)\,xy_{n+1}+n(n+1)y_n=0$$

77. If $y=x\log(x+1)$ prove that

$$y_n=\frac{(-1)^{n-1}\cdot(n-2)!\,(x+n)}{(x+1)^n} \qquad \text{[B.U. Feb., 1994]}$$

Let us choose $V=x$

$\therefore$ $V_1=1$

$V_2=0$ and $U=\log(x+1)$

Also $U_n=\dfrac{(-1)^n\cdot(n-1)!}{(x+1)^n}$

Now by Leibnitz's theorem, we get

$$D^n\,[x\log(x+1)]=x\cdot\frac{(-1)^n\,(n-1)!}{(x+1)^n}+n\cdot 1\cdot\frac{(-1)^{n-1}\,(n-2)!}{(x+1)^{n-1}}$$

$$=\frac{(-1)^{n-1}\,(n-2)!}{(x+1)^n}\,[-x\,(n-1)+n(x+1)]$$

$$=\frac{(-1)^{n-1}\,(n-2)!\,(x+n)}{(x+1)^n}$$

78. If $y^{1/m}+y^{-1/m}=2x$ prove that

$$(x^2-1)y_{n+2}+(2n+1)\,xy_{n+1}+(n^2-m^2)\,y_n=0 \qquad \text{[B.U. Feb., 1994] [B.U. Aug., 1994]}$$

Given: $y^{1/m}+y^{-1/m}=2x$

Put $y^{1/m}=z$, then we have

$$\left(z+\frac{1}{z}\right)=2x \quad \text{or} \quad z^2-2xz+1=0$$

$$\therefore \qquad z = \frac{2x \pm \sqrt{4x^2 - 4}}{2} = x \pm \sqrt{x^2 - 1}$$

i.e. $$y^{1/m} = x \pm \sqrt{x^2 - 1} \qquad \text{or} \qquad y = \left(x \pm \sqrt{x^2 - 1}\right)^m$$

If $$y = \left(x + \sqrt{x^2 - 1}\right)^m, \text{ then}$$

$$y_1 = m\left(x + \sqrt{x^2 - 1}\right)^{m-1}\left[1 + \frac{1}{2} \cdot \frac{1}{\sqrt{x^2 - 1}} \cdot 2x\right]$$

$$= m\left(x + \sqrt{x^2 - 1}\right)^{m-1}\left(\frac{x + \sqrt{x^2 - 1}}{\sqrt{x^2 - 1}}\right) = \frac{m\left(x + \sqrt{x^2 - 1}\right)^m}{\sqrt{(x^2 - 1)}} = \frac{my}{\sqrt{x^2 - 1}}$$

If $$y = \left(x - \sqrt{x^2 - 1}\right)^m$$

$$y_1 = m\left(x - \sqrt{x^2 - 1}\right)^{m-1}\left[1 - \frac{1}{2} \cdot \frac{1}{\sqrt{x^2 - 1}} \cdot 2x\right]$$

$$= m\left(x - \sqrt{x^2 - 1}\right)^{m-1}\left(\frac{\sqrt{x^2 - 1} - x}{\sqrt{x^2 - 1}}\right)$$

$$= -\frac{m\left(x - \sqrt{x^2 - 1}\right)^{m-1}\left[x - \sqrt{x^2 - 1}\right]}{\sqrt{x^2 - 1}}$$

$$= -\frac{m\left(x - \sqrt{x^2 - 1}\right)^m}{\sqrt{x^2 - 1}} = -\frac{my}{\sqrt{x^2 - 1}}$$

Squaring both sides, we have

$$(x^2 - 1)\, y_1^2 = m^2 y^2 \qquad [1]$$

(Differentiate again)

$$(x^2 - 1) \cdot 2y_1 y_2 + y_1^2 (2x) = m^2 \cdot 2y \cdot y_1$$

or $$(x^2 - 1)y_2 + xy_1 - m^2 y = 0 \qquad [2]$$

Applying Leibnitz's theorem term by term to differentiate n times, we get

$$D^n[(x^2 - 1)y_2] = y_{n+2}\,(x^2 - 1) + n \cdot y_{n+1} \cdot 2x + \frac{n\,(n-1)}{2!} \cdot y_n \cdot 2\Big]$$

$$D^n\,[xy_1] = [y_{n+1}\, x + n \cdot y_n \cdot 1]$$

$$D^n\,[m^2 y] = m^2 y_n$$

Adding, we get

$$(x^2 - 1)\, y_{n+2} + (2n + 1)\, xy_{n+1} + (n^2 - n + n - m^2)\, y_n = 0$$

or $$(x^2 - 1)\, y_{n+2} + (2n + 1)\, xy_{n+1} + (n^2 - m^2)\, y_n = 0.$$

79. If $x = \sin t$, $y = \sin pt$, show that:

$$(1 - x^2)\, y_{n+2} - (2n + 1)\, xy_{n+1} - (n^2 - p^2)\, y_n = 0$$ [V.T.U., J/F, 2004] [B.U. (MOD.Q.P-II)]

$$x = \sin t,\ y = \sin pt$$

$\therefore$ $$y = \sin (p \sin^{-1} x) \qquad (\because t = \sin^{-1} x)$$

Diff. w.r.t. x

$$y_1 = \cos (p \sin^{-1} x) \frac{p}{\sqrt{1 - x^2}}$$

$\therefore$ $$\sqrt{(1 - x^2)}\ y_1 = p \cos (p \sin^{-1} x)$$

Again diff. w.r.t. x,

$$\sqrt{(1 - x^2)}\ y_2 + y_1 \frac{(-2x)}{2\sqrt{1 - x^2}} = - p \sin (p \sin^{-1} x) \cdot \frac{p}{\sqrt{1 - x^2}}$$

i.e. $$(1 - x^2)\, y_2 - xy_1 = - p^2 y$$

i.e. $$(1 - x^2)\, y_2 - xy_1 + p^2 y = 0$$

Diff. n times by Leibnitz's theorem,

$$(1 - x^2)\, y_{n+2} + {}^nC_1 \cdot y_{n+1} (- 2x) + {}^nC_2 \cdot y_n (- 2) - y_{n+1} \cdot x - nC_1\, y_n (1) + p^2 y_n = 0$$

i.e. $$(1 - x^2)\, y_{n+2} - (2n + 1)\, xy_{n+1} - (n^2 - n + n - p^2)\, y_n = 0$$

i.e. $$(1 - x^2)\, y_{n+2} - (2n + 1)\, x\, y_{n+1} - (n^2 - p^2) y_n = 0$$

80. If $y = (\sin^{-1} x)^2$ prove that

$$(1 - x^2)\, y_{n+2} - (2n + 1)\, xy_{n+1} - n^2 y_n = 0$$

Hence find the value of y_n at $x = 0$ [B.U. M.Q.P-III]

Given: $$y = (\sin^{-1} x)^2 \qquad [1]$$

$\therefore$ $$y_1 = 2 \sin^{-1} x \cdot \frac{1}{\sqrt{1 - x^2}} \qquad [2]$$

or $$\sqrt{(1 - x^2)}\ y_1 = 2 \sin^{-1} x$$

$\Rightarrow$ $$(1 - x^2) \cdot y_1^2 = 4(\sin^{-1} x)^2$$

$\Rightarrow$ $$(1 - x^2) \cdot y_1^2 = 4y \qquad [\because y = (\sin^{-1} x)^2]$$

(Differentiate again w.r.t. x)

$$(1 - x^2)\, 2y_1 y_2 - 2xy_1^2 = 4y_1$$

$\Rightarrow$ $$(1 - x^2)\, y_2 - xy_1 - 2 = 0 \qquad [3]$$

Differentiate [1] n times using Leibnitz's theorem, we get

$$(1 - x^2)\, y_{n+2} - n\, (- 2x)\, y_{n+1} + \frac{n\,(n-1)}{2} \cdot (- 2)\, y_n - xy_{n+1} - ny_n = 0$$

or $$(1 - x^2) y_{n+2} - (2n + 1)\, xy_{n+1} - [n^2 - n + n]\, y_n = 0$$

or $$(1 - x^2)\, y_{n+2} - (2n + 1)\, xy_{n+1} - n^2 y_n = 0 \qquad [4]$$

To find y_n at $x = 0$

Putting $x = 0$ in [1], [2], [3] and [4].

$$y\,(0) = 0, \quad y_1\,(0) = 0, \quad y_2\,(0) = 2, \quad y_{n+2}\,(0) = n^2 y_n\,(0) \qquad [5]$$

Putting $n = 1, 2, 3, 4$; in [5]

$$y_3(0) = 1^2 \cdot y_1(0) = 0$$
$$y_4(0) = 2^2 \cdot y_2(0) = 2^2 \cdot 2$$
$$y_5(0) = 3^2 \cdot y_3(0) = 3^2 \cdot 0 = 0$$
$$y_6(0) = 4^2 \cdot y_4(0) = 4^2 \cdot 2^2 \cdot 2$$
$$\vdots \qquad \vdots \qquad \vdots$$

Hence, when n is odd, $y_n(0) = 0$ when n is even. $y_n(0) = (n-2)^2 \ldots 4^2 \cdot 2^2 \cdot 2$.

81. If $y = e^{m\cos^{-1}x}$, show that

$$(1 - x^2)\, y_{n+2} - (2n + 1)\, xy_{n+1} - (n^2 + m^2)\, y_n = 0$$

and calculate $y_n\,(0)$.

Solution: Let $y = e^{m\cos^{-1}x}$ [1]

$$y_1 = e^{m\cos^{-1}x} \cdot \frac{-m}{\sqrt{(1-x^2)}} = -\frac{my}{\sqrt{(1-x^2)}} \qquad [2]$$

Squaring and cross multiplying,

$$(1 - x^2)\, y_1^2 = m^2 y^2$$

(Diff. again)

$$(1 - x^2) \cdot 2y_1y_2 - 2xy_1^2 = m^2 \cdot 2yy_1$$

$$\Rightarrow \qquad (1 - x^2)y_2 - xy_1 = m^2 y \qquad [3]$$

Diff. 'n' times by Leibnitz's Theorem

$$\left[y_{n+2}\,(1-x^2) + ny_{n+1}\,(-2x) + \frac{n(n-1)}{1\cdot 2}\, y_n\,(-2)\right]$$
$$-\,[y_{n+1} \cdot x + ny_n \cdot 1] = m^2 y_n$$

or $$(1 - x^2)\, y_{n+2} - (2n + 1)\, xy_{n+1} = (m^2 + n^2)\, y_n \qquad [4]$$

Calculation of y_n (0)

Putting $x = 0$, in [1], [2], [3] and [4]

$$y(0) = e^{m\cos^{-1}0} = e^{m\frac{\pi}{2}}$$

$$y_1(0) = -\,my\,(0) = -\,m \cdot e^{m\frac{\pi}{2}}$$

$$y_2(0) = m^2 y(0) = m^2 e^{m\frac{\pi}{2}}$$

$$y_{n+2}(0) = (m^2 + n^2)\, y_n(0), \qquad [5]$$

Putting $n = 1, 2, 3, 4, \ldots$ in [5]

$$y_3(0) = (m^2 + 1^2)\, y_1(0) = -m(m^2 + 1^2)\, e^{m\frac{\pi}{2}}$$

$$y_4(0) = (m^2 + 2^2)\, y^2(0) = (m^2 + 2^2) \cdot m^2 e^{m\frac{\pi}{2}}$$

$$y_5(0) = (m^2 + 3^2)\, y_3(0) = (m^2 + 3^2)\left\{-\,m\,(m^2+1)^2 \cdot e^{m\frac{\pi}{2}}\right\}$$

$$y_6(0) = (m^2 + 4^2)\, y_4(0) = -\,(m^2 + 4^2)(m^2 + 3^2)\, m(m^2 + 1^2)\, e^{m\frac{\pi}{2}}$$

$$\vdots \qquad \vdots \qquad \vdots$$

Hence, when n is odd

$$y_n(0) = -\, m\,(m^2 + 1^2)(m^2 + 3^2) \ldots [m^2 + (n-2)^2]\, e^{m\frac{\pi}{2}}$$

$$y_n(0) = m^2(m^2 + 2^2)(m^2 + 4^2) \ldots [m^2 + (n-2)^2]\, e^{m\frac{\pi}{2}}$$

(when n is even)

82. If $x = \sin t$ and $y = \cos \lambda t$ then prove that

$$(1 - x^2)\, y_{n+2} - (2n + 1)\, xy_{n+1} - (n^2 - \lambda^2)\, y_n = 0$$

Solution: Since $x = x(t)$; $y = y(t)$

$$\therefore \qquad y_1 = \frac{dy}{dx} = \frac{\left(\frac{dy}{dt}\right)}{\left(\frac{dx}{dt}\right)} = -\,\lambda \sin \lambda t \cdot \frac{1}{\cos t} = -\,\lambda \cdot \frac{\sin \lambda t}{\cos t}$$

$$\therefore \qquad y_1^2 = \lambda^2 \cdot \frac{\sin^2 \lambda t}{\cos^2 t} = \lambda^2 \left[\frac{1 - \cos^2(\lambda t)}{1 - \sin^2 t}\right]$$

i.e., $$y_1^2 = \lambda^2 \left(\frac{1 - y^2}{1 - x^2}\right)$$

i.e., $(1 - x^2)\, y_1^2 - \lambda^2 (1 - y^2) = 0$

(Diff. once)

$$(1 - x^2)\, 2y_1y_2 - 2xy_1^2 - \lambda^2 (-2 \cdot yy_1) = 0$$

i.e., $(1 - x^2)\, y_2 - xy_1 + \lambda^2 y = 0$ [1]

Diff. term by term by using Leibnitz's theorem we get

$$\left[(1 - x^2) y_{n+2} + n \cdot (-2x)\, y_{n+1} + \left[\frac{n\,(n-1)}{1 \cdot 2}\right](-2)\, y_n\right] - \{xy_{n+1} + n \cdot 1 \cdot y_n\} + \lambda^2 y_n = 0$$

i.e., $$(1 - x^2)\, y_{n+2} - (2n + 1)\, xy_{n+1} - (n^2 - \lambda^2) y_n = 0$$

83. If $y = \sin \log (x^2 + 2x + 1)$ prove that

$$(x + 1)^2\, y_{n+2} + (2n + 1)(x + 1)\, y_{n+1} + (n^2 + 4) y_n = 0$$

Solution: Let $y = \sin \log (x^2 + 2x + 1) = \sin [\log (x + 1)^2] = \sin [2 \log (x + 1)]$

i.e., $\sin^{-1}(y) = 2 \log (x + 1)$

(Differentiate once)

$$\frac{1}{\sqrt{1 - y^2}} \cdot y_1 = 2 \cdot \frac{1}{x + 1}$$

$$\Rightarrow \qquad y_1(x + 1) = 2\sqrt{1 - y^2} \qquad \text{or} \qquad y_1^2 (x + 1)^2 = 4(1 - y^2)$$

PART-B

(Diff. again)

$$2y_1y_2(x+1)^2 + y_1^2 \cdot 2[x+1] = 4(-2y \cdot y_1)$$

or $$(x+1)^2y_2 + y_1(x+1) + 4y = 0 \qquad [1]$$

(on division by $2y_1$)

Applying Leibnitz's theorem term by term to get

$$\left[y_{n+2}(x+1)^2 + n \cdot y_{n+1} \cdot 2(x+1) + \left\{\frac{n(n-1)}{1 \cdot 2}\right\} \cdot y_n (2)\}\right]$$

$$+ [\{y_{n+1}(x+1) + n \cdot y_n \cdot 1\}] + 4y_n = 0$$

i.e., $$y_{n+2}(x+1)^2 + (2n+1)(x+1)y_{n+1} + (n^2+4)y_n = 0$$

84. If $y = e^{a\sin^{-1} x}$, show that

$$(1-x^2)y_{n+2} - (2n+1)xy_{n+1} - (a^2+n^2)y_n = 0$$

and find the value for $x = 0$.

Solution: Let $$y = e^{a\sin^{-1} x}; \qquad [1]$$

$$y_1 = e^{a\sin^{-1} x} \cdot \left[\frac{a}{\sqrt{1-x^2}}\right]$$

$$\therefore \quad \sqrt{(1-x^2)} \cdot y_1 = ae^{a\sin^{-1} x} = ay \qquad [2]$$

$$\therefore \quad \sqrt{(1-x^2)} \cdot y_2 + y_1 \cdot \frac{1}{2\sqrt{(1-x^2)}}(-2x) = y_1 a$$

i.e., $(1-x^2)y_2 - xy_1 = ay_1\sqrt{(1-x^2)} = a \cdot ay = a^2y$

(using [2])

i.e., $$(1-x^2)y_2 - xy_1 - a^2y = 0 \qquad [3]$$

Diff. n times by Leibnitz's theorem. We get

$$(1-x^2)y_{n+2} + n(-2x)y_{n+1} + \left[\frac{n(n-1)}{1 \cdot 2}\right](-2)y_n - x \cdot y_{n+1} + n(-1) \cdot y_n - a^2y_n = 0$$

$$(1-x^2)y_{n+2} - xy_{n+1}(2n+1) + y_n[-n^2 + n - n - a^2] = 0$$

or $$(1-x^2)y_{n+2} - (2n+1)xy_{n+1} - (n^2+a^2)y_n = 0 \qquad [4]$$

To find the value for x = 0

We have from [4]

$$(1-x^2)y_{n+2} - (2n+1)x\,y_{n+1} = (n^2+a^2)y_n$$

Putting $x = 0$ in [1], [2], [3] and [4]

$$y(0) = e^{m\sin-1(0)} = e^0 = 1$$

$$y_1(0) = my(0) = m \cdot 1 = m$$

$$y_2(0) = m^2y(0) = m^2 \cdot 1 = m^2$$

$$y_{n+2}(0) = (n^2+a^2)y_n(0) \qquad [5]$$

Putting $n = 1, 2, 3, 4, \ldots$ in [5]

$$y_3(0) = (a^2+1^2)y_1(0) = (a^2+1^2)m$$

$$y_4(0) = (a^2+2^2)y_2(0) = (a^2+2^2)m^2$$

$$y_5(0) = (a^2 + 3^2)\, y_3(0) = (a^2 + 3^2)\,(a^2 + 1^2)\, m$$
$$y_6(0) = (a^2 + 4^2)\, y_4(0) = (a^2 + 4^2)\,(a^2 + 2^2) m^2$$
$$\vdots \qquad \vdots \qquad \vdots$$

Hence when n is odd

$$y_n(0) = a[a^2 + 1^2]\,[a^2 + 3^2]\,[a^2 + (n-2)^2]$$

and when n is even

$$y_n(0) = a^2[a^2 + 2^2]\,[a^2 + 4^2]\,[a^2 + (n-2)^2]$$

EXERCISES

Find the nth derivatives of the following functions, using the Leibnitz's theorem

(I) (1) xe^{ax} (2) xe^{-x} (3) $(x^2 + 3x)\, e^{-2x}$

(4) $x^2 \sin(2x)$ (5) $x^2 \sin(3x)$ (6) $x \,.\, \log x$

(7) $x^3 \,.\, \log x$ (8) $x^3 \sin^3 x$

(II) (1) If $y = x^2 e^x$, P.T

$$y_n = \frac{1}{2}(n)\,(n-1)\, y_2 - n\,(n-2)\, y_1 + \frac{1}{2}(n-1)\cdot(n-2)\, y$$

(2) If $y = (1-x)^{-a} \,.\, e^{-ax}$, P.T.

$(1-x)\, y_{n+1} - (n + ax)\, y_n - nay_{n-1} = 0$

(3) If $y = (\cos^{-1} x)^2$, P.T.

$(1-x^2)\, y_{n+2} - (2n+1)\, xy_{n+1} - n^2 y_n = 0$

(4) If $y = \dfrac{\log x}{x^2}$, prove that

$x^2 y_{n+2} + (2n+5)\, xy_{n+1} + (n^2 + n)\, y_n = 0$

(5) If $y = \dfrac{\sin h^{-1}}{\sqrt{1+x^2}}$, prove that

$(1 + x^2)\, y_{n+2} + (2n+3)x\, y_{n+1} - (n+1)^2\, y_n = 0$

(6) If $y = \cos(m \sin^{-1} \sqrt{x})$, show that

$$\frac{y_{n+1}}{y_n} = \frac{4n^2 - m^2}{4n+2} \text{ at } x = 0$$

(7) If $y = \dfrac{x^5}{ax^2 + 2bx + c}$ show that

$(ax^2 + bx + c)y_n + 2n(ax + b)y_{n-1} + n(n-1)\, a\, y_{n-2} = 0$ if $n > 5$

(8) If $y = (\sin h^{-1} x)^2$, prove that

$(1 + x^2)\, y_{n+2} + (2n+1)\, xy_{n+1} + n^2 y_n = 0$

(9) If $y = (\sin^{-1} x)^2$, prove that

$(1 - x^2)\, y_{n+2} - (2n+1)\, xy_{n+1} - n^2 y_n = 0$

Hence find the value of y_n at $x = 0$.

(10) If $y = \tan^{-1} x$, prove that

$(1 + x^2)\, y_{n+1} + 2nxy_n + n(n-1)\, y_{n-1} = 0$

Hence find $y_n\,(0)$

(11) Find $y_n(0)$, when

$$y = \log\left(x + \sqrt{1+x^2}\right)$$

(12) Find $y_n(0)$ when $y = \sin h^{-1} n$

(13) If $y = \sin(m \sin^{-1} x)$,
find y_n at $x = 0$.

2.4. POLAR CURVES

1. *Polar coordinates in the Plane:* The system of coordinates in which a point is located by its distance from a fixed point and the angle that the line from this point to the given point makes with a fixed line, called the *polar axis*.

The fixed point O, in the figure is the POLE the distance $OP = r$, from the pole to the given point is the RADIUS VECTOR. The angle θ (taken positive when counter clockwise) is the polar angle or the vectorial angle.

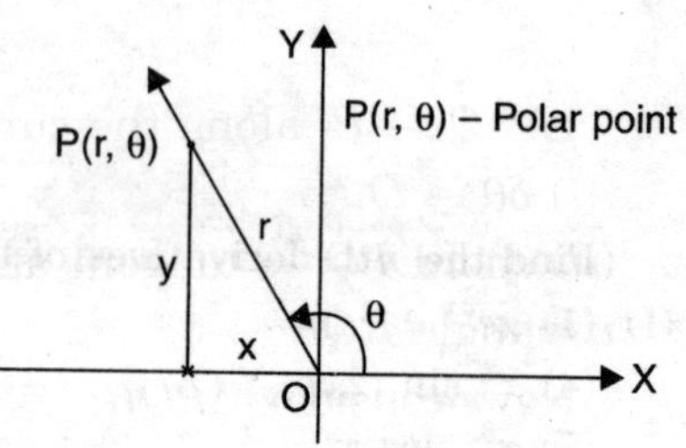

Fig. 2.1

2.5 ANGLE BETWEEN THE RADIUS VECTOR AND THE TANGENT

For any point (r, θ) of the curve $r = f(\theta)$, the angle ϕ between the radius vector and the tangent is given by

$$\boxed{\tan\phi = r\,\frac{d\theta}{dr}}$$ (VTU, August, 2000)

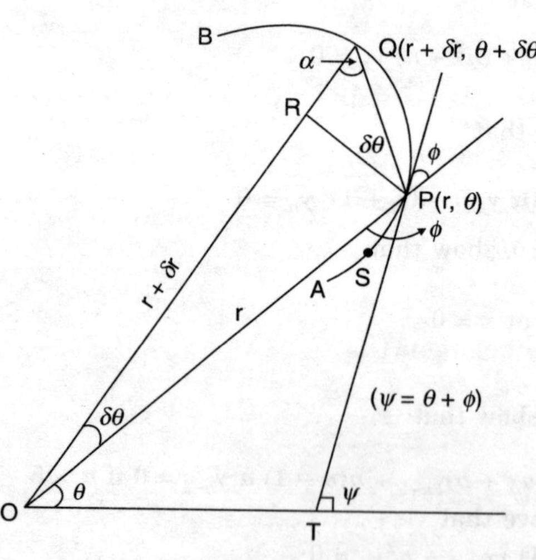

Fig. 2.2

Let $P = (r, \theta)$ and $Q\ (r + \delta r, \theta + \delta\theta)$ be two neighbouring points on the graph of the function $r = f(\theta)$.

Join PQ, OP, OQ

Let the tangent at P make an angle ϕ with the radius vector OP

From ΔOPR, $\quad \dfrac{PR}{OP} = \sin\delta\theta$

$$\frac{OR}{OP} = \cos \delta\theta$$

$$\Rightarrow \quad \left.\begin{aligned} PR &= r \sin \delta\theta \\ OR &= r \cos \delta\theta \end{aligned}\right\} \qquad (\because OP = r) \quad [1]$$

Now,
$$QR = OQ - OR = (r + \delta r) - r \cos \delta\theta;$$

$$= \delta r + r(1 - \cos \delta\theta); \qquad \left(\because \sin^2 \theta = \frac{1 - \cos 2\theta}{2}\right)$$

$$= \delta r + 2r \cdot \sin^2\left(\frac{\delta\theta}{2}\right); \qquad [2]$$

Let $Q \to P$, along the curve, then

(i) $\delta\theta \to O$,

(ii) chord $PQ \to$ tangent at P

(iii) angle $RQP \to \phi$

Now, consider ΔPQR,

$$\tan R\hat{Q}P = \frac{PR}{QR}$$

$$\lim_{Q \to P} \tan R\hat{Q}P = \lim_{Q \to P} \frac{PR}{QR}$$

i.e.,
$$\lim_{R\hat{Q}P \to \phi} \tan R\hat{Q}P = \lim_{\delta\theta \to O} \frac{r \sin \delta\theta}{\delta r + 2r \sin^2\left(\frac{\delta\theta}{2}\right)}$$

i.e.
$$\tan\phi = \lim_{\delta\theta \to O} \frac{r \cdot \frac{\sin \delta\theta}{\delta\theta}}{\frac{\delta r}{\delta\theta} + \frac{r \cdot \sin\frac{\delta\theta}{2}}{\frac{\delta\theta}{2}} \cdot \sin\frac{\delta\theta}{2}} = \frac{r(1)}{\frac{dr}{d\theta} + r \cdot 1 \cdot 0} \qquad \left(\because \lim_{\theta \to 0} \frac{\sin\theta}{\theta} = 1\right)$$

$$\therefore \quad \boxed{\tan \phi = r \cdot \left(\frac{d\theta}{dr}\right)} \qquad [3]$$

Note: (1) ϕ is the angle between the radius vector and the tangent.

Note (2): Relation between θ, ϕ, ψ: $\boxed{\psi = (\theta + \phi.)}$

$$\therefore \quad \tan\psi = \tan(\theta + \phi) = \frac{\tan\theta + \tan\phi}{1 - \tan\theta \tan\phi}$$

Knowing θ and ϕ, $\tan\psi$ can be calculated.

Angle of Intersection of Two Curves

Let the two curves given by the polar equations $r = f_1(\theta)$ and $r = f_2(\theta)$ intersect at P and ϕ_1, ϕ_2 be the angles between the common radius vector OP and the tangents PT_1, PT_2 to the two curves

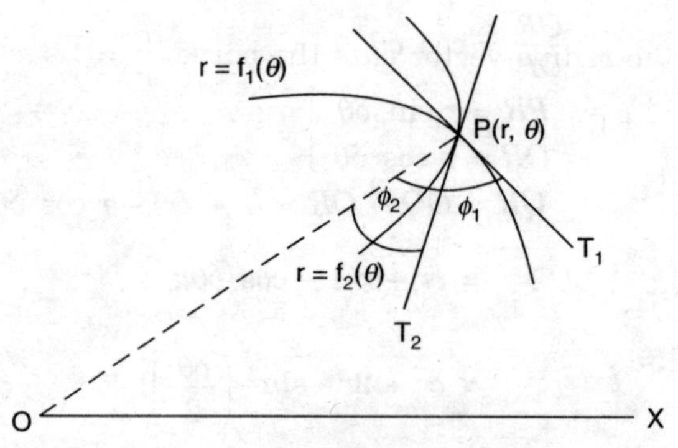

Fig. 2.3

Angle of intersection of two curves = angle between their tangents at a point of intersection.

$\therefore$ $\alpha = \phi_1 - \phi_2$ or $|\phi_1 - \phi_2|$

Note (I) For orthogonal intersection

$$\alpha = \frac{\pi}{2} \quad \therefore \quad \phi_1 = \frac{\pi}{2} + \phi_2$$

$$\tan \phi_1 = \tan\left(\frac{\pi}{2} + \phi_2\right) = -\cot \phi_2 = -\frac{1}{\tan \phi_2}$$

or $\tan \phi_1 \tan \phi_2 = -1.$

Note: (II) If $\alpha = 0$, then $\tan \phi_1 = \tan \phi_2$,

$\therefore$ the two curves touch if $\tan \phi_1 = \tan \phi_2$

A. Examples (85 – 103)

85. Show that the radius vector is inclined at a constant angle to the tangent at any point in the equiangular spiral $r = ae^{b\theta}$ or $(r = ae^{\theta \cot \alpha})$

Solution: We have $r = ae^{b\theta}$

Diff. w.r.t. θ,

$$\frac{dr}{d\theta} = ae^{b\theta}, b = rb$$

$$\tan \phi = r\frac{d\theta}{dr} = r \cdot \frac{1}{rb} = \frac{1}{b}$$

$$\Rightarrow \quad \phi = \tan^{-1}\left(\frac{1}{b}\right) = \text{constant}$$

OR

We have $r = ae^{\theta \cot \alpha}$

Diff. w.r.t. θ, we get

$$\frac{dr}{d\theta} = ae^{\theta \cot \alpha}(\cot \alpha) = r \cot \alpha$$

$$\therefore \quad \tan \phi = \frac{r}{\left(\frac{dr}{d\theta}\right)} = \frac{r}{r \cot \alpha} = \tan \alpha$$

$$\therefore \quad \phi = \alpha = \text{constant}$$

86. Find the angle at which the radius vector cuts the curve $\frac{l}{r} = 1 + e\cos\theta$.

Solution: **Given:** $\log l - \log r = \log(1 + e\cos\theta)$

Diff. w.r.t. θ, we get

$$0 - \frac{1}{r}\frac{dr}{d\theta} = \frac{-e\sin\theta}{1+e\cos\theta} \quad \text{or} \quad \frac{dr}{d\theta} = \frac{er\sin\theta}{1+e\cos\theta}$$

Now,
$$\tan\phi = \frac{r}{\left(\frac{dr}{d\theta}\right)} = \frac{r}{\frac{er\sin\theta}{(1+e\cos\theta)}} = \frac{1+e\sin\theta}{e\sin\theta}$$

$$\therefore \quad \phi = \tan^{-1}\left[\frac{1+e\cos\theta}{e\sin\theta}\right]$$

87. Find the angle between the radius vector and tangent of $\frac{2a}{r} = 1 - \cos\theta$.

Solution: **Given:**
$$\frac{2a}{r} = 1 - \cos\theta$$

$$\therefore \quad \log 2a - \log r = \log(1 - \cos\theta).$$

Differentiating w.r.t. θ,

$$0 - \frac{1}{r}\cdot\frac{dr}{d\theta} = \frac{1}{1-\cos\theta}(+\sin\theta)$$

$$\therefore \quad \frac{1}{r}\frac{dr}{d\theta} = \frac{-\sin\theta}{1-\cos\theta} = -\frac{2\sin(\theta/2)\cos(\theta/2)}{1-[1-2\sin^2(\theta/2)]}$$

$$= -\frac{\cos(\theta/2)}{\sin(\theta/2)} = -\cot\left(\frac{\theta}{2}\right)$$

$$\therefore \quad \tan\phi = r\frac{d\theta}{dr} = -\tan\left(\frac{\theta}{2}\right) = \tan\left(\pi - \frac{\theta}{2}\right)$$

$$\Rightarrow \quad \phi = \pi - \left(\frac{\theta}{2}\right)$$

88. Find the angle at which the radius vector cuts the curve $r^2 = a^2\cos 2\theta$.

Solution: **Given:** $r^2 = a^2\cos 2\theta$

$$\therefore \quad 2r\frac{dr}{d\theta} = -2a^2\sin 2\theta$$

$$\frac{dr}{d\theta} = \frac{-a^2\sin 2\theta}{r}$$

Now,
$$\tan\phi = r\frac{d\theta}{dr} = r\left(\frac{r}{-a^2\sin 2\theta}\right)$$

$$= \frac{a^2\cos 2\theta}{-a^2\sin 2\theta} = -\cot 2\theta = \tan\left(\frac{\pi}{2} + 2\theta\right)$$

$$\therefore \quad \boxed{\phi = \frac{\pi}{2} + 2\theta.}$$

PART-B

89. Find the slope of the curve

$$r = a \sin 2\theta$$

at $\theta = \dfrac{\pi}{4}$

$$r = a \sin 2\theta$$

$$\frac{dr}{d\theta} = 2a \cos 2\theta$$

$$\tan \phi = r\frac{d\theta}{dr} = \frac{a \sin 2\theta}{2a \cos 2\theta} = \frac{1}{2}\tan 2\theta$$

Let ψ be the angle made by the tangent with X-axis at $\theta = \dfrac{\pi}{4}$,

Then, solpe

$$= \tan \psi$$

$$= \frac{\tan \theta + \tan \phi}{1 - \tan \theta \tan \phi} = \frac{\tan \theta + \frac{1}{2}\tan 2\theta}{1 - \tan \theta \cdot \frac{1}{2}\tan 2\theta}$$

$$= \frac{\tan \theta + \dfrac{\tan \theta}{1 - \tan^2 \theta}}{1 - \dfrac{\tan^2 \theta}{1 - \tan^2 \theta}} = \frac{\tan \theta - \tan^3 \theta + \tan \theta}{1 - \tan^2 \theta - \tan^2 \theta}$$

$\therefore$ $$\text{Slope}\Big]_{\theta = \frac{\pi}{4}} = \tan \psi\Big]_{\theta = \frac{\pi}{4}}$$

and $$\tan \psi\Big]_{\theta = \frac{\pi}{4}} = \left[\frac{2\tan \theta - \tan^3 \theta}{1 - 2\tan^2 \theta}\right]_{\theta = \frac{\pi}{4}} = \frac{2\tan\left(\frac{\pi}{4}\right) - \tan^3\left(\frac{\pi}{4}\right)}{1 - 2\tan^2\left(\frac{\pi}{4}\right)} = \frac{2-1}{1-2} = -1$$

90. Find the value of ϕ for the curve $r^m = a^m (\cos m\theta - \sin m\theta)$ at the point $\theta = 0$.

Solution: **Given:** $$r^m = a^m (\cos m\theta - \sin m\theta)$$

taking logarithms, $$m \log r = m \log a + \log (\cos m\theta - \sin m\theta)$$

Diff. w.r.t. θ, $$\frac{m}{r} \cdot \frac{dr}{d\theta} = + \frac{-m \sin m\theta - m \cos m\theta}{\cos m\theta - \sin m\theta}$$

or $$\frac{1}{r} \cdot \frac{dr}{d\theta} = -\frac{\sin m\theta + \cos m\theta}{\cos m\theta - \sin m\theta}$$

$\therefore$ $$\tan \phi = r\frac{d\theta}{dr} = -\frac{\cos m\theta - \sin m\theta}{\sin m\theta + \cos m\theta}$$

At $\theta = 0$, $$\tan \phi = -\frac{1-0}{0+1} = -1 = \tan\frac{3\pi}{4}$$

$\Rightarrow$ $$\boxed{\phi = \frac{3\pi}{4}}$$

91. Prove that the tangent at any point (r, θ) on $r^2 = a^2 \sin 2\theta$ makes an angle 3θ with the initial line.

Solution: $r^2 = a^2 \sin 2\theta$ [1]

Diff. w.r.t. 'θ'

$$2r \cdot \frac{dr}{d\theta} = 2a^2 \cos 2\theta$$

$$\therefore \quad \frac{dr}{d\theta} = \frac{a^2}{r} \cos 2\theta \quad \text{or} \quad \frac{d\theta}{dr} = \frac{r}{a^2 \cos 2\theta}$$

$$\therefore \quad \tan\phi = r\frac{d\theta}{dr} = \frac{r^2}{a^2 \cos 2\theta} = \frac{a^2 \sin 2\theta}{a^2 \cos 2\theta} = \tan 2\theta$$

The angle which the tangent makes with the initial line = ψ

But $\quad \psi = \theta + \phi = \theta + 2\theta = 3\theta$

92. Find the angle of intersection of the curves

$r = \sin\theta + \cos\theta, \quad r = 2\sin\theta$ [VTU, March, 1999, VTU, Aug., 2002]

Solution: Equations of curves are

$$r = \sin\theta + \cos\theta \quad [1]$$

$$r = 2\sin\theta \quad [2]$$

Diff. [1] w.r.t. θ,

i.e. $(r = \sin\theta + \cos\theta)$, we get

$$\frac{dr}{d\theta} = \cos\theta - \sin\theta$$

$$\therefore \quad r\frac{d\theta}{dr} = \frac{\sin\theta + \cos\theta}{\cos\theta - \sin\theta} = \frac{\tan\theta + 1}{1 - \tan\theta}$$

$$\tan\phi_1 = \frac{\tan(\pi/4) + \tan\theta}{1 - \tan(\pi/4)\tan\theta} = \tan\left(\frac{\pi}{4} + \theta\right)$$

$$\Rightarrow \quad \phi_1 = \left(\frac{\pi}{4} + \theta\right) \quad [3]$$

ϕ_1 = angle between the radius vector and the first curve

|||ly, ϕ_2 = angle between the radius vector and the second curve

Since, 2nd curve is given by $r = 2\sin\theta$;

$$\therefore \quad \frac{dr}{d\theta} = 2\cos\theta$$

$$\therefore \quad r\frac{d\theta}{dr} = \frac{2\sin\theta}{2\cos\theta} = \tan\theta$$

$$\Rightarrow \quad \tan\phi_2 = \tan\theta$$

$$\Rightarrow \quad \phi_2 = \theta \quad [4]$$

$\therefore$ Angle of intersection of the curve is

$$|\phi_1 - \phi_2| = \left|\frac{\pi}{4} + \theta - \theta\right| = \frac{\pi}{4}$$

(They are not orthogonal)

93. Find the angle between the curves

$$r = a\,(1 + \sin\theta)$$

and $$r = a\,(1 - \sin\theta)$$ (VTU, March, 2000)

Solution: Equations of curves are given by

$$r = a(1 + \sin\theta) \quad [1]$$

$$r = a(1 - \sin\theta) \quad [2]$$

$$\Rightarrow \quad 1 + \sin\theta = 1 - \sin\theta \quad \Rightarrow \quad \sin\theta = 0$$

$$\therefore \quad \theta = 0 \quad \therefore \quad r = a$$

$\therefore$ The point of intersection of [1] and [2] is $(a, 0)$.

For first curve: [1]

$$\frac{dr}{d\theta} = a\cos\theta$$

$$\tan\phi = r\frac{d\theta}{dr} = a(1 + \sin\theta)\cdot\frac{1}{a\cos\theta} = \frac{1+\sin\theta}{\cos\theta}$$

$$\tan\phi_1 = [\tan\phi]_{\text{at }(a,\,0)}$$

$$= 1 \quad [3]$$

For the second curve: [2]

$$\frac{dr}{d\theta} = -a\cos\theta$$

$$\tan\phi = r\frac{d\theta}{dr} = a(1 - \sin\theta)\cdot\frac{1}{-a\cos\theta} = -\frac{1-\sin\theta}{\cos\theta}$$

$$\tan\phi_2 = [\tan\phi]_{\text{at }(a,\,0)}$$

$$= -1 \quad [4]$$

But, $\tan\phi_1 \tan\phi_2 = -1$

$\therefore$ Angle of intersection of two curve is $\frac{\pi}{2}$.

94. Find the angle of intersection of the curve

$$r = \frac{a\theta}{1+\theta} \quad \text{and} \quad r = \frac{a}{1+\theta^2}$$ [VTU, March, 2001]

Solution: $$r = \frac{a\theta}{1+\theta} \quad [1]$$

$$r = \frac{a}{1+\theta^2} \quad [2]$$

Eliminating r from the eqns. $r = \frac{a\theta}{1+\theta}$, $r = \frac{a}{1+\theta^2}$

$$\frac{a\theta}{1+\theta} = \frac{a}{1+\theta^2}$$

which gives $\theta^3 = 1$ or $\boxed{\theta = 1}$

Hence $r = \frac{9}{2}$.

Point of intersection is $\left(\frac{9}{2}, 1\right)$

Again, $r = \dfrac{a\theta}{1+\theta} \quad \Rightarrow \quad \dfrac{dr}{d\theta} = \dfrac{a}{(1+\theta)^2}$

$$\tan\phi_1 = \frac{a\theta}{1+\theta} \times \frac{(1+\theta)^2}{a} = \theta(1+\theta) = 2 \text{ at } \theta = 1$$

For the curve, $r = \dfrac{a}{1+\theta^2}, \quad \dfrac{dr}{d\theta} = -\dfrac{2a\theta}{(1+\theta^2)^2}$

$$\tan\phi_2 = \frac{a}{1+\theta^2} \times \frac{(1+\theta^2)^2}{-2a\theta} = -\frac{(1+\theta^2)}{2\theta} = -1 \text{ at } \theta = 1.$$

Now $\tan\phi = |\tan(\phi_1 - \phi_2)|$

$$= \left|\frac{\tan\phi_1 - \tan\phi_2}{1+\tan\phi_1 \cdot \tan\phi_2}\right| = \left|\frac{2-(-1)}{1+2(-1)}\right| = |-3| = 3$$

$\Rightarrow \quad \phi = \tan^{-1}(3).$

95. Find the angle of intersection of the curves:

$$r = a\log\theta, \quad r = \frac{a}{\log\theta}$$ (VTU, August, 2001)

Solution: **Given:** $r = a\log\theta$ and $r = \dfrac{a}{\log\theta}$

$\therefore \quad a\log\theta = \dfrac{a}{\log\theta}$

(Thus 'r' is automatically eliminated)

$\Rightarrow \quad (\log\theta)^2 = 1 \quad \Rightarrow \quad \log\theta = 1 \quad \Rightarrow \quad \boxed{\theta = e}$

Hence, $r = a\log e = a$

$\therefore$ The point of intersection is (a, e)

For the curve, $r = a\log\theta$,

$$\frac{dr}{d\theta} = \frac{a}{\theta}$$

Hence, $\tan\phi_1 = r\dfrac{d\theta}{dr} = (a\log\theta)\left(\dfrac{\theta}{a}\right) = \theta\log\theta.$

At (a, e), $\tan\phi_1 = e$

$\therefore \quad \phi_1 = \tan^{-1}(e)$

For the curve, $r = \dfrac{a}{\log\theta}, \quad \dfrac{dr}{d\theta} = \dfrac{-a}{(\log\theta)^2} \cdot \dfrac{1}{\theta}$

$$\therefore \quad \tan\phi_2 = r\frac{d\theta}{dr} = \frac{a}{\log\theta}\left\{\frac{-\theta(\log\theta)^2}{a}\right\} = -\theta\log\theta$$

At (a,e), $\tan\phi_2 = -e\log e = -e$

Hence $\phi_2 = \tan^{-1}(-e)$

$\therefore$ Required angle $= |\phi_1 - \phi_2| = |\tan^{-1}(e) - \tan^{-1}(-e)| = 2\tan^{-1}(e)$

$$= \tan^{-1}\frac{2e}{1-e^2}$$

PART-B

96. Find the angle of intersection of the two curves

$$r = a\,(1 - \cos\,\theta)$$

and $$r = a\,(1 + \cos\,\theta)$$ (VTU, Adv. Math.-1, 2002)

Solution: Given the two curves:

$$r = a\,(1 - \cos\,\theta) \qquad [1]$$

$$r = a\,(1 + \cos\,\theta) \qquad [2]$$

For the curve [1]

$$\frac{dr}{d\theta} = a\sin\theta;$$

$$\therefore \quad r\,\frac{d\theta}{dr} = a(1 - \cos\theta)\cdot\frac{1}{a\sin\theta} = \frac{2\sin^2\frac{\theta}{2}}{2\sin\frac{\theta}{2}\cos\frac{\theta}{2}} = \tan\left(\frac{\theta}{2}\right)$$

$$\therefore \quad \tan\phi_1 = r\,\frac{d\theta}{dr} = \tan\left(\frac{\theta}{2}\right) \qquad [3]$$

For the second curve [2]

$$\frac{dr}{d\theta} = -\,a\sin\theta$$

$$\therefore \quad r\,\frac{d\theta}{dr} = a(1 + \cos\theta)\cdot\frac{1}{-\,a\sin\theta} = -\frac{2\cos^2\frac{\theta}{2}}{2\sin\frac{\theta}{2}\cos\frac{\theta}{2}} = -\cot\left(\frac{\theta}{2}\right)$$

$$\therefore \quad \tan\phi_2 = -\cot\left(\frac{\theta}{2}\right) \qquad [4]$$

$$\therefore \quad \tan\phi_1\cdot\tan\phi_2 = \tan\left(\frac{\theta}{2}\right)\cdot\left[-\cot\frac{\theta}{2}\right] = -1$$

Hence the two curves cut orthogonally.

97. Find the angle between the curves

$$r = a\theta \quad \text{and} \quad r\theta = a$$ (B.U. Jan., 1993)

Solution: **Method (I)**

Given: $$r = a\theta \qquad [1]$$

$$r\theta = a \qquad [2]$$

Eliminating r, we have

$$a\theta^2 = a \quad \Rightarrow \quad \theta^2 = 1 \quad \Rightarrow \quad \theta = \pm 1$$

$\therefore$ Points of intersection are:

$(a, 1)$ and $(-\,a, 1)$

$$r = a\theta$$

$$\therefore \quad \frac{dr}{d\theta} = a$$

Now, $$\tan\phi_1 = r\,\frac{d\theta}{dr} = \frac{r}{\left(\frac{dr}{d\theta}\right)} = \left(\frac{r}{a}\right)$$

$$\boxed{\tan \phi_1 = \frac{r}{a}} \quad [3]$$

$\therefore$ $\tan, \phi_1 = [\tan \phi]$ at $(a, 1)$

$$= \left[\frac{r}{a}\right] \text{ at } (a, 1) = \left[\frac{a\theta}{a}\right] \text{ at } (a, 1) = 1$$

$\therefore$ $\tan \phi_1 = 1 \quad \Rightarrow \quad \phi_1 = \left(\frac{\pi}{4}\right)$

Again, $r\theta = a \quad \Rightarrow \quad r = \frac{a}{\theta}$

$\Rightarrow$ $\frac{dr}{d\theta} = -\frac{a}{\theta^2}$

Now, $\tan \phi_2 = r\frac{d\theta}{dr}$

or $= \frac{r}{\left(\frac{dr}{d\theta}\right)} = \frac{r}{\left(\frac{a}{\theta^2}\right)} = \frac{-r\theta^2}{a}$

$\Rightarrow$ $\left[\tan \theta_2 \Big|_{a,1}^{=} - \frac{a}{a} = -1\right]$

$\therefore$ $\tan \phi_2 = -1 \quad \Rightarrow \quad \phi_2 = \frac{3\pi}{4}$

$\therefore$ Angle of intersection

$$= (\phi_1 - \phi_2) = \left|\left(\frac{\pi}{4} - \frac{3\pi}{4}\right)\right| = \frac{\pi}{2}$$

Method (II):

Since $r = a\theta$

and $r\theta = a$ are the given curves.

$\therefore$ from $r = a\theta$, $\frac{dr}{d\theta} = a$

$\therefore$ $\tan \phi_1 = r\frac{d\theta}{dr}$ or $\tan \phi_1 = \frac{a\theta}{a} = \theta$ $\quad (\because \ r = a\theta)$

Also, $r\theta = a$ or $r = \frac{a}{\theta}$ or $\frac{dr}{d\theta} = \frac{-a}{\theta^2}$

$\therefore$ $\tan \phi_2 = r\frac{d\theta}{dr}$ or $\tan \phi_2 = -\frac{r\theta^2}{a}$

or $\tan \phi_2 = -\frac{a \cdot \theta}{a} = -\theta$ $\quad (\because \ r\theta = a)$

Now $\left(\begin{array}{l} r = a\theta\,; r\theta = a \\ \therefore a\theta^2 = a\,; \theta^2 = 1 \end{array}\right)$.

$$\tan \phi_1 \cdot \tan \phi_2 = (\theta) \cdot (-\theta) = -\theta^2 = -1$$

Thus, the curves cut orthogonally.

98. Find the angle between the curves

$$r = ae^{\theta},\ re^{\theta} = b$$ (B.U. Aug., 1996)

Solution: Given equations are $r = ae^{\theta}$

$$re^{\theta} = b$$

Eliminate r between them, we get

$$ae^{2\theta} = b$$

or $$e^{\theta} = \sqrt{\frac{b}{a}} \quad \Rightarrow \quad \theta = \log\sqrt{\frac{b}{a}}$$

Diff. the first equation $r = ae^{\theta}$, we get

$$\frac{dr}{d\theta} = ae^{\theta} = r$$

i.e. $$\tan\phi_1 = \frac{rd\theta}{dr} = \frac{r}{r} = 1$$

$\therefore$ $$\phi_1 = \tan^{-1}(1) = \frac{\pi}{4}$$

Second curve is $re^{\theta} = b$

$\Rightarrow$ $$\log r + \theta = \log b$$ (Differentiate once w.r.t. θ)

$$\frac{1}{r}\left(\frac{dr}{d\theta}\right) = -1$$

i.e., $$\tan\phi_2 = \frac{rd\theta}{dr} = -1; \qquad \phi_2 = \left(-\frac{\pi}{4}\right)$$

$\therefore$ Required angle of intersection

$$= |\phi_1 - \phi_2| = \left|\frac{\pi}{4} - \left(\frac{-\pi}{4}\right)\right| = \left(\frac{\pi}{2}\right).$$

99. Find the angle of intersection of the parabolas

$$r = \frac{a}{1+\cos\theta} \quad \text{and} \quad r = \frac{b}{1-\cos\theta}$$

Solution: Given equations are: $r = \dfrac{a}{1+\cos\theta}$; $r = \dfrac{b}{1-\cos\theta}$

For the first curve:

We have $$\log r = \log a - \log(1+\cos\theta)$$

$\therefore$ $$\frac{1}{r}\cdot\frac{dr}{d\theta} = 0 - \frac{-\sin\theta}{1+\cos\theta} = \frac{2\sin\left(\frac{\theta}{2}\right)\cos\left(\frac{\theta}{2}\right)}{2\cos^2\left(\frac{\theta}{2}\right)} = \tan\left(\frac{\theta}{2}\right)$$

$$\tan\phi_1 = \frac{rd\theta}{dr} = \frac{1}{\tan\left(\frac{\theta}{2}\right)} = \cot\left(\frac{\theta}{2}\right) \quad [1]$$

For the second curve:

We have, $$\log r = \log b - \log(1-\cos\theta)$$

$$\therefore \quad \frac{1}{r} \cdot \frac{dr}{d\theta} = 0 - \frac{-\sin\theta}{1-\cos\theta} = -\frac{2\sin\left(\frac{\theta}{2}\right)\cos\left(\frac{\theta}{2}\right)}{2\sin^2\left(\frac{\theta}{2}\right)} = -\cot\left(\frac{\theta}{2}\right)$$

$$\tan\phi_2 = \frac{rd\theta}{dr} = -\frac{1}{\cot\left(\frac{\theta}{2}\right)} \quad [2]$$

$\therefore \quad \tan\phi_1 \cdot \tan\phi_2 = -1$

$\Rightarrow$ Curves cut orthogonally.

100. Prove that the spirals

$$r^n = a^n \cos n\theta \quad \text{and} \quad r^n = b^n \sin n\theta$$

intersect orthogonally.

Solution: $r^n = a^n \cos n\theta$ [1]

Given: $r^n = b^n \sin n\theta$ [2]

For the curve [1]: We have

$$n \log r = n \log a + \log \cos n\theta$$

$$\therefore \quad n \cdot \frac{1}{r} \cdot \frac{dr}{d\theta} = 0 + \frac{-n\sin n\theta}{\cos n\theta} \quad \text{or} \quad \frac{1}{r} \cdot \frac{dr}{d\theta} = -\tan n\theta$$

$$\tan\phi_1 = \frac{rd\theta}{dr} = -\cot n\theta \quad [3]$$

For the curve [2]: We have

$$n \log r = n \log b + \log \sin n\theta$$

$$\therefore \quad \frac{n}{r} \cdot \frac{dr}{d\theta} = 0 + \frac{n\cos n\theta}{\sin n\theta} \quad \text{or} \quad \frac{1}{r} \cdot \frac{dr}{d\theta} = \cot n\theta$$

$$\tan\phi_2 = \frac{rd\theta}{dr} = \tan(n\theta) \quad [4]$$

$\therefore \quad \tan\phi_1 \tan\phi_2 = -\cot n\theta. \tan n\theta = -1$

$\therefore$ The two curves cut orthogonally.

Note: Another form of the problem: show that the curves $r^m = a^m \cos m\theta$ and $r^m = a^m \sin m\theta$ intersect orthogonally.

101. Find the angle of intersection of the curves:

$$r = 2a\cos\theta$$
$$r = 2a\sin\theta$$

Solution: $r = 2a\cos\theta$ [1]

$r = 2a\sin\theta$ [2]

$\therefore \quad 2a\cos\theta = 2a\sin\theta$

$\therefore \quad \cos\theta = \sin\theta$

or $\quad \tan\theta = 1 = \tan\left(\frac{\pi}{4}\right) \quad \Rightarrow \quad \boxed{\theta = \frac{\pi}{4}}$

$\therefore$ **From [2]**

$$r = 2a\sin\left(\frac{\pi}{4}\right) = \frac{2a}{\sqrt{2}} = \sqrt{2}a$$

$\therefore$ Point of intersection: is $\left(\sqrt{2}a, \dfrac{\pi}{4}\right)$

For the curve [1]

$$\frac{dr}{d\theta} = -2a \sin\theta$$

$$\tan\theta = \frac{rd\theta}{dr} = \frac{2a\cos\theta}{-2a\sin\theta}$$

$$\therefore \quad \tan\phi = -\cot\theta$$

Now,
$$\tan\phi_1 = [\tan\phi]_{\text{at}\left(\sqrt{2}a, \frac{\pi}{4}\right)}$$

$$= -\cot\left(\frac{\pi}{4}\right) = \tan\left(\frac{\pi}{2} + \frac{\pi}{4}\right) = \tan\left(\frac{3\pi}{4}\right)$$

$$\therefore \quad \phi_1 = \left(\frac{\pi}{2} + \frac{\pi}{4}\right) = \frac{3\pi}{4} \qquad [1]$$

For the curve [2]

$$\frac{dr}{d\theta} = 2a\cos\theta$$

$$\tan\phi = r\frac{d\theta}{dr} = \frac{2a\sin\theta}{2a\cos\theta} = \tan\theta$$

$$\tan\phi_2 = [\tan\phi] \text{ at } \left(\sqrt{2}a, \frac{\pi}{4}\right) = \tan\left(\frac{\pi}{4}\right)$$

$$\therefore \quad \phi_2 = \left(\frac{\pi}{4}\right) \qquad [2]$$

$\therefore$ Angle of intersection of two curves is ϕ, where

$$\phi = |\phi_1 - \phi_2| = \left|\frac{3\pi}{4} - \frac{\pi}{4}\right| = \frac{\pi}{2}$$

102. Find the angle of intersection of the curves:

$$r = a(1 - \cos\theta) \qquad [1]$$

$$r = 2a\cos\theta \qquad [2]$$

Solution:

For curve [1]

$$\tan\phi_1 = r\frac{d\theta}{dr} = a(1-\cos\theta)\cdot\frac{1}{a\sin\theta} = \frac{1-\cos\theta}{\sin\theta}$$

$$= \frac{2\sin^2\dfrac{\theta}{2}}{2\sin\dfrac{\theta}{2}\cos\dfrac{\theta}{2}} = \tan\frac{\theta}{2}$$

$$\Rightarrow \qquad \phi_1 = \frac{\theta}{2}$$

For the curve [2]

$$\tan \phi_2 = r\,\frac{d\theta}{dr} = \frac{2a\cos\theta}{-2a\sin\theta}$$

$$= -\cot\theta = \tan\left(\frac{\pi}{2} + \theta\right)$$

$$\Rightarrow \qquad \phi_2 = \left(\frac{\pi}{2}\right) + \theta$$

$$\therefore \qquad |\phi_1 - \phi_2| = \left|\frac{\theta}{2} - \left(\frac{\pi}{2} + \theta\right)\right| = \frac{\pi}{2} + \frac{\theta}{2} \qquad [3]$$

Point of intersection

$$a(1 - \cos\theta) = 2a\cos\theta$$

$$\Rightarrow \qquad \cos\theta = \frac{1}{3}$$

$$\Rightarrow \qquad \theta = \cos^{-1}\left(\frac{1}{3}\right).$$

Hence the equation [3] is

$$|\phi_1 - \phi_2| = \frac{\pi}{2} + \frac{1}{2}\cos^{-1}\left(\frac{1}{3}\right)$$

103. Prove that the following parrs of curves intersect orthogonally

$$r = a\sec^2\left(\frac{\theta}{2}\right) \qquad [1]$$

$$r = b\,\mathrm{cosec}^2\left(\frac{\theta}{2}\right) \qquad [2]$$

For curve [1], **we have**

$$\tan \phi_1 = r\,\frac{d\theta}{dr} = \frac{a\sec^2\left(\frac{\theta}{2}\right)}{a\sec^2\left(\frac{\theta}{2}\right)\tan\left(\frac{\theta}{2}\right)} = \cot\left(\frac{\theta}{2}\right)$$

$$= \tan\left(\frac{\pi}{2} - \frac{\theta}{2}\right)$$

PART-B

$$\therefore \qquad \phi_1 = \left(\frac{\pi}{2} - \frac{\theta}{2}\right)$$

For the curve [2], we have

$$\tan \phi_2 = r\,\frac{d\theta}{dr} = \frac{b \operatorname{cosec}^2\left(\frac{\theta}{2}\right)}{b \operatorname{cosec}^2\left(\frac{\theta}{2}\right)\cot\left(\frac{\theta}{2}\right)} = -\tan\left(\frac{\theta}{2}\right)$$

$$\Rightarrow \qquad \phi_2 = \left(-\frac{\theta}{2}\right)$$

$$\therefore \qquad |\phi_1 - \phi_2| = \frac{\pi}{2}$$

$\Rightarrow$ the curves cut orthogonally.

EXERCISES

I. Find the angle of intersection of the curve

(1) $r = 6 \cos \theta$ and $r = 2(1 + \cos \theta)$

(2) $a^2 = r^2 \sin 2\theta$ and $r^2 \cos 2\theta = b^2$

(3) $r^2 \sin 2\theta = 4$ and $r^2 = 16 \sin 2\theta$

$\left(Ans\text{: (1) } \frac{\pi}{6} \text{ (2) } \frac{\pi}{2} \text{ (3) } \frac{2\pi}{3}\right)$

II. (1) Show that the curves

$r^m = a^m \sec(m\theta + \alpha)$ and

$r^m = b^m \sec(m\theta + \beta)$

intersect at an angle which is independent of a and b.

(2) Find the value of ϕ for the curve $a\theta = \sqrt{r^2 - a^2} - a \cos^{-1}\frac{a}{r}$

$\left(Ans\text{: } \phi = \tan^{-1}\left(\frac{\sqrt{r^2 - a^2}}{a}\right)\right)$.

(3) Show that the circle

$r = b$ cuts the curve

$r^2 = a^2 \cos 2\theta + b^2$ at an angle $\tan^{-1}\left(\frac{a^2}{b^2}\right)$

(4) Show that the curves

$r^2 = 16 \sin 2\theta$ and $r^2 \sin 2\theta = 4$

intersect at angle $\frac{2\pi}{3}$

(5) Show that at any point (r, θ) the tangent to the curve $r^n = a^n \sin n\theta$ makes an angle $(n + 1)\theta$ with the initial line

(6) Prove that the curves

$r^2 = a^2 \cos 2\theta$ and $r = a(1 + \cos \theta)$

intersect at an angle $3 \sin^{-1}\left(\frac{3}{4}\right)^{1/4}$.

(7) Find the value of ϕ for the curve

$$r^m = a^m(\cos m\theta - \sin m\theta)$$

at the point $\theta = 0$

(8) Find the angle between the radius vector and the tangent in each of the following curves:

(*i*) $r = a\,(1 + \sin\theta)$ at $\theta = \frac{\pi}{6}$ (*ii*) $r = a \operatorname{cosec}^2\left(\frac{\theta}{2}\right)$ at $\theta = \frac{\pi}{2}$

(*iii*) $r^2 = a^2 \cos 2\theta$ at $\theta = \frac{\pi}{6}$ (*iv*) $r = a\,(1 + \cos\theta)$ at $\theta = \frac{\pi}{2}$

LENGTH OF PERPENDICULAR FROM POLE TO THE TANGENT (PEDAL)

From the pole O, draw $ON \perp r$ to the tangent at any point $P\,(r\,\theta)$ on the curve $r = f(\theta) \cdot N$ being the foot of the perpendicular.

Let $ON = p$ (length of perpendicular from O)

From rt. $\lfloor d$ ΔOPN, we have

$$ON = OP \sin\phi$$

$$\therefore \quad \boxed{p = r\sin\phi} \quad [1]$$

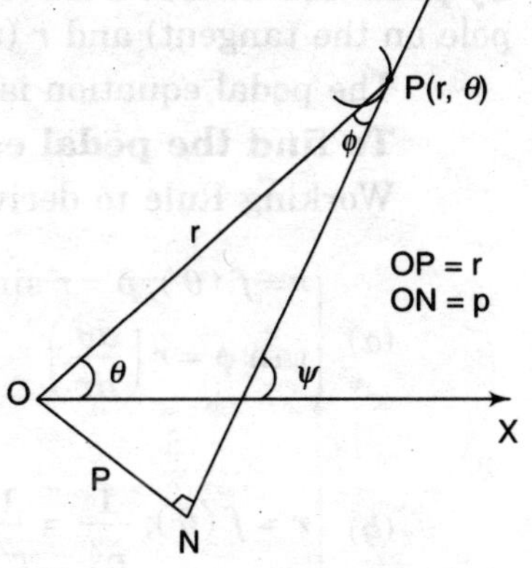

If p denotes the length of the perpendicular from the pole on the tangent at any point (r, θ) of a curve $r = f(\theta)$, we have

(*a*) $$\frac{1}{p^2} = \frac{1}{r^2} + \frac{1}{r^4}\left(\frac{dr}{d\theta}\right)^2$$

(*b*) $$\frac{1}{p^2} = u^2 + \left(\frac{du}{d\theta}\right)^2$$

where $$u = \frac{1}{r}$$

Solution:

$$\because \quad p = r\sin\phi$$

$$\therefore \quad \frac{1}{p^2} = \frac{1}{r^2\sin^2\phi} = \frac{1}{r^2}\operatorname{cosec}^2\phi = \frac{1}{r^2}(1 + \cot^2\phi)$$

$$= \frac{1}{r^2}\left[1 + \frac{1}{r^2}\left(\frac{dr}{d\theta}\right)^2\right] \qquad \left(\because \tan\phi = r\frac{d\theta}{dr}\right)$$

$$\therefore \quad \boxed{\frac{1}{p^2} = \frac{1}{r^2} + \frac{1}{r^4}\cdot\left(\frac{dr}{d\theta}\right)^2} \quad (2)$$

(*b*) Now $$\frac{1}{p^2} = \frac{1}{r^2} + \frac{1}{r^4}\left(\frac{dr}{d\theta}\right)^2 \quad [2]$$

Put $r = \frac{1}{u}$ $\therefore$ $\frac{dr}{d\theta} = -\frac{1}{u^2} \cdot \frac{du}{d\theta}$

$\therefore$ From [2],

$$\frac{1}{p^2} = u^2 + u^4 \cdot \left[-\frac{1}{u^2} \cdot \frac{du}{d\theta}\right]^2 = u^2 + \left(\frac{du}{d\theta}\right)^2$$

$\therefore$
$$\boxed{\frac{1}{p^2} = u^2 + (u^1)^2}$$

where $u^1 = \left(\frac{du}{d\theta}\right)$.

2.6 PEDAL EQUATIONS OR ($p - r$) EQUATIONS (POLAR CURVES ONLY)

By pedal equation of a curve we mean to find the relation between p (perpendicular from the pole on the tangent) and r (the radius vector) *i.e.*, $F(P_1\ r) = 0$

The pedal equation is also sometimes called the $(P - r)$ equation of the curve.

To find the pedal equation of a curve whose polar equation is given.

Working Rule to derive the pedal equation:

(*a*) $\begin{cases} r = f(\theta);\ p = r \sin\phi \\ \tan\phi = r\left(\dfrac{d\theta}{dr}\right) \end{cases}$ (ELIMINATE θ and ϕ)

(*b*) $\left\{ r = f(\theta);\ \dfrac{1}{p^2} = \dfrac{1}{r^2} + \dfrac{1}{r^4}\left(\dfrac{dr}{d\theta}\right)^2 \right.$ (ELIMINATE θ)

(*c*) $\left\{ \left(\dfrac{1}{r}\right) = u = f(\theta);\ \dfrac{1}{p^2} = (u^2 + u'^2). \right.$ (ELIMINATE θ)

Examples (104 – 122)

Problems based on finding the pedal equations of polar curve only

104. For the cardioid $r = a(1 - \cos\theta)$, find ϕ and the pedal equation.

Solution: Let $r = a(1 - \cos\theta)$ [1]

Diff. $\frac{dr}{d\theta} = a \sin\theta = \tan\phi = r\frac{d\theta}{dr} = a(1 - \cos\theta) \cdot \frac{1}{a \sin\theta}$

$$= \frac{2\sin^2\left(\frac{\theta}{2}\right)}{2\sin\left(\frac{\theta}{2}\right)\cos\left(\frac{\theta}{2}\right)} = \tan\left(\frac{\theta}{2}\right)$$

$\therefore \qquad \phi = \dfrac{\theta}{2}$ [2]

Now, since $p = r \sin \phi = r \sin\left(\dfrac{\theta}{2}\right)$

$\therefore \qquad \sin\left(\dfrac{\theta}{2}\right) = \dfrac{p}{r}$

$\therefore$ from [1], $r = a \cdot 2 \sin^2 \dfrac{\theta}{2} = 2a \cdot \dfrac{p^2}{r^2}$

$\therefore$ $2ap^2 = r^3$ is the required pedal equation.

105. Find the pedal equation of the curve

$$r = a\,(1 + \cos\,\theta)$$ [VTU, Aug., 1999]

Solution: **Given:** $r = a\,(1 + \cos\,\theta)$ [1]

$\therefore \qquad \dfrac{dr}{d\theta} = -\,a \sin\theta$

Hence, $\tan\,\phi = \dfrac{r}{\left(\dfrac{dr}{d\theta}\right)} = \dfrac{a\,(1+\cos\theta)}{-a \sin\theta} = -\cot\left(\dfrac{\theta}{2}\right) = \tan\left(\dfrac{\pi}{2} + \dfrac{\theta}{2}\right)$

$\Rightarrow \qquad \phi = \dfrac{\pi}{2} + \dfrac{\theta}{2}$ [2]

Also, we have $p = r \sin\,\phi = r \sin\left(\dfrac{\pi}{2} + \dfrac{\theta}{2}\right) = r \cos\left(\dfrac{\theta}{2}\right)$

i.e., $p = r \cos\left(\dfrac{\theta}{2}\right)$ [3]

Eliminate θ from [1] and [3]

From [1], we have $r = a\,(1 + \cos\,\theta) = 2a \cos^2\left(\dfrac{\theta}{2}\right) = 2a\left(\dfrac{p}{r}\right)^2$

i.e., $\boxed{r^3 = 2ap^2}$

is $(p - r)$ equation.

106. Find the pedal equations of the curve

$$r^m = a^m \cos\, m\theta$$ [VTU (Aug., 1999, 2000, Mar., 2001)]

Solution: **Given:** $r^m = a^m \cos\, m\theta$ [1]

$\therefore \qquad m \log r = m \log a + \log \cos m\theta$

Diff. w.r.t. 'θ', we get

$$\frac{m}{r}\,\frac{dr}{d\theta} = \left(0 - \frac{m \sin m\theta}{\cos m\theta}\right)$$

i.e., $$\frac{dr}{d\theta} = -r\tan m\theta$$

$\therefore$ $$\tan\theta = \frac{r}{\left(\frac{dr}{d\theta}\right)} = \frac{r}{-r\tan m\theta} = -\cot m\theta = \tan\left(\frac{\pi}{2}+m\theta\right)$$

Thus, $$\phi = \frac{\pi}{2} + m\theta \quad [2]$$

Since $$p = r\sin\phi$$

$$= r\sin\left(\frac{\pi}{2}+m\theta\right) = r\cos m\theta \quad [3]$$

We eliminate θ from [1] and [3]

from [1], $$r^m = a^m\cos m\theta = a^m\cdot\frac{p}{r} \quad \text{[using (3)]}$$

i.e., $$r^{m+1} = a^m p,$$

Which is the pedal eqn. of the curve.

107. Find the pedal equation of the curve

$$r^n = a^n\sin n\theta \quad \text{[B.U. 1995, B.U. 1965]}$$

Solution:

We have $$r^n = a^n\sin(n\,\theta) \quad [1]$$

Diff. w.r.t. a,

$$nr^{n-1}\frac{dr}{d\theta} = na^n\cos n\theta$$

i.e., $$\frac{dr}{d\theta} = \frac{r^{n-1}}{a^n\cos n\theta} \quad [2]$$

We know that $$\tan\phi = r\cdot\frac{r^{n-1}}{a^n\cos n\theta}$$

$$= \frac{r^n}{a^n\cos n\theta} = \frac{a^n\sin n\theta}{a^n\cos n\theta} = \tan n\theta$$

$\therefore$ $$\phi = n\theta \quad [3]$$

Also, we have

$$p = r\sin\phi = r\sin n\theta = r\cdot\frac{r^n}{a^n} \quad \text{[from (1)]}$$

$\therefore$ $$p = \frac{r^{n+1}}{a^n}$$

or $$pa^n = r^{n+1},$$ which is the required pedal equation.

108. Obtain the pedal equation of the curve

$$r \sin n\theta = a$$

Solution:

Differentiating

$$r \sin n\theta = a \text{ w.r.t. } \theta,$$

$$\frac{dr}{d\theta} \cdot \sin n\theta + nr \cos n\theta = 0$$

or $$\frac{1}{r} \cdot \frac{dr}{d\theta} = -n \cdot \frac{\cos n\theta}{\sin n\theta} = -n \cot n\theta$$

$\therefore$ $$\cot \phi = -n \cot n\theta$$

Now, $$p = r \sin \theta = \frac{r}{\operatorname{cosec} \phi}$$

$$p = \frac{r}{\operatorname{cosec} \phi} = \frac{r}{\sqrt{1 + \cot^2 \phi}} = \frac{r}{\sqrt{1 + n^2 \cot^2 n\theta}}$$

$$= \frac{r}{\sqrt{1 + n^2 \cdot \dfrac{\cos^2 n\theta}{\sin^2 n\theta}}} = \frac{r \sin n\theta}{\sqrt{\sin^2 n\theta + n^2 (1 - \sin^2 n\theta)}}$$

$$= \frac{a}{\sqrt{\dfrac{a^2}{r^2} + n^2 \left(1 - \dfrac{a^2}{r^2}\right)}} \qquad (\because \ r \sin n\theta = a)$$

$\therefore$ $$p = \frac{ar}{\sqrt{a^2 + n^2 (r^2 - a^2)}}$$

109. Find the pedal equation of the curve

$$r = ae^{m}\theta$$ [B.U. 1985, 1958]

Solution:

Diff. $$r = ae^{m\theta} \text{ w.r.t. } \theta$$

$$\frac{dr}{d\theta} = ame^{m\theta} = rm;$$

$\therefore$ $$\frac{1}{r} \frac{dr}{d\theta} = m;$$

or $$\cot \phi = m.$$

Now, $$p = r \sin \phi$$

$$= \frac{r}{\operatorname{cosec} \phi} = \frac{r}{\sqrt{1 + \cot^2 \phi}} = \frac{r}{\sqrt{1 + m^2}}$$

$$\therefore \qquad r^2 = (1 + m^2)p^2$$

110. Find the pedal equation of the curve

$$r = a\theta.$$ [M.U. 1973]

Solution: **Given:** $r = a\theta$

Method (I)

Differentiating w.r.t. θ, we get

$$\frac{dr}{d\theta} = a; \quad \text{or} \quad \frac{d\theta}{dr} = \frac{1}{a}$$

But $$\tan\phi = r\frac{d\theta}{dr} = \frac{r}{a}$$

Also $$p = r\sin\phi$$

$$= r \cdot \frac{\sin\phi}{\cos\phi\sec\phi}$$ (Note this step)

$$= \frac{r\tan\phi}{\sqrt{1+\tan^2\phi}}$$

or $$p = \frac{\frac{r^2}{a}}{\sqrt{1+\frac{r^2}{a^2}}} = \frac{r^2}{\sqrt{r^2+a^2}}$$

or $$p^2(a^2 + r^2) = r^4$$

Which is the required pedal equation.

Method (II)

Since: $$r = a\theta$$

$$r' = \frac{dr}{d\theta} = a$$

$$\left(\frac{1}{p^2}\right) = \left(\frac{1}{r^2}\right) + \left(\frac{(r^1)^2}{r^4}\right) = \left(\frac{1}{r^2}\right) + \left(\frac{a^2}{r^4}\right) = \left[\frac{(r^2+a^2)}{r^4}\right]$$

$$\therefore \qquad p^2(r^2 + a^2) = r^4$$

This is the pedal equation.

111. Find the pedal equation of the curve

$$r = ae^{\theta\cot\alpha}$$

Solution: **Given:** $$r = ae^{\theta\cot\alpha}$$ [1]

Diff. w.r.t θ, we get

$$\frac{dr}{d\theta} = ae^{\theta\cot\alpha} \cdot \cot\alpha$$

or $$\frac{dr}{d\theta} = \frac{r}{\tan\alpha}$$ from [1]

We know that

$$\tan\phi = r\frac{d\theta}{dr} = \frac{r}{\left(\frac{dr}{d\theta}\right)} = \frac{r\tan\alpha}{r} = \tan\alpha$$

$$\phi = \alpha.$$

We also know that

$$p = \sin\phi$$

$$\therefore \quad p = r\sin\alpha$$

is the required pedal eqation.

112. Find the pedal equation of

$$2a = r(1 + \cos\theta)$$

Solution: **Given:** $$\frac{2a}{r} = 1 + \cos\theta \qquad [1]$$

Diff. w.r.t. θ, we get

$$-\frac{2a}{r^2}\cdot\frac{dr}{d\theta} = -\sin\theta;$$

$$\frac{d\theta}{dr} = \frac{2a}{r^2\sin\theta}$$

But $$\tan\theta = r\frac{d\theta}{dr} = \frac{2a}{r\sin\theta} = \frac{(1+\cos\theta)}{\sin\theta}$$ (using [1])

$$\tan\phi = \frac{2\cos^2\left(\frac{\theta}{2}\right)}{2\sin\left(\frac{\theta}{2}\right)\cos\left(\frac{\theta}{2}\right)} = \cot\left(\frac{\theta}{2}\right) = \tan\left(\frac{\pi}{2} - \frac{\theta}{2}\right)$$

$$\therefore \quad \phi = \left(\frac{\pi}{2} - \frac{\theta}{2}\right)$$

Also $$p = r\sin\phi = r\sin\left(\frac{\pi}{2} - \frac{\theta}{2}\right) = r\cos\left(\frac{\theta}{2}\right)$$

or $$p^2 = r^2\cos^2\left(\frac{\theta}{2}\right) = r^2\frac{(1+\cos\theta)}{2}$$

$$p^2 = r^2\cdot\frac{a}{r} = ar.$$

Hence, the required pedal equation is $p^2 = ar$.

PART-B

113. Show that the pedal equation of the curve

$$r\theta = a, \text{ is } p^2 (a^2 + r^2) = a^2 r^2.$$

Given: $r\theta = a;$

$$\therefore \quad r = \left(\frac{a}{\theta}\right) \qquad [1]$$

$$r' = \frac{dr}{d\theta} = a\left(-\frac{1}{\theta^2}\right) = -a\left(\frac{r^2}{a^2}\right) = -\frac{1}{a}r^2 \qquad [2]$$

$$\frac{1}{p^2} = \frac{1}{r^2} + \frac{1}{r^4}\left(\frac{dr}{d\theta}\right)^2 = \frac{1}{r^2} + \frac{1}{r^4}\left(\frac{-r^2}{a}\right)^2$$

$$= \frac{1}{r^2} + \frac{1}{r^4}\frac{r^4}{a^2} = \left(\frac{1}{r^2} + \frac{1}{a^2}\right) = \frac{(a^2 + r^2)}{a^2 r^2}$$

i.e., $a^2 r^2 = p^2 (r^2 + a^2)$

114. Find the pedal equation of the curves

(*a*) $r = a(1 - \cos\theta)$ (*b*) $\left(\dfrac{2a}{r}\right) = (1 - \cos\theta)$ (*c*) $\left(\dfrac{l}{r}\right) = (1 + e\cos\theta)$

Solution:

(*a*) Let $r = a(1 - \cos\theta)$

$$\therefore \quad r^1 = \frac{dr}{d\theta} = a(\sin\theta) \qquad [1]$$

$$\therefore \quad \tan\phi = \left(\frac{r}{r^1}\right) = \frac{a\left[2\sin^2\left(\frac{\theta}{2}\right)\right]}{\left[a \cdot 2\sin\left(\frac{\theta}{2}\right)\cos\frac{\theta}{2}\right]}$$

$$\therefore \quad \tan\phi = \tan\frac{\theta}{2}$$

Hence, $$\phi = \left(\frac{\theta}{2}\right) \qquad [2]$$

$$p = r\sin\phi = r\sin\left(\frac{\theta}{2}\right)$$

$$\therefore \quad p^2 = r^2\sin^2\left(\frac{\theta}{2}\right) = r^2\left[\frac{(1-\cos\theta)}{2}\right]; \qquad \text{(using [1])}$$

$$= \left(\frac{r^2}{2}\right)\left(\frac{r}{a}\right) = \left(\frac{r^3}{2a}\right)$$

$\therefore$ Pedal Equation is

$$2ap^2 = r^3$$

(*b*) **Given:** $\left(\frac{2a}{r}\right) = (1 - \cos\theta)$

$$\therefore \quad 2au = (1 - \cos\theta) \qquad \left(\because\ r = \left(\frac{1}{u}\right)\right)$$

$$2au' = \left(2a \cdot \frac{du}{d\theta}\right) = [0 - (-\sin\theta)] = \sin\theta$$

$$2au = (1 - \cos\theta);\ \text{and}\ 2au^1 = \sin\theta$$

$\therefore$ Squaring and adding, we get

$$(4\,a^2u^2) + (4a^2\,u'^2)$$
$$= (1 - \cos\theta)^2 + \sin^2\theta$$
$$= (1 - 2\cos\theta) + \cos^2\theta + (\sin^2\theta)$$
$$= (1 + 1 - 2\cos\theta) = 2\,(1 - \cos\theta)$$

i.e., $$4a^2\,(u^2 + u'^2) = 2\,(1 - \cos\theta)$$

But, $$\left(\frac{1}{p^2}\right) = [(u^2) + (u')^2]$$

So, we get

$$4a^2\left(\frac{1}{p^2}\right) = 4\,a^2\,(u^2 + u'^2) = 2\,(1 - \cos\theta)$$

$$= 2\,(2au) = \left(\frac{4a}{r}\right), \qquad \text{[using (1)]}$$

$$\therefore \quad \left(\frac{4a^2}{p^2}\right) = \left(\frac{4a}{r}\right)$$

$\boxed{p^2 = ar}$ is the pedal equation.

(*c*) It is more convenient to use $u = \left(\frac{1}{r}\right)$

$$\therefore \quad lu = (1 + e\cos\theta);$$
$$lu' = -\,e\sin\theta \qquad [1]$$

Squaring and adding, we get

$$(lu)^2 + (lu')^2$$
$$= (1 + e\cos\theta)^2 + (-\,e\sin\theta)^2$$
$$= (1 + 2e\cos\theta + e^2\cos^2\theta) + e^2\sin^2\theta = (1 + 2e\cos\theta + e^2)$$

i.e., $l^2 (u^2 + u'^2) = (1 + e^2) + 2 [e \cos \theta]$ [using (1)]

i.e., $$\left(\frac{l^2}{p^2}\right) = (1 + e^2) + 2\left[\left(\frac{1}{r}\right) - 1\right] = 1 + e^2 + 2\left(\frac{l}{r}\right) - 2$$

$\therefore$ pedal equation

$$\left(\frac{l^2}{p^2}\right) = \left(\frac{2l}{r}\right) + (e^2 - 1)$$

115. **Find the pedal equation of the curve $a^2 = r^2 \cos 2\theta$.**

Given: $a^2 = r^2 \cos 2\theta$ [1]

Taking logarithms, we get

$$2 \log a = 2 \log r + \log \cos 2\theta$$

(Differentiating once)

$$0 = \frac{2}{r} \cdot \frac{dr}{d\theta} - \frac{2 \sin 2\theta}{\cos 2\theta}$$

or $$\frac{1}{r} \cdot \frac{dr}{d\theta} = \tan (2\theta)$$

$\therefore$ $$\tan \phi = r \frac{d\theta}{dr} = \cot (2\theta) = \tan\left(\frac{\pi}{2} - 2\theta\right)$$

$\therefore$ $$\phi = \left(\frac{\pi}{2} - 2\theta\right)$$

But $$p = r \sin \phi = r \sin\left(\frac{\pi}{2} - 2\theta\right) = r \cos 2\theta$$

or $$\cos 2\theta = \frac{p}{r}$$

$\therefore$ From [1], $$a^2 = r^2 \cdot \frac{p}{r}$$

or $$a^2 = pr$$

which is the required pedal equation.

116. **Find the pedal equation of:**

(*a*) $r^2 = a^2 \cos 2\theta$; (*b*) $r^3 = a^3 \cos 3\theta$ (*c*) $r^4 = a^4 \cos 4\theta$

Solution: (*a*)

Here $r^2 = a^2 \cos 2\theta$ [1]

$\therefore$ $$2 \log r = 2 \log a + \log \cos 2\theta$$

$$\therefore \quad 2 \cdot \left(\frac{1}{r}\right)\left(\frac{dr}{d\theta}\right) = 0 + \left(\frac{1}{\cos 2\theta}\right)(-2 \sin 2\theta)$$

i.e.,

$$\left(\frac{1}{r}\right)\left(\frac{dr}{d\theta}\right) = -\tan 2\theta$$

or $\cot \phi = -\tan 2\theta$ [2]

Hence, $\phi = \left[\left(\frac{\pi}{2}\right) + 2\theta\right]$ [3]

$$p = r \sin \phi = r \cdot \sin \left[\left(\frac{\pi}{2}\right) + 2\theta\right] = r \cos 2\theta$$

i.e., $p = r\left(\frac{r^2}{a^2}\right)$, (using (1))

Pedal equation is: $p \ a^2 = r^3$

(*b*) **Given:** $r^3 = a^3 \cos 2\theta$ (1)

$$3 \log r = 3 \log a + \log \cos 3\theta$$

$$\therefore \quad 3 . \left(\frac{1}{r}\right)\left(\frac{dr}{d\theta}\right) = 0 + \left(\frac{1}{\cos 3\theta}\right)(-3 \sin 3\theta)$$

i.e., $\left(\frac{1}{r}\right)\left(\frac{dr}{d\theta}\right) = -\tan 3\theta$ (2)

Hence $\phi = \left[\left(\frac{\pi}{2}\right) + 3\theta\right]$

$$p = r \sin \phi = r \sin \left[\left(\frac{\pi}{2}\right) + 3\theta\right] = r \cos 3\theta$$

i.e., $p = r\left(\frac{r^3}{a^3}\right)$, (using (1))

$P\ a^3 = r^4$ is the $(p - r)$ equation

(*c*) $r^4 = a^4 \cos 4\theta$

Proceed exactly as in (*a*) and (*b*) and the pedal equation is given by $pa^4 = r^5$

117. Find the pedal equation of

$$r = a \text{ sech } n\theta \text{ (Spiral)}$$

Solution: **Given:** $r = a \text{ sech } n\theta$

Diff. w.r.t. θ $\quad \frac{dr}{d\theta} = -an \text{ sech } n\theta. \tanh n\theta$

$$= -nr \cdot \sqrt{1 - \text{sech}^2 \ n\theta} = -nr\sqrt{1 - \frac{r^2}{a^2}}$$

We know that

$$\frac{1}{p^2} = \frac{1}{r^2} + \frac{1}{r^4}\left(\frac{dr}{d\theta}\right)^2$$

$$\therefore \quad \frac{1}{p^2} = \frac{1}{r^2} + \frac{1}{r^4}\cdot n^2r^2\left(1-\frac{r^2}{a^2}\right)$$

i.e., $$\frac{1}{p^2} = \frac{1}{r^2} + \frac{n^2}{r^2}\left(1-\frac{r^2}{a^2}\right) = \frac{n^2+1}{r^2} - \frac{n^2}{a^2}$$

Which is of the form

$$\frac{1}{p^2} = \frac{A}{r^2} + B$$

or $$p^{-2} = Ar^{-2} + B$$

$$\left(A = n^2+1, B = -\frac{n^2}{a^2}\right)$$

118. Find the pedal equation of

$$\frac{a(e^2-1)}{r} = 1 + e\cos\theta$$

Solution: **Given:** $$\frac{a(e^2-1)}{r} = 1 + e\cos\theta \quad [1]$$

Compare with the conic:

$$\frac{l}{r} = 1 + e\cos\theta \quad [2]$$

We find: $$l = a(e^2-1) \quad [3]$$

Now, differentiate [2], w.r.t. θ,

We get $$-\frac{l}{r^2}\cdot\frac{dr}{d\theta} = -e\sin\theta$$

$$\therefore \quad \frac{dr}{d\theta} = \frac{r^2 e\sin\theta}{l}$$

Since $$\frac{1}{p^2} = \frac{1}{r^2} + \frac{1}{r^4}\cdot\left(\frac{dr}{d\theta}\right)^2$$

$$\therefore \quad \frac{1}{p^2} = \frac{1}{r^2} + \frac{1}{r^4}\,\frac{r^4 e^2\sin^2\theta}{l^2}$$

$$= \frac{1}{r^2} + \frac{e^2 \sin^2 \theta}{l^2} = \frac{1}{l^2}\left[\frac{l^2}{r^2} + e^2 (1 - \cos^2 \theta)\right]$$

$$= \frac{1}{l^2}\left[\frac{l^2}{r^2} + e^2\left\{1 - \frac{1}{e^2}\left(\frac{l}{r} - 1\right)^2\right\}\right] \qquad \left(\because \frac{l}{r} = 1 + e\cos\theta\right)$$

$$= \frac{1}{l^2}\left[\frac{l^2}{r^2} + e^2 - \left(\frac{l^2}{r^2} - \frac{2l}{r} + 1\right)\right]$$

or
$$\frac{1}{p^2} = \frac{1}{l^2}\left[\frac{2l}{r} - 1 + e^2\right]$$

Which is the required pedal equation.

There $\quad l = a(e^2 - 1)$

119. Find the pedal equation of the curve

$$r^m = a^m \sin m\theta + b^m \cos m\theta \qquad [1]$$

Put $\quad a^m = k^m \cos \theta$

and $\quad b^m = k^m \sin \theta$

So that $\quad k^m = \sqrt{a^{2m} + b^{2m}}$

$\therefore$ **[1] becomes**

$$r^m = k^m \sin(m\theta + \theta) \qquad [2]$$

Differentiating

$$mr^{m-1}\frac{dr}{d\theta} = k^m \cdot m \cdot \cos(m\theta + \theta)$$

$$\therefore \quad \tan\theta = r\frac{d\theta}{dr} = \frac{r^m}{k^m \cos(m\theta + \theta)}$$

$$= \frac{k^m \sin(m\theta + \theta)}{k^m \cos(m\theta + \theta)} = \tan(m\theta + \theta)$$

Thus $\quad p = r\sin\phi = r\sin(m\theta + \theta)$

or $\quad p = r \cdot \frac{r^m}{k^m}$ **from [2]**

i.e., $\quad r^{m+1} = pk^m$

i.e., $\quad r^{m+1} = pk^m = p\sqrt{a^{2m} + b^{2m}}$

PART-B

120. Establish the pedal equation of the curve

$r^n = a^n \sin n\theta + b^n \cos n\theta$ in the form $p^2 (a^{2n} + b^{2n}) = r^{2n+2}$

(V.T.U. F/M, 2005)

Solution: **Given :** $r^n = a^n \sin n\theta + b^n \cos n\theta$ [1]

$\Rightarrow$ $n \log r = \log (a^n \sin n\theta + b^n \cos n\theta)$.

Differentiating w.r.t. θ

$$\frac{n}{r}\frac{dr}{d\theta} = \frac{na^n \cos n\theta - nb^n \sin n\theta}{a^n \sin n\theta + b^n \cos n\theta} \sin \theta$$

$$\cot \phi = \frac{a^n \cos n\theta - b^n \sin n\theta}{a^n \sin n\theta + b^n \cos n\theta}$$

Since, $p = r \sin \phi$. Squaring and taking the reciprocal,

$$\frac{1}{p^2} = \frac{1}{r^2} \operatorname{cosec}^2 \phi \quad \text{or} \quad \frac{1}{p^2} = \frac{1}{r^2} (1 + \cot^2 \phi)$$

$$\therefore \quad \frac{1}{p^2} = \frac{1}{r^2} \left\{1 + \frac{(a^n \cos n\theta - b^n \sin n\theta)^2}{(a^n \sin n\theta + b^n \cos n\theta)^2}\right\}$$

$$\frac{1}{p^2} = \frac{1}{r^2} \left\{\frac{(a^n \sin n\theta + b^n \cos n\theta)^2 + (a^n \cos n\theta - b^n \sin n\theta)^2}{(a^n \sin n\theta + b^n \cos n\theta)^2}\right\}$$

$$\frac{1}{p^2} = \frac{1}{r^2} \left\{\frac{a^{2n} (\sin^2 n\theta + \cos^2 n\theta)^2 + b^{2n} (\cos^2 n\theta + \sin^2 n\theta)}{(a^n \sin n\theta + b^n \cos n\theta)^2}\right\}$$

$$\frac{1}{p^2} = \frac{1}{r^2} \cdot \frac{a^{2n} + b^{2n}}{(a^n \sin n\theta + b^n \cos n\theta)^2}$$

$$\frac{1}{p^2} = \frac{1}{r^2} \cdot \frac{a^{2n} + b^{2n}}{(r^n)^2} \qquad \text{(Using [1])}$$

$p^2 (a^{2n} + b^{2n}) = r^{2n+2}$ is the required pedal equation.

121. Find $P - v$ equation of

$$r\left(1 - \sin\frac{\theta}{2}\right)^2 = a$$

Solution: **Given:** $r\left(1 - \sin\frac{\theta}{2}\right)^2 = a$ [1]

Diff. w.r.t. r

$$\left(1 - \sin\frac{\theta}{2}\right)^2 + r \cdot 2\left(1 - \sin\frac{\theta}{2}\right)\left(\frac{-1}{2}\cos\frac{\theta}{2}\right) \cdot \frac{d\theta}{dr} = 0$$

$$\therefore \quad \tan \phi = r\frac{d\theta}{dr};$$

$$= \frac{1-\sin\left(\frac{\theta}{2}\right)}{\cos\frac{\theta}{2}} = \frac{\sin^2\frac{\theta}{4}+\cos^2\frac{\theta}{4}-2\sin\frac{\theta}{4}\cos\frac{\theta}{4}}{\cos^2\left(\frac{\theta}{4}\right)-\sin^2\left(\frac{\theta}{4}\right)}$$

$$= \frac{\left(\cos\frac{\theta}{4}-\sin\frac{\theta}{4}\right)^2}{\left(\cos\frac{\theta}{4}-\sin\frac{\theta}{4}\right)\left(\cos\frac{\theta}{4}+\sin\frac{\theta}{4}\right)}$$

$$= \frac{\cos\left(\frac{\theta}{4}\right)-\sin\left(\frac{\theta}{4}\right)}{\cos\left(\frac{\theta}{4}\right)+\sin\left(\frac{\theta}{4}\right)} = \frac{1-\tan\left(\frac{\theta}{4}\right)}{1+\tan\left(\frac{\theta}{4}\right)} \quad \left(\div N^r \text{ and } D^r \text{ by } \cos\left(\frac{\theta}{4}\right)\right)$$

$$= \frac{\tan\frac{\pi}{4}-\tan\left(\frac{\theta}{4}\right)}{1+\tan\left(\frac{\pi}{4}\right)\tan\left(\frac{\theta}{4}\right)} = \tan\left(\frac{\pi}{4}-\frac{\theta}{4}\right)$$

or $$\phi = \left(\frac{\pi}{4}-\frac{\theta}{4}\right)$$

Thus $$p = r\sin\phi = r\sin\left(\frac{\pi}{4}-\frac{\theta}{4}\right)$$

or $$p^2 = \frac{r^2}{2}\cdot 2\sin^2\left(\frac{\pi}{4}-\frac{\theta}{4}\right) = \frac{r^2}{2}\cdot\left\{1-\cos\left(\frac{\pi}{2}-\frac{\theta}{2}\right)\right\}$$

$$= \frac{r^2}{2}\left[1-\sin\left(\frac{\theta}{2}\right)\right]$$

Squaring again

$$4p^4 = r^4, \left(1-\sin\frac{\theta}{2}\right)^2 = r^4, \left(\frac{a}{r}\right) \qquad (\because \text{ From (1)})$$

or $ar^3 = 4p^4$ **is the required pedal equation.**

122. Find the pedal equation of the curve

$$r = 2\,(1+\cos\theta) \qquad \text{(VTU, Aug., 1999)}$$

Solution: $$r = 2\,(1+\cos\theta) \qquad [1]$$

$$\frac{dr}{d\theta} = -2\sin\theta$$

$$\therefore \qquad r\,\frac{d\theta}{dr} = \frac{2(1+\cos\theta)}{-2\sin\theta} = \frac{2\cos^2\dfrac{\theta}{2}}{2\sin\dfrac{\theta}{2}\cos\dfrac{\theta}{2}} = -\cot\left(\frac{\theta}{2}\right)$$

$$\tan\phi = \tan\left(\frac{\pi}{2}+\frac{\theta}{2}\right)$$

$$\Rightarrow \qquad \phi = \left(\frac{\pi}{2}+\frac{\theta}{2}\right)$$

Again $$p = \sin\phi = r\sin\left(\frac{\pi}{2}+\frac{\theta}{2}\right) = r\cos\frac{\theta}{2}$$

$$\Rightarrow \qquad p = r\cos\frac{\theta}{2}$$

$$\Rightarrow \qquad \frac{p}{r} = \cos\frac{\theta}{2} \qquad [3]$$

Now [1] $\Rightarrow$ $$r = 4\cos^2\frac{\theta}{2}$$

$$r = 4\cdot\frac{p^2}{r^2} \qquad \text{(using [2] and [3])}$$

$$\Rightarrow \qquad 4p^2 = r^3$$

which is the pedal equation.

EXERCISES

I. Find the pedal equations of the following curves:

(i) $r^2 = a^2 \sec 2\theta$ (*Ans.* $pr = a^2$)

(ii) $r^m = \sin m\theta$ (*Ans.* $r^{m+1} = a^m p$)

(iii) $r = a + b\cos\theta$ $\left(\textit{Ans. } r^4 = \dfrac{(b^2 - a^2 + 2ar)}{p^2}\right)$

(iv) $r = ae^{\theta\cot\alpha}$ (*Ans.* $p = r\sin\alpha$)

(v) $r = a\theta$ (*Ans.* $p^2(r^2 + a^2) = r^4$)

(vi) $r^2 = a^2\cos 2\theta$ (*Ans.* $pa^2 = r^3$)

(vii) $r^3 = a^3\cos 3\theta$ (*Ans.* $pa^3 = r^3$)

(viii) $r^2 = a^2\sin 2\theta$ (*Ans.* $pa^2 = r^3$)

II. (1) Prove that pedal equation of the cardioide

$$r = a(1-\cos\theta) \text{ is } p^2 = \frac{r^3}{2a}$$

(2) Show that the pedal equation of the curve $r \sin \theta + a = 0$ is $p^2 = a^2$

(3) Show that the $(p - r)$ equation of the curve $r^2 \sin 2\theta + a^2 = 0$ is

$$\frac{1}{p^2} = \frac{1}{a^2} + \frac{1}{b^2} - \frac{r^2}{a^2 b^2}$$

2.7 PARTIAL DIFFERENTIATION: EULER'S THEOREM

If $u = f(x, y)$; or $u = f(x, y, z)$, and this can be further generalised, are functions of more than one variable then the result of differentiating u w.r.t. x (keeping the other independent variables unchanged) gives rise to the partial differential coefficient of u, w.r.t. x. Similarly, the partial *d.c.* of u, w.r.t. y and so on.

Partial Derivatives:

Let $u = f(x, y)$. The derivative of u w.r.t. x if it exists when x alone varies (*i.e.*, y remains constant) is called the partial derivative of u w.r.t. x and is denoted by the symbol $\frac{\partial u}{\partial x}$ sometimes it is also denoted by the symbol u_x.

Hence,
$$\frac{\partial u}{\partial x} = \lim_{\delta x \to 0} \frac{f(x + \delta x, y) - f(x, y)}{\delta x} = u_x$$

Similarly when y alone varies and x remains constant then P.D. of u w.r.t. y denoted by $\frac{\partial u}{\partial y}$ or sometimes denoted by u_y

Hence
$$\frac{\partial u}{\partial y} = \lim_{\delta y \to 0} \frac{f(x, y + \delta y) - f(x, y)}{\delta y} = u_y$$

For example,

(*i*) If
$$u = f(x, y, z) = 3x^2 y^3 z + 4xy^2 - yz^3$$

$$\frac{\partial u}{\partial x} = 6xy^3 z + 4y^2$$

$$\frac{\partial u}{\partial y} = 9x^2 y^2 z + 8xy - z^3$$

$$\frac{\partial u}{\partial z} = 3x^2 y^3 - 3yz^2$$

(*ii*) If
$$f(x, y) = x^y$$

Then,
$$\frac{\partial f}{\partial x} = yx^{y-1} \quad \text{and} \quad \frac{\partial f}{\partial y} = x^y \log x$$

Partial Derivatives of Higher Order

1. The second order partial derivatives are denoted as

(*i*) $$\frac{\partial}{\partial x}\left(\frac{\partial u}{\partial x}\right) = \frac{\partial^2 u}{\partial x^2} = u_{xx}$$

PART-B

(ii) $\dfrac{\partial}{\partial y}\left(\dfrac{\partial u}{\partial y}\right) = \left(\dfrac{\partial^2 u}{\partial y^2}\right) = u_{yy}$

(iii) $\dfrac{\partial}{\partial y}\left(\dfrac{\partial u}{\partial x}\right) = \dfrac{\partial^2 u}{\partial y\, \partial x} = u_{yx}$

(iv) $\dfrac{\partial}{\partial x}\left(\dfrac{\partial u}{\partial y}\right) = \left(\dfrac{\partial^2 u}{\partial x\, \partial y}\right) = u_{xy}$

2. In practically all the cases that are usually dealt with, $\dfrac{\partial^2 u}{\partial x\, \partial y} = \dfrac{\partial^2 u}{\partial y\, \partial x}$

3. If $u = f(x, y, z, ...,)$, then

$$\left(\frac{du}{dt}\right) = \left(\frac{\partial u}{\partial x}\cdot\frac{dx}{dt} + \frac{\partial u}{\partial y}\cdot\frac{dy}{dt} + \frac{\partial u}{\partial z}\cdot\frac{dz}{dt} + ...\right)$$

$$du = \left(\frac{\partial u}{\partial x}\right) dx + \left(\frac{\partial u}{\partial y}\right) dy + ... + ...$$

Note (1) If $u = u_1 + u_2 + ...$

where u_1, u_2 ... are functions of x, y

Then $\dfrac{\partial u}{\partial x} = \dfrac{\partial u_1}{\partial x} + \dfrac{\partial u_2}{\partial x} + \dfrac{\partial u_3}{\partial x} + ...$

Note (2) If u a function of t,

where t is a function of x and y,

then $\dfrac{\partial u}{\partial x} = \dfrac{du}{dt}\cdot\dfrac{\partial t}{\partial x}$

and $\dfrac{\partial u}{\partial y} = \dfrac{du}{dt}\cdot\dfrac{\partial t}{\partial y}$.

4. In the same way higher order partial differential coefficients are defined.

Thus $\dfrac{\partial}{\partial x}\left(\dfrac{\partial^n u}{\partial x^n}\right) = \dfrac{\partial^{n+1} u}{\partial x^{n+1}}$

$$\frac{\partial}{\partial y}\left(\frac{\partial^n u}{\partial x^n}\right) = \frac{\partial^{n+1} u}{\partial y\, \partial x^n}$$

$$\frac{\partial}{\partial x}\left(\frac{\partial^n u}{\partial y^n}\right) = \frac{\partial^{n+1} u}{\partial x\, \partial y^{n+1}}$$

$$\frac{\partial}{\partial y}\left(\frac{\partial^n u}{\partial y^n}\right) = \frac{\partial^{n+1} u}{\partial y^{n+1}}$$

Worked Examples (123 – 145)

123. If $z = e^{ax+by} f(ax - by)$

Prove that $b \dfrac{\partial z}{\partial x} + a \dfrac{\partial z}{\partial y} = 2abz.$

[VTU, March, 2000; VTU, Adv. Maths-1, Aug. 2002]

Solution:

$$z = e^{ax+by} f(ax - by)$$

$$\therefore \quad \frac{\partial z}{\partial x} = e^{ax+by} f^1 (ax - by).\, a + f(ax - by)\, e^{ax+by}$$

So that $$b\frac{\partial z}{\partial x} = abe^{ax+by} [f^1 (ax - by) + f(ax - by)] \quad [1]$$

$$\frac{\partial z}{\partial y} = e^{ax+by} f^1 (ax - by) \cdot (-b) + f(ax - by) \cdot e^{ax+by} \cdot b$$

So that $$a\frac{\partial z}{\partial y} = abc^{ax+by} \;[-f^1 (ax - by) + f(ax - by)] \quad [2]$$

Adding [1] and [2] we get,

$$b\frac{\partial z}{\partial x} + a\frac{\partial z}{\partial y} = 2ab\, e^{ax+by} f(ax - by) = 2abz.$$

Hence proved.

124. If $$f = \frac{1}{x^2} + \frac{1}{xy} + \frac{\log x - \log y}{x^2 + y^2},$$

Prove that $x\dfrac{\partial f}{\partial x} + y\dfrac{\partial f}{\partial x} + 2f = 0$ [VTU, August, 1999]

Solution:

$$\frac{\partial f}{\partial x} = \frac{-2}{x^3} - \frac{1}{x^2 y} + \frac{(x^2 + y^2)\frac{1}{x} - (\log x - \log y)\cdot 2x}{(x^2 + y^2)^2}$$

$$\frac{\partial f}{\partial y} = 0 - \frac{1}{xy^2} + \frac{(x^2 - y^2)\left(-\frac{1}{y}\right) - (\log x - \log y)\cdot 2y}{(x^2 + y^2)^2}$$

$$\therefore \quad x\frac{\partial f}{\partial x} + y\frac{\partial f}{\partial y} + 2f$$

$$= \frac{2}{x^2} - \frac{1}{xy} + \frac{(x^2 + y^2) - 2x^2 (\log x - \log y)}{(x^2 + y^2)^2}$$

$$- \frac{1}{xy} + \frac{-(x^2 + y^2) - 2y^2 (\log x - \log y)}{(x^2 + y^2)}$$

$$+ \frac{2}{x^2} + \frac{2}{xy} + \frac{2\log x - 2\log y}{(x^2 + y^2)}$$

$$= \frac{-2(\log x - \log y)(x^2 + y^2)}{(x^2 + y^2)^2} + \frac{2\log x - 2\log y}{x^2 + y^2} = 0$$

125. If $u = A \cos(x + at) + B \sin(x - at)$,

Show that $\left(\dfrac{\partial^2 u}{\partial t^2}\right) = a^2 \left(\dfrac{\partial^2 u}{\partial x^2}\right)$

Solution:

$$u = A \cos(x + at) + B \sin(x - at)$$

$$\therefore \quad \frac{\partial u}{\partial x} = A[-\sin(x + at) \cdot 1] + B[\cos(x - at) \cdot 1]$$

$$\frac{\partial^2 u}{\partial x^2} = -A \cos(x + at) \cdot 1 + B[-\sin(x - at) \cdot 1]$$

$$= -[A \cos(x + at) + B \sin(x - at)] = -u \quad [1]$$

$$\frac{\partial u}{\partial t} = A[-\sin(x + at) \cdot a] + B \cos(x - at) \cdot (-a)$$

$$= -a[A \sin(x + at) + B \cos(x - at)]$$

$$\frac{\partial^2 u}{\partial t^2} = -a\{[A \cos(x + at)\, a] + B[-a \sin(x - at)(-a)]\}$$

$$= -a^2[A \cos(x + at) + B \sin(x - at)] = -a^2 u \quad [2]$$

Clearly $\left(\dfrac{\partial^2 u}{\partial t^2}\right) = -a^2 u = a^2 \left(\dfrac{\partial^2 u}{\partial x^2}\right)$

126. Establish the result

$\left(\dfrac{\partial^2 u}{\partial x\, \partial y}\right) = \left(\dfrac{\partial^2 u}{\partial y\, \partial x}\right)$ for the following functions:

(*a*) $u = x^y$; (*b*) $u = \tan^{-1}\left(\dfrac{x}{y}\right)$; (*c*) $u = e^{ax} \sin by$; (*d*) $u = x \tan y + y \tan x$;

(*e*) $u = \log \dfrac{x^2 + y^2}{xy}$

(*a*) **Given:**

$$u = x^y \quad [1]$$

$$\therefore \quad \frac{\partial u}{\partial x} = y x^{y-1} \quad [2]$$

and $$\frac{\partial}{\partial y}\left[\frac{\partial u}{\partial x}\right] = \frac{\partial}{\partial y}[y x^{y-1}]$$

$$= 1 \cdot x^{y-1} + y x^{y-1} \cdot \log x$$

$$= x^{y-1}(1 + y \log x) \quad [3]$$

Again, diff. [1] w.r.t. y
(keeping x as constant)

$$\frac{\partial u}{\partial y} = x^y \log x \qquad [4]$$

Diff. [4] partially w.r.t. x
(keeping y as constant)

$$\frac{\partial}{\partial x}\left[\frac{\partial u}{\partial y}\right] = \frac{\partial}{\partial x}[x^y \log x]$$

$$= yx^{y-1} \cdot \log x + \frac{1}{x} \cdot x^y = y \cdot x^{y-1} \log x + x^{y-1}$$

$$= x^{y-1}[1 + y \log x] \qquad [5]$$

From [3] and [5],

$$\boxed{\frac{\partial^2 u}{\partial y\, \partial x} = \frac{\partial^2 u}{\partial x\, \partial y}}$$

(*b*) **Given:** $u = \tan^{-1}\left(\frac{x}{y}\right)$ [1]

(Diff. partially w.r.t. x), we get

$$\frac{\partial u}{\partial x} = \frac{1}{x^2 + y^2} \cdot \frac{\partial}{\partial x}\left(\frac{x}{y}\right) \qquad [2]$$

$$= \frac{y^2}{x^2 + y^2} \cdot \frac{1}{y} = \frac{y}{x^2 + y^2}$$

Diff. [2] w.r.t. 'y', we get

$$\frac{\partial^2 u}{\partial y\, \partial x} = \frac{\partial}{\partial y}\left[\frac{\partial u}{\partial x}\right] = \frac{\partial}{\partial y}\left[\frac{y}{x^2 + y^2}\right]$$

$$= \frac{(x^2 + y^2) \cdot 1 - y \cdot 2y}{(x^2 + y^2)^2} = \frac{x^2 - y^2}{(x^2 + y^2)^2} \qquad [3]$$

Diff. [1] partially w.r.t. y, we get

$$\frac{\partial u}{\partial y} = \frac{1}{1 + \left(\frac{x}{y}\right)^2} \frac{\partial}{\partial y}\left(\frac{x}{y}\right) = \frac{y^2}{x^2 + y^2} \cdot \left(\frac{-x}{y^2}\right) = -\frac{x}{x^2 + y^2}$$

i.e., $$\frac{\partial u}{\partial y} = -\frac{x}{x^2 + y^2} \qquad [4]$$

PART-B

Diff. [4] partially w.r.t. 'x', we get,

$$\frac{\partial^2 u}{\partial x\,\partial y} = -\frac{(x^2+y^2)\cdot 1 - x\cdot 2x}{(x^2+y^2)^2} = \frac{x^2-y^2}{(x^2+y^2)^2} \quad [5]$$

∴ From [3] and [5], we obtain

$$\boxed{\frac{\partial^2 u}{\partial x\,\partial y} = \frac{\partial^2 u}{\partial y\,\partial x}}$$

(*c*) **Given:** $u = e^{ax}\sin by$

$$\frac{\partial u}{\partial x} = ae^{ax}\sin by;$$

$$\frac{\partial u}{\partial y} = be^{ax}\cos by;$$

$$\frac{\partial^2 u}{\partial y\,\partial x} = abe^{ax}\cos by$$

Hence $\dfrac{\partial^2 u}{\partial x\,\partial y} = abe^{ax}\cos by$

(*d*) $u = x\tan y + y\tan x$ [1]

Diff. [1] partially w.r.t. x,

$$\frac{\partial u}{\partial x} = \tan y - y\sec^2 x \quad [2]$$

Diff. [2] partially w.r.t. y

$$\frac{\partial^2 u}{\partial y\,\partial x} = \frac{\partial}{\partial y}\left[\frac{\partial u}{\partial x}\right] = \frac{\partial}{\partial y}[\tan y + y\sec^2 x]$$

i.e.,

$$\frac{\partial^2 u}{\partial y\,\partial x} = \sec^2 y + \sec^2 x \quad [3]$$

Diff. [1] partially w.r.t. y

$$\frac{\partial u}{\partial y} = x\sec^2 y + \tan x \quad [4]$$

Diff. [4] partially w.r.t. x

$$\frac{\partial^2 u}{\partial x\,\partial y} = \frac{\partial}{\partial x}\left[\frac{\partial u}{\partial y}\right] = \frac{\partial}{\partial x}[x\sec^2 y + \tan x]$$

$$= \sec^2 y + \sec^2 x \quad [5]$$

From [3] and [5],

$$\frac{\partial^2 u}{\partial y\,\partial x} = \frac{\partial^2 u}{\partial x\,\partial y}$$

(*e*) **Given:** $$u = \log \frac{x^2 + y^2}{xy} = \log (x^2 + y^2) - \log x - \log y$$

Hence, $$\frac{\partial u}{\partial x} = \frac{2x}{x^2 + y^2} - \frac{1}{x}$$

$$\frac{\partial u}{\partial y} = \frac{2y}{x^2 + y^2} - \frac{1}{y}$$

$$\frac{\partial^2 u}{\partial y\, \partial x} = 2x \cdot \frac{-2y}{(x^2 + y^2)^2} - 0 = \frac{-4xy}{(x^2 + y^2)^2} \qquad [1]$$

$$\frac{\partial^2 u}{\partial x\, \partial y} = 2y \cdot \frac{-2y}{(x^2 + y^2)^2} - 0 = -\frac{4xy}{(x^2 + y^2)^2} \qquad [2]$$

From [1] and [2], we get

$$\frac{\partial^2 u}{\partial x\, \partial y} = \frac{\partial^2 u}{\partial y\, \partial x}$$

127. If $u = x^2 \tan^{-1}\left(\frac{y}{x}\right) - y^2 \tan^{-1}\left(\frac{x}{y}\right)$

Show that $$\frac{\partial^2 u}{\partial x\, \partial y} = \frac{x^2 - y^2}{x^2 + y^2}$$

Solution: Diff. the given equation partially w.r.t. '*y*', we get

$$\frac{\partial u}{\partial y} = \left[x^2 \cdot \frac{1}{1 + \left(\frac{y}{x}\right)^2} \left(\frac{1}{x}\right) \right] - \left[y^2 \cdot \frac{1}{1 + \left(\frac{x}{y}\right)^2} \cdot \frac{-x}{y^2} + \tan^{-1}\left(\frac{x}{y}\right) \cdot 2y \right]$$

$$= \frac{x^3}{x^2 + y^2} + \frac{xy^2}{x^2 + y^2} - 2y \tan^{-1}\left(\frac{x}{y}\right)$$

$$= \frac{x(x^2 + y^2)}{x^2 y^2} - 2y \tan^{-1}\left(\frac{x}{y}\right) = x - 2y \tan^{-1}\left(\frac{x}{y}\right) \qquad [1]$$

Diff. [1] partially w.r.t. '*x*', we get

$$\frac{\partial^2 u}{\partial x\, \partial y} = 1 - 2y \cdot \frac{1}{1 + \left(\frac{x}{y}\right)^2} \cdot \frac{1}{y} = 1 - \frac{2y^2}{x^2 + y^2}$$

$$= \frac{x^2 - y^2}{x^2 + y^2}. \qquad \text{(Reqd. Result)}$$

128. If $u = e^{x^3 + y^3}$,

Show that $x \dfrac{\partial u}{\partial x} + y \dfrac{\partial u}{\partial y} = 3u \log u$ [B.U., July, 1993]

Solution:

$$u = e^{x^3+y^3}$$

$$\Rightarrow \quad \frac{\partial u}{\partial x} = e^{x^3+y^3} . 3x^2 = 3ux^2$$

and

$$\frac{\partial u}{\partial y} = e^{x^3+y^3} \cdot 3y^2 = 3\,uy^2$$

Also, $x \dfrac{\partial u}{\partial x} + y \dfrac{\partial u}{\partial y} = 3u\,(x^3 + y^3) = 3u \log u.$ $\left[\begin{array}{l} \because \quad u = e^{x^3+y^3} \\ \Rightarrow x^3 + y^3 = \log u \end{array}\right]$

129. If $u = \dfrac{1}{\sqrt{x^2 + y^2 + z^2}}$

Show that $\dfrac{\partial^2 u}{\partial x^2} + \dfrac{\partial^2 u}{\partial y^2} + \dfrac{\partial^2 u}{\partial z^2} = 0$ [B.U., Jan., 1993]

Solution: Let $x^2 + y^2 + z^2 = r^2$ [1]

$$\therefore \quad u = \frac{1}{r} \qquad [2]$$

$$\left.\begin{aligned} \frac{\partial u}{\partial x} &= -\frac{1}{r^2}\frac{\partial r}{\partial x} \\ \frac{\partial u}{\partial y} &= -\frac{1}{r^2}\frac{\partial r}{\partial y} \\ \frac{\partial u}{\partial z} &= -\frac{1}{r^2}\frac{\partial r}{\partial z} \end{aligned}\right\} \qquad [1]$$

Now, $x^2 + y^2 + z^2 = r^2$

$$\therefore \quad \left.\begin{aligned} 2r \cdot \frac{\partial r}{\partial x} &= 2x\,; \\ 2r \cdot \frac{\partial r}{\partial y} &= 2y\,; \\ 2r \cdot \frac{\partial r}{\partial z} &= 2z\,; \end{aligned}\right\} \qquad [2]$$

$$\Rightarrow \quad \boxed{\frac{\partial r}{\partial x} = \frac{x}{r}\,;\; \frac{\partial r}{\partial y} = \frac{y}{r}\,;\; \frac{\partial r}{\partial z} = \frac{z}{r}} \qquad [3]$$

$\therefore \qquad \left.\begin{aligned} \frac{\partial u}{\partial x} &= -\frac{x}{r^3}; \\ \frac{\partial u}{\partial z} &= -\frac{z}{r^3}; \\ \frac{\partial u}{\partial z} &= -\frac{z}{r^3}; \end{aligned}\right\}$ [4]

Now, $\frac{\partial u}{\partial x} = -\frac{x}{r^3}$

$\Rightarrow \qquad \frac{\partial^2 u}{\partial x^2} = -\frac{1}{r^3} + \frac{3x}{r^4}\frac{\partial r}{\partial x} = -\frac{1}{r^3} + \frac{3x}{r^4}\left[\frac{x}{r}\right]$

$\Rightarrow \qquad \frac{\partial^2 u}{\partial x^2} = -\frac{1}{r^5}[r^2 - 3x^2]$ [5]

Similarly, $\frac{\partial^2 u}{\partial y^2} = -\frac{1}{r^5}[r^2 - 3y^2]$ [6]

and $\frac{\partial^2 u}{\partial z^2} = -\frac{1}{r^5}[r^2 - 3z^2]$ [7]

$\therefore$ [5] + [6] + [7] gives

$$\frac{\partial^2 u}{\partial x^2} + \frac{\partial^2 u}{\partial y^2} + \frac{\partial^2 u}{\partial z^2}$$

$$= -\frac{1}{r^5}[3r^2 - 3(x^2 + y^2 + z^2)]$$

$$= -\frac{1}{r^5}[3r^2 - 3r^2] = 0 \qquad (\because x^2 + y^2 + z^2 = r^2).$$

130. If $u = e^x (x \cos y - y \sin y)$

Show that $\frac{\partial^2 u}{\partial x^2} + \frac{\partial^2 u}{\partial y^2} = 0$ [B.U., Feb., 1994]

Solution: $u = e^x (x \cos y - y \sin y)$

$\therefore \qquad \frac{\partial u}{\partial x} = e^x (\cos y) + e^x (x \cos y - y \sin y)$

$= e^x [(x + 1) \cos y - y \sin y]$

and $\frac{\partial^2 u}{\partial x^2} = e^x [\cos y] + e^x [(x + 1) \cos y - y \sin y]$

$= e^x [(x + 2) \cos y - y \sin y]$ [1]

PART-B

Since $$u = e^x (x \cos y - y \sin y)$$

$$\frac{\partial u}{\partial y} = e^x (-x \sin y - y \cos y - \sin y)$$

$$= -e^x [(x+1) \sin y + y \cos y]$$

$$\therefore \quad \frac{\partial^2 u}{\partial y^2} = -e^x [(x+1) \cos y - y \sin y + \cos y]$$

$$= -e^x [(x+2) \cos y - y \sin y] \qquad [2]$$

Adding [1] and [2], we get

$$\frac{\partial^2 u}{\partial x^2} + \frac{\partial^2 u}{\partial y^2} = 0.$$

131. If $x^x y^y z^z = c$ find $\dfrac{\partial^2 z}{\partial x\, \partial y}$ [B.U., Feb., 1994]

Solution: **Given:**

$$x^x y^y z^z = c$$

taking log on both sides

$$\Rightarrow \quad x \log x + y \log y + z \log z = \log c$$

$$\Rightarrow \quad z \log z = \log c - x \log x - y \log y \qquad [1]$$

(Differentiate partially w.r.t. x), we get

$$z \cdot \frac{1}{z} \cdot \frac{\partial z}{\partial x} + \log z \frac{\partial z}{\partial x} = -x \cdot \frac{1}{x} - \log x$$

$$\Rightarrow \quad (1 + \log z) \frac{\partial z}{\partial x} = -(1 + \log x)$$

$$\Rightarrow \quad \frac{\partial z}{\partial x} = -\frac{1+\log x}{1+\log z} \qquad [2]$$

Again Differentiate (1) Partially w.r.t. y, we get

$$\frac{\partial z}{\partial y} = -\frac{1+\log y}{1+\log z} \qquad [3]$$

Differentiate [3] partially w.r.t. x, we get

$$\frac{\partial^2 z}{\partial x\, \partial y} = -(1 + \log y) \left[\frac{-1}{(1+\log z)^2} \cdot \frac{1}{z} \cdot \frac{\partial z}{\partial x} \right]$$

$$= (1 + \log y) \left[\frac{1}{z(1+\log z)^2} \left(-\frac{1+\log x}{1+\log z} \right) \right]$$

$$\frac{\partial^2 z}{\partial x\, \partial y} = -\frac{(1+\log y)(1+\log x)}{z(1+\log z)^3}$$

132. If $v = e^{a\theta} \cos (a \log r)$
then show that

$$\frac{\partial^2 v}{\partial r^2} + \frac{1}{r}\frac{\partial v}{\partial r} + \frac{1}{r^2}\frac{\partial^2 v}{\partial \theta^2} = 0 \quad \text{(B.U., Ang., 1994)}$$

Solution: **Given:** $V = e^{a\theta} \cos (a \log r)$

$$\frac{\partial v}{\partial r} = -\, e^{a\theta} \sin (a \log r) \cdot \frac{a}{r}$$

or

$$\frac{\partial^2 v}{\partial r^2} = -\, e^{a\theta}\left[\sin (a \log r)\left(-\frac{a}{r^2}\right) + \cos (a \log r) \cdot \frac{a^2}{r^2}\right]$$

$$= -\frac{ae^{a\theta}}{r^2}\,[a \cos (a \log r) - \sin (a \log r)]$$

Diff. [1] partially w.r.t. 'θ' we get

$$\frac{\partial r}{\partial \theta} = ae^{a\theta} \cos (a \log r)$$

$$\therefore \quad \frac{\partial^2 r}{\partial \theta^2} = a^2 e^{a\theta} \cos (a \log r)$$

Now, $$\frac{\partial^2 v}{\partial r^2} + \frac{1}{r}\frac{\partial v}{\partial r} + \frac{1}{r^2}\frac{\partial^2 v}{\partial \theta^2} = -\frac{ae^{a\theta}}{r^2}\,[a \cos (a \log r) - \sin (a \log r)]$$

$$+ \frac{1}{r}\left[-e^{a\theta} \sin (a \log r) \cdot \frac{a}{r}\right] + \frac{1}{r^2}\,[a^2 e^{a\theta} \cos (a \log r)]$$

$$= \frac{ae^{a\theta}}{r^2}\,[-\, a \cos (a \log r) + \sin (a \log r) - \sin (a \log r) + a \cos (a \log r)] = 0.$$

133. If $u = \log (x^3 + y^3 + z^3 - 3xyz)$

Show that

(*i*) $$\frac{\partial u}{\partial x} + \frac{\partial u}{\partial y} + \frac{\partial u}{\partial z} = \frac{3}{x + y + z} \quad \text{[VTU, F/M, 2005]}$$

(*ii*) $$\left(\frac{\partial}{\partial x} + \frac{\partial}{\partial y} + \frac{\partial}{\partial z}\right)^2 u = \frac{-9}{(x + y + z)^2} \quad \text{(VTU, Aug., 2002)}$$

Solution: We know that

$$\boxed{x^3 + y^3 + z^3 - 3xyz = (x + y + z)\,(x^2 + y^2 + z^2 - xy - yz - zx)}$$

Given: $u = \log (x^3 + y^3 + z^3 - 2xyz)$

$$\therefore \quad \frac{\partial u}{\partial x} = \frac{3x^2 - 3yz}{x^3 + y^3 + z^3 - 3xyz},$$

$$\frac{\partial u}{\partial y} = \frac{3y^2 - 3zx}{x^3 + y^3 + z^3 - 3xyz},$$

$$\frac{\partial u}{\partial z} = \frac{3z^2 - 3xy}{x^3 + y^3 + z^3 - 3xyz},$$

Adding, we get

PART-B

$$\frac{\partial u}{\partial x}+\frac{\partial u}{\partial y}+\frac{\partial u}{\partial z}=\frac{3x^2-3zy}{x^3+y^3+z^3-3xyz}+\frac{3y^2-3zx}{x^3+y^3+z^3-3xyz}+\frac{3z^2-3xy}{x^3+y^3+z^3-3xyz}$$

$$=\frac{3(x^2+y^2+z^2-xy-yz-zx)}{(x+y+z)(x^2+y^2+z^2-xy-yz-zx)}$$

$$=\frac{3}{(x+y+z)}$$ [VTU, F/M, 2005] [1]

(ii) $$\left(\frac{\partial}{\partial x}+\frac{\partial}{\partial y}+\frac{\partial}{\partial z}\right)^2 u=\left(\frac{\partial}{\partial x}+\frac{\partial}{\partial y}+\frac{\partial}{\partial z}\right)\left(\frac{\partial u}{\partial x}+\frac{\partial u}{\partial y}+\frac{\partial u}{\partial z}\right)$$

$$=\left(\frac{\partial}{\partial x}+\frac{\partial}{\partial y}+\frac{\partial}{\partial z}\right)\left(\frac{3}{x+y+z}\right)$$ from [1]

$$=\frac{\partial}{\partial x}\left(\frac{3}{x+y+z}\right)+\frac{\partial}{\partial y}\left(\frac{3}{x+y+z}\right)+\frac{\partial}{\partial z}\left(\frac{3}{x+y+z}\right)$$

$$=3\left[\frac{-1}{(x+y+z)^2}+\frac{-1}{(x+y+z)^2}+\frac{-1}{(x+y+z)^2}\right]$$

$$=\frac{-9}{(x+y+z)^2}$$ [VTU, F/M, 2005] [2]

134. If $u=\tan^{-1}\dfrac{xy}{(1+x^2+y^2)^{1/2}}$

P.T. $$\frac{\partial^2 u}{\partial x\,\partial y}=\frac{1}{\sqrt{(1+x^2+y^2)^3}}$$

Solution: **Given:** $$u=\tan^{-1}\frac{xy}{(1+x^2+y^2)^{1/2}}$$

$$\frac{\partial u}{\partial y}=\frac{1}{1+\dfrac{x^2y^2}{(1+x^2+y^2)}}\times \text{d.c of }\frac{xy}{\sqrt{1+x^2+y^2}}\ \text{w.r.t.}\times y$$

$$=\frac{1+x^2+y^2}{1+x^2+y^2+x^2y^2}\cdot x\,\frac{\sqrt{1+x^2+y^2}\cdot 1-y\cdot\dfrac{1}{2\sqrt{1+x^2+y^2}}\cdot 2y}{(1+x^2+y^2}$$

$$=\frac{x}{(1+x^2)(1+y^2)}\cdot\frac{(1+x^2+y^2)-y^2}{\sqrt{1+x^2+y^2}}$$

$$=\frac{x}{(1+y^2)\sqrt{1+x^2+y^2}}$$ [1]

Now, Differentiate [1] partially w.r.t x, we get

$$\frac{\partial^2 u}{\partial x\, \partial y} = \frac{1}{(1+y^2)} \cdot \frac{\sqrt{1+x^2+y^2} \cdot 1 - x \dfrac{1}{2\sqrt{1+x^2+y^2}} \cdot 2x}{(1+x^2+y^2)}$$

or
$$\frac{\partial^2 u}{\partial x\, \partial y} = \frac{1}{(1+y^2)} \frac{(1+x^2+y^2)-x^2}{(1+x^2+y^2)^{3/2}} = \frac{1}{(1+x^2+y^2)^{3/2}}$$

135. If $u = \sin^{-1} \dfrac{\sqrt{x}-\sqrt{y}}{\sqrt{x}+\sqrt{y}}$

show that
$$\frac{\partial u}{\partial x} = -\frac{y}{x} \frac{\partial u}{\partial y}$$

Solution: **Given:**
$$u = \sin^{-1} \frac{\sqrt{x}-\sqrt{y}}{\sqrt{x}+\sqrt{y}}$$

$$\therefore \quad \sin u = \frac{\sqrt{x}-\sqrt{y}}{\sqrt{x}+\sqrt{y}} \qquad [1]$$

Diff. [1] partially w.r.t. x

$$\cos u \,.\, \frac{\partial u}{\partial x} = \frac{\left(\sqrt{x}+\sqrt{y}\right)\dfrac{1}{2\sqrt{x}} - \left(\sqrt{x}-\sqrt{y}\right)\cdot \dfrac{1}{2\sqrt{x}}}{\left(\sqrt{x}+\sqrt{y}\right)^2} = \frac{\sqrt{y}}{\sqrt{x}\left(\sqrt{x}+\sqrt{y}\right)^2}$$

Diff. [1] partially w.r.t. y

$$\cos u \cdot \frac{\partial u}{\partial y} = \frac{\left(\sqrt{x}+\sqrt{y}\right)\left(\dfrac{1}{2\sqrt{y}}\right) - \left(\sqrt{x}-\sqrt{y}\right)\dfrac{1}{2\sqrt{y}}}{\left(\sqrt{x}+\sqrt{y}\right)^2} = \frac{-\sqrt{x}}{\sqrt{y}\left(\sqrt{x}+\sqrt{y}\right)^2}$$

$$\text{Now, } \cos u \left[x \frac{\partial u}{\partial y} + y \frac{\partial u}{\partial y}\right] = \frac{\sqrt{xy}}{\left(\sqrt{x}+\sqrt{y}\right)^2} - \frac{\sqrt{xy}}{\left(\sqrt{x}+\sqrt{y}\right)^2} = 0$$

But $\cos u \neq 0$

$$\therefore \quad x \frac{\partial u}{\partial x} + y \frac{\partial u}{\partial y} = 0$$

i.e.,
$$x \frac{\partial u}{\partial x} = -y \frac{\partial u}{\partial y}$$

or $$\frac{\partial u}{\partial x} = -\frac{y}{x}\frac{\partial u}{\partial y}$$

136. If $u = \log(\tan x + \tan y + \tan z)$

Show that $\sin 2x \cdot \frac{\partial u}{\partial x} + \sin 2y \frac{\partial u}{\partial y} + \sin 2z \cdot \frac{\partial u}{\partial z} = 2$

Solution: $$u = \log(\tan x + \tan y + \tan z)$$

$$\frac{\partial u}{\partial x} = \frac{1}{\tan x + \tan y + \tan z} \cdot \sec^2 x$$

$$\frac{\partial u}{\partial y} = \frac{1}{\tan x + \tan y + \tan z} \cdot \sec^2 y$$

$$\frac{\partial u}{\partial z} = \frac{1}{\tan x + \tan y + \tan z} \cdot \sec^2 z$$

Now, $\sin 2x \frac{\partial u}{\partial x} + \sin 2y \frac{\partial u}{\partial y} + \sin 2z \frac{\partial u}{\partial z}$

$$= \frac{\sin 2x \sec^2 x + \sin 2y \sec^2 y + \sin 2z \sec^2 z}{\tan x + \tan y + \tan z}$$

$$= \frac{2(\tan x + \tan y + \tan z)}{(\tan x + \tan y + \tan z)} = 2.$$

137. If $V = r^m$ where $r^2 = x^2 + y^2 + z^2$

Show that $\frac{\partial^2 v}{\partial x^2} + \frac{\partial^2 v}{\partial y^2} + \frac{\partial^2 v}{\partial z^2} = m(m+1) r^{m-2}$

Solution: **Given:** $$V = r^m$$

$$\frac{\partial v}{\partial x} = mr^{m-1} \cdot \frac{\partial v}{\partial x}$$

But $$r^2 = x^2 + y^2 + z^2$$

$\therefore$ $$\frac{\partial v}{\partial x} = mr^{m-1}\left(\frac{x}{r}\right) = mx\, r^{m-2}$$

$\because$ $$\boxed{2r\frac{\partial v}{\partial x} = 2x}$$

$$\frac{\partial v}{\partial x} = \frac{x}{r}$$

|||ly $$\frac{\partial v}{\partial y} = \frac{y}{r};\ \frac{\partial v}{\partial z} = \frac{z}{r}$$

$$\frac{\partial^2 v}{\partial x^2} = m\left[r^{m-2} + x(m-2)\,r^{m-3}\,\frac{\partial v}{\partial x}\right]$$

$$= m\left[r^{m-2} + x(m-2)\,r^{m-3}\left(\frac{x}{r}\right)\right] = mr^{m-4}\,[r^2 + (m-2)\,x^2]$$

|||ly $$\frac{\partial^2 v}{\partial y^2} = mr^{m-4}\,[r^2 + (m-2)\,y^2]$$

and $$\frac{\partial^2 v}{\partial z^2} = mr^{m-4}\,[r^2 + (m-2)\,z^2]$$

$\therefore$ $$\frac{\partial^2 v}{\partial x^2} + \frac{\partial^2 v}{\partial y^2} + \frac{\partial^2 v}{\partial z^2}$$

$$= mr^{m-4}\,[3r^2 + (m-2)\,(x^2 + y^2 + z^2)]$$
$$= mr^{m-4}\,[3r^2 + (m-2)\,(r^2)]$$
$$= mr^{m-4}\,[(m+1)\,r^2] = m\,(m+1)\,r^{m-2}$$

138. If $r^2 = x^2 + y^2 + z^2$, prove that

$$\left(\frac{\partial^2}{\partial x^2} + \frac{\partial^2}{\partial y^2} + \frac{\partial^2}{\partial z^2}\right)\left(\frac{1}{r}\right) = 0$$

Solution: we have

$$\left(\frac{\partial^2}{\partial x^2} + \frac{\partial^2}{\partial y^2} + \frac{\partial^2}{\partial z^2}\right)\frac{1}{r} = \frac{\partial^2}{\partial x^2}\left(\frac{1}{r}\right) + \frac{\partial^2}{\partial y^2}\left(\frac{1}{r}\right) + \frac{\partial^2}{\partial z^2}\left(\frac{1}{r}\right)$$

$$r^2 = x^2 + y^2 + z^2;\; 2r\cdot\frac{\partial r}{\partial x} = 2x;\; \frac{\partial r}{\partial x} = \frac{x}{r}$$

|||ly $$\frac{\partial r}{\partial y} = \frac{y}{r};\; \frac{\partial r}{\partial z} = \frac{z}{r}$$

Now $$\frac{\partial}{\partial x}\left(\frac{1}{r}\right) = -\frac{1}{r^2}\frac{\partial r}{\partial x} = -\frac{1}{r^2}\left(\frac{x}{r}\right) = -\frac{x}{r^3}$$

Hence $$\frac{\partial^2}{\partial x^2}\left(\frac{1}{r}\right) = \frac{\partial}{\partial x}\left(\frac{\partial}{\partial x}\cdot\frac{1}{r}\right) = \frac{\partial}{\partial x}\left(-\frac{x}{r^3}\right)$$

$$= -\frac{r^3 - 3r^2 x\,\dfrac{\partial r}{\partial x}}{r^6} = -\frac{r^3 - 3r^2 x\left(\dfrac{x}{r}\right)}{r^6} = \frac{3x^2 - r^2}{r^5}$$

|||ly $$\frac{\partial^2}{\partial y^2}\left(\frac{1}{r}\right) = \frac{3y^2 - r^2}{r^5};$$

and $$\frac{\partial^2}{\partial z^2}\left(\frac{1}{r}\right) = \frac{3z^2 - r}{r^5}$$

Hence $$\frac{\partial^2\left(\frac{1}{r}\right)}{\partial x^2} + \frac{\partial^2\left(\frac{1}{r}\right)}{\partial y^2} + \frac{\partial^2\left(\frac{1}{r}\right)}{\partial z^2} = \frac{3(x^2 + y^2 + z^2) - 3r^2}{r^5} = 0.$$

139. **If $z\,(x + y) = x^2 + y^2$, prove that**

$$\left(\frac{\partial z}{\partial x} - \frac{\partial z}{\partial y}\right)^2 = 4\left(1 - \frac{\partial z}{\partial x} - \frac{\partial z}{\partial y}\right)$$

Solution: $$z = \frac{x^2 + y^2}{x + y}$$

$$\therefore \quad \frac{\partial z}{\partial x} = \frac{(x + y)\cdot 2x - (x^2 + y^2)}{(x + y)^2} = \frac{x^2 + 2xy - y^2}{(x + y)^2}$$

$$\frac{\partial z}{\partial y} = \frac{y^2 + 2xy - x^2}{(x + y)^2}$$

$$\therefore \quad \frac{\partial z}{\partial x} + \frac{\partial z}{\partial y} = \frac{4xy}{(x + y)^2}$$

$$\frac{\partial z}{\partial x} - \frac{\partial z}{\partial y} = \frac{2(x^2 - y^2)}{(x^2 + y^2)} = \frac{2(x - y)}{(x + y)} \qquad [1]$$

Now $$1 - \frac{\partial z}{\partial x} - \frac{\partial z}{\partial y} = 1 - \left(\frac{\partial z}{\partial x} + \frac{\partial z}{\partial y}\right) = 1 - \frac{4xy}{(x + y)^2} = \left(\frac{x - y}{x + y}\right)^2 \qquad [2]$$

Now $$\left(\frac{\partial z}{\partial x} - \frac{\partial z}{\partial y}\right)^2 = 4\left(\frac{x - y}{x + y}\right)^2 \qquad \text{(using [1])}$$

$$= 4\left(1 - \frac{\partial z}{\partial x} - \frac{\partial z}{\partial y}\right) \qquad \text{(using [2])}$$

140. **If $u = e^{xyz}$, find the value of $\dfrac{\partial^3 u}{\partial x\,\partial y\,\partial z}$**

Solution: $u = e^{xyz}$

$$\frac{\partial u}{\partial z} = e^{xyz}\,(xy)$$

$$\frac{\partial^2 u}{\partial y\,\partial z} = e^{xyz}\,(x) + e^{xyz}\,(xz)\,(xy) = e^{xyz}\,(x + x^2yz)$$

$$\frac{\partial^3 u}{\partial x\,\partial y\,\partial z} = e^{xyz}\,(1 + 2\,xyz) + e^{xyz}\,(yz)\cdot(x + x^2yz)$$

$$= e^{xyz}\,[1 + 2\,xyz + xyz + x^2y^2z^2] = e^{xyz}\,[1 + 3\,xyz + x^2y^2z^2].$$

141. If $z = (1 - 2xy + y^2)^{-1/2}$

Show that $\quad x\dfrac{\partial z}{\partial x} - y\dfrac{\partial z}{\partial y} = y^2z^3$

Solution: $\quad z = (1 - 2xy + y^2)^{-1/2}$

$$\frac{\partial z}{\partial x} = -\frac{1}{2}\,(1 - 2xy + y^2)^{-3/2}\,(-2y) = yz^3$$

$$\frac{\partial z}{\partial y} = -\frac{1}{2}\,(1 - 2xy + y^2)^{-3/2}\,(2y - 2x) = (x - y)z^3$$

Now $\quad x\dfrac{\partial z}{\partial x} - y\dfrac{\partial z}{\partial y} = xyz^3 - xyz^3 + y^2z^3$

$\therefore \quad x\dfrac{\partial z}{\partial x} - y\dfrac{\partial z}{\partial y} = y^2z^3$

142. If $u = (1 - 2xy + y^2)^{-1/2}$

Show that $\quad \dfrac{\partial}{\partial x}\left[(1 - x^2)\dfrac{\partial u}{\partial x}\right] + \dfrac{\partial}{\partial y}\left[y^2\dfrac{\partial u}{\partial y}\right] = 0$

Solution: We have from the above example [143], above,

$$x\frac{\partial z}{\partial x} - y\frac{\partial z}{\partial y} = y^2z^3 \qquad [1]$$

Again, $\quad \dfrac{\partial z}{\partial x} = yz^3$

$$\Rightarrow \quad \frac{\partial^2 z}{\partial x^2} = 3yz^2\,\frac{\partial z}{\partial x} = 3y^2z \qquad [2]$$

Now, $\quad \dfrac{\partial z}{\partial y} = (x - y)\,z^3$

$$\Rightarrow \quad \frac{\partial^2 z}{\partial y^2} = 3z^2 (x - y) \frac{\partial z}{\partial y} - z^3$$

$$\Rightarrow \quad \frac{\partial^2 z}{\partial y^2} = 3z^5 (x - y)^2 - z^3 \qquad [3]$$

Consider $\frac{\partial}{\partial x}\left[(1 - x^2)\frac{\partial u}{\partial x}\right] + \frac{\partial}{\partial y}\left[y^2 \frac{\partial u}{\partial y}\right]$

$$= (1 - x^2) \frac{\partial^2 u}{\partial x^2} + (-2x) \frac{\partial u}{\partial x} + y^2 \frac{\partial^2 u}{\partial y^2} + 2y \frac{\partial u}{\partial y}$$

$$= (1 - x^2) \frac{\partial^2 u}{\partial x^2} + y^2 \frac{\partial^2 u}{\partial y^2} - 2\left(x \frac{\partial u}{\partial x} - y \frac{\partial u}{\partial y}\right)$$

$$= (1 - x^2)\, 3y^2 z^5 + y^2 [3z^5 (x - y)^2 - z^3] - 2y^2 z^3$$

(From [1], [2] and [3])

$$= 3y^2 z^5 [1 - x^2 + (x - y)^2] - 3y^2 z^3 = 3y^2 z^5 [1 - xy + y^2] - 3y^2 z^3$$

$$= 3y^2 z^5 \cdot z^{-2} - 3y^2 z^3 = 3y^2 z^3 - 3y^2 z^3 = 0$$

Hence the result.

143. If $x = r \cos \theta$; $y = r \sin \theta$, prove the following:

(*a*) $$\left(\frac{\partial r}{\partial x}\right)^2 + \left(\frac{\partial r}{\partial y}\right)^2 = 1$$

(*b*) $$\frac{\partial^2 r}{\partial x^2} + \frac{\partial^2 r}{\partial y^2} = \frac{1}{r}$$

(*c*) $$\frac{\partial^2 r}{\partial x^2} \cdot \frac{\partial^2 r}{\partial y^2} = \left(\frac{\partial^2 r}{\partial x \, \partial y}\right)^2$$

(*d*) $$\frac{\partial^2 \theta}{\partial x^2} + \frac{\partial^2 \theta}{\partial y^2} = 0$$

(*a*) $x = r \cos \theta$; $y = r \sin \theta$

Here $x^2 + y^2 = r^2$ [1]

Diff. [1] partially w.r.t. 'x' we get

$$2x = 2r \frac{\partial r}{\partial x};\ \frac{\partial r}{\partial x} = \left(\frac{x}{r}\right) \qquad [2]$$

Diff. [1] partially w.r.t. 'y' we get

$$2y = 2r \frac{\partial r}{\partial y};\ \frac{\partial r}{\partial y} = \frac{y}{r} \qquad [3]$$

Hence, $\left(\frac{\partial r}{\partial x}\right)^2 + \left(\frac{\partial r}{\partial y}\right)^2$

$$\frac{x^2}{r^2} + \frac{y^2}{r^2} = \frac{x^2 + y^2}{r^2} = \frac{r^2}{r^2} = 1.$$

(*b*) From [2], we get

$$\frac{\partial^2 r}{\partial x^2} = \frac{\partial}{\partial x}\left(\frac{x}{r}\right) = \frac{r \cdot 1 - x\frac{\partial r}{\partial x}}{r^2} = \frac{r - x\left(\frac{x}{r}\right)}{r^2} = \frac{r^2 - x^2}{r^3} = \frac{y^2}{r^3} \qquad [4]$$

From [3], we get

$$\frac{\partial^2 r}{\partial y^2} = \frac{\partial}{\partial y}\left(\frac{y}{r}\right) = \frac{r \cdot 1 - y\frac{\partial r}{\partial y}}{r^2}$$

$$= \frac{r - y\left(\frac{y}{r}\right)}{r^2} = \frac{r^2 - y^2}{r^3} = \frac{x^2}{r^3} \qquad [5]$$

(*c*) Diff. [3] partially w.r.t. '*x*', we obtain:

$$\frac{\partial^2 r}{\partial x\,\partial y} = \frac{\partial}{\partial x}\left(\frac{y}{r}\right) = \frac{-y}{r^2}\frac{\partial r}{\partial x} = -\frac{xy}{r^3} \qquad [6]$$

Hence, $$\frac{\partial^2 r}{\partial x^2} \cdot \frac{\partial^2 r}{\partial y^2} = \frac{y^2}{r^3} \cdot \frac{x^2}{r^3} = \left(\frac{-xy}{r^3}\right)^2 = \left(\frac{\partial^2 r}{\partial x\,\partial y}\right)^2$$

(*d*) Since
$$x = r\cos\theta;$$
$$y = r\sin\theta;$$
$$\frac{y}{x} = \tan\theta;\ \theta = \tan^{-1}\left(\frac{y}{x}\right)$$

$$\therefore \quad \frac{\partial\theta}{\partial x} = \frac{1}{1+\left(\frac{y}{x}\right)^2} \cdot \frac{-1}{x^2} = \frac{-y}{x^2 + y^2}$$

$$\frac{\partial^2\theta}{\partial x^2} = -y\left[\frac{-1}{(x^2 + y^2)^2}\right] \cdot 2x = \frac{2xy}{(x^2 + y^2)^2} \qquad [7]$$

$$\frac{\partial\theta}{\partial y} = \frac{1}{1+\left(\frac{y}{x}\right)^2} \cdot \frac{1}{x} = \frac{x}{x^2 + y^2}$$

$$\frac{\partial^2 \theta}{\partial y^2} = x\left[\frac{-1}{(x^2+y^2)^2}\right] \cdot 2y = -\frac{2xy}{(x^2+y^2)^2} \quad [8]$$

Adding [7] and [8], we get

$$\frac{\partial^2 \theta}{\partial x^2} + \frac{\partial^2 \theta}{\partial y^2} = 0$$

EXERCISES

1. If $z^3 - 3yz - 3x = 0$, show that

$$z\frac{\partial z}{\partial x} = \frac{\partial z}{\partial y};$$

$$z\left[\frac{\partial^2 z}{\partial x\,\partial y} + \left(\frac{\partial z}{\partial x}\right)^2\right] = \frac{\partial^2 z}{\partial y^2}$$

2. If $f(x, y) = x^3y, - xy^3$, find $\left[\frac{1}{\frac{\partial t}{\partial x}} + \frac{1}{\frac{\partial t}{\partial y}}\right]_{\substack{x=1\\y=2}}$ $\left(Ans. -\frac{13}{22}\right)$

3. If $z = y\,f(x^2 - y^2)$, show that $y\frac{\partial z}{\partial x} + x\frac{\partial z}{\partial y} = \frac{xz}{y}$

4. If $z = \log(e^x + e^y)$

Show that $rt - s^2 = 0$.

5. If $u = \log(\tan x + \tan y)$, prove that

$$\sin 2x\,\frac{\partial u}{\partial x} + \sin 2y\,\frac{\partial u}{\partial y} = 2$$

6. If $z = x^y + y^x$, verify that

$$\frac{\partial^2 z}{\partial x\,\partial y} = \frac{\partial^2 z}{\partial y\,\partial x}$$

7. If $V = (x^2 - y^2) \times f(x, y)$, prove that

$$\frac{\partial^2 r}{\partial x^2} + \frac{\partial^2 r}{\partial y^2} = (x^4 - y^4)\,f^{11}(x, y)$$

8. If $ux + vy = 0$, and $\frac{u}{x} + \frac{v}{y} = 1$,

Show that $\frac{\partial u}{\partial x} - \frac{\partial v}{\partial y} = \frac{x^2 + y^2}{y^2 - x^2}$.

9. If $\theta = t^n e^{-\frac{r^2}{4t}}$, find what value of n will make

$$\frac{1}{r^2}\frac{\partial}{\partial r}\left(r^2\frac{\partial \theta}{\partial r}\right) = \frac{\partial \theta}{\partial t}$$

10. If $u = \log(x^2 + y^2)$,

Show that $\left[\left(\frac{\partial^2 u}{\partial x^2}\right) + \left(\frac{\partial^2 u}{\partial y^2}\right)\right] = 0$

11. If $u = a\tan^{-1}\left(\frac{y}{x}\right)$,

Show that $\left[x\frac{\partial u}{\partial x} + y\frac{\partial u}{\partial y}\right] = 0$

12. If $u = x\tan u$, show that

$$\left[\left(\frac{\partial^2 u}{\partial x^2}\right) + \left(\frac{\partial^2 u}{\partial y^2}\right)\right] = 0$$

13. If $u = (x^3 + y^3 + z^3 + 3xyz)$,

Show that $\left[x\left(\frac{\partial u}{\partial x}\right) + y\left(\frac{\partial u}{\partial y}\right) + z\left(\frac{\partial u}{\partial z}\right)\right] = 3u.$

14. If $u = \left(\frac{1}{\sqrt{t}}\right)e^{-x^2/4kt}$, then $\left(\frac{\partial^2 u}{\partial x^2}\right) = \left(\frac{1}{k}\right)\left(\frac{\partial u}{\partial t}\right)$

15. If $u = \cos h\,(x^2 - y^2).\cos(2xy)$,

Show that $\left[\left(\frac{\partial^2 u}{\partial x^2}\right) + \left(\frac{\partial^2 u}{\partial y^2}\right)\right] = 0$

16. If $u = \left[\sin(xy) + \cos\left(\frac{x}{y}\right)\right]$

then show that $\left[x^2\left(\frac{\partial^2 u}{\partial x^2}\right) - y^2\left(\frac{\partial^2 u}{\partial y^2}\right)\right] + \left[x\left(\frac{\partial u}{\partial x}\right) - y\left(\frac{\partial u}{\partial y}\right)\right] = 0$

17. If $u = \log[x^3 + y^3 - x^2y - xy^2]$

S.T. $\frac{\partial^2 u}{\partial x^2} + 2\frac{\partial^2 u}{\partial x\,\partial y} + \frac{\partial^2 u}{\partial y^2} = -4(x + y)^{-2}$

PART-B

HOMOGENEOUS FUNCTIONS:

Definition

If $f(\lambda x, \lambda y) = \lambda^n f(x, y)$ then $f(x, y)$ is said to be a homogeneous function of degree n and is denoted by $H(n)$.

Examples:

1. $f(x, y) = \dfrac{x^3 + y^3}{x - y}$

$$f(\lambda x, \lambda y) = \frac{\lambda^3 x^3 + \lambda^3 y^3}{\lambda x - \lambda y} = \lambda^2 \cdot \frac{x^3 + y^3}{x - y} = \lambda^2 f(x, y)$$

$\therefore$ $f(x, y)$ is a Homogeneous function of degree 2, *i.e.* $H(2)$.

2. $f(x, y) = \tan^{-1}\left[\dfrac{(x^3 + y^3)}{(x - y)}\right]$ is not Homogeneous; since $\tan^{-1}(\lambda^2 v) \neq \lambda^2 (\tan^{-1} v)$.

3. $f(x, y) = \tan^{-1}\left[\dfrac{(2x + 3y)}{x - y}\right]$ is a $H(0)$.

4. $f(x, y) = a_0 x^n + a_1 x^{n-1} y + a_2 x^{n-2} y^2 + \ldots + \ldots + a_n y^n$

$$= x^n \left[a_0 + a_1\left(\frac{y}{x}\right) + a_2\left(\frac{y^2}{x^2}\right) + \ldots + a_n\left(\frac{y^n}{x^n}\right)\right] = x^n \phi\left(\frac{y}{x}\right)$$

This $\Rightarrow$ that Homogeneous function $f(x, y)$ of degree n in x and y, *i.e.* $H(n)$ in x and y is expressed as:

$$f(x, y) = x^n \phi\left(\frac{y}{x}\right)$$

We can also denote $f(x, y)$ as $f(x, y) = y^n\, \psi\left(\dfrac{y}{x}\right)$.

2.8 EULER'S THEOREM

Statement:

If $f(x, y)$ is a Homogeneous function of degree n in x and y then $x\dfrac{\partial f}{\partial x} + y\dfrac{\partial f}{\partial y} = nf$

Proof: By data, $f(x, y)$ being a Homogeneous function of degree n in x and y, we can express it as

$$f(x, y) = x^n \phi\left(\frac{y}{x}\right)$$

$$\frac{\partial f}{\partial x} = nx^{n-1}\, \phi\left(\frac{y}{x}\right) + x^n\, \phi^1\left(\frac{y}{x}\right) \cdot \left(\frac{-y}{x^2}\right)$$

$$\frac{\partial f}{\partial y} = x^n \phi^1 \left(\frac{y}{x}\right) \cdot \frac{1}{x}$$

Now, $$x \frac{\partial f}{\partial x} + y \frac{\partial f}{\partial y} = nx^n \phi \left(\frac{y}{x}\right) - x^{n-1} y \cdot \phi \left(\frac{y}{x}\right) + x^{n-1} y \cdot \phi \left(\frac{y}{x}\right)$$

$$= nx^n \phi \left(\frac{y}{x}\right) = nf$$

$$\therefore \quad \boxed{x \frac{\partial f}{\partial x} + y \frac{\partial f}{\partial y} = nf}$$

EXTENSION OF EULER'S THEOREM

If $f(x, y)$ is a Homogeneous function of degree n in x and y, then

$$x^2 \frac{\partial^2 f}{\partial x^2} + 2xy \frac{\partial^2 f}{\partial x \, \partial y} + y^2 \frac{\partial^2 f}{\partial y^2} = n (n-1) f.$$ **[VTU, August, 2001]**

[VTU, F/M, 2005]

Proof: **Since $f(x, y)$ is a $H(n)$, from Euler's theorem, we have**

$$x \frac{\partial f}{\partial x} + y \frac{\partial f}{\partial y} = nf \qquad [1]$$

Diff. [1] partially w.r.t. x

$$x \frac{\partial^2 f}{\partial x^2} + \frac{\partial f}{\partial x} + y \frac{\partial^2 f}{\partial y^2} = n \frac{\partial f}{\partial x} \qquad [2]$$

Diff. [1] partially w.r.t. y

$$x \frac{\partial^2 f}{\partial y \, \partial x} + y \frac{\partial^2 f}{\partial y^2} + \frac{\partial f}{\partial y} = n \frac{\partial f}{\partial y} \qquad [3]$$

$$[2]\, x + [3]\, y = x^2 \frac{\partial^2 f}{\partial x^2} + xy \left(\frac{\partial^2 f}{\partial x \, \partial y} + \frac{\partial^2 f}{\partial y \, \partial x}\right) + y^2 \frac{\partial^2 f}{\partial y^2} + \left(x \frac{\partial f}{\partial x} + y \frac{\partial f}{\partial y}\right)$$

$$= n \left(x \frac{\partial f}{\partial x} + y \frac{\partial f}{\partial y}\right)$$

But $$\frac{\partial^2 f}{\partial x \, \partial y} = \frac{\partial^2 f}{\partial y \, \partial x}$$

$$\therefore \quad x^2 \frac{\partial^2 f}{\partial x^2} + 2xy \frac{\partial^2 f}{\partial x \, \partial y} + y^2 \frac{\partial^2 f}{\partial y^2} = n \cdot nf - n \cdot f = n (n-1) f$$

PART-B

Examples: (146-170)

144. Verify Euler's Theorem for

$$f(x, y) = x^4 \cdot \log \frac{y}{x}$$

Solution: Here $f(x, y)$ is a Homogeneous function of degree 4 in x and y *i.e.* H (4)

$$f(x, y) = x^4 \cdot \log \frac{y}{x} = x^4 (\log y - \log x)$$

$$\frac{\partial f}{\partial x} = 4x^3 (\log y - \log x) + x^4 \left(-\frac{1}{x}\right)$$

$$\frac{\partial f}{\partial y} = x^4 \cdot \frac{1}{y}$$

$$\therefore \quad x \frac{\partial f}{\partial x} + y \frac{\partial f}{\partial y} = 4x^4 (\log y - \log x) = 4f(x, y)$$

Thus, the theorem is verified.

145. Verify Euler's theorem

for $$f(x, y) = \frac{1}{x^2 + xy + y^2}$$

Solution: Since $f(x, y)$ is H (2)

$$f(x, y) = (x^2 + xy + y^2)^{-1}$$

$$\frac{\partial f}{\partial x} = -(x^2 + xy + y^2)^{-2} (2x + y)$$

$$\frac{\partial f}{\partial y} = -(x^2 + xy + y^2)^{-2} (x + 2y)$$

$$\therefore \quad x \frac{\partial f}{\partial x} + y \frac{\partial f}{\partial y} = -\frac{1}{(x^2 + xy + y^2)^2} [2x^2 + xy + xy + 2y^2]$$

$$= -2 \cdot \frac{(x^2 + xy + y^2)}{(x^2 + xy + y^2)^2} = -2 (x^2 + y^2 + z^2)^{-1} = -2f(x, y).$$

146. If $u = \sin^{-1}\left(\frac{x^2 + y^2}{x + y}\right)$

show that $x \frac{\partial f}{\partial x} + y \frac{\partial f}{\partial y} = \tan u$

Solution:

$$u = \sin^{-1}\left(\frac{x^2+y^2}{x+y}\right)$$

is not a Homogeneous function of x and y

But, $$\sin u = \frac{x^2+y^2}{x+y} = z \qquad [1]$$

$$= \frac{x^2\left[1+\left(\frac{y}{x}\right)^2\right]}{x\left[1+\frac{y}{x}\right]} = x^1\phi\left(\frac{y}{x}\right)$$

⇒ sin u is a Homogeneous. of degree 1 in x and y *i.e.* H (1)

(It should be noted that, here u is not a homogeneous function but sin u is a Homogeneous function of degree 1 *i.e.* H (1). Therefore, Euler's theorem can be applied for sin u and not for u)

$$\therefore \quad x\frac{\partial z}{\partial x} + y\frac{\partial z}{\partial y} = 1.z = z$$

since, z = sin u (using [1])

$$\frac{\partial z}{\partial x} = \cos u \cdot \frac{\partial u}{\partial x}\frac{\partial z}{\partial y} = \cos u \cdot \frac{\partial u}{\partial y}$$

$$\therefore \quad x\frac{\partial z}{\partial x} + y\frac{\partial z}{\partial y} = \cos u\left[x\frac{\partial u}{\partial x} + y\frac{\partial u}{\partial y}\right]$$

i.e., $$z = \cos u\left[x\frac{\partial u}{\partial x} + y\frac{\partial u}{\partial y}\right]$$

$$\therefore \quad x\frac{\partial u}{\partial x} + y\frac{\partial u}{\partial y} = \frac{z}{\cos u} = \frac{\sin u}{\cos u} = \tan u$$

$$\therefore \quad x\frac{\partial u}{\partial x} + y\frac{\partial u}{\partial y} = \tan u$$

147. If $u = \tan^{-1}\left(\frac{y^2}{x}\right)$, **prove that**

$$x^2\frac{\partial^2 u}{\partial x^2} + 2xy\frac{\partial^2 u}{\partial x\,\partial y} + y^2\frac{\partial^2 u}{\partial y^2} = -\sin^2 u \sin 2u$$ [VTU, March, 2001]

PART-B

Solution: Let $f = \tan u = \dfrac{y^2}{x} = x\left(\dfrac{y^2}{x^2}\right)$

Hence f x a H (1).

Hence by Euler's theorem

$$x\frac{\partial f}{\partial x} + y\frac{\partial f}{\partial y} = f$$

i.e., $$x \cdot \sec^2 u\frac{\partial u}{\partial x} + y\sec^2 u\frac{\partial u}{\partial y} = \tan u$$

i.e., $$x\frac{\partial u}{\partial x} + y\frac{\partial u}{\partial y} = \sin u\cos u = \frac{1}{2}\sin 2u \qquad [1]$$

Diff. [1] partially w.r.t. 'x'

$$x\frac{\partial^2 u}{\partial x^2} + \frac{\partial u}{\partial x} + y\frac{\partial^2 u}{\partial x\,\partial y} = \cos 2u \cdot \frac{\partial u}{\partial x}$$

$$x^2\frac{\partial^2 u}{\partial x^2} + xy\frac{\partial^2 u}{\partial x\,\partial y} = (\cos 2u - 1)\,x\frac{\partial u}{\partial x} \qquad [2]$$

Perpendicularly Diff. [1] partially w.r.t. 'y', we get

$$xy\frac{\partial^2 u}{\partial y\,\partial x} + y^2\frac{\partial^2 u}{\partial y^2} = (\cos 2u - 1)\,y\frac{\partial u}{\partial y}. \qquad [3]$$

Adding [2] and [3], we get

$$x^2\frac{\partial^2 u}{\partial x^2} + 2xy\frac{\partial^2 u}{\partial x\,\partial y} + y^2\frac{\partial^2 u}{\partial y^2}$$

$$= (\cos 2u - 1)\left(x\frac{\partial u}{\partial x} + y\frac{\partial u}{\partial y}\right)$$

$$= \frac{1}{2}(\cos 2u - 1)\sin 2u, \qquad \text{(using [1])}$$

$$= \sin^2 u \sin 2u$$

148. If $u = \dfrac{x}{y+z} + \dfrac{y}{z+x} + \dfrac{z}{x+y}$

prove that $x\dfrac{\partial u}{\partial x} + y\dfrac{\partial u}{\partial y} + z\dfrac{\partial u}{\partial z} = 0$ [VTU, August, 2000]

Let $$u = (x, y, z) = \frac{x}{y+z} + \frac{y}{z+x} + \frac{z}{x+y},$$

$\therefore \qquad u(\lambda x, \lambda y, \lambda z) = \dfrac{\lambda x}{\lambda y + \lambda z} + \dfrac{\lambda y}{\lambda z + \lambda x} + \dfrac{\lambda z}{\lambda x + \lambda y}$

$$= \frac{x}{y+z} + \frac{y}{z+x} + \frac{z}{x+y} = \lambda^{\circ} \cdot u(x, y, z)$$

$\therefore$ **u is a $H(0)$**

Hence, by Euler's theorem

$$x\frac{\partial u}{\partial x} + y\frac{\partial u}{\partial y} + z\frac{\partial u}{\partial z} = 0$$

149. If $u = \sin^{-1} \dfrac{xy}{x^3 + y^3}$,

prove that $\qquad x\dfrac{\partial u}{\partial x} + y\dfrac{\partial u}{\partial y} = -\tan u$

Solution: **Given:** $\qquad u = \sin^{-1} \dfrac{xy}{x^3 + y^3}$

$\Rightarrow \qquad \sin u = \dfrac{xy}{x^3 + y^3} = V,$ (say) $\qquad$ [1]

Since $\qquad V = \dfrac{x^2}{x^3}\left[\dfrac{\left(\dfrac{y}{x}\right)}{1 + \left(\dfrac{y}{x}\right)^3}\right] = x^{-1}\phi\left(\dfrac{y}{x}\right),$

V is *a* $H(-1)$

$\therefore$ By Euler's theorem

$$x\frac{\partial v}{\partial x} + y\frac{\partial v}{\partial y} = -V, \qquad [2]$$

Since $V = \sin u$,

$$\frac{\partial v}{\partial x} = \cos u \cdot \frac{\partial u}{\partial x};$$

$$\frac{\partial v}{\partial y} = \cos u \cdot \frac{\partial u}{\partial y}$$

From [2], we get

$$x \cos u \frac{\partial u}{\partial x} + y \cos u \frac{\partial u}{\partial y} = -\sin u$$

PART-B

i.e., $$x\frac{\partial u}{\partial x}+y\frac{\partial u}{\partial y} = -\tan u$$

150. If $u = \log\dfrac{x^5+y^5}{x^3+y^3}$,

prove that $$x\frac{\partial u}{\partial x}+y\frac{\partial u}{\partial y} = 2$$

Solution: We have $$e^u = \frac{x^5+y^5}{x^3+y^3} = V \text{ (say)}$$

Since V is a H (2)

$\therefore$ **By Euler's theorem, we have**

$$x\frac{\partial v}{\partial x}+y\frac{\partial v}{\partial y} = 2V$$

i.e., $$x\frac{\partial}{\partial x}(e^u)+y\frac{\partial}{\partial y}(e^u) = 2e^u$$

i.e., $$xe^u\frac{\partial u}{\partial x}+ye^u\frac{\partial u}{\partial y} = 2e^u$$

i.e., $$x\frac{\partial v}{\partial x}+y\frac{\partial v}{\partial y} = 2.$$

151. If $u = \cot^{-1}\left(\dfrac{x+y}{\sqrt{x}+\sqrt{y}}\right)$ prove that $x\dfrac{\partial u}{\partial x}+y\dfrac{\partial u}{\partial y} = -\left(\dfrac{1}{4}\right)\sin 2u$

Solution: **Given:** $$u = \cot^{-1}\left(\frac{x+y}{\sqrt{x}+\sqrt{y}}\right)$$

$\Rightarrow$ $$\cot u = \frac{x+y}{\sqrt{x}+\sqrt{y}} = V, \text{ (say)}$$

Since V is a $H\left(\dfrac{1}{2}\right)$.

$\therefore$ By Euler's theorem

$$x\frac{\partial f}{\partial x}+y\frac{\partial f}{\partial y} = \frac{1}{2}V$$

i.e., $$x\frac{\partial}{\partial x}(\cot u)+y\frac{\partial}{\partial y}(\cot u) = \frac{1}{2}\cot u$$

i.e., $$x(-\operatorname{cosec}^2 u)\frac{\partial u}{\partial x} + y(-\operatorname{cosec}^2 u)\frac{\partial u}{\partial y} = \frac{1}{2}\cot u$$

or $$x\frac{\partial u}{\partial x} + y\frac{\partial u}{\partial y} = \frac{1}{2}(-\sin^2 u)\cot u$$

$$= -\frac{1}{2}\sin u\cos u = -\frac{1}{4}\sin 2u.$$

152. If $u = \cos^{-1}\dfrac{x^3+y^3-z^3}{\sqrt{x^4+y^4-z^4}}$

prove that $$x\frac{\partial u}{\partial x} + y\frac{\partial u}{\partial y} + z\frac{\partial u}{\partial z} = -\cot u$$

Solution:

Let $$V = \frac{x^3+y^3-z^3}{\sqrt{x^4+y^4-z^4}}$$

$\Rightarrow$ $\cos u = V$

Since V is a $H(1)$

$\therefore$ By Euler's theorem, we have

$$x\frac{\partial v}{\partial x} + y\frac{\partial v}{\partial y} + z\frac{\partial v}{\partial z} = V$$

i.e., $$x\frac{\partial}{\partial x}(\cos u) + y\frac{\partial}{\partial y}(\cos u) + z\frac{\partial}{\partial z}(\cos u) = \cos u$$

or $$x(-\sin u)\frac{\partial u}{\partial x} + y(-\sin u)\frac{\partial u}{\partial y} + z(-\sin u)\frac{\partial u}{\partial z} = \cos u$$

or $$x\frac{\partial u}{\partial x} + y\frac{\partial u}{\partial y} + z\frac{\partial u}{\partial z} = -\cot u,$$ thus, the result follows.

153. If $u = \tan^{-1}\dfrac{x^3+y^3}{x-y}$, prove that

$$x^2\frac{\partial^2 u}{\partial x^2} + 2xy\frac{\partial^2 u}{\partial x\,\partial y} + y^2\frac{\partial^2 u}{\partial y^2} = (1-4\sin^2 u)\sin 2u$$

Solution: **Given:** $$u = \tan^{-1}\left(\frac{x^3+y^3}{x-y}\right)$$

PART-B

$$\Rightarrow \qquad \tan u = \frac{x^3 + y^3}{x - y} = V$$

$$\Rightarrow \qquad V = \tan u \text{ is a } H(2)$$

$\therefore$ By Euler's theorem, we have

$$x \frac{\partial}{\partial x}(\tan u) + y \frac{\partial}{\partial y}(\tan u) = 2 \tan u$$

$$\Rightarrow \qquad x \cdot \sec^2 u \cdot \frac{\partial u}{\partial x} + y \sec^2 u \cdot \frac{\partial u}{\partial y} = 2 \tan u$$

or
$$x \frac{\partial z}{\partial x} + y \frac{\partial u}{\partial y} = \frac{2 \sin u}{\cos u} \cos^2 u = 2 \sin u \cos u$$

$$\therefore \qquad x \frac{\partial u}{\partial x} + y \frac{\partial u}{\partial y} = \sin 2u \qquad [1]$$

Diff. [1] partially w.r.t. x,

$$x \frac{\partial^2 u}{\partial x^2} + 1 \cdot \frac{\partial u}{\partial x} + y \frac{\partial^2 u}{\partial x \, \partial y} = 2 \cos 2u \cdot \frac{\partial u}{\partial x}$$

or
$$x \frac{\partial^2 u}{\partial x^2} + y \frac{\partial^2 u}{\partial x \, \partial y} = (2 \cos 2u - 1) \frac{\partial u}{\partial x} \qquad [2]$$

Diff. [1] -partially w.r.t. y,

$$x \frac{\partial^2 u}{\partial y \, \partial x} + y \frac{\partial^2 u}{\partial y^2} + 1 \cdot \frac{\partial u}{\partial y} = 2 \cos 2u \cdot \frac{\partial u}{\partial y}$$

or
$$x \frac{\partial^2 u}{\partial y \, \partial x} + y \frac{\partial^2 u}{\partial y^2} = (2 \cos 2u - 1) \frac{\partial u}{\partial y} \qquad [3]$$

[2] x + [3] y gives

$$x^2 \frac{\partial^2 u}{\partial x^2} + 2xy \frac{\partial^2 u}{\partial x \, \partial y} + y^2 \frac{\partial^2 u}{\partial y^2}$$

$$= (2 \cos 2u - 1) \left(x \frac{\partial u}{\partial x} + y \frac{\partial u}{\partial y} \right)$$

$$= [2(1 - 2 \sin^2 u) - 1] \sin 2u \qquad \text{(using [1])}$$

$$= [1 - 4 \sin^2 u] \sin 2u$$

154. If $u = \sin^{-1} \dfrac{x + y}{\sqrt{x} + \sqrt{y}}$, prove that

$$x^2 \frac{\partial^2 u}{\partial x^2} + 2xy \frac{\partial^2 u}{\partial x \, \partial y} + y^2 \frac{\partial^2 u}{\partial y^2} = -\frac{\sin u \cos 2u}{4 \cos^3 u}$$

Solution:

$$u = \sin^{-1} \frac{x+y}{\sqrt{x}+\sqrt{y}}$$

$\therefore$ $$\sin u = \frac{x+y}{\sqrt{x}+\sqrt{y}} \text{ is a } H\left(\frac{1}{2}\right)$$

$\therefore$ By Euler's theorem, we have

$$x \frac{\partial}{\partial x} (\sin u) + y \frac{\partial}{\partial y} (\sin u) = \frac{1}{2} \sin u$$

$\Rightarrow$ $$x \cos u \cdot \frac{\partial u}{\partial x} + y \cos u \cdot \frac{\partial u}{\partial y} = \frac{1}{2} \sin u$$

$\therefore$ $$x \frac{\partial u}{\partial x} + y \frac{\partial u}{\partial y} = \frac{1}{2} \tan u \qquad [1]$$

Diff. [1] w.r.t. x partially

$$x \frac{\partial^2 u}{\partial x^2} + 1 \cdot \frac{\partial u}{\partial x} + y \frac{\partial^2 u}{\partial x\, \partial y} = \frac{1}{2} \sec^2 u \frac{\partial u}{\partial x}$$

or $$x \frac{\partial^2 u}{\partial x^2} + y \frac{\partial^2 u}{\partial x\, \partial y} = \left(\frac{1}{2} \sec^2 u - 1\right) \frac{\partial u}{\partial x} \qquad [2]$$

Diff. [1] partially w.r.t. y

$$x \frac{\partial^2 u}{\partial x\, \partial y} + y \frac{\partial^2 u}{\partial y^2} + 1 \cdot \frac{\partial u}{\partial y} = \left(\frac{1}{2} \sec^2 u\right) \cdot \frac{\partial u}{\partial y}$$

or $$x \frac{\partial^2 u}{\partial y\, \partial x} + y \frac{\partial^2 u}{\partial y^2} = \left(\frac{1}{2} \sec^2 u - 1\right) \cdot \frac{\partial u}{\partial y} \qquad [3]$$

[2] x + [3] y gives

$$x^2 \frac{\partial^2 u}{\partial x^2} + 2xy \frac{\partial^2 u}{\partial x\, \partial y} + y^2 \frac{\partial^2 u}{\partial y^2}$$

$$= \left(\frac{1}{2} \sec^2 u - 1\right)\left(x \frac{\partial u}{\partial x} + y \frac{\partial u}{\partial y}\right)$$

$$= \left(\frac{1}{2 \cos^2 u} - 1\right) \cdot \frac{1}{2} \tan u$$

$$= -\frac{2 \cos^2 u - 1}{2 \cos^2 u} \cdot \frac{\sin u}{2 \cos u} = -\frac{\cos 2u \sin u}{4 \cos^3 u}$$

155. *Deduction:* **(From Euler's theorem.)**

If z is a homo. Function of x, y of degree n and $z = f(u)$, then

$$x\frac{\partial u}{\partial x} + y\frac{\partial u}{\partial y} = n\,\frac{f(u)}{f'(u)}.$$

Proof: Given: z is a $H(n)$, we have by Euler's theorem, $x\dfrac{\partial z}{\partial x} + y\dfrac{\partial z}{\partial y} = nz$ [1]

Since $z = f(u)$, $\therefore \dfrac{\partial z}{\partial x} = f'(u)\dfrac{\partial u}{\partial x}$

and $\dfrac{\partial z}{\partial y} = f'(u)\dfrac{\partial u}{\partial y}$

Substituting in [1], we get

$$x \cdot f'(u)\frac{\partial u}{\partial x} + y\,f'(u)\frac{\partial u}{\partial y} = n\,f(u)$$

$$\Rightarrow \qquad x\frac{\partial u}{\partial x} + y\frac{\partial u}{\partial y} = n\,\frac{f(u)}{f'(u)}$$

156. If $\sin V = \dfrac{x+2y+3z}{\sqrt{x^8+y^8+z^8}}$, show that

$$x\frac{\partial v}{\partial x} + y\frac{\partial v}{\partial y} + z\frac{\partial v}{\partial z} + 3\tan V = 0$$

Solution:

$$\sin V = \frac{x+2y+3z}{\sqrt{x^8+y^8+z^8}} = x^{-3}\cdot f\left(\frac{y}{x}, \frac{z}{x}\right)$$

$$= x^{-3}\left[\frac{x+2y+3z}{\sqrt{x^8+y^8+z^8}}\right]$$

$\Rightarrow$ $\sin V$ is a $H(-3)$ in x, y and z.

$\therefore$ By Euler's theorem

$$x\frac{\partial}{\partial x}(\sin V) + y\frac{\partial}{\partial y}(\sin V) + z\frac{\partial}{\partial z}(\sin V) = -3\sin V$$

$$\Rightarrow \qquad x\cos V\cdot\frac{\partial v}{\partial x} + y\cos V\frac{\partial v}{\partial y} + z\cos V\frac{\partial v}{\partial z} = -3\sin V$$

or

$$x\frac{\partial v}{\partial x} + y\frac{\partial v}{\partial y} + z\frac{\partial v}{\partial z} = -3\tan V$$

or $$x\frac{\partial v}{\partial x}+y\frac{\partial v}{\partial y}+z\frac{\partial v}{\partial z}+3\tan V=0$$

157. If $u=\tan^{-1}\frac{x^2+y^2}{\sqrt{x}+\sqrt{y}}$, prove that

$$x\frac{\partial u}{\partial x}+y\frac{\partial u}{\partial y}=\frac{3}{4}\sin 2u$$

Solution: we have $$\tan u=\frac{x^2+y^2}{\sqrt{x}+\sqrt{y}}=V\text{ (say)} \quad [1]$$

V is a $H\left(\frac{3}{2}\right)$.

Hence by Euler's theorem,

$$x\frac{\partial v}{\partial x}+y\frac{\partial v}{\partial y}=\frac{3}{2}V \quad [2]$$

Since $V=\tan u$, $\frac{\partial v}{\partial x}=\sec^2 u\cdot\frac{\partial u}{\partial x}$

$$\frac{\partial v}{\partial y}=\sec^2 u\frac{\partial u}{\partial y}$$

Hence [2] gives

$$x\sec^2 u\frac{\partial u}{\partial x}+y\sec^2 u\frac{\partial u}{\partial y}=\frac{3}{2}\tan u$$

i.e., $$x\frac{\partial u}{\partial x}+y\frac{\partial u}{\partial y}=\frac{3}{2}\tan u\cos^2 u$$

$$=\frac{3}{2}\sin u\cos u=\frac{3}{4}\sin 2u$$

158. If $u=\sin^{-1}\left(\frac{ax^3+by^3+cz^3}{\sqrt{x^2+y^2+z^2}}\right)$

Prove that $x\frac{\partial u}{\partial x}+y\frac{\partial u}{\partial y}+z\frac{\partial u}{\partial z}=2\tan u$

Solution:

Since, $$\sin u=\frac{ax^3+by^3+cz^3}{\sqrt{x^2+y^2+z^2}}=f(x,y,z)$$

$$=f\text{ (say)}$$

$$f=\sin u\text{ is a }H(2)$$

PART-B

Hence by Euler's theorem, we have

$$x\frac{\partial f}{\partial x} + y\frac{\partial f}{\partial y} + z\frac{\partial f}{\partial z} = 2f$$

i.e.,
$$x\,\frac{\partial}{\partial x}\,(\sin u) + y\,\frac{\partial}{\partial y}\,(\sin u) + z\,\frac{\partial}{\partial z}\,(\sin u) = 2\sin u$$

$$x\cos u\,\frac{\partial u}{\partial x} + y\cos u\,\frac{\partial u}{\partial y} + 2\cos u\,\frac{\partial u}{\partial z} = 2\sin u$$

or
$$x\,\frac{\partial u}{\partial x} + y\frac{\partial u}{\partial y} + z\frac{\partial u}{\partial z} = 2\tan u$$

159. If $u = x\,\phi\left(\frac{y}{x}\right) + \psi\left(\frac{y}{x}\right)$,

prove that $x^2\,\frac{\partial^2 u}{\partial x^2} + 2xy\,\frac{\partial^2 u}{\partial x \partial y} + y^2\,\frac{\partial^2 u}{\partial y^2} = 0$

Solution:

Let
$$U = x\,\phi\left(\frac{y}{x}\right) + \psi\left(\frac{y}{x}\right) = V + W$$

Where
$$V = x\,\phi\left(\frac{y}{x}\right) \text{ is a } H\,(1).$$

$$W = \psi\left(\frac{y}{x}\right) = x^0\,\psi\left(\frac{y}{x}\right) \text{ is a } H\,(0).$$

160. If u is a $H\,(n)$,

Then
$$x^2\,\frac{\partial^2 z}{\partial x^2} + 2xy\,\frac{\partial^2 z}{\partial x \partial y} + y^2\,\frac{\partial^2 z}{\partial y^2} = n\,(n-1)z$$

$\Rightarrow$
$$x^2\,\frac{\partial^2 v}{\partial x^2} + 2xy\,\frac{\partial^2 v}{\partial x \partial y} + y^2\,\frac{\partial^2 v}{\partial y^2} = 1\,(1-1) = 0$$

and
$$x^2\,\frac{\partial^2 w}{\partial x^2} + 2xy\,\frac{\partial^2 w}{\partial x \partial y} + y^2\,\frac{\partial^2 w}{\partial y^2} = 0\,(0-1)\,w = 0$$

$$\therefore\quad x^2\,\frac{\partial^2 u}{\partial x^2} + 2xy\,\frac{\partial^2 u}{\partial x \partial y} + y^2\,\frac{\partial^2 u}{\partial y^2}$$

$$= x^2\,\frac{\partial^2}{\partial x^2}\,(V+W) + 2xy\,\frac{\partial^2}{\partial x \partial y}\,(V+W) + y^2\,\frac{\partial^2}{\partial y^2}\,(V+W)$$

$$= \left(x^2 \frac{\partial^2 v}{\partial x^2} + 2xy \frac{\partial^2 v}{\partial x \partial y} + y^2 \frac{\partial^2 v}{\partial y^2}\right) + \left(x^2 \frac{\partial^2 w}{\partial x^2} + 2xy \frac{\partial^2 w}{\partial x \partial y} + y^2 \frac{\partial^2 w}{\partial y^2}\right)$$

$$= 0 + 0 = 0$$

161. If $u = x^2 (y - z) + y^2 (z - x) + z^2 (x - y)$, then

$$\left(x \frac{\partial u}{\partial x} + y \frac{\partial u}{\partial y} + z \frac{\partial u}{\partial z}\right) = 3u.$$

$$U = f(x, y, z) = [x^2 (y - z) + y^2 (z - x) + z^2 (x - y)]$$

$\therefore$ $f(\lambda x, \lambda y, \lambda z)$

$$= [\lambda^2 x^2 (\lambda y - \lambda z) + \lambda^2 y^2 (\lambda z - \lambda x) + \lambda^2 z^2 (\lambda x - \lambda y)].$$

$$= \lambda^3 [x^2 (y - z) + y^2 (z - x) + z^2 (x - y)] = \lambda^3 \cdot u$$

$\therefore$ $u = H(3)$

Hence, by Euler's theorem, we get $(n = 3)$

$$\left[x\left(\frac{\partial u}{\partial x}\right) + y\left(\frac{\partial u}{\partial y}\right) + z\left(\frac{\partial u}{\partial z}\right)\right] = nu = 3u$$

162. If $\sin u = \left[\frac{x^2 y^2}{(x + y)}\right]$, then show that

$$\left[x\left(\frac{\partial u}{\partial x}\right) + y\left(\frac{\partial u}{\partial y}\right)\right] = 3 \tan u$$

Solution: Let $V(x, y) = \left[\frac{x^2 y^2}{(x + y)}\right]$

$\therefore$ $V(\lambda x, \lambda y) = \frac{[\lambda^2 x^2 \cdot \lambda^2 y^2]}{(\lambda x + \lambda y)} = \lambda^3 \left[\frac{x^2 y^2}{(x + y)}\right] = \lambda^3 [V(x, y)]$

$\therefore$ $V = H(3)$ [1]

Hence, by Euler's theorem,

$$\left[x\left(\frac{\partial v}{\partial x}\right) + y\left(\frac{\partial v}{\partial y}\right)\right] = 3V \quad [2]$$

Now $\sin u = V$ [3]

$\therefore$ $\cos u \left(\frac{\partial u}{\partial x}\right) = \left(\frac{\partial v}{\partial x}\right);$

$\therefore$ $\cos u \left[x\left(\frac{\partial u}{\partial x}\right)\right] = x\left(\frac{\partial v}{\partial x}\right)$

and $$\cos u \left[\left(\frac{\partial u}{\partial y}\right)\right] = \left(\frac{\partial v}{\partial y}\right)$$

$$\therefore \qquad \cos u \left[y\left(\frac{\partial u}{\partial y}\right)\right] = y\left(\frac{\partial v}{\partial y}\right)$$

$$\therefore \qquad \cos u \left[x\left(\frac{\partial u}{\partial x}\right) + y\left(\frac{\partial u}{\partial y}\right)\right] = \left[x\left(\frac{\partial v}{\partial x}\right) + y\left(\frac{\partial v}{\partial y}\right)\right] = 3V = 3\,(\sin u)$$

$$\therefore \qquad \left[x\left(\frac{\partial u}{\partial x}\right) + y\left(\frac{\partial u}{\partial y}\right)\right] = \left[\frac{3\,(\sin u)}{(\cos u)}\right] = 3\tan u$$

163. If $u = \sec^{-1}\left(\dfrac{x^3 + y^3}{x + y}\right)$,

Prove that $$x\frac{\partial u}{\partial x} + y\frac{\partial u}{\partial y} = 2\cot u$$

Solution: **Given:** $$u = \sec^{-1}\left[\frac{x^3 + y^3}{x + y}\right]$$

or $$\sec u = \frac{x^3 + y^3}{x + y} = V$$

$\therefore$ $\sec u = V$ is a H (2).

$\therefore$ By Euler's theorem, we have

$$x\frac{\partial}{\partial x}(\sec u) + y\frac{\partial}{\partial y}(\sec u) = 2\sec u$$

or $$x\sec u\tan u\frac{\partial u}{\partial x} + y\sec u\tan u\frac{\partial u}{\partial y} = 2\sec u$$

$$x\frac{\partial u}{\partial x} + y\frac{\partial u}{\partial y} = 2\cot u$$

164. If $z = x f\left(\dfrac{y}{x}\right) + g\left(\dfrac{y}{x}\right)$ prove that

$$x^2\frac{\partial^2 z}{\partial x^2} + 2xy\frac{\partial^2 z}{\partial x \partial y} + y^2\frac{\partial^2 z}{\partial y^2} = 0$$

Solution: Put $$u = xf\left(\frac{y}{x}\right); v = g\left(\frac{y}{x}\right) = x^0\, g\left(\frac{y}{x}\right)$$

$\therefore \qquad z = u + v$ [1]

u is a H (1); By Euler's theorem

$$x\frac{\partial u}{\partial x} + y\frac{\partial u}{\partial y} = 1.u = u$$ [2]

since v is H (0); By Euler's theorem we have

$$x\frac{\partial v}{\partial x} + y\frac{\partial v}{\partial y} = 0 \cdot v = 0\cdot$$ [3]

[2] + [3] gives

$$x\frac{\partial}{\partial x}(u + v) + y\frac{\partial u}{\partial y}(u + v) = u$$

$$\Rightarrow \qquad x\frac{\partial z}{\partial x} + y\frac{\partial z}{\partial y} = u$$ [4]

(using [1])

Differentiating equation [4] partially *w.r.t* x, we get

$$x\frac{\partial^2 z}{\partial x^2} + 1.\frac{\partial z}{\partial x} + y\frac{\partial^2 z}{\partial x \partial y} = \frac{\partial u}{\partial x}$$

$$\Rightarrow \qquad x\frac{\partial^2 z}{\partial x^2} + y\frac{\partial^2 z}{\partial x \partial y} = \frac{\partial u}{\partial x} - \frac{\partial z}{\partial x}$$ [5]

Diff. equation [4] partially *w.r.t* y, we get

$$x\frac{\partial^2 z}{\partial y \partial x} + y\frac{\partial^2 z}{\partial y^2} + 1.\frac{\partial z}{\partial y} = \frac{\partial u}{\partial y}$$

$$\Rightarrow \qquad x\frac{\partial^2 z}{\partial y \partial x} + y\frac{\partial^2 z}{\partial y^2} = \frac{\partial u}{\partial y} - \frac{\partial z}{\partial y}$$ [6]

[5] x + [6] y, we get

$$x^2\frac{\partial^2 z}{\partial x^2} + 2xy\frac{\partial^2 z}{\partial x \partial y} + y^2\frac{\partial^2 z}{\partial y^2}$$

$$= \left(x\frac{\partial u}{\partial x} + y\frac{\partial u}{\partial y}\right) - \left(x\frac{\partial z}{\partial x} + y\frac{\partial z}{\partial y}\right) = u - u = 0$$

$$\therefore \qquad x^2\frac{\partial^2 z}{\partial x^2} + 2xy\frac{\partial^2 z}{\partial x \partial y} + y^2\frac{\partial^2 z}{\partial y^2} = 0$$

165. If $u = \sec^{-1}\dfrac{x^2 + y^2}{x - y}$, prove that

$$x\frac{\partial u}{\partial x} + y\frac{\partial u}{\partial y} = \cot u$$

Solution: **Given:**

$$u = \sec^{-1} \frac{x^2 + y^2}{x - y}$$

$$\Rightarrow \qquad \sec u = \frac{x^2 + y^2}{x - y} \text{ is a } H\,(1)$$

$\therefore$ By Euler's theorem for sec u, we have

$$x \frac{\partial}{\partial x} (\sec u) + y \frac{\partial}{\partial x} (\sec u) = 1 \cdot \sec u$$

or
$$x \sec u \tan u \frac{\partial u}{\partial x} + y \sec u \tan u \frac{\partial u}{\partial y} = \sec u$$

$\therefore$
$$\sec u \tan u \left(x \frac{\partial u}{\partial x} + y \frac{\partial u}{\partial y} \right) = \sec u$$

or
$$x \frac{\partial u}{\partial x} + y \frac{\partial u}{\partial y} = \frac{\sec u}{\sec u \tan u} = \frac{1}{\tan u} = \cot u$$

166. If $u = \dfrac{xy}{\sqrt{x} + \sqrt{y}}$, show that

$$x \frac{\partial^2 u}{\partial x^2} + y \frac{\partial^2 u}{\partial x \partial y} = \frac{1}{2} \frac{\partial u}{\partial x}.$$

Solution: We can write

$$u = \frac{xy}{\sqrt{x} + \sqrt{y}} = \frac{x^2 \left(\dfrac{y}{x} \right)}{\sqrt{x} \left[1 + \sqrt{\dfrac{y}{x}} \right]} = x^{3/2} \cdot \frac{\left(\dfrac{y}{x} \right)}{1 + \sqrt{\dfrac{y}{x}}}.$$

Thus u is $H\left(\dfrac{3}{2}\right)$; Hence

Euler's theorem can be applied to u.

i.e.,
$$x \frac{\partial u}{\partial x} + y \frac{\partial u}{\partial y} = \left(\frac{3}{2} \right) u$$

Diff. partially *w.r.t.* x,

$$\left(x \frac{\partial^2 u}{\partial x^2} + 1 \cdot \frac{\partial u}{\partial x} \right) + y \frac{\partial^2 u}{\partial x \partial y} = \frac{3}{2} \cdot \frac{\partial u}{\partial x}$$

$$\Rightarrow \quad x\frac{\partial^2 u}{\partial x^2} + y\frac{\partial^2 u}{\partial x \partial y} = \frac{3}{2}\cdot\frac{\partial u}{\partial x} - \frac{\partial u}{\partial x} = \frac{1}{2}\frac{\partial u}{\partial x}.$$

Thus $x\dfrac{\partial^2 u}{\partial x^2} + y\dfrac{\partial^2 u}{\partial x \partial y} = \dfrac{1}{2}\dfrac{\partial u}{\partial x}$, as required

167. If $x^x y^y z^z = k$, show that

at $x = y = z$,

$$\frac{\partial^2 z}{\partial x \partial y} = -(x \log ex)^{-1}$$

Solution: Taking logarithm of both sides of the given function, we get

$$x \cdot \log x + y \log y + z \cdot \log z = \log k.$$

Differentiate partially w.r.t. y, we get

$$0 + \left(y \cdot \frac{1}{y} + 1 \cdot \log y\right) + \left(z \cdot \frac{1}{z} + 1 \cdot \log z \frac{\partial z}{\partial y}\right) = 0$$

$$\therefore \quad \frac{\partial z}{\partial y} = -\frac{(1 + \log y)}{(1 + \log z)} \qquad [1]$$

$$|||^{ly} \quad \frac{\partial z}{\partial x} = -\frac{(1 + \log x)}{(1 + \log z)} \qquad [2]$$

Now, differentiate [1] partially w.r.t. x, we get

$$\frac{\partial^2 z}{\partial x \partial y} = -(1 + \log y) \cdot \frac{-1}{(1 + \log z)^2} \cdot \frac{1}{z} \cdot \frac{\partial z}{\partial x}$$

$$= \frac{(1 + \log y)}{z(1 + \log z)^2} \times \frac{(1 + \log x)}{(1 + \log z)} \text{ from [2]}$$

Now put $x = y = z$

$$\therefore \quad \frac{\partial^2 z}{\partial x \partial y} = -\frac{(1 + \log x)^2}{x(1 + \log x)^3} = \frac{-1}{x(\log e + \log x)}$$

$$= -\frac{-1}{x \log ex} = -(x \log ex)^{-1}.$$

Note: A similar problem may be stated as :

168. If $x^x y^y z^z = 108$

To find $\dfrac{\partial^2 z}{\partial x \partial y}$ at $x = 1, y = 2, z = 3$.

Solution: Take log on both sides and get

$$x \log x + y \log y + z \log z = \log 108$$

PART-B

Differentiate partially *w.r.t.* y, we get

$$y\left(\frac{1}{y}\right) + \log y + \log z\left(\frac{\partial z}{\partial y}\right) + z\left(\frac{1}{z}\right)\frac{\partial z}{\partial y} = 0$$

$$\therefore \qquad \frac{\partial z}{\partial y} = -\left(\frac{1+\log y}{1+\log z}\right) \qquad [1]$$

$$|||^{ly} \qquad \frac{\partial z}{\partial x} = -\left(\frac{1+\log x}{1+\log y}\right) \qquad [2]$$

Again Differentiate [1] partially w.r.t x, we get

$$\frac{\partial^2 z}{\partial x \partial y} = \frac{(1+\log y)}{(1+\log z)^2}\left(\frac{1}{z}\right)\left(\frac{\partial z}{\partial x}\right) = -\frac{(1+\log x)(1+\log y)}{z\,(1+\log z)^3}$$

(using [2])

$$\therefore \qquad \left[\frac{\partial^2 z}{\partial x \partial y}\right]_{at\,(1,2,3)} = -\frac{1}{3}\left(\frac{1+\log 2}{(1+\log 3)^2}\right)$$

169. Transform the equation

$$\frac{\partial^2 u}{\partial x^2} + \frac{\partial^2 u}{\partial y^2} = 0 \text{ into}$$

Polar coordinates

Solution: Put $\quad x = r\cos\theta;$

$\quad y = r\sin\theta$

$$\therefore \qquad \frac{\partial u}{\partial x} = \frac{\partial u}{\partial r}\cdot\frac{\partial v}{\partial x} + \frac{\partial u}{\partial \theta}\cdot\frac{\partial \theta}{\partial x}$$

$$= \frac{\partial u}{\partial r}\cdot\cos\theta - \frac{\partial u}{\partial \theta}\frac{\sin\theta}{r} = p, \text{ (say)}$$

We can similarly obtain

$$\boxed{\frac{\partial u}{\partial y} = \frac{\partial u}{\partial v}\sin\theta + \frac{\partial u}{\partial \theta}\frac{\cos\theta}{r}}$$

It follows that

$$\frac{\partial^2 u}{\partial x^2} = \frac{\partial}{\partial x}\left(\frac{\partial u}{\partial x}\right) = \frac{\partial p}{\partial x} = \cos\theta\cdot\frac{\partial p}{\partial r} - \frac{\sin\theta}{r}\frac{\partial p}{\partial \theta}.$$

$$= \cos\theta\frac{\partial}{\partial r}\left(\cos\theta\frac{\partial u}{\partial r} - \frac{\sin\theta}{r}\frac{\partial u}{\partial \theta}\right) - \frac{\sin\theta}{r}\frac{\partial}{\partial \theta}\left(\cos\theta\frac{\partial u}{\partial r} - \frac{\sin\theta}{r}\frac{\partial u}{\partial \theta}\right)$$

$$= \cos^2\theta \frac{\partial^2 u}{\partial r^2} - \frac{2\sin\theta\cos\theta}{r} \cdot \frac{\partial^2 u}{\partial r \partial \theta} + \frac{\sin^2\theta}{r^2} \frac{\partial^2 u}{\partial \theta^2}$$

$$+ \frac{\sin^2\theta}{r} \frac{\partial u}{\partial r} + \frac{2\sin\theta\cos\theta}{r^2} \cdot \frac{\partial u}{\partial \theta}$$

Similarly,

$$\frac{\partial^2 u}{\partial y^2} = \sin^2\theta \frac{\partial^2 u}{\partial r^2} + \frac{2\sin\theta\cos\theta}{r} \cdot \frac{\partial^2 u}{\partial r \partial \theta}$$

$$+ \frac{\cos^2\theta}{r^2} \cdot \frac{\partial^2 u}{\partial \theta^2} + \frac{\cos^2\theta}{r} \frac{\partial u}{\partial r} - \frac{2\sin\theta\cos\theta}{r^2} \cdot \frac{\partial u}{\partial \theta}$$

Adding, we get

$$\frac{\partial^2 u}{\partial x^2} + \frac{\partial^2 u}{\partial y^2} = \frac{\partial^2 u}{\partial r^2} + \frac{1}{r}\frac{\partial u}{\partial r} + \frac{1}{r^2}\frac{\partial^2 u}{\partial \theta^2}$$

Hence the transformed equation is

$$\frac{\partial^2 u}{\partial r^2} + \frac{1}{r}\frac{\partial u}{\partial r} + \frac{1}{r^2}\frac{\partial^2 u}{\partial \theta^2} = 0.$$

170. Prove that

$$\frac{\partial^2 u}{\partial x^2} + \frac{\partial^2 u}{\partial y^2} = \frac{\partial^2 u}{\partial \xi^2} + \frac{\partial^2 u}{\partial \eta^2},$$

where

$$x = \xi\cos\alpha - \eta\sin\alpha$$
$$y = \xi\sin\alpha + \eta\cos\alpha.$$

Solution: Here

$$x = \xi\cos\alpha - \eta\sin\alpha$$
$$y = \xi\sin\alpha + \eta\cos\alpha,$$

$$\therefore \quad \frac{\partial u}{\partial \xi} = \frac{\partial u}{\partial x}\frac{\partial x}{\partial \xi} + \frac{\partial u}{\partial y} \cdot \frac{\partial y}{\partial \xi} = \frac{\partial u}{\partial x}\cos\alpha + \frac{\partial u}{\partial y}\sin\alpha$$

or

$$\frac{\partial}{\partial \xi} = \left(\frac{\partial}{\partial x}\cos\alpha + \frac{\partial}{\partial y}\sin\alpha\right) \qquad [1]$$

$$\frac{\partial u}{\partial \eta} = \frac{\partial u}{\partial x}\frac{\partial x}{\partial \eta} + \frac{\partial u}{\partial y}\frac{\partial y}{\partial \eta}$$

$$= \frac{\partial u}{\partial x}(-\sin\alpha) + \frac{\partial u}{\partial y}(\cos\alpha)$$

$$\frac{\partial^2 u}{\partial \xi^2} = \frac{\partial}{\partial \xi}\left(\frac{\partial u}{\partial \xi}\right)$$

$$= \left(\cos\alpha\frac{\partial}{\partial x} + \sin\alpha\frac{\partial}{\partial y}\right)\frac{\partial u}{\partial \xi} \quad \text{by[1]}$$

PART-B

$$= \left(\cos\alpha \cdot \frac{\partial}{\partial x} + \sin\alpha \frac{\partial}{\partial y}\right)\left(\cos\alpha \frac{\partial u}{\partial x} + \sin\alpha \frac{\partial u}{\partial y}\right)$$

$$= \cos^2\alpha \cdot \frac{\partial^2 u}{\partial x^2} + 2\cos\alpha \sin\alpha \frac{\partial^2 u}{\partial x \partial y} + \sin^2\alpha \frac{\partial^2 u}{\partial y^2}$$

Similarly, $$\frac{\partial^2 u}{\partial \eta^2} = \left(-\sin\alpha \frac{\partial}{\partial x} + \cos\alpha \frac{\partial}{\partial y}\right)^2 u$$

$$= \sin^2\alpha \frac{\partial^2 u}{\partial x^2} - 2\sin\alpha\cos\alpha \frac{\partial^2 u}{\partial x \partial y} + \cos^2\alpha \frac{\partial^2 u}{\partial y^2}$$

$$\therefore \quad \frac{\partial^2 u}{\partial \xi^2} + \frac{\partial^2 u}{\partial \eta^2} = \frac{\partial^2 u}{\partial x^2} + \frac{\partial^2 u}{\partial y^2}$$

171. If $z = f(x, y)$, where
$x = e^u \cos v;\ y = e^u \sin v$,
Show that

$$\left(\frac{\partial z}{\partial u}\right)^2 + \left(\frac{\partial z}{\partial r}\right)^2 = e^{2u}\left\{\left(\frac{\partial z}{\partial x}\right)^2 + \left(\frac{\partial z}{\partial y}\right)^2\right\}$$

Solution:

Hence $x = e^u \cos v;\ y = e^u \sin v$,

$$\therefore \quad \frac{\partial z}{\partial u} = \frac{\partial z}{\partial x} \cdot \frac{\partial x}{\partial u} + \frac{\partial z}{\partial y} \cdot \frac{\partial y}{\partial u} = \frac{\partial z}{\partial x} \cdot e^u \cos v + \frac{\partial z}{\partial y} \cdot e^u \sin v$$

$$\frac{\partial z}{\partial v} = \frac{\partial z}{\partial x} \cdot \frac{\partial x}{\partial v} + \frac{\partial z}{\partial y} \cdot \frac{\partial y}{\partial v} = \frac{\partial z}{\partial x}(-e^u \sin v) + \frac{\partial z}{\partial y}(e^u \cos v)$$

Squaring and adding, we get

$$\left(\frac{\partial z}{\partial u}\right)^2 + \left(\frac{\partial z}{\partial v}\right)^2 = e^{2u}\left\{\left(\frac{\partial z}{\partial x}\right)^2 + \left(\frac{\partial z}{\partial y}\right)^2\right\}$$

172. If $V = f(r)$ where $r^2 = x^2 + y^2 + z^2$,
Show that

$$\frac{\partial^2 V}{\partial x^2} + \frac{\partial^2 V}{\partial y^2} + \frac{\partial^2 V}{\partial z^2} = \frac{d^2 V}{dr^2} + \frac{2}{r}\frac{dV}{dr}.$$

Solution: We are given $V = f(r)$, where

$$r^2 = x^2 + y^2 + z^2;$$

$$2r \frac{\partial r}{\partial x} = 2x + 0 + 0$$

or $$\frac{\partial r}{\partial x} = \frac{x}{r}; \frac{\partial V}{\partial x} = \frac{dV}{dr} \cdot \frac{\partial r}{\partial x} = \left(\frac{x}{r}\right)\frac{\partial V}{\partial r}$$

$$\frac{\partial^2 V}{\partial x^2} = \frac{\partial}{\partial x}\left[\frac{x}{r}\frac{dV}{dr}\right] = \frac{1}{r}\frac{dV}{dr} + x\frac{\partial}{\partial x}\left[\frac{1}{r}\frac{dV}{dr}\right]$$

$$= \frac{1}{r} \cdot \frac{dV}{dr} + x\frac{d}{dr}\left[\frac{1}{r} \cdot \frac{dV}{dr}\right] \cdot \frac{\partial r}{\partial x}$$

$$= \frac{1}{r} \cdot \frac{dV}{dr} + \frac{x^2}{r}\left[\frac{1}{r}\frac{d^2V}{dr^2} - \frac{1}{r^2}\frac{dV}{dr}\right]$$

Writing two similar expressions for $\frac{\partial^2 V}{\partial y^2}$ and $\frac{\partial^2 V}{\partial z^2}$ and adding, we get

$$\frac{\partial^2 V}{\partial x^2} + \frac{\partial^2 V}{\partial y^2} + \frac{\partial^2 V}{\partial z^2} = \frac{3}{r}\frac{dV}{dr} + \frac{(x^2+y^2+z^2)}{r^2}\left[\frac{d^2V}{dr^2} - \frac{1}{r}\frac{dV}{dr}\right]$$

$$= \frac{d^2V}{dr^2} + \frac{2}{r}\frac{dV}{dr}$$

or $$x\frac{\partial u}{\partial x} + y\frac{\partial u}{\partial y} = 2\cot u.$$

173. If $u = e^{xy}$, prove that $\frac{\partial^2 u}{\partial x^2} + \frac{\partial^2 u}{\partial y^2} = \frac{1}{u}\left[\left(\frac{\partial u}{\partial x}\right)^2 + \left(\frac{\partial u}{\partial y}\right)^2\right]$

Solution: $u = e^{xy}$; $\frac{\partial u}{\partial x} = e^{xy} \cdot y$ and $\frac{\partial u}{\partial y} = e^{xy} \cdot x$

Consider $$\frac{1}{u}\left[\left(\frac{\partial u}{\partial x}\right)^2 + \left(\frac{\partial u}{\partial y}\right)^2\right] = \frac{1}{e^{xy}}[y^2 e^{2xy} + x^2 e^{2xy}]$$

$$= (x^2 + y^2)\, e^{xy} \qquad [1]$$

Also, $$\frac{\partial^2 u}{\partial x^2} = e^{xy} \cdot y^2 \text{ and } \frac{\partial^2 u}{\partial y^2} = e^{xy} \cdot x^2$$

$$\Rightarrow \quad \frac{\partial^2 u}{\partial x^2} + \frac{\partial^2 u}{\partial y^2} = (x^2 + y^2)\, e^{xy} \qquad [2]$$

Form [1] and [2], we find that

$$\frac{\partial^2 u}{\partial x^2} + \frac{\partial^2 u}{\partial y^2} = \frac{1}{u}\left[\left(\frac{\partial u}{\partial x}\right)^2 + \left(\frac{\partial u}{\partial y}\right)^2\right].$$

PART-B

EXERCISES

1. Verify Euler's theorem on Homogenous functions for

$$u = \frac{x+y+z}{\sqrt{x}+\sqrt{y}+\sqrt{z}}$$ (*Ans.* 0)

2. If $u = \tan \dfrac{x^2+y^2+z^2}{xy+yz+zx}$, evaluate $x\dfrac{\partial u}{\partial x} + y\dfrac{\partial u}{\partial y} + z\dfrac{\partial u}{\partial z}$.

3. Verify Euler's Theorem for the function $u = \sin^{-1}\left(\dfrac{x}{y}\right) + \tan^{-1}\left(\dfrac{y}{x}\right)$.

4. If V is a $H(n)$, prove that $\dfrac{\partial v}{\partial x}, \dfrac{\partial v}{\partial y}$ are each $a\ H(n-1)$. If $V = f(x, y)\ X = \dfrac{\partial V}{\partial x}$, $Y = \dfrac{\partial V}{\partial y}$, show that $X\dfrac{\partial V}{\partial X} + Y\dfrac{\partial V}{\partial Y} = \dfrac{n}{n-1}V$.

5. If $V = \dfrac{x^3 y^3}{x^3+y^3}$, show that $x\cdot\dfrac{\partial v}{\partial x} + y\cdot\dfrac{\partial v}{\partial y} = 3V$.

6. If $u = \sin^{-1}\left[\left(x^2+y^2\right)^{1/5}\right]$, then show that

$$\left[x\left(\frac{\partial u}{\partial x}\right) + y\left(\frac{\partial u}{\partial y}\right)\right] = \left(\frac{2}{5}\right)\tan u$$

7. If $u = \log\left(\dfrac{x^4+y^4}{x+y}\right)$ show that $x\dfrac{\partial u}{\partial x} + y\dfrac{\partial u}{\partial y} = 3$.

8. If $u = \dfrac{xy}{(x+y)}$, then prove that $\left[x^2\left(\dfrac{\partial^2 u}{\partial x^2}\right) + 2xy\left(\dfrac{\partial^2 u}{\partial x\partial y}\right) + y^2\left(\dfrac{\partial^2 u}{\partial y^2}\right)\right] = 0$.

9. If $u = \left(\dfrac{z}{x^2+y^2}\right)^{1/2}$. Prove that $x\dfrac{\partial u}{\partial x} + y\dfrac{\partial u}{\partial y} + z\dfrac{\partial u}{\partial z} = -\dfrac{1}{2}u$

10. If $u = \dfrac{x+y}{x^2+y^2}$, applying Euler's theorem *P.T.* u is a $H(-1)$ in u and y

11. If $u = (x+y)\,f\left(\dfrac{y}{x}\right)$, prove that $x^2\dfrac{\partial^2 u}{\partial x^2} + 2xy\dfrac{\partial^2 u}{\partial x\partial y} + y^2\dfrac{\partial^2 u}{\partial y^2} = 0$

12. If $u = \text{cosec}^{-1}\left(\dfrac{x^{1/2}+y^{1/2}}{x^{1/3}+y^{1/3}}\right)$. *P.T.*

$$x^2\frac{\partial^2 u}{\partial x^2} + 2xy\frac{\partial^2 u}{\partial x\partial y} + y^2\frac{\partial^2 u}{\partial y^2} = \frac{1}{12}\left(1+\frac{1}{12}\sec^2 u\right)\tan u.$$

2.9 TOTAL DIFFERENTIATION

If $u = f(x, y)$ is a function of two variables x and y, where $x = x(t)$; $y = y(t)$, *i.e.*, x and y both are function of a single variables 't'.

Then $\left(\dfrac{du}{dt}\right)$ is called the total differenatial coeff (or the exact differential) and is given by the formula:

$$\frac{du}{dt} = \frac{\partial u}{\partial x}\cdot\frac{dx}{dt} + \frac{\partial u}{\partial y}\cdot\frac{dy}{dt} \quad [1]$$

or

$$du = \frac{\partial u}{\partial x}\cdot dx + \frac{\partial u}{\partial y}\cdot dy \quad [2]$$

If we put $t = x$ then [1] gives

$$\frac{du}{dx} = \frac{\partial u}{\partial x} + \frac{\partial u}{\partial y}\cdot\frac{dy}{dx} \quad [3]$$

Similarly, if $u = f(x, y, z)$,

Then

$$\frac{du}{dt} = \frac{\partial u}{\partial x}\cdot\frac{dx}{dt} + \frac{\partial u}{\partial y}\cdot\frac{dy}{dt} + \frac{\partial y}{\partial z}\cdot\frac{dz}{dt} \quad [5]$$

or

$$du = \frac{\partial u}{\partial x}dx + \frac{\partial u}{\partial y}dy + \frac{\partial u}{\partial z}dz \quad [6]$$

2.10 DIFFERENTIATION OF COMPOSITE FUNCTIONS

We note that $\dfrac{du}{dt}$ is called the total derivative of u w.r.t. 't'.

Now, if z is a function of two variables u and v *i.e.*, $z = z(u, v)$ where u and v are themselves are functions of x and y then the partial derivatives $\left(\dfrac{\partial z}{\partial x}\right)$ and $\left(\dfrac{\partial z}{\partial y}\right)$ are given by:

$$\frac{\partial z}{\partial x} = \frac{\partial z}{\partial u}\cdot\frac{\partial u}{\partial x} + \frac{\partial z}{\partial v}\cdot\frac{\partial v}{\partial x}, \quad [1]$$

$$\frac{\partial z}{\partial y} = \frac{\partial z}{\partial u}\cdot\frac{\partial u}{\partial y} + \frac{\partial z}{\partial v}\cdot\frac{\partial v}{\partial y} \quad [2]$$

Examples (176 – 195)

174. If $x = v + u$, $y = uv$ and z is a function of x and y, show that

$$u\frac{\partial z}{\partial u} + v\frac{\partial z}{\partial v} = x\frac{\partial z}{\partial x} + 2y\frac{\partial z}{\partial y}$$

Solution: Hence z is a composite function of u and v.

$$\therefore \quad \frac{\partial z}{\partial u} = \frac{\partial z}{\partial x}\cdot\frac{\partial x}{\partial u} + \frac{\partial z}{\partial y}\cdot\frac{\partial y}{\partial u} = \frac{\partial z}{\partial x}\cdot 1 + \frac{\partial z}{\partial y}\cdot v$$

and

$$\frac{\partial z}{\partial v} = \frac{\partial z}{\partial u}\cdot\frac{\partial x}{\partial v} + \frac{\partial z}{\partial y}\cdot\frac{\partial y}{\partial v} = \frac{\partial z}{\partial x}\cdot 1 + \frac{\partial z}{\partial y}\cdot u \qquad [2]$$

[1] u + [2] v gives

$$u\frac{\partial z}{\partial u} + v\frac{\partial z}{\partial v} = (u + v)\frac{\partial z}{\partial x} + 2uv\frac{\partial z}{\partial y} = x\frac{\partial z}{\partial x} + 2y\frac{\partial z}{\partial y}$$

175. Find $\frac{du}{dt}$ when $u = xy^2 + x^2y$; $x = at^2$; $y = 2at$

Solution: Here, u is a composite function of t

$$\therefore \quad \frac{du}{dt} = \frac{\partial u}{\partial x}\cdot\frac{dx}{dt} + \frac{\partial u}{\partial y}\cdot\frac{dy}{dt} = (y^2 + 2xy)(2at) + (2xy + x^2)(2a)$$

$$= (4a^2t^2 + 4a^2t^3)\,2at + (4a^2t^3 + a^2t^4)\,2a = 16a^3t^3 + 10a^3t^4.$$

176. If $z = f(x, y)$

and $x = u + v$

and $y = u - v$

Show that $\frac{\partial z}{\partial u} + \frac{\partial z}{\partial v} = 2\frac{\partial z}{\partial x}$

Solution: z is a composite function of u and v

$$\therefore \quad \frac{\partial z}{\partial u} = \frac{\partial z}{\partial x}\cdot\frac{\partial x}{\partial u} + \frac{\partial z}{\partial y}\,\frac{\partial y}{\partial u} = \frac{\partial z}{\partial x} + \frac{\partial z}{\partial y}$$

and

$$\frac{\partial z}{\partial v} = \frac{\partial z}{\partial x}\frac{\partial x}{\partial v} + \frac{\partial z}{\partial y}\frac{\partial y}{\partial v} = \frac{\partial z}{\partial x} - \frac{\partial z}{\partial y}$$

$$\therefore \quad \frac{\partial z}{\partial u} + \frac{\partial z}{\partial v} = 2\frac{\partial z}{\partial x}$$

177. Find $\frac{du}{dt}$ when

$$u = x^2y^2 + x^3y$$

where $x = 2t^2$ and $y = 4t$

Solution: $$\frac{du}{dt} = \frac{\partial u}{\partial x}\cdot\frac{dx}{dt} + \frac{\partial u}{\partial y}\cdot\frac{dy}{dt} = (2xy^2 + 3x^2y)(4t) + (2yx^2 + x^3)(4)$$

$$= [2\cdot 2t^2\cdot 16t^2 + 3\cdot 4t^4\cdot 4t](4t) + [2\cdot(4t)(4t^4) + 8t^6](4)$$

$$= 64t^4(4t) + 48t^5(4t) + (32t^5)(4) + 32t^6 = 384t^5 + 224t^6$$

178. It $z = e^{ax+by} f(ax - \text{by})$,

Show that $b\dfrac{\partial z}{\partial x} + a\dfrac{\partial z}{\partial y} = 2abz$

Solution: $z = e^{ax+by} f(ax - \text{by})$

$\Rightarrow$ $z = e^u f(v)$,

where $u = ax + \text{by}$,

$v = ax - \text{by}$

$\Rightarrow$ **z is a composite function of x and y**

$$\therefore \quad \frac{\partial z}{\partial x} = \frac{\partial z}{\partial u}\cdot\frac{\partial u}{\partial x} + \frac{\partial z}{\partial v}\cdot\frac{\partial v}{\partial x} = e^u f(v)\cdot a + e^u f'(v)\cdot a$$

$$= az + ae^u f'(v) \qquad [1]$$

and $$\frac{\partial z}{\partial y} = \frac{\partial z}{\partial u}\cdot\frac{\partial u}{\partial y} + \frac{\partial z}{\partial v}\cdot\frac{\partial v}{\partial y} = e^u f(v)\cdot b + e^u f'(v)\cdot(-b)$$

$$= bz - be^u f'(v) \qquad [2]$$

[1] b + [2] a gives

$$b\frac{\partial z}{\partial x} + a\frac{\partial z}{\partial y} = abz + abe^u f'(v) + abz - abe^u f'(v) = 2abz$$

179. Find the total derivative of u w.r.t. t

When $u = e^x \sin y$, when $x = \log t$, $y = t^2$

Also verify by direct calculations.

Solution: Total derivative of u w.r.t. t

$$\frac{du}{dt} = \frac{\partial u}{\partial x}\cdot\frac{dx}{dt} + \frac{\partial u}{\partial y}\cdot\frac{dy}{dt} = e^x \sin y\cdot\frac{1}{t} + e^x\cos y\cdot 2t$$

$$= t\sin t^2\cdot\frac{1}{t} + t\cos t^2\cdot 2t \qquad \left[\begin{aligned}\because x &= \log t;\\ \therefore e^x &= t\end{aligned}\right]$$

$$= \sin t^2 + 2t^2\cos t^2$$

Also $u = e^x \sin y = t\sin t^2$

$$\therefore \quad \frac{du}{dt} = 1\cdot\sin t^2 + t\cos t^2\cdot 2t = \sin t^2 + 2t^2\cos t^2$$

Hence the verification.

180. If $u = \sin^{-1}(x, y)$;

$x = 3t$;

$y = 4t^3$

Show that $\dfrac{du}{dt} = \dfrac{3}{\sqrt{1-t^2}}$

Solution: Here u is a composite function of t.

$$\frac{du}{dt} = \frac{\partial u}{\partial x}\cdot\frac{dx}{dt} + \frac{\partial u}{\partial y}\cdot\frac{dy}{dt}$$

$$= \frac{1}{\sqrt{1-(x-y)^2}}\cdot(1)\cdot(3) + \frac{1}{\sqrt{1-(x-y)^2}}\cdot(-1)\cdot 12t^2$$

$$= \frac{3-12t^2}{\sqrt{1-x^2+2xy-y^2}}$$

$$= \frac{3(1-4t^2)}{\sqrt{1-9t^2+24t^4-16t^2}} = \frac{3(1-4t^2)}{\sqrt{(1-t^2)(1-8t^4+16t^4)}}$$

$$= \frac{3(1-4t^2)}{\sqrt{(1-t^2)(1-4t^2)^2}} = \frac{3}{\sqrt{1-t^2}}.$$

181. If $z = f(x, y)$ and $u = ax + by$, $v = ay - bx$, prove that

$$\frac{\partial^2 z}{\partial x^2} + \frac{\partial^2 z}{\partial y^2} = (a^2 + b^2)\left(\frac{\partial^2 z}{\partial u^2} + \frac{\partial^2 z}{\partial v^2}\right)$$

Solution: We have

$$z = f(u, v),\ u = ax + by$$
$$v = ay - bx.$$

$$\frac{\partial u}{\partial x} = a;\quad \frac{\partial u}{\partial y} = b$$

$$\frac{\partial v}{\partial x} = -b,\quad \frac{\partial v}{\partial y} = a$$

Now,
$$\frac{\partial z}{\partial x} = \frac{\partial z}{\partial u}\cdot\frac{\partial u}{\partial x} + \frac{\partial z}{\partial v}\cdot\frac{\partial v}{\partial x}$$

$\therefore$
$$\frac{\partial z}{\partial x} = a\frac{\partial z}{\partial u} - b\frac{\partial z}{\partial v}$$

or
$$\frac{\partial}{\partial x} = a\frac{\partial}{\partial u} - b\frac{\partial}{\partial v}$$

$$\frac{\partial^2 z}{\partial x^2} = \frac{\partial}{\partial x}\left(\frac{\partial z}{\partial x}\right) = \left(a\frac{\partial}{\partial u} - b\frac{\partial}{\partial v}\right)\left(a\frac{\partial z}{\partial u} - b\frac{\partial z}{\partial v}\right)$$

$$= a^2 \frac{\partial}{\partial u}\left(\frac{\partial z}{\partial u}\right) - ab\frac{\partial}{\partial u}\left(\frac{\partial z}{\partial v}\right) - ab\frac{\partial}{\partial v}\left(\frac{\partial z}{\partial u}\right) + b\frac{2\partial}{\partial v}\left(\frac{\partial z}{\partial v}\right)$$

$$= a^2 \frac{\partial^2 z}{\partial u^2} - ab\frac{\partial^2 z}{\partial u \partial v} - ab\frac{\partial^2 z}{\partial v \partial u} + b^2 \frac{\partial^2 z}{\partial v^2}$$

$$= a^2 \frac{\partial^2 z}{\partial u^2} - 2ab\frac{\partial^2 z}{\partial u \partial v} + b^2 \frac{\partial^2 z}{\partial v^2} \qquad [1]$$

Similarly, $\frac{\partial z}{\partial y} = \frac{\partial z}{\partial u} \cdot \frac{\partial u}{\partial y} + \frac{\partial z}{\partial v} \cdot \frac{\partial v}{\partial y}$

$\therefore$ $\frac{\partial z}{\partial y} = \frac{\partial z}{\partial u} \cdot (b) + \frac{\partial z}{\partial v}(a)$

or $\frac{\partial}{\partial y} = b\frac{\partial}{\partial u} + a\frac{\partial}{\partial v}$

Now $\frac{\partial^2 z}{\partial y^2} = \frac{\partial}{\partial y}\left(\frac{\partial z}{\partial y}\right) = \left(b\frac{\partial}{\partial u} + a\frac{\partial}{\partial v}\right)\left(b\frac{\partial z}{\partial u} + a\frac{\partial z}{\partial v}\right)$

$$= b^2 \frac{\partial^2 z}{\partial u^2} + ab\frac{\partial^2 z}{\partial v \partial u} + ab\frac{\partial^2 z}{\partial u \partial v} + a^2 \frac{\partial^2 z}{\partial v^2}$$

$$= b^2 \frac{\partial^2 z}{\partial u^2} + 2ab\frac{\partial^2 z}{\partial u \partial v} + a^2 \frac{\partial^2 z}{\partial v^2} \qquad [2]$$

Now adding [1] and [2], we get

$$\frac{\partial^2 z}{\partial x^2} + \frac{\partial^2 z}{\partial y^2} = (a^2 + b^2)\frac{\partial^2 z}{\partial u^2} + (a^2 + b^2)\frac{\partial^2 z}{\partial v^2}$$

or $$\frac{\partial^2 z}{\partial x^2} + \frac{\partial^2 z}{\partial y^2} = (a^2 + b^2)\left(\frac{\partial^2 z}{\partial u^2} + \frac{\partial^2 z}{\partial v^2}\right)$$

182. If $u = f(y - z, z - x, x - y)$,

Porve that $\frac{\partial u}{\partial x} + \frac{\partial u}{\partial y} + \frac{\partial u}{\partial z} = 0$

Solution: Let $X = y - z;$

$Y = z - x$

$Z = x - y$

$\therefore$ $u = f(X, Y, Z)$

where X, Y, Z are function of x, y, z.

$\therefore$ u is a composite function of three variables x, y, z.

$$\therefore \quad \frac{\partial u}{\partial x} = \frac{\partial u}{\partial X}\frac{\partial X}{\partial x} + \frac{\partial u}{\partial Y}\frac{\partial Y}{\partial x} + \frac{\partial u}{\partial Z}\frac{\partial Z}{\partial x}$$

$$= \frac{\partial u}{\partial X}(0) + \frac{\partial u}{\partial Y}(-1) + \frac{\partial u}{\partial Z}(1)$$

$$\frac{\partial u}{\partial x} = -\frac{\partial u}{\partial y} + \frac{\partial u}{\partial z} \qquad [1]$$

Similarly, $$\frac{\partial u}{\partial y} = \frac{\partial u}{\partial x} - \frac{\partial u}{\partial z} \qquad [2]$$

and $$\frac{\partial u}{\partial z} = -\frac{\partial u}{\partial x} + \frac{\partial u}{\partial y} \qquad [3]$$

[1] + [2] + [3] gives

$$\frac{\partial u}{\partial x} + \frac{\partial u}{\partial y} + \frac{\partial u}{\partial z} = 0$$

183. If $z = f(x, y)$ where

$$x = e^u + e^{-v},\ y = e^{-u} - e^v$$

Show that $$\frac{\partial z}{\partial u} - \frac{\partial z}{\partial v} = x\frac{\partial z}{\partial x} - y\frac{\partial z}{\partial y}$$

Solution: $$\frac{\partial x}{\partial u} = e^u;\quad \frac{\partial x}{\partial v} = -e^{-v},$$

$$\frac{\partial y}{\partial u} = -e^{-u};\quad \frac{\partial y}{\partial v} = -e^v$$

We have $$\frac{\partial z}{\partial u} = \frac{\partial z}{\partial x}\cdot\frac{\partial x}{\partial u} + \frac{\partial z}{\partial y}\frac{\partial y}{\partial u} = \frac{\partial z}{\partial x}(e^u) + \frac{\partial z}{\partial y}(-e^{-u}) \qquad [1]$$

$$\frac{\partial z}{\partial v} = \frac{\partial z}{\partial x}\cdot\frac{\partial x}{\partial v} + \frac{\partial z}{\partial y}\cdot\frac{\partial y}{\partial v} = \frac{\partial z}{\partial x}(-e^{-v}) + \frac{\partial z}{\partial y}(-e^v) \qquad [2]$$

Subtracting [2] from [1], we get

$$\frac{\partial z}{\partial u} - \frac{\partial z}{\partial v} = (e^u + e^{-v})\frac{\partial z}{\partial x} - (e^{-u} - e^v)\frac{\partial z}{\partial y} = x\frac{\partial z}{\partial x} - y\frac{\partial z}{\partial y}$$

184. If $z = f(u, v)$ where $u = x^2 - y^2$, $v = 2xy$

Prove that

(*i*) $$x\frac{\partial z}{\partial x} - y\frac{\partial z}{\partial y} = 2(x^2 + y^2)\frac{\partial z}{\partial x}$$

(*ii*) $$\left(\frac{\partial z}{\partial x}\right)^2 + \left(\frac{\partial z}{\partial y}\right)^2 = 4(x^2 + y^2)\left[\left(\frac{\partial z}{\partial u}\right)^2 + \left(\frac{\partial z}{\partial v}\right)^2\right]$$

Solution: **Given:** $u = x^2 - y^2;\ v = 2xy$

$$\therefore \quad \frac{\partial u}{\partial x} = 2x;\ \frac{\partial u}{\partial y} = -2y;\ \frac{\partial v}{\partial x} = 2y,\ \frac{\partial v}{\partial y} = 2x$$

We have,
$$\frac{\partial z}{\partial x} = \frac{\partial z}{\partial u}\cdot\frac{\partial u}{\partial x} + \frac{\partial z}{\partial v}\cdot\frac{\partial v}{\partial x} = 2x\frac{\partial z}{\partial u} + 2y\frac{\partial z}{\partial v} \qquad [1]$$

$$\frac{\partial z}{\partial y} = \frac{\partial z}{\partial u}\cdot\frac{\partial u}{\partial y} + \frac{\partial z}{\partial v}\cdot\frac{\partial v}{\partial y} = -2y\frac{\partial z}{\partial u} + 2x\frac{\partial z}{\partial v} \qquad [2]$$

Hence (*i*) $x\dfrac{\partial z}{\partial x} - y\dfrac{\partial z}{\partial y} = 2x^2\dfrac{\partial z}{\partial u} + 2xy\dfrac{\partial z}{\partial v} - \left[-2y^2\dfrac{\partial z}{\partial u} + 2xy\dfrac{\partial z}{\partial v}\right]$

$$= 2(x^2 + y^2)\frac{\partial z}{\partial u}$$

(*ii*) squaring and adding [1] and [2], we get

$$\left(\frac{\partial z}{\partial x}\right)^2 + \left(\frac{\partial z}{\partial y}\right)^2 = 4\left[\left(x\frac{\partial z}{\partial u} + y\frac{\partial z}{\partial v}\right)^2 + \left(-y\frac{\partial z}{\partial u} + x\frac{\partial z}{\partial v}\right)^2\right]$$

$$= 4(x^2 + y^2)\left[\left(\frac{\partial z}{\partial u}\right)^2 + \left(\frac{\partial z}{\partial v}\right)^2\right]$$

185. **If u is a homogeneous functions of x, y, z of order n, prove that**

$$x\frac{\partial u}{\partial x} + y\frac{\partial u}{\partial y} + z\frac{\partial u}{\partial z} = nu$$

Solution: Since
$$u = x^n f\left(\frac{y}{x}, \frac{z}{x}\right) = x^n f(u, v),$$

where
$$\frac{y}{x} = u,\ \frac{z}{x} = v$$

Now, $f(u, v)$ is a composite function of x, y, z

$$\therefore \quad \frac{\partial u}{\partial x} = nx^{n-1} f(u, v) + x^n\left(\frac{\partial f}{\partial u}\cdot\frac{\partial u}{\partial x} + \frac{\partial f}{\partial v}\cdot\frac{\partial v}{\partial x}\right)$$

$$\left(\frac{\partial u}{\partial x} = -\frac{y}{x^2};\ \frac{\partial v}{\partial x} = -\frac{z}{x^2}\right)$$

$$\therefore \quad \frac{\partial u}{\partial x} = nx^{n-1} f(u,v) - x^{n-2}\left(y\frac{\partial f}{\partial u} + z\frac{\partial f}{\partial x}\right) \qquad [1]$$

Also,
$$\frac{\partial u}{\partial y} = x^n\left[\frac{\partial f}{\partial u}\cdot\frac{\partial u}{\partial y} + \frac{\partial f}{\partial v}\cdot\frac{\partial v}{\partial y}\right]$$

$$= x^{n-1}\left(\frac{\partial f}{\partial u}\right)\cdot\left(\because \frac{\partial u}{\partial y}=\frac{1}{x},\ \frac{\partial v}{\partial y}=0\right)$$

Similarly, $$\frac{\partial u}{\partial z} = x^{n-1}\frac{\partial f}{\partial v} \qquad [3]$$

Now [1] x + [2] y + [3] z gives

$$x\frac{\partial u}{\partial x}+y\frac{\partial u}{\partial y}+z\frac{\partial u}{\partial z} = nx^n f(u, v) = n\cdot u$$

Note: (The above result is Euler's theorem for a homogeneous function of three independent variables)

186. Find the total derivative of u w.r.t. t when

$$u = \cos h\left(\frac{y}{x}\right), \text{ where } x = t^2,\ y = e^t.$$

Solution:

Since, $$u = \cos h\left(\frac{y}{x}\right)$$

$$\therefore \quad \frac{\partial u}{\partial x} = \sin h\left(\frac{y}{x}\right)\left(-\frac{y}{x^2}\right) = -\frac{y}{x^2}\sin h\left(\frac{y}{x}\right)$$

$$\frac{\partial u}{\partial y} = \frac{1}{x}\sinh\left(\frac{y}{x}\right)$$

$$\frac{dx}{dt} = 2t;\quad \frac{dy}{dt} = e^t$$

$$\therefore \quad \frac{du}{dt} = \frac{\partial u}{\partial x}\cdot\frac{dx}{dt}+\frac{\partial u}{\partial y}\cdot\frac{dy}{dt}$$

$$= \left(-\frac{y}{x^2}\sin h\left(\frac{y}{x}\right)\right)(2t)+\left(\frac{1}{x}\sin h\left(\frac{y}{x}\right)\right)\cdot e^t$$

$$= \frac{1}{x^2}\sin h\left(\frac{y}{x}\right)[xe^t - 2yt].$$

187. If $u = f(r, s)$,

$r = x + y$

$s = x - y$, show that

$$\frac{\partial u}{\partial x}+\frac{\partial u}{\partial y} = 2\frac{\partial u}{\partial r}$$

Solution: Here u is a composite function of x and y

$$\therefore \quad \frac{\partial u}{\partial x} = \frac{\partial u}{\partial r} \cdot \frac{\partial r}{\partial x} + \frac{\partial u}{\partial s} \cdot \frac{\partial s}{\partial x} = \frac{\partial u}{\partial r} \cdot (1) + \frac{\partial u}{\partial s} \cdot (1)$$

$$= \frac{\partial u}{\partial r} + \frac{\partial u}{\partial s}$$

and
$$\frac{\partial u}{\partial y} = \frac{\partial u}{\partial r} \cdot \frac{\partial r}{\partial y} + \frac{\partial u}{\partial s} \cdot \frac{\partial s}{\partial y} = \frac{\partial u}{\partial r} \cdot (1) + \frac{\partial u}{\partial s}(-1) = \frac{\partial u}{\partial r} - \frac{\partial u}{\partial s}$$

$$\therefore \quad \frac{\partial u}{\partial x} + \frac{\partial u}{\partial y} = 2\frac{\partial u}{\partial r}$$

188. If $u = u\left(\dfrac{y-x}{xy}, \dfrac{z-x}{xz}\right)$,

Find the value of

$$x^2 \frac{\partial u}{\partial x} + y^2 \frac{\partial u}{\partial y} + z^2 \frac{\partial u}{\partial z}$$

(VTU, August, 2001)

Solution: **Given:**
$$u = u\left(\frac{y-x}{xy}, \frac{z-x}{xz}\right) = u\left(\frac{1}{x} - \frac{1}{y}, \frac{1}{x} - \frac{1}{z}\right) = u(p, q)$$

where
$$p = \left(\frac{1}{x} - \frac{1}{y}\right); q = \left(\frac{1}{x} - \frac{1}{z}\right)$$

Now
$$\frac{\partial u}{\partial x} = \frac{\partial u}{\partial p} \cdot \frac{\partial p}{\partial x} + \frac{\partial u}{\partial q} \cdot \frac{\partial q}{\partial x} = \frac{\partial u}{\partial p}\left(-\frac{1}{x^2}\right) + \frac{\partial u}{\partial q}\left(-\frac{1}{x^2}\right)$$

So that
$$x^2 \frac{\partial u}{\partial x} = -\frac{\partial u}{\partial p} - \frac{\partial u}{\partial q} \qquad [1]$$

$$\frac{\partial u}{\partial y} = \frac{\partial u}{\partial p} \cdot \frac{\partial p}{\partial y} + \frac{\partial u}{\partial q} \cdot \frac{\partial q}{\partial y} = \frac{\partial u}{\partial p}\left(\frac{1}{y^2}\right) + \frac{\partial u}{\partial q} \cdot 0$$

i.e.,
$$y^2 \frac{\partial u}{\partial y} = \frac{\partial u}{\partial p} \qquad [2]$$

and
$$\frac{\partial u}{\partial z} = \frac{\partial u}{\partial p} \cdot \frac{\partial p}{\partial z} + \frac{\partial u}{\partial q} \cdot \frac{\partial q}{\partial z} = \frac{\partial u}{\partial p} \cdot 0 + \frac{\partial u}{\partial q} \cdot \frac{1}{z^2}$$

i.e.,
$$z^2 \frac{\partial u}{\partial z} = \frac{\partial u}{\partial q} \qquad [3]$$

Adding [1], [2] and [3], we get

$$x^2 \frac{\partial u}{\partial x} + y^2 \frac{\partial u}{\partial y} + z^2 \frac{\partial u}{\partial z} = 0$$

189. If $z = \sqrt{x^2 + y^2}$ and $x^3 + y^3 + 3axy = 5a^2$

find the value of $\frac{dz}{dx}$, when $x = a\ y = a$.

Solution: **Given:** $z = \sqrt{x^2 + y^2}$

i.e, $z = z\ (x, y) =$ a function of x and y

Again $x^3 + y^3 + 3axy = 5a^2$ defines y as a function of x.

Thus z is a composite function of x

$$\therefore \quad \frac{dz}{dx} = \frac{\partial z}{\partial x} + \frac{\partial z}{\partial y} \cdot \frac{dy}{dx}$$

Now
$$\frac{\partial z}{\partial x} = \frac{2x}{2\sqrt{x^2 + y^2}} = \frac{x}{\sqrt{x^2 + y^2}}$$

$$\frac{\partial z}{\partial y} = \frac{2y}{2\sqrt{x^2 + y^2}} = \frac{y}{\sqrt{x^2 + y^2}}$$

Again let $\phi\ (x, y) = x^3 + y^3 + 3axy - 5a^2 = 0$

is an implicit function

$$\therefore \quad \frac{dy}{dx} = -\left(\frac{\frac{\partial f}{\partial x}}{\frac{\partial f}{\partial y}}\right) = -\frac{3x^2 + 3ay}{3y^2 + 3ax} = -\left[\frac{x^2 + ay}{y^2 + ax}\right]$$

$$\therefore \quad \frac{dz}{dx} = \frac{x}{\sqrt{x^2 + y^2}} - \frac{y}{\sqrt{x^2 + y^2}} \cdot \left[\frac{x^2 + ay}{y^2 + ax}\right]$$

$$\therefore \quad \left(\frac{dz}{dx}\right)_{at\,(a,a)} = \frac{a}{\sqrt{2a^2}} - \frac{a}{\sqrt{2a^2}}\left[\frac{a^2 + a^2}{a^2 + a^2}\right] = \frac{1}{\sqrt{2}} - \frac{1}{\sqrt{2}} = 0$$

190. If
$$z = \log\,(u^2 + v),$$
$$u = e^{x^2 + y^2},$$
$$v = x^2 + y$$

find $\frac{\partial z}{\partial x}, \frac{\partial z}{\partial y}$ at $x = 1,\ y = 0$

Solution: Hence

$u = e$ and $v = 1$

at $x = 1$ and $y = 0$

$$\frac{\partial z}{\partial x} = \frac{\partial z}{\partial u} \cdot \frac{\partial u}{\partial x} + \frac{\partial z}{\partial v} \cdot \frac{\partial v}{\partial x}$$

$$= \frac{1}{u^2 + v} \cdot 2u \cdot e^{x^2+y^2} \cdot 2x + \frac{1}{u^2 + v} \cdot 1 \cdot 2x$$

$$\therefore \quad \left[\frac{\partial z}{\partial x}\right]_{(1,0)} = \frac{2(2e^2 + 1)}{e^2 + 1}.$$

$$\left(\frac{\partial z}{\partial y}\right) = \frac{\partial z}{\partial u} \cdot \frac{\partial u}{\partial y} + \frac{\partial z}{\partial v} \cdot \frac{\partial v}{\partial y} = \frac{1}{u^2 + v} \cdot 2u \cdot e^{x^2+y^2} \cdot 2y + \frac{1}{u^2 + v} \cdot 1$$

$$= \frac{1}{e^2 + 1} \quad (\text{at } x = 1,\ y = 0)$$

191. If $z = f(x, y)$ where $x = e^u \cos v$ and $y = e^u \sin v$

Show that $y \dfrac{\partial z}{\partial u} + x \dfrac{\partial z}{\partial v} = e^{2u} \dfrac{\partial z}{\partial y}$

Solution: $x = e^u \cos v$; $y = e^u \sin v$

$$\frac{\partial z}{\partial u} = \frac{\partial z}{\partial x} \cdot \frac{\partial x}{\partial u} + \frac{\partial z}{\partial y} \cdot \frac{\partial y}{\partial u} = \frac{\partial z}{\partial x} \cdot e^u \cos v + \frac{\partial z}{\partial y} \cdot e^u \sin v$$

$$= x \frac{\partial z}{\partial x} + y \frac{\partial z}{\partial y}$$

$$\therefore \quad y \frac{\partial z}{\partial u} = xy \frac{\partial z}{\partial x} + y^2 \frac{\partial z}{\partial y} \qquad [1]$$

and

$$\frac{\partial z}{\partial v} = \frac{\partial z}{\partial x} \frac{\partial x}{\partial v} + \frac{\partial z}{\partial y} \frac{\partial y}{\partial v} = \frac{\partial z}{\partial x} (-e^u \sin v) + \frac{\partial z}{\partial y} (e^u \cos v)$$

$$= -y \frac{\partial z}{\partial x} + x \frac{\partial z}{\partial y}$$

$$\therefore \quad x \frac{\partial z}{\partial v} = -xy \frac{\partial z}{\partial x} + x^2 \frac{\partial z}{\partial y} \qquad [2]$$

on adding [1] and [2], we get

$$y \frac{\partial z}{\partial u} + x \frac{\partial z}{\partial v} = (x^2 + y^2) \frac{\partial z}{\partial y} = (e^{2u} \cos^2 v + e^{2u} \sin^2 v) \frac{\partial z}{\partial y}$$

$$= e^{2u} (\cos^2 v + \sin^2 v) \frac{\partial z}{\partial y} = e^{2u} \frac{\partial z}{\partial y}$$

192. If $u = x^2 + y^2$, where $x = e^t \cos t$ and $y = e^t \sin t$, show that

$$\frac{du}{dt} = 2e^{2t}.$$

PART-B

Solution: $u = x^2 + y^2$; $x = e^t \cos t$ and $y = e^t \sin t$

$$\frac{du}{dt} = \frac{\partial u}{\partial x} \cdot \frac{dx}{dt} + \frac{\partial u}{\partial y} \cdot \frac{dy}{dt}$$

(u is a composite function of t)

$$\frac{\partial u}{\partial x} = 2x; \quad \frac{\partial u}{\partial y} = 2y; \quad \frac{dx}{dt} = e^t (\cos t - \sin t);$$

$$\frac{dy}{dt} = e^t (\sin t + \cos t);$$

$$\therefore \quad \frac{du}{dt} = 2xe^t [(\cos t - \sin t) + 2ye^t (\sin t + \cos t)$$

$$= 2e^t [(x + y) \cos t + (y - x) \sin t]$$

$$= 2e^t [(\cos t + \sin t) e^t \cos t + (\sin t - \cos t) \cdot e^t \sin t] = 2e^t$$

EXERCISES

1. If $\frac{\partial u}{\partial x} = \frac{\partial v}{\partial y}, \frac{\partial u}{\partial y} = -\frac{\partial v}{\partial x}$ and $x = r \cos \theta$, $y = r \sin \theta$

   Prove that $\frac{\partial u}{\partial r} = \frac{1}{r}\frac{\partial v}{\partial \theta}, \frac{\partial v}{\partial r} = -\frac{1}{r}\frac{\partial u}{\partial \theta}$

2. If $x = u + v + w$, $y = uv + vw + wu$, $z = uvw$ and $f = f(x, y, z)$, prove that

   $$u \cdot \frac{\partial f}{\partial u} + v \cdot \frac{\partial f}{\partial v} + w \frac{\partial f}{\partial w} = w \frac{\partial f}{\partial x} + 2y \frac{\partial f}{\partial y} + 3z \frac{\partial f}{\partial z}$$

3. A function $f(x, y)$ is rewritten in terms of new variables $u = e^x \cos y$, $v = e^x \sin y$ show that

   (*i*) $\frac{\partial f}{\partial x} = u \frac{\partial f}{\partial u} + v \frac{\partial f}{\partial v}$

   and (*ii*) $\frac{\partial f}{\partial y} = -v \frac{\partial f}{\partial u} + u \frac{\partial f}{\partial v}$

4. Let $w = \frac{x + y}{x - y} + \sin z$ where $x = \sin t^2$, $y = \cos t^2$ and $z = t^2$.

   Find $\frac{dw}{dt}$

5. If $w = f(u, v)$, where $u = x + y$ and $v = x - y$

   Show that $\frac{\partial w}{\partial x} + \frac{\partial w}{\partial y} = 2 \frac{\partial w}{\partial x}$

6. Find $\frac{du}{dt}$ if

(a) $u = x^2 + y^2 + z^2$ where $x = e^t$, $y = e^t \sin t$, $z = e^t \cos t$

(b) $u = xy + yz + zx$ where $x = t \cos t$, $y = t \sin t$, $z = t$, at $t = \frac{\pi}{4}$

7. If $x^2 = au + bv$, $y = au - br$, prove that

$$\left(\frac{\partial u}{\partial x}\right)_y \cdot \left(\frac{\partial x}{\partial u}\right)_v = \frac{1}{2} = \left(\frac{\partial v}{\partial y}\right)_x \cdot \left(\frac{\partial y}{\partial v}\right)_u$$

8. By changing the independent variables x and t to u and v by means of the relationships

$$u = x - at, \; v = x + at$$

Show that $a^2 \frac{\partial^2 y}{\partial x^2} - \frac{\partial^2 y}{\partial t^2} = 4a^2 \cdot \frac{\partial^2 y}{\partial u \partial v}$

9. If $u = x \cos \frac{y}{z}$, $x = 3r^2 + 2s$, $y = 4r - 2s^3$, $z = 2r^2 - 3s^2$ find $\frac{\partial u}{\partial v}, \frac{\partial u}{\partial s}$.

10. If $u = z \,.\, \sin^{-1} \frac{y}{x}$, where $x = 3r^2 + 2s$, $y = 4r - 2s^3$, $z = 2r^2 - 3s^2$, find $\frac{\partial u}{\partial r}, \frac{\partial u}{\partial s}$.

11. Find $\frac{\partial w}{\partial v}$ when $u = 0$, $v = 0$

if $w = (x^2 + y - 2)^4 + (x - y + 2)^3$.

$x = u - 2v + 1$ and $y = 2u + v - 2$ **(*Ans.* 99)**

12. Let $w = f(v, z)$, and $r = x + 2t$, $z = x - 2t$

Assuming sufficient differentiability of function w, show that

$$4 \cdot \frac{\partial^2 w}{\partial x^2} - \frac{\partial^2 w}{\partial t^2} = 16 \cdot \frac{\partial^2 w}{\partial v \partial z}.$$

13. Given: $w = x + 2y + z^2$,

$x = r/s$, $y = r^2 + e^s$

and $z = 2r$, show that

$$r \frac{\partial w}{\partial r} + s \frac{\partial w}{\partial s} = 12r^2 + 2se^s.$$

Differentiation of Implicit Functions (Imp. Deductions)

If $u = f(x, y)$, then $\frac{du}{dt} = \frac{\partial f}{\partial x} \cdot \frac{dx}{dt} + \frac{\partial f}{\partial y} \cdot \frac{dy}{dt}$ [1]

where $x = x(t)$, $y = y(t)$, and if $t = x$, then $\frac{dx}{dt} = \frac{dx}{dx} = 1$.

Substituting in [1], we obtain:

$$\frac{du}{dx} = \frac{\partial f}{\partial x} + \frac{\partial f}{\partial y} \cdot \frac{dy}{dx} \quad [2]$$

Cor. Let the implicit function be: $f(x, y) = 0$

or $$\frac{du}{dx} = 0$$

Substituting in [2], we have

$$0 = \frac{\partial f}{\partial x} + \frac{\partial f}{\partial y} \cdot \frac{dy}{dx}; \frac{\partial f}{\partial y} \cdot \frac{dy}{dx} = -\frac{\partial f}{\partial x}$$

or $$\boxed{\left(\frac{\partial y}{\partial x}\right) = \frac{-\left(\frac{\partial f}{\partial x}\right)}{\left(\frac{\partial f}{\partial y}\right)}}$$ **is an important result.** [3]

Note: An implicit function can be written as $u(x, y) = c$, a function of x and y (c is a constant)

$$\therefore \quad \frac{du}{dx} = 0$$

To find $\frac{d^2 y}{dx^2}$; we have from equation [3] that

$$\frac{dy}{dx} = \frac{-\left(\frac{\partial f}{\partial x}\right)}{\left(\frac{\partial f}{\partial y}\right)} = \left(-\frac{f_x}{f_y}\right)$$

$$\therefore \quad \frac{d^2 y}{dx^2} = -\left[\frac{f_y \frac{d}{dx}(f_x) - f_x \frac{d}{dx}(f_y)}{f_y^2}\right],$$

(using quotient rule of differentiation)

$$= \frac{-f_y\left[\frac{\partial}{\partial x}(f_x) + \frac{\partial}{\partial y}(f_x)\frac{dy}{dx}\right] + f_x\left[\frac{\partial}{\partial x}(f_y) + \frac{\partial}{\partial y}(f_y)\frac{dy}{dx}\right]}{f_y^2}$$

$$= \frac{-f_y\left[f_{xx} + f_{yx}\left(-\frac{f_x}{f_y}\right)\right] + f_x\left[f_{xy} + f_{yy}\left(\frac{-f_x}{f_y}\right)\right]}{f_y^2}$$

$$= \frac{-f_y f_{xx} + f_{yx} f_x + f_{xy} f_x - \frac{f_x^2 f_{yy}}{f_y}}{f_y^2}$$

$$= \frac{-f_y^2 f_{xx} + 2f_{xy} f_x f_y - f_x^2 f_{yy}}{f_y^3}$$

$$\therefore \quad \boxed{\frac{d^2 y}{dx^2} = \frac{2f_{xy} f_x f_y - f_y^2 f_{xx} - f_x^2 f_{yy}}{f_y^3}}$$

Examples (193 – 209)

193. If $u = f(x, y)$ and $x = r \cos \theta$; $y = r \sin \theta$ prove that

$$\frac{\partial^2 u}{\partial x^2} + \frac{\partial^2 u}{\partial y^2} = \frac{\partial^2 u}{\partial r^2} + \frac{1}{r}\frac{\partial u}{\partial r} + \frac{1}{r^2}\frac{\partial^2 u}{\partial \theta^2}.$$

Solution: put $\left(x = r\cos\theta;\ y = r\sin\theta;\ r^2 = x^2 + y^2;\ \tan\theta = \frac{y}{x} \right)$

$$\therefore \quad 2r \cdot \frac{\partial r}{\partial x} = 2x; \quad \frac{\partial r}{\partial x} = \frac{x}{r} = \cos\theta;$$

and $\quad \sec^2\theta \dfrac{\partial\theta}{\partial x} = -\dfrac{y}{x^2} \quad \therefore \quad \dfrac{\partial\theta}{\partial x} = -\dfrac{y}{r^2} \qquad \left(\because \dfrac{\partial\theta}{\partial x} = -\dfrac{y}{x^2}\cos^2\theta \right)$

or $\quad \dfrac{\partial\theta}{\partial x} = -\dfrac{y}{r^2} = -\dfrac{\sin\theta}{r}$

$u = f(x, y)$, where $x = r\cos\theta$; $y = r\sin\theta$

$$\therefore \quad \frac{\partial u}{\partial x} = \frac{\partial u}{\partial r}\frac{\partial r}{\partial x} + \frac{\partial u}{\partial\theta}\cdot\frac{\partial\theta}{\partial x} = \frac{\partial u}{\partial r}(\cos\theta) + \frac{\partial u}{\partial\theta}\left(-\frac{\sin\theta}{r}\right)$$

$$= \left(\cos\theta \frac{\partial}{\partial r} - \frac{\sin\theta}{r}\frac{\partial}{\partial\theta}\right) u$$

$$\Rightarrow \quad \frac{\partial}{\partial x} = \left(\cos\theta\frac{\partial}{\partial r} - \frac{\sin\theta}{r}\frac{\partial}{\partial\theta}\right) \qquad [1]$$

$$\therefore \quad \frac{\partial^2 u}{\partial x^2} = \frac{\partial u}{\partial x^2}\frac{\partial}{\partial x}\left(\frac{\partial u}{\partial x}\right) = \left(\cos\theta\frac{\partial}{\partial r} - \frac{\sin\theta}{r}\frac{\partial}{\partial\theta}\right)\left(\cos\theta\frac{\partial u}{\partial r} - \frac{\sin\theta}{r}\frac{\partial u}{\partial\theta}\right)$$

on simplication and assuming that $\boxed{\dfrac{\partial^2 u}{\partial r\partial\theta} = \dfrac{\partial^2 u}{\partial\theta\partial r}}$

$$= \cos^2\theta\frac{\partial^2 u}{\partial r^2} - \frac{2\sin\theta\cos\theta}{r}\cdot\frac{\partial^2 u}{\partial r\partial\theta} + \frac{\sin^2\theta}{r^2}\frac{\partial^2 u}{\partial\theta^2}$$

PART-B

To find $\frac{\text{‘}\partial\text{’}}{\partial y}$ we change θ to $\left(\frac{\pi}{2} - \theta\right)$ in [1],

Hence $$\frac{\partial}{\partial y} = \left(\sin\theta \frac{\partial}{\partial r} + \frac{\cos\theta}{r}\frac{\partial}{\partial\theta}\right)$$

$$\left(\because \sin\left(\frac{\pi}{2} - \theta\right) = \cos\theta; \cos\left(\frac{\pi}{2} - \theta\right) = \sin\theta\right)$$

|||[ly] $\frac{\partial^2 u}{\partial y^2}$ can be found from $\frac{\partial^2 u}{\partial x^2}$ by changing θ to $\left(\frac{\pi}{2} - \theta\right)$. This gives

$$\frac{\partial^2 u}{\partial y^2} = \sin^2\theta \frac{\partial^2 u}{\partial r^2} + \frac{2\sin\theta\cos\theta}{r}\cdot\frac{\partial^2 u}{\partial r\partial\theta} + \frac{\cos^2\theta}{r^2}\frac{\partial^2 u}{\partial\theta^2}$$

$$+ \frac{\cos^2\theta}{r}\frac{\partial u}{\partial r} - \frac{2\sin\theta\cos\theta}{r^2}\cdot\frac{\partial u}{\partial\theta}$$

$$\therefore \quad \frac{\partial^2 u}{\partial x^2} + \frac{\partial^2 u}{\partial y^2} = \frac{\partial^2 u}{\partial r^2} + \frac{1}{r}\frac{\partial u}{\partial r} + \frac{1}{r^2}\frac{\partial^2 u}{\partial\theta^2}$$

194. If $u = x^2 + y^2 + z^2$; $x = e^t$; $y = e^t \sin t$;
$z = e^t \cos t$, then

prove that $\frac{du}{dt} = 4e^{2t}$

Solution: We have

$$\frac{du}{dt} = \left(\frac{\partial u}{\partial x}\cdot\frac{dx}{dt} + \frac{\partial u}{\partial y}\cdot\frac{dy}{dt} + \frac{\partial u}{\partial z}\cdot\frac{dz}{dt}\right)$$

$$= 2xe^t + 2y\,(e^t \cos t + \sin t\, e^t) + 2z\,(-e^t \sin t + \cos t\, e^t)$$

$$= 2\,e^t \cdot e^t + 2e^t \sin t\,(\cos t + \sin t)e^t + 2e^t \cos t\,(-e^t \sin t + \cos t\, e^t)e^t$$

$$= 2e^{2t}\,[1 + \sin t \cos t + \sin^2 t - \sin t \cos t + \cos^2 t] = 4\,e^{2t}$$

195. If $V = f\left(\frac{x}{z}, \frac{y}{z}\right)$, prove that

$$x\frac{\partial v}{\partial x} + y\frac{\partial v}{\partial y} + z\frac{\partial v}{\partial z} = 0$$

Solution: Let $P = \frac{x}{z}$; $q = \frac{y}{z}$

Then $$V = f(p, q)$$

Hence $$\frac{\partial V}{\partial x} = \frac{\partial V}{\partial p}\cdot\frac{\partial p}{\partial x} + \frac{\partial V}{\partial q}\cdot\frac{\partial q}{\partial x} = \frac{\partial V}{\partial p}\left(\frac{1}{z}\right) + \frac{\partial V}{\partial q}(0) = \frac{1}{z}\frac{\partial V}{\partial p}$$

$$\frac{\partial V}{\partial y} = \frac{\partial V}{\partial p}\cdot\frac{\partial p}{\partial y} + \frac{\partial V}{\partial q}\cdot\frac{\partial q}{\partial y} = \frac{\partial V}{\partial p}(0) + \frac{\partial V}{\partial q}\left(\frac{1}{z}\right) = \frac{1}{z}\frac{\partial V}{\partial q}$$

$$\frac{\partial V}{\partial z} = \frac{\partial V}{\partial p}\frac{\partial p}{\partial z} + \frac{\partial V}{\partial q}\cdot\frac{\partial q}{\partial z} = \frac{\partial V}{\partial p}\left(-\frac{x}{z^2}\right) + \frac{\partial V}{\partial q}\left(-\frac{y}{z^2}\right)$$

$$\therefore\ x\frac{\partial V}{\partial x} + y\frac{\partial V}{\partial y} + z\frac{\partial V}{\partial z} = \frac{x}{z}\frac{\partial V}{\partial p} + \frac{y}{z}\frac{\partial V}{\partial q} - \frac{x}{z}\frac{\partial V}{\partial p} - \frac{y}{z}\frac{\partial V}{\partial q} = 0$$

196. Find $\frac{du}{dx}$, given $u = e^{x^2+y^2}$,

where $$\frac{x^2}{a^2} + \frac{y^2}{b^2} = 1$$

Since u is $u(x, y)$ and x and y are connected by a relation.

$\therefore$ y is an implicit function of x in this case.

u is therefore a composite function of x

By total derivative, we have

$$\frac{du}{dx} = \frac{\partial u}{\partial x}\cdot\frac{dx}{dx} + \frac{\partial u}{\partial y}\cdot\frac{dy}{dx}$$

$$= \frac{\partial u}{\partial x}\cdot 1 + \frac{\partial u}{\partial y}\cdot\frac{dy}{dx}$$

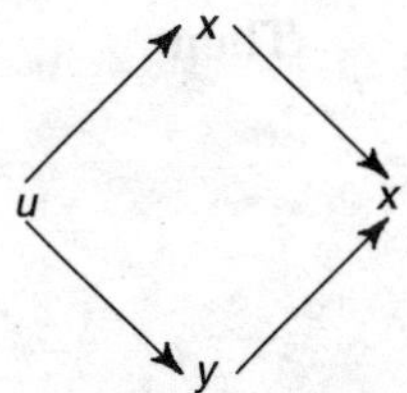

Fig. 2.5

But $$u = e^{x^2+y^2}$$

$\Rightarrow$ $$\frac{\partial u}{\partial x} = e^{x^2+y^2}\ .\ 2x$$

$$\frac{\partial u}{\partial y} = e^{x^2+y^2}\ .\ 2y$$

Also $$f = \frac{x^2}{a^2} + \frac{y^2}{b^2} - 1 = 0$$

$\therefore$ We have by implicit function formula

$$\frac{dy}{dx} = -\frac{\left(\frac{\partial f}{\partial x}\right)}{\left(\frac{\partial f}{\partial y}\right)} = -\frac{2x/a^2}{2y/b^2} = -\frac{b^2x}{a^2y}$$

Substituting in [1]

$$\frac{du}{dx} = e^{x^2+y^2}\cdot 2x + e^{x^2+y^2}\cdot 2y\left(\frac{-b^2x}{a^2y}\right) = 2e^{x^2+y^2}\left(x - \frac{b^2x}{a^2}\right)$$

Hence $$\frac{du}{dx} = \frac{2x}{a^2}\cdot e^{x^2+y^2}\cdot(a^2 - b^2)$$

197. Find $\frac{dy}{dx}$ given

$$xe^{-y} - 2ye^{x} = 1$$

Solution:

Hence $u = xe^{-y} - 2ye^{x} - 1 = 0$ is an implicit relation.

$$\therefore \quad \frac{dy}{dx} = -\frac{\left(\frac{\partial u}{\partial x}\right)}{\left(\frac{\partial u}{\partial y}\right)} = \frac{-(e^{-y} - 2ye^{x})}{(-xe^{-y} - 2e^{x})}$$

198. If $x^y + y^x = c$ find $\frac{dy}{dx}$

Solution: Let $u = x^y + y^x - c = 0$

Then $$\frac{dy}{dx} = -\frac{\left(\frac{\partial u}{\partial x}\right)}{\left(\frac{\partial u}{\partial y}\right)} = -\frac{(yx^{y-1} + y^x \log y)}{(x^y \log x + xy^{x-1})}$$

$$\therefore \quad \frac{dy}{dx} = -\frac{y \log y + yx^{y-1}}{x^y \log x + xy^{x-1}}$$

199. If $u = x \log(xy)$, where $x^3 + y^3 + 3xy = 1$

find $\frac{dy}{dx}$ and hence find $\frac{du}{dx}$

Here $u = f(x, y)$

and $\phi(x, y) = x^3 + y^3 + 3xy - 1 = 0$

is an implicit function.

$$\therefore \quad \frac{dy}{dx} = -\frac{\frac{\partial \phi}{\partial x}}{\frac{\partial \phi}{\partial y}} = -\frac{3x^2 + 3y}{3y^2 + 3x} = -\frac{x^2 + y}{y^2 + x}$$

Now, $$\frac{du}{dx} = \frac{\partial u}{\partial x}\cdot\frac{dx}{dx} + \frac{\partial u}{\partial y}\cdot\frac{dy}{dx} = \frac{\partial u}{\partial x}\cdot 1 + \frac{\partial u}{\partial y}\cdot\frac{dy}{dx}$$

$$= \left\{1\cdot\log(xy) + x\,\frac{1}{xy}\cdot y\right\} + \left\{x\cdot\frac{1}{xy}\cdot x\right\}\cdot\left\{-\frac{x^2+y}{y^2+x}\right\}$$

$$= [\log(xy) + 1] - \frac{x}{y}\cdot\frac{x^2+y}{y^2+x}$$

200. If $ax^2 + by^2 - 1 = 0$ find $\dfrac{dy}{dx}$

Since $u = ax^2 + by^2 - 1 = 0$ is an implicit function.

Hence $$\frac{dy}{dx} = \frac{-\left(\frac{\partial u}{\partial x}\right)}{\left(\frac{\partial u}{\partial y}\right)} = -\frac{2ax}{2by} = -\frac{ax}{by}$$

201. $x^3 + y^3 - 3axy = 0;$

find $\dfrac{dy}{dx}$ and $\dfrac{d^2y}{dx^2}$

Solution: Let $$f(x, y) = x^3 + y^3 - 3axy = 0$$

Let $$p = \frac{\partial f}{\partial x} = 3x^2 - 3ay$$

$$q = \frac{\partial f}{\partial y} = 3y^2 - 3ax$$

$$r = \frac{\partial^2 f}{\partial x^2} = 6x,$$

$$s = \frac{\partial^2 f}{\partial y\partial x} = -3a,$$

$$t = \frac{\partial^2 f}{\partial y^2} = 6y$$

$$\frac{dy}{dx} = -\frac{\left(\frac{\partial f}{\partial x}\right)}{\left(\frac{\partial f}{\partial y}\right)} = -\frac{3x^2 - 3ay}{3y^2 - 3ax} = \frac{ay - x^2}{y^2 - ax}$$

PART-B

$$\frac{d^2y}{dx^2} = -\frac{q^2r - 2pqs + p^2t}{q^3}$$

$$= \frac{-(3y^2 - 3ax)^2\, 6x - 2(3x^2 - 3ay)(3y^2 - 3ax)\cdot(-3a) + (3x^2 - 3ay)^2 \cdot 6y}{(3y^2 - 3ax)^3}$$

$$= \frac{-2x(y^2 - ax)^2 + 2a(x^2 - ay)(y^2 - ax) + 2y(x^2 - ay)^2}{(y^2 - ax)^3}$$

$$= \frac{2a^3xy}{(ax - y^2)^3}$$

202. If $x^3 + 3x^2y + 6xy^2 + y^3 = 1$, find $\frac{dy}{dx}$

Solution:

Let $f(x, y) = x^3 + 3x^2y + 6xy^2 + y^3 - 1 = 0$

is an implicit function.

$$\frac{\partial f}{\partial x} = 3x^2 + 6xy + 6y^2;$$

$$\frac{\partial f}{\partial y} = 3x^2 + 12xy + 3y^2$$

$$\frac{dy}{dx} = -\frac{\left(\frac{\partial f}{\partial x}\right)}{\left(\frac{\partial f}{\partial y}\right)} = -\frac{3x^2 + 6xy + 6y^2}{3x^2 + 12xy + 3y^2} = -\frac{x^2 + 2xy + 2y^2}{x^2 + 4xy + y^2}$$

203. If $f(x, y) = 0$ and $\phi(y, z) = 0$,

Show that

$$\frac{\partial f}{\partial y}\cdot\frac{\partial \phi}{\partial z}\cdot\frac{\partial z}{\partial x} = \frac{\partial f}{\partial x}\cdot\frac{\partial \phi}{\partial y}$$

Solution: $f(x, y) = 0$ [1]

$\phi(y, z) = 0$ [2]

Diff. (1) *w.r.t.* x we get

$$0 = \frac{\partial f}{\partial x} + \frac{\partial f}{\partial y}\cdot\frac{dy}{dx}$$

or $$\frac{dy}{dx} = -\frac{\left(\frac{\partial f}{\partial x}\right)}{\left(\frac{\partial f}{\partial y}\right)} \qquad [3]$$

Diff. (2) *w.r.t. y*, we get

$$0 = \frac{\partial \phi}{\partial y} + \frac{\partial \phi}{\partial z} \cdot \frac{dz}{dy}$$

or $$\frac{dz}{dy} = -\frac{\left(\frac{\partial \phi}{\partial y}\right)}{\left(\frac{\partial \phi}{\partial z}\right)} \qquad [4]$$

Multiply [3] and [4], we get

$$\frac{dy}{dx} \times \frac{dz}{dy} = \left[\frac{-\left(\frac{\partial f}{\partial x}\right)}{\left(\frac{\partial f}{\partial y}\right)}\right]\left[\frac{-\left(\frac{\partial \phi}{\partial y}\right)}{\left(\frac{\partial \phi}{\partial z}\right)}\right]$$

$$\frac{dz}{dx} = \frac{\left(\frac{\partial f}{\partial x}\right) \times \left(\frac{\partial \phi}{\partial y}\right)}{\left(\frac{\partial f}{\partial y}\right)\left(\frac{\partial \phi}{\partial z}\right)}$$

or $$\frac{\partial f}{\partial y} \cdot \frac{\partial \phi}{\partial z} \cdot \frac{dz}{dx} = \frac{\partial f}{\partial x} \cdot \frac{\partial \phi}{\partial y}$$

204. Find $\frac{dy}{dx}$ when $(\cos x)^y = (\sin y)^x$

Solution:

Let $$f(x, y) = (\cos x)^y - (\sin y)^x = 0$$
is an implicit function.

$$\left.\begin{aligned} \frac{\partial f}{\partial x} &= y (\cos x)^{y-1} (-\sin x) - (\sin y)^x \log \sin y \\ &= -[y \sin x (\cos x)^{y-1} + (\sin y)^x \log \sin y] \\ \frac{\partial f}{\partial y} &= (\cos x)^y \log \cos x - x (\sin y)^{x-1} \cos y \end{aligned}\right] \qquad [1]$$

$$\left(\frac{dy}{dx}\right) = -\frac{\left(\frac{\partial f}{\partial x}\right)}{\left(\frac{\partial f}{\partial y}\right)}$$

$$\frac{dy}{dx} = \frac{y \sin x\,(\cos x)^{y-1} + (\sin y)^x \log \sin y}{(\cos x)^y \log \cos x - x\,(\sin y)^{x-1} \cos y}$$

In [1], Put $(\cos x)^y$ for $(\sin y)^x$

$$\frac{dy}{dx} = \frac{y \sin x\,(\cos x)^{y-1} + (\cos x)^y \log \sin y}{(\cos x)^y \log \cos x - \frac{x\,(\cos x)^y}{\sin y} \cdot \cos y}$$

$$= \frac{(\cos x)^y [y \tan x + \log \sin y]}{(\cos x)^y [\log \cos x - x \cot y]} = \frac{y \tan x + \log \sin y}{\log \cos x - x \cot y} \qquad [2]$$

205. If $u = \sin(x^2 + y^2)$, where $a^2x^2 + b^2y^2 = c^2$, find $\frac{du}{dx}$

Solution: **Given:** $u = \sin(x^2 + y^2)$

$$\frac{\partial u}{\partial x} = 2x \cos(x^2 + y^2);$$

$$\frac{\partial u}{\partial y} = 2y \cos(x^2 + y^2)$$

Hence $$\frac{du}{dx} = \frac{\partial u}{\partial x} + \frac{\partial u}{\partial y} \cdot \frac{dy}{dx}$$

$$= 2x \cos(x^2 + y^2) + 2y \cos(x^2 + y^2) \cdot \frac{dy}{dx} \qquad [1]$$

If $z(x, y) = z = a^2x^2 + b^2y^2 - c^2 = 0$, defines an implicit function.

Hence $$\frac{\partial z}{\partial x} = 2a^2x;$$

$$\frac{\partial z}{\partial y} = 2b^2y$$

So that $$\frac{dy}{dx} = -\frac{\left(\frac{\partial z}{\partial x}\right)}{\left(\frac{\partial z}{\partial y}\right)} = -\frac{2a^2 x}{2b^2 y} = \frac{-a^2 x}{b^2 y}$$

From [1], we get

$$\frac{du}{dx} = 2x \cos(x^2 + y^2) + 2y \cos(x^2 + y^2) . \left(\frac{-a^2 x}{b^2 y} \right)$$

$$= 2x \cos(x^2 + y^2) \left[1 - \frac{a^2}{b^2} \right] = 2 \frac{(b^2 - a^2)}{b^2} x \cos(x^2 + y^2).$$

Special cases:

If $x = f(u, v)$, $y = \phi(u, v)$, then we have the values for $\frac{\partial u}{\partial x}, \frac{\partial u}{\partial y}, \frac{\partial v}{\partial x}, \frac{\partial v}{\partial y}$

(i) $$\frac{\partial u}{\partial x} = \frac{\left(\frac{\partial \phi}{\partial v}\right)}{\frac{\partial f}{\partial u} \cdot \frac{\partial \phi}{\partial v} - \frac{\partial f}{\partial v} \cdot \frac{\partial \phi}{\partial u}}$$

(ii) $$\frac{\partial u}{\partial y} = \frac{-\left(\frac{\partial f}{\partial v}\right)}{\frac{\partial f}{\partial u} \cdot \frac{\partial \phi}{\partial v} - \frac{\partial f}{\partial v} \cdot \frac{\partial \phi}{\partial u}}$$

(iii) $$\frac{\partial v}{\partial x} = \frac{-\left(\frac{\partial \phi}{\partial u}\right)}{\frac{\partial f}{\partial u} \cdot \frac{\partial \phi}{\partial v} - \frac{\partial f}{\partial v} \cdot \frac{\partial \phi}{\partial u}}$$ and

(iv) $$\frac{\partial v}{\partial y} = \frac{\left(\frac{\partial f}{\partial u}\right)}{\frac{\partial f}{\partial u} \cdot \frac{\partial \phi}{\partial v} - \frac{\partial f}{\partial v} \cdot \frac{\partial \phi}{\partial u}}$$

206. If $x = u^2 - v^2$ and $y = uv$, find $\frac{\partial u}{\partial x}, \frac{\partial u}{\partial y}, \frac{\partial v}{\partial x}$ and $\frac{\partial v}{\partial y}$

Solution: **Given:**

$$x = u^2 - v^2 \qquad [1]$$

$$y = uv \qquad [2]$$

$$\frac{\partial x}{\partial u} = 2u, \quad \frac{\partial y}{\partial u} = v, \quad \frac{\partial x}{\partial v} = -2v,$$

$$\frac{\partial y}{\partial v} = u$$

Diff. [1] w.r.t. x (partially), we get

$$1 = 2u \frac{\partial u}{\partial x} - 2v \frac{\partial v}{\partial x} \qquad [3]$$

Diff. [2] w.r.t. x, we get

$$0 = v \frac{\partial u}{\partial x} + u \frac{\partial v}{\partial x} \qquad [4]$$

by solving [3] and [4], we get

$$\frac{\partial u}{\partial x} = \frac{u}{2u^2 + 2v^2}, \quad \frac{\partial v}{\partial x} = -\frac{v}{2u^2 + 2v^2}$$

Diff. [1] and [2], *w.r.t.* y, we get

$$0 = 2u\frac{\partial u}{\partial y} - 2v\frac{\partial v}{\partial y} \qquad [5]$$

$$1 = v\frac{\partial u}{\partial y} + u\frac{\partial v}{\partial y} \qquad [6]$$

on solving [5] and [6], we get

$$\frac{\partial v}{\partial y} = \frac{u}{u^2+v^2}, \frac{\partial u}{\partial y} = \frac{v}{u^2+v^2}$$

EXERCISES

1. If $y^3 - 3ax^2 + x^3 = 0$ then prove that

$$\frac{d^2y}{dx^2} = -\frac{2a^2x^2}{y^5}$$

2. If $x^3 + 3x^2y + 6xy^2 + y^3 = 1$, find $\frac{dy}{dx}$ $\left(Ans: -\frac{x^2+2xy+2y^2}{x^2+4xy+y^2}\right)$

3. Find $\frac{d^2y}{dx^2}$ if $ax^2 + 2hxy + by^2 = 1$

4. If $u = x^2y + y^2z + z^2x$ and if z is defined implicitly as a function of x and y by the equation $x^2 + yz + z^3 = 0$

find $\frac{\partial u}{\partial x}$, where u is considered as a function of x and y alone.

5. Find $\frac{d^2y}{dx^2}$ if $x^5 + y^5 = 5a^3xy$

6. Find p and q, if $x = \sqrt{a}\,[\sin u + \cos v]$

$$y = \sqrt{a}\,(\cos u - \sin v),\ z = 1 + \sin(u - v).$$

2.11 JACOBIANS

Let us suppose that u and v are function of two independent variables x and y, and let the function be given by the equations (transformations).

$x = x(u, v)$ and $y = y(u, v)$, where the functions x and y have continuous first order partial derivatives.

Definition:

The Jacobian J of the function u and v *w.r.t.* x and y is defined as

$$J = \begin{vmatrix} \frac{\partial u}{\partial x} & \frac{\partial u}{\partial y} \\ \frac{\partial v}{\partial x} & \frac{\partial v}{\partial y} \end{vmatrix}$$

The Jacobian J is also denoted by $\dfrac{\partial(u,v)}{\partial(x,y)}$ · or $J(u, v)$.

If u, v and w are function of three independent variables x, y and z then the Jacobian J of the functions is defined as

$$J = \frac{\partial(u,v,w)}{\partial(x,y,z)} = \begin{vmatrix} \dfrac{\partial u}{\partial x} & \dfrac{\partial u}{\partial y} & \dfrac{\partial u}{\partial z} \\ \dfrac{\partial v}{\partial x} & \dfrac{\partial v}{\partial y} & \dfrac{\partial v}{\partial z} \\ \dfrac{\partial w}{\partial x} & \dfrac{\partial w}{\partial y} & \dfrac{\partial w}{\partial z} \end{vmatrix}$$

Also, the Jacobian of n functions of n independent variables can be defined in a similar way.

(Jacobian plays an important role in evaluating multiple integrals by transforming one coordinate system to another).

Properties

P_1: Let $x = x(u, v)$ and $y = y(u, v)$.

If $J = \dfrac{\partial(x,y)}{\partial(u,v)}$ and $J' = \dfrac{\partial(u,v)}{\partial(x,y)}$ then

$JJ^1 = 1.$

Proof: We have $x = x(u, v)$

$$y = y(u, v) \qquad [1]$$

We can also express u and v in terms of x and y, *i.e.*, $u = u(x, y)$ and $v = v(x, y)$

On differentiating [1] *w.r.t* x and y we get

$$\frac{\partial x}{\partial x} = 1; \frac{\partial y}{\partial y} = 1; \frac{\partial x}{\partial y} = 0 \text{ and } \frac{\partial y}{\partial x} = 0.$$

Now,

$$\frac{\partial x}{\partial x} = 1 \Rightarrow \frac{\partial x}{\partial u}\cdot\frac{\partial u}{x} + \frac{\partial x}{\partial v}\cdot\frac{\partial v}{\partial x} = 1$$

$$\frac{\partial x}{\partial y} = 0 \Rightarrow \frac{\partial x}{\partial u}\cdot\frac{\partial u}{\partial y} + \frac{\partial x}{\partial v}\cdot\frac{\partial v}{\partial y} = 0$$

$$\frac{\partial y}{\partial x} = 0 \Rightarrow \frac{\partial y}{\partial u}\cdot\frac{\partial u}{\partial x} + \frac{\partial y}{\partial v}\cdot\frac{\partial v}{\partial x} = 0$$

$$\frac{\partial y}{\partial y} = 1 \Rightarrow \frac{\partial y}{\partial u}\cdot\frac{\partial u}{\partial y} + \frac{\partial y}{\partial v}\cdot\frac{\partial v}{\partial y} = 1$$

Now, consider

$$JJ' = \frac{\partial(x,y)}{\partial(u,v)} \times \frac{\partial(u,v)}{\partial(x,y)} = \begin{vmatrix} \dfrac{\partial x}{\partial u} & \dfrac{\partial x}{\partial v} \\ \dfrac{\partial y}{\partial u} & \dfrac{\partial y}{\partial v} \end{vmatrix} \times \begin{vmatrix} \dfrac{\partial u}{\partial x} & \dfrac{\partial u}{\partial y} \\ \dfrac{\partial v}{\partial x} & \dfrac{\partial v}{\partial y} \end{vmatrix}$$

$$= \begin{vmatrix} \frac{\partial x}{\partial u}\cdot\frac{\partial u}{\partial x}+\frac{\partial x}{\partial v}\cdot\frac{\partial v}{\partial x} & \frac{\partial x}{\partial u}\cdot\frac{\partial u}{\partial y}+\frac{\partial x}{\partial v}\cdot\frac{\partial v}{\partial y} \\ \frac{\partial y}{\partial u}\cdot\frac{\partial u}{\partial x}+\frac{\partial y}{\partial v}\cdot\frac{\partial v}{\partial x} & \frac{\partial y}{\partial u}\cdot\frac{\partial u}{\partial y}+\frac{\partial y}{\partial v}\cdot\frac{\partial v}{\partial y} \end{vmatrix} = \begin{vmatrix} 1 & 0 \\ 0 & 1 \end{vmatrix} = 1$$

$\therefore \quad JJ' = 1$

P_2 : If u and v are the functions of r and s where r and s are functions of x and y, then

$$\frac{\partial (u,v)}{\partial (r,s)} \times \frac{\partial (r,s)}{\partial (x,y)} = \frac{\partial (u,v)}{\partial (x,y)}$$

Proof:

$$\frac{\partial (u,v)}{\partial (r,s)} \times \frac{\partial (r,s)}{\partial (x,y)} = \begin{vmatrix} \frac{\partial u}{\partial r} & \frac{\partial u}{\partial s} \\ \frac{\partial v}{\partial r} & \frac{\partial v}{\partial s} \end{vmatrix} \times \begin{vmatrix} \frac{\partial r}{\partial x} & \frac{\partial r}{\partial y} \\ \frac{\partial s}{\partial x} & \frac{\partial s}{\partial y} \end{vmatrix}$$

$$= \begin{vmatrix} \frac{\partial u}{\partial r}\cdot\frac{\partial r}{\partial x}+\frac{\partial u}{\partial s}\cdot\frac{\partial s}{\partial x} & \frac{\partial u}{\partial r}\cdot\frac{\partial r}{\partial y}+\frac{\partial u}{\partial s}\cdot\frac{\partial s}{\partial y} \\ \frac{\partial v}{\partial r}\cdot\frac{\partial r}{\partial x}+\frac{\partial v}{\partial s}\cdot\frac{\partial s}{\partial x} & \frac{\partial v}{\partial r}\cdot\frac{\partial v}{\partial y}+\frac{\partial v}{\partial s}\cdot\frac{\partial s}{\partial y} \end{vmatrix}$$

$$= \begin{vmatrix} \frac{\partial u}{\partial x} & \frac{\partial u}{\partial y} \\ \frac{\partial v}{\partial x} & \frac{\partial v}{\partial y} \end{vmatrix} = \frac{\partial (u,v)}{\partial (x,y)}.$$

P_3: (A property related to functional dependency)

If u and v are functions of two independent variables x and y, then x and y can be expressed in terms of u and v

iff $J = \frac{\partial (u,v)}{\partial (x,y)}$, $\neq 0$, identically otherwise they are dependent

Similarly if (u,v,w) are functions of 3 independent variables x, y, z, they are independent

iff $\quad \frac{\partial (u,v,w)}{\partial (x,y,z)} \neq 0$

(not indentically zero).

Worked Examples (210-240)

207. If $x = r\cos\theta$; $y = r\sin\theta$, find $\frac{\partial (x,y)}{\partial (r,\theta)}$ and $\frac{\partial (r,\theta)}{\partial (x,y)}$ and verify $\frac{\partial (x,y)}{\partial (r,\theta)}\cdot\frac{\partial (r,\theta)}{\partial (x,y)}$ 1 or JJ' = 1. [VTU, August, 1999]

Solution:

$$\left(\begin{array}{l} x = r\cos\theta;\ \dfrac{\partial x}{\partial r} = \cos\theta;\ \dfrac{\partial x}{\partial \theta} = -r\sin\theta \\ y = r\sin\theta;\ \dfrac{\partial y}{\partial x} = \sin\theta;\ \dfrac{\partial y}{\partial \theta} = r\cos\theta \end{array}\right)$$

Now $$J = \frac{\partial(x, y)}{\partial(r, \theta)} = \begin{vmatrix} \dfrac{\partial x}{\partial r} & \dfrac{\partial x}{\partial \theta} \\ \dfrac{\partial y}{\partial r} & \dfrac{\partial y}{\partial \theta} \end{vmatrix} \quad \text{(By definition)}$$

$$\therefore \quad J = \frac{\partial(x, y)}{\partial(r, \theta)} = \begin{vmatrix} \cos\theta & -r\sin\theta \\ \sin\theta & r\cos\theta \end{vmatrix} \begin{array}{l} = r\cos^2\theta - (-r\sin^2\theta) \\ = r(\cos^2\theta + \sin^2\theta) = r \end{array}$$

$$J' = \frac{\partial(r, \theta)}{\partial(x, y)} = \begin{vmatrix} \dfrac{\partial r}{\partial x} & \dfrac{\partial r}{\partial y} \\ \dfrac{\partial \theta}{\partial x} & \dfrac{\partial \theta}{\partial y} \end{vmatrix} = \begin{vmatrix} \cos\theta & \sin\theta \\ \dfrac{-\sin\theta}{r} & \dfrac{\cos\theta}{r} \end{vmatrix} = \frac{1}{r}(\cos^2\theta + \sin^2\theta) = \frac{1}{r}$$

Hence, $$\frac{\partial(x, y)}{\partial(r, \theta)} \times \frac{\partial(r, \theta)}{\partial(x, y)} = r \times \frac{1}{r} = 1.$$

208. If $x = u(1 - v),\ y = uv,$

evaluate $J = \dfrac{\partial(x, y)}{\partial(u, v)},\ J' = \dfrac{\partial(u, v)}{\partial(x, y)}$

and verify $JJ' = 1$ [VTU, August, 2000]

Solution:

$$J = \frac{\partial(x, y)}{\partial(u, v)} = \begin{vmatrix} \dfrac{\partial x}{\partial u} & \dfrac{\partial x}{\partial v} \\ \dfrac{\partial y}{\partial u} & \dfrac{\partial y}{\partial v} \end{vmatrix} = \begin{vmatrix} 1 - v & -u \\ v & u \end{vmatrix}$$

$$= u(1 - v) + uv = u.$$

Given: $x = u(1 - v)$

$y = uv$

Adding $x + y = u$

or $u = x + y$, But $v = \dfrac{y}{u}$ $\therefore$ $v = \dfrac{y}{x + y}$ [1]

Differentiating [1] partially w.r.t x and y

separately, $\therefore$ $\dfrac{\partial u}{\partial x} = 1$ and $\dfrac{\partial u}{\partial y} = 1.$

$$\frac{\partial v}{\partial x} = \frac{-y}{(x+y)^2} = \frac{-y}{u^2};\ \frac{\partial v}{\partial y} = \frac{x+y-y}{(x+y)^2} = \frac{x}{u^2}$$

Again,
$$J' = \frac{\partial(u,v)}{\partial(x,y)} = \begin{vmatrix} \frac{\partial u}{\partial x} & \frac{\partial u}{\partial y} \\ \frac{\partial v}{\partial x} & \frac{\partial v}{\partial y} \end{vmatrix} = \begin{vmatrix} 1 & 1 \\ \frac{-y}{u^2} & \frac{x}{u^2} \end{vmatrix}$$

or
$$J' = \frac{x}{u^2} + \frac{y}{u^2} = \frac{x+y}{u^2} = \frac{u}{u^2} = \frac{1}{u}$$

Hence
$$JJ' = u \times \frac{1}{u} = 1$$

209. If $x = e^v \sec u,\ y = e^v \tan u$, find $J = \frac{\partial(x,y)}{\partial(u,v)}$

$J' = \frac{\partial(u,v)}{\partial(x,y)}$ and verify $JJ' = 1$. [B.U. July, 1993]

Solution:

$$\left(\begin{array}{l} x = e^v \sec u;\ \frac{\partial x}{\partial u} = e^v \sec u \tan u;\ \frac{\partial y}{\partial u} = e^v \sec^2 u \\ y = e^v \tan u;\ \frac{\partial x}{\partial v} = e^v \sec u;\ \frac{\partial y}{\partial v} = e^v \tan u \end{array} \right) \quad [1]$$

Hence,
$$\frac{\partial(x,y)}{\partial(u,v)} = \begin{vmatrix} \frac{\partial x}{\partial u} & \frac{\partial x}{\partial v} \\ \frac{\partial y}{\partial u} & \frac{\partial y}{\partial v} \end{vmatrix} = \begin{vmatrix} e^v \sec u \tan u & e^v \sec u \\ e^v \sec^2 u & e^v \tan u \end{vmatrix}$$

$$= e^{2v} \sec u\ (\tan^2 u - \sec^2 u) = -e^{2v} \sec u \quad [2]$$

From [1],
$$\frac{y}{x} = \frac{\tan u}{\sec u} = \sin u \Rightarrow \quad u = \sin^{-1}\left(\frac{y}{x}\right)$$

$$(x^2 - y^2) = e^{2v} (\sec^2 u - \tan^2 u) = e^{2v} \Rightarrow V = \frac{1}{2} \log (x^2 - y^2)$$

Hence
$$u = \sin^{-1}\left(\frac{y}{x}\right);\ V = \frac{1}{2} \log (x^2 - y^2) \quad [3]$$

$\therefore$
$$\frac{\partial u}{\partial x} = \frac{1}{\sqrt{1 - \frac{y^2}{x^2}}}\left(\frac{-y}{x}\right) = \frac{-y}{x\sqrt{x^2 - y^2}}$$

$$\frac{\partial u}{\partial y} = \frac{1}{\sqrt{1-\frac{y^2}{x^2}}}\left(\frac{1}{x}\right) = \frac{1}{\sqrt{x^2-y^2}}$$

$$\frac{\partial V}{\partial x} = \frac{x}{(x^2-y^2)}; \frac{\partial V}{\partial y} = \frac{-y}{(x^2-y^2)}$$

$$\therefore \quad J' = \frac{\partial(u,v)}{\partial(x,y)} = \begin{vmatrix} \frac{\partial u}{\partial x} & \frac{\partial u}{\partial y} \\ \frac{\partial v}{\partial x} & \frac{\partial v}{\partial y} \end{vmatrix} = \begin{vmatrix} \frac{-y}{x\sqrt{x^2-y^2}} & \frac{1}{\sqrt{x^2-y^2}} \\ \frac{x}{x^2-y^2} & \frac{-y}{x^2-y^2} \end{vmatrix}$$

$$= \frac{1}{(x^2-y^2)\sqrt{x^2-y^2}}\left(\frac{y^2}{x} - x\right) = \frac{y^2-x^2}{x(x^2-y^2)\left(\sqrt{x^2-y^2}\right)}$$

$$= -\frac{1}{x\sqrt{(x^2-y^2)}} = -\frac{1}{e^v \sec u \cdot e^v} = -\frac{1}{e^{2v} \cdot \sec u} \quad [4]$$

$$\therefore \quad JJ' = -e^{2v} \sec u \left(-\frac{1}{e^{2v} \sec u}\right) = 1$$

Hence verified

210. **If $x + y + z = u, y + z = v, z = uvw$**

evaluate $\frac{\partial(x,y,z)}{\partial(u,v,w)}$ **[VTU, March, 2000]**

Solution: **First we shall express x, y, z in terms of u, v, w**

$$x + v = u \;\Rightarrow\; x = u - v; \; y + uvw = v$$

or $$y = v - uvw$$

$$\therefore \quad (x = u - v; \; y = v - uvw; \; z = uvw)$$

$$\frac{\partial x}{\partial u} = 1; \; \frac{\partial x}{\partial v} = -1; \; \frac{\partial x}{\partial w} = 0;$$

$$\frac{\partial y}{\partial u} = -vw; \; \frac{\partial y}{\partial v} = 1 - uw; \; \frac{\partial y}{\partial w} = -uv.$$

$$\frac{\partial z}{\partial u} = vw; \; \frac{\partial z}{\partial v} = uw; \; \frac{\partial z}{\partial w} = uv;$$

PART-B

Hence $$J = \frac{\partial(x,y,z)}{\partial(u,v,w)} = \begin{vmatrix} 1 & -1 & 0 \\ -vw & 1-uw & -uv \\ vw & uw & uv \end{vmatrix} = uv$$

211. If $x = r \sin\theta \cos\phi$; $y = r \sin\theta \sin\phi$ and $z = r\cos\theta$;

Show that $J\left(\dfrac{x, y, z}{r, \theta, \phi}\right) = r^2 \sin\theta$ [B.U. Feb., 1994]

Solution:
$$\frac{\partial x}{\partial r} = \sin\theta\cos\phi;\ \frac{\partial y}{\partial r} = \sin\theta\sin\phi;\ \frac{\partial z}{\partial r} = \cos\theta$$

$$\frac{\partial x}{\partial \theta} = r\cos\theta\cos\phi;\ \frac{\partial y}{\partial \theta} = r\cos\theta\sin\phi;\ \frac{\partial z}{\partial \theta} = -r\sin\theta$$

$$\frac{\partial x}{\partial \phi} = -r\sin\theta\sin\phi;\ \frac{\partial y}{\partial \phi} = r\sin\theta\cos\phi;\ \frac{\partial z}{\partial \phi} = 0.$$

Now, $$\frac{\partial(x,y,z)}{\partial(r,\theta,\phi)} = \begin{vmatrix} \sin\theta\cos\phi & r\cos\theta\cos\phi & -r\sin\theta\sin\phi \\ \sin\theta\sin\phi & r\cos\theta\sin\phi & r\sin\theta\cos\phi \\ \cos\theta & -r\sin\theta & 0 \end{vmatrix}$$

$$= (\sin\theta\cos\phi)(r^2\sin^2\theta\cos\phi) + (r\cos\theta\cos\phi)(r\sin\theta\cos\theta\cos\phi)$$
$$+ (r\sin\theta\sin\phi)(r\sin^2\theta\sin\phi + r\cos^2\theta\sin\phi)$$
$$= r^2\sin\theta\cos^2\phi(\sin^2\theta + \cos^2\theta) + r^2\sin\theta\sin^2\phi(\sin^2\theta + \cos^2\theta)$$
$$= r^2\sin\theta\cos^2\phi + r^2\sin\theta\sin^2\phi$$
$$= r^2\sin\theta(\cos^2\phi + \sin^2\theta) = r^2\sin\theta$$

212. If $u = x + 3y^2 - z^3$; $v = 4x^2yz$; $w = 2z^2 - xy$

evaluate $\dfrac{\partial(u,v,w)}{\partial(x,y,z)}$ at $(1, -1, 0)$ [VTU, August, 2001]

Solution:
$$\left[\frac{\partial v}{\partial x}\right]_{(1,-1,0)} = 1;\ \left[\frac{\partial u}{\partial y}\right]_{(1,-1,0)} = 6y = -6,\ \left[\frac{\partial u}{\partial z}\right]_{(1,-1,0)} = 0 = -3z^2$$

$$\left[\frac{\partial v}{\partial x}\right]_{(1,-1,0)} = 8xyz = 0;\ \left[\frac{\partial v}{\partial y}\right]_{(1,-1,0)} = 4x^2z = 0;\ \left[\frac{\partial v}{\partial z}\right]_{(1,-1,0)}$$
$$= 4xy^2 = -4$$

$$\left.\frac{\partial w}{\partial x}\right]_{(1,-1,0)} = -y = +1;\ \left.\frac{\partial w}{\partial y}\right]_{(1,-1,0)} = -x = -1;\ \left[\frac{\partial w}{\partial z}\right]_{(1,-1,0)} = 4z = 0$$

Hence at (1, –1, 0)

$$\frac{\partial(u,v,w)}{\partial(x,y,z)} = \begin{vmatrix} \frac{\partial u}{\partial x} & \frac{\partial u}{\partial y} & \frac{\partial u}{\partial z} \\ \frac{\partial v}{\partial x} & \frac{\partial v}{\partial y} & \frac{\partial v}{\partial z} \\ \frac{\partial w}{\partial x} & \frac{\partial w}{\partial y} & \frac{\partial w}{\partial z} \end{vmatrix} = \begin{vmatrix} 1 & -6 & 0 \\ 0 & 0 & -4 \\ 1 & -1 & 0 \end{vmatrix}$$

$$= 1\,(0 - 4) + 6\,(0 + 4) + 0 = 20.$$

213. Evaluate $\frac{\partial(u,v,w)}{\partial(x,y,z)}$, where

$$u = x^2 + y^2 + z^2,\ V = x + y + z,\ w = xy + yz + zx$$

[VTU, March, 2001]

Solution:

$$u = x^2 + y^2 + z^2;\ v = x + y + z;\ w = xy + yz + zx$$

$$\Rightarrow \quad \frac{\partial u}{\partial x} = 2x;\ \frac{\partial v}{\partial x} = 1;\ \frac{\partial w}{\partial x} = y + z$$

$$\frac{\partial u}{\partial y} = 2y;\ \frac{\partial v}{\partial y} = 1;\ \frac{\partial w}{\partial y} = x + z$$

$$\therefore \quad \frac{\partial u}{\partial z} = 2z;\ \frac{\partial v}{\partial z} = 1;\ \frac{\partial w}{\partial z} = x + y$$

$$\frac{\partial(u,v,w)}{\partial(x,y,z)} = \begin{vmatrix} \frac{\partial u}{\partial x} & \frac{\partial u}{\partial y} & \frac{\partial u}{\partial z} \\ \frac{\partial v}{\partial x} & \frac{\partial v}{\partial y} & \frac{\partial v}{\partial z} \\ \frac{\partial w}{\partial x} & \frac{\partial w}{\partial y} & \frac{\partial w}{\partial z} \end{vmatrix}$$

$$\therefore \quad J = \frac{\partial(u,v,w)}{\partial(x,y,z)} = \begin{vmatrix} 2x & 2y & 2z \\ 1 & 1 & 1 \\ y+z & z+x & x+y \end{vmatrix}$$

Applying

$(R_1 \to R_1 + 2\,R_3)$

$$\Rightarrow \quad J = \begin{vmatrix} 2(x+y+z) & 2(y+z+x) & 2(z+x+y) \\ 1 & 1 & 1 \\ y+z & z+x & x+y \end{vmatrix} = 0$$

($\because$ Elements of 1st and 2nd rows are proportional)

PART-B

Hence $\dfrac{\partial(u,v,w)}{\partial(x,y,z)} = 0$

$\Rightarrow$ u, v and w are functionally dependent.

214. If $u = x^2 - y^2$; $v = 2xy$ and $x = r\cos\theta$; $y = r\sin\theta$ determine the value of the Jacobian $\dfrac{\partial(u,v)}{\partial(r,\theta)}$. [VTU, March, 1999]

Solution:

We know that

$$\frac{\partial(u,v)}{\partial(r,\theta)} = \frac{\partial(u,v)}{\partial(x,y)} \times \frac{\partial(x,y)}{\partial(r,\theta)}$$

$$= \begin{vmatrix} \dfrac{\partial u}{\partial x} & \dfrac{\partial u}{\partial y} \\ \dfrac{\partial v}{\partial x} & \dfrac{\partial v}{\partial y} \end{vmatrix} \times \begin{vmatrix} \dfrac{\partial x}{\partial r} & \dfrac{\partial x}{\partial \theta} \\ \dfrac{\partial y}{\partial r} & \dfrac{\partial y}{\partial \theta} \end{vmatrix} = \begin{vmatrix} 2x & -2y \\ 2y & 2x \end{vmatrix} \times \begin{vmatrix} \cos\theta & -r\sin\theta \\ \sin\theta & r\cos\theta \end{vmatrix}$$

$$= (4x^2 + 4y^2)\,(r\cos^2\theta + r\sin^2\theta)$$

$$= 4\,(x^2 + y^2)\,r\,(\cos^2\theta + \sin^2\theta)$$

$$= 4\,r^2 \cdot r = 4r^3. \qquad (\because\ x^2 + y^2 = r^2)$$

215. If $x = u\,(1 + v)$ and $y = v\,(1 + u)$,

show that $\dfrac{\partial(x,y)}{\partial(u,v)} = 1 + u + v$

Solution: **Given:**

$$\left(\begin{array}{ll} x = u + uv; & y = v + uv \\ \dfrac{\partial x}{\partial u} = 1 + v; & \dfrac{\partial x}{\partial v} = u \\ \dfrac{\partial y}{\partial u} = v; & \dfrac{\partial y}{\partial v} = u + 1 \end{array}\right)$$

$$\therefore \quad \frac{\partial(x,y)}{\partial(u,v)} = \begin{vmatrix} \dfrac{\partial x}{\partial u} & \dfrac{\partial x}{\partial v} \\ \dfrac{\partial y}{\partial u} & \dfrac{\partial y}{\partial v} \end{vmatrix} = \begin{vmatrix} 1+v & u \\ v & 1+u \end{vmatrix}$$

$$= (1 + v)\,(1 + u) - uv = 1 + v + u + uv - uv$$

$$= 1 + u + v.$$

216. If $u = \dfrac{y^2}{2x}$; $v = \dfrac{x^2 + y^2}{2x}$; compute $\dfrac{\partial(u,v)}{\partial(x,y)}$

Solution: $u = \dfrac{y^2}{2x}$; $\therefore \dfrac{\partial u}{\partial x} = -\dfrac{y^2}{2x^2}$

$$\frac{\partial u}{\partial y} = \frac{2y}{2x} = \frac{y}{x}$$

$$v = \frac{x^2 + y^2}{2x}; \quad \therefore \frac{\partial v}{\partial x} = \frac{1}{2} - \frac{y^2}{2x^2}, \frac{\partial v}{\partial y} = \frac{2y}{2x} = \frac{y}{x}$$

Now
$$\frac{\partial(u,v)}{\partial(x,y)} = \begin{vmatrix} \dfrac{\partial u}{\partial x} & \dfrac{\partial u}{\partial y} \\ \dfrac{\partial v}{\partial x} & \dfrac{\partial v}{\partial y} \end{vmatrix} = \begin{vmatrix} \dfrac{-y^2}{2x} & \dfrac{y}{x} \\ \dfrac{1}{2} - \dfrac{y^2}{2x} & \dfrac{y}{x} \end{vmatrix}$$

$$= -\frac{y^2}{2x^2} - \frac{y}{2x} + \frac{y^3}{2x^2} = -\frac{y}{2x}$$

217. If $x = a^u \cos v$ and $y = a^u \sin v$ show that $JJ' = 1$. [VTU, F/M, 2005]

Solution: **Given:** $x = a^u \cos v$; $y = a^u \sin v$

$$\frac{\partial x}{\partial u} = a^u \log a \cos v = x \log a, \quad \frac{\partial y}{\partial u} = a^u \log a \sin v = y \log a$$

$$\frac{\partial x}{\partial v} = -a^u \sin v = -y, \quad \frac{\partial y}{\partial v} = a^u \cos v = x$$

$$J = \frac{\partial(x,y)}{\partial(u,v)} = \begin{vmatrix} \dfrac{\partial x}{\partial u} & \dfrac{\partial x}{\partial v} \\ \dfrac{\partial y}{\partial u} & \dfrac{\partial y}{\partial v} \end{vmatrix} = \begin{vmatrix} x \log a & -y \\ y \log a & x \end{vmatrix} = \log a\,[x^2 + y^2]$$

Since $x^2 + y^2 = (a^u)^2 = a^{2u} \Rightarrow \log(x^2 + y^2) = 2u \log a$

$\therefore$
$$u = \frac{\log(x^2 + y^2)}{2 \log a} \quad \text{and} \quad \frac{y}{x} = \tan V, \text{ or } V = \tan^{-1}(y/x)$$

Hence,
$$\frac{\partial u}{\partial x} = \frac{x}{\log a\,[(x^2 + y^2)]}; \frac{\partial v}{\partial y} = \frac{y}{\log a\,[(x^2 + y^2)]}$$

and
$$\frac{\partial v}{\partial x} = \frac{-y}{x^2 + y^2}, \frac{\partial v}{\partial y} = \frac{x}{x^2 + y^2}$$

$\therefore$
$$J' = \frac{\partial(v,v)}{\partial(x,y)} = \begin{vmatrix} \dfrac{\partial u}{\partial x} & \dfrac{\partial u}{\partial y} \\ \dfrac{\partial v}{\partial x} & \dfrac{\partial v}{\partial y} \end{vmatrix} = \begin{vmatrix} \dfrac{x}{\log a\,(x^2 + y^2)} & \dfrac{y}{\log a\,(x^2 + y^2)} \\ \dfrac{-y}{x^2 + y^2} & \dfrac{x}{x^2 + y^2} \end{vmatrix}$$

$$= \frac{1}{\log a\,(x^2+y^2)^2}\begin{vmatrix} x & y \\ -y & x \end{vmatrix} = \frac{x^2+y^2}{(\log a)(x^2+y^2)^2}$$

or, $$J' = \frac{1}{\log a\,(x^2+y^2)}\ ; \quad \therefore \quad JJ' = \frac{\log a\,(x^2+y^2)}{\log a\,(x^2+y^2)} = 1$$

$\Rightarrow$ $JJ' = 1$ is verified.

218. If $x = u\cos v$, $y = u\sin v$

prove that $\dfrac{\partial(x,y)}{\partial(u,v)} \times \dfrac{\partial(u,v)}{\partial(x,y)} = 1$

Solution:

Since $$x = u\cos v \qquad [1]$$
$$y = u\sin v \qquad [2]$$

$$\therefore \quad \frac{\partial(x,y)}{\partial(u,v)} = \begin{vmatrix} \dfrac{\partial x}{\partial u} & \dfrac{\partial x}{\partial v} \\ \dfrac{\partial y}{\partial u} & \dfrac{\partial y}{\partial v} \end{vmatrix} = \begin{vmatrix} \cos v & -u\sin v \\ \sin v & u\cos v \end{vmatrix} = u\cos^2 v + u\sin^2 v$$

$$= u(\cos^2 v + \sin^2 v) = u.$$

We now find from [1] and [2]

that $$u = \sqrt{x^2+y^2}\ ; \ v = \tan^{-1}\left(\frac{y}{x}\right)$$

$$\frac{\partial(u,v)}{\partial(x,y)} = \begin{vmatrix} \dfrac{\partial u}{\partial x} & \dfrac{\partial u}{\partial y} \\ \dfrac{\partial v}{\partial x} & \dfrac{\partial v}{\partial y} \end{vmatrix} = \begin{vmatrix} \dfrac{x}{\sqrt{x^2+y^2}} & \dfrac{y}{\sqrt{x^2+y^2}} \\ \dfrac{-y}{(x^2+y^2)} & \dfrac{x}{(x^2+y^2)} \end{vmatrix}$$

$$= \frac{(x^2+y^2)}{(x^2+y^2)^{3/2}} = \frac{1}{\sqrt{(x^2+y^2}} = \frac{1}{u}$$

$$\therefore \quad \frac{\partial(x,y)}{\partial(u,v)} \times \frac{\partial(u,v)}{\partial(x,y)} = u\left(\frac{1}{u}\right) = 1.$$

219. If $$u = x + y + z \qquad [1]$$
$$uv = y + z \qquad [2]$$
$$uvw = z \qquad [3]$$

show that $\dfrac{\partial(x,y,z)}{\partial(u,v,w)} = u^2 v.$ [VTU, MQP-2]

Solution: We shall express x, y, z in terms of u, v, w

From [2], we get $y = uv - z$

$\Rightarrow$ $y = uv - uvw = uv\,(1 - w)$

From [1], we get $u = x + (y + z)$

$\Rightarrow$ $u = x + uv$

$\Rightarrow$ $x = u\,(1 - v)$

From [3], we get $z = uvw$

$$\therefore \quad \frac{\partial\,(x, y, z)}{\partial\,(u, v, w)} = \begin{vmatrix} \frac{\partial x}{\partial u} & \frac{\partial x}{\partial v} & \frac{\partial x}{\partial w} \\ \frac{\partial y}{\partial u} & \frac{\partial y}{\partial v} & \frac{\partial y}{\partial w} \\ \frac{\partial z}{\partial u} & \frac{\partial z}{\partial v} & \frac{\partial z}{\partial w} \end{vmatrix} = \begin{vmatrix} 1-v & -u & 0 \\ v(1-w) & u(1-w) & -uv \\ vw & uw & uv \end{vmatrix}$$

$$= (1 - v)\,[u^2v\,(1 - w) + u^2vw] + u\,[uv^2\,(1 - w) + uv^2w]$$

$$= u^2v - u^2v^2 + u^2v^2 = u^2v.$$

220. If $x = r\cos\theta$; $y = r\sin\theta$; then

$$\text{P.T.} \quad \frac{\partial(x, y)}{\partial(r, \theta)} = r.$$

Solution: **Given:** $\begin{pmatrix} x = r\cos\theta; \\ y = r\sin\theta; \end{pmatrix}$

$$\therefore \quad \frac{\partial(x, y)}{\partial(r, \theta)} = \begin{vmatrix} \frac{\partial x}{\partial r} & \frac{\partial x}{\partial \theta} \\ \frac{\partial y}{\partial r} & \frac{\partial y}{\partial \theta} \end{vmatrix} = \begin{vmatrix} \cos\theta & -r\sin\theta \\ \sin\theta & r\cos\theta \end{vmatrix}$$

$$= r\cos^2\theta + r\sin^2\theta = r\,(1) = r$$

221. If $u = \dfrac{y^2}{x}$; $v = \dfrac{x^2}{y}$ find $\dfrac{\partial(x, y)}{\partial(u, v)}$

Solution:

$$\text{Now,} \quad \frac{\partial(u, v)}{\partial(x, y)} = \begin{vmatrix} \frac{\partial u}{\partial x} & \frac{\partial u}{\partial y} \\ \frac{\partial v}{\partial x} & \frac{\partial v}{\partial y} \end{vmatrix} = \begin{vmatrix} \frac{-y^2}{x^2} & \frac{2y}{x} \\ \frac{2x}{y} & \frac{-x^2}{y^2} \end{vmatrix} = -3$$

$$\text{Hence} \quad \frac{\partial\,(x, y)}{\partial\,(u, v)} = -\frac{1}{3}$$

PART-B

222. If $x + y = u;\ 2x - 3y = v$

Find $\dfrac{\partial(x, y)}{\partial(u, v)}$

Solution: We shall express u, v in terms of x, y

$$\therefore \quad x = \frac{1}{5}(3u + v);$$

$$y = \frac{1}{5}(2u - v).$$

$$\therefore \quad \frac{\partial(x, y)}{\partial(u, v)} = \begin{vmatrix} \frac{3}{5} & \frac{1}{5} \\ \frac{2}{5} & \frac{-1}{5} \end{vmatrix} = -\frac{1}{5}$$

Note: Since, here $u = u(x, y)$ and $v = v(x, y)$. The Jocobian J' of u and v w.r.t x and y can be written as

$$J' = \frac{\partial(u, v)}{\partial(x, y)} = \begin{vmatrix} \frac{\partial u}{\partial x} & \frac{\partial u}{\partial y} \\ \frac{\partial v}{\partial x} & \frac{\partial v}{\partial y} \end{vmatrix} = \begin{vmatrix} 1 & 1 \\ 2 & -3 \end{vmatrix} = -5$$

223. If $x = a \cos h\xi \cos \eta$

$y = a \sin h\xi \sin \eta$

Show that $\dfrac{\partial(x, y)}{\partial(\xi, \eta)} = \dfrac{1}{2} a^2 [\cos 2\xi - \cos 2\eta]$

Solution: We know that

$$\sin h^2 x = \frac{\cos h\, 2x - 1}{2}; \quad \cos h^2 x = \frac{\cos h\, 2x + 1}{2}$$

and $$\cos 2\theta = \cos^2\theta - \sin^2\theta$$

Now,

Given: $x = a \cos h\xi \cos \eta$

$y = a \sin h\xi \sin \eta$

$$\therefore \quad \frac{\partial(x, y)}{\partial(\xi, \eta)} = \begin{vmatrix} \frac{\partial x}{\partial \xi} & \frac{\partial x}{\partial \eta} \\ \frac{\partial y}{\partial \xi} & \frac{\partial y}{\partial \eta} \end{vmatrix} = \begin{vmatrix} a \sin h\xi \cos \eta & -a \cos h\xi \sin \eta \\ a \sin h\xi \cos \eta & a \sin h\xi \cos \eta \end{vmatrix}$$

$$= a^2 [\sin h^2\xi \cos^2 \eta + \cos h^2\xi \sin^2 \eta]$$

$$= \frac{1}{2} a^2 [\{(\cos h\, 2\xi - 1) \cos^2\eta\} + (\cos h\, 2\xi + 1) \sin^2\eta)\}]$$

$$= \frac{1}{2} a^2 [\cos h\ 2\xi - \cos 2\eta] \qquad (\because \cos 2\eta = (\cos^2\eta - \sin^2\eta)$$

224. If $u = \dfrac{yz}{x}$; $v = \dfrac{zx}{y}$; $w = \dfrac{xy}{z}$

prove that $\dfrac{\partial(u, v, w)}{\partial(x, y, z)} = 4.$

Solution:

$$\frac{\partial(u,v,w)}{\partial(x,y,z)} = \begin{vmatrix} \frac{\partial u}{\partial x} & \frac{\partial u}{\partial y} & \frac{\partial u}{\partial z} \\ \frac{\partial v}{\partial x} & \frac{\partial v}{\partial y} & \frac{\partial v}{\partial z} \\ \frac{\partial w}{\partial x} & \frac{\partial w}{\partial y} & \frac{\partial w}{\partial z} \end{vmatrix} = \begin{vmatrix} -\frac{yz}{x^2} & \frac{z}{x} & \frac{y}{x} \\ \frac{z}{y} & -\frac{zx}{y^2} & \frac{x}{y} \\ \frac{y}{z} & \frac{x}{z} & -\frac{xy}{z^2} \end{vmatrix}$$

$$= \frac{1}{xyz} \begin{vmatrix} -\frac{yz}{x} & z & y \\ z & -\frac{zx}{y} & x \\ y & x & -\frac{xy}{z} \end{vmatrix}$$

$$= \frac{1}{xyz} \left[-\frac{yz}{x}(x^2 - x^2) - z(-xy - xy) + y(xz + xz) \right]$$

$$= \frac{1}{xyz} [0 + 2xyz + 2xyz] = 4.$$

225. If $x = \dfrac{1}{2}(u + v)$; $y = \dfrac{1}{2}(u - v)$

evaluate $\dfrac{\partial(x, y)}{\partial(u, v)}$

Solution:

$$\begin{pmatrix} x = \frac{1}{2}(u+v)\,; \frac{\partial x}{\partial u} = \frac{1}{2}\,; \frac{\partial x}{\partial v} = \frac{1}{2} \\ y = \frac{1}{2}(u-v)\,; \frac{\partial y}{\partial u} = \frac{1}{2} \text{ and } \frac{\partial y}{\partial v} = -\frac{1}{2} \end{pmatrix}$$

$$\therefore \qquad J = \frac{\partial(x,y)}{\partial(u,v)} = \begin{vmatrix} \frac{\partial x}{\partial u} & \frac{\partial x}{\partial v} \\ \frac{\partial y}{\partial u} & \frac{\partial y}{\partial v} \end{vmatrix} = \begin{vmatrix} \frac{1}{2} & \frac{1}{2} \\ \frac{1}{2} & -\frac{1}{2} \end{vmatrix} = -\frac{1}{2}$$

PART-B

226. If $x = \cos u;\ y = \cos v \sin u;$
$z = \cos w \sin v \sin u$

Show that $\dfrac{\partial(x, y, z)}{\partial(u, v, w)} = -\sin^3 u \sin^2 v \sin w.$

Solution:

$$x = \cos u;\ \frac{\partial x}{\partial u} = -\sin u;\ \frac{\partial x}{\partial v} = 0;\ \frac{\partial x}{\partial w} = 0$$

$$y = \cos v \sin u;\ \frac{\partial y}{\partial u} = \cos v \cos u;\ \frac{\partial y}{\partial v} = -\sin v \sin u;\ \frac{\partial y}{\partial w} = 0$$

Again; $$z = \cos w \sin v \sin u$$

$$\frac{\partial z}{\partial u} = \cos w \sin v \cos u;\ \frac{\partial z}{\partial v} = \cos w \cos v \sin u$$

$$\frac{\partial z}{\partial w} = -\sin w \sin v \sin u$$

$$\frac{\partial(x, y, z)}{\partial(u, v, w)} = \begin{vmatrix} -\sin u & 0 & 0 \\ \cos v \cos u & -\sin v \sin u & 0 \\ \cos w \sin v \cos u & \cos w \cos v \sin u & -\sin w \sin v \sin u \end{vmatrix}$$

$$= -\sin^3 u \sin^2 v \sin w.$$

227. If $F = xu + v - y;$
$G = u^2 + vy + w;$
$H = zu - v + vw$

find $\dfrac{\partial(F, G, H)}{\partial(u, w, v)}$

Solution:

$$\frac{\partial F}{\partial u} = x;\ \frac{\partial F}{\partial w} = 0;\ \frac{\partial F}{\partial v} = 1$$

$$\frac{\partial G}{\partial u} = 2u;\ \frac{\partial G}{\partial w} = 1;\ \frac{\partial G}{\partial v} = y$$

$$\frac{\partial H}{\partial u} = z;\ \frac{\partial H}{\partial w} = v;\ \frac{\partial H}{\partial v} = -1 + w$$

Now $$\frac{\partial(F, G, H)}{\partial(u, w, v)} = \begin{vmatrix} \frac{\partial F}{\partial u} & \frac{\partial F}{\partial w} & \frac{\partial F}{\partial v} \\ \frac{\partial G}{\partial u} & \frac{\partial G}{\partial w} & \frac{\partial G}{\partial v} \\ \frac{\partial H}{\partial u} & \frac{\partial H}{\partial w} & \frac{\partial H}{\partial v} \end{vmatrix} = \begin{vmatrix} x & 0 & 1 \\ 2u & 1 & y \\ z & v & u-1 \end{vmatrix} = x\begin{vmatrix} 1 & y \\ v & u-1 \end{vmatrix} + 1\begin{vmatrix} 2u & 1 \\ z & v \end{vmatrix}$$

$$= x(u - 1 - vy) + 2uv - z$$

228. If u and v are functions of x and y and $u = f(v)$, show that

$$\begin{vmatrix} \frac{\partial u}{\partial x} & \frac{\partial v}{\partial x} \\ \frac{\partial u}{\partial y} & \frac{\partial v}{\partial y} \end{vmatrix} = 0$$

Verify this result in the case of $u = \cos(x^2 + 2y)$; $v = x^4 + 4x^2y + 4y^2$

Solution: $u = f(v)$; $\therefore \quad \frac{\partial u}{\partial x} = f'(v)\frac{\partial v}{\partial x}$ [1]

and $\quad f'(v)\frac{\partial v}{\partial y} = \frac{\partial u}{\partial y}$ [2]

Multiplying [1] and [2], we get

$$f'(v)\frac{\partial u}{\partial x} \times \frac{\partial v}{\partial y} = f'(v) \cdot \frac{\partial v}{\partial x} \times \frac{\partial u}{\partial y}$$

or

$$\frac{\partial u}{\partial x} \cdot \frac{\partial v}{\partial y} - \frac{\partial v}{\partial x}\frac{\partial u}{\partial y} = 0$$

$$\begin{vmatrix} \frac{\partial u}{\partial x} & \frac{\partial v}{\partial x} \\ \frac{\partial v}{\partial y} & \frac{\partial v}{\partial y} \end{vmatrix} = 0$$

Now,

$$\frac{\partial u}{\partial x} = -2x\sin(x^2 + 2y),$$

$$\frac{\partial u}{\partial y} = -2\sin(x^2 + y^2)$$

$$\frac{\partial v}{\partial x} = 4x^2 + 8xy;$$

$$\frac{\partial v}{\partial y} = 4x^2 + 8y$$

We find that $\begin{vmatrix} -2x\sin(x^2+2y) & 4x^3+8xy \\ -2\sin(x^2+2y) & 4x^2+8y \end{vmatrix} = 0$

Here $\quad u = \cos(x^2 + 2y) = \cos\sqrt{v}$

229. If $u = xyz$; $v = (xy + yz + zx)$; $w = x + y + z$ compute the Jacobian.

Solution:

$$J = \frac{\partial(u, v, w)}{\partial(x, y, z)} = \begin{vmatrix} \frac{\partial u}{\partial x} & \frac{\partial u}{\partial y} & \frac{\partial u}{\partial z} \\ \frac{\partial v}{\partial x} & \frac{\partial v}{\partial y} & \frac{\partial v}{\partial z} \\ \frac{\partial w}{\partial x} & \frac{\partial w}{\partial y} & \frac{\partial w}{\partial z} \end{vmatrix} = \begin{vmatrix} yz & xz & xy \\ y+z & x+z & y+x \\ 1 & 1 & 1 \end{vmatrix}$$

Applying

$$\xrightarrow[C_2 \to C_2 - C_3]{C_1 \to C_1 - C_3} \begin{vmatrix} (yz - xy) & (xz - xy) & xy \\ (z - x) & (z - y) & x + y \\ 0 & 0 & 1 \end{vmatrix}$$

$$= (z - x)\,((z - y) \begin{vmatrix} y & x & xy \\ 1 & 1 & x + y \\ 0 & 0 & 1 \end{vmatrix} \begin{matrix} = (z - x)(z - y)(y - x) \\ = (x - y)(y - z)(z - x) \end{matrix}$$

230. If $x = \sin\theta\sqrt{1 - c^2 \sin^2\phi}$

$y = \cos\theta \cos\phi$, show that

$$\frac{\partial(x, y)}{\partial(\theta, \phi)} = -\sin\phi \frac{(1 - c^2)\cos^2\theta + c^2 + \cos^2\phi}{\sqrt{1 - c^2 \sin^2\phi}}$$

Solution:

$$\frac{\partial x}{\partial\theta} = \cos\theta\sqrt{(1 - c^2 \sin^2\phi)}$$

$$\frac{\partial y}{\partial\theta} = \cos\phi\,(-\sin\theta) = -\sin\theta\cos\phi$$

$$\frac{\partial x}{\partial\phi} = \frac{1}{2} \cdot \frac{\sin\theta\,(-2c^2 \sin\phi\cos\phi)}{\sqrt{(1 - c^2 \sin^2\phi)}}$$

$$\frac{\partial y}{\partial\phi} = \cos\theta\,(-\sin\phi) = -\cos\theta\sin\phi$$

Now

$$\frac{\partial(x, y)}{\partial(\theta, \phi)} = \begin{vmatrix} \dfrac{\partial x}{\partial\theta} & \dfrac{\partial x}{\partial\phi} \\ \dfrac{\partial y}{\partial\theta} & \dfrac{\partial y}{\partial\phi} \end{vmatrix} = \begin{vmatrix} \cos\theta\sqrt{(1 - c^2 \sin^2\phi)} & \dfrac{1}{2} \cdot \dfrac{\sin\theta\,(-2c^2 \sin\phi\cos\phi}{\sqrt{(1 - c^2 \sin\phi)}} \\ -\sin\theta\cos\phi & -\cos\theta\sin\phi \end{vmatrix}$$

$$= -\sin\phi\cos^2\theta\,\sqrt{(1 - c^2 \sin^2\phi)} - \frac{c^2 \sin^2\theta\sin\phi\cos^2\phi}{\sqrt{(1 - c^2 \sin^2\phi)}}$$

$$= -\frac{\sin\phi\cos^2\theta\,(1 - c^2 \sin^2\phi) + c^2 \sin^2\theta\sin\phi\cos^2\phi}{\sqrt{(1 - c^2 \sin^2\phi)}}$$

$$= -\frac{\sin\phi\,[\cos^2\theta\,\{1 - c^2\,(1 - \cos^2\phi)\} + c^2\,(1 - \cos^2\theta)\cos^2\phi]}{\sqrt{(1 - c^2 \sin^2\phi)}}$$

231. If $x = u(1-v)$;
$y = uv(1-v)$;
$z = uvw$

then find $J = \dfrac{\partial(x, y, z)}{\partial(u, v, w)}$

Solution: We have

$$J = \frac{\partial(x,y,z)}{\partial(u,v,w)} = \begin{vmatrix} \frac{\partial x}{\partial u} & \frac{\partial x}{\partial v} & \frac{\partial x}{\partial w} \\ \frac{\partial y}{\partial u} & \frac{\partial y}{\partial v} & \frac{\partial y}{\partial w} \\ \frac{\partial z}{\partial u} & \frac{\partial z}{\partial v} & \frac{\partial z}{\partial w} \end{vmatrix}$$

$$= \begin{vmatrix} (1-v) & -u & 0 \\ v(1-v) & u(1-2v) & 0 \\ vw & uw & uv \end{vmatrix} \text{ Expend from 3rd column}$$

$$= uv\,[u(1-2v)(1-v) + uv(1-v)] = u_v^2\,(1-v^2).$$

232. If $x = ar\cos\theta\sin\phi$;
$y = br\sin\theta\sin\phi$;
$z = cr\cos\theta$; find $\dfrac{\partial(x, y, z)}{\partial(r, \theta, \phi)}$

Solution: We have $J = \dfrac{\partial(x, y, z)}{\partial(r, \theta, \phi)}$

$$= \begin{vmatrix} a\cos\theta\sin\phi & -ar\sin\theta\sin\phi & ar\cos\theta\cos\phi \\ b\sin\theta\sin\phi & br\cos\theta\sin\phi & br\sin\theta\cos\phi \\ c\cos\theta & -cr\sin\theta & 0 \end{vmatrix}$$

$$= a\cos\theta\cdot\sin\phi\,[bcr^2\sin^2\theta\cdot\cos\phi]$$
$$+ ar\sin\theta\sin\phi\,[-bcr\sin\theta\cos\theta\cos\phi]$$
$$+ ar\cos\theta\cdot\cos\phi\,[-bcr\sin^2\theta\cdot\sin\phi - bcr\cos^2\theta\cdot\sin\phi]$$
$$= abcr^2\,[\sin^2\theta\cos\theta\cdot\sin\phi\cos\phi - \sin^2\theta\cdot\cos\theta\cdot\sin\phi\cos\phi - \sin\phi]$$
$$= -abcr^2\sin\phi$$

233. If $x = z\sec\theta\cdot\cos\phi$;
$y = z\sec\theta\sin\phi$;
$z = z$ then find $J = \dfrac{\partial(x, y, z)}{\partial(\theta, \phi, z)}$

Solution:
$$J = \frac{\partial(x, y, z)}{\partial(\theta, \phi, z)}$$

$$= \begin{vmatrix} z \sec\theta \cdot \tan\theta \cos\phi & -z \sec\theta \sin\phi & \sec\theta \sin\phi \\ z \sec\theta \cdot \tan\theta \sin\phi & z \sec\theta \cdot \cos\phi & \sec\theta \cdot \sin\phi \\ 0 & 0 & 1 \end{vmatrix}$$

$$= z^2 \sec^2\theta \cdot \tan\theta \cdot \cos^2\phi + z^2 \sec^2\theta \tan\theta \sin^2\phi$$
$$= z^2 \sec^2\theta \cdot \tan\theta$$

234. Show that $JJ' = 1$ for

(*a*) $x = uv;\ y = \frac{u}{v}$

(*b*) $x = v^2 + w^2;\ y = w^2 + u^2;\ z = u^2 + v^2$

(*c*) $x + y = 2e^{\theta} \cos\phi;\ x - y = 2ie^{\theta} \sin\phi$

Solution: (*a*) [to find $\frac{\partial(u,v)}{\partial(x,y)}$ we have to express u, v in terms of x, y]

We have $x = uv;\ y = \frac{u}{v}$

$$\therefore \quad J = \frac{\partial(x,y)}{\partial(u,v)} = \begin{vmatrix} v & u \\ \frac{1}{v} & -\frac{u}{v^2} \end{vmatrix} = -\frac{u}{v} - \frac{u}{v} = -\frac{2u}{v} \qquad [1]$$

since $x = uv,\ y = \frac{u}{v}$

$\therefore \quad u^2 = xy \qquad (\therefore\ u = \sqrt{xy} = x^{1/2}\, y^{1/2})$

and $v^2 = \frac{x}{y} \qquad \left(\therefore\ v = \sqrt{\frac{x}{y}} = x^{1/2} \cdot y^{-1/2}\right)$

$$J' = \frac{\partial(u,v)}{\partial(x,y)} = \begin{vmatrix} \frac{1}{2}x^{-1/2} \cdot y^{1/2} & \frac{1}{2}x^{1/2} \cdot y^{-3/2} \\ \frac{1}{2}x^{-1/2} \cdot y^{-1/2} & -\frac{1}{2}x^{1/2} \cdot y^{-3/2} \end{vmatrix} = \frac{1}{4}[-y^{-1} - y^{-1}] = -\frac{1}{2y}$$

[2]

$\therefore$ From [1] and [2],

$$JJ' = \left(-\frac{2u}{v}\right)\left(-\frac{1}{2y}\right) = 1 \qquad \left(\because\ y = \frac{u}{v}\right)$$

Solution (*b*): ($x = v^2 + w^2;\ y = w^2 + u^2;\ z = u^2 + v^2$) [1]

$$\therefore \quad J = \frac{\partial(x,y,z)}{\partial(u,v,w)} = \begin{vmatrix} 0 & 2v & 2w \\ 2u & 0 & 2w \\ 2u & 2v & 0 \end{vmatrix}$$

$$= -2v\,[-4uw] + 2w\,[4uv] = 16\,uvw \qquad [2]$$

since $x = v^2 + w^2,\ y = w^2 + u^2,\ z = u^2 + v^2$

$$\therefore \quad u^2 + v^2 + w^2 = \frac{1}{2}(x + y + z) = \frac{x}{2} + \frac{y}{2} + \frac{z}{2}$$

$$u^2 = -\frac{x}{2} + \frac{y}{2} + \frac{z}{2};\ v^2 = \frac{x}{2} - \frac{y}{2} + \frac{z}{2};$$

$$w^2 = \frac{x}{2} + \frac{y}{2} - \frac{z}{2}$$

$$\therefore \quad J' = \frac{\partial(u,v,w)}{\partial(x,y,z)} = \begin{vmatrix} -\frac{1}{4}u & \frac{1}{4}u & \frac{1}{4}u \\ \frac{1}{4}v & -\frac{1}{4}v & \frac{1}{4}v \\ \frac{1}{4}w & \frac{1}{4}w & -\frac{1}{4}w \end{vmatrix}$$

$$= \frac{1}{64\,uvw}\begin{vmatrix} -1 & 1 & 1 \\ 1 & -1 & 1 \\ 1 & 1 & -1 \end{vmatrix} = \frac{4}{64\,uvw} = \frac{1}{16\,uvw} \qquad [2]$$

From [1] and [2], we have

$$JJ' = (16\,uvw)\left(\frac{1}{16\,urw}\right) = 1.$$

Solution (c): **Given:**

$$x + y = 2e^{\theta}\cos\phi$$

$$x - y = 2ie^{\theta}\sin\phi,$$

Adding and subtracting

$$x = e^{\theta}(\cos\phi + i\sin\phi) = e^{\theta}\cdot e^{i\phi}$$

$$y = e^{\theta}(\cos\phi - i\sin\phi) = e^{\theta}\cdot e^{-i\phi}$$

$$\therefore \quad J = \frac{\partial(x,y)}{\partial(\theta,\phi)} = \begin{vmatrix} e^{\theta}\cdot e^{i\phi} & e^{\theta}\cdot i\cdot e^{i\phi} \\ e^{\theta}\cdot e^{-i\phi} & -ie^{\theta}\cdot e^{i\phi} \end{vmatrix}$$

$$= -ie^{2\theta} - ie^{2\theta} = -2ie^{2\theta} \qquad [1]$$

Again, we have

$$e^{2\theta} = x\cdot y \text{ and } \frac{x}{y} = e^{2i\phi} \qquad [2]$$

$$J' = \frac{\partial(\theta,\phi)}{\partial(x,y)} = \begin{vmatrix} \frac{y}{2e^{2\theta}} & \frac{x}{2e^{2\theta}} \\ \frac{1}{2iye^{2\phi},e^{2i\theta}} & \frac{-x}{2iy^2e^{2i\phi}} \end{vmatrix}$$

$$= \frac{-x}{4iye^{2\theta}\cdot e^{2i\phi}} - \frac{x}{4iye^{2\theta}\cdot e^{2i\phi}} = \frac{-x}{2iye^{2\theta}\cdot e^{2i\phi}} \qquad [3]$$

From [1] and [3],

$$JJ' = (-2ie^{2\theta})\left(\frac{-x}{2iye^{2\theta}\cdot e^{2i\phi}}\right)$$

$$= \frac{x}{y}\cdot\frac{1}{e^{2i\phi}} = (e^{2i\phi})\left(\frac{1}{e^{2i\phi}}\right) \quad \text{(using [2])}$$

$$= 1$$

$$\therefore \quad JJ' = 1$$

235. If $x^2 + y^2 - v^2 + u^2 = 0;$
$uv + xy = 0$

then find $\dfrac{\partial(u,v)}{\partial(x,y)}$

(An Implicit functional relation)

Solution:

Let $f_1 = x^2 + y^2 + u^2 - v^2 = 0$

$f_2 = uv + xy = 0$

Then to evaluate Jacobian from implicit functional relation, we apply the formula:

$$\frac{\partial(u,v)}{\partial(x,y)} = (-1)^2 \frac{\dfrac{\partial(f_1,f_2)}{\partial(x,y)}}{\dfrac{\partial(f_1,f_2)}{\partial(u,v)}} \quad [1]$$

Now, $$\frac{\partial(f_1,f_2)}{\partial(x,y)} = \begin{vmatrix} 2x & 2y \\ y & x \end{vmatrix} = 2(x^2 - y^2) \quad [2]$$

And $$\frac{\partial(f_1,f_2)}{\partial(u,v)} = \begin{vmatrix} 2u & -2v \\ v & u \end{vmatrix} = +\,2u^2 + 2v^2 = 2\,(u^2 + v^2) \quad [3]$$

$\therefore$ From [1],

$$\frac{\partial(u,v)}{\partial(x,y)} = \frac{2(x^2 - y^2)}{2(u^2 + v^2)} = \frac{(x^2 - y^2)}{(u^2 + v^2)}$$

236. Verify whether the following functions are functionally dependent and so find the relation between them.

(1) $$u = \frac{x+y}{1-xy},\ v = \tan^{-1} x + \tan^{-1} y$$

(2) $$u = \frac{x-y}{x+y}, v = \frac{x+y}{x}$$

(3) $u = \dfrac{x-y}{x+z}, v = \dfrac{x+z}{y+z}$

(4) $u = x + y + z,\ V = x^2 + y^2 + z^2 - 2xy - 2yz - 2zx$

$w = x^3 + y^3 + z^3 - 3xyz$

(5) $u = x + y - z,\ V = x - y + z,\ w = x^2 + y^2 + z^2 - 2yz$

Solution: (1) Since $u = \dfrac{x+y}{1-xy}$

$$\therefore \quad \frac{\partial u}{\partial x} = \frac{(1-xy)-(-y)(x+y)}{(1-xy)^2} = \frac{(1+y^2)}{(1-xy)^2}$$

$$\frac{\partial u}{\partial y} = \frac{(1+x^2)}{(1-xy)^2}$$

Again, $V = \tan^{-1}(x) + \tan^{-1}(y)$

$$\therefore \quad \frac{\partial v}{\partial x} = \frac{1}{1+x^2}; \frac{\partial v}{\partial y} = \frac{1}{1+y^2}$$

Hence
$$J = \frac{\partial(u,v)}{\partial(x,y)} \begin{vmatrix} \dfrac{(1+y^2)}{(1-xy)^2} & \dfrac{(1+x^2)}{(1-xy)^2} \\ \dfrac{1}{(1+x^2)} & \dfrac{1}{(1+y^2)} \end{vmatrix} = 0$$

Thus, u, v are functionally related (or functionally dependent)

To find the relation:

We notice that

$$v = \tan^{-1}(x) + \tan^{-1}(y) = \tan^{-1}\frac{(x+y)}{(1-xy)} = \tan^{-1}(u)$$

$\therefore$ $u = \tan v$ is the relation.

Solution: (2): **Given:** $u = \dfrac{x-y}{x+y}$ and $v = 1 + \dfrac{y}{x}$

we first find $\dfrac{\partial u}{\partial x}, \dfrac{\partial u}{\partial y}$

since $u = \dfrac{x-y}{x+y}$

$$\frac{\partial u}{\partial x} = \frac{(x+y)\cdot(1)-(x-1)(1)}{(x+y)^2} = \frac{2y}{(x+y)^2}$$

and $$\frac{\partial u}{\partial y} = \frac{-2x}{(x+y)^2}$$

$\therefore$ $$J = \frac{\partial(u,v)}{\partial(x,y)} = \begin{vmatrix} \frac{\partial u}{\partial x} & \frac{\partial u}{\partial y} \\ \frac{\partial v}{\partial x} & \frac{\partial v}{\partial y} \end{vmatrix} = \begin{vmatrix} \frac{2y}{(x+y)^2} & \frac{-2x}{(x+y)^2} \\ \frac{-y}{x^2} & \frac{1}{x} \end{vmatrix}$$

$$= \frac{2y}{x(x+y)^2} \begin{vmatrix} 1 & -x \\ -\frac{1}{x} & 1 \end{vmatrix} = 0$$

$\Rightarrow$ **u and v are functionally related.**

To find the relation:

Since $$\frac{y}{x} = v - 1;\ u = \frac{x-y}{x+y} = \frac{1-\frac{y}{x}}{1+\frac{y}{x}}$$

$\therefore$ $$u = \frac{1-(v-1)}{v}$$

$\Rightarrow$ $$uv + v = 2$$

$\Rightarrow$ $$v\,(u+1) = 2$$

Solution (3): **Given:** $$u = \frac{x-y}{x+z};\ V = \frac{x+z}{y+z}$$

$\therefore$ (*i*) $$\frac{\partial(u,v)}{\partial(x,y)} = \begin{vmatrix} \frac{(y+z)}{(x+z)^2} & \frac{-1}{(x+z)} \\ \frac{1}{y+z} & -\frac{(x+z)}{(y+z)^2} \end{vmatrix}$$

$$= \frac{-1}{(x+z)(y+z)} + \frac{1}{(x+z)(y+z)} = 0$$

(*ii*) $$\frac{\partial(u,v)}{\partial(y,z)} = \begin{vmatrix} -\frac{1}{(x+z)} & -\frac{(x-y)}{(x+z)^2} \\ -\frac{(x+z)}{(y+z)^2} & -\frac{(x-y)}{(y+z)^2} \end{vmatrix} = 0$$

(iii) $$\frac{\partial(u,v)}{\partial(z,x)} = \begin{vmatrix} -\dfrac{(x-y)}{(x+z)^2} & \dfrac{(y+z)}{(x+z)^2} \\ -\dfrac{(x-y)}{(y+z)^2} & \dfrac{1}{(y+z)} \end{vmatrix} = 0$$

⇒ u and v are functionally dependent (or related)

To find the relation

$$u = \frac{x-y}{x+z} \text{ and } \frac{1}{v} = \frac{y+z}{x+z}$$

Now, $$u + \frac{1}{v} = \frac{(x-y)}{(x+z)} + \frac{(y+z)}{(x+z)} = \frac{x+z}{x+z} = 1$$

∴ $$u + \frac{1}{v} = 1.$$

⇒ $uv + 1 = v$ is the relation.

Solution: (4) $u = x + y + z$; $v = x^2 + y^2 + z^2 - 2xy - 2yz - 2zx$

$$w = x^3 + y^3 + z^3 - 3xyz$$

∴ $$J = \frac{\partial(u,v,w)}{\partial(x,y,z)}$$

$$= \begin{vmatrix} 1 & 1 & 1 \\ 2x - 2y - 2z & 2y - 2x - 2z & 2z - 2y - 2x \\ 3x^2 - 3yz & 3y^2 - 3xz & 3z^2 - 3xy \end{vmatrix}$$

Applying

$$\xrightarrow[c_2 \to c_2 - c_3]{c_1 \to c_1 - c_2}$$

$$= 6 \begin{vmatrix} 1 & 1 & 1 \\ x - y & y - z & z - y - x \\ x^2 - y^2 - yz + zx & y^2 - z^2 - xz + xy & z^2 - xy \end{vmatrix}$$

$$= 6(x-y)(y-z) \begin{vmatrix} 1 & 1 & 1 \\ 1 & 1 & z - y - x \\ (x+y+z) & (x+y+z) & z^2 - xy \end{vmatrix} = 0$$

∴ u, v, w are functionally related.

To find the relation: we have

$$u^2 + v = (x + y + z)^2 + (x^2 + y^2 + z^2 - 2xy - 2yz - 2zx)$$
$$= 2(x^2 + y^2 + z^2)$$

$$\therefore \qquad (x^2 + y^2 + z^2) = \frac{1}{2}(u^2 + v) \qquad [1]$$

Again $\qquad u^2 - v = (x + y + z)^2 - (x^2 + y^2 + z^2 - 2xy - 2yz - 2zx)$

$$= 4(xy + yz + zx)$$

$$\therefore \qquad (xy + yz + zx) = \frac{(u^2 - v)}{4} \qquad [2]$$

and $\qquad w = x^3 + y^3 + z^3 - 3xyz$

$$= (x + y + z)(x^2 + y^2 + z^2 - xy - yz - zx)$$

(Formula)

$$= u\left[\frac{1}{2}(u^2 + v) - \left(\frac{u^2 - v}{4}\right)\right] = \frac{u}{4}[2u^2 + 2v - u^2 + v]$$

$$w = \frac{u}{4}(u^2 + 3v) \text{ is the relation.}$$

Solution (5): $\qquad u = x + y - z;$

$$v = x - y + z,$$

$$w = x^2 + y^2 + z^2 - 2yz$$

If $\dfrac{\partial(u, v, w)}{\partial(x, y, z)} = 0$, then the functions are functionally dependent.

$$\frac{\partial(u, v, w)}{\partial(x, y, z)} = \begin{vmatrix} \frac{\partial u}{\partial x} & \frac{\partial u}{\partial y} & \frac{\partial u}{\partial z} \\ \frac{\partial v}{\partial x} & \frac{\partial v}{\partial y} & \frac{\partial v}{\partial z} \\ \frac{\partial w}{\partial x} & \frac{\partial w}{\partial y} & \frac{\partial w}{\partial z} \end{vmatrix}$$

$$= \begin{vmatrix} 1 & 1 & -1 \\ 1 & -1 & 1 \\ 2x & 2y - 2z & 2z - 2y \end{vmatrix} = 0$$

(since $c_2 = -c_3$)

i.e., (column 2) = – (column 3)

Hence the functions are functionally dependent.

To find the relation:

We get, $\qquad u + v = 2x$ and $u - v = 2(y - z)$

$$w = x^2 + (y - z)^2$$

$$= \left(\frac{u + v}{2}\right)^2 + \left(\frac{u - v}{2}\right)^2 = \frac{(u^2 + v^2)}{2}$$

EXERCISES

I. In each of the following cases find $J_1 = \dfrac{\partial(x,y)}{\partial(u,v)}$ and $J_2 = \dfrac{\partial(u,v)}{\partial(x,y)}$

Also verify $J_1 J_2 = 1$.

(a) $x = u^2 - 2v$ and $y = u + v$ (Ans. $J_1 = 2(u + 1)$

(b) $x = u^2 - v^2$ and $y = 2uv$ (Ans. $J_1 = 4(u^2 + v^2)$

(c) $x = u(1 - v)$ and $y = uv$ $\left(\text{Ans. } J_1 = -\dfrac{v}{2u}\right)$

(d) $u = x^2 - 2y$ and $v = x + y$ (Ans. $J_2 = 2(x + 1)$

(e) $u = x^2 - y^2$ and $v = 2xy$ (Ans. $J_2 = 4(x^2 + y^2)$

(f) $u = \dfrac{y^2}{2x}$ and $v = \dfrac{x^2 + y^2}{2x}$ $\left(\text{Ans. } J_2 = -\dfrac{y}{2x}\right)$

II. For the following functions verify.

that $J_1 J_2 = 1$

(a) $u = x - y$ and $v = xy$

(b) $u = x - y$ and $v = y^2 + 2x$

(c) $u = x(1 + y)$ and $v = xy$

(d) $u = x - y$ and $v = y^2 + 2x$

(e) $u = \sqrt{x^2 + y^2}$ and $v = \sqrt{x^2 - y^2}$

III. 1. If $F = xu + v - y$,

$G = u^2 + vy + w$

and $H = zu - v + vw$ prove that

$$\frac{\partial(F,G,H)}{\partial(u,w,v)} = xw - x - xyv + 2uv - z$$

2. If $u = x^2 - 2y^2$ and $v = 2x^2 - y^2$

where $x = r\cos\theta$ and

$y = r\sin\theta$.

prove that $\dfrac{\partial(u,v)}{\partial(r,\theta)} = 6r^3 \sin 2\theta$

3. Prove that the functions $x = r(w - u)$, $y = w(u - v)$ and $z = (v - w)$ are functionally dependent.

4. If x, y, z are connected by the relation $f(x, y, z) = 0$, show that $\left[\dfrac{\partial y}{\partial x}\right]_z = \dfrac{\partial(y,z)}{\partial(x,z)}$.

5. If $u = \dfrac{2yz}{x}, v = \dfrac{3zx}{y}, w = \dfrac{4xy}{z}$, show that $\dfrac{\partial(x,y,z)}{\partial(u,v,w)} = \dfrac{1}{96}$

6. If $X = \dfrac{u^2}{v}, Y = \dfrac{v^2}{u}$ show that $\dfrac{\partial(u,v)}{\partial(X,Y)} = \dfrac{1}{3}$.

PART-B

7. If $x = u,\ y = u \tan v,\ z = w$, show that $\dfrac{\partial(u,v,w)}{\partial(x,y,z)} = \dfrac{\cos^2 v}{u}$

8. If $u = y + z,\ v = x + 2z^2,\ w = x - 4yz - 2y^2$

   Find $\dfrac{\partial(u,v,w)}{\partial(x,y,z)}$. Are u, v and w functionally related?

   If so, find the relationship (*Ans:* O; $w = v - u^2$)

9. If $u = \dfrac{x}{(1-r^2)^{1/2}},\ v = \dfrac{y}{(1-r^2)^{1/2}},\ w = \dfrac{z}{(1-r^2)^{1/2}}$ and $r^2 = x^2 + y^2 + z^2$ then show that

   $$\frac{\partial(u,v,w)}{\partial(x,y,z)} = \frac{1}{(1-r^2)^{5/2}}$$

10. If $x = \sqrt{uw}\ ;\ y = \sqrt{vw}\ ;\ z = \sqrt{uv}$ and $u = r \sin\theta \times \cos\phi$

    $v = r\sin\theta \cdot \sin\phi,\ w = r\cos\theta$ then find $\dfrac{\partial(x,y,z)}{\partial(v,\theta,\phi)}$ (*Ans:* $r^2 \sin\theta$)

2.12 ERRORS AND APPROXIMATIONS

From differential calculus, we know that if $y = f(x)$ is a function of x, then

$$\frac{dy}{dx} = \left[\lim_{\delta x \to 0} \frac{\delta y}{\delta x}\right]$$

since $\dfrac{\delta y}{\delta x} = \dfrac{dy}{dx}$ (where δx is sufficiently small)

$$\Rightarrow \qquad \delta y = \left(\frac{dy}{dx}\right)\delta x$$

or

$$\delta y = f'(x)\, dx \text{ (approximately)}$$

(*i*) δx is called absolute error in x

(*ii*) $\dfrac{\delta x}{x}$ is called relative error in x

(*iii*) $\dfrac{\delta x}{x} \times 100$ is called the percentage error in x.

Further, Let $z = f(x, y)$ and [1]

if δx, δy are small increments in x and y respectively and δz, the corresponding increment in z, then

$$z + \delta z = f(x + \delta x,\ y + \delta y) \qquad [2]$$

Subtracting [1] from [2], we get

$$\delta z = f(x + \delta x,\ y + \delta y) - f(x, y)$$

$$= f(x, y) + \delta x\,\frac{\partial f}{\partial x} + \delta y\,\frac{\partial f}{\partial y} + \ldots - f(x, y)$$

$$= \delta x \frac{\partial f}{\partial x} + \delta y \frac{\partial f}{\partial y} \text{ (approximately)}$$

(neglecting higher power of δx, δy)

$$\therefore \qquad \delta z = \frac{\partial f}{\partial x} \cdot \delta x + \frac{\partial f}{\partial y} \cdot \delta y \text{ (approx.)}$$

If δx and δy are small changes (or errors) in x and y respectively, then an approximate change (in error) in z is δz.

Replacing δx, δy, δz by dx, dy, dz respectively, we have

$$\boxed{dz = \frac{\partial f}{\partial x} dx + \frac{\partial f}{\partial y} \cdot dy}$$

Worked Examples (241-261)

237. A triangle ABC is inscribed in a fixed circle. If the sides of the triangle vary prove that

$$\frac{\delta a}{\cos A} + \frac{\delta b}{\cos B} + \frac{\delta c}{\cos C} = 0$$ [VTU, August, 2002]

[VTU, August, 2001]

Solution: The circum radius R of a ΔABC is given by

$$R = \frac{a}{2 \sin A} = \frac{b}{2 \sin B} = \frac{c}{2 \sin C}$$

Now $a = 2R \sin A$ (R is constant)

Taking differentials, $\delta a = 2R \cos A dA$

or
$$\frac{\delta a}{\cos A} = 2RdA$$

$|||^{ly}$,
$$\frac{\delta b}{\cos B} = 2RdB \text{ and } \frac{\delta c}{\cos C} = 2RdC$$

Adding,
$$\frac{\delta a}{\cos A} + \frac{\delta b}{\cos B} + \frac{\delta c}{\cos C}$$

$$= 2R\,(dA + dB + dC) \qquad [1]$$

Also $\qquad A + B + C = \pi$

Taking differentials, $\delta A + \delta B + \delta C = 0$

$$\therefore \text{ From [1], } \frac{\delta a}{\cos A} + \frac{\delta b}{\cos B} + \frac{\delta c}{\cos C} = 0.$$

238. Find the percentage error in the evaluation of the area of the ellipse $\frac{x^2}{a^2} + \frac{y^2}{b^2} = 1$ if an error of + 1 percent is made while measuring the semi-major and semi-minor axis.

(VTU, August, 2000)

Solution:

We know that the error in the function of two variables, $f(x, y) = u$ is approximated by the formula:

$$\delta u = \frac{\partial u}{\partial x}\,\delta x + \frac{\partial u}{\partial y}\,\delta y$$

Also, we know that the area A of an ellipse with semi-major axis a and semi-minor axis b is given by $A = \pi ab$.

We are given: $100 \times \dfrac{\delta a}{a} = 1$

And $100 \times \dfrac{\delta b}{b} = 1$

Now, $\delta A = \dfrac{\partial A}{\partial a}\,\delta a + \dfrac{\partial A}{\partial b}\,\delta b = \pi b\,\delta a + \pi a\,\delta b$

(Error Relation)

($\div\, A = \pi ab$) we get

$$\frac{\delta A}{A} = \frac{\delta a}{a} + \frac{\delta b}{b}$$

or $$100 \times \frac{\delta A}{A} = 100 \times \frac{\delta a}{a} + 100 \times \frac{\delta b}{b} = 1 + 1 = 2.$$

239. The diameter and the altitude of a can in the form of right circular cylinder are found to be 4.5 cms and 8.25 cms respectively. The possible error in each measurement is 0.1 cms. Find the appoximate error in the volume and lateral surface. (VTU, March, 1999)

Solution: Diameter of cylindrical can. (D) = 4.5 cms. ($\therefore\ r = 2.25$ cms)

(r = radius) Height of cylindrical can.

$$h = 8.25 \text{ cms.}$$

$$\delta D = \delta h = 0.1, \quad \Rightarrow \quad 2\delta r = 0.1$$

or $$\delta r = 0.05$$

Volume of the can $= V = \pi r^2 h = \dfrac{\pi}{4} D^2 h$ [1]

$$\therefore \quad \delta V = \pi\delta\,(r^2 h)$$

$$= \pi\,(r^2 \delta h + h \cdot 2r\delta r)$$

$$= \pi\,[(2.25)^2\,(0.1) + 2 \times 8.25 \times 2.25 \times 0.05] = 2.36\,\pi$$

and Lateral surface of the can

$$= S = \pi Dh = 2\pi rh$$ [2]

Differentiating [2]

$$\delta S = 2\pi\,(r\delta h + h\delta r)$$

$$= 2\pi\,(2.25 \times 0.1 + 8.25 \times 0.05) = 1.275\pi$$

240. If the deflection at the centre of a rod of length l and diameter d supported at its end and loaded at the centre with a weight w is proportional to $\frac{wl^3}{d^4}$. Determine the percentage increase in the deflection of w, l and d are increased by 5%, 4% and 3% respectively.

(VTU, August, 1999)

Solution:

Let f be the deflection of the rod.

Given: $f \propto \frac{wl^3}{d^4}$

$$\Rightarrow \quad f = k\left(\frac{wl^3}{d^4}\right) \quad [1]$$

(k = constant of proportionality)

Also, We are given; $\frac{\delta w}{w} \times 100 = 5$

$$\frac{\delta l}{l} \times 100 = 4$$

$$\frac{\delta d}{d} \times 100 = 3$$

Taking log. on both sides of [1], we get.

$$\log f = \log k + \log w + 3 \log l - 4 \log d \quad (2)$$

Differentiating [2], we get

$$\frac{df}{f} = \frac{dw}{w} + \frac{3dl}{l} - \frac{4d(d)}{d}$$

$$\therefore \quad \frac{df}{f} \times 100 = \frac{dw}{w} \times 100 + 3\frac{dl}{l} \times 100 - \frac{4d(d)}{d} \times 100$$

$\therefore$ The error relation is

$$\frac{\delta f}{f} \times 100 = \frac{\delta w}{w} \times 100 + 3\frac{\delta l}{l} \times 100 - \frac{4\delta d}{d} \times 100$$

$$= 5 + (3 \times 4) - (4 \times 3) = 5$$

241. In estimating the cost of a pile of bricks measured as $2m. \times 15 \times 1.2\ m.$ the tape is stretched 1% beyond the standard length. If the count is 450 bricks per cubic meters and the bricks cost is Rs. 1100 per thousand, find the approximate error in the cost.

(VTU, March, 2001)

Solution:

If x, y, z be the length, breath and height of the pile, then the volume is given by

$$V = xyz \quad [1]$$

$$= 2 \times 15 \times 1.2 = 36 \text{ m}^3$$

($\because$ by data x = 2m, y = 15m, z = 1.2m)

Also we note that

$$100 \times \frac{\delta x}{x} = 1; \ 100 \times \frac{\delta y}{y} = 1, \ 100 \times \frac{\delta z}{z} = 1$$

Take log. For [1], we get

$$\log V = \log x + \log y + \log z \qquad [2]$$

Differentiate [2], we get

$$\frac{\delta v}{v} = \frac{\delta x}{x} + \frac{\delta y}{y} + \frac{\delta z}{z}$$

i.e.,

$$100 \frac{\delta v}{v} = 100 \frac{\delta x}{x} + 100 \frac{\delta y}{y} + 100 \frac{\delta z}{z}$$

$$= 1 + 1 + 1 = 3$$

Thus, the percentage error in V

$$= 100 \frac{\delta v}{v} = 3$$

or

$$\delta v = \frac{3}{100} v$$

But $V = 36$ m.

Hence

$$\delta v = \frac{3}{100} \times 36 = 1.08 \text{ m}^3$$

$\therefore$ Error in number of bricks

$$= 1.08 \times 450 = 486.$$

Thus error in cost

$$= 486 \times \frac{1100}{1000} = 534.6 \text{ Rs.}$$

242. Compute an approximate value of $(1.04)^{3.01}$

Solution: Let $f(x, y) = x^y$

$$\Rightarrow \quad \frac{\partial f}{\partial x} = yx^{y-1}; \ \frac{\partial f}{\partial y} = x^y \cdot \log x$$

Let $x = 1, dx = .04, y = 3, dy = 0.1$ [1]

Now

$$df = \left(\frac{\partial f}{\partial x}\right) dx + \left(\frac{\partial f}{\partial y}\right) dy$$

$$= yx^{y-1} \, dx + x^y \log x \, dy$$

Substituting the values from [1], we get

$$df = 3.1 \, (0.04) + (1)^3 \, (\log 1) \, (.01) = 0.12$$

Now, $f(1, 3) = 1$

$$(1.04)^{3.01} = f(1, 3) + df = 1 + .12 = 1.12 \ \textit{Ans.}$$

243. If the kinetic energy $T = \frac{mv^2}{2}$, find approximately the change in T as m changes from 49 to 49.5 and v changes from 1600 to 1590.

Solution: **Given:** $T = \frac{1}{2} mv^2$; $m = 49$; $m + \delta m = 49.5$; $\therefore \delta m = 0.5$

$v = 1600$; $v + \delta v = 1590$; $\therefore \delta v = -10$

Hence,
$$\delta T = mv\,\delta v + \frac{v^2}{2}\,\delta m$$
$$= (49\,(1600)\,(-10) + \frac{(1600)^2}{2}\,(0.5)$$
$$= -784000 + 64000 = -144000$$

244. The period of a simple pendulum of length l is given

By
$$T = 2\pi\sqrt{\frac{l}{g}}$$

Find (*i*) the error

(*ii*) percent error (relative error)

made in computing T by using $l = 2ft$, $g = 32\ ft/\text{sec}^2$, if the true value are $l = 1.95\ ft$ and $g = 32.2\ ft/\text{sec}^2$.

Solution: Here
$$T = 2\pi\sqrt{(l/g)} \qquad [1]$$
$$\therefore \quad \delta T = 2\pi\,\frac{1}{2}\cdot\left(\frac{g}{l}\right)^{1/2}\left[\frac{gdl - ldg}{g^2}\right]$$
$$\delta T = \frac{22}{7}\left(\frac{32.2}{1.95}\right)^{1/2}\left[\frac{(32.2)(0.05)}{(32.2)^2} - (1.95)(-.2)\right]$$
$$= \frac{178.64}{7257.88} = 0.0246.$$

Taking log for eqn [1], we have $\log T = \log(2\pi) + \left(\frac{1}{2}\right)\log l - \left(\frac{1}{2}\right)\log g$.

$$\therefore \quad \frac{\delta T}{T} = 0 + \left(\frac{1}{2}\right)\frac{\delta l}{l} - \left(\frac{1}{2}\right)\frac{\delta g}{g}$$
$$= \frac{1}{2}\left(\frac{\delta l}{l} - \frac{\delta g}{g}\right) = \frac{1}{2}\left(\frac{0.05}{1.95} - \frac{-0.2}{32.2}\right)$$
$$= \frac{0.5}{32} \text{ (approx.).}$$

PART-B

$$\therefore \quad \frac{\delta T}{T} \times 100 = \frac{0.05 \times 100}{32} = \frac{25}{16} = 1{\cdot}5625\%$$

245. Find the possible relative error in computing the resistance r from the formula $\frac{1}{r} = \frac{1}{r_1} + \frac{1}{r_2}$ if r_1 and r_2 are both have error 2%.

Solution:
$$\frac{1}{r} = \frac{1}{r_1} + \frac{1}{r_2} \qquad [1]$$

$$\therefore \quad -\frac{1}{r^2}\,\delta r = \frac{1}{r_1^2}\,\delta r_1 - \frac{1}{r_2^2}\,\delta r_2$$

i.e.,
$$\frac{1}{r^2}\,\delta r = \frac{1}{r_1^2}\,\delta r_1 - \frac{1}{r_2^2}\,\delta v_2$$

i.e.,
$$\frac{1}{r}\cdot\frac{\delta r}{r} = \frac{1}{r_1}\cdot\frac{\delta r_1}{r_1} + \frac{1}{r_2}\cdot\frac{\delta r_2}{r_2}$$

Given $\frac{\delta r_1}{r_1} = \frac{\delta r_2}{r_2} = 2\%$ .We find

$$\frac{1}{r}\cdot\frac{\delta r}{r} = \frac{1}{r_1}\cdot\left(\frac{2}{100}\right) + \frac{1}{r_2}\left(\frac{2}{100}\right)$$

$$= \frac{2}{100}\left(\frac{1}{r_1} + \frac{1}{r_2}\right) = \frac{2}{100}\cdot\frac{1}{r}$$

$$\Rightarrow \quad \frac{\delta r}{r} = 2\%$$

246. Find the percentage error in calculating the area of a rectangle when an error of 2% is made in measuring its sides.

Solution: If 'a' and 'b' are the length, breadth of the rectangle respectively, then, area $A = ab$.

$$\therefore \quad \log A = \log a + \log b \qquad [1]$$

Diff [1],
$$\frac{dA}{A} = \frac{da}{a} + \frac{db}{b}$$

$$\therefore \quad \left(\frac{\delta A}{A}\right) \times 100 = \left(\frac{\delta a}{a}\right) \times 100 + \left(\frac{\delta b}{b}\right) \times 100$$

$$\left(\text{since } \frac{\delta A}{A} = b; \frac{\partial A}{\partial b} = a\right) = 2 + 2 = 4.$$

i.e., (4%) is the required percentage error in calculating the area.

247. What error in common logarithm of a number will be produced by an error of 1% in the number?

Solution: Let x be the number

Let $$y = \log_{10}^{x}$$

We are to find δy

$$\delta y = \frac{1}{x}\log_{10}^{e}\cdot\delta x$$

$$= \left(\frac{100\delta x}{x}\right)\times\frac{\log_{10}^{e}}{100}$$

putting $\dfrac{100\delta x}{x} = 1$, we get

$$\delta y = \frac{1\times\log_{10}^{e}}{100} = \frac{0.4343}{100} = 0{\cdot}004343$$

248. Prove that the error δA in the area A of an ellipse, due to small errors δa and δb in the calculated lengths of the semi-axes, a and b *i.e.* given by $\left(\dfrac{\delta A}{A}\right)=\left(\dfrac{\delta a}{a}\right)+\left(\dfrac{\delta b}{b}\right)$

Solution:

Area of the Ellipse $A = \pi ab$

$$\therefore \quad \log A = \log\pi + \log a + \log b$$

$$\therefore \quad \left(\frac{1}{A}\right)(\delta A) = 0 + \left(\frac{1}{a}\right)\delta a + \left(\frac{1}{b}\right)\delta b$$

$$\therefore \quad \left(\frac{\delta A}{A}\right)=\left(\frac{\delta a}{a}\right)+\left(\frac{\delta b}{b}\right)$$

249. If the percentage error in the radius of the sphere is 0.5%, find the percentage error in the volume and the surface area of the sphere.

Solution: If r, V and A denote the radius, volume and the surface area of the sphere respectively, then

$$V = \frac{4}{3}\pi r^3 \qquad [1]$$

$$\therefore \quad \delta v = \frac{4}{3}\pi\,3r^2\,\delta r = 4\pi r^2\,\delta r$$

i.e., $$\frac{\delta v}{v} = \frac{4\pi r^2\delta r}{(4/3)\pi r^3} = 3\frac{\delta r}{r}$$

[Given: $100\times\dfrac{\delta r}{r} = 0.5$]

$$\therefore \qquad 100 \times \frac{\delta v}{v} = 3\left(100 \times \frac{\delta r}{r}\right)$$

$$= 3\,(0.5) = 1.5$$

Again, $A = 4\pi\, r^2.$

$$\Rightarrow \qquad \delta A = 4\pi\, 2r\delta r = 8\pi r\delta r$$

i.e.
$$\frac{\delta A}{A} = \frac{8\pi r\, \delta r}{4\pi r^2} = 2\frac{\delta r}{r}$$

Hence
$$100 \times \frac{\delta A}{A} = 2\left(100 \times \frac{\delta r}{r}\right) = 2\,(0.5) = 1.$$

Thus, the required percentage errors in volume and surface area are 1·5% and 2% respectvely

250. If A be the area of a triangle, prove that the error in A resulting from a small error C is given by

$$\delta A = \frac{1}{4}\, A\, [s^{-1} + (s-a)^{-1} + (s-b)^{-1} - (s-c)^{-1}]\; \delta C.$$

Solution:

We know that

$$A = \sqrt{s\,(s-a)\,(s-b)\,(s-c)}$$

$$\Rightarrow \qquad \log A = \frac{1}{2}\,[\log s + \log\,(s-a) + \log\,(s-b) + \log\,(s-c)]$$

$$= f(s,\, c)\ (a,\, b\text{-constants})$$

Differentiate [1].

$$\frac{\delta A}{A} = \frac{1}{2}\left[\frac{\delta s}{s} + \frac{\delta s}{s-a} + \frac{\delta s}{s-b} + \frac{\delta s - \delta c}{s-c}\right] \qquad [1]$$

Now
$$s = \frac{a+b+c}{2}$$

$$\Rightarrow \qquad \delta s = \frac{1}{2}\,\delta c$$

$\therefore$ From [1],

$$\delta A = \frac{1}{2}A\left[\frac{\left(\frac{1}{2}\delta c\right)}{s} + \frac{\left(\frac{1}{2}\delta c\right)}{(s-a)} + \frac{\left(\frac{1}{2}\delta c\right)}{(s-b)} + \frac{\left(\frac{1}{2}\delta c\right)}{(s-c)}\right]$$

$$= \frac{1}{4}\,A\,[s^{-1} + (s-a)^{-1} + (s-b)^{-1} - (s-c)^{-1}]\;\delta c$$

251. The radius of a sphere is found to be 10 cm with a possible error of 0.02 cm. What is the relative error in computing the volume ?

Solution:

$$\text{Volume} \quad = V = \frac{4}{3}\pi r^3$$

$$\Rightarrow \quad \log r = \log\left(\frac{4}{3}\pi\right) + 3\log r$$

$$\text{Differentiate,} \quad \frac{dr}{r} = 0 + \frac{3dr}{r}$$

Here, $r = 10$ cm, $dr = 0.02$ cms

$$\therefore \quad \text{Relative errors in volume } \frac{dr}{r} = 3\left(\frac{0.02}{10}\right) = 0.006 \text{ c.c}$$

252. The focal length of a mirror is given by the formula

$\frac{1}{v} - \frac{1}{u} = \frac{2}{f}$. If equal errors δ

are made in the determination of u and v, show that the relative error in the focal length is given by $\left[\frac{1}{u} + \frac{1}{v}\right]\delta$.

Solution: **Given:** $\frac{1}{u} - \frac{1}{v} = \frac{2}{f}$

or $f = \frac{2uv}{u - v}$

Taking log on both sides, we get

$$\log f = \log 2 + \log u + \log v - \log (u - v) \qquad [1]$$

Differentiate [1],

$$\frac{df}{f} = \frac{du}{u} + \frac{dv}{v} - \frac{du - dv}{u - v}$$

$$\leq \text{Error Function } \frac{\Delta f}{f} = \frac{\Delta u}{u} + \frac{\Delta v}{v} - \frac{\Delta u - \Delta v}{u - v}$$

$$= \frac{\delta}{u} + \frac{\delta}{v} - \frac{\delta - \delta}{u - v} = \left[\frac{1}{u} + \frac{1}{v}\right]\delta.$$

253. The side of a triangle ABC is calculated from the sides b and c and the angle A. If small errors δc, δb, δA are made in the values of c, b and A respectively prove that the error δa in the calculated value of a is equal to

$$\cos B \cdot \delta c + \cos C\, \delta b + b \sin C\, \delta A$$

Solution: We know that from trigonometry, the cosine rule is given by

$$a^2 = b^2 + c^2 - 2bc \cos A \qquad [1]$$

Differentiating, [1],

$$2a\,da = 2bdb + 2cdc + 2bc \sin A\, dA - 2c \cos A\, db - 2b \cos A dc$$

or,
$$2a\delta a = 2b\delta b + 2c\delta c + 2\text{ bc} \sin A\, \delta A - 2c \cos A\, \delta b - 2\, b \cos A\, \delta c$$

$$\therefore \quad \delta a = \left[\frac{b - c \cos A}{a}\right]\delta b + \left[\frac{c - b \cos A}{a}\right]\delta c + \left[\frac{bc \sin A}{a}\right]\delta A$$

$$= \left[\frac{a \cos C}{a}\right]\delta b + \left[\frac{a \cos B}{a}\right]\delta c + b\left[\frac{2R \sin C \sin A}{2R \sin A}\right]\delta A$$

($\because b = c \cos A + a \cos B$)

$$\therefore \delta a = \cos C\, \delta b + \cos B\, \delta c + b \sin C\, \delta A$$

254. **In a plane triangle, if the sides *a*, *b* be constant, prove that the variation of its angles are given by the relations**

$$\frac{dA}{\sqrt{a^2 - b^2 \sin^2 A}} = \frac{dB}{\sqrt{b^2 - a^2 \sin^2 B}} = \frac{dC}{c},$$

the letters having their usual significance.

Solution: We know from trigonometry,

By sine rule,

$$\frac{a}{\sin A} = \frac{b}{\sin B}$$

$$\therefore \quad a \sin B = b \sin A \qquad [1]$$

Differentiate [1],

$$a \cos B \cdot dB = b \cos A \cdot dA$$

or
$$\frac{dA}{a \cos B} = \frac{dB}{b \cos A} = \frac{dA + dB}{a \cos B + b \cos A}$$

(by an algebraic principle)

Now
$$a \cos B = a\sqrt{1 - \sin^2 B}$$

$$= \sqrt{a^2 - a^2 \sin^2 B}$$

$$= \sqrt{a^2 - b^2 \sin^2 A} \qquad \text{[using (1)]}$$

$$a \cos B + b \cos A = C$$

(by projection formula)

and
$$A + B + C = \pi, \Rightarrow A + B = \pi - C$$

$$\therefore \quad dA + dB = -dc$$

$\therefore$ From [2], we have

$$\frac{dA}{\sqrt{a^2 - b^2 \sin^2 A}} = \frac{dB}{\sqrt{b^2 - a^2 \sin^2 B}} = \frac{-dC}{C}$$

255. If $ax^2 + by^2 + cz^2 = 1$
and $lx + my + nz = 0$,

prove that $$\frac{dx}{bny - cmz} = \frac{dy}{clz - anx} = \frac{dz}{amx - bly}$$

Solution: **Given:** $ax^2 + by^2 + cz^2 = 1$ [1]

and $lx + my + nz = 0$ [2]

Taking differentials in [1] and [2]

$$2axdx + 2bydy + 2\,czdz = 0$$

or $axdx + bydy + czdz = 0$ [3]

and $ldx + mdy + ndz = 0$ [4]

Solving for dx, dy, dz, by Cross-multiplication method

$$\frac{dx}{bny - cmz} = \frac{dy}{clz - anx} = \frac{dz}{amx - bly}$$

256. If $x^2 + y^2 + z^2 - 2xyz = 1$,

show that $$\frac{dx}{\sqrt{(1-x^2)}} + \frac{dy}{\sqrt{(1-y^2)}} + \frac{dz}{\sqrt{(1-z^2)}} = 0$$

Solution: By data, we have

$x^2 + y^2 + z^2 - 2xyz = 1$ [1]

Taking differentials, we get

$$2xdx + 2ydy + 2zdz - 2yzdx - 2zxdy - 2xydz = 0$$

$(x - yz)\,dx + (y - zx)\,dy + (z - xy)\,dz = 0$ [2]

Consider $(x - yz)^2$

$$= x^2 - 2xyz + y^2z^2$$
$$= 1 - y^2 - z^2 + y^2z^2 \quad \text{(by [1])}$$
$$= (1 - y^2) - z^2(1 - y^2)$$
$$= (1 - y^2)(1 - z^2)$$

$\Rightarrow$ $$x - yz = \sqrt{(1-y^2)(1-z^2)}$$

|||ly $$y - zx = \sqrt{(1-z^2)(1-x^2)}$$

and $$z - xy = \sqrt{(1-x^2)(1-y^2)}$$

$\therefore$ From [2], we have

$$\sqrt{(1-y^2)(1-z^2)}\,dx + \sqrt{(1-z^2)(1-x^2)}\,dy + \sqrt{(1-x^2)(1-y^2)}\,dz = 0$$

$\div \sqrt{(1-x^2)(1-y^2)(1-z^2)}$, we get

$$\frac{dx}{\sqrt{1-x^2}} + \frac{dy}{\sqrt{1-y^2}} + \frac{dz}{\sqrt{1-z^2}} = 0$$

PART-B

257. The current measured by a tangent galvanometer is given by the relation $c = k \tan\theta$ where θ is the angle of deflection. Show that the relative error in c due to a given error in θ is minimum when $\theta = 45°$. [VTU F/M, 2005]

Solution: **Given :** $c = k \tan\theta$; (k = constant)

Take log on both sides, we get

$$\log c = \log k + \log(\tan\theta)$$

$$\Rightarrow \quad \delta(\log c) = \delta(\log k) + \delta \log(\tan\theta)$$

$$\Rightarrow \quad \frac{1}{c}\,\delta c = 0 + \frac{\sec^2\theta}{\tan\theta}\,\delta\theta$$

$$\Rightarrow \quad \frac{\delta c}{c} = \frac{\cos\theta}{\sin\theta}\cdot\frac{1}{\cos^2\theta}\,\delta\theta \text{ or } \frac{\delta c}{c} = \frac{\delta\theta}{\sin\theta\cos\theta}$$

$$\Rightarrow \quad \frac{\delta c}{c} = \frac{2}{\sin 2\theta}\,d\theta$$

Now, $\dfrac{\delta c}{c}$ = relative error in c which when $\sin 2\theta$ is maximum, $\delta c/c$ is minimum. Its min. value is 1.

$$\therefore \quad \sin 2\theta = 1 \;\Rightarrow\; 2\theta = 90° \text{ or } \theta = 45°.$$

Thus the relative error in c is minimum when $\theta = 45°$.

EXERCISES

1. What error in the common logarithm of a number will be produced by an error of 1% in the number? (*Ans.* 0.01)
2. In calculating the volume of a right circular cone, errors in defect of 2% and 1% are made in measuring the height and radius of base respectively. Find the error in the calculated volume. (*Ans.* – 4%)
3. Find approximate value of $\sqrt{(0.98)^2 + (2.01)^2 + (1.94)^2}$ (*Ans.* 2.96)
4. Find the possible percentage error in computing the parallel resistance r of three resistance r_1, r_2, r_3 from the formula $\dfrac{1}{r} = \dfrac{1}{r_1} + \dfrac{1}{r_2} + \dfrac{1}{r_3}$

   If r_1, r_2, r_3 are each in error by plus 1.2% (*Ans:* 1.2%)
5. The time t of oscillations of a simple pendulum of length l is given by $t = 2\pi\sqrt{(l/g)}$ where l is the length of the pendulum and g is a constant.

(*a*) prove that if a clock pendulum $\frac{1}{2}$ meter long is shortened by one millimeter, the clock will go faster by 86.4 secs. every day (*Ans.* 86.4 seconds)

6. The area of a spherical balloon (r = 100) increases by 0·1%, find the percentage increase in the volume (*Ans.* 0.3)

7. If in a triangle, *ABC*, the side a and *A* remain constant and the other sides and angles are slightly varied show that $\frac{\delta b}{\cos B} + \frac{\delta c}{\cos C} = 0$

8. If the formula $T = 2\pi\sqrt{(l/g)}$ is used to calculate the value of g from observations of T and l. Find the percentage error in g due to an error of (0.2) percent in the value of T.

(*Ans.* 0.4 percent)

9. The error in the measurement of the diameter of a sphere is 1%. Find the percentage error in (*i*) the volume (*ii*) the surface area of the sphere $\left(Ans: \frac{3}{d}, \frac{2}{d}\right)$

10. In triangle *ABC* the side b and the angle B are fixed and the remaining elements undergo small changes. Show that Cos $A\delta c$ + cos $C\delta a = 0$

11. The dimensions of a cone are: radius 4cm. and height 6 cm. What is the error in the calculated volume of the cone if there is a shortage of 0.01 cm. per cm. in the measure used.

12. Find the percentage error in the evaluation of the area of an ellipse if – 2% error is made while measuring the major axis and + 1.5% error is made while measuring the minor axis

(*Ans.* – 0.5)

13. The diameter of a right circular cylinder is measured as 6.0 ± 0.01 cm. While its height is measured correctly. Find the resulting error in the volume per unit height. (*Ans.* ± 0.03)

PART-B

PART-C

3

Integral Calculus

REDUCTION FORMULAE FOR THE FUNCTIONS: $\sin^n x$, $\cos^n x$, $\tan^n x$, $\cot^n x$, $\sec^n x$, $\text{cosec}^n x$ AND $\sin^m x \cos^n x$ – EVALUATION OF THESE INTEGRALS WITH STANDARD LIMITS

Reduction formulae reduces a given integral to a known integration form by the repeated application of integration by parts. The relations or formulas expressing an integral as the sum of certain functions and a simpler integral are called reduction formulae. Such relations or formulas are most commonly derived by integration by parts. Some useful properties of definite integrals

1. $\int_0^a f(x)dx = \int_0^a f(a-x)dx$;

2. $\int_0^a f(x)dx = \begin{bmatrix} 2\int_0^{a/2} f(x)dx, \text{if } f(a-x)=f(x) \\ 0, \text{if } f(a-x)=-f(x) \end{bmatrix}$

3. $\int_{-a}^a f(x)dx = \begin{bmatrix} 2\int_0^a f(x)dx, \text{if } f(x) \text{ is even, } i.e., f(-x)=f(x) \\ 0, \text{if } f(x) \text{ is odd } i.e., f(-x)=-f(x) \end{bmatrix}$

3.1 REDUCTION FORMULAE FOR: $\int \sin^n x\, dx$ and $\int \cos^n x\, dx$

Denote $I_n = \int \sin^n x \cdot dx$ (VTU, MQP – I, VTU March, 2000)

(n is a +ve integer)

We express I_n as follows:

$$I_n = \int \sin^n x \cdot dx = \int \sin^{n-1} x \cdot \sin x\, dx$$

and integrate by parts, we get

$$I_n = \sin^{n-1} x \cdot (-\cos x) - \int(-\cos x)\cdot(n-1)\sin^{n-2} x \cdot \cos x \cdot dx$$
$$= -\sin^{n-1} x \cdot \cos x + \int (n-1)\cdot \cos^2 x \cdot \sin^{n-2} x \cdot dx$$
$$= -\sin^{n-1} x \cdot \cos x + (n-1)\int(1-\sin^2 x)\cdot \sin^{n-2} x \cdot dx$$

$$= -\sin^{n-1} x \cdot \cos x + (n-1) \int \sin^{n-2} x \, dx - (n-1) \cdot \int \sin^n x \, dx$$

$$= -\sin^{n-1} x \cdot \cos x + (n-1) I_{n-2} - (n-1) I_n$$

Combining I_n terms

$$n \cdot I_n = \sin^{n-1} x \cdot \cos x + (n-1) \cdot I_{n-2}$$

Hence $$I_n = -\frac{\sin^{n-1} x \cdot \cos x}{n} + \frac{n-1}{n} I_{n-2} \qquad [1]$$

Cor. (*i*) If $$I_n = \int_0^{\pi/2} \sin^n x \, dx,$$

Then $$I_n = \left[\frac{-\sin^{n-1} x \cdot \cos x}{n}\right]_0^{\pi/2} + \frac{n-1}{n} I_{n-2}$$

$$= 0 + \frac{n-1}{n} I_{n-2}$$

Thus, we have: $$\boxed{I_n = \frac{n-1}{n} I_{n-2}, \ (n > 1)} \qquad [2]$$

Replacing n by $(n-2)$, $(n-4)$ in equation [2], we find:

$$I_{n-2} = \frac{n-3}{n-2} I_{n-4},$$

$$I_{n-4} = \frac{n-5}{n-4} I_{n-6},$$

... ...

... ...

Hence, $$I_n = \frac{n-1}{n} I_{n-2} = \frac{n-1}{n} \cdot \frac{n-3}{n-1} I_{n-4}$$

$$= \frac{(n-1)(n-3)(n-5)}{n(n-2)(n-4)} I_{n-6},$$

Case (*i*): When '*n*' is odd the last integral will be of the form

$$\int_0^{\pi/2} \sin x \, dx = \left[-\cos x\right]_0^{\pi/2} = 1$$

and hence when n is odd, we have

$$I_n = \frac{(n-1)(n-3)(n-5) \ldots 4.2}{n(n-2)(n-3) \ldots . 5.3} \cdot \int_0^{\pi/2} \sin x \, dx$$

$$= \frac{(n-1)(n-3)(n-5) \ldots 4.2}{n(n-2)(n-4) \ldots 5.3} \cdot 1 \qquad [3]$$

Case (*ii*) When 'n' is even, then the last integral will be of the form:

$$\int_0^{\pi/2} \sin x\, dx = \int_0^{\pi/2} 1\, dx = \frac{1}{2}\frac{\pi}{2}$$

Hence $$I_n = \frac{(n-1)(n-3)(n-5)\ldots 3.1}{n(n-2)(n-3)\ldots 4.2} \times \frac{\pi}{2} \quad [4]$$

For example,

1. If $I_n = \int_0^{\pi/2} \sin^n x\, dx$, then

$$I_2 = \frac{1}{2}\cdot\frac{\pi}{2} = \frac{\pi}{4}, \qquad I_3 = \frac{2}{3}.1 = \frac{2}{3}$$

$$I_4 = \frac{3}{4}\cdot\frac{1}{2}\cdot\frac{\pi}{2} = \frac{3\pi}{16}, \qquad I_5 = \frac{4}{5}\cdot\frac{2}{3}\cdot 1 = \frac{8}{15}$$

$$I_6 = \frac{5}{6}\cdot\frac{3}{4}\cdot\frac{1}{2}\cdot\frac{\pi}{2} = \frac{5\pi}{32}, \qquad I_7 = \frac{6}{7}\cdot\frac{4}{5}\cdot\frac{2}{3}\cdot 1 = \frac{16}{35}.$$

Reduction formula for $\int \cos^n x\, dx$ (VTU, August, 2001)

Denote $$I_n = \int \cos^n x\, dx \quad [1]$$

or, $$I_n = \int \cos^{n-1} x \cdot \cos x\, dx$$

Integrate by parts, Integrate (cos x) first, we obtain:

$$I_n = \cos^{n-1} x \cdot \sin x - \int \sin x\,(n-1)\cos^{n-2} x \cdot (-\sin x)\cdot dx$$

$$= \cos^{n-1} x \cdot \sin x + (n-1)\int \cos^{n-2} x \cdot \sin^2 x\, dx$$

$$= \cos^{n-1} x \cdot \sin x + (n-1)\int \cos^{n-2} x\, dx - (n-1)\int \cos^n x\, dx$$

$(\because\ \sin^2 x = 1 - \cos^2 x)$

$$= \cos^{n-1} x \cdot \sin x + (n-1)\, I_{n-2} - (n-1)\, I_n \quad \text{(using [1])}$$

Grouping I_n terms.

$$n \cdot I_n = \cos^{n-1} x \cdot \sin x + (n-1)\, I_{n-2}$$

or, $$\boxed{I_n = \frac{1}{n}\cos^{n-1} x \cdot \sin x + \frac{(n-1)}{n}\, I_{n-2}} \quad [2]$$

(Reduction formula for the indefinite integral [1])

If $I_n = \int_0^{\pi/2} \cos^n x\, dx$, then from [2],

we get $$I_n = \left[\frac{1}{n}\cos^{n-1} x \cdot \sin x\right]_0^{\frac{\pi}{2}} + \frac{(n-1)}{(n)}\, I_{n-2}$$

$$I_n = 0 + \left(\frac{n-1}{n}\right) I_{n-2}, \qquad \left(\text{Since, } \cos\frac{\pi}{2} = 0, \sin o = 0\right)$$

or, $$I_n = \left[\frac{n-1}{n}\right] I_{n-2} \qquad [3]$$

This is the Reduction formula for the definite integral.

$$I_n = \int_0^{\pi/2} \cos^n x \, dx$$

Now, from [3], we have,

$$I_n = \left(\frac{n-1}{n}\right) I_{n-2} = \left(\frac{n-1}{n}\right)\left(\frac{n-3}{n-2}\right) I_{n-4}$$

$$= \left(\frac{n-1}{n}\right)\left(\frac{n-3}{n-2}\right)\left(\frac{n-5}{n-4}\right) I_{n-6} \text{ and so on.}$$

Finally, $$I_n = \left(\frac{n-1}{n}\right)\left(\frac{n-3}{n-2}\right)\left(\frac{n-5}{n-4}\right) \ldots\ldots k \qquad [4]$$

Where $k = I_1$, (according as n is odd)

and $k = I_0$, (according as n is even)

Again, $I_1 = \int_0^{\pi/2} \cos x \, dx = \left[\sin x\right]_0^{\frac{\pi}{2}} = 1$

and $I_0 = \int_0^{\pi/2} \cos^0 x \, dx = \int_0^{\pi/2} dx = \left[x\right]_0^{\frac{\pi}{2}} = \frac{\pi}{2}$

Eqn. [4] gives the reduction formula for $I_n = \int_0^{\pi/2} \cos^n x dx$, $k = 1$ or $\frac{\pi}{2}$ (according as n is odd or even)

$$\int_0^{\pi/2} \sin^n x \, dx = \int_0^{\pi/2} \sin^n \left(\frac{\pi}{2} - x\right) dx = \int_0^{\pi/2} \cos^n x \, dx$$

(This formula is called Walli's formula)

Note: (*i*) The above formula is applicable only when limits are 0 to $\frac{\pi}{2}$

(*ii*) In order to evaluate

$$\int_0^{\pi/2} \sin^n x \, dx \quad \text{or} \quad \int_0^{\pi/2} \cos^n x \, dx$$

PART-C

Start with $(n - 1)$ in the N^r and go on diminishing by 2 till you get either 2 or 1. Similarly start with n in the D^r and go on diminishing by 2 till you get either 2 or 1.

(*iii*) Multiply by $k = \frac{\pi}{2}$ in case n is even.

Worked Examples (1 – 30)

1. $$\int_0^{\pi/2} \sin^8 x\, dx = \int_0^{\pi/2} \cos^8 x\, dx$$

$$= \frac{7 \cdot 5 \cdot 3 \cdot 1}{8 \cdot 6 \cdot 4 \cdot 2} \cdot \left(\frac{\pi}{2}\right) = \frac{35}{256}\pi$$

(We have multiplied by $\left(\frac{\pi}{2} = k\right)$ because n is even).

2. (*a*) $$\int_0^{\pi/2} \sin^7 x\, dx = \frac{6 \cdot 4 \cdot 2}{7 \cdot 5 \cdot 3 \cdot 1} \cdot 1 = \frac{16}{35}$$

(Here we did not multiply by $\frac{\pi}{2}$ as n is odd *i.e.,* 7)

(*b*) $$\int_0^{\pi/2} \sin^7 \left(\frac{x}{2}\right) dx = \frac{32}{35}$$

put $\frac{x}{2} = t$

$\therefore$ $dx = 2dt$

Limits: $x = 0, t = 0$

$x = \frac{\pi}{2}, t = \frac{\pi}{4}$

$$\therefore \quad I = 2 \int_0^{\pi/4} \sin^7 (t)\, dt = 2 \cdot \frac{16}{35} = \left(\frac{32}{35}\right)$$

3. (*a*) $$\int_0^{\pi/4} 8 \cos^4 x \sin^4 x\, dx$$

Now, $\sin 2x = 2 \sin x \cos x$

$$\therefore \quad I = \frac{1}{2} \int_0^{\pi/4} (\sin 2x)^4\, dx$$

Now, put $2x = t$;

$$\therefore \quad dx = \frac{1}{2}\, dt$$

and when $x = \frac{\pi}{4}$,

$$t = \frac{\pi}{2}$$

and when $x = 0;\ t = 0$

$$\therefore \quad I = \frac{1}{4}\int_0^{\pi/2} \sin^4 t\, dt = \frac{1}{4}\left[\frac{3}{4}\cdot\frac{1}{2}\cdot\frac{\pi}{2}\right] = \frac{3\pi}{64}$$

(b) $\int_0^{\pi/8} \cos^3(4x)\, dx$

$$I = \frac{1}{4}\int_0^{\pi/2} \cos^3 t\, dt = \frac{1}{4}\cdot\frac{2}{3} = \frac{1}{6}$$

4. $I_6 = \int_0^{\pi/2} \sin^6 x\, dx = \frac{5\cdot 3\cdot 1}{6\cdot 4\cdot 2}\left(\frac{\pi}{2}\right) = \frac{5\pi}{32}$

5. $I_6 = \int_0^{\pi/2} \cos^6 x\, dx = \frac{5\cdot 3\cdot 1}{6\cdot 4\cdot 2}\cdot\frac{\pi}{2} = \frac{5\pi}{32}$

6. $I_5 = \int_0^{\pi/2} \cos^5 x\, dx = \frac{4\cdot 2}{5\cdot 3\cdot 1}\cdot 1 = \frac{8}{15}$

($k = 1$, $\because$ 5 is odd)

7. Prove that: (Imp.)

$$I_{2n} = \int_0^{\pi/2} \sin^{2n} x\, dx = \frac{(2n)!}{(2^n \cdot n!)^2}\left(\frac{\pi}{2}\right) = \int_0^{\pi/2} \cos^{2n} x\, dx$$

Solution: We know that

$$\int_0^{\pi/2} \sin^{2n} x\, dx = \left[\frac{2n-1}{2n}\cdot\frac{2n-3}{2n-2}\cdot\frac{2n-5}{2n-4} \cdots\cdots \frac{3}{4}\cdot\frac{1}{2}\right]\frac{\pi}{2}$$

(Wallis formula)

Multiplying N^r and D^r by
$[2n\,(2n-2)\,(2n-4)\ \ldots\ 4.2]$

$$I_{2n} = \frac{2n\,(2n-1)(2n-2)(2n-3)\ldots 4\cdot 3\cdot 2\cdot 1}{[2n\,(2n-2)(2n-4)\ldots 4\cdot 2]^2}$$

$$N^r = (2n)!$$

and D^r each term has got 2 as a factor and there are n factors

$$\therefore \quad D^r = [2^n\,\{n\,(n-1)\ .\ .\ .\ (n-2)\ 2.1\}]^2 = [2^n \cdot n!]^2$$

$$\therefore \quad I_{2n} = \frac{(2n)!}{(2^n\cdot n!)^2}\cdot\frac{\pi}{2} \quad \text{Proved.}$$

PART-C

8. Evaluate: $\int_0^1 \frac{x^9}{\sqrt{1-x^2}}\,dx$ (VTU, Aug., 2000)

Solution: Let $I = \int_0^1 \frac{x^9}{\sqrt{1-x^2}}\,dx$

put $x = \sin\theta;$

$dx = \cos\theta\,d\theta$

Also, $\theta = \sin^{-1}(x).$

Limits: $x = 0,\ \theta = 0$

$x = 1,\ \theta = \frac{\pi}{2}$

$$\therefore \quad I = \int_0^{\pi/2} \frac{\sin^9\theta\cos\theta\,d\theta}{\sqrt{1-\sin^2\theta}} = \int_0^{\pi/2}\sin^9\theta\,d\theta = \frac{8\cdot 6\cdot 4\cdot 2}{9\cdot 7\cdot 5\cdot 3\cdot 1}\cdot 1 = \frac{128}{315}$$

9. (Problems similar to Ex. 8)

Evaluate $\int_0^a \frac{x^{5/2}}{\sqrt{a-x}}\,dx$

Solution: Let $I = \int_0^a \frac{x^{5/2}}{\sqrt{a-x}}\,dx$

put $x = a\sin^2\theta;$

$(a-x) = a\cos^2\theta;$

$dx = 2a\sin\theta\cos\theta\,d\theta$

$(x = 0,\ \theta = 0);$

$\left[x = a, \sin^2\theta = 1, \theta = \frac{\pi}{2}\right]$

Hence
$$I = \int_0^a \frac{x^{5/2}}{\sqrt{a-x}}\,dx = \int_0^{\pi/2} \frac{(a\sin^2\theta)^{5/2}\cdot 2a\cdot\sin\theta\cdot\cos\theta\,d\theta}{\sqrt{(a\cos^2\theta)}}$$

$$= \int_0^{\pi/2} \frac{a^2\sin^5\theta\cdot 2a\sin\theta\cos\theta\,d\theta}{\cos\theta} = 2a^3\cdot\int_0^{\pi/2}\sin^6\theta\,d\theta$$

$$= 2a^3\left[\frac{5\cdot 3\cdot 1}{6\cdot 4\cdot 2}\left(\frac{\pi}{2}\right)\right] = \frac{5\pi a^3}{16}$$

10. Evaluate $\int_0^a \frac{\sqrt{x}\, dx}{\sqrt{(a-x)}}$

Solution:

$$\begin{bmatrix} x = a\sin^2\theta, (a-x) = a\cos^2\theta, dx = 2a\sin\theta\cos\theta\, d\theta \\ (x = 0, \sin\theta = 0, \theta = 0);\ x = a, \sin\theta = 1, \theta = \left(\frac{\pi}{2}\right) \end{bmatrix} \quad [1]$$

$$I = \int_0^a \frac{\sqrt{x}\, dx}{\sqrt{(a-x)}} = \int_0^{\pi/2} \frac{\sqrt{(a\sin^2\theta)}\, 2a\sin\theta\cos\theta\, d\theta}{\sqrt{(a\cos^2\theta)}}$$

$$= \int_0^{\pi/2} \frac{2a^2\sin^2\theta\cos\theta\, d\theta}{\cos\theta} = 2a\int_0^{\pi/2} \sin^2\theta\, d\theta$$

$$= 2a\left[\left(\frac{1}{2}\right)\left(\frac{\pi}{2}\right)\right] = \left(\frac{\pi a}{2}\right).$$

11. Evaluate:

$$\int_0^{\pi} x\sin^8 x\, dx$$

Solution: Let

$$I = \int_0^{\pi} x\sin^8 x\, dx$$

$$= \int_0^{\pi} (\pi - x)\sin^8(\pi - x)\, dx \qquad \left(\because \int_0^a f(x)\, dx = \int_0^a f(a-x)\, dx\right)$$

$$= \int_0^{\pi} (\pi - x)\sin^8 x\, dx \qquad (\because \sin^8(\pi - \theta) = \sin^8\theta)$$

$$= \pi\int_0^{\pi} \sin^8 x\, dx - \int_0^{\pi} x\sin^8 x\, dx$$

$$I = 2\pi\int_0^{\pi/2} \sin^8 x\, dx - I$$

$$\Rightarrow \qquad 2I = 2\pi\int_0^{\pi/2} \sin^8 x\, dx$$

$$= 2\pi \cdot \frac{(7\cdot 5\cdot 3\cdot 1)}{(8\cdot 6\cdot 4\cdot 2)} \cdot \left(\frac{\pi}{2}\right) = (\pi^2)\,\frac{35}{128}$$

$$\therefore \qquad I = \frac{35\pi^2}{256}$$

12. Evaluate:

$$\int_0^{\pi} x \sin^n x \, dx$$

Solution:

We know that $\int_0^b f(x)\,dx = \int_0^b f(b-x)\,dx$ [1]

and $\sin(\pi - x) = \sin x$ [2]

Again, $\int_0^{\pi} f(\sin x)\,dx = 2\int_0^{\pi/2} f(\sin x)\,dx$ [3]

Let $I = \int_0^{\pi} x \sin^n x \, dx,$

Hence by [1], we have

$$I = \int_0^{\pi} (\pi - x)\sin^n(\pi - x)\,dx$$

$$= \int_0^{\pi} (\pi - x)\sin^n x\,dx$$

$$\therefore \qquad 2I = \int_0^{\pi} (x\sin^n x)\,dx + \int_0^{\pi} (\pi - x)\sin^n x\,dx$$

$$= \int_0^{\pi} [x\sin^n x + (\pi - x)\sin^n x]\,dx$$

$$= \int_0^{\pi} \pi \sin^n x\,dx = \pi \cdot 2\int_0^{\pi/2} \sin^n x\,dx \qquad \text{(using [3])}$$

$$\therefore \qquad I = \pi \int_0^{\pi/2} \sin^n x\,dx$$

$$= (\pi)\,\frac{(n-1)(n-3)(n-5)\ldots}{n(n-2)(n-4)\ldots}\cdot k$$

$[k = 1 \text{ or } (\pi/2)]$

13. Evaluate

$$\int_0^{\pi/6} \sin^8 (3x)\, dx$$

put $3x = t$	Limits
$dx = \dfrac{dt}{3}$	$x = 0,\ t = 0$
	$x = \dfrac{\pi}{0},\ t = \dfrac{\pi}{2}$

Let $$I = \int_0^{\pi/6} \sin^8 (3x)\, dx$$

$$= \int_0^{\pi/2} \sin^8 (t) \cdot \frac{dt}{3} = \frac{1}{3} \int_0^{\pi/2} \sin^8 t\, dt$$

$$= \frac{1}{3}\left[\frac{7\cdot 5\cdot 3\cdot 1}{8\cdot 6\cdot 4\cdot 2} \cdot \frac{\pi}{2}\right] = \frac{35\pi}{768}$$

14. Evaluate:

$$\int_0^{2\pi} \sin^7 (x/4)\, dx$$

Solution: Let $$I = \int_0^{2\pi} \sin^7 (x/4)\, dx \qquad [1]$$

put $$\frac{x}{4} = t$$

$\therefore$ $$dx = 4dt$$

Limits: $x = 0,\ t = 0$

$$x = 2\pi,\ t = \frac{\pi}{2}$$

$$I = \int_0^{\pi/2} \sin^7 (t) \cdot 4\, dt = 4 \int_0^{\pi/2} \sin^7 t\, dt$$

$$= (4)\left[\frac{(6\cdot 4\cdot 2)}{(7\cdot 5\cdot 3\cdot 1)}\right] \cdot 1 = \frac{4\times 48}{105} = 4 \times \frac{16}{35} = \frac{64}{35}$$

15. Evaluate

$$\int_0^{a} \frac{x^4}{\sqrt{a^2 - x^2}}\, dx$$

Solution:

Put $x = a \sin \theta;$

$dx = a \cos \theta \, d\theta$

Limits: When $x = 0,\ \theta = \sin^{-1}(0) = 0$

When $x = a,\ \theta = \sin^{-1} \dfrac{a}{a} = \dfrac{\pi}{2}$

Let
$$I = \int_0^a \frac{x^4}{\sqrt{(a^2 - x^2)}} dx = \int_0^{\pi/2} \frac{a^4 \sin^4 \theta (a \cos \theta \, d\theta)}{\sqrt{a^2 - a^2 \sin^2 \theta}}$$

$$= a^4 \int_0^{\pi/2} \sin^4 \theta \, d\theta = a^4 \cdot \left[\frac{(3 \cdot 1)}{(4 \cdot 2)}\left(\frac{\pi}{2}\right)\right]$$

$$= \frac{3\pi a^4}{16}$$

16. Show that

$$\int_0^a \frac{x^n}{\sqrt{ax - x^2}} dx = \frac{1 \cdot 3 \cdot 5 \ldots (2n-1)}{2 \cdot 4 \cdot 6 \ldots 2n} \cdot \pi a^n$$

Solution: Let
$$I = \int_0^a \frac{x^n}{\sqrt{ax - x^2}} dx$$ [1]

put $x = a \sin^2 \theta;$

$dx = 2a \sin \theta \cos \theta \, d\theta$

Limits: When $x = 0,\ \theta = 0,$

When $x = a,\ \theta = \dfrac{\pi}{2}$

$\therefore$
$$I = \int_0^{\pi/2} \frac{(a \sin^2 \theta)^n}{\sqrt{a^2 \sin^2 \theta - a^2 \sin^4 \theta}} \cdot 2a \sin \theta \cos \theta \, d\theta$$

$$= \int_0^{\pi/2} \frac{a^n \sin^{2n} \theta}{a \sin \theta \cos \theta} 2a \sin \theta \cos \theta \, d\theta = 2a^n \int_0^{\pi/2} \sin^{2n} \theta \, d\theta$$

$$= (2a^n) \frac{(2n-1)(2n-3) \ldots 5 \cdot 3 \cdot 1}{2n(2n-2) \ldots 6 \cdot 4 \cdot 2} \left(\frac{\pi}{2}\right)$$

(2n is even)

$$= \frac{1 \cdot 3 \cdot 5 \ldots (2n-1)}{2 \cdot 4 \cdot 6 \ldots 2n} \pi a^n$$

17. Evaluate $\int \cos^5 x\,dx$, using reduction formula.

Solution: We know that

If $\quad I_n = \int \cos^n x\,dx$, then

$$I_n = \frac{\cos^{n-1} x \sin x}{n} + \frac{(n-1)}{n} I_{n-2} \quad [1]$$

For, $\quad (n = 5)$, we have

$$I_5 = \frac{\cos^4 x \sin x}{5} + \frac{4}{5} I_3$$

$$I_3 = \frac{\cos^2 x \sin x}{5} + \frac{2}{3} I_1$$

$$I_1 = \int \cos x\,dx = -\sin x$$

$$\therefore \quad \int \cos^5 x\,dx = \frac{\cos^4 x \sin x}{5} + \frac{4}{5}\left\{\frac{\cos^2 x \sin x}{3} + \frac{2}{3}(-\sin x)\right\}$$

$$= \sin x\left[\frac{\cos^4 x}{5} + \frac{4}{15}\cos^2 x - \frac{8}{15}\right]$$

18. Prove that

$$\int_0^a (a^2 - x^2)^n\,dx = \frac{2n(2n-2)....4\cdot 2}{(2n+1)(2n-1)....5\cdot 3\cdot 1} a^{2n+1}$$

Deduce that $\quad I_n = \frac{2n}{2n+1} a^2 I_{n-1}$

Solution:

Let $\quad I_n = \int_0^a (a^2 - x^2)^n\,dx, n > 0$

put $\quad x = a \sin \theta;$

$$dx = a \cos \theta\,d\theta$$

Limits: $\quad x = 0, \theta = 0$

$$x = a, \theta = \frac{\pi}{2}$$

$$\therefore \quad I_n = \int_0^{\pi/2} (a^2 - a^2 \sin^2 \theta)^n \cdot (a \cos \theta\,d\theta)$$

PART-C

$$= \int_0^{\pi/2} a^{2n+1} \cdot \cos^{2n+1}\theta \, d\theta$$

$$= a^{2n+1} \cdot \frac{2n \cdot 2n-2 \ldots . 4 \cdot 2}{(2n+1)(2n-1) \ldots . 5 \cdot 3 \cdot 1} \qquad [1]$$

Replace n by $n - 1$ in the above result.

$$I_{n-1} = a^{2n-1} \cdot \frac{2n-2 \cdot 2n-4 \ldots . 4 \cdot 2}{(2n-1)(2n-3) \ldots . 5 \cdot 3} \qquad [2]$$

$$\therefore \quad \frac{[1]}{[2]} \text{ gives } \frac{I_n}{I_{n-1}} = a^2 \cdot \frac{2n}{2n+1}$$

$$\Rightarrow \quad \boxed{I_n = \frac{2n}{2n+1} \; a^2 \cdot I_{n-1}}$$

19. If n is a +ve integer $P.T$

$$\int_0^{\infty} \frac{dx}{(x^2 + a^2)^n} = \frac{1}{a^{2n-1}} \cdot \frac{(2n-3)(2n-5) \ldots .1}{(2n-2)(2n-4) \ldots . 2} \cdot \frac{\pi}{2}$$

if $n > 1$,

Solution:

Put $x = a \tan \theta$;

$dx = a \sec^2 \theta \, d\theta$

Limits:	when $x = 0$, $\theta = 0$
	when $x = \infty$, $\theta = \frac{\pi}{2}$

$$\therefore \quad I = \int_0^{\pi/2} \frac{a \sec \theta \, d\theta}{(a^2 \tan^2 \theta + a^2)^n} = \int_0^{\pi/2} \frac{a \sec^2 \theta \, d\theta}{a^{2n} \sec^{2n} \theta}$$

$$= \frac{1}{a^{2n-1}} \int \cos^{2n-2} \theta \, d\theta$$

$$= \frac{1}{a^{2n-1}} \cdot \frac{(2n-3)(2n-5) \ldots . 5 \cdot 3 \cdot 1}{(2n-2)(2n-4) \ldots . 6 \cdot 4 \cdot 2} \left(\frac{\pi}{2}\right)$$

($\because n > 1$, $2n - 2$ is even)

Note: If $n = 1$, then

$$I = \int_0^{\infty} \frac{dx}{x^2 + a^2} = \left[\frac{1}{a} \tan^{-1}\left(\frac{x}{a}\right)\right]_0^{\infty} = \frac{\pi}{2a}$$

20. Evaluate

$$\int_0^{2a} \frac{x^{9/2}}{\sqrt{2a-x}}\, dx\cdot$$

Solution:

Put $x = 2a\sin^2\theta;$

$dx = 4a\sin\theta\cos\theta\, d\theta$

Limits: when $x = 0,\ \theta = 0$

when $x = 2a,\ \theta = \dfrac{\pi}{2}$

$$\therefore \quad I = \int_0^{\pi/2} \frac{(2a\sin^2\theta)^{9/2}}{\sqrt{2a-2a\sin^2\theta}}\, 4a\sin\theta\cos\theta\, d\theta$$

$$= 64\, a^5 \int_0^{\pi/2} \sin^{10}\theta\, d\theta$$

$$= 64a^5\cdot\frac{9}{10}\cdot\frac{7}{8}\cdot\frac{5}{6}\cdot\frac{3}{4}\cdot\frac{1}{2}\cdot\frac{\pi}{2} = \frac{63\pi a^5}{8}.$$

21. Evaluate

$$\int_0^{\pi} \frac{\sin^4\theta\,(1-\cos\theta)^{1/2}}{(1+\cos\theta)^2}\cdot d\theta$$

Solution:

Let $$I = \int_0^{\pi} \frac{[2\sin(\theta/2)\cos(\theta/2)]^4\,[2\sin^2(\theta/2)]^{1/2}}{[2\cos^2(\theta/2)]^2}\, d\theta$$

$$= 4\sqrt{2}\int_0^{\pi}\sin^5(\theta/2)\,d\theta = 4\sqrt{2}\int_0^{\pi}\sin^5 t\,(2dt)\ \left(\frac{\theta}{2}=t\right)$$

$$= 8\sqrt{2}\int_0^{\pi/2}\sin^5 t\, dt = 8\sqrt{2}\cdot\frac{4}{5}\cdot\frac{2}{3}\cdot 1 = \frac{64\sqrt{2}}{15}.$$

22. Evaluate $\displaystyle\int_0^{\pi} \frac{\sin^4\theta}{(1+\cos\theta)^2}\, d\theta$

Solution:

Let $$I = \frac{[2\sin(\theta/2)\cos(\theta/2)]^4}{[2\cos^2(\theta/2)]^2}\, d\theta$$

$$= \int_0^{\pi} \frac{16 \sin^4 (\theta/2) \cos^4 (\theta/2)}{4 \cos^4 (\theta/2)} d\theta$$

$$= 4 \int_0^{\pi} \sin^4 (\theta/2) d\theta = 4 \int_0^{\pi/2} \sin^4 t (2dt)$$

$$\left(\text{where } t = \frac{\theta}{2}\right)$$

Limits	$\theta = 0, t = 0$
	$\theta = \pi, t = \frac{\pi}{2}$

$$= 8 \int_0^{\pi/2} \sin^4 t \, dt = 8 \cdot \frac{3}{4} \cdot \frac{1}{2} \cdot \frac{\pi}{2} = \frac{3\pi}{2}$$

23. Evaluate $\int_0^a x^2 (a^2 - x^2)^{3/2} dx$

Solution: Let $I = \int_0^a x^2 (a^2 - x^2)^{3/2} dx$ [1]

put $x = a \sin \theta$

$dx = a \cos \theta \, d\theta$

Limits: when $x = 0, \theta = 0$

when $x = a, \theta = \frac{\pi}{2}$

$$(a^2 - x^2) = a^2 (1 - \sin^2 \theta) = a^2 \cos^2 \theta$$

$$\therefore \quad I = \int_0^{\pi/2} a^2 \sin^2 \theta \cdot a^3 \cos^3 \theta \cdot a \cos \theta \, d\theta$$

$$= a^6 \int_0^{\pi/2} \cos^4 \theta (1 - \cos^2 \theta) d\theta$$

$$= a^6 \left[\int_0^{\pi/2} \cos^4 \theta \, d\theta - \int_0^{\pi/2} \cos^6 \theta \, d\theta \right]$$

$$= a^6 \left[\frac{3 \cdot 1}{4 \cdot 2} \cdot \left(\frac{\pi}{2}\right) - \frac{5 \cdot 3 \cdot 1}{6 \cdot 4 \cdot 2} \left(\frac{\pi}{2}\right) \right]$$

$$= a^6 \frac{3\pi}{8} \left[1 - \frac{5}{6}\right] = \frac{a^6 \pi (3)}{8} \left(\frac{1}{6}\right)$$

$$\boxed{I = \frac{a^6 \pi}{16}}$$

24. Evaluate $\int_0^{\infty} \frac{x^2\, dx}{(1+x^2)^2}$

Solution: Let $I = \int_0^{\infty} \frac{x^2\, dx}{(1+x^2)^2}$ [1]

put $\begin{cases} x = \tan\theta, dx = \sec^2\theta\, d\theta, (1+x^2) = \sec^2\theta \\ [x = 0, \tan\theta = 0, \theta = 0]; \left[x = \infty, \tan\theta = \infty, \theta = \frac{\pi}{2}\right] \end{cases}$

$$\therefore \quad I = \int_0^{\infty} \frac{x^2\, dx}{(1+x^2)^2} = \int_0^{\pi/2} \frac{\tan^2\theta \sec^2\theta\, d\theta}{(\sec^2\theta)^2}$$

$$= \int_0^{\pi/2} \frac{\tan^2\theta}{\sec^2\theta}\, d\theta = \int_0^{\pi/2} \sin^2\theta\, d\theta = \frac{1}{2}\cdot\frac{\pi}{2} = \frac{\pi}{4}$$

25. Evaluate $\int_0^{2a} y dx$, where

$$y^2 = \left[\frac{x^3}{(2a-x)}\right]$$

Solution: Let $I = \int_0^{2a} \left[\frac{x^3}{2a-x}\right]^{1/2} \cdot dx$

put:
$$x = 2a \sin^2\theta$$
$$(2a - x) = 2a \cos^2\theta;$$
$$dx = 4a \sin\theta \cdot \cos\theta\, d\theta$$

$$(x = 0,\ \theta = 0);\ \left[x = 2a, \sin\theta = 1, \theta = \left(\frac{\pi}{2}\right)\right]$$

$$I = \int_0^{2a} y\, dx = \int_0^{2a} \left[\frac{x^3}{2a-x}\right]^{1/2} \cdot dx$$

$$= \int_0^{\pi/2} \frac{\sqrt{8a^3 \sin^6\theta}}{\sqrt{(2a\cos^2\theta)}} \cdot 4a \sin\theta \cdot \cos\theta\, d\theta$$

$$= \int_0^{\pi/2} \frac{2a \sin^3 \theta}{\cos \theta} 4a \sin \theta \cos \theta \, d\theta$$

$$= 8a^2 \int_0^{\pi/2} \sin^4 \theta \, d\theta = 8a^2 \frac{(3 \cdot 1)}{(4 \cdot 2)} \cdot \left(\frac{\pi}{2}\right) = \frac{3\pi a^2}{2}$$

26. Evaluate: $\int_0^3 \frac{x^{5/2} \, dx}{\sqrt{(3-x)}}$

Solution: $\begin{bmatrix} x = 3 \sin^2 \theta; dx = 6 \sin \theta \cos \theta \, d\theta; (3 - x) = 3 \cos^2 \theta \\ (x = 0, \sin \theta = 0, \theta = 0); x = 3, \sin \theta = 1, \theta = (\pi/2) \end{bmatrix}$

$$\therefore \qquad I = \int_0^3 \frac{x^{3/2}}{\sqrt{3-x}} \cdot dx = \int_0^{\pi/2} \frac{(3 \sin^2 \theta)^{3/2}}{\sqrt{3 \cos^2 \theta}} \cdot 6 \sin \theta \cos \theta \, d\theta$$

$$= \int_0^{\pi/2} \frac{3\sqrt{3} \sin^3 \theta}{\sqrt{3} \cos \theta} 6 \sin \theta \cos \theta \, d\theta$$

$$= 18 \int_0^{\pi/2} \sin^4 \theta \, d\theta = 18 \cdot \frac{3 \cdot 1}{4 \cdot 2} \cdot \frac{\pi}{2} = \frac{27}{8} \pi.$$

27. Evaluate

$$\int_0^1 x^6 \sin^{-1} x \, dx$$

Solution: Let $\qquad I = \int_0^1 x^6 \sin^{-1} x \, dx$

put $\qquad \{\sin^{-1} x = t; x = \sin t, dx = \cos t \, dt\}$

$$[x = 0, \sin^{-1}(0) = t = 0] \left[x = 1, \sin^{-1}(1) = \frac{\pi}{2} = t\right]$$

$$\therefore \qquad I = \int_0^{\pi/2} t \sin^6 t \cos t \, dt$$

$$= \left[t \cdot \frac{\sin t}{t}\right]_0^{\pi/2} - \int_0^{\pi/2} \cdot 1 \cdot \frac{\sin^7}{7} - dt$$

$$= \frac{\pi}{14} - \frac{1}{7} \cdot \frac{6 \cdot 4 \cdot 2}{7 \cdot 5 \cdot 3 \cdot 1} = \left(\frac{\pi}{14}\right) - \frac{16}{245}$$

28. Show that

$$\int_0^1 \frac{x^{2n-1}}{\sqrt{(1-x^2)}}\,dx = \frac{2\cdot 4\cdot 6\ldots 2n}{3\cdot 5\cdot 7\ldots(2n-1)}$$

Solution: Let $$I = \int_0^1 \frac{x^{2n-1}}{\sqrt{(1-x^2)}}\,dx \quad [1]$$

put $$\left\{\begin{array}{l} x=\sin\theta;(1-x^2)=\cos^2\theta \\ dx=\cos\theta\,d\theta;[x=0,\text{gives }\sin\theta=0,\theta=0 \\ x=1,\text{gives }\sin\theta=1,\theta=\pi/2 \end{array}\right\}$$

$$\therefore \quad I = \int_0^1 \frac{x^{2n-1}}{\sqrt{(1-x^2)}}\,dx$$

$$= \int_0^{\pi/2} \frac{\sin^{2n-1}\theta\cdot\cos\theta\,d\theta}{\cos\theta} = \int_0^{\pi/2}\sin^{2n-1}\theta\,d\theta$$

[Since $(2n-1)$ is odd, and $k = 1$]

$$= \frac{(2n-2)(2n-4)\ldots 4\cdot 2}{(2n-1)(2n-3)\ldots 5\cdot 3}\,.\,k$$

$$= \frac{2\cdot 4\cdot 6\ldots\ldots(2n-4)(2n-2)(2n)}{3\cdot 5\cdot 7\ldots(2n-3)(2n-1)}$$

29. Evaluate: $\displaystyle\int_0^1 \frac{x^{2n}}{\sqrt{(1-x^2)}}\,dx$

Solution: Let $$I = \int_0^1 \frac{x^{2n}}{\sqrt{(1-x^2)}}\,dx$$

put $$\left\{\begin{array}{l} x=\sin\theta;dx=\cos\theta\,d\theta;\sqrt{1-x^2}=\cos\theta \\ x=0,\sin\theta=0,\theta=0;x=1,\sin\theta=1,\theta=\dfrac{\pi}{2} \end{array}\right\}$$

$$\therefore \quad I = \int_0^1 \frac{x^{2n}}{\sqrt{(1-x^2)}}\,dx = \int_0^{\pi/2}\frac{(\sin\theta)^{2n}\cdot\cos\theta\,d\theta}{\cos\theta}$$

$$= \int_0^{\pi/2}\sin^{2n}\theta\,d\theta = \frac{(2n-1)(2n-3)\ldots\cdot 3\cdot 1}{2n\cdot(2n-2)\ldots 4\cdot 2}\cdot\frac{\pi}{2}$$

PART-C

30. Evaluate: $\int_0^\infty \frac{dx}{(1+x^2)^{n+\left(\frac{1}{2}\right)}}$

Solution: Let $$I = \int_0^\infty \frac{dx}{(1+x^2)^{\frac{2n+1}{2}}} \qquad [1]$$

put $\left\{\begin{array}{l} x = \tan\theta ; dx = \sec^2\theta\, d\theta; \\ \text{when } x = 0; \theta = 0 \\ \text{when } x = \infty, \tan\theta = \infty; \theta = \frac{\pi}{2} \end{array}\right\}$

$$\therefore \quad I = \int_0^{\pi/2} \frac{\sec^2\theta\, d\theta}{(\sec^2\theta)^{\frac{2n+1}{2}}} = \int_0^{\pi/2} \cos^{2n-1}\theta\, d\theta$$

$$= \frac{(2n-2)(2n-4)\ldots 4\cdot 2}{(2n-1)\cdot(2n-3)\ldots 5\cdot 3\cdot 1}$$

3.2 REDUCTION FORMULA FOR $\int \sin^m x \cos^n x\, dx$,

[(*m*, *n* are positive integers)] (VTU, March, 1999)

Let $$I(m, n) = \int \sin^m x \cdot \cos^n x\, dx \qquad [1]$$

$$= \int (\sin^m x \cos x)\cos^{n-1} x\, dx \qquad \text{(by parts)}$$

$$= \cos^{n-1} x \left(\frac{\sin^{m+1} x}{m+1}\right) - \int \frac{\sin^{m+1} x}{m+1}(n-1)\cos^{n-2} x\,(-\sin x)\cdot dx$$

$$= \cos^{n-1} x \left(\frac{\sin^{m+1} x}{m+1}\right) + \left(\frac{n-1}{m+1}\right)\int \sin^m x \cos^{n-2} x \sin^2 x\, dx$$

$$= \frac{\cos^{n-1} x \sin^{m+1} x}{m+1} + \left(\frac{n-1}{m+1}\right)\int \sin^m x \cos^{n-2} x\,(1-\cos^2 x)\, dx$$

$$= \frac{\cos^{n-1} x \sin^{m+1} x}{m+1} + \left(\frac{n-1}{m+1}\right)\int \sin^m x \cos^{n-2} x\, dx$$

$$-\left(\frac{n-1}{m+1}\right)\int \sin^m x \cos^n x\, dx$$

$$I(m, n) = \frac{\cos^{n-1} x \sin^{m+1} x}{m+1} + \left(\frac{n-1}{m+1}\right) I(m, n-2)$$

$$-\left(\frac{n-1}{m+1}\right) I(m, n) \qquad \text{(using [1])}$$

$$\therefore \quad I(m, n) = \frac{\cos^{n-1} x \sin^{m+1} x}{m+n} + \left(\frac{n-1}{m+1}\right) I_{m, n-2} \qquad [2]$$

is the required reduction formula for the indefinite integral $I(m, n)$ given in (1)

Evaluation of the integral: $\int_0^{\pi/2} \sin^m x \cos^n x \, dx$

Let $$I(m, n) = \int_0^{\pi/2} \sin^m x \cos^n x \, dx$$

$$= \int_0^{\pi/2} (\sin^m x \cos x) \cos^{n-1} x \, dx \qquad [1]$$

Integrate by parts, integrating ($\sin^m x \cos x$) first.

put $\sin x = t, \cos x \, dx = dt$ [2]

$$\int \sin^m x \cos x \, dx = \int t^m \, dt = \frac{t^{m+1}}{m+1} = \frac{\sin^{m+1} x}{(m+1)} \qquad [3]$$

Hence, $$I(m, n) = \int_0^{\pi/2} (\sin^m x \cos x) \cos^{n-1} x \, dx$$

$$= \left[\frac{\sin^{m+1} x \cos^{n-1} x}{m+1}\right]_0^{\pi/2}$$

$$- \int_0^{\pi/2} \frac{\sin^{m+1} x}{(m+1)} (n-1) \cos^{n-2} x (-\sin x) \, dx$$

$$I(m, n) = \frac{1.0 - 0.1}{(m+1)} + \frac{(n-1)}{(m+1)} \int_0^{\pi/2} \sin^{m+2} x \cos^{n-2} x \, dx$$

$$\therefore \quad (m+1) I(m, n) = (n-1) \int_0^{\pi/2} \sin^m x \sin^2 x \cos^{n-2} x \, dx$$

$$= (n-1) \int_0^{\pi/2} \sin^m x (1 - \cos^2 x) \cos^{n-2} x \, dx$$

PART-C

$$= (n-1)\int_0^{\pi/2} (\sin^m x \cos^{n-2} x - \sin^m x \cos^n x)\, dx$$

$$= (n-1)\int_0^{\pi/2} \sin^m x \cos^{n-2} x\, dx - (n-1)\int_0^{\pi/2} \sin^m x \cos^n x\, dx$$

$$(m+1)\, I(m, n) = (n-1)\, I(m, n-2) - (n-1)\, I(m, n) \qquad [4]$$

$$\therefore \quad (m+1)\, I(m, n) + (n-1)\, I(m, n) = (n-1)\, I(m, n-2)$$

$$I(m, n)(m+1+n-1) = (n-1)\, I(m, n-2)$$

$$I(m, n)(m+n) = (n-1)\, I(m, n-2)$$

$$\therefore \quad I(m, n) = \frac{(n-1)}{(m+n)}\, I(m, n-2) \qquad [5]$$

Note: (*i*) by taking ($\sin x$) and integrating by parts the integral

$$\int_0^{\pi/2} \sin^{m-1} x\, (\sin x \cdot \cos^n x)\, dx, \text{ we can}$$

get a reduction formula where m is reduced and n is retained.

i.e., $I(m, n) = [(m-1)/(m+n)]\, [I(m-2, n)]$.

Thus, we have,

Note: (*ii*) $$I(m, n) = \frac{(m-1)}{(m+n)}\, I(m-2, n)$$

$$= \frac{(m-1)}{(m+n)}\, I(m, n-2)$$

So, the index of ($\sin x$) or of ($\cos x$) can be reduced at choice.

Working Rule

$$I(m, n) = \frac{\{(m-1)(m-3)\ldots][(n-1)(n-3)\ldots]}{[(m+n)(m+n-2)(m+n-4)\ldots]}\, k;$$

where $k = (\pi/2)$, if both m and n are even: and $k = 1$ otherwise,

N^r = Start with $(m-1)$ as first factor. Subtract 2 from it and get the 2nd factor $(m-3)$ and continue like this till the last factor is either 2 or 1. Similarly start with $(n-1)$. Subtract 2 from it and get the second factor $(n-3)$ and continue till the last factor is either 2 or 1.

D^r = Start with $(m+n)$ as first factor subtract 2 from it and get the 2nd factor $(m+n-2)$; continue till the last factor is either 2 or 1.

Multiply by $\left[k = \left(\frac{\pi}{2}\right)\right]$ only when both m and n are even positive integers.

The formulae for

$$\int_0^{\pi/2} \sin^m x\,dx \text{ and } \int_0^{\pi/2} \cos^n x\,dx$$

are indicated in the above

(*i.e.*, when $m = 0$ or $n = 0$), is the some as that when m is an even number (or when n is even).

In other words, 0 is considered or treated as an even number.

For:
$$\int_0^{\pi/2} \sin^m x\,dx = \int_0^{\pi/2} \sin^m x \cdot \cos^0 x\,dx$$

$$\int_0^{\pi/2} \cos^n x\,dx = \int_0^{\pi/2} \sin^0 x \cdot \cos^n x\,dx$$

$$\int_0^{\pi/2} \sin^6 x\,dx = \int_0^{\pi/2} \sin^6 x \cos^0 x\,dx$$

($m = 6$, $n = 0$ (both even))

$\therefore \quad k = (\pi/2)$

Worked Examples (31 – 45)

31. $$\int_0^{\pi/2} \sin^6\theta \cos^5\theta\,d\theta = \frac{(5\cdot3\cdot1)(4\cdot2)}{(11\cdot9\cdot7\cdot5\cdot3\cdot1)} = \frac{8}{693}$$

32. $$\int_0^{\pi/2} \sin^2\theta \cos^4\theta\,d\theta = \frac{(1)(3)}{6\cdot4\cdot2}(k) = \frac{(1\cdot3)}{(6\cdot4\cdot2)}\cdot\left(\frac{\pi}{2}\right) = \frac{\pi}{32}$$

33. $$\int_0^{\pi/4} (\cos 2\theta)^{3/2} \cdot \cos\theta\,d\theta$$

Solution: Let
$$I = \int_0^{\pi/4} (1 - 2\sin^2\theta)^{3/2} \cdot \cos\theta\,d\theta$$

put and
$$\left\{\begin{array}{l} \sqrt{2}\sin\theta = \sin z;\ \sqrt{2}\cos\theta\,d\theta = \cos z\,dz \\ \text{when } \theta = \dfrac{\pi}{4}, \sin z = 1, : z = \dfrac{\pi}{2} \\ \theta = 0, \text{ then } \sin z = 0 \end{array}\right\}$$

$$\therefore \quad I = \int_0^{\pi/2} (1 - \sin^2 z)^{3/2} \cdot \frac{\cos z\,dz}{\sqrt{2}}$$

$$= \frac{1}{\sqrt{2}} \int_0^{\pi/2} \cos^4 z\,dz = \frac{1}{\sqrt{2}} \cdot \frac{(3\cdot1)}{(4\cdot2)} \cdot \frac{\pi}{2} = \frac{3\pi}{16\sqrt{2}}$$

PART-C

34. $\int_0^{\pi/6} \cos^4(3\phi)\cdot\sin^2(6\phi)\,d\phi$

put $\quad 3\phi = t;$

$\therefore \quad 3d\phi = dt : d\phi = \dfrac{dt}{3}$

Limit: $\quad \phi = 0;\ t = 0\ ;\ \phi = \dfrac{\pi}{6}\ ;\ t = \dfrac{\pi}{2}$

$$\therefore \quad I = \frac{1}{3}\int_0^{\pi/2} \cos^4 t \sin^2 2t\, dt = \frac{1}{3}\int_0^{\pi/2} \cos^4 t\,(2\sin t\cos t)^2\, dt$$

$$= \frac{4}{3}\int_0^{\pi/2} \cos^6 t \sin^2 t\, dt$$

$$= \frac{4}{3}\left[\frac{(5\cdot 3\cdot 1)(1)}{8\cdot 6\cdot 4\cdot 2}\left(\frac{\pi}{2}\right)\right] = \frac{5\pi}{192}$$

35. Evaluate:

$$\int_0^{\pi} \sin^6 x \cos^4 x\, dx$$ (VTU, March, 2001)

Solution: Let $\quad I = \int_0^{\pi} \sin^6 x \cos^4 x\, dx$ [1]

$$= 2\int_0^{\pi/2} \sin^6 x \cos^4 x\, dx$$

$$\left\{\therefore \int_0^a f(x)\,dx = \begin{cases} 2\int_0^{a/2} f(x)\,dx, \text{ if } f(a-x) = f(x) \\ 0, \text{ if } f(a-x) = -f(x) \end{cases}\right\}$$

Also, $\sin^6(\pi - x)\cos^4(\pi - x) = \sin^6 x \cos^4 x$

$$= 2\cdot\frac{(5\cdot 3\cdot 1)(3\cdot 1)}{10\cdot 8\cdot 6\cdot 4\cdot 2}\cdot(k),\ \left(k = \frac{\pi}{2}\right) = \frac{3k}{128} = \frac{3\pi}{256}$$

36. Using the reduction formula find the value of:

$$\int_0^1 x^2 (1 - x^2)^{3/2}\, dx$$ (VTU, March, 1999)

Solution: Let $\quad I = \int_0^1 x^2 (1 - x^2)^{3/2}\, dx,$ [1]

put $\begin{cases} x = \sin\theta; dx = \cos\theta; d\theta \\ \text{limits : when } x = 0; \theta = 0 \\ \text{when } x = 1; \sin\theta = 1, \theta = \dfrac{\pi}{2} \end{cases}$

$$\therefore \qquad I = \int_0^{\pi/2} \sin^2\theta\,(\cos^2\theta)^{3/2}\,\cos\theta\,d\theta$$

$$= \int_0^{\pi/2} \sin^2\theta\,\cos^4\theta\,d\theta = \frac{(1)(3\cdot 1)}{6\cdot 4\cdot 2}\left(\frac{\pi}{2}\right) = \frac{\pi}{32}$$

$$\left(k = \frac{\pi}{2}, \text{ since } m = 2, n = 4 \text{ both even}\right)$$

37. Evaluate

$$\int_0^{\pi/6} \sin^2(6\theta)\cos^4(3\theta)\,d\theta$$ (VTU, August, 1999)

Solution: Let
$$I = \int_0^{\pi/6} \sin^2(6\theta)\cos^4(3\theta)\,d\theta$$

$$= \int_0^{\pi/6} [2\sin 3\theta\cos 3\theta]^2 \cdot \cos^4(3\theta)\,d\theta$$

$$= 4\cdot \int_0^{\pi/6} \sin^2(3\theta)\cos^6(3\theta)\,d\theta$$

Put $\begin{cases} 3\theta = t; \theta = \dfrac{t}{3} \\ \theta = 0; t = 0 \\ \theta = \pi/6, t = \left(\dfrac{\pi}{2}\right) \end{cases}$

$$= 4\int_0^{\pi/2} \sin^2(t)\cos^6(t)\left(\frac{dt}{3}\right)$$

$$= \frac{4}{3}\int_0^{\pi/2} \sin^2 t\cos^6 t\,(dt)$$

$$= \frac{4}{3}\cdot\frac{(1)(5\cdot 3\cdot 1)}{8\cdot 6\cdot 4\cdot 2}\cdot k = \frac{5k}{96}$$

$$= \frac{5}{96}\left(\frac{\pi}{2}\right) = \frac{5\pi}{192} \quad k = \frac{\pi}{2}, \text{ since } \begin{matrix} m = 2 \\ n = 6 \end{matrix} \text{ (both even)}$$

PART-C

38. Evaluate $\int_0^\infty \frac{x^6}{(1+x^2)^{9/2}}\,dx$

Solution: Let $$I = \int_0^\infty \frac{x^6}{(1+x^2)^{9/2}}\,dx$$ [1]

Put $\left\{\begin{array}{l} x = \tan\theta;\, dx = \sec^2\theta\, d\theta \\ \text{limits: when } x = 0, \tan\theta = 0, \theta = 0 \\ \text{when } x = \infty, \tan\theta = \infty, \theta = \dfrac{\pi}{2} \end{array}\right.$

$$\therefore \quad I = \int_0^{\pi/2} \frac{\tan^6\theta}{(1+\tan^2\theta)^{9/2}} \sec^2\theta\, d\theta$$

$$= \int_0^{\pi/2} \frac{\tan^6\theta \sec^2\theta}{\sec^9\theta}\, d\theta = \int_0^{\pi/2} \sin^6\theta \cos\theta\, d\theta$$

$$= \left[\frac{\sin^7\theta}{7}\right]_0^{\pi/2} = \frac{1}{7}$$

39. Evaluate:

(a) $\int_0^1 x^2 (1-x)^{5/2}\, dx$ (b) $\int_0^1 x^3 (1-x)^{5/2}\, dx$

(c) $\int_0^1 x^{3/2} (1-x)^{3/2}\, dx$ (d) $\int_0^1 x^5 (1-x)^{3/2}\, dx$

Solution: (a) Let $$I = \int_0^1 x^2 (1-x)^{5/2}\, dx$$ [1]

put $x = \sin^2\theta;\ (1-x) = \cos^2\theta;$

$dx = 2\sin\theta\cos\theta\, d\theta$

$(x = 0,\ \theta = 0);\quad [x = 1;\ \sin\theta = 1,\ \theta = (\pi/2)]$

$$I = \int_0^1 x^2 (1-x)^{5/2}\, dx$$

$$= \int_0^{\pi/2} (\sin^4\theta)(\cos^2\theta)^{5/2} \cdot 2\sin\theta\cos\theta\, d\theta$$

$$= \int_0^{\pi/2} \sin^4 \theta \cdot \cos^5 \theta \cdot 2 \sin\theta \cdot \cos\theta \, d\theta$$

$$= 2 \int_0^{\pi/2} \sin^5 \theta \cos^6 \theta \, d\theta = 2 \cdot \frac{(4 \cdot 2)(5 \cdot 3 \cdot 1)}{11 \cdot 9 \cdot 7 \cdot 5 \cdot 3 \cdot 1} \cdot 1 = \frac{16}{693}$$

Solution: (*b*) Let $I = \int_0^1 x^3 (1-x)^{5/2} \, dx$

$$\therefore \quad I = \int_0^1 x^3 (1-x)^{5/2} \, dx; \quad (x = \sin^2 \theta)$$

$$= \int_0^{\pi/2} (\sin^6 \theta)(\cos^2 \theta)^{5/2} \cdot 2 \sin\theta \cos\theta \, d\theta$$

$$= 2 \int_0^{\pi/2} \sin^6 \theta \cdot \cos^5 \theta \sin\theta \cos\theta \, d\theta$$

$$= 2 \int_0^{\pi/2} \sin^7 \theta \cos^6 \theta \, d\theta$$

$$= 2 \cdot \frac{(6 \cdot 4 \cdot 2)(5 \cdot 3 \cdot 1)}{13 \cdot 11 \cdot 9 \cdot 7 \cdot 5 \cdot 3 \cdot 1} \cdot 1 = \frac{32}{3003}$$

Solution: (*c*) Let $I = \int_0^1 x^{3/2} (1-x)^{3/2} \, dx$

put $x = \sin^2 \theta; \quad dx = 2 \sin \theta \cos \theta \, d\theta$

Limits: $x = 0; \sin^2 \theta = 0; \theta = 0$

$x = 1, \sin^2 \theta = 1, \theta = \pi/2$

$$\therefore \quad I = \int_0^{\pi/2} (\sin^2 \theta)^{3/2} (\cos^2 \theta)^{3/2} \cdot 2 \sin\theta \cos\theta \, d\theta$$

$$= 2 \int_0^{\pi/2} \sin^4 \theta \cos^4 \theta \, d\theta = 2 \frac{(3 \cdot 1)(3 \cdot 1)}{8 \cdot 6 \cdot 4 \cdot 2} \cdot \frac{\pi}{2} = \frac{3\pi}{128}$$

Solution: (*d*) Let $I = \int_0^1 x^5 (1-x)^{3/2} \, dx$

Solution: $\begin{bmatrix} x = \sin^2 \theta; (1-x) = \cos^2 \theta; dx = 2 \sin\theta \cos\theta \, d\theta \\ (x = 0, \sin\theta = 0, \theta = 0); (x = 1, \sin\theta = 1, \theta = (\pi/2)) \end{bmatrix}$

PART-C

$$\therefore \quad I = \int_0^1 x^5 (1-x)^{3/2}\, dx$$

$$= \int_0^{\pi/2} (\sin^2\theta)^5 (\cos^2\theta)^{3/2} \cdot 2\sin\theta\cos\theta\, d\theta$$

$$= 2\int_0^{\pi/2} \sin^{11}\theta \cos^4\theta\, d\theta$$

$$= 2\,\frac{(10\cdot 8\cdot 6\cdot 4\cdot 2)(3\cdot 1)}{(15\cdot 13\cdot 11\cdot 9\cdot 7\cdot 5\cdot 3\cdot 1)}\cdot 1 = \frac{512}{45045}$$

40. Evaluate the following

(*a*) $\int_0^{2a} \left[\sqrt{2\,ax - x^2}\right] dx$ (*b*) $\int_0^{2a} x\sqrt{(2\,ax - x^2)}\, dx$

(*c*) $\int_0^{2a} x^2 \left[\sqrt{(2\,ax - x^2)}\right] dx$ (*d*) $\int_0^{2a} x^3 \left[\sqrt{(2\,ax - x^2)}\right] dx$

(VTU F/M, 2005)

Note: The standard substitution for the above (*a*), (*b*), (*c*), (*d*) is to put $x = 2a\sin^2\theta$ and adjust the limits.

Solution: (*a*)

Put
$$x = 2a\sin^2\theta;$$
$$dx = 2a\cdot 2\sin\theta\cdot\cos\theta\, d\theta$$
$$(2a - x) = (2a - 2a\sin^2\theta)$$
$$= (2a)(1-\sin^2\theta) = 2a\cos^2\theta$$

$$\begin{Bmatrix} (x = 0, \sin\theta = 0, \theta = 0), \\ x = 2a, 2a\sin^2\theta = 2a, \sin\theta = 1, \theta = (\pi/2) \end{Bmatrix}$$

$$\therefore \quad I = \int_0^{2a} \sqrt{2ax - x^2}\, dx = \int_0^{2a}\left[\sqrt{x(2a-x)}\right] dx$$

$$= \int_0^{\pi/2} \sqrt{[2a\sin^2\theta\cdot 2a\cos^2\theta]}\cdot 4a\sin\theta\cos\theta\, d\theta$$

$$= \int_0^{\pi/2} 2a\sin\theta\cos\theta\cdot 4a\sin\theta\cos\theta\, d\theta$$

$$= 8a^2\int_0^{\pi/2}\sin^2\theta\cos^2\theta\, d\theta = 8a^2\,\frac{(1)(1)}{4\cdot 2}\left(\frac{\pi}{2}\right) = \frac{\pi a^2}{2}$$

Solution: (*b*)

$$x = 2a\sin^2\theta;\quad dx = 4a\sin\theta\cdot\cos\theta\,d\theta$$

$$(2a - x) = 2a\cos^2\theta$$

$$(x = 0,\ \theta = 0)\ [x = 2a,\ \theta = (\pi/2)] \qquad [1]$$

$$\therefore \quad I = \int_0^{2a} x\sqrt{(2ax - x^2)\,dx} = \int_0^{2a} x\sqrt{[x(2a - x)]}\,dx$$

$$= \int_0^{2a} 2a\cdot\sin^2\theta\sqrt{[2a\sin^2\theta\cdot 2a\cos^2\theta]}\cdot 4a\sin\theta\cos\theta\,d\theta$$

$$= 16a^3\int_0^{\pi/2}\sin^4\theta\cos^2\theta\,d\theta = 16a^3\,\frac{(3\cdot 1)(1)}{6\cdot 4\cdot 2}\cdot\frac{\pi}{2} = \frac{\pi a^3}{2}$$

Exactly in a similar manner, we can evaluate (*c*) and (*d*).

Ans. for (*c*) $\cdot\ \dfrac{5\pi a^4}{8}$

Ans. for (*d*) $\cdot\ \dfrac{7\pi a^5}{8}$

41. (V. Imp.)

If '*n*' is a positive integer,

Prove that

$$\int_0^{2a} x^n\sqrt{[2ax - x^2]}\,dx = \frac{(2n+1)!\,a^{n+2}\,\pi}{(n+2)!\,n!\,2^n} \qquad \text{[VTU F/M, 2005]}$$

Solution: Put $(a - x) = a\cos\theta;$

$\Rightarrow$ $x = a(1 - \cos\theta) = 2a\cdot\sin^2(\theta/2).$

and $dx = a\sin\theta\,d\theta$

Limits:

when	$x = 0;\ \theta = 0$
when	$x = 2a;\ \theta = \pi$

$$\therefore \quad I = \int_0^{2a} x^n\sqrt{(2ax - x^2)}\,dx$$

$$= \int_0^{2a} x^n\sqrt{a^2 - (a - x)^2}\,dx$$

$$= \int_0^{\pi} 2^n\cdot a^n\cdot\sin^{2n}(\theta/2)(a\sin\theta)(a\sin\theta\,d\theta)$$

$$= (2a)^{n+2} \int_0^{\pi} \sin^{2n}\left(\frac{\theta}{2}\right)\sin^2\left(\frac{\theta}{2}\right)\cos^2\left(\frac{\theta}{2}\right)d\theta$$

$$= (2a)^{n+2} \int_0^{\pi} \sin^{2n+2}\left(\frac{\theta}{2}\right)\cos^2\left(\frac{\theta}{2}\right)d\theta \quad \left[\begin{array}{l}\text{put } \frac{\theta}{2} = \phi;\ \frac{d\theta}{2} = d\phi;\ d\theta = 2d\phi \\ \text{limits: when } \theta = 0;\ \phi = 0 \\ \text{and when } \theta = \pi;\ \phi = \frac{\pi}{2}\end{array}\right]$$

$$\Rightarrow \quad I = (2a)^{n+2} \cdot 2 \cdot \int_0^{\pi/2} \sin^{2n+2}\phi \cdot \cos^2\phi\, d\phi$$

$$= (2a)^{n+2} \cdot (2) \cdot \frac{(2n+1)(2n-1)\ldots 1}{(2n+4)(2n+2)\ldots 2} \cdot \left(\frac{\pi}{2}\right)$$

$$= (2a)^{n+2}\,(\pi)\,\frac{(2n+1)(2n-1)\ldots 2\cdot 1}{2^{n+2}\cdot(n+2)(n+1)\ldots 1}$$

$$= (a)^{n+2}\,(\pi)\,\frac{(2n+1)(2n)(2n-1)(2n-2)\ldots 2\cdot 1}{(n+2)!\,(2n)(2n-2)\ldots 2}$$

$$= a^{n+2}\,(\pi)\,\frac{(2n+1)!}{(n+2)!\,2^n\cdot n!} = \frac{(2n+1)!}{(n+2)!\,n!}\cdot\frac{a^{n+2}}{2^n}\cdot\pi$$

Hence proved.

42. Evaluate:

$$\int_0^{\pi} \sin^2\theta\,(1-\cos\theta)^3\,d\theta$$

Solution:

Let $$I = \int_0^{\pi} \sin^2\theta\,(1-\cos\theta)^3\,d\theta$$

$$= \int_0^{\pi} [2\sin(\theta/2)\cos(\theta/2)]^2\,[2\sin^2(\theta/2)]^3\,d\theta$$

$$= \int_0^{\pi} 4\sin^2(\theta/2)\cdot\cos^2(\theta/2)\,8\cdot\sin^6\theta\,d\theta$$

$$= 32\int_0^{\pi} [\sin^8(\theta/2)]\cdot[\cos^2(\theta/2)]\,d\theta \qquad [1]$$

$$\left[\begin{array}{l}\text{put } \theta = 2\phi;\ d\theta = 2d\phi; \\ (\theta = 0, \phi = 0) \\ [\theta = \pi, \phi = (\pi/2)\end{array}\right] \qquad [2]$$

$$\therefore \quad I = 32 \int_0^{\pi/2} \sin^8 \phi \cdot \cos^2 \phi \cdot 2d\phi$$

$$= 64 \int_0^{\pi/2} \sin^8 \phi \cdot \cos^2 \phi \, d\phi$$

$$= 64 \frac{(7 \cdot 5 \cdot 3 \cdot 1)(1)}{(10 \cdot 8 \cdot 6 \cdot 4 \cdot 2)} \cdot \frac{\pi}{2} = \frac{7\pi}{8}$$

43. Evaluate:

$$\int_a^b \sqrt{[(x-a)(b-x)]}\, dx$$

Solution: Let $I = \int_a^b \sqrt{[(x-a)(b-x)]}\, dx$

put $x = (a \cos^2 \theta + b \sin^2 \theta)$

$(x - a) = (b - a) \sin^2 \theta.$

$x = a; \theta = 0$

Also, $dx = 2(b - a) \sin \theta \cos \theta \, d\theta$

$(b - x) = (b - a) \cos^2 \theta$

$x = b, \quad \theta = (\pi/2)$

Note: $(x - a) = [(a \cos^2 \theta + b \sin^2 \theta) - a]$

$= b \sin^2 \theta - a [(1 - \cos^2 \theta)]$

$= (b \sin^2 \theta - a \sin^2 \theta) = (b - a)(\sin^2 \theta)$

Similarly $(b - x) = (b - a) \cos^2 \theta$

$$\therefore \quad I = \int_a^b \sqrt{[(x-a)(b-x)]}\, dx$$

$$= \int_0^{\pi/2} \sqrt{[(b-a)\sin^2 \theta \cdot (b-a)\cos^2 \theta]} \cdot 2(b-a) \sin\theta \cos\theta \, d\theta$$

$$= 2 \int_0^{\pi/2} (b-a)^2 \sin\theta \cos\theta \sin\theta \cos\theta \, d\theta$$

$$= 2(b-a)^2 \int_0^{\pi/2} \sin^2 \theta \cdot \cos^2 \theta \, d\theta$$

$$= 2(b-a)^2 \cdot \frac{(1)(1)}{(4 \cdot 2)} \cdot \left(\frac{\pi}{2}\right) = \frac{\pi (b-a)^2}{8}$$

44. *P.T* $\int_0^1 x^m (1-x)^n \, dx = \frac{m!\,n!}{(m+n+1)!}$

Solution: Let $I = \int_0^1 x^m (1-x)^n \, dx$

Put $x = \sin^2 \theta$; $dx = 2 \sin \theta \cos \theta \, d\theta$

Limits: when $x = 0$; $\theta = 0$

when $x = 1$, $\theta = \frac{\pi}{2}$

$$\therefore \quad \int_0^1 x^m (1-x)^n \, dx = 2 \int_0^{\pi/2} \sin^{(2m+1)} \theta \cos^{(2n+1)} \theta \, d\theta$$

$$= 2 \left[\frac{\{2m(2m-2) \ldots .\, 4 \cdot 2\}\{2n(2n-2) \ldots .\, 4 \cdot 2\}}{(2m+2n+2)(2m+2n) \ldots . 4 \cdot 2} \right]$$

$$= 2 \left[\frac{\{2^m \, m(m-1) \ldots 3 \cdot 2 \cdot 1\}\{2^n \, n(n-1) \ldots 3 \cdot 2 \cdot 1\}}{2^{m+n+1} (m+n+1)(m+n) \ldots . 3 \cdot 2 \cdot 1} \right]$$

$$= \frac{m!\,n!}{(m+n+1)!}$$

45. If $I_{(m,\,n)} = \int_0^{\pi/2} \sin^{2m} x \cos^{2n} x \, dx$,

S.T $(2m + 2n)\, I_{(m,\,n)} = (2n - 1)\, I_{(m,\,n-1)}$.

Evaluate $I_{(m,\,n)}$.

Solution: Compare with the reduction formula for $\int_0^{\pi/2} \sin^m x \cos^n x \, dx$

The only difference here is that the two indices are ($2m$ and $2n$) both even, we get

$$I_{(m,\,n)} = \frac{[(2m-1)(2m-3)\ldots][(2n-1)(2n-3)\ldots]}{(2m+2n)(2m+2n-2) \ldots .\, 4 \cdot 2} \cdot \frac{\pi}{2}$$

The same working rule holds, here also.

EXERCISES

1. Evaluate

$$\int_0^{\pi/6} \cos^2 (6\phi) \sin^4 (3\phi) \, d\phi \qquad \left(Ans.\ \frac{7\pi}{12} \right)$$

2. P.T $\int_0^{\pi/6} \sin^2(6\phi)\cos^5(3\phi)\,d\phi = \frac{64}{945}$

3. Evaluate $\int_0^4 x^3\sqrt{4x - x^2}\,dx$ (*Ans.* 28π)

4. Evaluate

(*i*) $\int_0^\infty \frac{x^5}{(1+x^2)^6}\,dx$ $\left(Ans.\ \frac{1}{60}\right)$ (*ii*) $\int_0^\infty \frac{x^3}{(x^2+4)^{3/2}}\,dx$ $\left(Ans.\ \frac{1}{2}\right)$

(*iii*) $\int_0^\infty \frac{x}{(x^6+1)^{7/2}}\,dx$ $\left(Ans.\ \frac{8}{45}\right)$ (*iv*) $\int_0^\infty \frac{x}{(x^2+a^2)^5}\,dx$ $\left(Ans.\ \frac{1}{12a^6}\right)$

(*v*) $\int_0^\infty \frac{x^2}{(x^2+1)^{1/2}}\,dx$ $\left(Ans.\ \frac{2}{15}\right)$ (*vi*) $\int_0^2 x^3\sqrt{2-x}$ $\left(Ans.\ \frac{512\sqrt{2}}{315}\right)$

(*vii*) $\int_0^\infty \frac{x\,dx}{(a+x)^5}$ $\left(Ans.\ \frac{1}{12a^3}\right)$

5. Evaluate

(*a*) $\int_0^3 x^2\sqrt{(3-x)}$ $\left(Ans.\ \frac{144\sqrt{3}}{35}\right)$

Hint. put $x = 3\sin^2\theta$

(*b*) $\int_0^3 \frac{x^{5/2}\,dx}{\sqrt{(3-x)}}$ $\left(Ans.\ \frac{27}{8}\pi\right)$

6. Evaluate

$$\int_a^b \sqrt{[(b-x)/(x-a)]}\,dx \qquad \left(Ans.\ 2(b-a)\frac{1}{2}\cdot\frac{\pi}{2} = \frac{\pi(b-a)}{2}\right)$$

7. If m, n are the positive integers,

Prove that $\int_a^b (x-a)^m (b-x)^n\,dx = (b-a)^{m+n+1}\frac{m!\,n!}{(m+n+1)!}$

3.3 REDUCTION FORMULAE FOR:

(*a*) $\int \tan^n x\,dx$ (*b*) $\int \cot^n x\,dx$

(*c*) $\int \sec^n x\,dx$ (VTU March, 2001) (*d*) $\int \text{cosec}^n x\,dx$ (VTU August, 1999)

PART-C

(*a*) Let $$I_n = \int \tan^n x\, dx \qquad [1]$$

$$= \int \tan^{n-2} x \tan^2 x\, dx = \int \tan^{n-2} x (\sec^2 x - 1)\, dx$$

$$= \int \tan^{n-2} x \sec^2 x\, dx - \int \tan^{n-2} x\, dx$$

$\therefore$ $$I_n = \int \tan^{n-2} x \sec^2 x - \int \tan^{n-2} x\, dx \qquad [2]$$

The second integral is (I_{n-2}) clearly.

To evaluate the first integral we have:

put $[\tan x = t, \sec^2 x \cdot dx = dt]$ [3]

$$\int \tan^{n-2} x \sec^2 x\, dx = \int t^{n-2}\, dt = \left[\frac{t^{n-1}}{(n-1)}\right] \qquad [4]$$

$\therefore$ $$I_n = \int \tan^{n-2} x \sec^2 x\, dx - \int \tan^{n-2} x\, dx = \left[\frac{(\tan^{n-1} x)}{(n-1)}\right] - T_{n-2} \qquad [5]$$

or, $$[I_n + I_{n-2}] = \left[\frac{(\tan^{n-1} x)}{(n-1)}\right] \qquad [6]$$

To evaluate: $\int_0^{\pi/4} \tan^n x\, dx$ (BUB, July, 1993)

If $$I_n = \int_0^{\pi/4} \tan^n x\, dx \text{, then}$$

$$(I_n + I_{n-2}) = \frac{1}{(n-1)}.$$

Since, $$I_n = \int_0^{\pi/4} \tan^n x\, dx = \int_0^{\pi/4} \tan^{n-2} x\, dx$$

$$= \int_0^{\pi/4} \tan^{n-2} x (\sec^2 x - 1)\, dx$$

$$= \int_0^{\pi/4} \tan^{n-2} x \sec^2 x\, dx - \int_0^{\pi/4} \tan^{n-2} x\, dx$$

$$I_n = \left[\frac{\tan^{n-1} x}{(n-1)}\right]_0^{\pi/4} - \int_0^{\pi/4} \tan^{n-2} x\, dx \qquad [7a]$$

$$= \left[\frac{1}{(n-1)}\right] - [1 - 0] - I_{n-2} = \left[\frac{1}{(n-1)}\right] - I_{n-2}$$

$$\therefore \qquad \boxed{I_n + I_{n-2} = \frac{1}{(n-1)}} \qquad [7b]$$

Examples (46 – 50)

46. If $I_n = \int_0^{\pi/4} \tan^n x\, dx$

Then $\qquad n\,(I_{n-1} + I_{n+1}) = 1$

In the result given by (7) replace n by $(n + 1)$: and so put $(n - 1)$ instead of $(n - 2)$, and n in the place $(n - 1)$

$$\therefore \qquad I_{n+1} + I_{n-1} = (1/n)$$

$$\Rightarrow \qquad n\,[I_{n-1} + I_{n+1}] = 1.$$

47. Evaluate $\int \tan^4 x\, dx$

Solution: In the reduction formula (7*a*),

put $\qquad n = 4$; we get

$$I_4 = [(\tan^3 x)/3] - I_2 \qquad [1]$$

$$n = 2;\ I_2 = \left[\frac{\tan x}{1}\right] - I_0$$

$$I_0 = \int (\tan x)^0\, dx = \int 1 dx = x \qquad [2]$$

$$\therefore \qquad I_4 = \left[\frac{\tan^3 x}{3}\right] - I_2 = \left[\frac{\tan^3 x}{3}\right] - \{[(\tan x)/1] - I_0\}$$

$$= \left[\frac{\tan^3 x}{3}\right] - [(\tan x)/1] + x$$

$$\therefore \qquad \int \tan^4 x\, dx = [(\tan^3 x)/3] - [(\tan x)/1] + x$$

48. $\int \tan^7 x\, dx = \dfrac{\tan^6 x}{6} - \dfrac{\tan^4 x}{4} + \dfrac{\tan^2 x}{2} - \log \sec x.$

49. If $I_n = \int_0^{\pi/3} \tan^n x\, dx$

Also *S.T.*

$$(n - 1)\,[I_n + I_{n-2}] = \left(\sqrt{3}\right)^{n-1}$$

PART-C

Solution:

$$I_n = \int_0^{\pi/3} \tan^n x\,dx = \left[\frac{\tan^{n-1} x}{n-1}\right]_0^{\pi/3} - \int_0^{\pi/3} \tan^{n-2} x\,dx$$

$$I_n = \frac{\left(\sqrt{3}\right)^{n-1}}{n-1} - I_{n-2}$$

$$\therefore \quad (n-1)\,[I_n + I_{n-2}] = \left(\sqrt{3}\right)^{n-1}$$

50. Evaluate:

(*a*) $\int \tan^5 x\,dx$ (*b*) $\int \tan^6 x\,dx$

Solution: (*a*) We know that

$$I_n + I_{n-2} = \frac{1}{(n-1)} \qquad [1]$$

put $n = 5$, and proceed, we get

$$I_5 = \left[\frac{\tan^4}{4}\right] - I_3; \qquad \text{[put } n = 3 \text{ and apply]}$$

$$= \left[\left(\frac{\tan^4 x}{4}\right)\right] - \left\{\left[\frac{(\tan^2 x)}{2}\right] - I_1\right\}$$

$$= \left[(\tan^4 x/4)\right] - \left[(\tan^2 x)/2\right] + \int \tan x\,dx$$

$$= \left[\frac{(\tan^4 x)}{4}\right] - \left[(\tan^2 x)/2\right] + \log \sec x$$

Solution: (*b*) Using the formula

$$I_n = \left[\frac{1}{(n-1)}\right] - I_{n-2}$$

put $n = 6$, $\quad I_6 = \left[\frac{1}{(6-1)}\right] - I_4 = (1/5) - I_4$

$n = 4$; $\quad I_4 = \left[\frac{1}{(n-1)}\right] - I_2 = \frac{1}{3} - I_2$

$n = 2$; $\quad I_2 = \left[\frac{1}{(2-1)}\right] - I_0 = 1 - I_0$

$$I_0 = \int 1 dx = \Big[x \Big]_0^{\pi/4} = (\pi/4)$$

$\therefore$

$$I_6 = (1/5) - I_4 = 1/5 - [(1/3) - I_2]$$

$$= (1/5) - (1/3) + (1 - I_0)$$

$$= \left[\left(\frac{1}{5}\right) - \left(\frac{1}{3}\right) + 1 - \left(\frac{\pi}{4}\right)\right] = \left[\left(\frac{13}{15}\right) - \left(\frac{\pi}{4}\right)\right] \quad [2]$$

(b) Reduction formula for $I_n = \int \cot^n x\, dx$ [1]

Let

$$I_n = \cot^{n-2} x \cdot \cot^2 x\, dx \quad [2]$$

$$= \int \cot^{n-2} x\, (\operatorname{cosec}^2 x - 1)\, dx$$

$$= \int \cot^{n-2} x \operatorname{cosec}^2 x\, dx - \int \cot^{n-2} x\, dx$$

$$= \frac{\cot^{n-1} x}{(n-1)} - I_{n-2} \quad [3]$$

This is the required reduction formula. We can also write [3] as

$$[I_n + I_{n-2}] = \frac{-\cot^{n-1} x}{(n-1)}; \ (n-1)\, [\, I_n + I_n - 2] + \cot^{n-1} x = 0, \textit{ i.e.,}$$

In particular,

If

$$I = \int_{\pi/4}^{\pi/2} \cot^n x\, dx, \text{ then} \quad [1]$$

[1] becomes

$$I_n = -\left[\frac{\cot^{n-1} x}{n-1}\right]_{\pi/4}^{\pi/2} - I_{n-2}$$

$$I_n = \frac{1}{(n-1)} - I_{n-2} \quad [2]$$

$$I_n = \frac{1}{(n-1)} - I_{n-2} \quad [3]$$

Successively reducing n by $n - 2$ is [3] we get

$$I_n = \frac{1}{n-1} - \frac{1}{n-3} + \frac{1}{n-5} - .. I \quad [4]$$

Where $I = I_1$ or $I = I_0$ according as n is odd or even.

Note: that

$$I_1 = \int_{\pi/4}^{\pi/2} \cot x\, dx = \Big[\log \sin x\Big]_{\pi/4}^{\pi/2}$$

$$= + \log \sqrt{2} = \frac{1}{2} \log 2$$

PART-C

$$I_0 = \int_{\pi/4}^{\pi/2} dx = \frac{\pi}{4}$$

Again, we know that

$$I_n = \frac{1}{(n-1)} - I_{n-2}$$

For example, put $n = 3$

Examples (51 – 53)

51.
$$I_3 = \frac{1}{2} - I_1 = \log \sqrt{2}$$

$$I_4 = \frac{1}{3} - \frac{1}{1} + I_0 = -\frac{2}{3} + \frac{\pi}{4}$$

$$I_5 = \frac{1}{4} - \frac{1}{2} + I_1 = -\frac{1}{4} + \log \sqrt{2}$$

$$I_6 = \frac{1}{5} - \frac{1}{3} + \frac{1}{1} - I_0 = \frac{13}{15} - \frac{\pi}{4} \text{ etc.}$$

52. Evaluate $\int \cot^5 x\, dx$

Solution:

We know that $\int \cot^n x\, dx = I_n = \left[\frac{-\left(\cot^{n-1} x\right)}{(n-1)}\right] - I_{n-2}$ [1]

put $\left.\begin{aligned} n = 5; 1_5 &= [-(\cot^4 x)/4] - I_3; \\ n = 3; I_3 &= [-(\cot^2 x)] - I_1 \end{aligned}\right\}$ [2]

$$I_1 = \int \cot x\, dx = (\log \sin x)$$

$$\therefore \quad I_5 = \left[-(\cot^4 x)/4\right] - I_3$$

$$= \left[\frac{-(\cot^4 x)}{4}\right] - \left[\frac{\cot^2 x}{2}\right] + I_1$$

$$= [-(1/4)\cot^4 x + (1/2)\cot^2 x + \log(\sin x)]$$

$$I_n = \frac{1}{(n-1)} - I_{n-2}$$

53. Evaluate

(*a*) $\int \cot^3 x\, dx$ (*b*) $\int \cot^6 x\, dx$

Solution: (*a*) We know that

$$\int \cot^n x\, dx = I_n = \frac{-\cot^{n-1} x}{(n-1)} - I_{n-2} \qquad [1]$$

put $n = 3$, we get

$$\int \cot^3 x\, dx = I_3 = -\frac{1}{2}\cot^2 x - I_1$$

$$= -\frac{1}{2}\cot^2 x - \int \cot x\, dx = -\frac{1}{2}\cot^2 x - \log \sin x$$

Solution:
(*b*) put $n = 6, 4, 2$
successively in (1), we get

$$\cot^6 x\, dx = I_6 = -\frac{1}{5}\cot^5 x - I_4$$

$$I_4 = \int \cot^4 x\, dx = -\frac{1}{3}\cot^3 x - I_2$$

$$\int \cot^2 x\, dx = I_2 = -\cot x - I_0$$

$$= -\cot x - \int dx = -\cot x - x$$

$$\Rightarrow \qquad \int \cot^6 x\, dx = I_6 = -\frac{1}{5}\cot^5 x + \frac{1}{3}\cot^3 x - \cot x - x$$

(c) Reduction formula for $\int \sec^n x\, dx$ (VTU, March, August, 2001)

Let $\qquad I_n = \int \sec^n x\, dx = \int \sec^{n-2} x \sec^2 x\, dx \qquad [1]$

Integrate by parts, taking the part ($\sec^2 x$) first

$$I_n = \int \sec^{n-2} x \cdot \sec^2 x\, dx$$

$$= \tan x \cdot \sec^{n-2} x - \int \tan x \cdot (n-2)\sec^{n-3} x \cdot \sec x \tan x \cdot dx$$

$$= \tan x \cdot \sec^{n-2} x - \int \tan x \cdot (n-2)\sec^{n-3} x \cdot \sec x \tan x \cdot dx$$

$$= \tan x \cdot \sec^{n-2} x - (n-2)\int \sec^{n-2} x \cdot \tan^2 x\, dx$$

$$= \tan x \cdot \sec^{n-2} x - (n-2)\int \sec^{n-2} x (\sec^2 x - 1) dx$$

$$= \tan x \cdot \sec^{n-2} x - (n-2)\int (\sec^n x - \sec^{n-2} x)\, dx$$

$$= \tan x \cdot \sec^{n-2} x - (n-2)\int \sec^n x\, dx + (n-2)\int \sec^{n-2} x\, dx$$

$$\therefore \qquad I_n = \tan x \cdot \sec^{n-2} x - (n-2) I_n + (n-2) I_{n-2} \qquad [2]$$

i.e., $[I_n + (n-2)\, I_n] = \tan x \cdot \sec^{n-2} x + (n-2)\, I_{n-2}$

i.e., $(n-1)\, I_n = \tan x \cdot \sec^{n-2} x + (n-2)\, I_{n-2}$

$$\therefore \quad \int \sec^n x\, dx = I_n = \frac{\tan x \cdot \sec^{n-2} x}{(n-1)} + \frac{(n-2)}{(n-1)} I_{n-2} \qquad [3]$$

To Evaluate:

$$I_n = \int_0^{\pi/4} \sec^n x\, dx \text{, Hence} \qquad \text{[VTU F/M, 2005]}$$

or otherwise evaluate I_4, I_3 and I_5

$$\text{Let} \quad I_n = \int_0^{\pi/4} \sec^n x\, dx \qquad [1]$$

$$= \left[\frac{\sec^{n-2} x \tan x}{n-1}\right]_0^{\pi/4} + \frac{(n-2)}{(n-1)} \int_0^{\pi/4} \sec^{n-2} x\, dx$$

(using formula [3])

$$= \frac{1}{(n-1)}\left[\tan\left(\frac{\pi}{4}\right)\sec^{n-2}\left(\frac{\pi}{4}\right)\right] + \frac{(n-2)}{(n-1)} I_{n-2}$$

$$= \frac{1}{(n-1)}\left[1 \cdot \left(\sqrt{2}\right)^{n-2}\right] + \frac{(n-2)}{(n-1)} I_{n-2}$$

$$= \frac{\left(\sqrt{2}\right)^{n-2}}{(n-1)} + \frac{(n-2)}{(n-1)} I_{n-2}$$

Putting $n = 4$, 3 or 5, in the formula above, we get I_4, I_3, I_5.

$$(i) \qquad I_4 = \frac{\left(\sqrt{2}\right)^{4-2}}{3} + \frac{2}{3} I_2 = \frac{2}{3} + \frac{2}{3} I_2$$

$$I_2 = \int_0^{\pi/4} \sec^2 x\, dx = \Big[\tan x\Big]_0^{\pi/4} = (1-0) = 1.$$

$$\therefore \qquad I_4 = \frac{2}{3} + \frac{2}{3} I_2 = \frac{2}{3} + \frac{2}{3} \cdot 1 = \frac{4}{3}$$

$$(ii) \qquad I_3 = \frac{\left(\sqrt{2}\right)^{3-2}}{2} + \frac{1}{2} I_1 = \frac{1}{\sqrt{2}} + \frac{1}{2} \int_0^{\pi/4} \sec x\, dx$$

$$= \frac{1}{\sqrt{2}} + \frac{1}{2}\Big[\log(\sec x + \tan x)\Big]_0^{\pi/4}$$

$$= \frac{1}{\sqrt{2}} + \frac{1}{2}\left[\log\left(\sec\frac{\pi}{4} + \tan\frac{\pi}{4}\right) - \log(\sec 0 + \tan 0)\right]$$

$$= \frac{1}{\sqrt{2}} + \frac{1}{2}\left[\log\left(\sqrt{2}+1\right) - \log(1+0)\right]$$

$$= \frac{1}{\sqrt{2}} + \frac{1}{2}\left[\log\left(\sqrt{2}+1\right)\right] = \left(1/\sqrt{2}\right) + (1/2)\log\left(\sqrt{2}+1\right)$$

(*iii*)
$$I_n = \frac{\tan x \sec^{n-2} x}{(n-1)} + \frac{(n-2)}{(n-1)} I_{n-2}$$

$$I_5 = \frac{\tan x \sec^3 x}{4} + \frac{3}{4} I_3$$

$$= \frac{\tan x \sec^3 x}{4} + \frac{3}{4}\left[\frac{1}{\sqrt{2}} + \frac{1}{2}\log\left(\sqrt{2}+1\right)\right] \quad \text{[using (b)]}$$

54. Evaluate:

(*a*) $\int \sec^3 x\, dx$ (*b*) $\int \sec^4 x\, dx$ (*c*) $\int \sec^5 x\, dx$

Solution: (*a*) We know that

$$\int \sec^n x\, dx = I_n = \frac{\tan x \cdot \sec^{n-2} x}{(n-1)} + \frac{(n-2)}{(n-1)} I_{n-2} \quad [1]$$

Put $n = 3$, in [1],

$$I_3 = \frac{\tan x \cdot \sec x}{2} + \frac{1}{2} I_1,$$

$$I_1 = \int \sec x\, dx = \log(\sec x + \tan x)$$

$$\therefore \quad I_3 = \frac{\tan x \sec x}{2} + \frac{1}{2}\log(\sec x + \tan x)$$

(*b*) $\int \sec^4 x\, dx$

put $n = 4$, in [1],

$$\therefore \quad \left.\begin{aligned} I_4 &= (1/3)\tan x \cdot \sec^2 x + (2/3) I_2 \\ I_2 &= \int \sec^2 x\, dx = (\tan x) \end{aligned}\right\} \quad [2]$$

$$\therefore \quad I_4 = (1/3)\tan x \cdot \sec^2 x + (2/3)\tan x \quad [3]$$

Solution: (*c*) Put $n = 5, 3$

$$\left.\begin{aligned} I_5 &= (1/4)\tan x \cdot \sec^3 x + (3/4) I_3 \\ I_3 &= (1/2)\tan x \cdot \sec x + (1/2) I_1 \end{aligned}\right\} \quad [1]$$

PART-C

$$I_1 = \int \sec x \, dx = \log(\sec x + \tan x) \qquad [2]$$

$\therefore$ $$I_5 = (1/4) \tan x \sec^3 x + (3/4) [(1/2) \tan x \sec x + (1/2) \log(\sec x + \tan x)]$$

$$= (1/4) \tan x \cdot \sec^2 x + (3/8) \tan x \cdot \sec x + (3/8) \log(\sec x + \tan x)$$

(*d*) Reduction formula for

$$\int \text{cosec}^n \, x \, dx, \qquad \text{(VTU, August, 1999)}$$

Let $$I_n = \int \text{cosec}^n \, x \, dx$$

$$= \int \text{cosec}^{n-2} x \cdot \text{cosec}^2 x \, dx \qquad [1]$$

Integrate by parts, integrating ($\text{cosec}^2 x$) first.

$$I_n = \left[-\cot x \cdot \text{cosec}^{n-2} x \right]$$

$$- \int -\cot x \cdot (n-2) \text{cosec}^{n-3} x \, (-\text{cosec}\, x \cot x) \, dx$$

$$= -(\cot x \cdot \text{cosec}^{n-2} x) - (n-2) \int \text{cosec}^2 x \cdot \cot^2 x \, dx$$

$$= -(\cot x \cdot \text{cosec}^{n-2} x) - (n-2) \int \text{cosec}^{n-2} x \cdot (\text{cosec}^2 x - 1) \, dx$$

$$= -(\cot x \cdot \text{cosec}^{n-2} x) - (n-2) \left[\int \text{cosec}^n x \, dx - \int \text{cosec}^{n-2} x \, dx \right]$$

$$= -(\cot x \cdot \text{cosec}^{n-2} x) - (n-2)(I_n - I_{n-2})$$

$\therefore$ $$I_n = -(\cot x \cdot \text{cosec}^{n-2} x) - (n-2) I_n + (n-2) I_{n-2}$$

$\Rightarrow$ $$I_n [1 + (n-2)] = -\cot x \cdot \text{cosec}^{n-2} x + (n-2) I_{n-2}.$$

$\therefore$ $$I_n = \frac{-\cot x \, \text{cosec}^{n-2} x}{(n-1)} + \frac{(n-2)}{(n-1)} I_{n-2} \qquad [2]$$

55. Evaluate $\int \text{cosec}^4 x \, dx$

Solution:

Put $n = 4$, in (2) we get

$$\left.\begin{aligned} I_4 &= \left[(-\cot x \cdot \text{cosec}^2 x)/3 \right] + (2/3) I_2 \\ I_2 &= \int \text{cosec}^2 x \, dx = -\cot x \end{aligned}\right\} \qquad [1]$$

$\therefore$ $$I_4 = (-1/3)(\cot x \cdot \text{cosec}^2 x) - (2/3) \cot x \qquad [2]$$

56. Evaluate $\int \text{cosec}^5 x \, dx$

Solution: Put $n = 5$ in (2), we obtain:

$$\int \text{cosec}^5 x \, dx = I_5$$

$$= \left(-\frac{1}{4} \right) (\cot x \cdot \text{cosec}^3 x) + \frac{3}{4} I_3$$

$$= \left(-\frac{1}{4}\right)(\cot x \cdot \operatorname{cosec}^3 x) + \frac{3}{4}\left[-\frac{1}{2}\cot x \operatorname{cosec} x + \frac{1}{2} I_1\right]$$

$$= \left(-\frac{1}{4}\right)(\cot x \cdot \operatorname{cosec}^3 x) - \frac{3}{8}\cot x \operatorname{cosec} x + \frac{3}{8}\int \operatorname{cosec} x\, dx$$

$$= \left(-\frac{1}{4}\right)(\cot x \cdot \operatorname{cosec}^3 x) - \frac{3}{8}(\cot x \cdot \operatorname{cosec} x) + \frac{3}{8}\log|\operatorname{cosec} x - \cot x|.$$

57. Evaluate

$$\int_{\pi/6}^{\pi/2} \operatorname{cosec}^5 \theta\, d\theta$$

Solution: We have from the result proved (in Ex. 56) above that

$$\int_{\pi/6}^{\pi/2} \operatorname{cosec}^5 \theta\, d\theta$$

$$= \left[-\frac{1}{4}\cot\theta \operatorname{cosec}^3\theta - \frac{3}{8}\cot\theta\operatorname{cosec}\theta + \frac{3}{8}\log|\operatorname{cosec}\theta - \cot\theta|\right]_{\pi/6}^{\pi/2}$$

$$= \frac{1}{4}\cot\frac{\pi}{6}\operatorname{cosec}^3\frac{\pi}{6} + \frac{3}{8}\cot\frac{\pi}{6}\operatorname{cosec}\frac{\pi}{6} - \frac{3}{8}\log\left(\operatorname{cosec}\frac{\pi}{6} - \cot\frac{\pi}{6}\right)$$

$$= 2\sqrt{3} + \frac{3}{4}\sqrt{3} - \frac{3}{8}\log\left(2-\sqrt{3}\right)$$

$$= \frac{11\sqrt{3}}{4} - \frac{3}{8}\log\left(2-\sqrt{3}\right).$$

3.4 TRACING OF STANDARD CURVES IN CARTESIAN FORM, PARAMETRIC FORM AND POLAR FORM

A. Rules for Tracing Cartesian Equations (VTU, Aug., 1999)

(*i*) Symmetry (mirror image)

If in the equation of the curve, y occurs in even powers only, then the curve is symmetrical about the x-axis that is,

$$f(x, y) = f(x, -y)$$

If in the equation of the curve, x occurs in even powers only, then the curve is symmetrical about the y-axis. (The stress on the word 'even and only even' should be observed, $x^2 + y^2 = ax$ is not symmetrical about y-axis, as it involves odd powers of x as well.

(*ii*) If the equation of the curve involves even and only even powers of x as well as of y, then the curve is symmetrical about both the axes *i.e.,* the circle $x^2 + y^2 = a^2$ or ellipse $\frac{x^2}{a^2} + \frac{y^2}{b^2} = 1$ are both symmetrical about the axes.

PART-C

(*iii*) Interchanging x and y leaves the equation unaltered, then there is symmetry about the line $y = x$

For example, $xy = c^2$

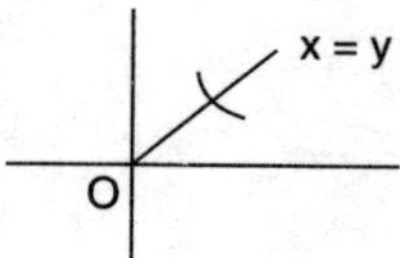

(*iv*) If the equation remains unaltered when x is replaced by $-y$ and y is replaced by $-x$, then there is symmetry about the line $y = -x$.

STANDARD POINTS

B. POINTS OF INTERSECTION

(*i*) Put $y = 0$ in the equation of the curve; we get the points of intersection of the curve and the x-axis.

(*ii*) Put $x = 0$ and we get the intersection of the curve and y-axis.

C. NATURE OF ORIGIN

(*a*) Find out whether the origin lies on the curve. [from (B)] If so, write down the equation of the tangent or tangents at the origin. This is given by equating to zero, the Lowest Degree Terms (L.D.T.) in the equation of the curve. (LDT = 0 gives T_0)

(*b*) Find out whether the tangents so obtained are:

(*i*) Real and distinct, if so, the origin is a node

(*ii*) Real and coincident; if so, the origin is a cusp

(*iii*) Imaginary; if so, the origin is a conjugate point.

D. ASYMPTOTES

An Asymptote is a tangent to the curve at infinity. A curve of nth degree has n (and only n) asymptotes.

(*i*) To find the equation of asymptotes parallel to x-axis (or y-axis) equate to zero, the coeffts of HDT in x (or y) *e.g.*, $y^2 (x + a) = x^2 (3a - x)$ Asymptotes $\|^{el}$ to y-axis is $x + a = 0$

E. REGION OF EXISTENCE

We find the points of the boundary (or region) in which the whole curve lies.

Worked Examples [58 – 71]

58. Trace the curve:

$$ay^2 = x^2 (a - x),\ a > 0 \qquad \text{(VTU, Sept., 1999)}$$

Solution:

1. Symmetry:

y is in even powers only, $\therefore$ there is symmetry about the x-axis.

2. Standard points: put $y = 0$,

we get $x^2 (a - x) = 0$

$\Rightarrow \quad x = 0,\ x = a$

put $x = 0$, we get $y^2 = 0$

$\Rightarrow \quad y = 0,\ y = 0.$

The curve passes through the origin. (Since there is no constant term)

3. T_0 (tangents at origin) are given by L.D.T. = 0

$\Rightarrow$ $ay^2 = ax^2$

$\Rightarrow$ $y^2 = x^2$

$\Rightarrow$ $y = \pm x.$

$\Rightarrow$ Two real and distinct tangents

$\Rightarrow$ origin is a node.

Nodal tangents being $y = + x$, $y = - x$

4. Asymptotes:

(*i*) Here n = degree of x in the eqn. of the curve = 3, there are 3 asymptotes.

(*ii*) H.D. power of y is y^2.

Its coeff. is a

$\therefore$ $a = 0$ is an asymptote at '∞'

(*iii*) Significant values of x are $x = 0$, $x = a$.

We tabulate the values of x and y

x	$y^2 = \left[\dfrac{x^2\,(a-x)}{a}\right]$	y
$x < 0$	= (+) (+) = (+)	real
$0 < x < a$	= (+) (+) = (+)	real
$x > a$	= (+) (–) = – ve	imaginary

The entire curve lies to the left of $x = a$.

The tangent at $(a, 0)$ is parallel to the y-axis.

The shape of the curve is as shown in Fig. 3.1

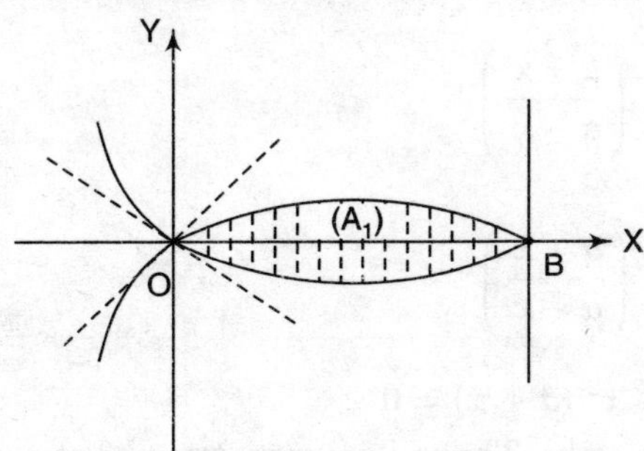

Fig. 3.1

59. Trace the curve: $y^2 = \dfrac{x^3}{(a-x)}$, $(a > 0)$

(CISSOID of DIOCLES) (VTU, Sept., 2000)

Solution:

(*i*) Since the equation has even powers of y, the curve is symmetrical about x-axis.

(*ii*) It passes through the origin (0, 0)

(*iii*) The tangents at origin are given by
$y^2 = 0$, real and coincident
so, origin is a CUSP.

(*iv*) The curve meets the x-axis and y-axis only at the origin. Equate to zero, (HDT in y to zero)
we get $x - a = 0$, $x = a$ is the asymptote parallel to y-axis

(*v*) From the given eqn:

$$y = \sqrt{\frac{x^3}{(a-x)}}, \ (x \neq a) \text{ (Taking positive sign)}$$

when x is –ve y is imaginary
∴ No portion of the curve lies to the left of y-axis.
Also, $x = a$ is an asymptote, the curve lies between y-axis and $x = a$ $(a > 0)$.

(*vi*)
$$y = \sqrt{\frac{x^3}{a-x}}$$

$$\therefore \quad \frac{dy}{dx} = \frac{\left(\frac{3}{2}a - x\right)\sqrt{x}}{(a-x)\sqrt{(a-x)}}$$

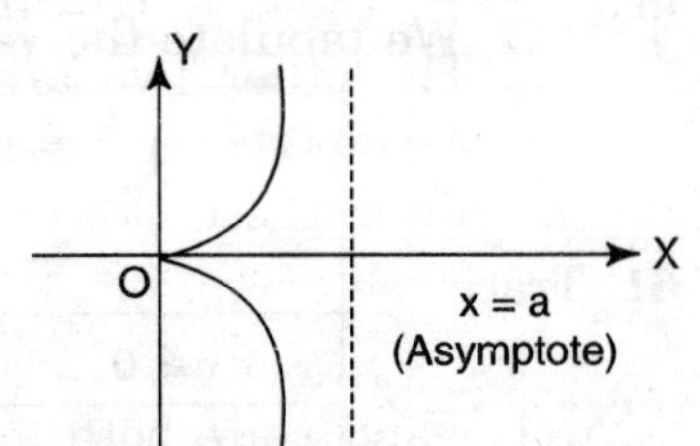

Fig. 3.2

We find $\frac{dy}{dx} > 0$

When $0 \leq x \leq a$.
⇒ The function is increasing when x increases from 0 to a.
From these discussions, the curve can be traced as shown in the Fig. 3.2.

60. Trace the cruve $y^2 = x^2\left(\frac{a+x}{a-x}\right)$ (VTU, March, 2000)

Solution: Given: $$y^2 = x^2\left(\frac{a+x}{a-x}\right)$$

or, $$y^2(a-x) - x^2(a+x) = 0$$

(1) y is in even powers only. There is symmetry about x-axis

(2) (*i*) $y = 0$ gives $x^2(a + x) = 0$
∴ $x = 0, 0, -a$
(*ii*) $x = 0$ gives, $y^2 = 0$

(3) The origin is clearly a point on the curve.

$$\therefore \quad T_0 = \text{Tangent at origin}$$
$$= \text{LDT} = 0$$

$$\Rightarrow \quad ay^2 - ax^2 = 0$$

$\Rightarrow \qquad y^2 = x^2$

$\Rightarrow \qquad y = \pm x.$

There are two real tangents. The origin is a node.

(4) (*i*) Here degree; $n = 3$

$\therefore$ There are 3 asymptotes,

(*ii*) y^3 is absent; Highest power of y is y^2; its coeff. is $(a - x)$

$\therefore \quad a - x = 0 \Rightarrow x = a$ is an asymptote

(*iii*) HDT = 0, give directions of Asymptotes

i.e., $\qquad - xy^2 - x^3 = 0$

$\Rightarrow \qquad x\,(x^2 + y^2) = 0.$

One asymptote is parallel to

$$x = 0$$

i.e., $\qquad x = a$

The other two are parallel to $x^2 + y^2 = 0$ and so are imaginary. The only real asymptote is $x = a$.

From these observations, we find the shape of the curve will be as shown.

Fig. 3.3 Stroiphoid

61. Trace the curve:

$$a^2y^4 = x^4\,(a^2 - x^2)$$

(1) y and x are both in even powers, so, the curve is symmetrical about both the axes.

(2) $\qquad y = 0$ gives $x^4 = 0$; $a^2 = x^2$

i.e., $\qquad x = 0,\ x = \pm a.$

(3) T_0 (tangent at the origin) is given by

LDT = 0 *i.e.*, $a^2y^4 - a^2x^4 = 0$

$\Rightarrow \qquad a^2\,(y^2 - x^2)\,(y^2 + x^2) = 0$

$\Rightarrow \qquad y^2 - x^2 = 0,$

$\Rightarrow \qquad y = \pm x$. The origin is a node.

(4) Asymptotes are at infinity

(5) $x^2 < a^2$ gives real values for y and $x^2 > a^2$ gives imaginary values.

The whole curve lies between $x = -a$ and $x = +a$.

(6) Tangents at $(a, 0)$ and $(-a, 0)$ are parallel to the y-axis

From these observations, we find the shape of the curve is as sketched in the Fig. 3.4.

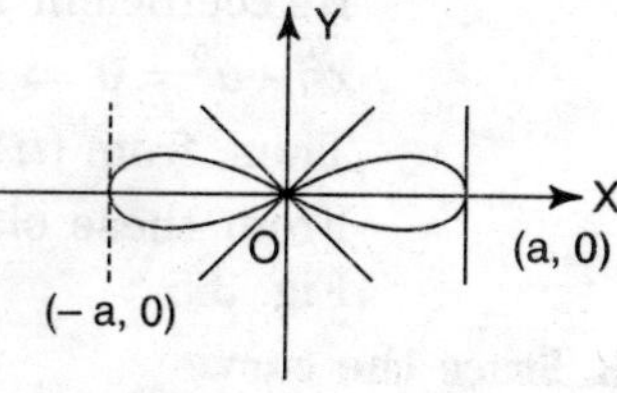

Fig. 3.4 Lemniscate

62. Trace the curve:

$$x^2y^2 = a^2\,(x^2 + y^2).$$

Solution: Given: $\qquad x^2y^2 = a^2x^2 + a^2y^2;$

$\therefore \qquad x^2\,(y^2 - a^2) = a^2y^2;$

$$\Rightarrow \qquad x^2 = \frac{a^2\,y^2}{(y^2 - a^2)}$$

Similarly $\quad y^2 (x - a)= a^2x^2$

$$\therefore \quad y^2 = \frac{a^2 x^2}{(x^2 - a^2)}$$

(1) *Symmetry:* Both x and y are in the even powers only. So, the curve is symmetrical about the x-axis and also the y-axis.

(2) *Standard points:* put $x = 0$,

We get $y^2 = 0$, put $y = 0$, we get $x = 0$.

(3) *Nature of origin:* T_0 (tangents at the origin) is given by

$$T_0 = \text{LDT} = 0$$

Since x^2y^2 is of the 4th degree; and a^2x^2, a^2y^2 are both of degree two.

so $\quad \text{LDT} = a^2 (x^2 + y^2)$

$\therefore \quad T_0 = a^2 (x^2 + y^2) = 0, y^2 = -x^2,$

$\Rightarrow \quad y = \pm ix$

(Tangents at the origin are imaginary)

$\Rightarrow$ The origin is a conjugate point.

(4) *Asymptotes:*

$$x^2y^2 - a^2x^2 - a^2y^2 = 0$$

(*i*) n = degree = 4;

(There are 4 asymptotes)

(*ii*) x^4 is absent;

the highest power of x is x^2

Its coefficient is $y^2 - a^2$

$\therefore \quad y^2 - a^2 = 0$

$\Rightarrow \quad y = \pm$ a yield two asymptotes.

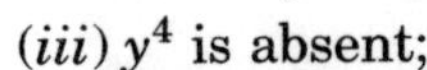

(*iii*) y^4 is absent;

the highest power of y is y^2

Its coefficient is $x^2 - a^2$

$x^2 - a^2 = 0 \Rightarrow x = \pm a$ yield two other asymptotes

Thus, from (*ii*) and (*iii*) we have all the 4 asymptotes.

From these observations, we find the shape of the curve is as sketched in the Fig. 3.5

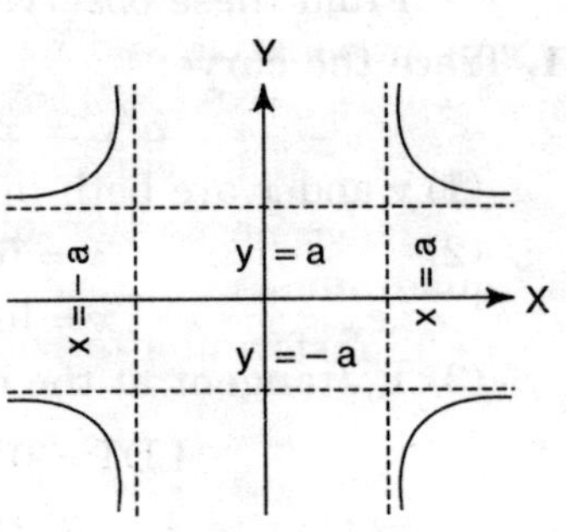

Fig. 3.5

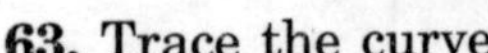

63. Trace the curve

Folium of Descarte:

$$x^3 + y^3 - 3axy = 0.$$

Solution:

(*i*) The equation remains unchanged when x and y are interchanged. Thus the curve is symmetrical about the line $y = x$

(*ii*) *Origin:* origin is a point on the curve.

T_0 (tangents at the origin) are obtained by

$$\text{LDT} = 0 \;\Rightarrow\; xy = 0 \;\Rightarrow\; x = 0, y = 0$$

yields real and distinct tangents

$\Rightarrow$ origin is a node.

The line $y = x$ meets the curve in the point $\left(\frac{3a}{2}, \frac{3a}{2}\right)$ and the tangent at this point is

$$x + y = 3a$$

(iii) Asymptotes:

(*a*) There are no parallel asymptotes to the curve.

(*b*) There is an oblique asymptote which is calculated as below:

put $x = 1$, $y = m$ in the HDT (*i.e.*, 3rd degree)

and equals to zero, we get

$$\phi_3(m) = 1 + m^3 = 0 \Rightarrow m = -1$$

Now $\phi_2(m) = -3am$

Putting $(x = 1)$, $(y = m)$ in the 2nd degree terms,

$$\therefore \quad c = \frac{\phi_2(m)}{\phi_3'(m)} = \frac{-3am}{3m^2} = \frac{+a}{m} = -a \qquad (\because m = -1)$$

$\therefore$ The asymptotes is: $y = mx + c$ or $y = -x - a$

or $$x + y + a = 0$$

(*iv*) If x and y are both negative, then LHS of eqn. is negative and RHS will be positive (*i.e.*, in $x^3 + y^3 = 3axy$)

$\therefore$ x and y cannot be both negative

Hence no portion of the curve lies in the 3rd quadrant.

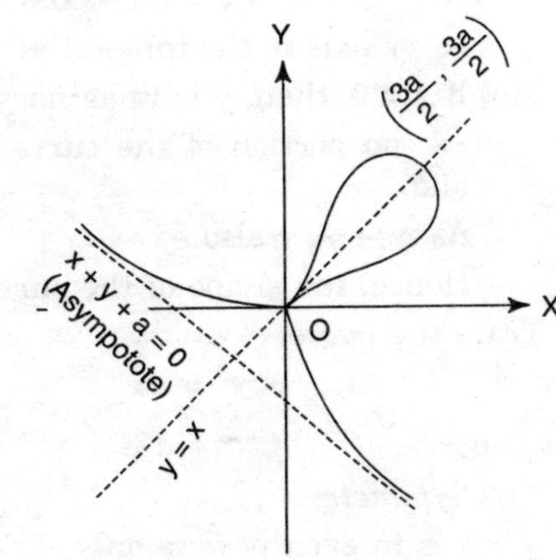

Fig. 3.6 Folium of Descarte

Transforming to polars, by putting

$$x = r\cos\theta;$$

$$y = r\sin\theta, \text{ we get}$$

$$r^3(\cos^3\theta + \sin^3\theta) = 3ar^2\sin\theta\cos\theta$$

$$\Rightarrow \quad r = \frac{3a\sin\theta\cos\theta}{(\cos^3\theta + \sin^3\theta)}$$

For $\left(0 \le \theta \le \frac{\pi}{4}\right)$ $\left(r \text{ is +ve and nonzero and increases from 0 to } \frac{3a}{\sqrt{2}}\right)$

For $\left(\frac{\pi}{4} \le \theta \le \frac{\pi}{2}\right)$ $\left(r \text{ decreases from } \frac{3a}{\sqrt{2}} \text{ to } a\right)$

There is a loop between

$$\therefore \quad \theta = 0 \text{ and } \theta = \frac{\pi}{2}$$

Now as $\theta \to \frac{3\pi}{4}$, $\cos^2\theta + \sin^2\theta \to 0$

$\therefore$ as $\theta \to \frac{3\pi}{4}$, $r \to \infty$.

$\therefore$ as θ increases from $\dfrac{\pi}{2}$ to $\dfrac{3\pi}{4}$,

r is –ve in this range and numerically increases from 0 to ∞.

Thus, the portion of the curve corresponding to this region will be in the fourth quadrant.

Hence, the shape of the curve is as shown in the Fig. 3.6.

64. Trace the curve:

$$y^2 = 4ax.$$

Solution:

(*i*) *Symmetry:* y is in even powers only. There is symmetry about x-axis.

(*ii*) Origin is clearly a point on the curve

($\because$ $x = 0, y = 0$ satisfies the curve $y^2 = 4ax$)

(*iii*) T_0 = (tangent at the origin)

$\Rightarrow$ LDT = 0

$\Rightarrow$ $4ax = 0$

$\Rightarrow$ $x = 0$ (y-axis)

$\Rightarrow$ y axis is the tangent at the origin.

(*iv*) If $x < 0$, then, y is imaginary.

$\Rightarrow$ no portion of the curve lies to the left of y-axis and

As $x \to \infty$, y also $\to \infty$.

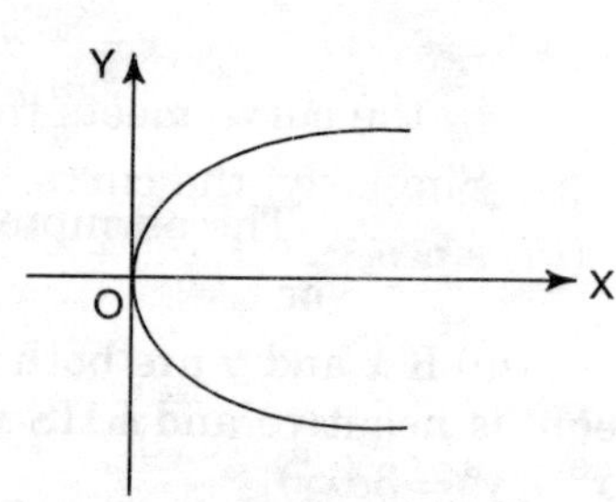

Fig. 3.7 Parabola

Hence, the shape of the curve (which is a parabola) is as sketched in the Fig. 13.7.

65. Trace the curve:

$$y^2 = ax^3$$

Solution:

(*i*) *Symmetry:*

y is in even powers only. There is symmetry about x-axis.

(*ii*) *Standard points:*

The curve passes through the origin (O, O) because $x = 0$, $y = 0$ satisfies the given equation.

(*iii*) T_0 (tangent at the origin) is obtained by making

LDT = 0

$\Rightarrow$ $y^2 = 0$

$\Rightarrow$ $y = 0$ (x-axis)

$\Rightarrow$ x-axis is the tangent at the origin. The curve has no asymptotes.

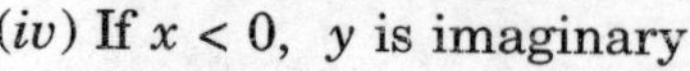

(*iv*) If $x < 0$, y is imaginary

$\Rightarrow$ no portion of the curve lies to the left of y-axis, as $x \to \infty$, $y \to \infty$.

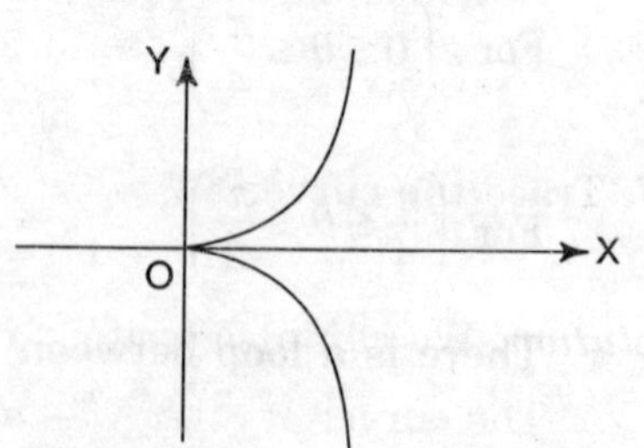

Fig. 3.8 Semi-cubical parabola

Since the x-axis is the tangent and the curve is symmetric about the x-axis the two branches of the curve lie on either side of the tangent.

Hence origin is a cusp.

Hence, the shape of the curve (which is a semi-cubical parabola) is as sketched in the Fig. 3.8

66. Trace the curve:

$$x^{2/3} + y^{2/3} = a^{2/3} \qquad \text{(Astroid)}$$ [VTU, F/M, 2005]

Solution:

(i) *Symmetry:*

The curve is symmetrical about both the axes.

(ii) *Standard points:*

Put $y = 0$, we get

$$x^{2/3} = a^{2/3}$$

$$\therefore \quad x^2 = a^2$$

$$\Rightarrow \quad x = \pm a.$$

$\Rightarrow$ the curve meets the x-axis at $(a, 0)$ and $(-a, 0)$.

Similarly, the curve meets the y-axis at $(0, a)$ and $(0, -a)$.

(iii) since, $x^{2/3} + y^{2/3} = a^{2/3}$

$$\Rightarrow \quad \left(\frac{y}{a}\right)^{2/3} + \left(\frac{x}{a}\right)^{2/3} = 1$$

$$\Rightarrow \quad \left(\frac{y}{a}\right)^{2/3} = 1 - \left(\frac{x}{a}\right)^{2/3}$$

If $|x| > a$, then $\left(\frac{y}{a}\right)^{2/3} < 0.$

and hence y is imaginary.

$\Rightarrow$ no portion of the curve lie beyond $x = \pm a$.

Similarly, the curve does not lie beyond

$$y = \pm a.$$

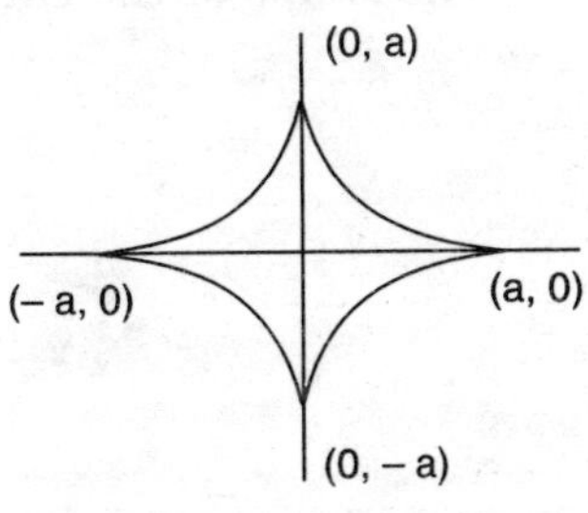

Fig. 3.9 Astroid

(iv) There are asymptotes. The shape of the curve is as sketched in Fig. 3.9.

Note: parametric equations of the curve are:

$$x = a\cos^3\theta;$$

$$y = a\sin^3\theta.$$

67. Trace the curve:

$$y^2(a^2 + x^2) = x^2(a^2 - x^2),$$ (VTU, August, 2001)

Solution: We can also write: $x^2(x^2 + y^2) = a^2(x^2 - y^2)$

The curve: $a^2y^2 + x^2y^2 - a^2x^2 + x^4 = 0$ passes through the origin, because $x = 0$, $y = 0$ satisfies the given equation. T_0 (tangents at the origin) are obtained by making LDT = 0,

1. *Symmetry:*

The powers of both x and y are even, the curve is symmetrical about both axes.

2. *Standard points:*

The curve passes through the origin (0, 0), because $x = 0$, $y = 0$ satisfies the given equation T_0 (tangents at the origin) are obtained

(since $a^2y^2 + x^2y^2 - a^2x^2 + x^4 = 0$)

$$\text{LDT} = 0 \Rightarrow a^2(y^2 - x^2) = 0$$

$\Rightarrow$ $y^2 = x^2$

$\Rightarrow$ $y = \pm x,$

$\Rightarrow$ real and distinct

$\Rightarrow$ origin is a node

3. The curve has no asymptotes (because coeff. of x^4 is 1 and that of y^2 is $a^2 + x^2$ cannot be equal to zero)

4. Solving for y, we have

$$y = \pm x\sqrt{\frac{a^2 - x^2}{a^2 + x^2}}$$

$\Rightarrow$ $a^2 - x^2 \geq 0$ (if y is real)

or $x^2 \leq a^2$ or $|x| \leq a$

The whole curve lies within the lines $x = \pm a$.
with a loop between (0 to a) and also between ($-a$ to O)

The shape of the curve is as sketched in Fig. 3.10

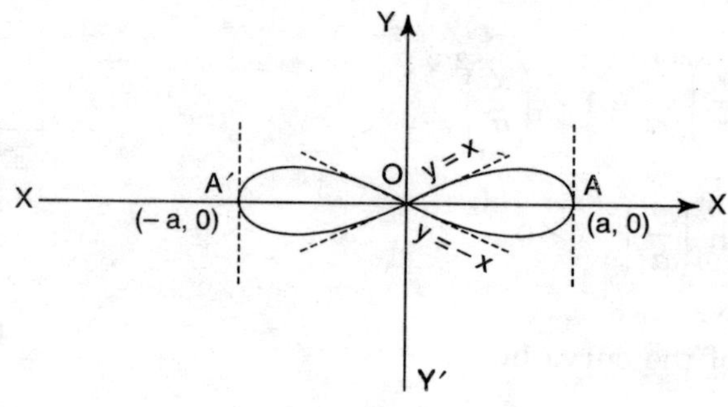

Fig. 3.10

68. Trace the curve

$$xy^2 = a^2(a - x)$$

Solution:

1. *Symmetry:* y is in even powers only, the curve is symmetrical about x-axis.

2. *Origin:* The curve does not pass through the origin and the curve meets x-axis in the point $A(a, 0)$ and it does not meet y-axis.

Shifting the origin to the point

$A\ (a, 0)$ by putting $x = X + a$,

$y = Y + 0$, the given equation transforms to

$$(X + a)Y^2 = a^2\,[a - (X + a)]$$

or $$(x + a)Y^2 + a^2X = 0$$

Equating to zero the LDT,

The tangent at the new origin is $X = 0$ *i.e.*, new y-axis.

Hence at the new origin *i.e.*, at $A\ (a, 0)$, the tangent is $\perp^r$ to x-axis.

3. *Asymptotes:* Coeff. of y^2 is equated to zero to get asymptotes parallel to y-axis *i.e.*, $x = 0$, (y-axis itself) and it has no other asymptote.

4. *Standard points:*

Given: $y = a(a - x)^{1/2} x^{-1/2}$

We obtain: y'

as $$y' = \frac{-a^2}{\left(2\sqrt{a-x}\right)x^{3/2}}$$

and $y' \neq 0$ for any real x

and $y' \to \infty$ when $x \to 0$ or a

when $x \to 0$, $y \to \infty$, showing that $x = 0$ is an asymptote. When $x \to a$, $y \to 0$. Thus the tangent $(a, 0)$ is $\perp^r$ to x-axis. Solving for y from the

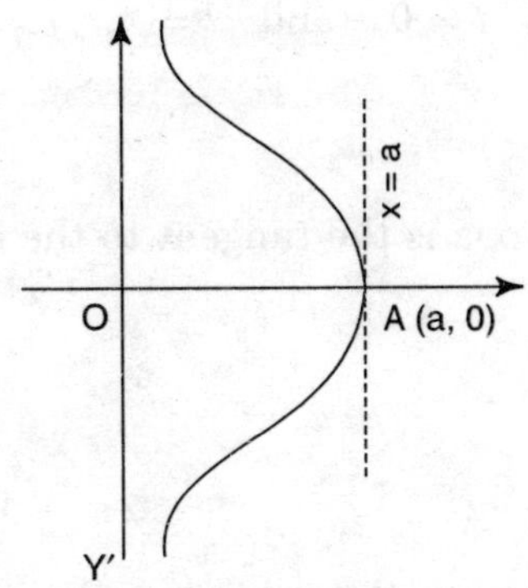

Fig. 3.11 Witch of Agensi

Given eqn. we get $$y = a\sqrt{\frac{a-x}{x}}$$

When $x < 0$ or $x > a$,

y is imaginary.

Thus, no portion of the curve lies to the left of y-axis and to the right of line $x = a$.

Hence the shape of the curve is as shown in the Fig. 3.11

69. Trace the curve. $x = a \cos^3 t$,

$y = a \sin t$. (VTU, M.Q. II)

Solution: We note the following:

1. *Symmetry:* The curve is symmetrical about the line $y = x$.

We have, $x = a \cos^3 t,\ \ y = a \sin^3 t$

When $x = 0$, $a \cos^3 t = 0$

$$\Rightarrow \quad \cos t = 0 = \cos\frac{\pi}{2} \text{ or } \cos\frac{3\pi}{2}$$

i.e., when $t = \frac{\pi}{2}$, and $t = \frac{3\pi}{2}$, the curve touches the y-axis at $y = a$, and $y = -a$

When $y = 0$, $a \sin^3 t = 0$

$$\Rightarrow \quad \sin t = 0 = \sin 0 \text{ or } \sin \pi$$

i.e. when $t = 0$ and π, the curve touches the x-axis at $x = a$ and $x = -a$

2. $|\cos t| \le 1$, $|\sin t| \le 1$, the curve lies within the circle, centre at the origin and radius a (*i.e.* $|x| = a|\cos t|^3$ and $|y| = a|\sin t|^3$, we have $|x| \le a$, $|y| \le a$)

PART-C

3. we have $$y' = \frac{dy}{dx} = \frac{\frac{dy}{dt}}{\frac{dx}{dt}} = \frac{3a \sin^2 t \cdot \cos t}{-3a \cos^2 t \cdot \sin t} = -\tan t$$

at $t = 0$, and $t = \pi$,

$$y' = \frac{dy}{dx} = 0$$

x-axis is the tangent to the curve at $A\ (a, 0)$ and $A'(-a, 0)$.

at $$t = \frac{\pi}{2}, \quad \text{and} \quad t = \frac{3\pi}{2}$$

$$y' = \frac{dy}{dx} = \infty$$

y-axis is the tangent to the curve at $B(0, a)$, and $B'(0, -a)$.

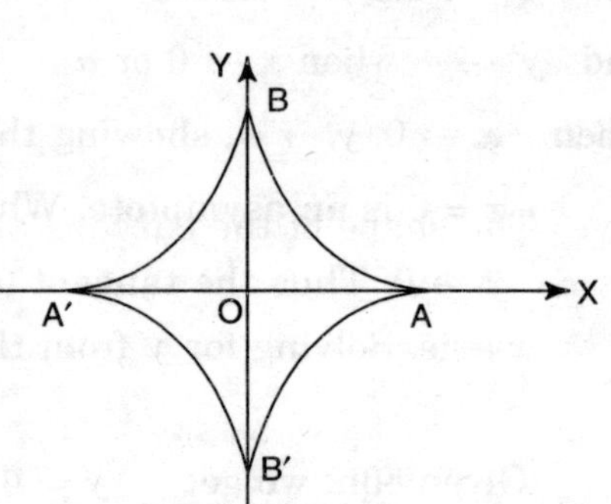

Fig. 3.12

Thus, the entire curve is enclosed within the circle of radius a and centred at (0, 0). As t increases from 0 to $\frac{\pi}{2}$, $y' = \frac{dy}{dx}$ is less than zero.

The shape of the curve is found as shown in Fig. 3.12

70. Trace the cycloid:

$$x = a\ (\theta - \sin \theta)$$
$$y = a\ (1 - \cos \theta), \qquad (0 \le \theta \le 2\pi)$$

Solution: The greatest value of y is $2a$ and the least value is 0.

$\Rightarrow$ The curve lies between $y = 0$ and $y = 2a$.

1. *Symmetry:*

$$x = a\ [-\theta - \sin(-\theta)]$$
$$\Rightarrow \quad x = a\ [-\theta + \sin \theta] = -a\ (\theta - \sin \theta)$$
$$\Rightarrow \quad x(-\theta) = -x(\theta).$$
$\therefore$ $x = a\ (\theta - \sin \theta)$ is an odd function of θ.

and $$y = a\ (1 - \cos \theta)$$
$\Rightarrow$ $x(-\theta) = x(\theta)$ an even function of θ.

$\therefore$ The curve is symmetrical about y-axis.

2. *Origin:*

(*i*) Put $x = 0$

$$\Rightarrow \quad \theta - \sin \theta = 0$$

or, $\theta = 0$ and then $y = 0$

$\Rightarrow$ The curve passes through the origin.

3. As θ varies from 0 to 2π, the changes in the values of x, y and $y' = \cot\left(\frac{\theta}{2}\right)$ are tabulated as follows:

θ	0	$\frac{\pi}{2}$	π	$\frac{3\pi}{2}$	2π
x	0	$a\left(\frac{\pi}{2}-1\right)$	$a\pi$	$a\left(\frac{3\pi}{2}+1\right)$	$2a\pi$
y	0	a	$2a$	a	0
$\frac{dy}{dx}$	∞	1	0	-1	∞

The shape of the curve is as shown in Fig. 3.13.

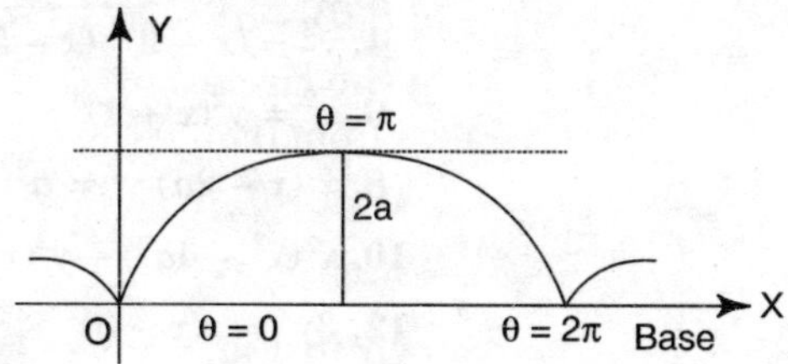

Fig. 3.13 Cycloid

71. Trace the curve:

$$y = c \cdot \cosh\left(\frac{x}{c}\right), x > 0$$

Solution:

1. *Symmetry:*

$$\cosh\left(\frac{-x}{c}\right) = \cosh\left(\frac{x}{c}\right)$$

$\Rightarrow$ The curve is symmetrical about the y-axis.

2. *Standard points:*

Put $x = 0$

$\Rightarrow$ $y = c$ $(\because \cosh 0 = 1)$

$y > 0$ for all $x > 0$

$\Rightarrow$ The curve lies above the x-axis and cuts the y-axis at $(0, c)$

3. Since, $$y = c\cosh\left(\frac{x}{c}\right) = c \cdot \frac{e^{x/c} + e^{-x/c}}{2}$$ [1]

Now, $$y' = \sinh\frac{x}{c} = \frac{e^{x/c} - e^{-x/c}}{2} > 0, \text{ for } x > 0, \text{ and } y' < 0 \text{ for } x < 0.$$

$\Rightarrow$ curve is increasing when $x > 0$
(*i.e.*, to the right of y-axis)
and decreasing when $x < 0$.
(*i.e.*, to the left of y-axis)

At $(0, c)$, $y' = 0$

$\Rightarrow$ the tangent to the curve at $(0, c)$ is parallel to the x-axis.

Thus, the shape of the curve is as sketched in the Fig. 3.14.

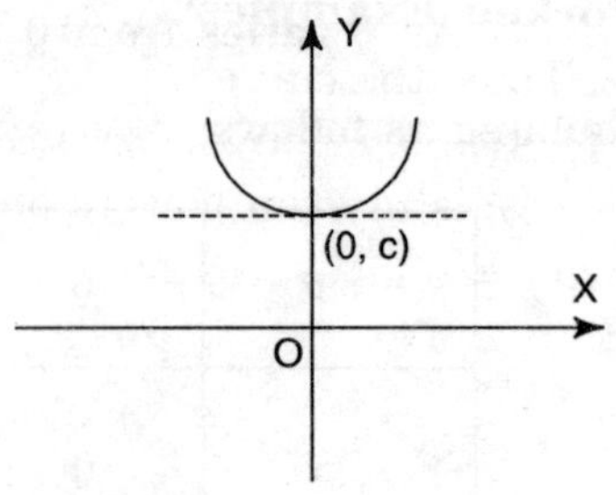

Fig. 3.14 Catenary

EXERCISES

Trace the following curves:

1. $a^2y^2 = x^3(2a - x)$
2. $a^2x^2 = y^3(2a - y)$
3. $y\,(x^2 + 4a^2) = 8a^3$
4. $y^2 = (x - 2)^2\,(x - 5)$
5. $9ay^2 = x(x - 3a)^2$
6. $x^2 = y^2(x + 1)^3$
7. $y^3 = x(x + a)$
8. $x\,(x - 2a)\,y^2 = a^2(x - a)\,(x - 3a)$
9. $(x^2 - 1)y^2 = x$
10. $x^2(x^2 - 4a^2) - y^2(x^2 - a^2) = 0$
11. $x^6 + y^6 = a^2x^2y^2$
12. $3y^2 = x(x - 3)^2$
13. $a^4y^2 = x^5(2a - x)$
14. $a^4y^2 = a^2x^4 - x^6$
15. $y^3 = a^2x - x^3$
16. $x^2y^2 = (a + y)^2\,(b^2 - y^2)$, $(b > a)$

3.5 POLAR EQUATIONS

Working Rule for tracing the curve whose equations are given in polar coordinates

(VTU, March, 1999)

1. Let $f(r, \theta) = 0$ be the given equation of the curve.
   (*a*) *Symmetry:* If $f(r, -\theta) = f(r, \theta)$,
   then, the curve is symmetrical about the initial line (in particular if θ occurs as $\cos\theta$ only in the given equation, we have $\cos(-\theta) = \cos\theta$, so the curve is symmetrical about the initial line (*b*) If the equation contains only even powers of r, the curve is symmetrical about the pole.
2. Put $r = 0$ and find whether the pole lies on the curve and if so for what values of θ.
3. Find ϕ using, $\tan\phi = r\left(\dfrac{d\theta}{dr}\right)$ and find particularly where $\phi = 0$ or $\left(\dfrac{\pi}{2}\right)$, usually, $\phi = \dfrac{\pi}{2}$

   when $\dfrac{dr}{d\theta} = 0$

   i.e., when r is a maximum or a minimum.
4. Find the region or regions in which real points on the curve are possible.

   For spirals and other systems of curves, slightly different methods are used.

Worked Examples

72. Trace the curve: $r^2 = a^2 \cos 2\theta$.

1. *Symmetry:* $\cos(-\theta) = \cos\theta$; there is symmetry about the initial line
r is in even powers only; there is symmetry about the pole.

2. $r = 0$ gives, $\cos 2\theta = 0$

$$\Rightarrow \quad 2\theta = (\pi/2)$$

$$\Rightarrow \quad \theta = (\pi/4) \qquad [1]$$

3. $$2r\left(\frac{dr}{d\theta}\right) = a^2(-2\sin 2\theta)$$

$$\therefore \quad \left(\frac{dr}{d\theta}\right) = -(a^2/r)\sin 2\theta$$

$$\tan\phi = \frac{r}{\left(\frac{dr}{d\theta}\right)} = \frac{r}{-(a^2/r)\sin 2\theta} = \frac{r^2}{-a^2\sin 2\theta} = \frac{a^2\cos 2\theta}{-a^2\sin^2\theta}$$

i.e., $$\tan\phi = -\cot(2\theta)$$

$$\Rightarrow \quad \phi = (\pi/2) + (2\theta) \qquad [2]$$

At $\quad \theta = 0,\ \phi = \pi/2;\ \theta = (\pi/4),\ \phi = \pi$ [3]

θ	2θ	$\cos 2\theta$	$r^2 = a^2\cos 2\theta$	r	ϕ
0	0	$\perp$	$= a^2$	a	$(\pi/2)$
0 to $(\pi/4)$	0 to $(\pi/2)$	1 to 0	positive	real	
$(\pi/4)$	$(\pi/2)$	0	0	0	π
$(\pi/4)$ to $(\pi/2)$	$(\pi/2)$ to π	(– ve)	(– ve)	imagi.	

Thus, real points on the curve are possible, when θ lies between 0 to $(\pi/4)$; hence by symmetry, it lies in the angle $(-\pi/4)$ to 0 (by symmetry about initial line) and in the opposite angle (symmetry about the pole).

Thus, the curve lies entirely within the angle $(-\pi/4)$ to $(\pi/4)$ and in the opposite angle.

The shape of the curve is as given in the rough sketch Fig. 3.15.

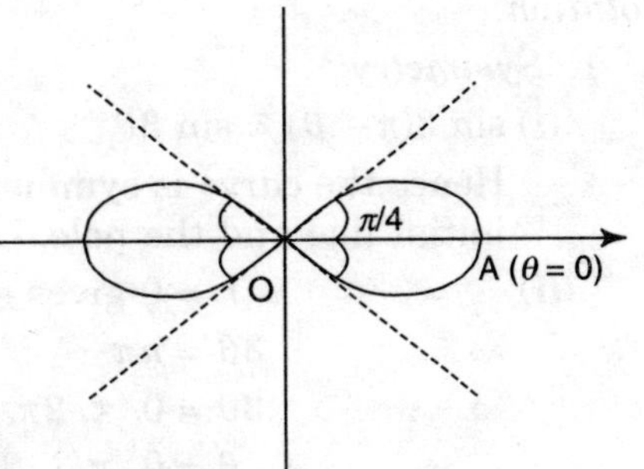

Fig. 3.15 Lemniscate

73. Trace the curve:

$$r = a(1 + \cos\theta)$$ (VTU, Model Q.P. 1)

Solution:

1. *Symmetry:*

(*i*) $\quad \cos(-\theta) = \cos\theta$.

Hence the curve is symmetrical about the initial line

PART-C

(*ii*) $r = 0$ gives $(1 + \cos\theta) = 0$

$\Rightarrow \quad \cos\theta = -1 = \cos\pi$

$\Rightarrow \quad (\theta = \pi)$

(*iii*) $$\tan\phi = \frac{r}{\left(\frac{dr}{d\theta}\right)} = \frac{a(1+\cos\theta)}{a(-\sin\theta)}$$

$$= \frac{2\cos^2\theta/2}{-2\sin\theta/2\cos\theta/2} = -\cot\theta/2.$$

$\therefore \quad \phi = (\pi/2 + \theta/2)$, since $|\cos\theta| \le 1$, $v \le 2a$

No portion of the curve lies to right side at $(2a, 0)$

(*vi*) We tabulate corresponding values of r and θ.

θ	0	increases	$\pi/2$	$\pi/2$ to π	π
$\cos\theta$	1	decreases	0	– ve	– 1
r	$2a$		a	decreases	0
ϕ	$\pi/2$		$\frac{3\pi}{4}$		π

Curve is symmetrical about the initial line.

Using these facts, the shape of the curve can be drawn and is as in the Fig. 3.16

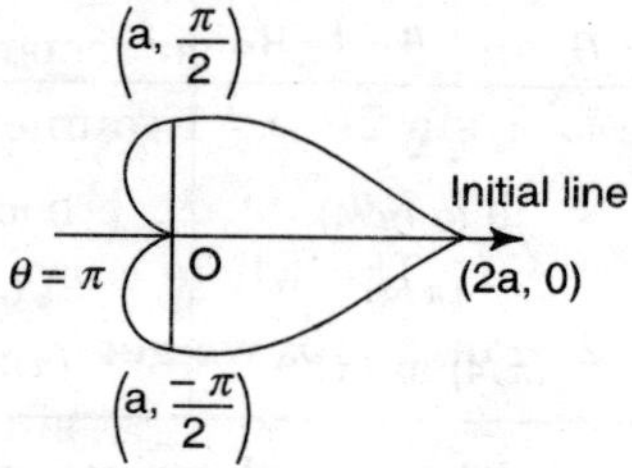

Fig. 3.16 Cardioide

74. Trace the curve: $r = a\sin 3\theta$, $a > 0$.

Solution:

1. *Symmetry:*

(*i*) $\sin 3(\pi - \theta) = \sin 3\theta$

Hence the curve is symmetrical about the line $\theta = \pi/2$, it is not symmetric about the initial line and the pole.

(*ii*) $r = 0$ gives $\sin 3\theta = 0$

$\Rightarrow \quad 3\theta = n\pi$

$\Rightarrow \quad 3\theta = 0, \pi, 2\pi, 3\pi, \ldots$ or

$\theta = 0, \pi/3, 2\pi/3, \pi, \ldots$

$\Rightarrow$ The curve passes through the pole and the tangents at the pole are $\theta = 0$, and $\theta = 2\pi/3$

$\theta = \pi/3$

($\because$ the other values of θ gives the same tangents)

(*iii*) since $|\sin 3\theta| \le 1$,

$\Rightarrow \quad |r| \le a$.

$\Rightarrow$ the entire curve lies within the circle $r = a$

2. **Standard points:**

θ	0	$\pi/6$	$\pi/3$	$\pi/2$	$2\pi/3$	$5\pi/6$	π	$4\pi/3$	$3\pi/2$	$5\pi/3$	2π
r	0	a	0	$-a$	0	a	0	0	a	0	0

With this information, we can draw the curve as shown in Fig. 3.17

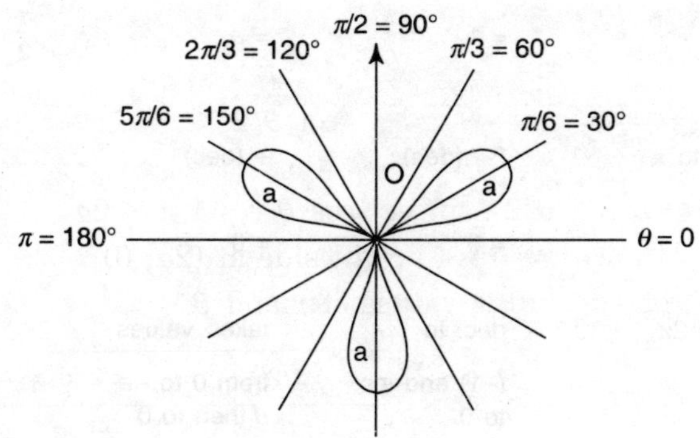

Fig. 3.17

75. Trace the curve:

$$r = a \sin 2\theta,$$

Solution: We have to take values from 0 to 2π for θ, and find the corresponding values for r.

1. $|\sin 2\theta| \le 1$ (numerically) $\forall\ \theta$

   $\therefore r \le a$

   $\Rightarrow$ The whole curve lies within a circle of radius a, and centre at the pole.

2. Put $r = 0$; we get

$$\sin 2\theta = 0$$

$$\therefore \quad 2\theta = (0, \pi, 2\pi, \ldots)$$

$$\Rightarrow \quad \theta\left(0, \frac{\pi}{2}, \pi, \frac{3\pi}{2} \text{ and } 2\pi\right) \qquad [1]$$

3. When $r = a$, $\sin 2\theta = 1$;

$$\therefore \quad 2\theta = \pi/2, \frac{3\pi}{2};$$

$$\therefore \quad r \text{ is a maximum at } \theta = \left(\frac{\pi}{4}\right), \left(\frac{3\pi}{4}\right), \left(\frac{5\pi}{4}\right) \text{ and } \left(\frac{7\pi}{4}\right) \qquad [2]$$

At these points $\quad \dfrac{dr}{d\theta} = 0; \tan\phi = \infty; \phi = \dfrac{\pi}{2}$ [3]

θ	2θ	$\sin 2\theta$	$r = a \sin 2\theta$	ϕ	curve
0	0	0	$= a \,.\, 0 = 0$	0	
0 to $\frac{\pi}{4}$	0 to $\frac{\pi}{2}$	(+) (inc.)	+ (inc.)	. . .	a loop of angle $\frac{\pi}{2}$
$\frac{\pi}{4}$	$\frac{\pi}{2}$	= 1	$= a$	$\frac{\pi}{2}$	with +ve values = (+)ve
$\frac{\pi}{4}$ to $\frac{\pi}{2}$	$\frac{\pi}{2}$ to π	(+) (dec)	+ (dec)	. . .	loop
$\frac{\pi}{2}$	π	= 0	= 0	. . .	
$\frac{\pi}{2}$ to π	π to 2π	dec. to (–1) and inc. to 0.	takes values from 0 to $-a$ θ then to 0		negative loop
π to $\frac{3\pi}{2}$	2π to 3π	(+)	+ 0 to a and back to 0		+ve loop
$\frac{3\pi}{2}$ to π	3π to 4π	(–)	0 to $-a$ and back to 0		–ve loop

Thus, the curve is seen to consist of 4 loops of angle $\left(\frac{\pi}{2}\right)$ each. With this information we can sketch the curve as shown in Fig. 3.18.

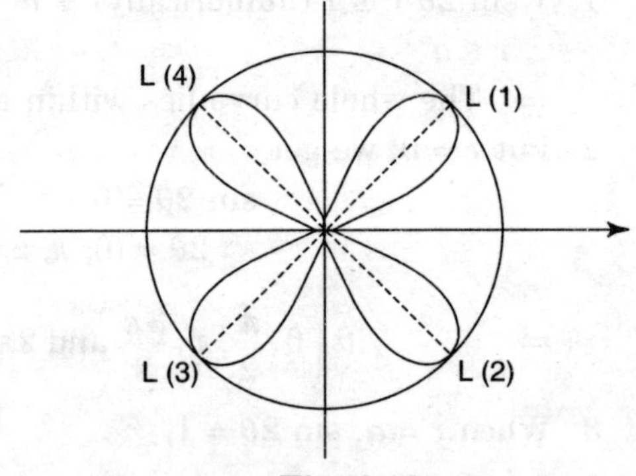

Fig. 3.18

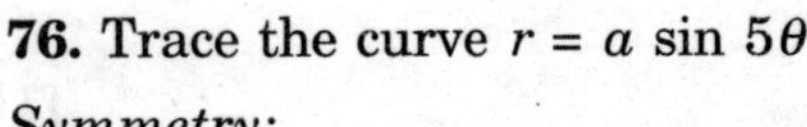

76. Trace the curve $r = a \sin 5\theta$

Symmetry:

(*i*) We have to take values 0 to 2π ($\sin 5\theta$ is a periodic function with period 2π) for θ and find the corresponding values of r.

(*ii*) $r = 0$ gives $\sin 5\theta = 0$

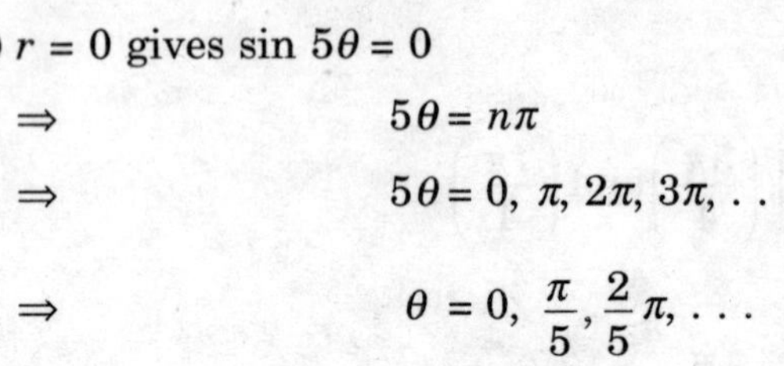

$$\Rightarrow \quad 5\theta = n\pi$$

$$\Rightarrow \quad 5\theta = 0, \pi, 2\pi, 3\pi, \ldots$$

$$\Rightarrow \quad \theta = 0, \frac{\pi}{5}, \frac{2}{5}\pi, \ldots$$

(*iii*) The maximum value of r is a

$\Rightarrow$ The whole curve lies within a circle of radius a, centre at the pole.

(*iv*) When $r = \pm a$, $\sin 5\theta = \pm 1$

The corresponding values of θ are

$$\frac{\pi}{10}, \frac{3\pi}{10}, \frac{5\pi}{10}, \frac{7\pi}{10}$$

(*v*) The curve passes through the pole and the tangents at the pole are:

$$\theta = \pi, \text{ and } \theta = 2\pi$$

(The other values of θ give the same tangents)

The following table gives the values of θ and corresponding values of r.

5θ	0	Intermediate values	$\frac{\pi}{2}$	Intermediate values	π	Intermediate values
θ	0	Intermediate values	$\frac{\pi}{10}$	Intermediate values	$\frac{2\pi}{10}$	Intermediate values
r	0	positive and Increasing	a	positive and decreasing	0	negative and decreasing
5θ	$\frac{3\pi}{2}$	Intermediate values	2π	Intermediate values	$\frac{5\pi}{2}$	Intermediate values
θ	$\frac{3\pi}{10}$	Intermediate values	$\frac{4\pi}{10}$	Intermediate values	$\frac{5\pi}{10}$	Intermediate values
r	$-a$	negative and increasing	0	positive and increasing	0	positive and decreasing
5θ	3π	Intermediate values	$\frac{7\pi}{2}$	Intermediate values	4π	Intermediate values
θ	$\frac{6\pi}{10}$	Intermediate values	$\frac{7\pi}{10}$	Intermediate values	$\frac{8\pi}{10}$	Intermediate values
r	0	negative and decreasing	$-a$	negative and decreasing	0	positive and increasing
5θ	$\frac{9\pi}{2}$	Intermediate values	5π			
θ	$\frac{9\pi}{10}$	Intermediate values	$\frac{10\pi}{10} = \pi$			
r	a	positive and increasing	4			

Thus, the curve is seen to consist of 5 loops $\left(\frac{\pi}{5}\right)$ each. With these information we can sketch the curve as shown in Fig. 3.19.

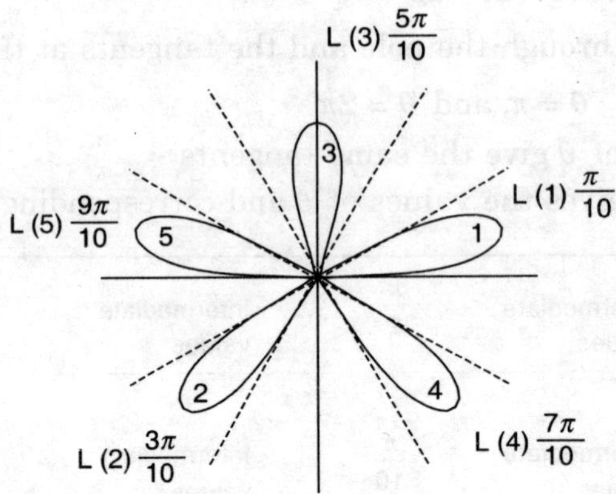

Fig. 3.19

77. Trace the curve $r = a \sin 6\theta$

Solution:

1. *Symmetry:* We have to take values from 0 to 2π for θ and find the corresponding values of r.

(*i*) $| \sin 6\theta | \leq 1, \forall \theta$

$$\therefore \quad r \leq a$$

$\Rightarrow$ The whole curve lies within a circle of radius a and centre at the pole.

(*ii*) Put $r = 0$, we get

$$\sin 6\theta = 0,$$

$$6\theta = n\pi,$$

$$6\theta = (0, \pi, 2\pi, 3\pi, \ldots)$$

$$\Rightarrow \quad \theta = 0, \frac{\pi}{6}, \frac{\pi}{3}, \frac{\pi}{2}, \ldots$$

(*iii*) When $r = a$, $\sin 6\theta = 1$

$$\Rightarrow \quad 6\theta = \frac{\pi}{2}, \frac{5\pi}{2}, \frac{9\pi}{2}, \ldots$$

$$\Rightarrow \quad \theta = \frac{\pi}{12}, \frac{5\pi}{12}, \frac{9\pi}{12}, \ldots$$

(*iv*) When $r = -a$,

$$\sin 6\theta = -1$$

i.e.,

$$6\theta = \frac{3\pi}{2}, \frac{7\pi}{2}, \ldots$$

$$\Rightarrow \quad \theta = \frac{3\pi}{12}, \frac{7\pi}{12}, \frac{11\pi}{12}, \ldots$$

The values of r for different values of θ are shown in the following table:

θ	0	$\frac{\pi}{12}$	$\frac{\pi}{6}$	$\frac{\pi}{4}$	$\frac{\pi}{3}$	$\frac{5\pi}{12}$	$\frac{\pi}{2}$	$\frac{7\pi}{12}$	$\frac{2\pi}{3}$
6θ	0	$\frac{\pi}{2}$	π	$\frac{3\pi}{2}$	2π	$\frac{5\pi}{2}$	3π	$\frac{7\pi}{2}$	4π
r	0	a	0	$-a$	0	a	0	$-a$	0
θ	$\frac{3\pi}{4}$	$\frac{5\pi}{6}$	$\frac{11\pi}{12}$	π	$\frac{13\pi}{12}$	$\frac{7\pi}{6}$	$\frac{15\pi}{12}$	$\frac{4\pi}{3}$	17π
6θ	$\frac{9\pi}{2}$	5π	$\frac{11\pi}{2}$	6π	$\frac{13\pi}{2}$	7π	$\frac{15\pi}{2}$	8π	$\frac{17\pi}{2}$
r	a	0	$-a$	0	a	0	$-a$	0	a
θ	$\frac{9\pi}{6}$	$\frac{19\pi}{12}$	$\frac{5\pi}{3}$	$\frac{21\pi}{12}$	$\frac{22\pi}{12}$	$\frac{23\pi}{12}$	2π		
6θ	9π	$\frac{19\pi}{2}$	10π	$\frac{21\pi}{2}$	11π	$\frac{23\pi}{2}$	12π		
r	0	$-a$	0	a	0	$-a$	0		

Thus, the curve is seen to consist of 12 similar loops and is sketched as shown in Fig. 3.20.

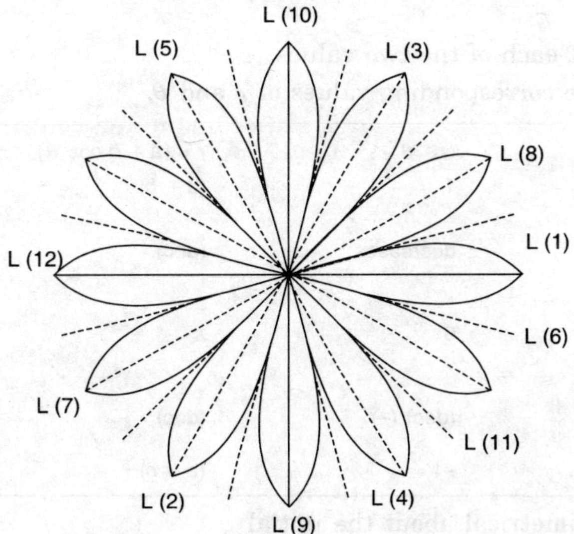

Fig. 3.20

78. Trace the curve $r = (a + b \cos \theta)$ considering the cases:

(i) $(a > b)$ *(ii)* $(a < b)$

(This curve is called Limacon).

Solution: **Case (i)** Let $a > b$

1. *Symmetry*

(*i*) $\cos(-\theta) = \cos\theta$

$\Rightarrow$ the curve is symmetrical about the initial line.

(*ii*) Put $r = 0$; we get $(b\cos\theta + a = 0)$

or $\cos\theta = -(a/b)$ which is impossible. ($\because a > b$) and $\cos\theta < 1$, (numerically)

$\therefore$ The curve does not pass through the pole.

(*iii*) When $\theta = \pi$, $\cos\theta = -1$ (least value)

and when $\theta = 0$ $\cos\theta = +1$ (greatest value)

The maximum value of r is

$$(a + b \cdot 1) = (a + b);$$

and when $\theta = \pi$, the min. value

is $[a + b(-1)] = (a - b) = +\text{ve}$ ($\because a > b$)

(*iv*) $$\left(\frac{dr}{d\theta}\right) = -b\sin\theta = 0,$$

$$\Rightarrow \qquad \theta = 0 \text{ or } \pi$$

$$\tan\phi = \left[r \div \left(\frac{dr}{d\theta}\right)\right]$$

$$= [(a \pm b)/0] = \infty$$

at $\theta = 0$, or π.

i.e., $\phi = (\pi/2)$ at each of the two values.

We tabulate corresponding values of r and θ.

θ	$\cos\theta$	$r = (a + b\cos\theta)$	ϕ
0	1	$(a + b)$	$(\pi/2)$
0 to $\frac{\pi}{2}$	decreases	(dec)	—
$\frac{\pi}{2}$	0	a	—
$\frac{\pi}{2}$ to π	(dec) (–)	(dec)	—
π	–1	$(a - b)$	$(\pi/2)$

The curve is symmetrical about the initial line.

Hence, the shape of the curve is roughly sketched in Fig. 3.21.

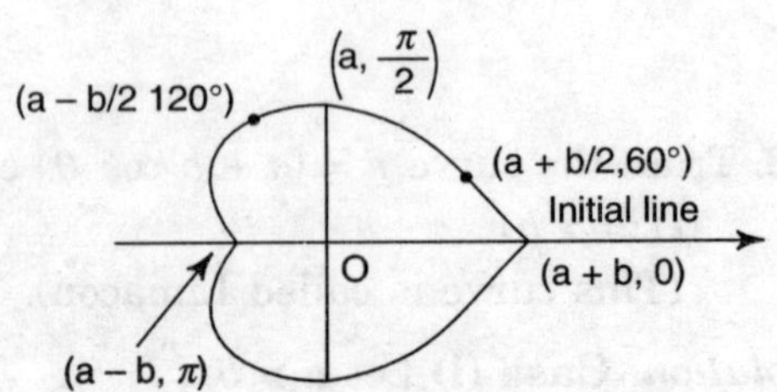

Fig. 3.21

Case (ii) Let $(a < b)$

1. The curve is symmetrical about the initial line.
2. Put $r = 0$, $\cos\theta = -(a/b)$ [1]

Since $(a < b)$, $(1 < 2)$, Let us take $a = 1$, $b = 2$

We get $\cos\theta = -\dfrac{1}{2}$ [using [1]]

$\Rightarrow \quad \theta = (2\pi/3)$

3. $\tan\phi = (r/r') = \dfrac{(1 + 2\cos\theta)}{(-2\sin\theta)} \left(r' = \dfrac{dr}{d\theta}\right)$ $(\because \; r = 1 + 2\cos\theta)$

$\therefore \quad \phi = (\pi/2)$ when $r' = \dfrac{dr}{d\theta} = 0$

i.e., when $\sin\theta = 0$

i.e., when $\theta = 0$ and $\theta = \pi$

We tabulate corresponding values of r and θ

θ	$\cos\theta$	$(1 + 2\cos\theta) = r$	ϕ
0	1	$r = 1 + 2 = 3$	$(\pi/2)$
0 to $\pi/2$	decreases	decreases	—
$\pi/2$	0	$r = 1 + 0 = 1$	—
$\pi/2$ to $2\pi/3$	(– ve)	r (+ve) dec.	—
$(2\pi/3)$	(–1/2)	0	—
$2\pi/3$ to π	(–)	$r = (-)$	—
π	–1	$(1 - 2) = -1$	$(\pi/2)$

Using these, we find that there is loop inside a bigger loop as is Fig. 3.22.

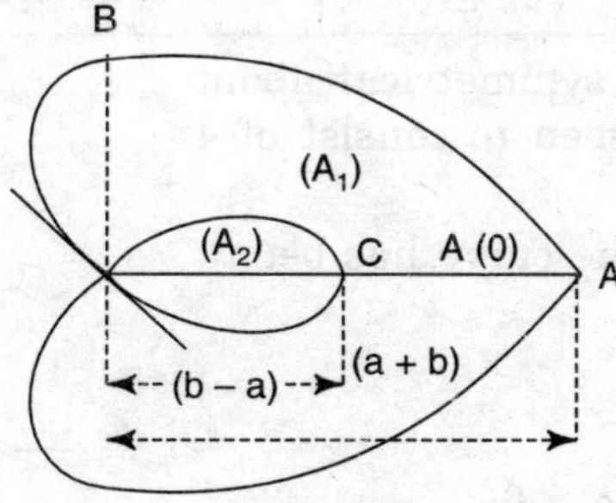

Fig. 3.22

79. Trace the curve

$$r = a\cos 2\theta,\; a > 0$$ [VTU (March, 1999, 2001]

Solution:

1. *Symmetry*:

(*i*) $\cos(-2\theta) = \cos 2\theta$

$\Rightarrow$ The curve is symmetrical about the initial line and the line $\theta = \dfrac{\pi}{2}$

PART-C

(*ii*) Put $r = 0,$

$\Rightarrow \quad \cos 2\theta = 0$

$$\Rightarrow \quad 2\theta = \frac{\pi}{2}, \frac{3\pi}{2}, \frac{5\pi}{2} \text{ or } \frac{7\pi}{2}$$

$$\Rightarrow \quad \theta = \frac{\pi}{4}, \frac{3\pi}{4}, \frac{5\pi}{4} \text{ or } \frac{7\pi}{4}$$

$\therefore$ The curve passes through the pole and the tangents at the pole are:

$$\theta = \frac{\pi}{4}, \quad \theta = \frac{3\pi}{4}, \quad \theta = \frac{5\pi}{4}, \quad \theta = \frac{7\pi}{4}$$

(the other values of θ give the same tangents)

(*iii*) $|\cos 2\theta| \le 1, \forall \theta$

$\therefore \quad r \le a$

$\Rightarrow$ The whole curve lies within a circle of radius a, and centre at the pole.

(*iv*) $$\frac{dr}{d\theta} = -a \sin 2\theta;$$

$$\frac{dr}{d\theta} = 0 \Rightarrow \sin 2\theta = 0$$

$$\Rightarrow \quad \tan \phi = \infty, \phi = \frac{\pi}{2}$$

(*v*) We tabulate the corresponding values of r and θ

θ	$r = a \cos 2\theta$
$\pi/4$ to $\pi/2$	increases from 0 to a
$\pi/2$ to $3\pi/4$	decreases from a to 0

$\therefore$ The curve has a loop symmetrical about the line $\theta = \pi/2$. The curve is seen to consist of 4 similar loops.

With these information, the curve has been sketched as shown in Fig. 3.23.

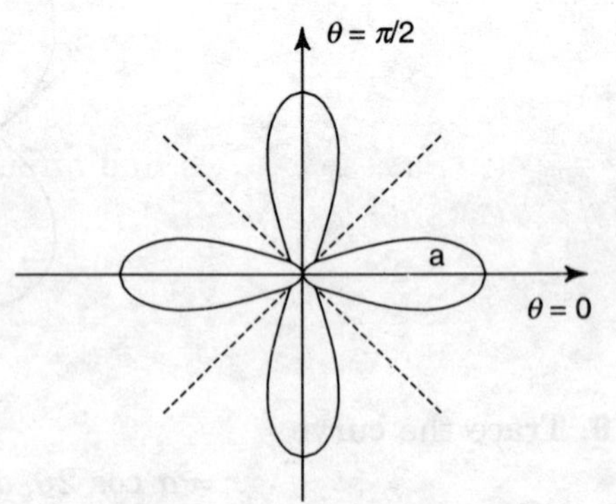

Fig. 3.23 Four leaved rose

80. Trace the curve: $r = ae^{\theta \cot \alpha}$, $a > 0$.

Solution:

1. *Symmetry:*

(*i*) The curve is not symmetrical.

(*ii*) The curve does not pass through the pole ($r \ne 0, \because e^{\theta \cdot \cos \alpha} \ne 0$)

(*iii*) $\theta = 0 \Rightarrow r = a$

$\Rightarrow$ The curve passes through the point $(a, 0)$

As θ increases, r increases from a to ∞. for –ve values of θ, r decreases, with θ.

(iv) $$\frac{dr}{d\theta} = ae^{\theta \cot \alpha} (\cot \alpha)$$

$$= r \cdot \cot \alpha$$

$\Rightarrow$ $$\tan \phi = r \frac{d\theta}{dr} = \frac{r}{r \cot \alpha}$$

$$= \tan \alpha$$

$\Rightarrow$ $$\phi = \alpha$$

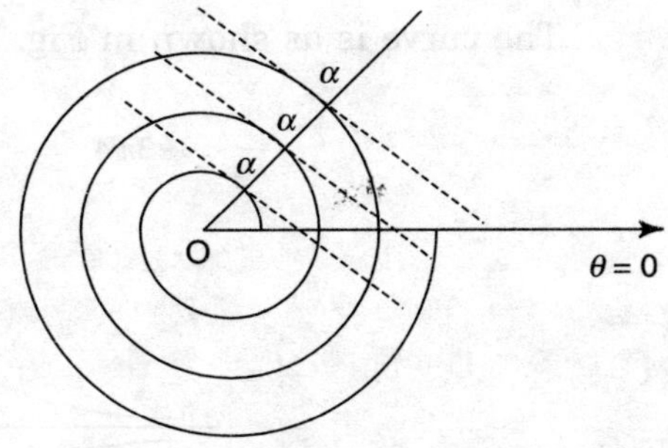

Fig. 3.24

$\Rightarrow$ The angle between the radius vector and the tangent is constant ($= \alpha$).

The curve is sketched roughly as shown in Fig. 3.24

(It is called Equiangular spiral)

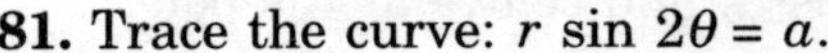

81. Trace the curve: $r \sin 2\theta = a$.

Solution:

(i) We can rewrite the given equation as: $r = \dfrac{a}{\sin 2\theta}$ [1]

Since $\sin 2\theta = 0$

$\Rightarrow$ $2\theta = n\pi$

$\Rightarrow$ $2\theta = 0, \pi, 2\pi, \ldots$

$\Rightarrow$ $\theta = 0, \dfrac{\pi}{2}, \pi, \dfrac{3\pi}{2} \ldots$

from Eqn. [1], we find the value of

$$r = \frac{a}{\sin 0} = \frac{a}{\sin 2 \cdot \left(\frac{\pi}{2}\right)} = \frac{a}{0} = \frac{a}{0} = \ldots = \infty$$

$\therefore$ $r = \infty$

(2) Since $r = \infty$, we find an asymptote in all directions.

(3) *Standard points*

Since $| \sin 2\theta | \leq 1$,

if, $\theta = \dfrac{\pi}{4}, \dfrac{5\pi}{4}$,

$\Rightarrow$ $r = a$ for these values of θ

and $\sin 2\theta = -1$,

if $\theta = \dfrac{3\pi}{4}, \dfrac{7\pi}{4}$

$\therefore$ $r = -a$ for these values of θ.

This $\Rightarrow$ that r increases or decreases} depending on sin 2θ.
The curve is as shown in Fig. 3.25.

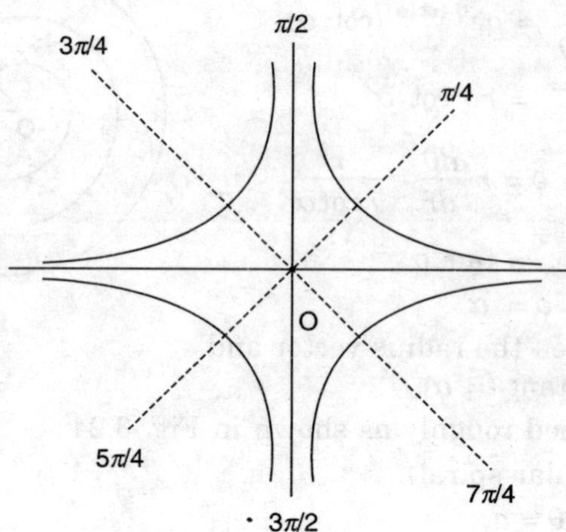

Fig. 3.25

82. Trace the curve:

$$r = a \cos 3\theta$$

Solution: $\boxed{r = a \cos 3\theta}$

(1) *Symmertry:*

$$\cos 3(-\theta) = \cos 3\theta.$$

$\Rightarrow$ The curve is symmetrical about the initial line.

(2) *Pole:*

put $r = 0$

$\Rightarrow$ $\cos 3\theta = 0$

$\Rightarrow$ $3\theta = (2n + 1)\dfrac{\pi}{2},$

$(n = 0, 1, 2, \ldots -1, -2, \ldots$
$= 0, \pm 1, \pm 2, \ldots \pm \ldots)$

$\Rightarrow$ $\theta = \pm\dfrac{\pi}{6}, \pm\dfrac{3\pi}{6}, \pm\dfrac{5\pi}{6}, \ldots$

$\Rightarrow$ origin (or pole) is a point on the curve.

T_0 (tangents at the origin) are obtained by:

$$\theta = \pm\frac{\pi}{6}, \theta = \pm\frac{\pi}{2} \text{ and } \theta = \pm\frac{5\pi}{6}$$

(The other values of θ give the same tangents)

(3) *Asymptote:*

Since $r \to \infty$, (for no value of θ)

$\Rightarrow$ The curve has no asymptote.

(4) *Standard points:*

$$| \cos 3\theta | \leq 1$$

$$\Rightarrow \quad | r | \leq a \ (i.e., \ r \not> a)$$

$\therefore$ The whole curve lies entirely within the circle

$$r = a$$

We tabulate the values of r with the corresponding value of θ

θ	0	$\frac{\pi}{6}$	$\frac{\pi}{3}$	$\frac{\pi}{2}$	$\frac{2\pi}{3}$	$\frac{5\pi}{6}$	π
r	a	0	$-a$	0	a	0	$-a$

Tracing the above values of θ we get the shape of the curve as shown in Fig. 3.26

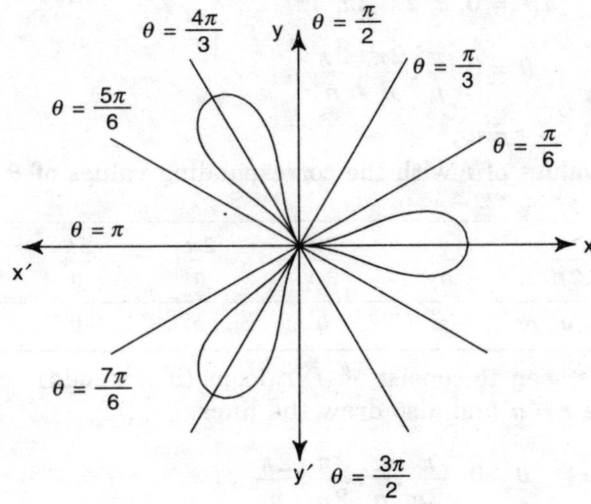

Fig. 3.26 3 leaved rose

83. Trace the curve:

$$r = a \cos (n\theta)$$

(1) *Symmetry:* $\cos [n (-\theta)] = \cos n\theta$.

$\Rightarrow$ The curve is symmetrical about the initial line.

(2) *Pole:* Put $r = 0$,

$$\Rightarrow \quad \cos n\theta = 0$$

$$\Rightarrow \quad n\theta = [2p + 1) \frac{\pi}{2}$$

(p is zero or any integer, positive or negative)

$$\Rightarrow \quad \theta = (2p + 1) \frac{\pi}{2n} = \frac{\pi}{2n}, \frac{3\pi}{2n}, \frac{5\pi}{2n}, \frac{7\pi}{2n}, \ldots$$

$\therefore$ pole is a point on the curve

T_0 (tangents at the poles) are given by

$$\theta = \frac{\pi}{2n}, \theta = \frac{3\pi}{2n}, \theta = \frac{5\pi}{2n}, \ldots$$

PART-C

3. *Asymptotes:* The curve has no asymptote.
4. Standard points:

(*i*) $| \cos n\,\theta | \leq 1$

$\Rightarrow \quad r \not> a$

$\Rightarrow$ curve lies entirely within the circle $r = a$

(*ii*) we have $\quad r' = \dfrac{dr}{d\theta} = -na \sin n\theta$

For max. value of r,

put, $\quad r' = \dfrac{dr}{d\theta} = 0$

$\Rightarrow \quad \sin n\,\theta = 0$

$\Rightarrow \quad n\theta = 0, \pi, 2\pi, 3\pi, 4\pi, \ldots$

$\Rightarrow \quad \theta = 0, \dfrac{\pi}{n}, \dfrac{2\pi}{n}, \dfrac{3\pi}{n}, \ldots$

$\Rightarrow$ max. value], $\quad r = a$

We tabulate the values of r with the corresponding values of θ

θ	0	$\frac{\pi}{2n}$	$\frac{\pi}{n}$	$\frac{3\pi}{2n}$	$\frac{2\pi}{n}$	$\frac{5\pi}{n}$	$\frac{3\pi}{n}$	...
r	a	0	a	0	a	0	a	...

Thus, the curve is seen to consist of (*i*) n loops (if n is odd) and (*ii*) $2n$ loops (if n is even) we draw the circle $r = a$ and also draw the lines.

$$\theta = 0, \frac{\pi}{2n}, \frac{\pi}{n}, \frac{3\pi}{2n}, \frac{2\pi}{n}, \ldots$$

Note: To trace the curve

$$r = a \sin (n\theta)$$

(*i*) put $\quad r = 0$, to get

$$\sin n\theta = 0, \pi, 2\pi, 3\pi, \ldots$$

$\Rightarrow \quad \theta = 0, \dfrac{\pi}{n}, \dfrac{2\pi}{n}, \dfrac{3\pi}{n}, \ldots$

Draw these lines $\quad \theta = 0, \theta = \dfrac{\pi}{n}, \theta = \dfrac{2\pi}{n},$

$$\theta = \frac{3\pi}{n}, \ldots$$

This will divide the circle $r = a$ into $2n$ equal parts.

The class of curves $r = a \sin n\theta$ are called rose petal curves. The curve $r = a \sin n\theta$ consist of $2n$ loops of angle $\left(\dfrac{\pi}{n}\right)$ each. When n is even $2n$ loops are distinct when n is odd, they coincide 2 by 2 into n distinct loops.

84. Trace the curve

(*i*) $r^2\ \theta = a^2$

(*ii*) $r\theta = a$ (Reciprocal spiral)

Solution: Given: $r^2\theta = a^2$

1. *Symmetry:*

$$(-r)^2\ \theta = a^2$$

$$\Rightarrow \quad r^2\theta = a^2$$

$\therefore$ The curve is symmetrical about the pole.

2. *Pole:* put $\quad r = 0;\ \theta = \dfrac{a^2}{r^2},$

$$\Rightarrow \quad \theta = \infty, \text{ when } r = 0$$

Thus, origin (or pole) is not a point on the curve.

3. *Asymptotes:*

since $\quad \theta \to 0 \Rightarrow r \to \infty.$

so, the curve has asymptotes which may be calculated as:

From the given curve, $r^2\theta = a^2$

we can write; $\quad \dfrac{1}{r} = \pm \dfrac{\sqrt{\theta}}{a} = f(\theta)$ (say)

$$f(\theta) = 0 \quad \text{gives } \theta = 0$$

and $\quad f'(\theta) = \pm \dfrac{1}{a} \cdot \dfrac{1}{2\sqrt{\theta}}$

$\therefore$ The asymptote is given by

$$r \sin(\theta - 0) = \frac{1}{f'(0)}$$

$$\Rightarrow \quad r \sin\theta = 0$$

$$\Rightarrow \quad \sin\theta = 0, \text{ or, } \theta = 0$$

Thus $\theta = 0$ is the only asymptote

4. Standard points

We tabulate the corresponding values of θ and r

$\theta = 0$	$\dfrac{\pi}{3}$	$\dfrac{\pi}{2}$	π
$r = \infty$	$\pm a\sqrt{\dfrac{3}{\pi}}$	$\pm a\sqrt{\dfrac{2}{\pi}}$	$\pm a\sqrt{\dfrac{1}{\pi}}$

As θ increases from 0 to ∞, r has equal and opposite values and numerically decreases from ∞ to 0.

PART-C

The curve is sketched as in Fig. 3.27

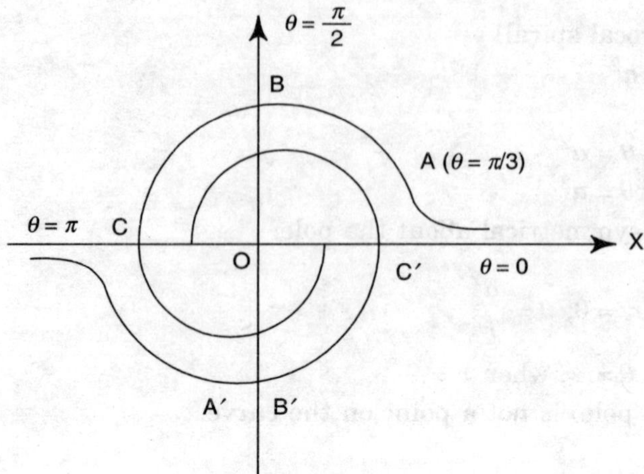

Fig. 3.27

(*ii*) To trace $r\,\theta = a$

1. *Symmetry:*

$$(-r)\,(-\theta) = a$$

$\Rightarrow r\theta = a$. Thus, the curve is symmetrical about the line $\theta = \dfrac{\pi}{2}$

$\left(\theta = \dfrac{\pi}{2}\ i.e., y-\text{axis}\right)$

2. *Pole:* $r \neq 0$ (for any finite value of θ)

$\therefore$ pole is not a point on the curve.

3. *Asymptotes:*

As $\quad \theta \to 0,\ r \to \infty$

The curve has an asymptote

From the given eqn:

$$r\theta = a$$

$$\frac{1}{r} = \frac{\theta}{a} = f(\theta)$$

$$f(\theta) = 0 \text{ gives } \theta = 0$$

$$\therefore \qquad f'(\theta) = \frac{1}{a},\ \therefore f'(0) = \frac{1}{a}$$

$\therefore$ The asymptote is given by

$$r\sin(\theta - 0) = \frac{1}{f'(0)}$$

or $$r\sin\theta = a$$

or $$y = a$$

$$\therefore \qquad r \sin \theta = a \cdot \frac{\sin\theta}{\theta} < a, \ (r\theta = a) \qquad \left(\because \ \frac{\sin\theta}{\theta} < 1\right)$$

∴ Thus, the whole curve lies entirely below the asymptote $r \sin \theta = a$

or $\qquad y = a$

4. *Standard points:*

We tabulate the corresponding values of θ and r.

θ	0	$\frac{\pi}{2}$	π	2π
r	∞	$\frac{2a}{\pi}$	$\frac{a}{\pi}$	$\frac{a}{2\pi}$

Thus, the shape of the curve is as shown in Fig. 3.28

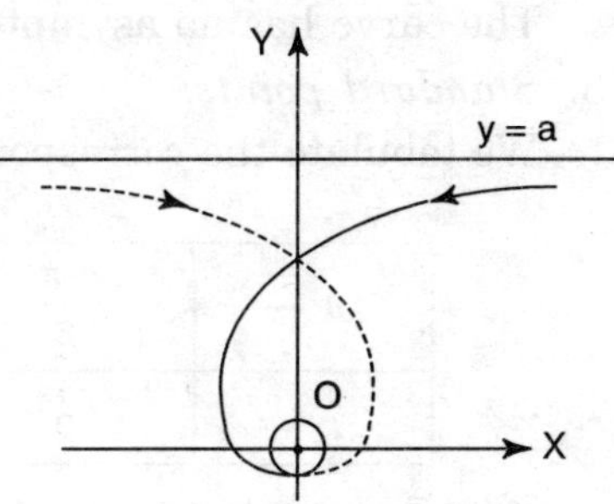

Fig. 3.28

85. Trace the curve :

$$r = a\,(1 + \sin\,\theta)$$

Solution:

1. Symmetry:

$$r = a\,[(1 + \sin\,(\pi - \theta)] \qquad (\because \ \sin\,(\pi - \theta) = \sin\,\theta)$$
$$= a\,(1 + \sin\,\theta)$$

∴ The curve is symmetrical about the line $\theta = \dfrac{\pi}{2}$

2. *Origin*: put $\qquad \theta = 0, \ r = a$

where $\qquad \theta = -\dfrac{\pi}{2}, \ r = 0$

∴ the line $\theta = \dfrac{\pi}{2}$ is tangent to curve at pole through' which the curve passes. Curve meets initial line at $(a, 0)$ and the line

$$\theta = \left(\frac{\pi}{2}\right), \text{ at } \left(2a, \frac{\pi}{2}\right)$$

3. $$\tan\phi = r\,\frac{d\theta}{dr} = \frac{a\,(1 + \sin\theta)}{a\cos\theta}$$

$$= \frac{(\sin\theta/2 + \cos\theta/2)^2}{\cos^2\theta/2 - \sin^2\theta/2} = \frac{\cos(\theta/2) + \sin\theta/2}{\cos\theta/2 - \sin\theta/2}$$

PART-C

$$= \frac{1+\tan\theta/2}{1-\tan\theta/2} = \tan\left(\frac{\pi}{4}+\frac{\theta}{2}\right)$$

$\therefore \qquad \phi = \frac{\pi}{4} + \theta/2$

when $\theta = 0$, $\phi = \frac{\pi}{4}$ when

$\theta = \frac{\pi}{2}, \phi = \frac{\pi}{2}$

i.e., at $\theta = \frac{\pi}{2}$, the tangent is $\perp r$ to the line $\theta = \frac{\pi}{2}$

4. *Asymptotes:*
The curve has no asymptotes because for any value of θ, r does not tend to ∞.
5. *Standard points:*
We tabulate the corresponding values of θ and r as below:

θ	$\frac{-\pi}{2}$	0	$\pi/2$	π
r	0	a	$2a$	a

since $|\sin\theta| \le 1$,
and we know that $r \le 2a$

$\therefore$ The whole curve lies within the circle $r = 2a$ and the shape of the curve is as shown in Fig. 3.29

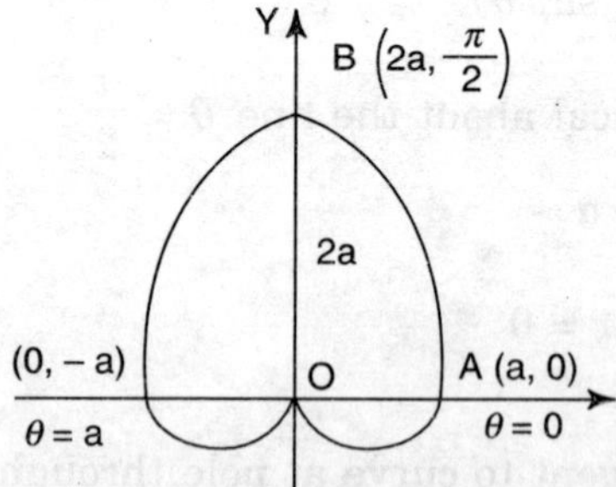

Fig. 3.29

EXERCISES

Trace the following curves:

(*i*) $r = a \cos 5\theta$ (Five leaved rose)
(*ii*) $r = a(1 - \sin\theta)$ (cardioids)
(*iii*) $r = 2a \cos\theta$ (circle)
(*iv*) $r = ae^{m\theta}$ $(a > 0, m > 0)$ (Equiangular spiral)

(v) $r = a \dfrac{\sin^2\theta}{\cos\theta}$ [Cissoid] (vi) $r = a(1 - \cos\theta)$

(vii) $r = (2 + 3\cos\theta)$ (viii) $r = a(1 + \sec\theta)$

Names of Important Curves and Their Cartesian, Parametric and Polar Equations.

1. Clssoid of Diocles

(a) $y^2(a - x) = x^3$

(b) $x = a\sin^2 t,\ y = a\dfrac{\sin^3 t}{\cos t}$

(c) $r = a\dfrac{\sin^2\theta}{\cos\theta}$

2. Folium of Decarte's

(a) $x^3 + y^3 = 3axy$

(b) $x = \dfrac{3at}{1+t^3},\ y = \dfrac{3at^2}{1+t^3}$

(c) $r = \dfrac{3a\sin\theta\cos\theta}{\sin^3\theta + \cos^3\theta}$

3. Cardioide

(a) $r = a(1 - \cos\theta)$, (Polar Equation)

(b) $r = a(1 + \cos\theta)$, (Polar Equation)

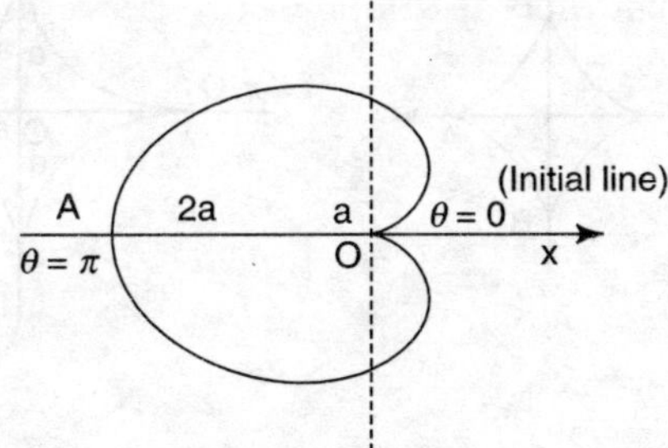

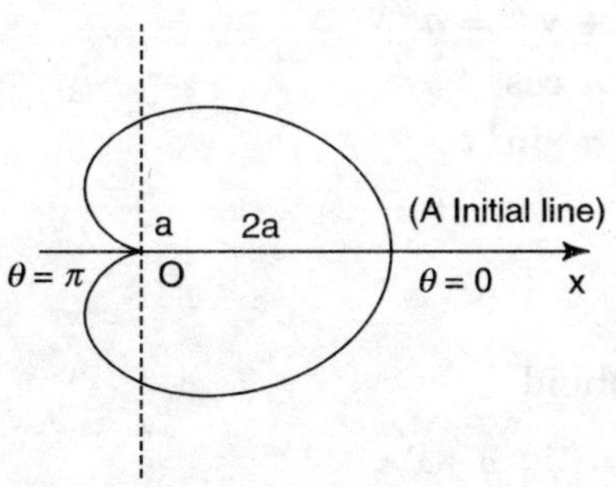

4. Limacon

(a) $r = a + b\cos\theta$ (Polar Equation) $(a > b)$

(b) $r = a + b\cos\theta$ (Polar Equation) $(a < b)$

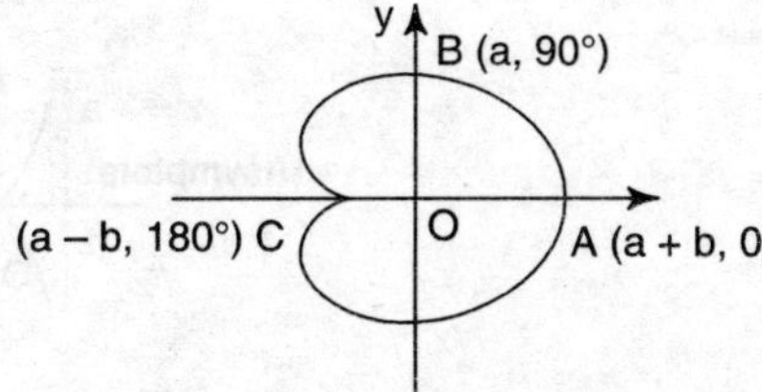

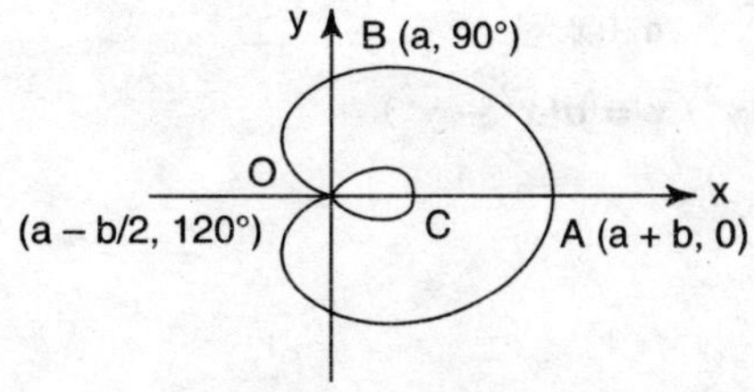

PART-C

5. **Lemniscate of Bernoulli**

(a) $(x^2 + y^2)^2 = a\,(x^2 - y^2)$, (Cartesian)

(b) $r^2 = a^2 \cos^2 \theta$ (Polar)

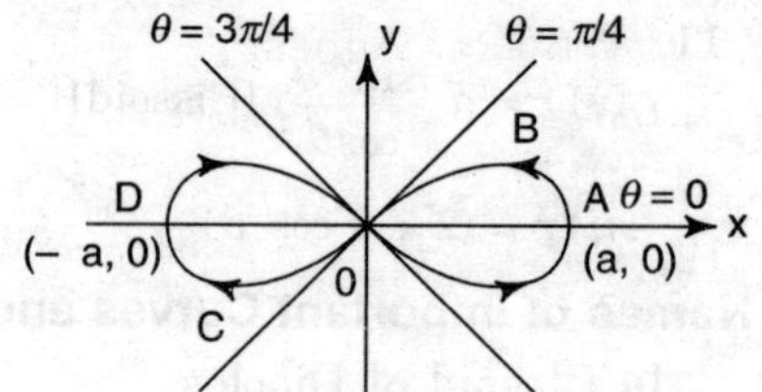

(Tangents at origin are y = ± x)

6. **Rose Lemniscate**

(a) $r^2 = a^2 \sin^2 \theta$

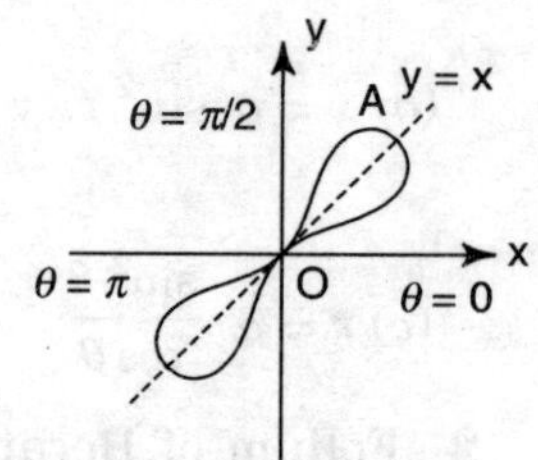

(Axes are tangents at origin)

7. **Hypocycloid**

(a) $\left(\dfrac{x}{a}\right)^{2/3} + \left(\dfrac{y}{b}\right)^{2/3} = 1$

(b) $x = a \cos^3 t$

$y = b \sin^3 t$

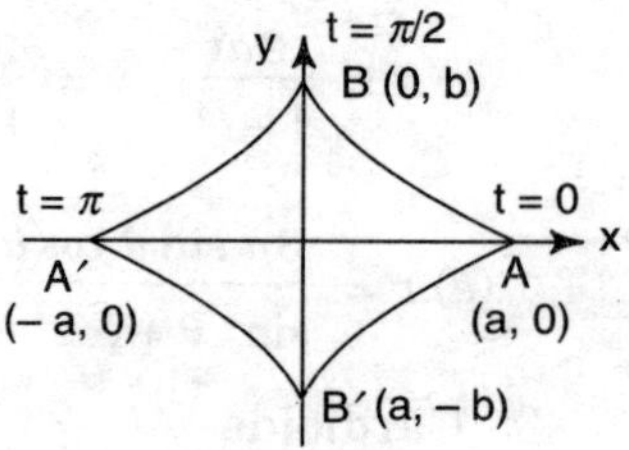

8. **Astroid**

(a) $x^{2/3} + y^{2/3} = a^{2/3}$

(b) $x = a \cos^3 t$

$y = a \sin^3 t$

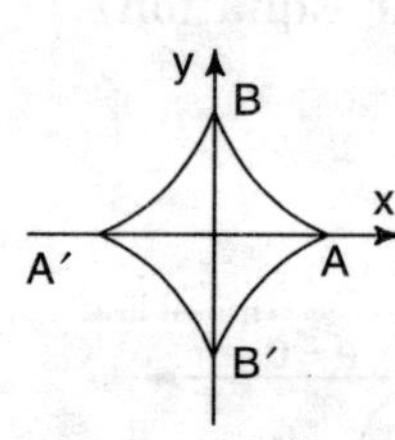

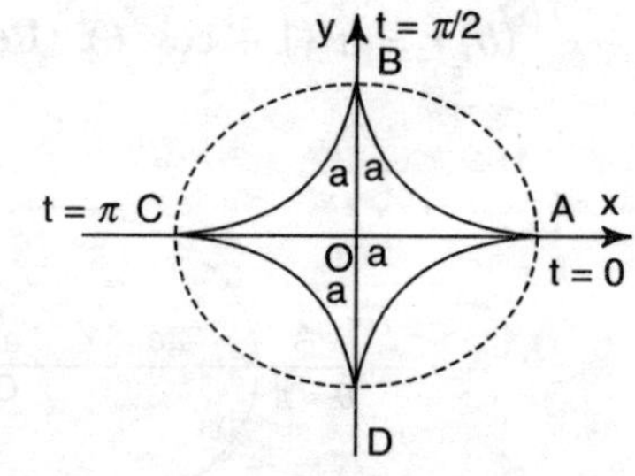

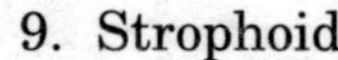

9. **Strophoid**

(a) $y^2 = x^2 \cdot \dfrac{a + x}{a - x}$

or $(x^2 + y^2)\,x = a\,(y^2 - x^2)$

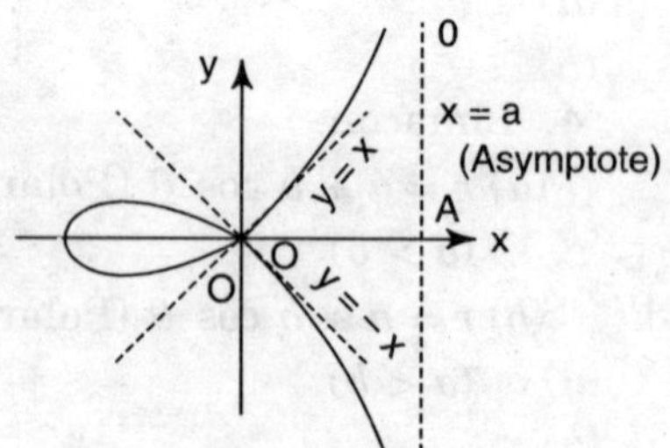

10. $y^2 = x^2 \cdot \dfrac{a - x}{a + x}$

or $(a^2 + y^2)\,x = a(x^2 - y^2)$

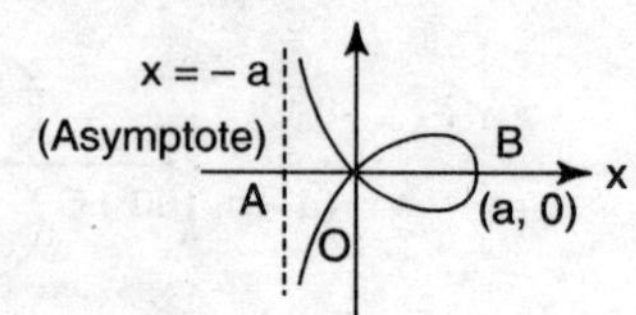

11. Witch of Agnesi

(a) $xy^2 = 4a^2 (2a - x)$

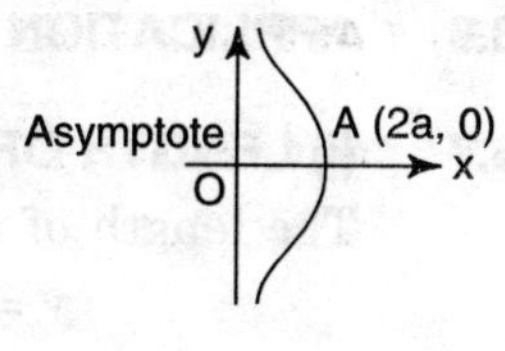

12. Catenary

(a) $y = c \cosh \left(\frac{x}{c}\right)$

or

$$y = \frac{c}{2} (e^{x/c} + e^{-x/c})$$

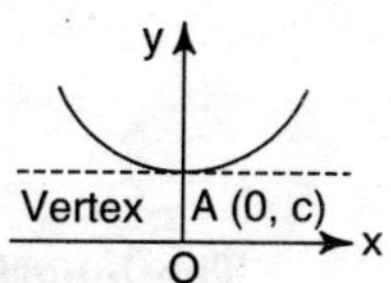

13. Tractrix

(a) $x = \sqrt{a^2 - y^2} + \frac{a}{2} \log \frac{a - \sqrt{a^2 - y^2}}{a + \sqrt{a^2 - b^2}}$

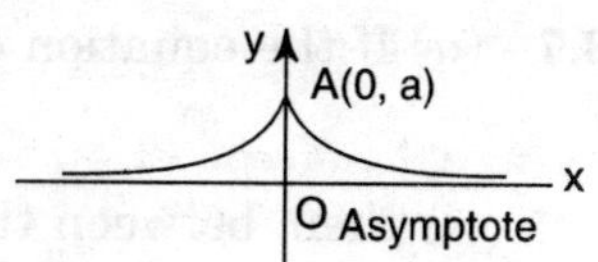

(b) $x = a\left(\cos t + \log \tan \frac{t}{2}\right),$

$y = a \sin t$

14. Cycloid

(a) $x = a (\theta - \sin \theta)$

$y = a (1 - \cos \theta)$

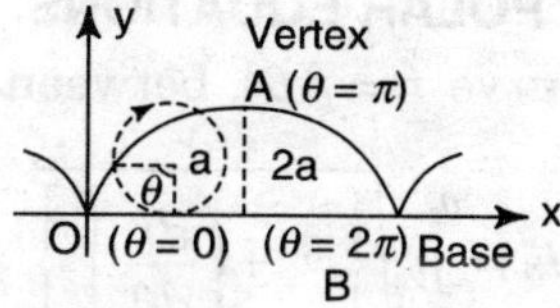

(b) $x = a (\theta + \sin \theta)$

$y = a (1 + \cos \theta)$

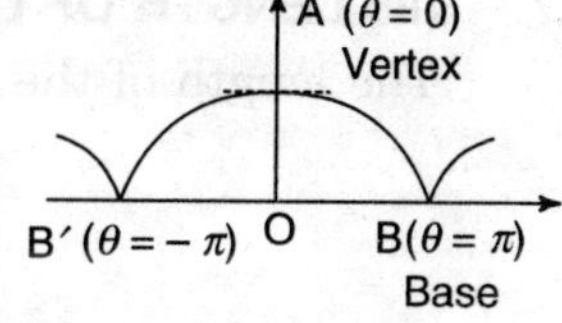

15. Inverted Cycloid

(a) $x = a (\theta + \sin \theta)$

$y = a (1 - \cos \theta)$

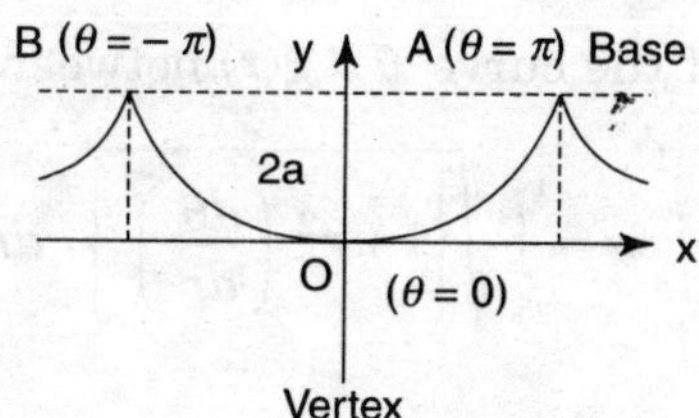

(b) $x = a (\theta - \sin \theta)$

$y = a (1 + \cos \theta)$

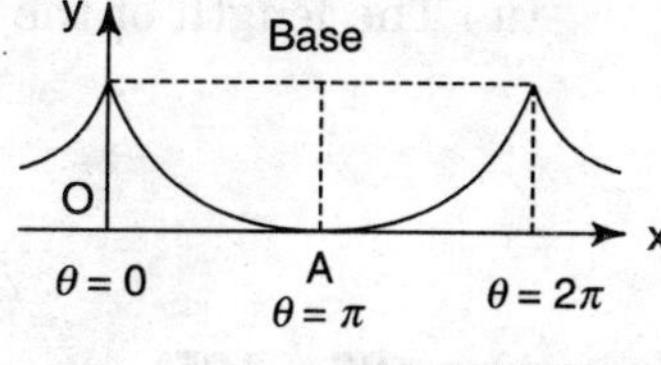

16. Cubical Parabola

(a) $y = ax^3$

(b) $x = t, y = at^3$

(c) $r^2 = \frac{\sin \theta}{a \cdot \cos^3 \theta}$

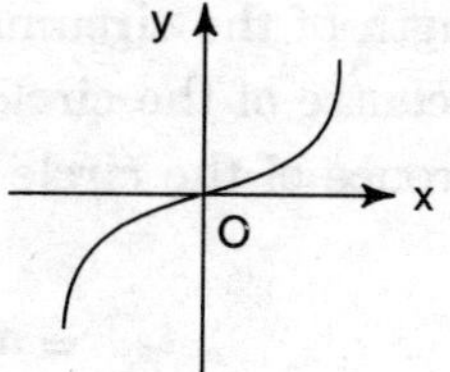

17. Semi-cubical Parabola

(a) $ay^2 = x^3$

(b) $x = at$

$y = a \cdot t^{3/2}$

(c) $r = a \cdot \frac{\sin^2 \theta}{\cos^3 \theta}$

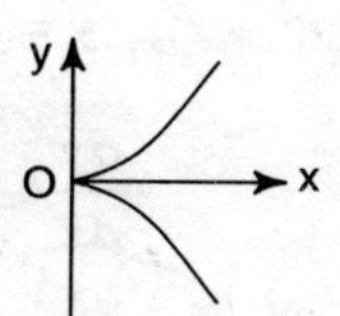

3.6 APPLICATION TO FIND AREA, LENGTH, VOLUME AND SURFACE AREA

3.7 (i) LENGTH OF PLANE CURVES (CARTESIAN)

The length of the arc of the curve

$y = f(x)$ between $x = a$ and $x = b$ is

$$S = \int_a^b ds = \int_a^b \sqrt{\left\{1 + \left(\frac{dy}{dx}\right)^2\right\}}\, dx$$

The length of the curve $x = g(y)$ between $y = c$ and $y = d$ is

$$S = \int_c^d ds = \int_c^d \sqrt{\left\{1 + \left(\frac{dx}{dy}\right)^2\right\}}\, dy$$

3.7 (*ii*) **If the equation of a curve given in parametric form: $x = x(t)$, $y = y(t)$**

then, between the points t_1 and t_2, we have, $S = \int_{t_1}^{t_2} \sqrt{\left(\frac{dx}{dt}\right)^2 + \left(\frac{dy}{dt}\right)^2}\, dt,$

where $(t_1 < t < t_2)$.

3.7 (iii) LENGTH OF CURVES OF POLAR EQUATIONS

The length of the arc of the curve $r = f(\theta)$, between $\theta = \theta_1$ and $\theta = \theta_2$ is

$$S = \int_{\theta_1}^{\theta_2} ds = \int_{\theta_1}^{\theta_2} \sqrt{\left\{r^2 + \left(\frac{dr}{d\theta}\right)^2\right\}} \cdot d\theta$$

(*iv*) The length of the arc of the curve $\theta = g(r)$ between $r = r_1$ and $r = r_2$ is:

$$S = \int_{v_1}^{v_2} ds = \int_{v_1}^{v_2} \sqrt{\left\{1 + r^2 \left(\frac{d\theta}{dr}\right)^2\right\}}\, dr$$

Examples (86 – 105)

86. Find the length of the circumference of circle of radius r by integration.

If 0(0, 0) is the centre of the circle, then $x^2 + y^2 = r^2$ is its equation.

Circumference of the circle = 4 × arc length in the first quardent

$$= 4\int_0^r \sqrt{1 + \left(\frac{dy}{dx}\right)^2}\, dx$$

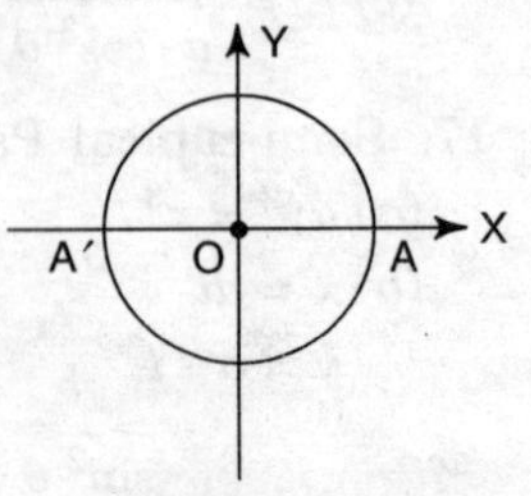

Fig. 3.30

Since $y^2 = r^2 - x^2$, $y = \sqrt{r^2 - x^2}$ in the I Quadrant

$$\therefore \quad \frac{dy}{dx} = \frac{1}{2\sqrt{r^2 - x^2}}(-2x) = \frac{-x}{\sqrt{r^2 - x^2}}$$

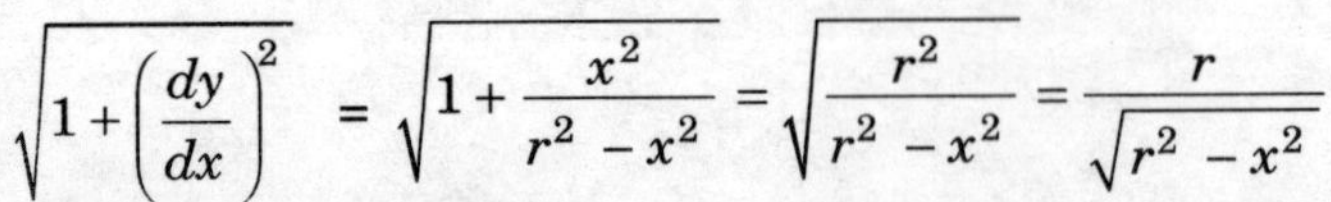

$$\sqrt{1 + \left(\frac{dy}{dx}\right)^2} = \sqrt{1 + \frac{x^2}{r^2 - x^2}} = \sqrt{\frac{r^2}{r^2 - x^2}} = \frac{r}{\sqrt{r^2 - x^2}}$$

$$\therefore \quad \text{Circumference} = 4\int_0^r \frac{r}{\sqrt{r^2 - x^2}}\,dx = 4r\left[\sin^{-1}\frac{x}{r}\right]_0^r$$

$$= 4r\,[\sin^{-1}(1) - \sin^{-1}(0)] = 4r\left(\frac{\pi}{2} - 0\right) = 2\pi r.$$

87. Find the complete perimeter of the curve $x^{2/3} + y^{2/3} = a^{2/3}$.

Since, $x = (a\cos^3 t),\ y = (a\sin^3 t)$

are the parametric equations of the curve.

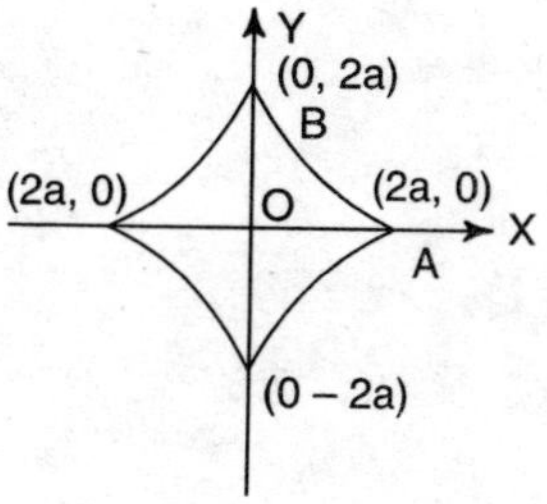

(0, 2a)
B
(2a, 0)
O
(2a, 0)
A
X
Y
(0 – 2a)

Fig. 3.31

$$\frac{dx}{dt} = -3a\cos^2 t\sin t$$

$$\frac{dy}{dt} = 3a\sin^2 t\cos t$$

$$\sqrt{\left(\frac{dx}{dt}\right)^2 + \left(\frac{dy}{dt}\right)^2}$$

$$= \sqrt{9a^2\cos^4 t\sin^2 t + 9a^2\sin^4 t\cos^2 t}$$

$$= \sqrt{9a^2\cos^2 t\sin^2 t} = 3a\cos t\sin t$$

$$= \frac{3}{2}\,a\sin 2t$$

By symmetry about x- and y-axis, we can write:

Perimeter = 4 × length of arc AB in the first quadrant which extends from $t = 0$ at A

to $\quad t = \dfrac{\pi}{2}$ at B

$$= 4\int_0^{\pi/2} \sqrt{\dot{x}^2 + \dot{y}^2}\,dt \quad \dot{x} = \frac{dx}{dt};\ \dot{y} = \frac{dy}{dt}$$

$$= 4\int_0^{\pi/2} \frac{3}{2}\,a\sin 2t\,dt = 6a\left[\frac{-\cos 2t}{2}\right]_0^{\pi/2}$$

$$= -3a(\cos\pi - \cos 0) = -3a(-1-1) = 6a \text{ units.}$$

88. Find the complete perimeter of the curve $y^2 = x^3$ from the origin to the point (4, 8)

Solution: Given: $y^2 = x^3$ (semi-cubical parabola)

$$\therefore \quad \frac{dy}{dx} = \frac{3x^2}{2y}$$

PART-C

$$\therefore \quad \sqrt{1+\left(\frac{dy}{dx}\right)^2} = \sqrt{1+\left(\frac{3x^2}{2y}\right)^2} = \sqrt{1+\frac{9x^4}{4x^3}}$$

$$= \sqrt{\left[1+\frac{9x}{4}\right]} = \frac{1}{2}(4+9x)^{1/2}$$

$\therefore$ **The required arc length**

$$l = \int_0^4 \sqrt{\left\{1+\left(\frac{dy}{dx}\right)^2\right\}}\,dx = \frac{1}{2}\int_0^4 (4+9x)^{1/2}\,dx$$

$$= \frac{1}{27}[(4+9x)^{3/2}]_0^4 = \frac{1}{27}\,[(40)^{3/2} - 4^{3/2}]$$

$$= \frac{8}{27}\left[10\sqrt{10}-1\right].$$

89. Find the length of an arc of the cycloid $x = a(t - \sin t)$
$y = a(1 - \cos t)$ **(VTU, F/M, 2005)**

Solution:

$$\frac{dx}{dt} = a(1 - \cos t)$$

$$\frac{dy}{dt} = a\,(0 + \sin t) = a \sin t$$

$$\sqrt{\left(\frac{dx}{dt}\right)^2 + \left(\frac{dy}{dt}\right)^2} = \sqrt{a^2\,(1-\cot)^2 + a^2 \sin^2 t}$$

$$= \sqrt{a^2\,(2 - 2\cos t)} = \sqrt{4a^2 \sin^2 \frac{t}{2}} = 2a \sin t/2$$

$$S = \int_0^{2\pi} 2a \sin t/2\, dt = 4a\left[-\cos\frac{t}{2}\right]_0^{2\pi}$$

$$= 4a\,[-\cos \pi + \cos 0] = 8a$$

90. Find the length of the curve
$r = a(1 + \cos \theta),\ a > 0.$ **(VTU, March, 1999)**

Solution:

$$S = 2\int_0^{\pi} \sqrt{\left\{r^2 + \left(\frac{dr}{d\theta}\right)^2\right\}} \cdot d\theta$$

$$= 2\int_0^{\pi} [a^2\,(1+\cos\theta)^2 + a^2 \sin^2\theta]^{1/2}\, d\theta$$

$$= 2a\ \sqrt{2}\int_0^{\pi} (1+\cos\theta)^{1/2}\, d\theta$$

$$= 2a\left(\sqrt{2}\right)\int_0^{\pi}\left(2\cdot\cos^2\left(\frac{\theta}{2}\right)\right)^{1/2} d\theta$$

$$= (2)\ a\ (2)\int_0^{\pi}\cos\frac{\theta}{2}\,d\theta = 4a\left[\sin\frac{\theta}{2}\right]_0^{\pi}\cdot 2$$

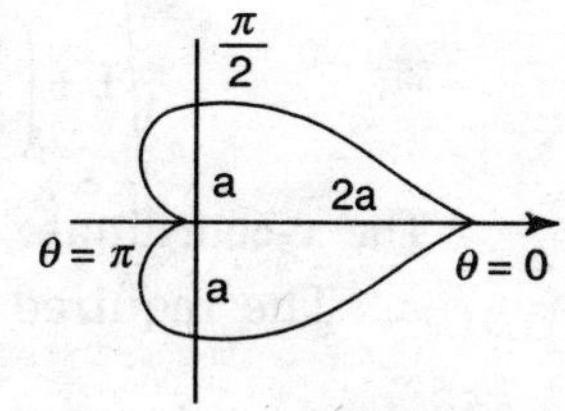

$$= 8a\left[\sin\frac{\pi}{2} - \sin 0\right]$$

$$= 8a\ [1-0] = 8a\ \textit{Ans.}$$

Fig. 3.32

91. **Prove that the length of the arc of the catenary $y = c \cosh (x/c)$ from the vertex to any point (x, y) is given by**

$$s^2 = y^2 - c^2$$

Solution:

Given: $y = c\cosh\left(\dfrac{x}{c}\right)$ [1]

Differentiate w.r.t. x, weget,

$$y' = c\cdot\sinh\left(\frac{x}{c}\right)\cdot\frac{1}{c} = \sinh\left(\frac{x}{c}\right)$$

$\therefore$ $$\sqrt{1+(y')^2} = \sqrt{1+\sinh^2\left(\frac{x}{c}\right)} = \cosh\left(\frac{x}{c}\right)$$

$\because$ $\cosh^2\theta - \sinh^2\theta = 1$

Vertex is $(0, c)$; $$s = \int_0^x \cosh\left(\frac{x}{c}\right)dx$$

$$= \left[\frac{\sinh(x/c)}{1/c}\right]_0^x = c\ \{\sinh(x/c) - \sinh(0)\}$$

$$= c\cdot\sinh(x/c) \qquad \because \sinh(0) = 0$$

Now $$s^2 = c^2\sinh^2(x/c) = c^2[\cosh^2(x/c) - 1] = c^2\cosh^2(x/c) - c^2$$

$$= y^2 - c^2 \qquad (\because y = \cosh(x/c))$$

92. **Determine the length of the parabola $x^2 = 4ay$ from its vertex to one extremity of the latus rectum.**

Solution:

Given: $x^2 = 4ay$

$\therefore$ $$\frac{dy}{dx} = (x/2a)$$

PART-C

$$\therefore \quad \sqrt{1+\left(\frac{dy}{dx}\right)^2} = \sqrt{\left[1+\frac{x^2}{4a^2}\right]} = \frac{1}{2a}\sqrt{(4a^2+x^2)}$$

The x-coordinate of one extremity of the latus rectums is $2a$.

$\therefore$ The required length l, where

$$l = \int_0^{2a} \sqrt{1+(y')^2}\, dx = \frac{1}{2a}\int_0^{2a} \sqrt{(4a^2+x^2)}\, dx$$

$$= \frac{1}{2a}\left[\frac{1}{2}x\sqrt{x^2+4a^2} + 2a^2 \sinh^{-1}(x/2a)\right]_0^{2a}$$

$$\left[\because \int \sqrt{a^2+x^2}\, dx = \frac{x}{2}\sqrt{x^2+a^2} + \frac{a^2}{2}\sinh^{-1}\left(\frac{x}{a}\right)\right]$$

$$= \frac{1}{2a}\left[2\sqrt{2}\, a^2 + 2a^2 \sinh^{-1}(1)\right]$$

$$= a\left[\sqrt{2} + \log\left(1+\sqrt{2}\right)\right] \qquad \left(\because \sinh^{-1}(x) = \log\left[x+\sqrt{(1+x^2)}\right]\right)$$

93. Find the length of the arc of the curve

$$x = e^{\theta}\left[\sin\frac{\theta}{2} + 2\cos\frac{\theta}{2}\right]$$

and
$$y = e^{\theta}\left[\cos\frac{\theta}{2} - 2\sin\frac{\theta}{2}\right]$$

from $\theta = 0$ to $\theta = \pi$

Solution: We know that

$$S = \int_0^{\pi} \sqrt{\left(\frac{dx}{d\theta}\right)^2 + \left(\frac{dy}{d\theta}\right)^2}\, d\theta \qquad [1]$$

Since
$$\frac{dx}{d\theta} = e^{\theta}\left[\sin\frac{\theta}{2} + 2\cos\frac{\theta}{2}\right] + e^{\theta}\left[\frac{1}{2}\cos\frac{\theta}{2} - \sin\frac{\theta}{2}\right] = \frac{5}{2}e^{\theta}\cdot\cos\frac{\theta}{2}$$

and similarly,
$$\frac{dy}{d\theta} = \frac{5}{2}e^{\theta}\sin\frac{\theta}{2}$$

$$\therefore \quad \sqrt{\left(\frac{dx}{d\theta}\right)^2 + \left(\frac{dy}{d\theta}\right)^2} = \sqrt{\frac{25}{4}e^{2\theta}\left(\cos^2\frac{\theta}{2} + \sin^2\frac{\theta}{2}\right)}$$

$\therefore$ from [1]

$$S = \int_0^{\pi} \frac{5}{2} e^{\theta} \cdot d\theta = \frac{5}{2}(e^{\theta})_0^{\pi} = \frac{5}{2}(e^{\pi} - 1)$$

94. Find the length of the arc of the equiangular spiral $r = (ae^{\theta \cot \alpha})$ between the points for which the radii vectors are r_1 and r_2.

Solution:

Here $$r = ae^{\theta \cot \alpha}$$

$$\therefore \quad \frac{dr}{d\theta} = a \cot \alpha\, e^{\theta \cot \alpha} = r \cot \alpha$$

Hence the required length

$$= \int_{v_1}^{v_2} \sqrt{\left\{ r^2 \left(\frac{d\theta}{dr} \right)^2 + 1 \right\}}\, dr$$

$$= \int_{v_1}^{v_2} \sqrt{\left(\tan^2 \alpha + 1 \right)}\, dr = \sec \alpha \left[r \right]_{r_1}^{r_2} = (r_2 - r_1) \sec \alpha$$

95. Show that the length of the curve $y = \log \sec x$ between the points where $x = 0$ and $x = \frac{\pi}{3}$ is $\log_e \left(2 + \sqrt{3}\right)$

Solution: $$y = \log \sec x$$

$$\therefore \quad \frac{dy}{dx} = \frac{1}{\sec x} \cdot \sec x \cdot \tan x = \tan x$$

$$\therefore \quad \frac{ds}{dx} = \sqrt{\left[1 + \left(\frac{dy}{dx} \right)^2 \right]} = \sqrt{1 + \tan^2 x} = \sec x$$

$\therefore$ required length $$S = \int_0^{\pi/2} \frac{ds}{dx} \cdot dx = \int_0^{\pi/2} \sec x \cdot dx$$

$$= \left[\log \tan \left(\frac{x}{2} + \frac{\pi}{4} \right) \right]_0^{\pi/3} = \log \tan \left(\frac{\pi}{6} + \frac{\pi}{4} \right) - \log \tan \frac{\pi}{4}$$

$$= \log \tan \frac{5\pi}{12} = \log \frac{1 + \frac{1}{\sqrt{3}}}{1 - \frac{1}{\sqrt{3}}} = \log \frac{\sqrt{3} + 1}{\sqrt{3} - 1}$$

$$= \log \frac{(\sqrt{3}+1)^2}{3-1} = \log(2+\sqrt{3})$$

96. Find the length of an arc of the parabola, $y = x^2$, measured from the vertex. Calculate the length of the arc to the point (1, 1) given $\log_e (2+\sqrt{5}) = 1 \cdot 45$.

Solution:

Given: $$y = x^2$$

$$\therefore \quad \frac{dy}{dx} = 2x$$

$$\therefore \text{ required length} \quad S = \int_0^x \sqrt{\left[1 + \left(\frac{dy}{dx}\right)^2\right]} \cdot dx$$

$$= \int_0^x \sqrt{(1+4x^2)}\, dx = 2\int_0^x \sqrt{\left(\frac{1}{4}+x^2\right)}\, dx$$

$$= a\,2\left[\frac{1}{2}x\sqrt{\left(\frac{1}{4}+x^2\right)} + \frac{1}{3}\cdot\frac{1}{4}\log\left(x + \sqrt{\frac{1}{4}+x^2}\right)\right]_0^x$$

$$= x\sqrt{\left(\frac{1}{4}+x^2\right)} + \frac{1}{4}\log\left[x + \sqrt{\left(\frac{1}{4}+x^2\right)}\right] - \frac{1}{4}\log\frac{1}{2}$$

Required length upto the point (1, 1)

$$S = 2\left[\frac{1}{2}x\sqrt{\frac{1}{4}+x^2} + \frac{1}{8}\log\left(x + \sqrt{\frac{1}{4}+x^2}\right)\right]_0^1$$

$$= \sqrt{\left(\frac{1}{4}+1\right)} + \frac{1}{4}\log\left[1 + \sqrt{\frac{1}{4}+1}\right] - \frac{1}{4}\log\frac{1}{2}$$

$$= \frac{\sqrt{5}}{2} + \frac{1}{4}\log\left(1+\frac{\sqrt{5}}{2}\right) - \frac{1}{4}\log\frac{1}{2}$$

$$= \frac{\sqrt{5}}{2} + \frac{1}{4}\log(2+\sqrt{5}) = \frac{2.24}{2} + \frac{1}{4}(1.45)$$

$$= 1.12 + .36 = 1.48$$

97. Prove that the length of the loop of the curve:

$$3ay^2 = x(x - a)^2 \text{ is } \frac{4a}{\sqrt{3}}$$

Solution:

Given: $3ay^2 = x(x - a)^2$ [1]

$$\therefore \quad 6ay\, y' = (x - a)^2 + 2x(x - a) = (x - a)(x - a + 2x)$$

$$= (x - a)(3x - a)$$

$$\therefore \quad y' = \frac{(x-a)(3x-a)}{6ay}$$

or

$$(y')^2 = \frac{(x-a)^2 (3x-a)^2}{36a^2 y^2}$$

$$= \frac{(x-a)^2 \cdot (3x-a)^2 \cdot 3a}{36a^2 \cdot x(x-a)^2} = \frac{(3x-a)^2}{12ax}$$

(using [1])

$\therefore$ required arc length

$$S = 2\int_0^a \sqrt{\left[1+(y')^2\right]}\, dx = 2\int_0^a \sqrt{\left[1+\frac{(3x-a)^2}{12ax}\right]}\, dx$$

$$= 2\int_0^a \sqrt{\left[\frac{12ax + 9x^2 + a^2 - 6ax}{12ax}\right]}\, dx = 2\int_0^a \frac{3x+a}{\sqrt{12ax}}\, dx$$

Put $x = t^2$, $dx = 2tdt$

$$\therefore \quad S = \frac{2}{2\sqrt{(3a)}} \int_0^{\sqrt{a}} \frac{3t^2 + a}{t} \cdot 2tdt$$

$$= \frac{2}{\sqrt{(3a)}} \int_0^{\sqrt{a}} (3t^2 + a)\, dt = \frac{2}{\sqrt{(3a)}} \left[3 \cdot \frac{t^3}{3} + at\right]_0^{\sqrt{a}}$$

$$= \frac{2}{\sqrt{(3a)}} \left[a\sqrt{a} + a\sqrt{a}\right] = \frac{4a}{\sqrt{3}}$$

98. Find the length of the curve

$$y = \log \frac{e^x - 1}{x^x + 1} \text{ from } x = 1 \text{ to } x = 2$$

Solution:

Given: $$y = \log \frac{e^x - 1}{x^x + 1}$$

$$\therefore \quad \frac{dy}{dx} = \frac{e^x - 1}{x^x + 1} \cdot \frac{e^x (e^x + 1) - e^x (e^x - 1)}{(e^x + 1)^2} = \frac{e^x (2)}{e^{2x} - 1} = \frac{2e^x}{e^{2x} - 1}$$

$\therefore$ Reqd. length

$$S = \int_1^2 \sqrt{\left[1 + \frac{4e^{2x}}{(e^{2x} - 1)^2}\right]}\, dx = \int_1^2 \frac{\sqrt{\left[(e^{2x} - 1)^2 + 4e^{2x}\right]}}{e^{2x} - 1} dx$$

$$= \int_1^2 \frac{e^{2x} + 1}{e^{2x} - 1} dx = \int_1^2 \frac{e^x + e^{-x}}{e^x - e^{-x}} dx$$

$$= [\log (e^x - e^{-x})]_1^2 = \log (e^2 - e^{-2}) - \log (e - e^{-1})$$

$$= \log (e + e^{-1})$$

99. Find the length of the arc of the cardioid $r = a\,(1 - \cos\,\theta)$ between points whose vertical angles arc θ_1 and θ_2.

Solutions:

Given: $$r = a(1 - \cos\,\theta)$$

$$\frac{dr}{d\theta} = a \sin\,\theta$$

$\therefore$ Reqd. length $$S = \int_{\theta_1}^{\theta_2} \sqrt{\left[r^2 + \left(\frac{dr}{d\theta}\right)^2\right]}\, d\theta$$

$$= \int_{\theta_1}^{\theta_2} \sqrt{\left\{a^2\,(1 - \cos\theta)^2 + a^2 \sin^2\theta\right\}}\, d\theta = a \int_{\theta_1}^{\theta_2} 2 \sin\theta/2\, d\theta$$

$$= 4a \left[-\cos\frac{\theta}{2}\right]_{\theta_1}^{\theta_2} = 4a \left[\cos\frac{\theta_1}{2} - \cos\frac{\theta_2}{2}\right]$$

100. Find the length of the arc of the spiral $r = a\theta$ between the points whose radii vectors are r_1 and r_2.

Solution:

Given: $$r = a\theta;$$

$$\therefore \quad 1 = a\frac{d\theta}{dr}$$

or $$\frac{d\theta}{dr} = \frac{1}{a}$$

$$\therefore \quad S = \int_{v_1}^{v_2} \sqrt{\left\{r^2 \left(\frac{d\theta}{dr}\right)^2 + 1\right\}}\, dr = \int_{v_1}^{v_2} \sqrt{\left\{r^2 \cdot \frac{1}{a^2} + 1\right\}}\, dr$$

$$= \frac{1}{a} \int_{v_1}^{v_2} \sqrt{(r^2 + a^2)}\, dr$$

$$= \frac{1}{a}\left[\frac{1}{2} r \cdot \sqrt{(r^2 + a^2)} + \frac{1}{2} a^2 \log\left\{r + \sqrt{r^2 + a^2}\right\}\right]_{r_1}^{r_2}$$

$$= f\left(\frac{r_2}{a}\right) - f\left(\frac{r_1}{a}\right)$$

Where $$f(\theta) = \frac{1}{2} a\left[\theta \sqrt{(1+\theta^2)} + \log\left(\theta + \sqrt{1+\theta^2}\right)\right]$$

101. **Show that the perimeter of the curve**

$r = 5(1 - \cos\theta)$ is 40 units **(VTU, MQ P.I)**

Solution:

The eqn is given by: $x = a(\theta - \sin\theta)$

$y = a(1 - \cos\theta)$

Now, the perimeter 'l' is

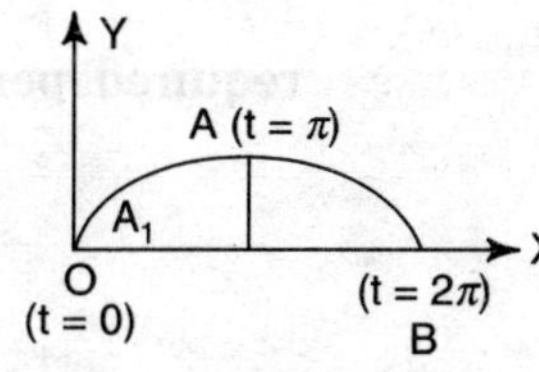

Fig. 3.33

given by: $$l = 2\int_0^{\pi} \sqrt{\left\{r^2 + \left(\frac{dr}{d\theta}\right)^2\right\}}\, d\theta$$

$$= 2\int_0^{\pi} \sqrt{25(1-\cos\theta)^2 + 25\sin^2\theta} \cdot d\theta$$

$$= 2.5 \int_0^{\pi} \sqrt{1 - 2\cos\theta + \cos^2\theta + \sin^2\theta} \cdot d\theta$$

$$= 10 \int_0^{\pi} \sqrt{2(1-\cos\theta)} \cdot d\theta$$

$$= 10 \int_0^{\pi} \sqrt{2 \cdot 2 \cdot \sin^2\theta/2} \cdot d\theta$$

$$= 10 \times 2 \int_0^{\pi} \sin\theta/2\, d\theta = 10 \times 2\left[\frac{-\cos\theta/2}{(1/2)}\right]_0^{\pi}$$

$$= -20\left[2\left(\cos\frac{\pi}{2} - \cos 0\right)\right]$$

$$= -20\ [2\ (0 - 1) = 40 \text{ units.}$$

102. Find the perimeter of the curve $r = a \sin^3\left(\frac{\theta}{3}\right)$ (VTU, Sept-1999)

Solution: We tabulate the values of θ and r in order to find the limits of integration.

θ	0	$\pi/2$	π	$3\pi/2$	2π	$5\pi/2$	3π
r	0	$a/8$	$\frac{3\cdot\sqrt{3}\cdot a}{8}$	a	$\frac{3\cdot\sqrt{3}\cdot a}{8}$	$a/8$	0

For a loop, θ varies from 0 to 3π

Also, $$\frac{dr}{d\theta} = 3a \sin^2(\theta/3)\cos(\theta/3)\left(\frac{1}{3}\right)$$

$$= a \sin^2(\theta/3)\cos(\theta/3)$$

Now, $$\left\{r^2 + \left(\frac{dr}{d\theta}\right)^2\right\}^{1/2} = \{a^2 \sin^6(\theta/3) + a^2 \sin^4(\theta/3)\cos^2(\theta/3)\}^{1/2}$$

$$= a \sin^2(\theta/3)\{\sin^2(\theta/3) + \cos^2(\theta/3)\}^{1/2} = a \sin^2(\theta/3).$$

$\therefore$ required perimeter

$$= \int_0^{3\pi}\left\{r^2 + \left(\frac{dr}{d\theta}\right)^2\right\}^{1/2} d\theta$$

$$= \int_0^{3\pi} a \sin^2(\theta/3)\, d\theta = \frac{a}{2}\int_0^{3\pi}\{1 - \cos(2\theta/3)\}\, d\theta$$

$$= \frac{a}{2}\left[\theta - \frac{3}{2}\sin(2\theta/3)\right]_0^{3\pi} = \frac{3a\pi}{2}$$

103. Find the length of the curve $y^2 = (2x - 1)^3$ cut-off by the line $x = 4$. (VTU, Sept, 2000)

Solution: **Given:** $y^2 = (2x - 1)^3;\ x = 4$

$\Rightarrow$ $y^2 = 7^3$ or $y = \pm 7\sqrt{7}$

Length, $$l = \int_{-7\sqrt{7}}^{7\sqrt{7}} \sqrt{1 + \left(\frac{dx}{dy}\right)^2}\, dy \qquad [1]$$

Now, we have $y^2 = (2x - 1)^3$

Diff. w.r.t. y, we get

$$2y = 3(2x-1)^2 \cdot 2\,\frac{dx}{dy}$$

$$\therefore \quad \frac{dx}{dy} = \frac{y}{3(2x-1)^2} = \frac{y}{3\cdot y^{4/3}} = \frac{1}{3y^{1/3}}$$

Again, $1+\left(\dfrac{dx}{dy}\right)^2 = 1+\dfrac{1}{9y^{2/3}} = \dfrac{9y^{2/3}+1}{9y^{2/3}}$

$$\therefore \quad \sqrt{\left\{1+\left(\frac{dx}{dy}\right)^2\right\}} = \frac{\sqrt{9y^{2/3}+1}}{3y^{1/3}}$$

From [1], $\quad l = \displaystyle\int_{-7\sqrt{7}}^{7\sqrt{7}} \frac{\sqrt{9y^{2/3}+1}}{3y^{1/3}}\,dy$

Put $y^{2/3} = t$

$$\Rightarrow \quad \frac{2}{3}y^{-1/3}\,dy = dt \quad \text{or} \quad \frac{dy}{3y^{1/3}} = \frac{dt}{2}$$

Also, t-varies from -7 to 7 as y varies from $[-(7)^{3/2})]$ to $(7)^{3/2}$

$$\therefore \quad l = \int_{-7}^{7}\sqrt{9t+1}\,\frac{dt}{2}$$

Since $t = y^{2/3}$ remain unaltered as y changes to $-y$ we can as well take it for 't' also

Hence $\quad l = 2\displaystyle\int_0^7 \sqrt{9t+1}\,\frac{dt}{2}$

$$= \int_0^7 (9t+1)^{1/2}\,dt = \frac{1}{9}\cdot\frac{2}{3}\left[(9t+1)^{3/2}\right]_0^7$$

$$= \frac{2}{27}\left[64^{3/2}-1\right] = \frac{2}{27}(512-1) = \frac{1022}{7}$$

104. Find the length of the cardioid $r = a\,(1-\cos\theta)$ lying outside the circle $r = a\cos\theta$.

Solution:

$$\left.\begin{aligned} r &= a\,(1-\cos\theta) \\ r &= a\cos\theta \end{aligned}\right\} \qquad [1]$$

$$\Rightarrow \quad a\cos\theta = a\,(1-\cos\theta)$$

$$\Rightarrow \quad 2\cos\theta = 1, \quad \Rightarrow \quad \cos\theta = \frac{1}{2}$$

$$\Rightarrow \quad \theta = \pi/3$$

PART-C

Hence required length is arc of the cardioid between the limits $\pi/3$ to π (for upper portion)

$$\therefore \quad \text{Total arc} = 2\int_{\pi/3}^{\pi} 2a \sin \theta/2 \, d\theta = 4a\left(-2\cos\theta/2\right)_{\pi/3}^{\pi}$$

$$= -8a\left[\cos\frac{\pi}{2} - \cos\frac{\pi}{6}\right] = 8a \cdot \frac{\sqrt{3}}{2} = 4a\sqrt{3}.$$

105. Show that the length of an arc of the curve $x^2 = a^2(1 - e^{y/a})$ measured from origin to any point is $a \log \dfrac{a+x}{a-x} - x$.

Solution: **Given:**

$$e^{y/a} = \frac{a^2 - x^2}{a^2}$$

$$\Rightarrow \quad y = a\left[\log(a^2 - x^2) - \log a^2\right]$$

$$\therefore \quad \frac{dy}{dx} = \frac{-2ax}{a^2 - x^2}$$

Hence

$$\sqrt{\left[1 + (y')^2\right]} = \frac{a^2 + x^2}{a^2 - x^2} = -1 + \frac{2a^2}{a^2 - x^2}$$

$$\therefore \quad S = \left(-x + 2a^2 \frac{1}{2a} \log \frac{a+x}{a-x}\right)_0^x$$

$$S = a \log \frac{a+x}{a-x} - x$$

EXERCISES

1. Find the length of the hypo-cycloid

$$\frac{x^{2/3}}{a^{2/3}} + \frac{y^{2/3}}{b^{2/3}} = 1 \qquad \left(\textit{Ans.}\ 4\frac{(a^2 + ab + b^2)}{(a+b)}\right)$$

2. Show that the perimeter of loop of the curve $9ay^2 = (x - 2a)(x - 5a)^2$ is $4\sqrt{3}$.
3. Find the perimeter of one loop of the cure $r = a\cos^2(\theta/3)$.
4. Find the arc length of:

   (*i*) $x^2 = 4ay$ cut off by the latus rectum

   (*ii*) $y^2 = 12x$ from the vertex to one end of the latus rectum

   (*iii*) $ay^2 = x^3$, from (0, 0) to (a, a)

   (*iv*) $x = e^t \sin t$, $y = e^t \cos t$
   from $t = 0$ to $t = \pi/2$

(v) $x = a(\cos t + t \sin t)$
$y = a(\sin t - t \cos t)$
from $t = 0$ to $t = \alpha$

5. Find the length of the arc of the curve $y = ae^x$ from the point $(0, a)$ to the point (x_1, y_1).

3.8 AREAS AND VOLUMES

I. Areas: (Cartesian equations)

The area of plane figure bounded by the curve $y = f(x)$, the x-axis and the two ordinates $x = a$, $x = b$ is equal to:

(a) $A = \int_a^b y dx$ or $A = \int_a^b f(x)\, dx$

(b) If the area between the curve and the x-axis be requied, we proceed as:

(i) Put $y = 0$; we get intersections of the curve with the x-axis.

$\therefore \quad x = a, x = b$

(ii) The area is $\quad A = \int_a^b y\, dx$

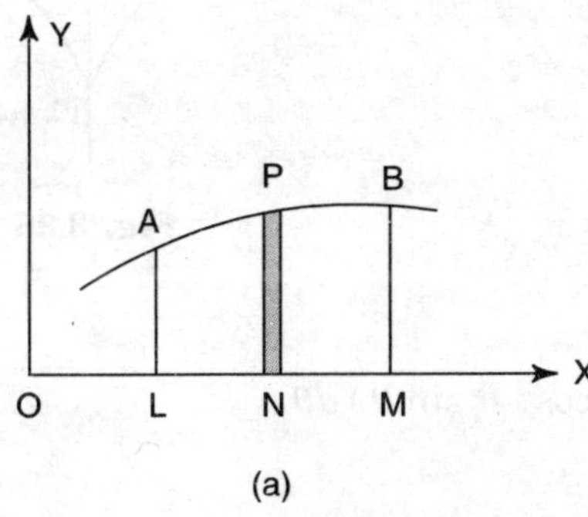

(a)

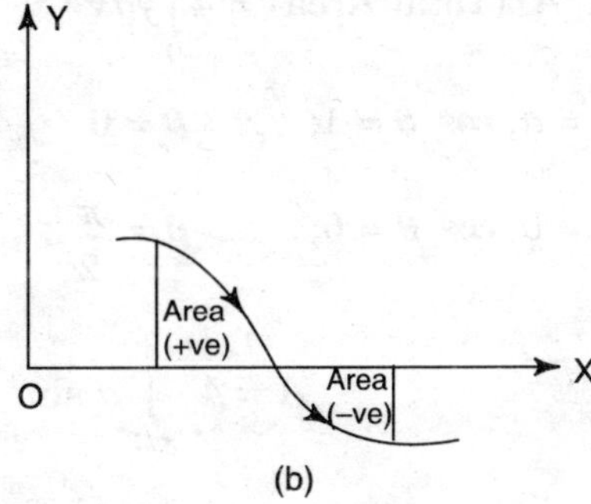

(b)

Fig. 3.34

(c) If the area between two curves $y = f(x)$ and $y = \phi(x)$ be required, we proceed as:

(i) Find the points of intersection by solving for x and y from the two given equations; we get $x = a$ and $x = b$. (Fig. 3.35)

(ii) For the first curve,

$$A_1 = \int_a^b y\, dx = \int_a^b f(x)\, dx$$

For the second curve,

$$A_2 = \int_a^b y\, dx = \int_a^b \phi(x)\, dx$$

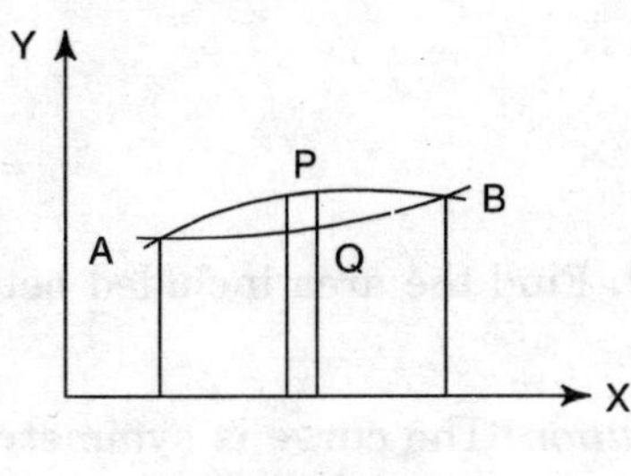

Fig. 3.35

(iii) The area required is the area between the two;

i.e., $\quad A = |A_1 - A_2|$

(iv) Sometimes, it is more convenient to use the area between the curve and the y-axis. Then, we use

PART-C

$A = \int f(y)\,dy$, between appropriate limit. (polar equations)

(*v*) If the equation of the curve is in the polar form: $r = f(\theta)$, then the area bounded by the curve and the lines $\theta = \theta_1$, $\theta = \theta_2$ $(\theta_1 < \theta_2)$, is given by the formula

$$\text{Area,} \qquad A = \int_{\theta=\theta_1}^{\theta_2} \frac{1}{2} r^2\, d\theta$$

Examples (106 – 120)

106. Find the total area of the Astroid:

$$x^{2/3} + y^{2/3} = a^{2/3}$$

Solution:

The curve astroid is as shown in the figure, it is symmetrical in all the 4 quadrants.

Hence, $\quad$ Area = $4AOB$

The parametric equations are:

$$x = a\cos^3\theta;\ y = a\sin^3\theta.$$

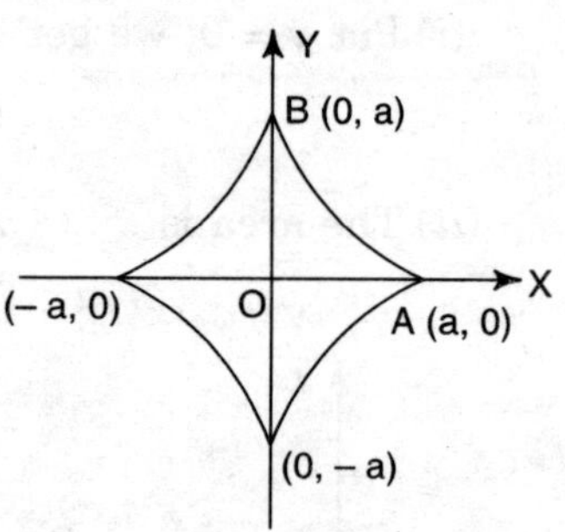

Fig. 3.36

$$\therefore \qquad A, \text{(Total Area)} = 4\int_0^a y\,dx$$

when $\quad x = a$, $\cos\theta = 1$; $\therefore \quad \theta = 0$

when $\quad x = 0$, $\cos\theta = 0$; $\therefore \quad \theta = \dfrac{\pi}{2}$

$$\therefore \qquad A = 4\int_{\pi/2}^{0} a\sin^3\theta\,(-3a\cos^2\theta\sin\theta)\,d\theta$$

$$= 12a^2 \int_0^{\pi/2} \sin^4\theta\cos^2\theta\,d\theta$$

$$= 12a^2\, \frac{3\cdot 1}{6\cdot 4\cdot 2}\cdot\frac{\pi}{2} \qquad \text{(by reduction formula)}$$

$$= \frac{3\pi a^2}{8} \text{ sq. units}$$

107. Find the area included between the curve $y^2(a - x) = x^3$ and its asymptote.

(VTU, Aug. 2001)

Solution: The curve is symmetrical about x-axis and passes through the origin. The asymptote is $x = a$. Requd. Area, $A = 2\int_0^a y\,dx$

$$= 2\int_0^a \frac{x^{3/2}}{\sqrt{a - x}}\,dx$$

Put $x = a \sin^2 \theta$

$dx = 2a \sin \theta \cdot \cos \theta \, d\theta;$

When $x = a, \theta = \pi/2,$

When $x = 0, \theta = 0$

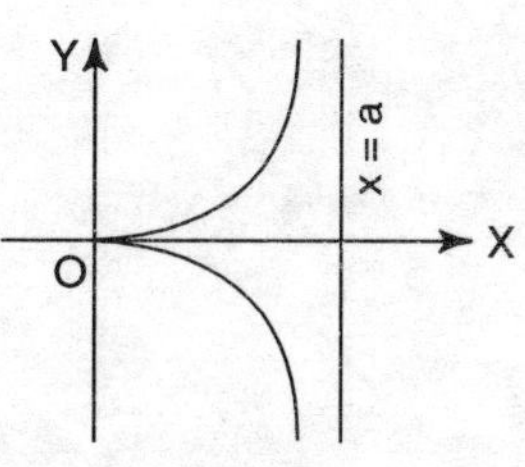

Fig. 3.37

$$\text{Area} = 2 \int_0^{\pi/2} \frac{a^{3/2} \sin^3 \theta}{\sqrt{a - a \sin^2 \theta}} \cdot 2a \sin \theta \cos \theta \, d\theta$$

$$= 4a^2 \int_0^{\pi/2} \sin^4 \theta \, d\theta$$

$$= 4a^2 \left[\frac{4-1}{4} \frac{4-3}{4-2} \cdot \frac{\pi}{2} \right] = 4a^2 \cdot \frac{3}{4} \cdot \frac{1}{2} \cdot \frac{\pi}{2} = \frac{3}{4} \pi a^2.$$

108. Find the area between the x-axis and the curve $y = (2x^2 - 8x + 6)$.

Solution: In order to find the points of intersection between the curve and the x-axis,
We put $y = 0$

$\therefore \quad (2x^2 - 8x + 6) = 0$

$\Rightarrow \quad (x^2 - 4x + 3) = 0$

$\Rightarrow \quad (x - 1)(x - 3) = 0$

$\Rightarrow \quad x = 1, x = 3$ [1]

$$\therefore \quad \text{Area, } A = \int_1^3 y \, dx = \int_1^3 (2x^2 - 8x + 6) \, dx \quad [2]$$

$$= [2(x^3/3) - 8(x^2/2) + 6x]_1^3$$

$$= [2(27/3) - 4(3^2) + 6(3)] - [2(1/3) - 4(1^2) + 6]$$

$$= \{0 - 8/3\} = -8/3$$

Note: (The part of the curve lies below the x-axis and so we get the negative sign). The area $A = 8/3$ (numerically)

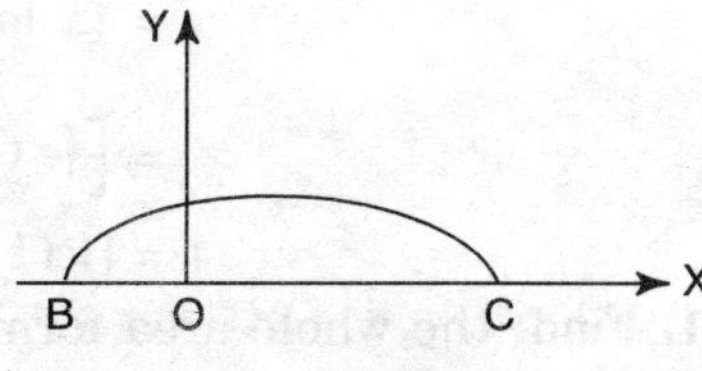

Fig. 3.38

109. Find the area between the curve $y = x^4/(x + 2)$, the x-axis and the ordinates $x = 1$, and $x = -1$.

Solution: Area required $= \int_{-1}^{+1} y \, dx = \int_{-1}^{+1} \left[\frac{x^4}{(x+2)} \right] dx$ [1]

$$\text{Now,} \quad \int \frac{x^4 \, dx}{(x+2)} = \int \frac{(x^4 - 16) + 16}{(x+2)} \, dx$$

$$= \int \frac{(x^2 + 4)(x^2 - 4) + 16}{(x+2)} \, dx$$

$$= \int \frac{(x^2 + 4)(x - 2)(x + 2) + 16}{(x+2)} \, dx$$

PART-C

$$= \int \left\{(x^2+4)(x-2)+\frac{16}{x+2}\right\} dx$$

$$= \int [x^3 - 2x^2 + 4x - 8 + 16/(x+2)]\, dx$$

$$= \frac{1}{4}x^4 - \frac{2}{3}x^3 + \frac{4}{2}x^2 - 8x + 16\log(x+2) \qquad [2]$$

$$\therefore \quad \text{Area} = \left[\left(x^4/4\right) - 2x^3/3 + 2x^2 - 8x + 16\log(x+2)\right]_{-1}^{+1}$$

$$= \left(\frac{1}{4}\right)(1-1) - \left(\frac{2}{3}\right)(1+1) + 2(1-1) - 8(1+1) + 16[\log 3 - \log 1]$$

$$= \left(-\frac{2}{3}\right)(2) - 8(2) + 16\log 3 = 16\log 3 - \frac{52}{3}.$$

110. Find the area between the x-axis, the curve $y = \log x$ and the line $x = e$

Solution: put $y = 0$;

$\therefore \quad \log x = 0$

$\Rightarrow \quad x = e^0 = 1$

Thus, we have; $x = 1$ and from the given line $x = e$

$$\therefore \text{ Area, } \quad A = \int_1^e y\, dx = \int_1^e (\log x)\, dx = \int_1^e \log x \cdot 1 \cdot dx \qquad \text{(by parts)}$$

$$= \left[x \cdot \log x - \int x \cdot (1/x)\, dx\right]_1^e$$

$$= (x\log x - x)_1^e = \{x[(\log x) - 1]\}_1^e$$

$$= \left[\{e(\log e - 1)\} - 1\{\log 1 - 1\}\right]$$

$$= [\{e(1-1) - (0-1)\}] = 1 \text{ sq. unit}$$

111. Find the whole area formed by the loop $a^2y^2 = x^2(a^2 - x^2)$

Solution: **Given:** $a^2y^2 = x^2(a^2 - x^2)$

or, $\quad y = \dfrac{x}{a}\sqrt{a^2 - x^2}$

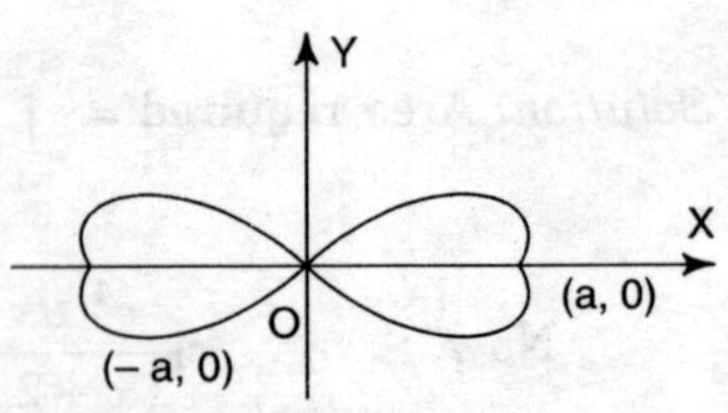

Fig. 3.39

x-varies from 0 to a

$$\left(\because \text{ if put } y = 0, \text{ implies } \frac{x}{a}\sqrt{a^2 - x^2} = 0 \Rightarrow x = 0, x = a\right)$$

$\therefore \quad$ Whole area = 4 (Area of half of a loop)

$$= 4\int_0^a y\,dx = 4\int_0^a \frac{x}{a}\sqrt{(a^2 - x^2)}\,dx$$

Put $x = a\sin\theta$; $dx = a\cos\theta\,d\theta$

$$\therefore \text{ required area} = \frac{4}{a}\int_0^{\pi/2} a\sin\theta\cdot a\cos\theta\cdot a\cos\theta\,d\theta$$

$$= 4a^2\int_0^{\pi/2}\sin\theta\cdot\cos^2\theta\,d\theta = 4a^2\cdot\left[\int_0^{\pi/2}(\sin\theta - \sin^3\theta)\right]d\theta$$

$$= 4a^2\left[(-\cos\theta)_0^{\pi/2} - \frac{(2)}{3\cdot 1}\cdot 1\right] \qquad \text{(by reduction formula)}$$

$$= -4a^2\left[(0-1)+\frac{2}{3}\right] = -4a^2\left[\frac{2}{3}-1\right] = -4a^2\left[-\frac{1}{3}\right]$$

$$= \frac{4}{3}a^2 \text{ sq. units}$$

112. Find the area between the curve $x = a(\theta - \sin\theta)$
$y = a(1 - \cos\theta)$ and its base. (VTU, March, 2000)

Solution:

If $\theta = 0$, then $x = 0$, $y = 0$
If $\theta = 2\pi$, then $x = 2a\pi$, $y = 0$

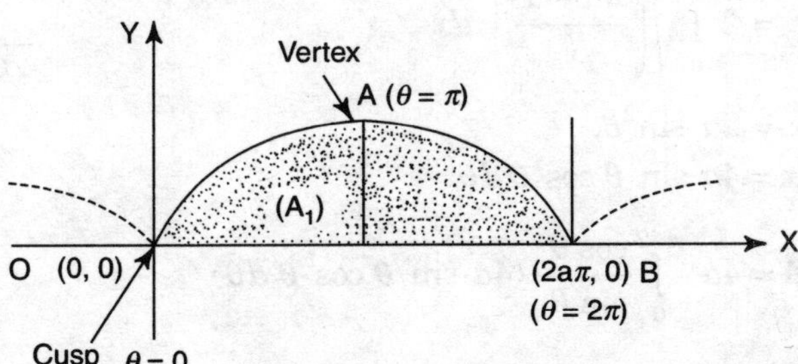

Fig. 3.40

The x-axis is along the base line, and the origin is at a cusp.

Limits for θ:

$\theta = 0$ at O
$\theta = \pi$ at A
$\theta = 2\pi$ at B

By symmetry, the area between the curve and the base, is twice that between the curve from cusp to vertex and the base. At the cusp, $\theta = 0$, and at the vertex $\theta = \pi$

$$\therefore \quad A = 2A_1 = 2\int_0^{\pi} y\left(\frac{dx}{d\theta}\right)d\theta = 2\int_0^{\pi} a\,(1-\cos\theta)\cdot a\,(1-\cos\theta)\,d\theta$$

$$= 2a^2\int_0^{\pi}(1-\cos\theta)^2\,d\theta = 2a^2\int_0^{\pi}[2\sin^2(\theta/2)]^2\,d\theta$$

$$= 8a^2\int_0^{\pi}\sin^4(\theta/2)\,d\theta$$

put $\theta = 2\phi,\ d\theta = 2d\phi$ $\quad [(\theta = 0,\ \phi = 0);\ (\theta = \pi,\ \phi = (\pi/2)]$

$$= 8a^2\int_0^{\pi/2}\sin^4\phi\cdot 2d\phi$$

$$= 16a^2\cdot\left[\frac{3\cdot 1}{(4\cdot 2)}\right]\left(\frac{\pi}{2}\right) = (3\pi a^2).$$

113. Find the area bounded by the witch of Agnesi

$y^2 = 4a^2\,(2a - x)/x$ and its asymptote. (VTU, March, 2001)

Solution: The curve is obviously symmetrical about axis of x cutting it at $(2a, 0)$, tangent at which is y-axis and asymptote is $x = 0$ (*i.e.,* y-axis itself)
(obtained on equaling to zero the coefft of y, HDT namely of y^2)

$$\therefore \quad A = 2\,A_1 = 2\int_0^{2a} y\,dx$$

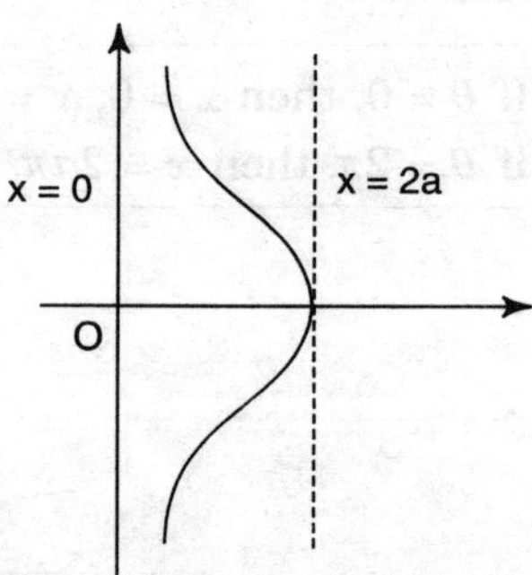

Fig. 3.41

$$= 2\int_0^{2a}\sqrt{\left(\frac{2a-x}{x}\right)}\,dx$$

put $x = 2a\sin^2\theta,$

$\therefore \quad dx = 4a\sin\theta\cos\theta\,d\theta$

$$A = 4a\int_0^{\pi/2}\frac{\cos\theta}{\sin\theta}\,(4a\sin\theta\cos\theta\,d\theta)$$

$$= 16a^2\int_0^{\pi/2}\cos^2\theta\,d\theta = 16a^2\cdot\frac{1}{2}\cdot\frac{\pi}{2} = 4\pi a^2.$$

114. Find the area of the ellipse

(*i*) $\dfrac{x^2}{a^2} + \dfrac{y^2}{b^2} = 1$ (*ii*) $x^2 + y^2 = a^2$

Solution: (*i*) The ellipse is symmetrical about both the axes and so the whole area of the ellipse = 4 × area of *OAB*

$$A = 4A_1$$

(For the area *OAB*, x-axes from $x = 0$, to $x = a$).

$$A = 4\int_0^a y\,dx \qquad (1)$$

$$\left(\frac{y^2}{b^2} = 1 - \frac{x^2}{a^2} = \frac{a^2 - x^2}{a^2}\right)$$

or $$y = \pm\,\frac{b}{a}\sqrt{a^2 - x^2}\,.$$

$$\left(y = \frac{b}{a}\sqrt{a^2 - x^2}\right)\text{ (numerically)}$$

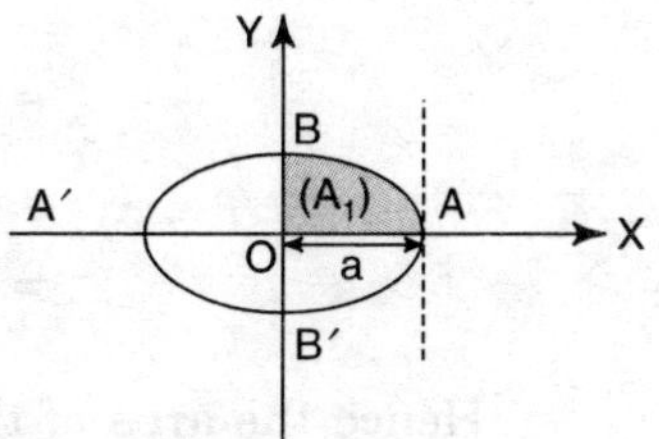

Fig. 3.42

$$= 4\int_0^a \frac{b}{a}\sqrt{a^2 - x^2}\;dx = \frac{4b}{a}\int_0^a \sqrt{a^2 - x^2}\;dx$$

$$A = \frac{4b}{a}\left[\frac{x}{2}\sqrt{a^2 - x^2} + \frac{a^2}{2}\sin^{-1}\frac{x}{a}\right]_0^a$$

$$= \frac{4b}{a}\left[0 + \frac{a^2}{2}\sin^{-1}(1) - 0 - 0\right]$$

$$= \frac{4b}{a}\left[\frac{a^2}{2}\sin^{-1}(1)\right] = \frac{4b}{a}\left(\frac{a^2}{2}\cdot\frac{\pi}{2}\right)$$

$$= 4\,\frac{ba\,\pi}{4} = \pi ab \text{ sq. units.}$$

(*ii*) The curve is symmetrical about both the axes. The whole area is 4 times the area in the first quadrant.

$$\therefore \qquad A = 4A_1 \qquad [1]$$

$$A_1 = \int_0^a y\,dx$$

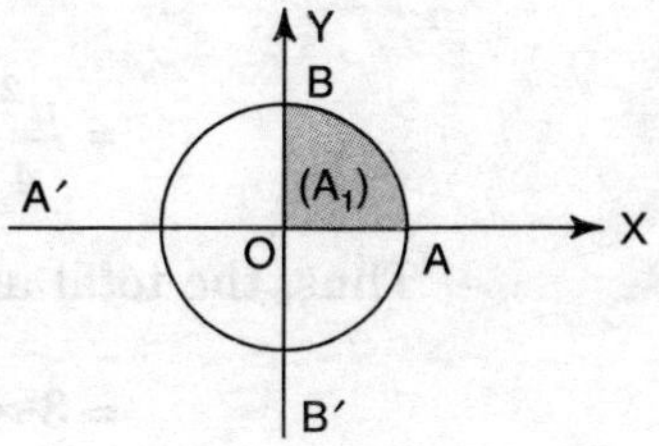

$$\left(\begin{aligned} &\because x^2 + y^2 = a^2 \\ &y^2 = a^2 - x^2 \\ &y = \sqrt{a^2 x^2} \quad \text{(numerically)} \end{aligned}\right)$$

$$= \int_0^a \sqrt{a^2 - x^2}\;dx$$

Fig. 3.43

$$= \left[\frac{x}{2}\sqrt{a^2 - x^2} + \frac{a^2}{2}\sin^{-1}\frac{x}{a}\right]_0^a$$

$$= 0 + \frac{a^2}{2} \sin^{-1}(1) - 0 - \frac{a^2}{2} \sin^{-1}(0)$$

$$= \frac{a^2}{2} \cdot \sin^{-1} \qquad (1)$$

$$= \frac{a^2}{2} \cdot \frac{\pi}{2} = \frac{\pi a^2}{4} = A_1$$

Hence the area of the circle by using [1]

$$A = 4\left[\frac{\pi a^2}{4}\right] = \pi a^2. \text{ sq. units.}$$

115. Find the area of a loop of each of the following curves

(i) $r = a \sin 3\theta$ (3 leaved rose)

(ii) $r = a \cos 2\theta$ (4 leaved rose)

Solution:

(i) Given $r = a \sin 3\theta$ [1]

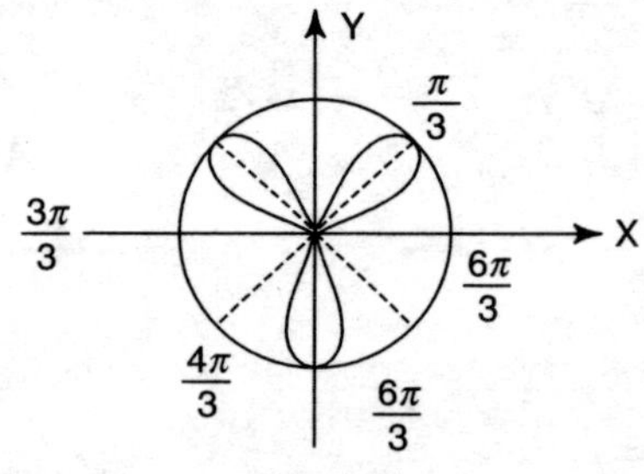

Fig. 3.44

For a loop, the limits for θ are

$$\theta = 0, \sin 3\theta = 0$$

$$\Rightarrow \quad 3\theta = 0, \pi$$

$$\therefore \quad \theta = 0, \frac{\pi}{3}$$

Thus, a loop is obtained when θ varies from 0 to $\frac{\pi}{3}$.

The required area, (of a loop) is:

$$A = \int_0^{\pi/3} \frac{1}{2} r^2 \, d\theta = \frac{1}{2} \int_0^{\pi/3} a^2 \sin^2(3\theta)\, d\theta$$

$$= \frac{a^2}{2} \int_0^{\pi/3} \left(\frac{1 - \cos 6\theta}{2}\right) d\theta$$

$$= \frac{a^2}{4}\left[\theta - \frac{\sin 6\theta}{6}\right]_0^{\pi/3} = \frac{a^2}{4}\left[\frac{\pi}{3} + 0 - 0\right] = \frac{\pi a^2}{12}.$$

Thus, the total area = 3 × area of one loop

$$= 3 \times \frac{\pi a^2}{12} = \frac{1}{4} \pi a^2$$

(ii) $r = a \cos 2\theta$ (1)

The curve is symmetrical about the initial line and consists of 4 loops. For a loop, putting $r = 0$,

$$\Rightarrow \quad \cos 2\theta = 0 \Rightarrow 2\theta = -\frac{\pi}{2}, \frac{\pi}{2}$$

$$\Rightarrow \qquad \theta = -\frac{\pi}{4}, \frac{\pi}{4}$$

Thus, one loop of the curve is obtained as θ varies from $-\frac{\pi}{4}$ to $\frac{\pi}{4}$.

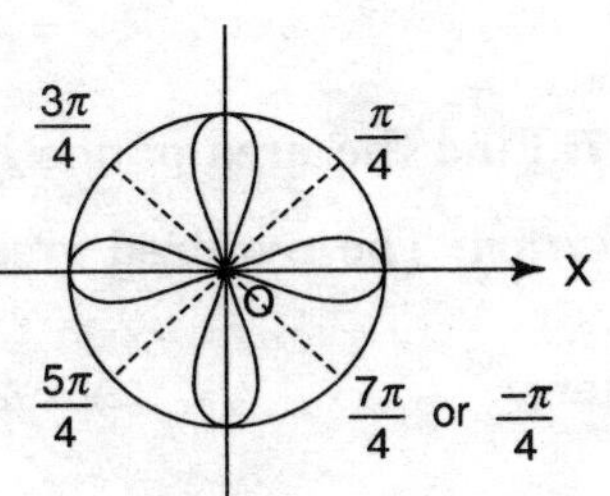

Fig. 3.45

$\therefore$ Area of one loop = A;

$$A = \int_{-\pi/4}^{\pi/4} \frac{1}{2} r^2 \, d\theta = \frac{1}{2} \int_{-\pi/4}^{\pi/4} a^2 \cos^2 2\theta \, d\theta$$

$$= \frac{a^2}{2} \cdot \int_{-\pi/4}^{\pi/4} \left(\frac{1 + \cos 4\theta}{2} \right) d\theta = \frac{a^2}{2} \left[\theta + \frac{\sin 4\theta}{4} \right]_{-\pi/4}^{\pi/4}$$

$$= \frac{a^2}{4} \left[\frac{\pi}{4} + \frac{1}{4} \sin \pi - \left(-\frac{\pi}{4} \right) - \frac{1}{4} \sin(-\pi) \right]$$

$$= \frac{a^2}{4} \cdot \left(\frac{\pi}{2} + 0 \right) = \frac{\pi a^2}{8}$$

Hence the total area $= 4 \times \frac{\pi a^2}{8} = \frac{\pi a^2}{2}$.

116. Find the area between the curve $x^2 y^2 = a^2 (y^2 - x^2)$ and its asymptote.

Solution: The required area A is 4 times the area (A_1) in the first Quadrant, *i.e.*,

$$A = 4A_1 = 4 \int_0^a y \, dx$$

$$= 4 \int_0^a \frac{ax}{\sqrt{a^2 - x^2}} \, dx$$

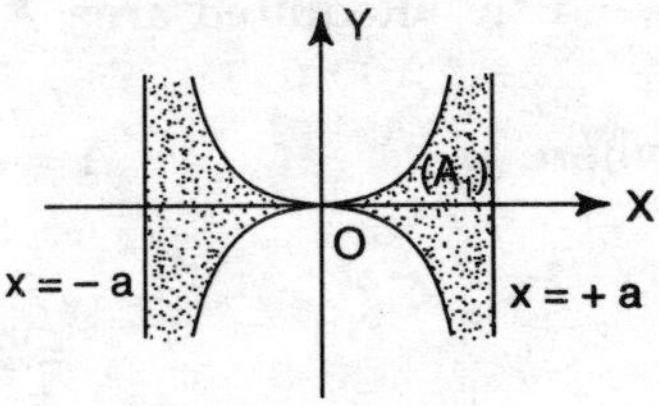

Fig. 3.46

$$\left[\begin{array}{l} \because a^2 y^2 - a^2 x^2 = x^2 y^2 \\ \Rightarrow a^2 y^2 - x^2 y^2 = a^2 x^2 \\ \Rightarrow (a^2 - x^2) y^2 = a^2 x^2 \\ \Rightarrow y^2 = \dfrac{a^2 x^2}{(a^2 - x^2)} \\ \Rightarrow y = \dfrac{ax}{\sqrt{a^2 - x^2}} \qquad \text{(numerically)} \end{array} \right]$$

$$\therefore \qquad A = 4a \int_0^a \frac{x}{\sqrt{a^2 - x^2}} \, dx = -4a \left[\sqrt{a^2 - x^2} \right]_0^a$$

$$= -4a\left(-\sqrt{a^2}\right) = 4a^2, \text{ sq. units.}$$

117. Find the area of one loop of the curve $a^2y^2 = x^2(a^2 - x^2)$

Solution: The required area is A

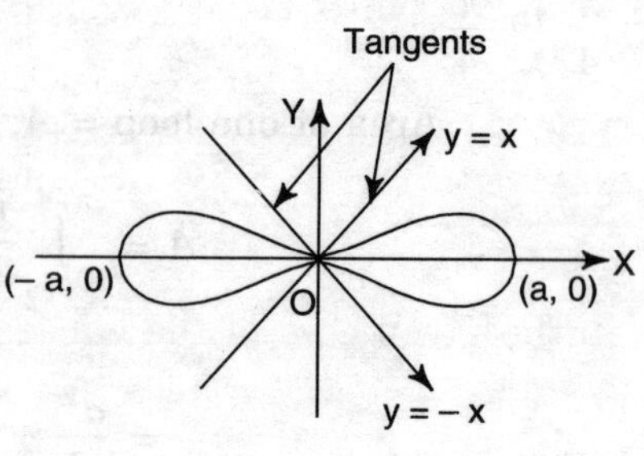

Fig. 3.47

where $$A = 2A_1 = 2\int_0^a y\,dx$$

$$= 2\int_0^a \left(\frac{x}{a}\right)\cdot\sqrt{a^2 - x^2}\cdot dx$$

$$= -\frac{2}{3a}\left[(a^2 - x^2)^{3/2}\right]_0^a$$

$$= \left(\frac{2}{3}\right)a^2; \text{ sq. units}$$

118. Show that the area included between the cardioids $r = a\,(1 + \cos\theta)$ and $r = a\,(1 - \cos\theta)$ is $\frac{1}{2}a^2\,(3\pi - 8)$.

Solution: First we find the points of intersection on the curve. From the given curves, we have,

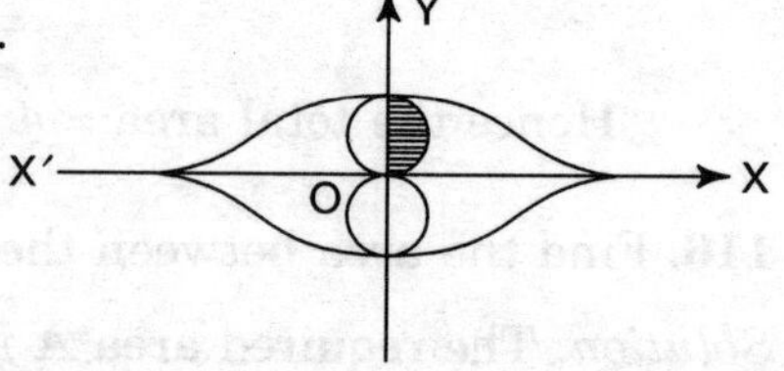

Fig. 3.48

From $$1 + \cos\theta = 1 - \cos\theta$$

$\Rightarrow$ $$2\cos\theta = 0$$

$\Rightarrow$ $$\cos\theta = 0$$

$\Rightarrow$ $$\theta = \pi/2$$

$\therefore$ Required area A,

where $$A = 4\int_0^{\pi/2} \frac{1}{2}r^2\,d\theta, \ [r = a\,(1 - \cos\theta)]$$

$$= 2\int_0^{\pi/2} a^2\,(1 - \cos\theta)^2\,d\theta$$

$$= 2\int_0^{\pi/2} a^2\left[1 - 2\cos\theta + \cos^2\theta\right]d\theta$$

$$= a^2\int_0^{\pi/2} \{2 - 4\cos\theta + 2\cos^2\theta\}\,d\theta$$

$$= a^2\int_0^{\pi/2} \{2 - 4\cos\theta + 1 + \cos^2\theta\}\,d\theta$$

$$= a^2\int_0^{\pi/2} \{3 - 4\cos\theta + \cos 2\theta\}\,d\theta$$

$$= a^2\left[3\theta - 4\sin\theta + \frac{\sin 2\theta}{2}\right]_0^{\pi/2}$$

$$= a^2\left[\frac{3\pi}{2} - 4\right] = \frac{a^2}{2}\ (3\pi - 8).$$

119. Find the area of the loop of the folium $x^3 + y^3 = 3axy$

Solution: Put $x = r\cos\theta$;

$y = r\sin\theta$

in the given eqn. of the curve, we get

$$r^3(\cos^3\theta + \sin^3\theta) - 3ar^2\sin\theta\cos\theta = 0$$

$$\Rightarrow \qquad r = \frac{3a\sin\theta\cos\theta}{\cos^3\theta + \sin^3\theta} \qquad [1]$$

Limits:

Put $r = 0$, then $\sin\theta\cos\theta = 0$ or $\sin 2\theta = 0$

$\Rightarrow \theta = 0$ and $\theta = \pi/2$.

Thus, the loop passes through (0, 0), and varies from 0 to $\dfrac{\pi}{2}$.

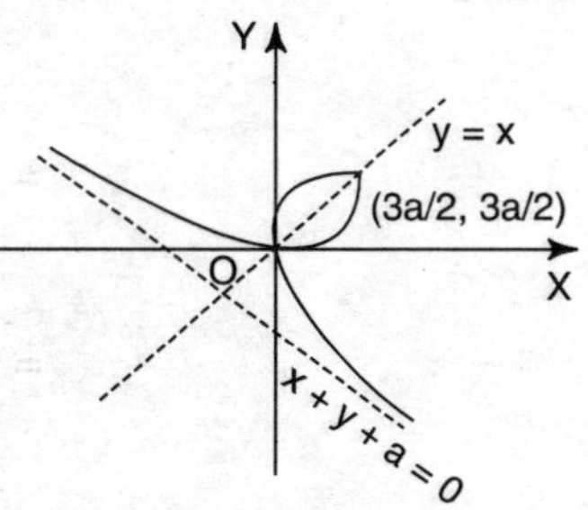

Fig. 3.49

$\therefore$ The required area is A, where

$$A = \frac{1}{2}\int_0^{\pi/2} r^2\,d\theta$$

$$= \frac{9a^2}{2}\int_0^{\pi/2}\frac{\sin^2\theta\cos^2\theta}{(\sin^3\theta + \cos^3\theta)^2}\,d\theta$$

$$(\div \cos^6\theta) \quad \Rightarrow A = \frac{9a^2}{2}\int_0^{\pi/2}\frac{\tan^2\theta\sec^2\theta}{(1+\tan^3\theta)^2}\,d\theta$$

Put, $1 + \tan^3\theta = t$;

$\therefore$ $3\tan^2\theta\sec^2\theta\,d\theta = dt$

Limits:	$\theta = 0, t = 1,$
	$\theta = \dfrac{\pi}{2}, t = \infty$

$$\therefore \quad A = \frac{3a^2}{2}\int_0^{\infty}\frac{dt}{t^2}$$

$$\therefore \qquad A = -\frac{3a^2}{2}\left[\frac{1}{t}\right]_1^{\infty}$$

$$= -\frac{3a^2}{2}(0 - 1) = \frac{3a^2}{2} \text{ sq. units}$$

120. Find the area lying between the parabola $y = 4x - x^2$ and the line $y = x$.

Solution: Points of intersection

Given equations are

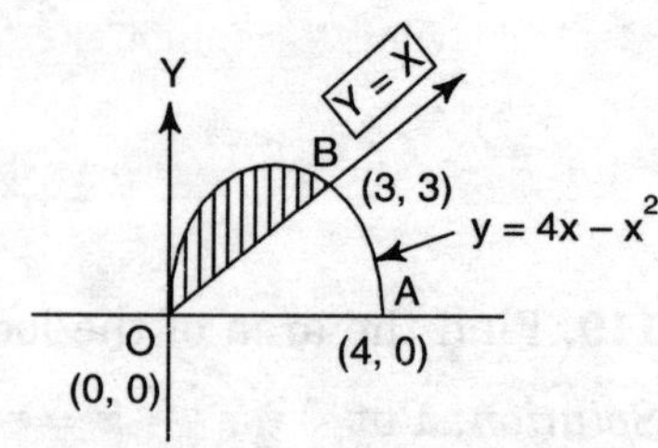

Fig. 3.50

$$y = 4x - x^2 \quad \text{and } y = x$$

$$\therefore \quad x = 4x - x^2$$

or $$x^2 + x - 4x = 0$$

$$x(x - 3) = 0$$

$$x = 0,\ x = 3.$$

where $x = 0,\ y = 0$

where $x = 3,\ y = 3,$

Thus, the line meets the parabola (1) in the points $O(0, 0)$ and $B(3, 3)$

$\therefore$ The required area is A

where $$A = \int_0^3 (y \text{ of upper curve} - y \text{ of lower curve})\, dx$$

$$= \int_0^3 [y \text{ of parabola} - y \text{ of line}]\, dx$$

$$= \int_0^3 [(4x - x^2) - x]\, dx$$

$$= \int_0^3 (3x - x^2)\, dx = \left[\frac{3x^2}{2} - \frac{x^3}{3}\right]_0^3$$

$$= \frac{27}{2} - \frac{27}{3} = \frac{9}{2} = 4\frac{1}{2} \text{ sq. units.}$$

EXERCISES

1. Find the area of the curve: $a^2x^2 = y^3(2a - y)$ (*Ans.* πa^2)
2. Find the areas enclosed by the curves

(1) $x = a[\theta + \sin\theta)]$,
$y = a[1 - \cos\theta)]$
and x-axis

(2) $y^2 = 4ax$
and $x^2 = 4by$

(3) $x^2 + y^2 = 1$
and $(x - 1)^2 + y^2 = 1$

(4) $y = \sin x$, $y = \cos x$

between $x = 0$ and $x = \dfrac{\pi}{4}$.

3. Evaluate the are of the region bounded by the curve $y = 2\sqrt{1-x^2}$ and the x-axis, after drawing the rough sketch of the same.

(*Ans.* π)

4. Find the area of the triangular region whose sides have the eqns.

$y = 2x + 1$

$y = 3x + 1$

And $x = 4$. (*Ans.* 8)

5. Show that the area common to the circles $r^2 = 2a^2$ and $r = 2a \cos \theta$ is $a^2 (\pi - 1)$

6. Find the area of a loop of each of the following curves:

(*i*) $r = a \sin 2\theta$

(*ii*) $r = a \cos 3\theta$

(*iii*) $r = a \cos 4\theta$

7. Find the area outside the circle $r = 2a \cos \theta$ and inside the cordiod $r = a (1 + \cos \theta)$

(*Ans.* πa^2)

3.9 SURFACE AREAS

(i) Surface area for Cartesian equation

The surface area of the solid generated by the revolution about x-axis by the area bounded by the curve $y = f(x)$, the x-axis and the ordinates at $x = a$ and $x = b$ is given by

$$S = 2\pi \int_a^b y\, ds, \text{ (if about } x\text{-axis)}$$

Similarly,

$$S = 2\pi \int_e^d x\, ds \text{ (if about } y\text{-axis)}$$

(ii) Surface area for polar Equation

The surface area of the solid generated by the revolution about the initial line by the area bounded by the curve $r = f(\theta)$ and radii vectors at $\theta = \theta_1$ and $\theta = \theta_2$ is:

$$S = \int_{\theta=\theta_1}^{\theta_2} r \sin\theta \cdot \frac{ds}{d\theta}\, d\theta .$$

or,

$$S = 2\pi \int_{\theta_1}^{\theta_2} y\, ds$$

where $y = r \sin \theta$, and $(ds)^2 = (dr)^2 + (rd\ \theta)^2$

Worked Examples (121 – 136) (Related to Surface areas)

121. Find the surface area of solid generated when one arch of cycloid $x = a\ (\theta + \sin\ \theta)$, $y = a\ (1 - \cos\ \theta)$ revolves about x-axis. (VTU, F/M, 2005; VTU, Sept. 2000)

Solution:

$$x = a(\theta + \sin\ \theta) \quad [1]$$

$$y = a(1 - \cos\ \theta) \quad [2]$$

$$\therefore \quad \frac{dx}{d\theta} = a(1 + \cos\ \theta); \quad \frac{dy}{d\theta} = a \sin\ \theta$$

PART-C

Now $(ds)^2 = (dx)^2 + (dy)^2$

$$\therefore \quad \left(\frac{ds}{d\theta}\right)^2 = \left(\frac{dx}{d\theta}\right)^2 + \left(\frac{dy}{d\theta}\right)^2 = a^2(1 + \cos\theta)^2 + a^2\sin^2\theta$$

$$= a^2[1 + 2\cos\theta + \cos^2\theta] + a^2\sin^2\theta$$

$$= a^2[1 + 2\cos\theta] + a^2 = 2a^2[1 + \cos\theta]$$

$$= 2a^2[2\cos^2\theta/2] = 4a^2\cos^2\theta/2$$

$$\therefore \quad \left(\frac{ds}{d\theta}\right) = 2a\cos\theta/2 \qquad [1]$$

Now, surface area is denoted by S, where

$$S = 2s_1 = 2\int 2\pi\, y\, ds = 2\cdot 2\pi\int_0^{\pi} y\left(\frac{ds}{d\theta}\right)d\theta$$

$$= 4\,\pi\int_0^{\pi} a(1-\cos\theta)\cdot 2a\cos\theta/2\, d\theta \qquad [2] \quad \text{(using [1])}$$

$$= 8\,\pi\, a^2\int_0^{\pi}\cdot 2\sin^2(\theta/2)\cdot\cos(\theta/2)\, d\theta, \text{ [put } \theta = 2\phi]$$

$$= 16\,\pi\, a^2\int_0^{\pi}\sin^2\phi\cos\phi\cdot 2d\phi$$

$$= 32\,\pi\, a^2\left[\frac{1}{3\cdot 1}\right] = \frac{32\,\pi a^2}{3}.$$

122. Find the surface area of solid obtained by the revolution of one loop of the curve $r^2 = a^2\cos^2\theta$ about the initial line (VTU, March, 2001)

Solution: The required surface area is twice the surface area of the solid generated by the upper part of one loop,

θ – varies from 0 to $\pi/4$.

$\therefore$ The required surface area is S, where

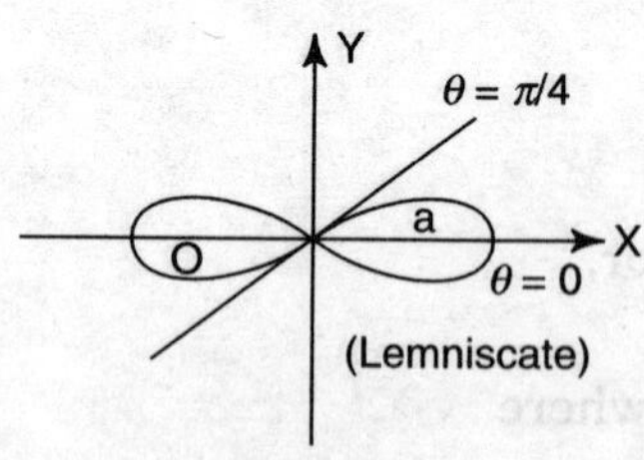

Fig. 3.52

$$S = 2S_1 = 2\int_0^{\pi/4} 2\pi\cdot y\cdot ds$$

$$= 2\int_0^{\pi/4} 2\pi y\cdot\frac{ds}{d\theta}d\theta$$

$$= 2\int_0^{\pi/4} 2\pi\,(r\sin\theta)\,\frac{ds}{d\theta}\cdot d\theta \qquad [1]$$

($\because$ $y = r\sin\theta$ comes from the formula and $(ds)^2 = (dr)^2 + (rd\,\theta)^2$)

Now $r^2 = a^2\cos^2\theta$

or $r = a\sqrt{\cos 2\theta}$

$$\therefore \quad \frac{dr}{d\theta} = a \cdot \frac{1}{2\sqrt{\cos 2\theta}}(-2\sin 2\theta) = -\frac{a\sin 2\theta}{\sqrt{\cos 2\theta}}$$

$$\frac{ds}{d\theta} = \left[r^2 + \left(\frac{dr}{d\theta}\right)^2\right]^{1/2} = \left\{a^2\cos 2\theta + \frac{a^2\sin^2 2\theta}{\cos 2\theta}\right\}^{1/2}$$

$$= \left\{\frac{a^2}{\cos 2\theta}\right\}^{1/2} = \frac{a}{\sqrt{\cos 2\theta}}$$

$\therefore$ From [1], we obtain

$$S = 4\pi \int_0^{\pi/4} a\sqrt{\cos 2\theta} \cdot \sin\theta \cdot \frac{a}{\sqrt{\cos 2\theta}}\, d\theta$$

$$= 4\pi a^2 \left[-\cos\theta\right]_0^{\pi/4} = 4\pi a^2\left[1 - \frac{1}{\sqrt{2}}\right]$$

123. Find the area of surface of revolution about x-axis of the astroid:

$$x^{2/3} + y^{2/3} = a^{2/3}$$

(or in parametric form: $x = a\cos^3\theta,\ y = a\sin^3\theta$)

Solution: By symmetry, the required surface area, S where

$$S = 2s_1 = 2\int_{x=0}^{a} 2\pi\, y\, ds$$

$$= 2\int_{x=0}^{\pi/2} 2\pi a\sin^3\theta \cdot \frac{ds}{d\theta}\, d\theta$$

Since $x = a\cos^3\theta$;

$$\therefore \quad \frac{dx}{d\theta} = -3a\cos^2\theta\sin\theta$$

$$y = a\sin^3\theta$$

$$\therefore \quad \frac{dy}{d\theta} = 3a\sin^2\theta\cos\theta$$

$$\therefore \quad \frac{ds}{d\theta} = \sqrt{\left(\frac{dx}{d\theta}\right)^2 + \left(\frac{dy}{d\theta}\right)^2}$$

$$= \sqrt{9a^2\cos^4\theta\sin^2\theta + 9a^2\sin^4\theta\cos^2\theta}$$

$$= 3a\cos\theta\sin\theta$$

$$\therefore \quad S = 2\int_0^{\pi/2} 2\pi \cdot a\sin^3\theta \cdot 3a\cos\theta \sin\theta\, d\theta$$

$$= 12a^2 \int_0^{\pi/2} \sin^4\theta \cos\theta\, d\theta$$

Put $\sin\theta = t$

$\Rightarrow \cos\theta\, d\theta = dt$

When $\theta = 0,\ t = 0$

and when $\theta = \frac{\pi}{2},\ t = 1$

$$\therefore \quad S = 12\pi a^2 \int_0^1 t^4\, dt = 12\pi a^2 \left[\frac{t^5}{5}\right]_0^1 = 12\pi a^2\left[\frac{1}{5} - 0\right]$$

$$= \frac{12}{5}\pi a^2 \text{ sq. units.}$$

124. Find the surface area when the portion of the parabola $x^2 = 4ay$ cut off by the latus rectum rotates about the y-axis.

Solution: The curve is symmetrical about y-axis,
The surface area S, where S is given by

$$S = \int_0^a 2\pi\, x ds$$

where
$$ds = \sqrt{1 + \left(\frac{dx}{dy}\right)^2}\, dy = \sqrt{1 + \left(\frac{4a^2}{x^2}\right)}\, dy = \frac{\sqrt{(y+a)}}{\sqrt{y}}\, dy$$

$$\therefore \quad S = 2\pi \int_0^a \frac{x\sqrt{y+a}}{\sqrt{y}}\, dy = 4\pi\sqrt{a}\int_0^a \sqrt{(a+y)}\, dy$$

$$= \frac{8}{3}\pi\sqrt{a}\left[(a+y)^{3/2}\right]_0^a = \frac{8}{3}\pi a^2\left(2\sqrt{2} - 1\right).$$

125. Find the area of the surface of revolution formed by revolving the loop of the curve $9ay^2 = x(3a - x)^2$ about x-axis.

Solution: The required surface area, is given by S,

where
$$S = \int_0^{3a} 2\pi y \cdot \frac{ds}{dx}\, dx$$

Now, $18\, ay \dfrac{dy}{dx} = 1 \cdot (3a - x)^2 + x \cdot 2\,(3a - x) \cdot (-1)$

$$= (3a - x) \cdot 3\,(a - x)$$

$$\therefore \quad \frac{dy}{dx} = \frac{(3a-x)(a-x)}{6ay} = \frac{a-x}{6a}\sqrt{\left(\frac{9a}{x}\right)} = \frac{a-x}{2\sqrt{ax}}$$

$$\therefore \quad \frac{ds}{dx} = \sqrt{\left\{1+\left(\frac{dy}{dx}\right)^2\right\}} = \sqrt{\left\{1+\frac{(a-x)^2}{4ax}\right\}} = \frac{a+x}{2\sqrt{(ax)}}$$

$$\therefore \quad s = \int_0^{3a} 2\pi \cdot \sqrt{(x)}\,\frac{(3a-x)}{3\sqrt{a}}\,\frac{(a+x)}{2\sqrt{a}\sqrt{x}}\,dx$$

$$= \frac{\pi}{3a}\int_0^{3a}(3a-x)(a+x)\,dx = \frac{\pi}{3a}(3a^2+2ax-x^2)\,dx$$

$$= \frac{\pi}{3a}\cdot 9a^3 = 3\pi a^2$$

126. Find the area of the surface of revolution formed by revolving the curve $r = 2a \cos \theta$ about the initial line.

Solution: The given curve is a circle as shown in Fig. 3.53 whose diameter is the initial line and of length $2a$.

$\therefore$ $A\ (2a, 0)$

$\therefore$ The surface area is demoted by S,

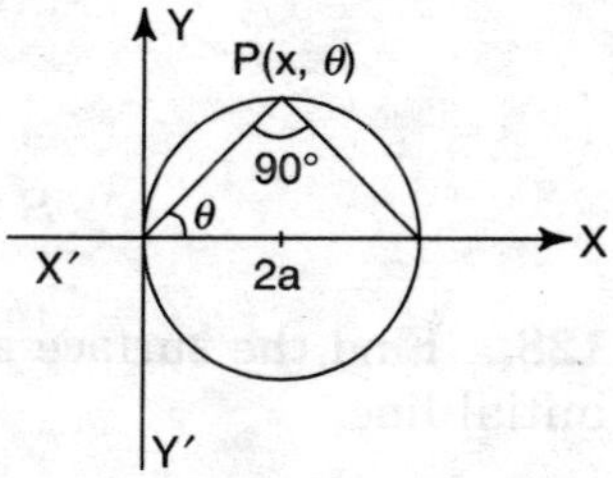

where $$S = 2\pi\int_0^{\pi/2} y\,ds = 2\pi\int_0^{\pi/2} y\cdot\frac{ds}{d\theta}\,d\theta$$

$$= 2\pi\int_0^{\pi/2} r\sin\theta\,(2a\,d\theta)$$

$$= 2\pi\int_0^{\pi/2} 2a\cos\theta\sin\theta\,(2a\,d\theta)$$

$$= 8\pi a^2\int_0^{\pi/2}\sin\theta\cdot\cos\theta\,d\theta$$

$$= 8\pi a^2\left[\frac{\sin^2\theta}{2}\right]_0^{\pi/2} = 4\pi a^2 \text{ sq. units.}$$

Fig. 3.53

127. Find the surface area of the solid generated by revoling the cycloid

$x = a\ (t + \sin t)$

$y = a\ (1 + \cos t)$ about its base (VTU, MQP. I)

Solution:

$x = a\ (t + \sin t)$

$y = a\ (1 + \cos t)$

The x-axis is along the base, the y-axis passes through the vertex A. The limits for 't' are:

$$t = -\pi \text{ at } B'$$
$$t = 0 \text{ at } A$$
$$t = \pi \text{ at } B$$

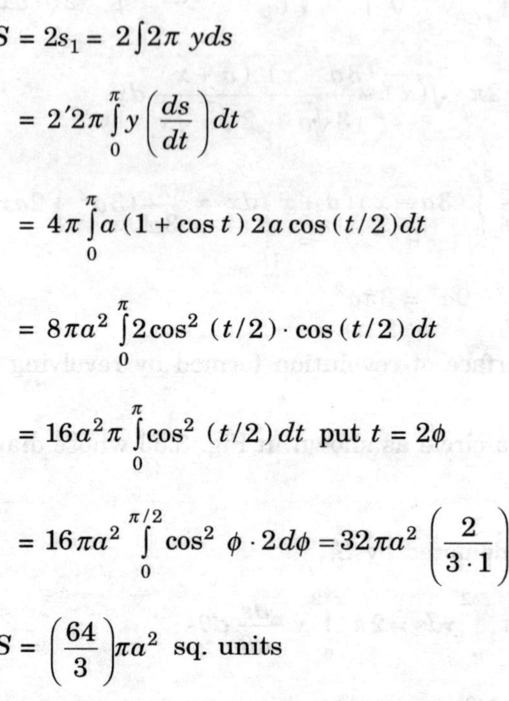

Fig. 3.54

Surface area:

$$S = 2s_1 = 2\int 2\pi \, yds$$

$$= 2'2\pi \int_0^{\pi} y\left(\frac{ds}{dt}\right)dt$$

$$= 4\pi \int_0^{\pi} a\,(1+\cos t)\,2a\cos(t/2)\,dt$$

$$= 8\pi a^2 \int_0^{\pi} 2\cos^2(t/2)\cdot\cos(t/2)\,dt$$

$$= 16a^2\pi \int_0^{\pi} \cos^2(t/2)\,dt \quad \text{put } t = 2\phi$$

$$= 16\pi a^2 \int_0^{\pi/2} \cos^2\phi \cdot 2d\phi = 32\pi a^2\left(\frac{2}{3\cdot 1}\right)$$

$$S = \left(\frac{64}{3}\right)\pi a^2 \text{ sq. units}$$

128. Find the surface area generated by revolving the cardiode $r = a\,(1 + \cos\,\theta)$ about the initial line. (VTU, March, Aug., 2000)

Solution: The cardiod is symmetrical about the initial line $\theta = 0$. The required surface area is denoted by S.

where $$S = 2\pi \int_0^{\pi} r\cdot\sin\theta\cdot\frac{ds}{d\theta}\,d\theta$$

$$= 2\pi \int_0^{\pi} a\,(1+\cos\theta)\cdot\sin\theta\cdot 2a\cos(\theta/2)\,d\theta$$

$$= 4\pi a^2 \int_0^{\pi} 2\cos^2(\theta/2)\cdot 2\sin(\theta/2)\cos(\theta/2)\cdot\cos(\theta/2)\,d\theta$$

$$= 16\pi a^2 \int_0^{\pi} \cos^4(\theta/2)\sin(\theta/2)\,d\theta$$

$$= 16\pi a^2 \int_0^{\pi/2} \cos^4\phi\,\sin\phi\cdot 2d\phi, \text{ where} \qquad (\phi = \theta/2)$$

$$= 32\pi a^2 \left[\frac{-\cos^5\phi}{5}\right]_0^{\pi/2} = \frac{32}{5}\pi a^2 \text{ sq. units}$$

129. **Find the surface area generated by revolving the curve $x = y^3$ between the points with ordinates $y = 0$ and $y = 1$, about the y-axis**

Solution: The required surface area is:

$$S = \int_{y=0}^{1} 2\pi x ds = 2\pi\int_0^1 x\frac{ds}{dy}dy = 2\pi\int_0^1 \left\{1+\left(\frac{dx}{dy}\right)^2\right\}^{1/2} dy$$

$$= 2\pi\int_0^1 y^3\sqrt{1+9y^4}\,dy = \frac{\pi}{18}\int_0^a \sqrt{1+t}\,dt$$

where $\quad t = 9y^4 = \frac{\pi}{27}[10^{3/2} - 1]$

130. **Show that the area of the surface generated by revolving the ellipse $\left(\frac{x^2}{a^2}\right)+\left(\frac{y^2}{b^2}\right) = 1$ about the x-axis is**

$$\left\{\frac{(2\pi ab)}{e}\left[e\sqrt{1-e^2} + \sin^{-1} e\right]\right\}$$

where e is the eccentricity

Solution: $\quad x = a\cos t,$

$y = b\sin t$

are the parametric equations of the ellipse.

The surface area S, where

$$S = 2S_1 = 2\int 2\pi y ds = 4\pi\int_0^{\pi/2} y\cdot\frac{ds}{dt}dt$$

$$= 4\pi\int_0^{\pi/2} b\sin t\cdot\sqrt{a^2\sin^2 t + b^2\cos^2 t}\,dt \qquad (\text{since } b^2 = a^2 - a^2e^2)$$

$$= 4\pi\int_0^{\pi/2} \sin t\sqrt{(a^2 - a^2e^2\cos^2 t)}\,dt$$

put $e\cos t = z \quad \therefore \quad -e\sin t\,dt = dz$

Limits: when $t = 0, z = e$
when $t = \frac{\pi}{2}, z = 0$

PART-C

$$= 4\pi ab \int_e^0 \sqrt{(1-z^2)}\left(-\frac{dz}{e}\right)$$

$$= \frac{4\pi ab}{e}\left[\frac{z}{2}\sqrt{(1-z^2)} + \frac{1}{2}\cdot \sin^{-1}(e)\right]_0^e$$

$$= \frac{2\pi ab}{e}\left[e\sqrt{(1-e^2)} + \sin^{-1}(e)\right]$$

[Now, area of the ellipse = $A = \pi ab$]

$$\therefore \qquad S = 2A\left[\sqrt{(1-e^2)} + \frac{1}{e}\sin^{-1}(e)\right]$$

131. Show that the area generated by revolving the curve

$$x = a\cos t + \frac{1}{2}\ a \cdot \log \tan^2\left(\frac{t}{2}\right)$$

$$y = a\sin t;\ \left(0 \le t \le \frac{\pi}{2}\right)$$

about its asymplote is $4\pi a^2$.

Solution: **Given:** $x = a\cos t + \frac{1}{2}\ a \cdot 2\log\tan\left(\frac{t}{2}\right)$

$$\therefore \qquad \frac{dx}{dt} = -a\sin t + a\cdot \frac{1}{\tan\left(\frac{t}{2}\right)}\cdot \sec^2\left(\frac{t}{2}\right)\cdot\frac{1}{2}$$

$$= -a\sin t + \frac{a}{2\sin\frac{t}{2}\cdot\cos\frac{t}{2}}$$

$$= \frac{a}{\sin t}\left[1-\sin^2 t\right] = \frac{a\cos^2 t}{\sin t}$$ [1]

$$\frac{dy}{dt} = a\cos t;$$

$$\therefore \qquad \frac{ds}{dt} = \sqrt{\left(\frac{dx}{dt}\right)^2 + \left(\frac{dy}{dt}\right)^2}$$

or, $$\frac{ds}{dt} = a\cos t\sqrt{\left(\frac{\cos^2 t}{\sin^2 t}+1\right)} = \frac{a\cos t}{\sin t}$$

$$V = 2\int_0^{\pi/2} \pi\, y^2 \frac{dx}{dt}\, dt = 2\int_0^{\pi/2} \pi a^2 \cdot \sin^2 t \cdot a \cdot \frac{\cos^2 t}{\sin t}\, dt\,; \qquad \text{(using [1])}$$

$$= 2a^3\pi \int_0^{\pi/2} \sin t \cos^2 t\, dt = -2\pi a^3 \left[\frac{\cos^3 t}{3}\right]_0^{\pi/2} = \frac{2}{3}\pi a^3$$

= **half the volume of the sphere of radius 'a'**

$$\therefore \qquad S = 2\int_0^{\pi/2} 2\pi y \frac{dS}{dt}\, dt = 2 \cdot 2\pi \int_0^{\pi/2} a \sin t \cdot a \frac{\cos t}{\sin t}\, dt$$

$$= 4\pi a^2 [\sin t]_0^{\pi/2} = 4\pi a^2$$

= **surface area of the sphere of radius 'a'**

132. Find the surface area of the solid generated by the revolution about the x-axis of the loop of the curve $x = t^2$, $y = t - \frac{1}{3}t^3$.

Solution: **Given:**

$$x = t^2 \qquad [1]$$

$$y = t - \frac{1}{3}t^3 \qquad [2]$$

Eliminating 't' between [1] and [2], we get

$$y^2 = t^2 - \frac{2}{3}\cdot t^4 + \frac{1}{9}t^6 = x - \frac{2}{3}\cdot x^2 + \frac{1}{9}x^3$$

$$= x\left[1 - \frac{2}{3}x + \frac{1}{9}x^2\right] = x\left[1 - \frac{1}{3}x\right]^2$$

$$\therefore \qquad y^2 = x\left(1 - \frac{x}{3}\right)^2$$

Limits: when $y = 0$, we have $x = 0$, $x = 3$.

and when $x = 0$, $t = 0$

when $x = 3$, $t = \sqrt{3}$

The loop is symmetrical about axis of x. The required surface area is S, where

$$S = 2\pi \int_0^{\sqrt{3}} y \frac{ds}{dt}\, dt = 2\pi \int_0^{\sqrt{3}} \left(t - \frac{1}{3}t^3\right)\sqrt{\left[(2t)^2 + (1-t^2)^2\right]} \cdot dt$$

$$= 2\pi \int_0^{\sqrt{3}} \left(t - \frac{1}{3}t^3\right)(1 + t^2)\, dt$$

$$= 2\pi \int_0^{\sqrt{3}} \left(t + t^3 - \frac{1}{3}t^3 - \frac{1}{3}t^5\right) dt$$

$$= 2\pi \int_0^{\sqrt{3}} \left[t + \frac{2}{3}t^3 - \frac{1}{3}t^5 \right] dt = 2\pi \left[\frac{t^2}{2} + \frac{t^4}{6} - \frac{t^6}{18} \right]_0^{\sqrt{3}}$$

$$= 2\pi \left[\frac{2}{3} + \frac{9}{6} - \frac{27}{18} \right] = 3\pi .$$

133. Find the surface area generated by the curve $y^2 = 4ax$ between the vertex and latus rectum.

Solution:

Let $A = (0, 0); L = (a, 2a)$

$y^2 = 4ax$ (Given)

$\therefore$ The required surface area is S, where

$$S = \int_0^a 2\pi y \cdot \frac{dS}{dx}\, dx \qquad [1]$$

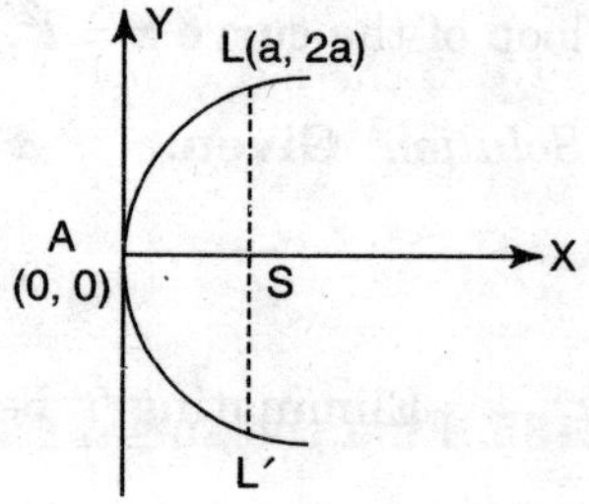

Y L(a, 2a)
A (0, 0) S X
L′

$\left(y^2 = 4ax \,;\, y = 2\sqrt{a}\sqrt{x} \right)$

$$\therefore \quad y^1 = \frac{\sqrt{a}}{\sqrt{x}} ; \frac{dS}{dx} = \sqrt{1 + \left(\frac{a}{x} \right)}$$

$$S = \int_0^a 2\pi \cdot 2\sqrt{a} \cdot \sqrt{x} \sqrt{\left(1 + \frac{a}{x} \right)}\, dx$$

$$= 4\pi\sqrt{a} \int \sqrt{(x + a)}\, dx = 4\pi\sqrt{a} \left[\frac{2}{3}(x + a)^{3/2} \right]_0^a$$

$$= \frac{8\pi}{3} a^2 [2\sqrt{2} - 1].$$

134. Find the surface area of the solid generated by revolving a catenary $y = c \cosh\left(\frac{x}{c} \right)$ from the vertex to any point (x, y), about the x-axis

Solution: The required surface area is

$$S = 2\pi \cdot \int_0^c y \cdot \cosh\left(\frac{x}{c} \right) dx \qquad [1]$$

$$\left(\because y = c \cosh\left(\frac{x}{c} \right); y' = c \sinh\left(\frac{x}{c} \right) \cdot \frac{1}{c} = \sinh\left(\frac{x}{c} \right) \right)$$

$$\therefore \qquad \sqrt{1+(y')^2} = \sqrt{1+\sinh^2\left(\frac{x}{c}\right)} = \sqrt{\cosh^2\left(\frac{x}{c}\right)} = \cosh\left(\frac{x}{c}\right)$$

Using [1], we have

$$S = 2\pi \int_0^c c\cosh\left(\frac{x}{c}\right)\cosh\left(\frac{x}{c}\right)dx$$

$$= 2\pi c \int_0^c \cosh^2\left(\frac{x}{c}\right)dx = 2\pi c \int_0^c \left[\frac{1+\cosh\left(\frac{2x}{c}\right)}{2}\right]dx$$

$$= \frac{2\pi c'}{2}\left[x + \frac{\sinh\left(\frac{2x}{c}\right)}{\left(\frac{2}{c}\right)}\right]_0^c = \pi c \left[c + \frac{c}{2}\sinh 2 - 0\right]$$

$$= \frac{\pi c^2}{2}\ [2 + \sinh 2]$$

135. Find the surface area of the sphere of radius 'a'

Solution: The required surface area is twice the surface generated by the revolution of the quadrant AOB about x-axis.

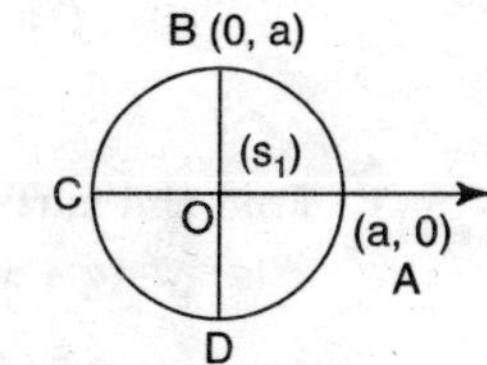

Fig. 3.55

$$\therefore \qquad S = 2S_1 = 2\int_0^a 2\pi y \frac{ds}{dx}\,dx$$

$$\left(x^2 + y^2 = a^2 \therefore y' = -\frac{x}{y}, \left(\frac{ds}{dx}\right)^2 = 1 + (y')^2 = \frac{x^2+y^2}{y^2} = \frac{a^2}{y^2}\right)$$

$$S = 4\pi \int_0^a y \cdot \frac{a}{y}\,dx = 4\pi a\,[x]_0^a = 4\pi a^2$$

136. The part of the parabola $y^2 = 4ax$; bounded by latus rectum revolves about the tangent at the vertex, find the surface of the reel thus generated.

Solution: The required surface area is s, where, $s = 2s_1$

or $$S = 2\int_0^a 2\pi x \frac{ds}{dx}\,dx$$

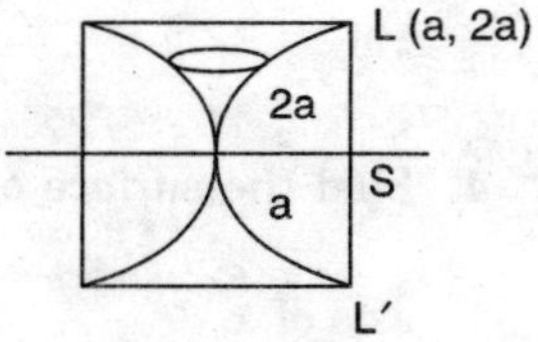

$$\left(y^2 = 4ax,\ y = 2\sqrt{a}\sqrt{x};\quad \therefore y' = \frac{\sqrt{a}}{\sqrt{x}}\right)$$

Fig. 3.56

$$\therefore \qquad S = 4\pi\int_0^a x\sqrt{\left(1+\frac{a}{x}\right)}\,dx = 4\pi\int_0^a \sqrt{(x^2+ax)}\,dx$$

$$= 4\pi\int_0^a \sqrt{\left\{\left(x+\frac{a}{2}\right)^2-\left(\frac{a}{2}\right)^2\right\}}\,dx$$

$$= 4\pi\left[\left(\frac{x+\frac{a}{2}}{2}\right)\sqrt{x^2+ax}-\frac{1}{2}\left(\frac{a}{2}\right)^2\log\left\{\left(x+\frac{a}{2}\right)+\sqrt{x^2+ax}\right\}\right]_0^a$$

$$= 4\pi\left[\frac{1}{2}\cdot\frac{3a}{2}\cdot a\sqrt{2}-\frac{a^2}{1}\left\{\log\left(\frac{3a}{2}+a\sqrt{2}\right)-\log\frac{a}{2}\right\}\right]$$

$$= \pi a^2\left[3\sqrt{2}-\frac{1}{2}\log\left(3+2\sqrt{2}\right)\right] = \pi a^2\left[3\sqrt{2}-\frac{1}{2}\log\left(\sqrt{2}+1\right)^2\right]$$

$$= \pi a^2\left[3\sqrt{2}-\log\left(\sqrt{2}+1\right)\right]$$

EXERCISES

1. Find the surface area generated by the revolution of the following curves about x-axis

(a) $y^2(a+x) = x^2(a-x)$ (*Ans.* $\pi a^2(8\log 2-3)$

(b) $y^2 = x^2(a-x)$ $\left(Ans.\ \frac{\pi a^4}{12}\right)$

(c) $y^2(a-x) = x^2(a+x)$ $\left(Ans.\ 2\pi a^3\left(\log 2-\frac{2}{3}\right)\right)$

2. Find the area of the surface formed by the revolution of ellipse $x^2+4y^2 = 16$ about its major axis $\left(Ans.\ 8\pi\left[1+\frac{4}{3\sqrt{3}}\cdot 1\right]\right)$

3. Find the area of surface generated by revolving about y-axis, the arc $x = y^3$ from $y = 0$ to $y = 2$ $\left(Ans.\ \frac{\pi}{27}\left[(145)^{5/2}-1\right]\right)$

4. Find the surface of the solid generated by the revolution of $x = a\cos^3 t$, $y = a\sin^3 t$ about the axis of x $\left(Ans.\ \frac{12}{5}\pi a^2\right)$

5. Find the surface area generated by revolving the loop of the curve $3ay^2 = x(x - a)^2$ for which $0 \le x \le a$ about x-axis $\left(Ans. \dfrac{\pi}{12} a^4\right)$

6. Find the surface of the cone generated by the revolution of the line segment $y = 2x$ from $x = 0$ to $x = 2$

(*i*) about x-axis

(*ii*) about y-axis $\left(Ans. 8\pi\sqrt{5}; 4\pi\sqrt{5}\right)$

7. Find the surface area obtained by revolving the loop of the curve $8a^2y^2 = a^2x^2 - x^4$ about x-axis

8. Prove that the surface area of a solid generated by revolving the curve given

$$x = a\,(2\cos\theta - \cos 2\theta);$$
$$y = a\,(2\sin\theta - \sin 2\theta)$$

about x-axis is $\dfrac{128}{5}\pi a^2$.

9. Find the area of the surface formed by rotating $y^2 = x^2$ from $x = 0$ to $x = 4$ about the y-axis. (*Ans.* 262.2)

3.10 VOLUME OF SOLID OF REVOLUTION OF PLANE CURVES

(*a*) Cartesian Equations:

(*i*) The volume of solid formed by the revolution of the curve $y = f(x)$ about the x-axis is V, where

$$V = \pi \int_a^b y^2\, dx$$

(*ii*) If it is about the y-axis

$$V = \pi \int_a^d x^2\, dy,$$

where $x = f(y)$

Note: (*i*) Here, the curve does not intersect (or cut) x-axis and y-axis respectively.

(*ii*) When the curve is symmetrical about x-axis, or y-axis, only one portion will be rotated to generate the solid.

(*b*) *Polar equations:*

If the equation of the curve is: $r = f(\theta)$ then the volume generated by an arc of the curve between

$\theta = \alpha$, $\theta = \beta$ about the lines

$\theta = 0$ (x-axis)

PART-C

and $\theta = \dfrac{\pi}{2}$ (y-axis) are given by

(*i*) about the initial line

i.e., $\theta = 0$

is
$$V = \int_{\theta=\alpha}^{\beta} \frac{2}{3}\pi r^3 \sin\theta \, d\theta$$

(*ii*) about the line $\theta = \dfrac{\pi}{2}$, the volume V is

$$V = \int_{\theta=\alpha}^{\beta} \frac{2}{3}\pi r^3 \cos\theta \, d\theta.$$

Worked Examples (137 – 152)

137. Find the volume and the surface area of the solid generated by revolving the astroid: $x^{2/3} + y^{2/3} = a^{2/3}$ about the x-axis (VTU. August, 2001)

Solution: Since the volume required is obtained by revolving the curve from $x = 0$ to $x = a$ about x-axis and taking twice the result (by symmetry)

$$\therefore \quad V = 2\int_0^a \pi y^2 \, dx$$

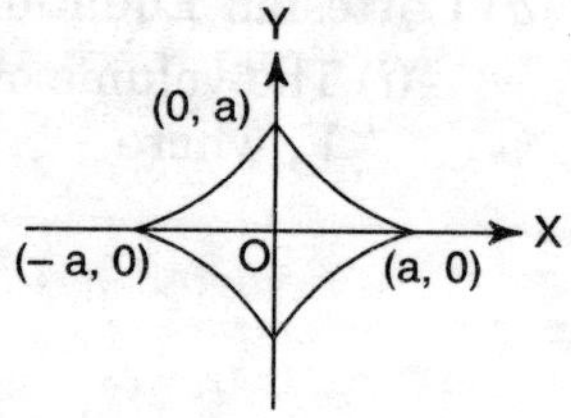

Since, $x = a\cos^3\theta;\ y = a\sin^3\theta$

$\therefore$ $dx = -3a\cos^2\theta\sin\theta\,d\theta$

where $x = 0,\ a\cos^3\theta = 0$

$\therefore$ $\theta = \dfrac{\pi}{2}$

where $x = a,\ a\cos^3\theta = a$

$\therefore$ $\theta = 0$

Fig. 3.57

Thus,
$$V = 2\int_{\pi/2}^{0} \pi(a\sin^3\theta)^2 \cdot (-3a\cos^2\theta\sin\theta)\cdot d\theta$$

$$= 6\pi a^3 \int_0^{\pi/2} \sin^7\theta\cos^2\theta\,d\theta,$$

$$= 6\pi a^3 \cdot \frac{6\cdot4\cdot2\cdot1}{9\cdot7\cdot5\cdot3\cdot1} \qquad [1]$$

$$= 32\pi \cdot \frac{a^3}{105} \text{ C. units}$$

and the surface area S, (already found in example (123) on surface areas)

$$S = 2\int_{\theta=0}^{\pi/2} 2\pi y\, ds = 4\pi \int_{\theta=0}^{\pi/2} y \cdot \frac{ds}{d\theta}\, d\theta \qquad (i)$$

where $\frac{ds}{d\theta} = \left[\left(\frac{dx}{d\theta}\right)^2 + \left(\frac{dy}{d\theta}\right)^2\right]^{1/2}$

$= 3a \sin\theta \cos\theta$ (after a minor simplification)

Hence $S = 4\pi \int_0^{\pi/2} a \sin^3\theta\,(3a \sin\theta \cos\theta)\, d\theta$

$= \frac{12\pi a^2}{5}$ Sq. units

138. Find the volume and surface area of the solid formed by revolution of one arc of the cycloid, $x = a(\theta + \sin\theta)$, $y = a(1 - \cos\theta)$ about the tangent at the vertex

(VTU, March, '99, Aug., 2000)

Solution: The required volume is given by

$$V = 2\int_0^{\pi} \pi y^2 \frac{dx}{d\theta}\, d\theta$$

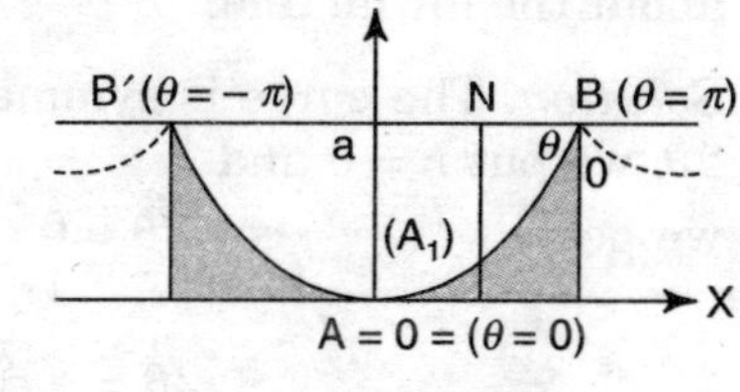

Fig. 3.58

From the figure, we find:

$x = a(\theta + \sin\theta)$

$y = a(1 - \cos\theta)$

Inverted cycloid:

$\therefore \quad \frac{dx}{d\theta} = a(1 + \cos\theta)$

The x-axis is the tangent at the vertex and the vertex A is at the origin. The base line is not at the x-axis.

Limits for θ: θ-Varies

$\theta = 0$; at O from $-\pi$ to π

$\theta = \pi$; at B

$\theta = -\pi$; at B'

Volume: $V = \int_{-\pi}^{\pi} \pi y^2 \frac{dx}{d\theta} \cdot d\theta = \pi \int_{-\pi}^{\pi} a^2 (1 - \cos\theta)^2\, a(1 + \cos\theta)\, d\theta$

$$= \pi a^3 \int_{-\pi}^{\pi} \left[2\sin^2\left(\frac{\theta}{2}\right)\right]^2 \left[2\cos^2\frac{\theta}{2}\right] d\theta$$

$$= 8\pi a^3 \int_{-\pi}^{\pi} \sin^4\left(\frac{\theta}{2}\right)\cos^2\left(\frac{\theta}{2}\right) d\theta$$

PART-C

$$= (16\pi a^3) \int_{-\pi/2}^{\pi/2} \sin^4 \phi \cos^2 \phi \, d\phi \,; \left(\phi = \frac{\theta}{2}\right)$$

$$= 32\pi a^3 \int_0^{\pi/2} \sin^4 \phi \cos^2 \phi \, d\phi = 32\pi a^3 \frac{(3 \cdot 1)(1)}{6 \cdot 4 \cdot 2} \cdot \frac{\pi}{2}$$

$$= \pi^2 a^3, \text{ c. units}$$

Surface area: $S = 2S_1 = 2\int_0^{\pi} 2\pi y \frac{dS}{d\theta} \, d\theta = 4\pi \int_0^{\pi} a(1 - \cos\theta) 2a \cos\frac{\theta}{2} \, d\theta$

$$= 16\pi a^2 \int_0^{\pi} \sin^2 \frac{\theta}{2} \cos\frac{\theta}{2} \, d\theta = 16\pi a^2 \left[2 \cdot \frac{\sin^3 \theta/2}{3}\right]_0^{\pi}$$

$$= \frac{32}{3} \pi a^2 \text{ Sq. units}$$

139. Find the volume of the solid generated by revolving one loop of the Lemniscate $r^2 = a^2 \cos 2\theta$ about the initial line.

Solution: The curve is symmetrical about the initial line. For a loop of the curve $r^2 = a^2 \cos 2\theta$, we put $r = 0$ and

we get $\cos 2\theta = 0$

$$\Rightarrow \quad 2\theta = \pm \frac{\pi}{2};$$

$$\Rightarrow \quad \theta = \pm \frac{\pi}{4}$$

$\Rightarrow \quad \theta = -\frac{\pi}{4}$ and $\theta = +\frac{\pi}{4}$ are the limits for θ, and we get a loop.

(It is enough if we revolve the part from $\theta = 0$ to $\theta = \pi/4$) about the initial line to get the volume of revolution of the loop.

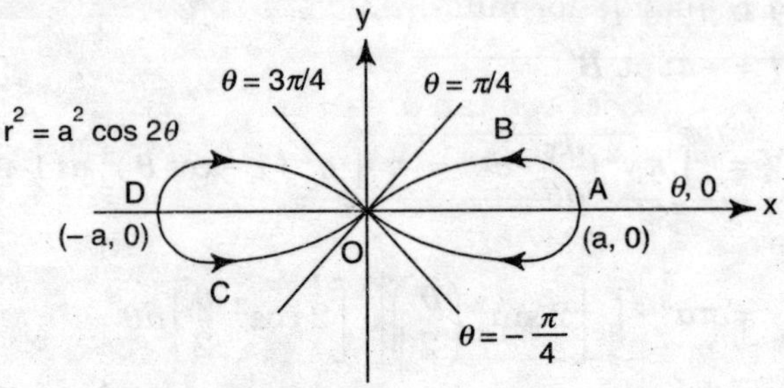

Fig. 3.59 Tangents at origin are $y = \pm x$

$$V = \int_0^{\pi/4} \frac{2}{3} \cdot \pi r^3 \sin\theta \, d\theta$$

$$= \int_0^{\pi/4} \frac{2}{3} \pi a^3 (\cos 2\theta)^{3/2} \sin\theta \, d\theta \qquad \left(\because r = a\sqrt{\cos 2\theta}\right)$$

$$= \frac{2\pi a^3}{3} \int_0^{\pi/4} (2\cos^2\theta - 1)^{3/2} \cdot \sin\theta \, d\theta$$

$$\left[\begin{array}{l}\text{Put } \sqrt{2}\cos\theta = \sec\phi \; ; \therefore \; -\sqrt{2}\sin\theta \, d\theta = \sec\phi\tan\phi \, d\phi \\ \therefore \; \sin\theta \, . d\theta = -\frac{1}{\sqrt{2}} \sec\phi\tan\phi \, d\phi\end{array}\right]$$

$$\left\{\begin{array}{l}\text{Limits : when } \theta = 0, \sec\phi = \sqrt{2}\cos 0 = \sqrt{2.1} = \sqrt{2} \\ \Rightarrow \; \phi = \left(\frac{\pi}{4}\right) \\ \text{when } \theta = \frac{\pi}{4}, \sec\phi = \sqrt{2}\cos\frac{\pi}{4} = \sqrt{2} \, . \frac{1}{\sqrt{2}} = 1 \\ \Rightarrow \; \phi = 0\end{array}\right.$$

$$\therefore \quad V = \frac{2\pi a^3}{3} \int_0^{\pi/4} (\sec^2\phi - 1)^{3/2} \, . \left[-\frac{1}{\sqrt{2}} \sec\phi\tan\phi\right] d\phi$$

$$\Rightarrow \quad V = \frac{\sqrt{2}\pi a^3}{3} \int_0^{\pi/4} \tan^4\phi \sec\phi \, d\phi$$

$$\Rightarrow \quad V = \frac{\sqrt{2}\pi a^3}{3} \int_0^{\pi/4} (\sec^2\phi - 1)^2 \, . \sec\phi \, d\phi$$

$$\Rightarrow \quad V = \frac{\sqrt{2}\pi a^3}{3} \int_0^{\pi/4} (\sec^5\phi - 2\sec^3\phi + \sec\phi) \, d\phi \qquad [1]$$

Now, by using the reduction formula :

$$\boxed{\int \sec^n\phi \, d\phi = \frac{\sec^{n-2}\phi\tan\phi}{(n-1)} + \frac{n-2}{n-1} \int \sec^{n-2}\phi \, d\phi}$$

$\therefore$ We have ; (here $n = 5$)

$$\int_0^{\pi/4} \sec^5\phi \, d\phi = \left[\frac{\sec^3\phi\tan\phi}{4}\right]_0^{\pi/4} + \frac{3}{4} \int_0^{\pi/4} \sec^{n-2}\phi \, d\phi$$

$$= \frac{\sqrt{2}}{2} + \frac{3}{4}\left\{\left[\frac{\sec\phi\tan\phi}{2}\right]_0^{\pi/4} + \frac{1}{2}\int_0^{\pi/4} \sec\phi \, d\phi\right\}$$

$$= \frac{\sqrt{2}}{2} + \frac{3}{4}\left\{\frac{\sqrt{2}}{2} + \frac{1}{2}[\log(\sec\phi + \tan\phi)]_0^{\pi/4}\right\}$$

$$= \frac{\sqrt{2}}{2} + \frac{3\sqrt{2}}{8} + \frac{3}{8}\log(\sqrt{2}+1)$$

$$\int_0^{\pi/4} \sec^3\phi\, d\phi = \left[\frac{\sec\phi\tan\phi}{2}\right]_0^{\pi/4} + \frac{1}{2}\int_0^{\pi/4}\sec\phi\, d\phi$$

$$= \frac{\sqrt{2}}{2} + \frac{1}{2}\log(\sqrt{2}+1)$$

and $\int_0^{\pi/4} \sec\phi\, d\phi = \log(\sqrt{2}+1),$

Hence, the required volume from [1] is:

$$= \frac{\sqrt{2}\pi a^3}{3}\left[\frac{7\sqrt{2}}{8} + \frac{3}{8}\log(\sqrt{2}+1) - 2\left\{\frac{\sqrt{2}}{2} + \frac{1}{2}\log(\sqrt{2}+1)\right\} + \log(\sqrt{2}+1)\right]$$

or, $$\boxed{V = \frac{\pi a^3\sqrt{2}}{24}[3\log(\sqrt{2}+1) - \sqrt{2}] \text{ C.C.}}$$

Method: (2)

Given: $r^2 = a^2\cos 2\theta$

We can write $r^4 = a^2 \,.\, r^2 \,.\, (\cos^2\theta - \sin^2\theta)$

Changing to cartesians, the equation becomes

$$(x^2 + y^2)^2 = a^2(x^2 - y^2)$$

$$\Rightarrow \quad y^4 + y^2(2x^2 + a^2) + x^4 - a^2x^2 = 0$$

Solving for y^2, we find :

$$y^2 = \frac{[-(2x^2 + a^2) \pm \sqrt{\{(2x^2+a^2) - 4(x^4 - a^2x^2)\}}]}{2}$$

Neglecting the – ve sign because y^2 cannot be negative, we have

$$y^2 = \frac{-(2x^2 + a^2) + \sqrt{(8a^2x^2 + a^4)}}{2}$$

or, $$y^2 = \frac{-(2x^2 + a^2) + 2\sqrt{2a}\sqrt{\left(x^2 + \frac{1}{8}a^4\right)}}{2}$$

Now, for one loop of the given curve x-varies from 0 to a.

$\therefore$ The required volume $= \pi\int_0^a y^2\, dx$

$$= \frac{\pi}{2}\int_0^a \left[-2x^2 - a^2 + 2\sqrt{2a}\sqrt{\left(x^2 + \frac{1}{8}a^2\right)}\right] dx$$

$$= \frac{\pi}{2}\left[-\frac{2}{3}x^3 - a^2 x + 2\sqrt{2a} \cdot \frac{x}{2}\sqrt{\left(x^2 + \frac{1}{8}a^2\right)}\right.$$

$$\left. + 2\sqrt{2a} \cdot \frac{1}{16}a^2 \log\left\{x + \sqrt{\left(x^2 + \frac{1}{8}a^2\right)}\right\}\right]_0^a$$

$$= \frac{\pi}{2}\left[-\frac{2}{3}a^3 - a^3 + 2\sqrt{2a} \cdot \frac{a}{2} \cdot \frac{3a}{2\sqrt{2}} + \frac{1}{8}\sqrt{2}\,a^3\left\{\log\left(a + \frac{3a}{2\sqrt{2}}\right)\right.\right.$$

$$\left.\left. - \log\frac{a}{2\sqrt{2}}\right\}\right]$$

$$= \frac{\pi}{2}\left[-\frac{5}{2}a^3 + \frac{3}{2}a^3 + \frac{1}{8}\sqrt{2}\,a^3 \log\left\{\frac{a(2\sqrt{2}+3)}{2\sqrt{2}} \cdot \frac{2\sqrt{2}}{a}\right\}\right]$$

$$= \frac{\pi}{2}\left[-\frac{1}{6}a^3 + \frac{1}{8}\sqrt{2}\,a^3 \log\left(2\sqrt{2}+3\right)\right]$$

$$= \frac{\pi}{2}\left[-\frac{1}{6}a^3 + \frac{1}{8}\sqrt{2}\,a^3 \log\left(\sqrt{2}+1\right)^2\right]$$

$$= \frac{\pi a^3}{2}\left[2 \cdot \frac{1}{8}\sqrt{2}\,\log\left(\sqrt{2}+1\right) - \frac{1}{6}\right]$$

$$= \frac{\pi a^3}{2}\left[\frac{1}{4} \cdot \sqrt{2}\,\log\left(\sqrt{2}+1\right) - \frac{1}{6}\right]$$

$$= \frac{\pi a^3}{24}\left[3\sqrt{2}\,\log\left(\sqrt{2}+1\right) - 2\right]$$

$$= \frac{\pi a^3\sqrt{2}}{24}\left[3 \cdot \log\left(\sqrt{2}+1\right) - \sqrt{2}\right]$$

or, $$V = \frac{\pi a^3\sqrt{2}}{24}\left[3\log\left(\sqrt{2}+1\right) - \sqrt{2}\right] \text{ C.C.}$$

140. If the curve $(a-x)\,y^2 = a^2 x$ revolves about its asymptote, find the volume formed.

Solution: **y occurs in even power only so, the equation is symmetrical about x-axis**

PART-C

Writing the equation as

$$y^2 = \frac{a^2 x}{a - x},$$

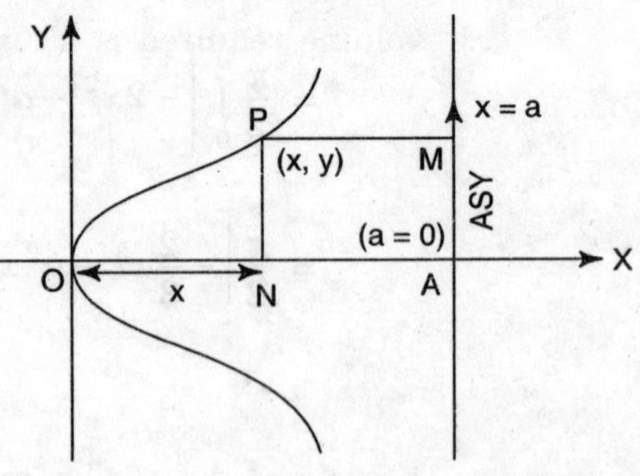

Fig. 3.60

we find:

when, $x = a, y^2 \to \infty$.

$\Rightarrow$ The line $x = a$ is an asymptote. From the figure PM is $\perp r$ to asymptote.

and $PN \perp r\ OX$

$\therefore$ Required volume, $V = 2\int_0^\infty \pi x^2 dy$

$$V = 2\pi \int_0^\infty (PM)^2\, dy \qquad (PM = NA = OA - ON = a - x)$$

Hence, $$V = 2\pi \int_0^\infty (a - x)^2\, dy$$

(But from the given curve, we have

$$x(a^2 + y^2) = ay^2$$

or, $$x = \left(\frac{ay^2}{a^2 + y^2}\right)$$

$\therefore$ $$V = 2\pi \int_0^\infty \left(a - \frac{ay^2}{a^2 + y^2}\right)^2 dy = 2\pi a^6 \int_0^\infty \frac{dy}{(a^2 + y^2)^2}$$

$$= 2\pi a^3 \cdot \int_0^{\pi/2} \cos^2\theta\, d\theta \qquad \text{(on putting } y = a\tan\theta)$$

or, $$V = 2\pi a^3 \cdot \frac{1}{2} \cdot \left(\frac{\pi}{2}\right) = \frac{\pi^2 a^3}{2} \text{ C·units}$$

141. Find the volume of the solid generated by the revolution of $r = 2a\cos\theta$ about the initial line.

Solution: Given: $r = 2a\cos\theta$ [1]

The curve is symmetrical about the initial line, since it is a circle passing through the pole.

For the upper half of the circle, θ varies from 0 to $\frac{\pi}{2}$

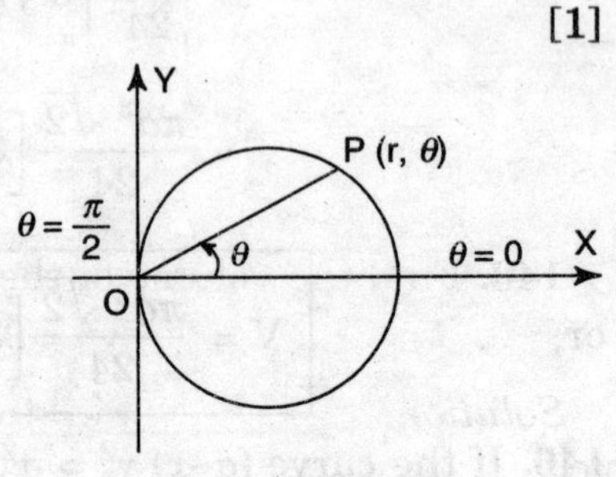

Fig. 3.61

$$\left(\begin{array}{l}\because \text{when } \theta = 0,\ r = 2a, \\ \text{when } \theta = \frac{\pi}{2}, r = 0\end{array}\right)$$

$\therefore$ volume required is V, where

$$V = \int_0^{\pi/2} \frac{2}{3}\pi r^3 \sin\theta\, d\theta = \frac{2}{3}\pi \int_0^{\pi/2} (2a\cos\theta)^3 \sin\theta\, d\theta$$

$$= \frac{16}{3}\pi a^3 \int_0^{\pi/2} \cos^3\theta \sin\theta\, d\theta = -\frac{16}{3}\pi a^3 \left[\frac{\cos^4\theta}{4}\right]_0^{\frac{\pi}{2}}$$

$$= -\frac{4}{3}\cdot \pi a^3(0-1) = \frac{4}{3}\pi a^3 \text{ C}\cdot\text{units}$$

142. Show that the volume of the solid formed by the revolution of the curve $r = a + b\cos\theta$, $(a > b)$ about the initial line is $\frac{4}{3}\pi a(a^2 + b^2)$.

Solution: **Given:** $r = a + b\cos\theta$; $(a > b)$

Since, $\cos(-\theta) = \cos\theta$,

$\Rightarrow$ The curve is symmetrical about the initial line and for the upper half of the curve θ varies from 0 to π,

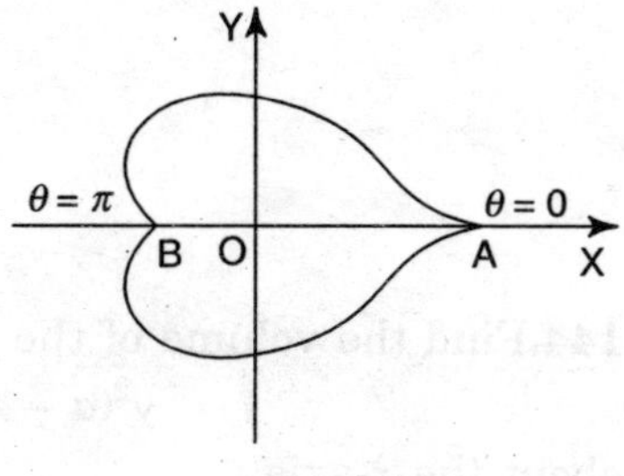

Fig. 3.62

$\therefore$ Volume,
$$V = \int_0^{\pi} \frac{2}{3}\pi r^3 \sin\theta\, d\theta$$

$$= \frac{2}{3}\pi \int_0^{\pi} (a + b\cos\theta)^3 \sin\theta\, d\theta$$

$$= -\frac{2}{3}\frac{\pi}{b} \int_0^{\pi} (a + b\cos\theta)^3 (-b\sin\theta)\, d\theta$$

$$= -\frac{2}{3}\frac{\pi}{b}\left[\frac{(a+b\cos\theta)^4}{4}\right]_0^{\pi} = -\frac{2}{3}\frac{\pi}{b}\left[\frac{(a-b)^4}{4} - \frac{(a+b)^4}{4}\right]$$

$$= \frac{\pi}{6b}\left[(a+b)^4 - (a-b)^4\right]$$

$$= \frac{\pi}{6b}\left[(a+b)^2 + (a-b)^2\right]\left[(a+b)^2 - (a-b)^2\right]$$

$$= \frac{\pi}{6b}\cdot 2(a^2 + b^2)\cdot 4ab = \frac{4}{3}\pi a(a^2 + b^2) \text{ C}\cdot\text{units} \quad [2]$$

143. Find the volume of the solid generated by the revolution of the plane area bounded by $y^2 = 9x$ and $y = 3x$ about the x-axis.

Solution:
$$y^2 = 9x \quad [1]$$
$$y = 3x \quad [2]$$

Solving (1) and (2), we get

$$9x^2 = 9x$$
$$9x^2 - 9x = 0$$
$$\Rightarrow \quad 9x(x-1) = 0$$
$$\therefore \quad x = 0, 1$$
$$x = 0, \text{ from [2] } y = 0$$
$$x = 1, \text{ from [2], } y = 3.$$

$\therefore$ The parabola and the line cut at $O(0, 0)$ and $A(1, 3)$.

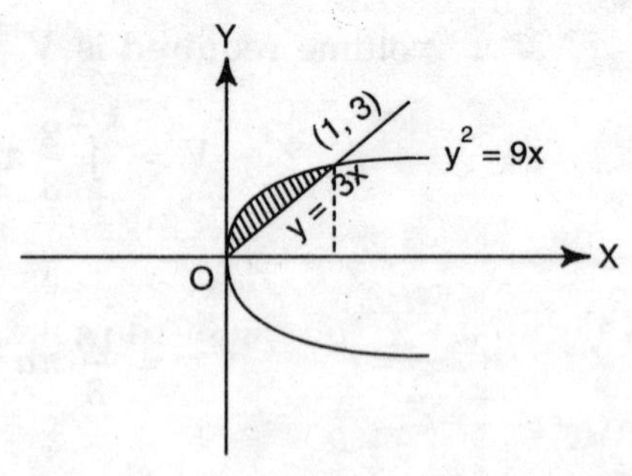

Fig. 3.63

Required volume is V, where

$$V = \int_0^1 \pi\left(y_1^2 - y_2^2\right) dx$$

$$= \int_0^1 \pi\ (9x - 9x^2) dx \qquad (y_1\text{-parabola, } y_2\text{-st. line})$$

$$= \pi\left[\frac{9x^2}{2} - 3x^3\right]_0^1 = \pi\left(\frac{9-6}{2}\right) = \frac{3\pi}{2} \text{ C} \cdot \text{C.}$$

144. Find the volume of the solid formed by revoling the loop of the curve

$$y^2(a - x) = x^2(a + x)$$

about the x-axis.

Solution: y is in even powers only. Therefore the curve is symmetrical about the x-axis For the upper half of the loop, x-varies from $-a$ to 0.

$\therefore$ Required volume is:

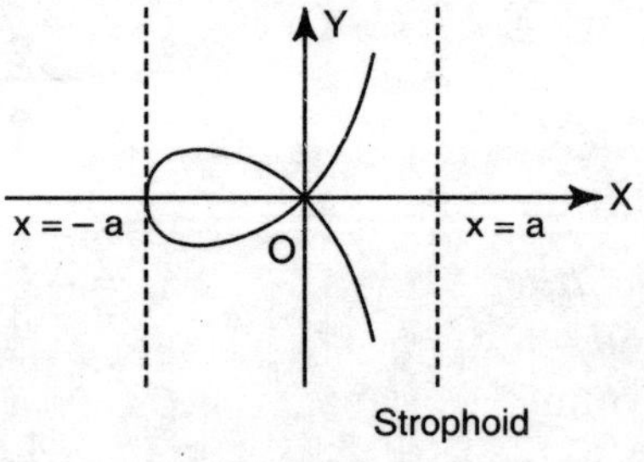

$$V = \int_{-a}^{0} \pi y^2\, dx$$

$$= \pi \int_{-a}^{0} \frac{x^2\,(a+x)}{(a-x)} dx$$

$$= \pi \int_{-a}^{0} \frac{ax^2 + x^3}{a - x} dx$$

$$= \pi \int_{-a}^{0} \left[-x^2 - 2ax - 2a^2 + \frac{2a^3}{a-x}\right] dx$$

$$= \pi\left[-\frac{x^3}{3} - 2a\frac{x^2}{2} - 2a^2x - 2a^3 \log(a - x)\right]_{-a}^{0}$$

$$= \pi\left[-2a^3 \log a - \left(\frac{a^3}{3} - a^3 + 2a^3 \log 2a\right)\right]$$

Strophoid

Fig. 3.64

$$= \pi \left[2a^3 \ (\log 2a - \log a) - \frac{4}{3} a^3 \right]$$

$$= 2\pi a^3 \left(\log 2 - \frac{2}{3} \right) \text{ C·C.}$$

145. **Find the volume of the solid formed when the area between the curve**

$$y = \frac{x}{(1 + x^2)}$$ **and the x-axis ($x > 0$) is revolved about the x-axis.**

Solution: The curve stretches from $x = 0$ to $x = \infty$ on the positive side of the axis. The volume generated is therefore equal to V, where $V = \pi \int y^2 \, dx$

$$= \pi \int [x^2/(1 + x^2)^2] \, dx$$

(from $x = 0$, to $x = \infty$)

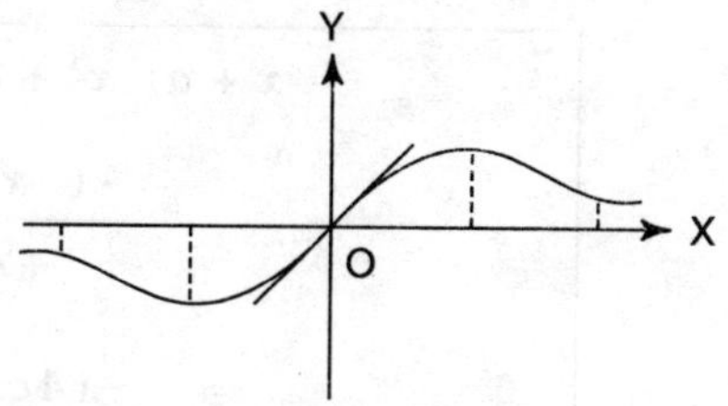

Fig. 3.65

put

$$x = \tan \theta;$$
$$dx = \sec^2 \theta \, d\theta;$$
$$1 + x^2 = \sec^2 \theta$$

($x = 0$, gives $\theta = 0$)

($x = \infty$ gives $\tan \theta = \infty$, $\theta = \pi/2$)

$$\therefore \quad V = \pi \int_0^{\infty} \frac{x^2 \, dx}{(1 + x^2)^2} = \pi \int_0^{\pi/2} \frac{\tan^2 \theta \cdot \sec^2 \theta \, d\theta}{(\sec^2 \theta)^2}$$

$$= \pi \int_0^{\pi/2} \sin^2 \theta \, d\theta = (1/2)(\pi/2)\pi = \frac{\pi^2}{4}.$$

146. **Find the volume generated by rotating a parabola lying between its vertex and latus rectum about x-axis**

Solution: **Given:** $y^2 = 4ax$

LSL' is the latus rectum at the focus s (a, 0) and A (0, 0) is the Vertex. The curve is symmetrical about x-axis, as it contains even powers of y only.

$\therefore$ Required volume

$$V = \int_0^a \pi y^2 \, dx$$

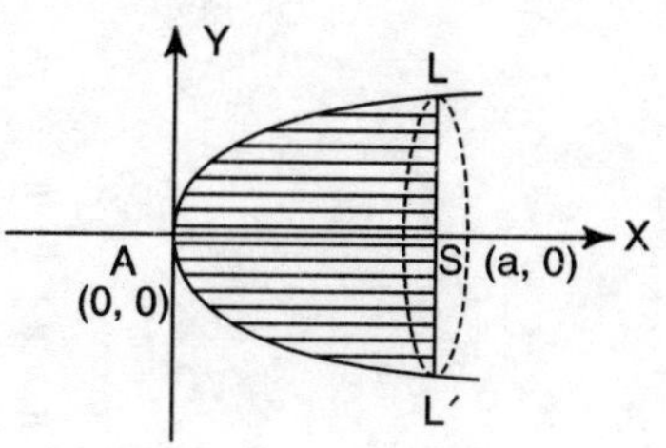

Fig. 3.66

$$= \pi \int_0^a 4ax \, dx$$

(as x-axis varies between $x = 0$ and $x = a$)

$$= 4a\pi \left[\frac{x^2}{2} \right]_0^a$$

$$= 2a\pi (a^2 - 0) = 2\pi a^3$$

PART-C

147. Find the volume generated by revolving the loop of the curve $y^2 (a + x) = x^2 (3a - x)$ for which $0 \le x \le 3a$, about the x-axis.

Solution: Since x-varies from 0 to $3a$

The required volume is V,

where $$V = \int_0^{3a} \pi y^2 \, dx = \pi \int_0^{3a} \frac{x^2 (3a - x)}{a + x} dx$$

$$= \pi \int_0^{3a} \left[\frac{-x^3 + 3ax^2}{(x + a)} \right] dx \quad [1]$$

Actual division in an Elegant form.

$$x + a)-x^3 + 3ax^2 \quad (-x^2 + 4ax - 4a^2$$

$$\frac{-(-x^3 - ax^2)}{+4ax^2}$$

$$\frac{-(4ax^2 + 4a^2 x)}{-4a^2 x}$$

$$\frac{-(-4a^2 x - 4a^3)}{+4a^3}$$

$\therefore$ Quotient $= -x^2 + 4ax - 4a^2$

Remainder $= 4a^3$

$$\therefore \quad \frac{-x^3 + 3ax^2}{x + a} = -x^2 + 4ax - 4a^2 + \frac{4a^3}{x + a}$$

Eqn. [1] becomes:

$$A = \pi \int_0^{3a} \left[-x^2 + 4ax - 4a^2 + \frac{4a^3}{x + a} \right] dx$$

$$= \pi \left[-\frac{x^3}{3} + 2ax^2 - 4a^2 x + 4a^3 x + 4a^3 \log (x + a) \right]_0^{3a}$$

$$= \pi \{[-9a^3 + 18a^3 - 12a^3 + 4a^3 \log 4a] - [4a^3 \log a]\}$$

$$= \pi a^3 [4 \log 4 - 3]$$

$$(\because \quad 4a^3 \log 4a = 4a^3 \log 4 + 4a^3 \log a)$$

148. Find the volume and surface area generated by revolving the cardioide $r = a (1 + \cos \theta)$ about the initial line (VTU, March, Aug., 2000 ; VTU, F/M, 2005)

Solution: **Given:** $r = a (1 + \cos \theta)$

when $\theta = 0, r = 0$

when $\theta = \pi, r = 2a$

Limits of θ are 0 to π

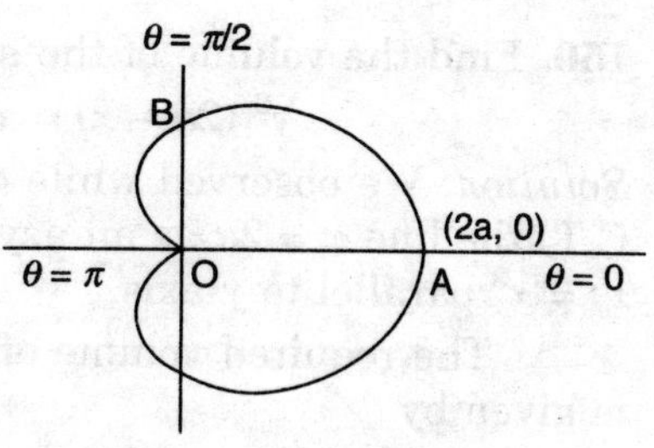

Fig. 3.67 *Cardioide*

$$\therefore \quad \text{Volume:} \quad V = \int_0^{\pi} \frac{2}{3} \pi r^3 \sin\theta \, d\theta$$

$$= \int_0^{\pi} \frac{2}{3} \pi a^3 (1 + \cos\theta)^3 \cdot \sin\theta \, d\theta$$

$$= \frac{2}{3} \pi a^3 \left[-\frac{(1+\cos\theta)^4}{4} \right]_0^{\pi}$$

$$= -\frac{1}{6} \pi a^3 (0 - 16) = \frac{8}{3} \pi a^3$$

For surface area:
(Refer solved example on S.A)

and we have $\quad S = \dfrac{32}{5} \pi a^2$ Sq. units

149. Prove that the volume of the solid generated by the revolution of the curve $y = \dfrac{a^3}{a^2 + x^2}$ about its asymptote is $\dfrac{\pi^2 a^3}{2}$ (VTU, August, 1999)

Solution: The given curve is symmetrical about y-axis and cutting in at the point $(0, a)$ tangent at which is parallel to the axis of x, the asymptote is given by equating the coefft. of HDT of x to zero.

$\Rightarrow \quad y = 0$ (*i.e.*, x-axis itself)

Volume: $\quad V = 2v_1$

Limits: $\quad x = 0$ to $x = \infty$

(in 1st Quadrant)

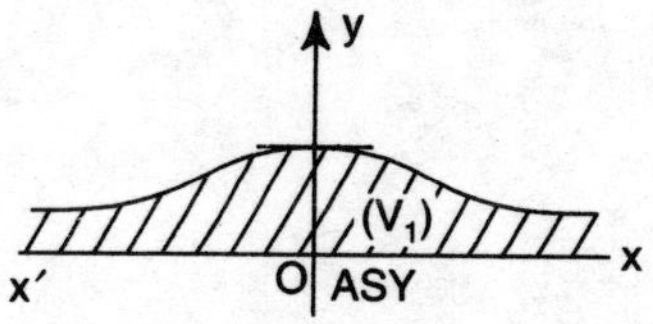

Fig. 3.68

$$\therefore \quad V = 2 \int_0^{\infty} \pi y^2 \, dx$$

$$= 2\pi \int_0^{\infty} \frac{a^6}{(a^2 + x^2)^2} \, dx$$

put $x = a \tan\theta$; and θ varies from 0 to $\dfrac{\pi}{2}$

$$\therefore \quad V = 2\pi a^6 \int_0^{\pi} \frac{a \sec^2\theta \, d\theta}{(a^2 + a^2 \tan^2\theta)^2} = 2\pi a^3 \int_0^{\pi/2} \cos^2\theta \, d\theta$$

$$= 2\pi a^3 \left(\frac{\pi}{4}\right) = \frac{\pi^2 a^3}{2} \text{ c. units.}$$

150. Find the volume of the solid obtained by rotating the CISSOID: $y^2(2a - x) = x^3$ about its asymptote. (VTU, March, 2001)

Solution: We observed while curve tracing that (See example on C.T) the line $x = 2a$ as an asymptote to the given CISSOID: $y^2(2a - x) = x^3$ parallel to y-axis.

The required volume of resolution about y-axis (its asymptote) is given by

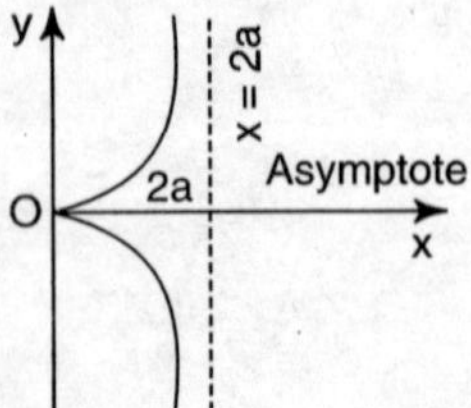

Fig. 3.69 *Cissoid*

$$V = 2v, = 2\int \pi x^2\, dy;$$

where
$$y = \frac{(2a - x)^{3/2}}{\sqrt{x}}$$

$$\left(\begin{array}{ll} \text{put } x = 2a\sin^2\theta; & y = \dfrac{(2a - 2a\sin^2\theta)^{3/2}}{\sqrt{2a\sin^2\theta}} = \dfrac{2a\cos^3\theta}{\sin\theta} \\ x = 0;\ \theta = 0; & \\ x = 2a;\ \theta = \dfrac{\pi}{2};\quad \therefore & dy = \dfrac{-2a\cos^2\theta\,(3\sin^2\theta + \cos^2\theta)}{\sin^2\theta}\, d\theta \end{array}\right)$$

$$\therefore \quad V = -2\int_0^{\frac{\pi}{2}} (\pi)(4a^2\sin^4\theta)\,\frac{(2a\cos^2\theta\,[3\sin^2\theta + \cos^2\theta])}{\sin^2\theta}\, d\theta$$

$$= -16\pi a^3 \int_0^{\frac{\pi}{2}} (3\cos^2\theta\sin^4\theta + \sin^2\theta\cos^4\theta)\, d\theta$$

$$\left[\text{Since } \int_0^{\frac{\pi}{2}} \sin^2\theta\cos^4\theta\, d\theta = \int_0^{\frac{\pi}{2}} \cos^2\theta\sin^4\theta\, d\theta = \frac{13\cdot 3\cdot 1}{6\cdot 4\cdot 2}\cdot\frac{\pi}{2}\cdot\frac{\pi}{32}.\right]$$

$$\therefore \quad |V| = 16\pi a^3\left[3\frac{(\pi)}{32} + \frac{(\pi)}{32}\right] = 2\pi^2 a^3 \quad \text{(taking numerical value)}$$

or,
$$V = 2\pi^2 a^3 \text{ C.C.}$$

151. Find the volume of the solid obtained by revolving the ellipse $\dfrac{x^2}{a^2} + \dfrac{y^2}{b^2} = 1$ about the x-axis.

Solution: The required volume is given by

$$V = \pi \int_a^b y^2 dx$$

$$\Rightarrow \qquad V = \pi \int_{-a}^{a} \left(\frac{b^2}{a^2}\right)(a^2 - x^2)\, dx$$

$$= \frac{2\pi b^2}{a^2} \int_0^a (a^2 - x^2)\, dx = \frac{2\pi b^2}{a^2}\left[a^2 x - \frac{x^3}{3}\right]_0^a$$

$$= \frac{4}{3} \pi a b^2 \text{ C.C.}$$

152. Find the volume bounded by the curves $y = xe^x$, $x = 1$ and $y = 0$ revolving about the x-axis.

Solution:

Volume:
$$V = \int_0^1 \pi y^2 dx = \pi \int_0^1 x^2 e^{2x}\, dx$$

$$= \pi \left[\left\{(x^2)\left(\frac{e^{2x}}{2}\right)\right\} - \frac{1}{2} x e^{2x} + \frac{1}{4} e^{2x}\right]_0^1 = \frac{\pi}{4}\,[e^2 - 1] \text{ C.C.}$$

(by parts)

EXERCISES

1. Find the volume of the solid generated by the revolution of the curve $y = \sqrt{1 + x^2}$ between $x = 0$ and $x = 4$ about the x-axis. $\left(\textit{Ans.}\ \frac{76\pi}{3}\right)$

2. Find the volume generated by revolving the loops of the following curve about the x-axis $y^2 = x^2 (1 - x^2)$. $\left(\textit{Ans.}\ \frac{\pi^2}{4}\right)$

3. Show that the volume of revolution of

(*i*) The portion of the parabola $y^2 = 4ax$ from $x = a$ to $x = 2a$ about x-axis is $6\pi a^3$.

(*ii*) The loop of the curve $2ay^2 = x(x - a)^2$ about x-axis is $\frac{\pi a^3}{24}$.

(*iii*) The area enclosed by the parabola $y^2 = 4ax$ and its latus rectum about y-axis is $\frac{16\pi a^3}{5}$.

4. Find the volume of the solid of revolution obtained by rotating the area included between the curves $y^2 = x^3$ and $x^2 = y^3$ about the x-axis $\left(\textit{Ans.}\ \frac{5}{28}\pi\right)$

PART-C

5. Find the volume of the solid generated by revolving the cycloid

$$x = a\,(\theta + \sin\theta),$$
$$y = a\,(1 + \cos\theta) \text{ about its base.}$$

(*Ans.* $5\pi^2 a^3$)

6. Find the volume when the loop of the curve $y^2 = x\,(2x - 1)^2$ revolves about the x-axis

$\left(Ans.\ \dfrac{\pi}{48}\right)$

7. Show that the volume of revolution, about the x-axis, generated by the parabola $y^2 = 4ax$ from $x = 0$ to $x = ah$ is $2\,\pi\, ah^2$.

8. Show that the volume generated by revolving the curve $y = x \sin x$,

$0 \le x \le \cdot\, 2\pi$, about the x-axis is $\dfrac{\pi^2\,(8\pi^2 - 3)}{6}$.

PART-D

4 Differential Equations

4.1 SOLUTIONS OF FIRST ORDER AND FIRST DEGREE EQUATIONS

An equation of the form

$$F\left[x, y, \frac{dy}{dx}\right] = 0 \qquad (1)$$

in which x is the independent variable and y is the dependent variable is called differential equation of the first order and first degree.

In the equation of the first order and first degree, the derivative $\left(\frac{dy}{dx}\right)$ occurs in the first degree in $F\left[x, y, \frac{dy}{dx}\right] = 0$.

So, it can be separated out, and written.

$$f_2(x, y)\left(\frac{dy}{dx}\right) + f_1(x, y) = 0;$$

or $\qquad f_1(x, y)\, dx + f_2(x, y)\, dy = 0.$

More usually, $f_1(x, y)$ is put equal to $P(x, y)$ or P

and $\qquad f_2(x, y) = Q(x, y) = Q.$

Thus, an equation of First order and First degree is written in the standard form:

$$P\, dx + Q dy = 0$$

If there exists a relation

$$\phi(x, y, c) = 0 \qquad (2)$$

where c is an arb. constant, which statisfies (1) then the relation (2) is called the general solution of the differential equation (1). For a particular value of c, (2) is called a particular solution

Standard methods of solving the differential equation (1), that is;

$$F\left[x, y, \frac{dy}{dx}\right] = 0$$

4.2 VARIABLES SEPARABLE

(i) Solution by the method of separation of variables

If $P = f(x)$ = a function of x only, and $Q = \phi(y)$ = a function of y only, we have the equation $Pdx + Qdy = 0$,

$$f(x)\,dx + \phi(y)\,dy = 0 \tag{1}$$

Evidently, the variables are separated, and the solution is got by direct integration.

Solution:

$$\int f(x)\,dx + \int \phi(y)\,dy = \text{(const.)}.$$

e.g., consider $\cos x\,dx + \sec^2 y\,dy = 0$. Here, the variables are separated.

So, integrating,

$$\int \cos x\,dx + \int \sec^2 y dy = \text{(const.)}$$

i.e.,
$$\sin x + \tan y = \text{(const.)}$$

A. Examples (Variables separable: 1–41)

1. Solve: $$y - x\frac{dy}{dx} = y^2 + \frac{dy}{dx}$$ (VTU, March, 2001)

Solution:

Given:
$$y - x\frac{dy}{dx} = y^2 + \frac{dy}{dx}$$

or
$$(1 + x)\frac{dy}{dx} = y\,(1 - y)$$

i.e.,
$$\frac{dy}{y\,(1-y)} = \frac{dx}{1+x}$$

on integrating,
$$\int\frac{dy}{y(1-y)} = \int\frac{dx}{1+x} + c$$

i.e.,
$$\int\left\{\frac{1}{y} + \frac{1}{1-y}\right\} dy = \log\,(1 + x) + c$$

i.e.,
$$\log y - \log\,(1 - y) = \log\,(1 + x) + \log c_1$$

i.e.,
$$\frac{y}{1-y} = c_1(1 + x)$$

($c = \log c_1$) is the requd. general solution.

2. Solve: $$e^y\left[\frac{dy}{dx} - 1\right] = e^x$$ (VTU, Aug. 2000)

Solution:

We can write the given equation:

$$\frac{dy}{dx} - 1 = e^{x-y}$$

Put $x - y = u$

$$\Rightarrow \quad 1 - \frac{dy}{dx} = \frac{du}{dx} \text{ (on differentiation)}$$

or $$\frac{dy}{dx} - 1 = -\frac{du}{dx}$$

Hence G.E. becomes: $-\dfrac{du}{dx} = e^u$

i.e., $$-e^{-u} \cdot du = dx$$

on integrating, we get

$$e^{-u} = x + c$$

i.e., $$e^{-x+y} = x + c$$

is the solution

3. Solve: $$\frac{dy}{dx} = \frac{x(2\log x + 1)}{\sin y + y\cos y}$$ (VTU, March, 2000)

Solution: $$(\sin y + y\cos y)\,dy = x(2\log x + 1)\,dx$$

Integrating, $\int \sin y\,dy + \int y\cos y\,dy = \int x(2\log x + 1)\,dx$ + const.

i.e., $$-\cos y + y\sin y - \int \sin y dy = (2\log x + 1)\frac{x^2}{2}$$

$$-\int \frac{x^2}{2}\left(\frac{2}{x} + 0\right)dx + \text{const}$$

i.e., $$-\cos y + y\sin y + \cos y$$

$$= \frac{x^2}{2}(2\log x + 1) - \frac{x^2}{2} + c$$

i.e., $$y\sin y = x^2\log x + c$$

is the requd. solution.

4. Solve: $$\frac{dy}{dx} = \cos(x + y + 1)$$ [VTU, Aug. 1999]

Solution:

(Put $x + y + 1 = z$)

$$\therefore \quad 1 + \frac{dy}{dx} = \frac{dz}{dx} \quad \therefore \quad \frac{dy}{dx} = \left(\frac{dz}{dx} - 1\right)$$

$\therefore$ Given equation becomes

$$\left(\frac{dz}{dx} - 1\right) = \cos z$$

or $$\frac{dz}{dx} = 1 + \cos z \quad \text{or} \quad \frac{dz}{1+\cos z} = dx$$

Method (i) from this step:

or $$\int dx + c = \int \frac{dz}{1+\cos z}$$

$$(x + c) = \int \frac{dz}{2\cos^2 z/2} = \frac{1}{2}\int \sec^2 \frac{z}{2}\, dz$$

$$= \frac{1}{2}\,\frac{\tan z/2}{1/2} = \tan\frac{z}{2} \quad (x + c) = \tan\frac{(x+y+1)}{2}$$

or

Method (ii)

We have $$\frac{dz}{1+\cos z} = dx$$

or $$\frac{(1-\cos z)\,dz}{1-\cos^2 z} = dx$$

$\Rightarrow$ $$\frac{(1-\cos z)\,dz}{\sin^2 z} = dx$$

or $[\text{cosec}^2 z - \text{cosec}\, z \cdot \cot z]\, dz = dx$

Integrating, we get,

$$\int \text{cosec}^2 z\, dz - \int \text{cosec}\, z \cdot \cot z\, dz = \int dx$$

or $$-\cot z + \text{cosec}\, z = x + c$$

or $$-\cot(x + y + 1) + \text{cosec}\,(x + y + z) = x + c$$

$\because$ $$\int \text{cosec}^2 x dx = -\cot x$$
$$\int \text{cosec}\, x \cot x\, dx = -\text{cosec}\, x$$

or $\text{cosec}\,(x + y + 1) - \cot(x + y + 1) = x + c$ is the required solution

5. Solve: $\dfrac{dy}{dx} = \sin(x + y)$ [B.U. Jan. 1993]

Solution: put $x + y = u$

$\Rightarrow$ $$1 + \frac{dy}{dx} = \frac{du}{dx}$$

$\therefore$ The given equation becomes

$$\frac{du}{dx} - 1 = \sin u$$

$$\Rightarrow \qquad \frac{du}{dx} = 1 + \sin u$$

$$\Rightarrow \qquad \frac{du}{1+\sin u} = dx$$

$$\Rightarrow \qquad \int \frac{du}{1+\sin u} = \int dx + c$$

$$\Rightarrow \qquad \int \frac{1-\sin u}{1-\sin^2 u}\, du = x + c$$

$$\Rightarrow \qquad \int \frac{1-\sin u}{\cos^2 u}\, du = x + c$$

$$\Rightarrow \qquad \int \sec^2 u du - \int \tan u \sec u\, du = x + c$$

$$\Rightarrow \qquad \tan u - \sec u = x + c$$

or $$\tan(x+y)\, u - \sec(x+y) = x + c$$

is the solution.

6. Solve: $\dfrac{dy}{dx} = xe^{y-x^2}$ [VTU, July, 2002]

Solution: The given equation can be written as:

$$\frac{dy}{dx} = x \cdot e^{y} \cdot e^{-x^2}$$

or $$e^{-y}\, dy = xe^{-x^2}\, dx$$

(Variables are separated)

So, the solution is got by direct Integration.

i.e., $$\int e^{-y}\, dy = \int xe^{-x^2}\, dx + c_1$$

i.e., $$-e^{-y} = \left[\int xe^{-x^2}\, dx\right] + c_1$$

Put $x^2 = z$; $$xdx = \frac{dz}{2}$$

$$\therefore \qquad -e^{-y} = \int e^{-x^2}\, (xdx) + c_1 = \int e^{-z} \cdot \frac{dz}{2} + c_1$$

or $$-e^{-y} = \frac{1}{2}\int e^{-z}\, dz + c_1$$

or $$-e^{-y} = \frac{1}{2}(-e^{-z}) + c_1$$

or $$\frac{e^{-z}}{2} - e^{-y} = c_1$$

or $$\frac{e^{-x^2}}{2} - e^{-y} = c_1$$

or $$e^{-x^2} - 2e^{-y} = 2c_1 = \text{const.}$$

or $$e^{-x^2} - 2e^{-y} = c \qquad (2c_1 = c)$$

is the required solution

7. Solve: $e^y\left(\frac{dy}{dx}+1\right) = e^x$ [VTU July/Aug., 2002]

Solution: The given equation can be written as:

$$\frac{dy}{dx} + 1 = e^{x-y}$$

Put $x + y = z$; $(y = z - x)$

$$\therefore \qquad 1 + \frac{dy}{dx} = \frac{dz}{dx}$$

$$\therefore \qquad \frac{dz}{dx} = e^{x-z+x} = e^{2x-z} = \frac{e^{2x}}{e^z}$$

or $$e^z dz = e^{2x} dx$$

We have $$e^z dz = e^{2x}\, dx$$

(V.S. Form)

$$\therefore \qquad \int e^z dz = \int e^{2x}\, dx + k$$

or $$e^z = \frac{1}{2} e^{2x} + k$$

or $$e^{x+y} = \frac{1}{2} e^{2x} + k$$

or $$\boxed{e^y = \frac{1}{2} e^x + ke^{-x}}$$

is the solution.

8. Solve: $x^4 \frac{dy}{dx} + x^3 y + \text{cosec}\,(xy) = 0$ [B.U. July, 1993]

Solution: The given equation can be written as

$$x^3\left[x\frac{dy}{dx} + y\right] + \text{cosec}\,(xy) = 0$$

Put $xy = v$

$$\therefore \qquad x\frac{dy}{dx} + y = \frac{dv}{dx}$$

$\therefore$ G.E. becomes $x^3 \dfrac{dv}{dx} + \text{cosec } v = 0;$

Separating the variables and Integrating,

$$x^3 \frac{dv}{dx} = -\text{ cosec } v$$

or
$$\frac{dv}{\text{cosec } v} = -\frac{dx}{x^3}$$

$$\int \sin v\, dv + \int x^{-3}\, dx = c - \cos v + \frac{x^{-2}}{-2} = c$$

or
$$x^{-2} + 2\cos(xy) = k, \ (k = 2c)$$

is the reqd. solution.

9. Solve: $\dfrac{dy}{dx} = \dfrac{1}{\cos(x+y)}$ [B.U. MQP-I]

Solution: Put $x + y = v$

$$\therefore \qquad 1 + \frac{dy}{dx} = \frac{dv}{dx}$$

$\therefore$ the equation becomes

$$\frac{dy}{dx} - 1 = \frac{1}{\cos v}$$

$$\Rightarrow \qquad \frac{dv}{dx} = \frac{1}{\cos v} + 1$$

or
$$\frac{dv}{dx} = \frac{1+\cos v}{\cos v}$$

or
$$\frac{\cos v}{1+\cos v}\, dv = dx$$

Integrating, we get,

$$\int \frac{\cos v}{1+\cos v}\, dv = \int dx + c$$

$$\int \frac{\cos^2 v/2 - \sin^2 v/2}{2\cos^2 v/2}\, dv = (x + c)$$

$$\Rightarrow \qquad \frac{1}{2}\int dv - \frac{1}{2}\int \tan^2 \frac{v}{2}\, dv = x + c$$

PART-D

$$\Rightarrow \quad \frac{1}{2}v - \frac{1}{2}\int\left(\sec^2\frac{v}{2} - 1\right)dv = x + c$$

$$\Rightarrow \quad \frac{1}{2}v - \tan\frac{v}{2} + \frac{1}{2}v = (x + c)$$

or
$$v - \tan\frac{v}{2} = x + c$$

or
$$(x + y) - \tan\left(\frac{x + y}{2}\right) = x + c$$

or
$$y - \tan\left(\frac{x + y}{2}\right) = c$$

is the solution

10. Solve: $\dfrac{dy}{dx} = \sqrt{y - x}$ [M.U. 1983]

Solution: Put $y - x = z^2$

$$\therefore \quad \frac{dy}{dx} - 1 = 2z\frac{dz}{dx}$$

$\therefore$ G. E. becomes $\quad 2z\dfrac{dz}{dx} + 1 = z$

$$2z\frac{dz}{dx} = z - 1$$

or
$$\frac{zdz}{z - 1} = \frac{dx}{z} \quad \text{or} \quad \frac{z - 1 + 1}{z - 1}\,dz = \frac{dx}{2}$$

or
$$\left(1 + \frac{1}{z - 1}\right)dz = \frac{dx}{2} \quad \text{or} \quad dz + \frac{dz}{z - 1} = \frac{dx}{2}$$

Integrating, we get

$$z + \log(z - 1) = \frac{1}{2}x + k$$

or
$$\sqrt{y - x} + \log\left(\sqrt{y - x} - 1\right) = \frac{1}{2}x + k$$

is the solution

11. Solve: $(x + 2y)(dx - dy) = (dx + dy)$ [VTU, Sept., 2000]

Solution: The given equation can be written as:

$$\frac{dy}{dx} = \frac{x - 2y - 1}{x - 2y + 1} \tag{1}$$

put $x - 2y = z$

$$\therefore \qquad 1 - 2\frac{dy}{dx} = \frac{dz}{dx}$$

or
$$\frac{dy}{dx} = \frac{1}{2}\left[1 - \frac{dz}{dx}\right]$$

$\therefore$ (1) becomes
$$\frac{1}{2}\left[1 - \frac{dz}{dx}\right] = \frac{z-1}{z+1}$$

i.e.,
$$\frac{dz}{dx} = 1 - \frac{2(z-1)}{z+1} = \frac{-z+3}{z+1}$$

or
$$\left[\frac{z+1}{z-3}\right] dz = -dx$$

or
$$\left[1 + \frac{4}{z-3}\right] dz = -dx$$

on integrating,

$$z + 4\log(z-3) = -x + c$$

i.e., $\quad x - 2y + 4\log(x - 2y - 3) = -x + c$

i.e., $\quad 2(x - y) + 4\log(x - 2y - 3) = c$

is the requd solution

12. Solve: $\dfrac{dy}{dx} = \left(\dfrac{y}{x}\right) + \tan\left(\dfrac{y}{x}\right)$

Solution: Put $z = \dfrac{y}{x}$

$$\Rightarrow \qquad xz = y$$

$$\therefore \qquad x\frac{dz}{dx} + z = \frac{dy}{dx}$$

$\therefore$ G.E. becomes
$$x\left(\frac{dz}{dx}\right) + z = \tan z + z$$

$$\Rightarrow \qquad x\left(\frac{dz}{dx}\right) = \tan z = \frac{1}{\cot z}$$

$$\Rightarrow \qquad \frac{dx}{x} = \cot z \, dz$$

(Variable separables form)

$\therefore$ Solution is got by direct integration.

$$\therefore \qquad \log \sin z = \log x + \log c$$

Hence $$\sin z = cx$$

$\therefore$ $$\sin\left(\frac{y}{x}\right) = cx$$

13. Solve: $\dfrac{dy}{dx} + \dfrac{y^2+y+1}{x^2+x+1} = 0$ [B.E. 1975]

Solution: $$\frac{dy}{y^2+y+1} + \frac{dx}{x^2+x+1} = 0$$

i.e., $$\frac{dy}{\left(y+\frac{1}{2}\right)^2 + \left(\frac{\sqrt{3}}{2}\right)^2} + \frac{dx}{\left(x+\frac{1}{2}\right)^2 + \left(\frac{\sqrt{3}}{2}\right)^2} = 0$$

$\therefore$ $$\int \frac{dy}{\left(y+\frac{1}{2}\right)^2 + \left(\frac{\sqrt{3}}{2}\right)^2} + \int \frac{dx}{\left(x+\frac{1}{2}\right)^2 + \left(\frac{\sqrt{3}}{2}\right)^2} = c$$

$$\frac{1}{\left(\frac{\sqrt{3}}{2}\right)} \tan^{-1} \frac{\left(y+\frac{1}{2}\right)}{\left(\frac{\sqrt{3}}{2}\right)} + \frac{2}{\sqrt{3}} \tan^{-1}\left(\frac{2x+1}{\sqrt{3}}\right) = c$$

$$\frac{2}{\sqrt{3}} \tan^{-1}\left(\frac{2y+1}{\sqrt{3}}\right) + \frac{2}{\sqrt{3}} \tan^{-1}\left(\frac{2x+1}{\sqrt{3}}\right) = c$$

Solve: $xdy + ydx + 4\sqrt{(1-x^2y^2)}\, dx = 0$ [B.E. 1985]

Solution: We can write the G.E as

$$d(xy) + 4\sqrt{(1-x^2y^2)}\, dx = 0$$

i.e., $$\frac{d(xy)}{\sqrt{(1-x^2y^2)}} + 4dx = 0$$

Integrating, $$\int \frac{d(xy)}{\sqrt{(1-x^2y^2)}} + 4\int dx = c$$

$\sin^{-1}(xy) + 4x = c$ is the solution.

14. Solve: $(x+y)^2 \dfrac{dy}{dx} = a^2$

Solution: put $(x + y) = z$

$$\therefore \quad 1 + \frac{dy}{dx} = \frac{dz}{dx} \text{ i.e., } \frac{dy}{dx} = \frac{dz}{dx} - 1$$

$\therefore$ The G.E becomes

$$z^2 \left(\frac{dz}{dx} - 1\right) = a^2 \quad \frac{dz}{dx} - 1 = \frac{a^2}{z^2}$$

$$\frac{dz}{dx} = \frac{a^2}{z^2} + 1 \quad \frac{dz}{dx} = \frac{a^2 + z^2}{z^2}$$

$$\frac{z^2}{(a^2 + z^2)} dz = dx$$

Integrating, $$\int \frac{z^2}{a^2 + z^2} dz = x + c$$

i.e., $$\int \left(\frac{z^2 + a^2 - a^2}{z^2 + a^2}\right) - a^2 dz = x + c$$

$$\int \left(1 - \frac{a^2}{z^2 + a^2}\right) dz = x + c$$

$$\Rightarrow \quad z - a^2 \cdot \frac{1}{a} \cdot \tan^{-1}\left(\frac{z}{a}\right) = x + c$$

or $$(x + y) - a \tan^{-1} \frac{(x + y)}{a} = x + c$$

i.e., $$y - a \tan^{-1} \frac{(x + y)}{a} = c$$

is the solution

15. Solve: $[\cos x \tan y + \cos (x + y)]\, dx + [\sin x \sec^2 y + \cos (x + y)\, dy] = 0$

[VTU March-1999]

Solution: We can write the G.E. as

$$\cos x.\ \tan y\, dx + \cos (x + y)\, dx + \sin x \sec^2 y\, dy + \cos (x + y)\, dy = 0$$

$$\cos x \tan y dx + \sin x \sec^2 y\, dy + \cos (x + y)\, (dx + dy) = 0$$

$$d\, (\sin x \tan y) + d\, (\sin (x + y)) = 0$$

Integrating, we get $$\int d\, (\sin x \tan y) + \int d\, (\sin (x + y) = c$$

$\sin \tan y + \sin (x + y) = c$ is the solution

PART-D

16. Solve: $(x + y)^2 \left[x \dfrac{dy}{dx} + y\right] = xy\left[1 + \dfrac{dy}{dx}\right]$ [VTU, Aug. 1999]

Solution: Put $x + y = u;\ xy = v$

$$1 + \frac{dy}{dx} = \frac{du}{dx};\ x\frac{dy}{dx} + y = \frac{dv}{dx}$$

$\therefore$ G.E. becomes: $(u)^2 \dfrac{dv}{dx} = v\dfrac{du}{dx}$ or $\dfrac{dv}{v} = \dfrac{du}{u^2}$

$$\Rightarrow \quad \log v = -u^{-1} + c$$

$$\Rightarrow \quad \log v = \frac{-1}{u} + c$$

or $$\log (xy) = \frac{-1}{(x+y)} + c$$

is the solution

17. Solve: $\sqrt{(1+x^2)(1+y^2)}\ dx + xydy = 0$

$$\left(\div x\sqrt{1+y^2}\right)$$

$$\therefore \quad \frac{\sqrt{(1+x^2)}}{x}dx + \frac{y}{\sqrt{(1+y^2)}}dy = 0$$

$$\therefore \quad \int \frac{\sqrt{(1+x^2)}}{x}dx + \int \frac{y}{\sqrt{(1+y^2)}}dy = \text{const.}$$

$$\left(\sqrt{1+x^2}\right) - \sin h^{-1}\left(\frac{1}{x}\right) + \sqrt{(1+y^2)} = c$$

$$\therefore \quad \int \frac{\sqrt{1+x^2}}{x}dx = \int \frac{1+x^2}{x\sqrt{1+x^2}}dx$$

$$= \int \frac{dx}{x\sqrt{1+x^2}} + \int \frac{x}{\sqrt{1+x^2}}dx$$

$$= \int \frac{-\frac{1}{t^2}dt}{\frac{1}{t}\sqrt{1+\frac{1}{t^2}}} + \frac{1}{2}\int (1+x^2)^{-1/2} \cdot 2xdx \qquad \left(t = \frac{1}{x}\right)$$

$$= -\int \frac{dt}{\sqrt{t^2+1}} + \frac{1}{2} \cdot \frac{(1+x^2)^{1/2}}{\frac{1}{2}}$$

$$= -\sin h^{-1}(t) + \sqrt{1+x^2}.$$

18. Solve $\frac{dy}{dx} = x \tan(y - x) + 1$ [V.T.U. F/M 2005]

Solution. Given $\frac{dy}{dx} = x \tan(y - x) + 1$

Put $u = y - x \Rightarrow \frac{du}{dx} = \frac{dy}{dx} - 1$

or $$\frac{du}{dx} + 1 = \frac{dy}{dx}$$

$\therefore$ G.E. becomes $$\frac{du}{dx} + 1 = \tan u + 1$$

or $$\frac{du}{dx} = x \tan u \quad \text{or} \quad \frac{du}{\tan u} = x dx$$

$\therefore$ $$\int \cot u \, du - \int x \, dx = \text{constant.}$$

$\Rightarrow$ $$\log \sin u - \frac{x^2}{2} = c$$

$\therefore$ $$\boxed{\log \sin (y - x) - \frac{x^2}{2} = c},$$

is the complete solution.

19. Solve : $(1 + x)\frac{dy}{dx} + 1 = 2e^{-y}$

Solution: $$(1 + x)\frac{dy}{dx} = (2e^{-y} - 1)$$

$\therefore$ $$\frac{dy}{(2e^{-y} - 1)} = \frac{dx}{(x+1)}$$

or $$\frac{e^y}{(2 - e^y)} = \frac{dx}{(x+1)}$$

or $$\frac{e^y}{(e^y - 2)} dy + \frac{dx}{(x+1)} = 0 \qquad \left(\because \int \frac{f'(y)}{f(y)} dy = \log f(y)\right)$$

PART-D

V.S. Form, integrate

$$\log(e^y - 2) + \log(x + 1) = \log c$$

$$\therefore \quad (e^y - 2)(x + 1) = c$$

20. Solve: $3e^x \tan y \cdot dx + (1 + e^x)\sec^2 y \cdot dy = 0$

given $y = \frac{\pi}{4}$ when $x = 0$

Solution: The given equation is

$$\frac{3e^x}{1+e^x}dx + \frac{\sec^2 y}{\tan y}dy = 0 \qquad \text{(V.S. Form, Integrate)}$$

$$3\log(1 + e^x) + \log \tan y = \log c$$

$$\therefore \quad (1 + e^x)^3 \tan y = c$$

for $\quad x = 0, y = \frac{\pi}{4}$

$$\therefore \quad (1 + 1)^3 \tan\left(\frac{\pi}{4}\right) = c$$

$$\therefore \quad c = 8$$

$\therefore$ Particular solution is

$$(1 + ex)^3 \cdot \tan y = 8$$

21. Find the curve which satisfies the differential equation

$$(y^2 + 1)\, xdx + (x^2 + 1)\, ydy = 0$$

and passes through the origin.

Solution:

Given: $\quad (y^2 + 1)\, xdx + (x^2 + 1)\, ydy = 0 \qquad [1]$

$\left[\div[(x^2+1)(y^2+1)]\right]$ we get

$$\frac{x}{(x^2+1)}dx + \frac{y}{(y^2+1)}dy = 0 \qquad [2]$$

$$\therefore \quad \int\frac{x}{(x^2+1)}dx + \int\frac{y}{(y^2+1)}dy = \left(\frac{c}{2}\right)$$

$$\left(\frac{1}{2}\right)\log(x^2 + 1) + \left(\frac{1}{2}\right)\log(y^2 + 1) = \left(\frac{c}{2}\right)$$

$$[\log(x^2 + 1) + \log(y^2 + 1)] = c = \log(k)$$

$$\log[(x^2 + 1)(y^2 + 1)] = \log(k)$$

$$\therefore \quad (x^2 + 1)(y^2 + 1) = k \qquad [3]$$

Since the curve passes through the origin (0, 0)

Hence, $\quad (0 + 1)(0 + 1) = k;$

$\Rightarrow \quad k = 1$

$$\therefore \quad (x^2 + 1)(y^2 + 1) = 1$$

or $$x^2y^2 + (x^2 + y^2) = 0 \text{ is the solution.}$$ [4]

22. $\sec^2 x \cdot \tan y \, dx + \sec^2 y \cdot \tan x \cdot dy = 0.$

Solution: $(\div \tan x \cdot \tan y)$, we get

$$\frac{\sec^2 x}{\tan x} + \frac{\sec^2 y}{\tan y} = 0$$

$$\therefore \quad \int \frac{\sec^2 x}{\tan x} dx + \int \frac{\sec^2 y}{\tan y} dy = \text{const.}$$

i.e., $$\log \tan x + \log \tan y = \text{const.}$$

i.e., $$\log (\tan x \cdot \tan y) = \text{const.}$$

$$\therefore \quad \tan x \cdot \tan y = \text{const.}$$

23. Solve: $(e^x + 1)\, y dy = (y + 1)\, e^x dx$

Solution:

Given: $$(e^x + 1)\, y dy - (y + 1)\, e^x dx = 0$$ [1]

$[\div (e^x + 1)(y + 1)]$

$$\therefore \quad \frac{(e^x + 1)\, y dy}{(e^x + 1)(y + 1)} - \frac{(y + 1)\, e^x \, dx}{(e^x + 1)(y + 1)} = 0$$

$$\frac{y dy}{(y + 1)} - \frac{e^x \, dx}{(1 + e^x)} = 0$$

or $$\frac{y dy}{(y + 1)} - \frac{dt}{(t + 1)} = 0$$

$$\left[1 - \frac{1}{(1 + y)}\right] dy - \frac{1}{(t + 1)} dt = 0$$

So, $$\int \left[1 - \frac{1}{(1 + y)}\right] dy - \int \frac{dt}{(t + 1)} = c$$ [2]

i.e., $$y - \log (1 + y) - \log (1 + t) = (\text{const.})$$

or $$y - \log [(1 + y)(1 + t)] = \text{const} = c$$

or $$y - \log [(1 + y)(1 + e^x)] = c$$ [3]

24. Solve: $\left(\dfrac{dy}{dx}\right) = e^{x-y} + x^2 e^{-y}$

Solution: Multiply be e^y; we get

$$e^y dy = e^x dx + x^2 dx$$ [V.S. Form]

$$\therefore \quad \int e^y dy = \int e^x dx + \int x^2 dx + \text{const.}$$

i.e., $$e^y = e^x + (x^3/3) + \text{const.}$$

$$\therefore \quad x^3 + 3\,(e^x - e^y) = \text{const.}$$

25. $\frac{dy}{dx} = e^{x+y} + x^2e^y$ (A Similar problem)

Solution:
$$\frac{dy}{dx} = e^x \cdot e^y + x^2e^y = e^y(e^x + x^2)$$

$$\therefore \quad e^{-y} \cdot \frac{dy}{dx} = (x^2 + e^x)$$

or
$$e^{-y}dy = (x^2 + e^x)\,dx$$

$$\therefore \quad (-e^{-y}) = [(x^3/3) + e^x] + c$$

is the required solution.

26. Solve: $(x+y)^2\left[x\frac{dx}{dy} + y\right] = xy\left[1 + \frac{dy}{dx}\right]$ [VTU, August, 1999]

Solution:

Put $(x + y) = u$; $xy = v$

$$\Rightarrow \quad 1 + \frac{dy}{dx} = \frac{du}{dx};\ x\frac{dy}{dx} + y = \frac{dv}{dx}$$

$$(u^2)\frac{dv}{dx} = v\frac{du}{dx} \quad \text{or} \quad \frac{dv}{v} = \frac{du}{u^2}$$

$$\Rightarrow \quad \log(xy) = \frac{-1}{x+y} + c$$

27. $(x + y + 1)\frac{dy}{dx} = 1.$

Solution: put $x + y + 1 = z;$

Then
$$1 + \frac{dy}{dx} = \frac{dz}{dx}$$

$\therefore$ Given equation becomes

$$z\left(\frac{dz}{dx} - 1\right) = 1$$

or
$$\frac{dz}{dx} - 1 = \frac{1}{z} \quad \text{or} \quad \frac{dz}{dx} = 1 + \frac{1}{z} = \frac{z+1}{z}$$

Separating the variables

$$\frac{z}{1+z}\,dz = dx \text{ or, } \left(1 - \frac{1}{z+1}\right)dz = dx$$

Integrating, we get

$$z\text{-}\log(z + 1) = x + c,$$

or $\quad (x + y + 1) - \log (x + y + 1 + 1) = x + c_1$

or, $\quad y - \log (x + y + 2) = c_1 - 1$

or, $\quad y - \log (x + y + 2) = c$ is the requd. solution

28. $\sin^{-1}\left(\dfrac{dy}{dx}\right) = x + y$

$$\Rightarrow \quad \frac{dy}{dx} = \sin (x + y)$$

Put $x + y = z$; then

$$1 + \frac{dy}{dx} = \frac{dz}{dx}$$

$\therefore$ Given eqn. becomes $\quad \dfrac{dz}{dx} - 1 = \sin z$

or $$\frac{dz}{dx} = 1 + \sin z;$$

$$\Rightarrow \quad \frac{dz}{1+\sin z} = dx$$

Integrate $$\int \frac{dz}{1+\sin z} = x + c;$$

or $$\int \frac{1-\sin z}{1-\sin^2 z}\, dz = x + c$$

or $$\int \frac{1-\sin z}{\cos^2 z}\, dz = x + c;$$

$$\int (\sec^2 z - \sec z \tan z)\, dz = x + c$$

or $\quad \tan z - \sec z = x + c;\quad$ where $z = x + y$

29. Solve: $\dfrac{dy}{dx} = 1 + 6xe^{x-y}$ [B.U. Jan. 1993]

Solution: put $\quad x - y = z; \quad \therefore \quad \dfrac{dy}{dx} = \left(1 - \dfrac{dz}{dx}\right)$

G.E. becomes $$\left(1 - \frac{dz}{dx}\right) = 1 + 6 \cdot x \cdot e^z$$

$$\therefore \quad \frac{dz}{dx} = -\,6 \cdot x \cdot e^z$$

PART-D

Separating the variables

$$6x \cdot dx = -\frac{dz}{e^z}$$

Integrating $\quad 6\int x dx = -\int e^{-z}\, dz + \text{const.}$

i.e., $$6 \cdot \frac{x^2}{2} = \frac{e^{-z}}{-1} + c$$

or $$3x^2 = e^{-z} + c$$

i.e., $$3x^2 = e^{-(x-y)} + c$$

30. $\dfrac{x\,dx + y\,dy}{x\,dy - y\,dx} = \sqrt{\left(\dfrac{a^2 - x^2 - y^2}{x^2 + y^2}\right)}$

Solution: Put $x = r\cos\theta$;

$$\frac{dx}{d\theta} = \frac{dr}{d\theta}\sin\theta + r\cos\theta,$$

$$\frac{dy}{dx} = \frac{dy}{d\theta} \div \frac{dx}{d\theta}$$

Also $\quad x^2 + y^2 = r^2$

The equation reduces to

$$\frac{r\cos\theta\left(\dfrac{dr}{d\theta}\cdot\cos\theta - r\sin\theta\right) + r\sin\theta\left(\dfrac{dr}{d\theta}\sin\theta + r\cos\theta\right)}{r\cos\theta\left(\dfrac{dr}{d\theta}\sin\theta + r\cos\theta\right) - r\sin\theta\left(\dfrac{dr}{d\theta}\cdot\cos\theta - r\sin\theta\right)}$$

$$= \sqrt{\left(\frac{a^2 - r^2}{r^2}\right)} \qquad \frac{r\dfrac{dr}{d\theta}(\cos^2\theta + \sin^2\theta)}{r^2(\cos^2\theta + \sin^2\theta)} = \frac{\sqrt{(a^2 - r^2)}}{r}$$

or $$\frac{dr}{d\theta} = \sqrt{(a^2 - r^2)}$$

or $$\frac{dr}{\sqrt{(a^2 - r^2)}} = d\theta$$

or $$\sin^{-1}\left(\frac{r}{a}\right) = \theta + c$$

$\therefore$ $$r = a\sin(\theta + c)$$

where $$r = \sqrt{(x^2 + y^2)} \quad \text{and} \quad \theta = \tan^{-1}\left(\frac{y}{x}\right).$$

31. $\left(\frac{x+y-a}{x+y-b}\right)\frac{dy}{dx} = \frac{x+y+a}{x+y+b}$

Solution: Put $(x + y) = v$;

$$\therefore \quad 1 + \frac{dy}{dx} = \frac{dv}{dx}$$

$$\frac{dv}{dx} - 1 = \frac{(v+a)(v-b)}{(v-a)(v+b)}$$

or
$$\frac{dv}{dx} = \frac{(v+a)(v-b)}{(v-a)(v+b)} + 1 = 2\frac{(v^2-ab)}{v^2+(b-a)v-ab}$$

$$\therefore \quad \frac{v^2+(b-a)v-ab}{v^2-ab}\,dv = 2dx$$

or
$$\left[1+\frac{(b-a)v}{v^2-ab}\right]dv = 2dx,$$

Integrating, we get

$$v + \frac{(b-a)}{2}\int\frac{2v}{v^2-ab}\,dv = 2x + c$$

or
$$v + \frac{1}{2}(b-a)\log(v^2-ab) = 2x + c$$

$$x + y + \frac{1}{2}(b-a)\log[(x+y)^2-ab] = 2x + c$$

$$\frac{1}{2}(b-a)\log[(x+y)^2-ab] = x - y + c$$

32. Solve: $(xy^2 + x)\,dx + y\,(x^2 + 1)\,dy = 0.$

Solution: Since, $(xy^2 + x) = x\,(y^2 + 1)$

$\therefore$ The G.E. becomes:

$$x\,(y^2 + 1)\,dx + y\,(x^2 + 1)\,dy = 0$$

(V.S. Form)

$[\div (y^2 + 1)(x^2 + 1)]$, we get

$$\therefore \quad \frac{x\,dx}{(x^2+1)} + y\frac{dy}{(y^2+1)} = 0$$

$$\therefore \quad \int\frac{2xdx}{(x^2+1)} + \int\left(\frac{2ydy}{(y^2+1)}\right) = 2\log c$$

$$\log (x^2 + 1) + \log (y^2 + 1) = 2 \log c$$

$$\log [(x^2 + 1)(y^2 + 1)] = \log c^2$$

$$\therefore \quad (x^2 + 1)(y^2 + 1) = c^2;$$

33. Solve: $\dfrac{dy}{dx} = \dfrac{\sin x + x \cos x}{y(2 \log y + 1)}$

Solution: Separating the variables,

$$(\sin x + x \cos x)\, dx = y(2 \log y + 1)\, dy$$

$$\Rightarrow \quad \int (\sin x + x \cos x)\, dx = \int (2 \log y + 1)\, y dy + c$$

$$\int x \cos x dx = (x)(\sin x) - \int 1 \cdot \sin x\, dx = x \sin x + \cos x$$

$$\int (2 \log y + 1)\, y\, dy = (2 \log y + 1)\frac{y^2}{2} - \int 2 \cdot \frac{1}{2} \cdot \frac{y^2}{2}\, dy \text{ by parts}$$

$$= y^2 \log y + \frac{y^2}{2} - \frac{y^2}{2} = y^2 \log y$$

$\therefore$ the solution

$$-\cos x + x \sin x + \cos x = y^2 \log y + c$$

34. Solve: $\dfrac{dy}{dx} = \sin (x + y) + \cos (x + y)$

Solution: Let $x + y = u$

$$\Rightarrow \quad \frac{dy}{dx} = \left(\frac{du}{dx} - 1\right)$$

$\therefore$ the given equation reduces to

$$\frac{du}{dx} = 1 + \sin u + \cos u$$

$$\Rightarrow \quad \frac{du}{dx} = 2 \sin \frac{u}{2} \cos \frac{u}{2} + 2 \cos^2 \frac{u}{2}$$

$$\Rightarrow \quad \frac{du}{dx} = 2 \cos^2 \frac{u}{2}\left\{\tan \frac{u}{2} + 1\right\}$$

Separating the variables, and integrating,

$$\int \frac{du}{2 \cos^2 \frac{u}{2}\left[\tan \frac{u}{2} + 1\right]} = \int dx$$

$$\Rightarrow \quad \log \left[1 + \tan \frac{(x + y)}{2}\right] = x + c$$

35.
$$(x^2 - yx^2)\frac{dy}{dx} + (y^2 + y^2 \cdot x) = 0$$

$$x^2(1 - y)\frac{dy}{dx} + y^2(1 + x) = 0$$

$$x^2(1 - y)\frac{dy}{dx} + y^2(1 + x) = 0$$

or
$$\left(\frac{1-y}{y^2}\right)dy + \left(\frac{1+x}{x^2}\right)dx = 0$$

or
$$\left(\frac{1}{y^2} - \frac{1}{y}\right)dy + \left(\frac{1}{x^2} + \frac{1}{x}\right)dx = 0$$

Integrate
$$\log x - \log y - \left(\frac{1}{x} + \frac{1}{y}\right) = c = \log k$$

$$\log\frac{x}{ky} = \frac{x+y}{xy}$$

36. $\cot x \cot y\, dx + dy = 0$

$(\div \cot y)$

$$\cot x + \left(\frac{1}{\cot y}\right)dy = 0$$

$$\cot x\, dx + \tan y dy = 0$$

$$\int \cot x\, dx + \int \tan y\, dy = c$$

$$\log(\sin x) + \log(\sec y) = c$$

$$\log \sin x + \log \sec y = (\text{const.})$$

37. $y f(xy)\, dx + x\phi(xy)\, dy = 0$ (A standard form of equation)

Solution: Since (xy) occurs as a single unit

Put $xy = u$; $\therefore$ $y = \frac{u}{x}$,

$$dy = \frac{x\,du - u\,dx}{x^2}$$

Given:
$$yf(xy)\, dx + x\,\phi(xy)\, dy = 0$$

$\therefore$
$$\frac{u}{x} f(u)\, dx + x\phi(u)\frac{(x\,du - u\,dx)}{x^2} = 0$$

$$uf(u)\, dx + x\phi(u)\, dx - x\phi(u)\, dx = 0$$

i.e.,
$$u\,[f(u) - \phi(u)]\, dx + x\phi(u)\, du = 0$$

PART-D

$$\therefore \quad \frac{dx}{x} + \frac{\phi(u)\,du}{u[f(u)-\phi(u)]} = 0$$

(V.S. Form)

$$\therefore \quad \int \frac{dx}{x} + \int \frac{\phi(u)}{u[f(u)-\phi(u)]}\,du = \text{(const.)}$$

38. Solve: $(1 - xy + x^2y^2)\,dx + (x^3y - x^2)\,dy = 0$

Solution: Multiply by y

$$[1 - (xy) + (xy)^2]\,ydx + (x^3y^2 - x^2y)\,dy = 0$$
$$[1 - (xy) + (xy)^2]\,ydx + x\,[x^2y^2 - xy]\,dy = 0$$
$$[1 - (xy) + (xy)^2]\,ydx + x\,[(xy)^2 - (xy)]\,dy = 0$$

This is the std form.

put $xy = u;\ y = \dfrac{u}{x},\qquad dy = \dfrac{(x\,du - u\,dx)}{x^2}$

$$\therefore \quad (1 - u + u^2)\,\frac{u}{x}\,dx + x\,(u^2 - u)\,\frac{(x\,du - u\,dx)}{x^2} = 0$$

i.e.,
$$(1 - u + u^2)\,udx + u\,(u - 1)\,(xdu - udx) = 0$$

or
$$dx\,(1 - u + u^2) + (u - 1)\,xdu - (u - 1)\,udx = 0$$
$$dx\,[1 - u + u^2 - u^2 + u] + (u - 1)\,xdu = 0$$
$$dx\,(1) + (u - 1)\,xdu = 0$$
$$dx\,(1) + (u - 1)\,xdu = 0$$
$$\left(\frac{dx}{x}\right) + (u - 1)\,du = 0$$

$$\therefore \quad \int \frac{dx}{x} + \int (u-1)\,du = c$$

$$\log x + \left[\frac{(u-1)^2}{2}\right] = c$$

or,
$$2\log x + (xy - 1)^2 = \text{(const.)}$$

39. $\left(2 + 2x^2\sqrt{y}\right)ydx + \left(x^3\sqrt{y} + 2x\right)dy = 0$

Solution: Rearranging, we get

$$2ydx + 2x^2y^{3/2}\,dx + x^3y^{1/2}\,dy + 2xdy = 0$$
$$2(ydx + xdy) + [2x^2y^{3/2}\,dx + x^3\,y^{1/2}\,dy] = 0$$
$$2(ydx + xdy) + (2/3)\,[3x^2dxy^{3/2} + x^3\,(3/2)y^{1/2}\,dy] = 0$$
$$2(ydx + xdy) + (2/3)\,[y^{3/2}\,d\,(x^3) + x^3d\,(y^{3/2}) = 0$$
$$2d\,(xy) + (2/3)\,d\,[x^3y^{3/2}] = 0$$

$$\therefore \quad 2xy + (2/3)\,(x^3y^{3/2}) = \text{const.} = (2k/3)\ \text{(say)}$$

$$\therefore \quad 3xy + x^3y^{3/2} = k$$

40. $\dfrac{dy}{dx} = \dfrac{2}{x+2y-3}$

Solution: Put $x + 2y - 3 = z$, then

$$1 + 2 \cdot \frac{dy}{dx} = \frac{dz}{dx} \qquad \left(\because \frac{dy}{dx} = \frac{1}{2}\left(\frac{dz}{dx} - 1\right)\right)$$

$\therefore$ G.E. becomes $\quad \dfrac{1}{2}\left(\dfrac{dz}{dx} - 1\right) = \dfrac{2}{z}$

or $\quad \dfrac{dz}{dx} - 1 = \dfrac{4}{z};$

or $\quad \dfrac{dz}{dx} = 1 + \dfrac{4}{z} = \dfrac{z+4}{z}$

(V.S. Form) $\quad \dfrac{z}{z+4}\, dz = dx$

or $\quad \left(1 - \dfrac{4}{z+4}\right) dz = dx$

Integrating, both sides,

$$z - 4 \log (z + 4) = x + c_1$$

or $\quad (x + 2y - 3) - 4 \log (x + 2y - 3 + 4) = x + 1,$

or $\quad 2y - 4 \log (x + 2y + 1) = c_1 + 3$

or $\quad 2y - 4 \log (x + 2y + 1) = 2c$

or $\quad y - 2 \log (x + 2y + 1) = c$

is the reqd. eqn.

41. $\dfrac{dy}{dx} = \dfrac{1}{x-y} + 1$

Solution: Put $x - y = z$

then $\quad 1 - \dfrac{dy}{dx} = \dfrac{dz}{dx}$ or $\dfrac{dy}{dx} = 1 - \dfrac{dz}{dx}$

$\therefore$ G.E. becomes

$$1 - \frac{dz}{dx} = \frac{1}{z} + 1 \quad \text{or} \quad -z dx = dx$$

Integrating both sides,

$$-\frac{z^2}{2} = x + c_1$$

or, $\quad 2x + (x - y)^2 + 2c_1 = 0.$

PART-D

EXERCISES

1. Solve the following Differential equations

(1) $(1 + x^2)\, ydx + (y^2 - 1)\, xdy = 0$

(2) $(1 + y^2)\, dx - xy\, (1 + x^2)\, dy = 0$

(3) $x\sqrt{(1-y^2)}\, dx + \sqrt{(x^2+1)}\, dx = 0$

(4) $y\,(x + 1)\, dy + x\,(y - 1)\, dx = 0$

(5) $\dfrac{xy \log x}{\log y} = \dfrac{dy}{dx} = 1$

(6) $\dfrac{x\,(1 + 2\log x)}{(\sin y + y\cos y)}$

(7) $\dfrac{dy}{dx} = \dfrac{\cos x - x\sin x}{y\,[2\log y + 1]}$

(8) $\dfrac{dy}{dx} + \tan x \tan y = 0$

(9) $e^x \cot y dx + (e^x - 1)\, \text{cosec}^2 y\, dy = 0$

(10) $y^2 \sec^2 \sqrt{x} - 2\sqrt{x}\, e^{1/y} \left(\dfrac{dy}{dx}\right) = 0$

4.3 HOMOGENEOUS AND NON-HOMOGENEOUS DIFFERENTIAL EQUATIONS

A function $f(x, y)$ is said to be a homogeneous function of degree n if it can be expressed as $x^n \phi\left(\dfrac{y}{x}\right)$ where $\phi\left(\dfrac{y}{x}\right)$ is a function of $\left(\dfrac{y}{x}\right)$.

The equation
$$\frac{dy}{dx} = \frac{f(x, y)}{g(x, y)} \qquad [1]$$

is said to be a Homogeneous differential equation if both $f(x, y)$ and $g(x, y)$ are homogeneous functions of the same degree.

$$\frac{dy}{dx} = \frac{x^n\, \phi_1\left(\dfrac{y}{x}\right)}{x^n\, \phi_2\left(\dfrac{y}{x}\right)} = \frac{\phi_1\left(\dfrac{y}{x}\right)}{\phi_2\left(\dfrac{y}{x}\right)}$$

or
$$\frac{dy}{dx} = F\left(\frac{y}{x}\right) \qquad [2]$$

Working Rule

This equation can be changed into a differential equation with separable variable by means of the substitution $y = vx$

Then
$$\frac{dy}{dx} = v + x\frac{dv}{dx}$$

$\therefore$ (2) becomes

$$v + x\frac{dv}{dx} = F(v)$$

$$\therefore \quad x\frac{dv}{dx} = F(v) - v$$

separating the variables, we get

$$\int \frac{dv}{F(v)-v} = \int \frac{dx}{x} + c$$

In this solution putting $v = \frac{y}{x}$, we get the solution of the given equaiton.

42. Solve:

$$\left(x \tan \frac{y}{x} - y \sec^2 \frac{y}{x}\right) dx + x \sec^2 \frac{y}{x}\, dy = 0$$ [VTU, March-1999]

Solution:

The given equation be written as

$$\frac{dy}{dx} = \frac{\left(\frac{y}{x}\right)\sec^2\left(\frac{y}{x}\right) - \tan\left(\frac{y}{x}\right)}{\sec^2\left(\frac{y}{x}\right)}$$ [1]

Clearly *RH* is a fucntion of $\left(\frac{y}{x}\right)$.

This a Homo. D.E.

$\therefore$ Put $y = vx$, or $\left(\frac{y}{x}\right) = v$ is (1)

we get $$v + x\frac{dv}{dx} = \frac{v \sec^2 v - \tan v}{\sec^2 v} = v - \frac{\tan v}{\sec^2 v}$$

i.e., $$x\frac{dv}{dx} = -\frac{\tan v}{\sec^2 v}$$

i.e., $$\frac{\sec^2 v}{\tan v}\, dv = -\frac{dx}{x}$$

(This is V.S. Form)

Hence, on integrating we get

$$\log \tan v = -\log x + \log c$$

i.e., $$x \tan v = c$$

or $$x \tan\left(\frac{y}{x}\right) = c$$

is the C.S.

PART-D

43. Solve: $\frac{dy}{dx} = \frac{y}{x} + \sin\frac{y}{x}$ [VTU, Aug., 2000]

Solution: Put $\frac{y}{x} = v$ or $y = vx$ so that

$$\frac{dy}{dx} = v + x\frac{dv}{dx}$$

Given equation reduces to

$$v + x\frac{dv}{dx} = v + \sin v$$

i.e.,

$$x\frac{dv}{dx} = \sin v$$

or

$$\text{cosec } v\, dv = \frac{dx}{x}$$

on integrating we get

$$\log\tan\left(\frac{v}{2}\right) = \log x + \log c$$

i.e.,

$$\tan\frac{v}{2x} = cx, \text{ where } v = \left(\frac{y}{x}\right)$$

is the complete solution.

44. Solve: $(x + y)\,dx + (y - x)\,dy = 0$ [VTU, August, 2001]

Solution:

Given equation:

$$(x + y)\,dx + (y - x)\,dy = 0 \quad [1]$$

Put $y = px; (x + px)\,dx + (px - x)\,(pdx + xdp) = 0$

$$x\,[(1 + p)\,dx + (p - 1)]\,(pdx + xdp) = 0$$

$$dx\,(1 + p + p^2 - p) + (p - 1)\,x\,dp = 0$$

$$dx\,(1 + p^2) + (p - 1)\,xdp = 0$$

$$\frac{dx}{x} + \frac{(p-1)}{(1+p^2)}\,dp = 0 \quad [2]$$

$\therefore$

$$\int\frac{dx}{x} + \int\frac{(p-1)}{(p^2+1)}\,dp = c;$$

$$\log x + \left[\int\left\{\frac{p}{(p^2+1)} - \frac{1}{(p^2+1)}\right\}dp\right]$$

or

$$\log x + \frac{1}{2}\log(p^2 + 1) - \tan^{-1}(p) = \log k$$

or $$\log x^2 + \log (p^2 + 1) - 2 \tan^{-1} (p)$$

or $$= 2 \log k = c$$

or $$\log [x^2 (p^2 + 1)] - 2 \tan^{-1} (p) = c$$

or $$\tan^{-1} \left(\frac{y}{x}\right) - \frac{1}{2} \log \left[(x^2) \left(\frac{y^2}{x^2} + 1 \right) \right] = c$$

or $$\tan^{-1} \left(\frac{y}{x}\right) - \frac{1}{2} \log (x^2 + y^2) = c$$

45. Solve: $x^2 y dx - (x^3 + y^3) dy = 0$ (VTU, March 2000]

Solution:

Given equation: $x^2 y dx = (x^3 + y^3) dy$

Put $y = vx$ and $dy = (vdx + xdv)$

we get, $$x^2 \cdot vx \cdot dx = (x^3 + v^3 x^3) \cdot (vdx + xdv)$$

$$x^3 \cdot [vdx - (1 + v^3) (vdx + x \cdot dv)] = 0$$

i.e., $$dx (v - v - v^4) - (1 + v^3) \cdot x \cdot dv = 0$$

$\therefore$ $$\frac{dx}{x} + \left[(v^3 + 1)/v^4 \right] dv = 0$$

i.e., $$\frac{dx}{x} + \left[\frac{1}{v} + \frac{1}{v^4} \right] dv = 0$$

i.e., $$\int \left(\frac{dx}{x} \right) + \int \left[\frac{1}{v} + \frac{1}{v^4} \right] dv = \text{const.}$$

i.e., $$\log x + \log v - \frac{1}{3v^3} = c;$$

Put $v = \dfrac{y}{x}$ $\log vx - \dfrac{x^3}{3y^3} = \text{const};$

$$\log y - \frac{x^3}{3y^3} = \text{const.}$$

$$\log y - \log a = \frac{x^3}{3y^3};$$

(Putting $c = \log a$)

$$\log \left(\frac{y}{a} \right) = \frac{x^3}{3y^3};$$

$\therefore$ $$y = ae^{x^3/3y^3}$$

PART-D

46. (a similar problem)

Solve: $$xy^2dy = (x^3 + y^3)\, dx$$

Put $y = vx$; $dy = vdx + xdv$

and compare with the above Example (45)

x and y are interchanged

$$x \cdot v^2x^2\,(vdx + xdv) = (x^3 + v^3x^3)\, dx$$

or $$x^3 \cdot v^2\,(vdx + xdv) = x^3\,(1 + v^3)\, dx$$

or $$x^3 \cdot v^3dx + x^4 \cdot v^2\, dv = x^3\, dx + v^3x^3dx$$

or $$v^2\, dv = dx\left(\frac{x^3}{x^4}\right) = \frac{dx}{x}$$

or $$\frac{dx}{x} = v^2dv$$

$\therefore$ $$\int \frac{dx}{x} - \int v^2dv = c$$

$$\log x - \frac{v^3}{3} = c \log x - \frac{y^3}{3x^3} = c$$

or $$\log x - \log a = \frac{y^3}{3x^3}$$

or $$\log\left(\frac{x}{a}\right) = \frac{y^3}{3x^3}$$

or $$\frac{x}{a} = e^{y^3/3x^3}$$

or $$x = ae^{y^3/3x^3}$$

47. Solve: $(x^2 + 2xy - y^2)\, dx + (y^2 + 2xy - x^2)\, dy = 0$

Solution: $$\frac{dy}{dx} = \frac{y^2 - 2xy - x^2}{y^2 + 2xy - x^2}$$

Put $y = vx$; $\dfrac{dy}{dx} = v + x\dfrac{dv}{dx}$

$\therefore$ $$v + x\frac{dv}{dx} = \frac{v^2x^2 - 2x\,(vx) - x^2}{v^2x^2 + 2x\,(vx) - x^2} = \frac{v^2 - 2v - 1}{v^2 + 2v - 1}$$

$\therefore$ $$x\frac{dv}{dx} = \frac{v^2 - 2v - 1}{v^2 + 2v - 1} - v$$

$$= \frac{+v^2 - 2v - 1 - v^3 - 2v^2 + v}{v^2 + 2v - 1} = \frac{-v^3 - v^2 - v - 1}{v^2 + 2v - 1}$$

Separating the variables,

we get $$\frac{(v^2 + 2v - 1)}{v^3 + v^2 + v + 1} dv = -\frac{dx}{x}$$

i.e., $$\frac{(v^2 + 2v - 1)}{(v + 1)(v^2 + 1)} dv = -\frac{dx}{x}$$

Integrating,

$$\int \frac{v^2 + 2v - 1}{(v + 1)(v^2 + 1)} dv = -\int \frac{dx}{x} + \text{const.}$$

i.e., $$\int \frac{dv}{v + 1} - \int \frac{2v}{v^2 + 1} dv = \int \frac{dx}{x} + \text{const.}$$

i.e., $$\log (v + 1) - \log (v^2 + 1) = \log x + \log c$$

i.e., $$\log \frac{v + 1}{(v^2 + 1)} = \log cx$$

$$\therefore \quad \frac{v + 1}{v^2 + 1} = cx \qquad \frac{\frac{y}{x} + 1}{\frac{y^2}{x^2} + 1} = cx$$

i.e., $$\frac{y + x}{x} \cdot \frac{x^2}{y^2 + x^2} = cx$$

$(y + x) = c\,(x^2 - y^2)$ is the required solution.

48. Solve: $(1 + e^{x/y})\, dx + e^{x/y}\left(1 - \frac{x}{y}\right) dy = 0$ [VTU, March, 2001]

Solution:

Given equation: $$\frac{dx}{dy} = -\frac{e^{x/y}(1 - x/y)}{1 + e^{x/y}} \qquad [1]$$

since RHS is only a function of (x/y), given equation is homogeneous, substituting $x = vy$;

so that $$\frac{dx}{dy} = v + y\frac{dv}{dy}$$

Then (1), reduces to

$$v + y\frac{dv}{dy} = -\frac{e^v(1 - v)}{1 + e^v}$$

i.e., $$y\frac{dv}{dy} = -\frac{e^v(1-v)}{1+e^v} - v$$

$$y\frac{dv}{dy} = -\frac{v+e^v}{1+e^v}$$

i.e., $$\frac{1+e^v}{v+e^v}\,dv = -\frac{dy}{y}$$

On integrating, we get

$$\log(v+e^v) = -\log y + \log c$$

or $$y(v+e^v) = c$$

since $v = x/y$, this yields

$$x + ye^{x/y} = c$$

is the required complete solution.

49. Solve: $(x^2+y^2)\,dx = 2xy\,dy$ [B.U. Feb., 1983]

Solution: $$\frac{dy}{dx} = \frac{x^2+y^2}{2xy}$$

(This is a homogeneous $d \cdot e$)

Put $y = vx$; $\dfrac{dy}{dx} = v + x\dfrac{dv}{dx}$

$$\therefore \quad v + x\frac{dv}{dx} = \frac{x^2+y^2x^2}{2x\cdot vx} = \frac{x^2(1+v^2)}{2x^2v}$$

i.e., $$x\cdot\frac{dv}{dx} = \frac{1+v^2}{2v} - v$$

$$x\frac{dv}{dx} = \frac{1+v^2-2v^2}{2v} = \frac{1-v^2}{2v}$$

Separating the variables

$$\frac{2v\,dv}{1-v^2} = \frac{dx}{x}$$

Integrating we get

$$-\log(1-v^2) = \log x + \log c$$

i.e., $$\log\left(\frac{1}{1-v^2}\right) = \log cx$$

i.e., $$\frac{1}{1-v^2} = cx; \quad \text{put } v = \frac{y}{x}$$

i.e., $$\frac{1}{1-\frac{y^2}{x^2}} = cx$$

i.e, $$\frac{x^2}{x^2 - y^2} = cx$$

or $c(x^2 - y^2) = x$ **is the required solution.**

50. Solve: $x\dfrac{dy}{dx} - y = \sqrt{x^2 + y^2}$

Solution: $$x\frac{dy}{dx} = y + \sqrt{(x^2 + y^2)}$$

$$\therefore \quad \frac{dy}{dx} = \frac{y + \sqrt{x^2 + y^2}}{x}$$

(Homogeneous)

Put $y = vx$, $\dfrac{dy}{dx} = v + x\dfrac{dv}{dx}$

$$\therefore \quad v + x\frac{dv}{dx} = \frac{vx + \sqrt{x^2 + v^2x^2}}{x} = v + \sqrt{1+v^2}$$

$$\therefore \quad x\frac{dv}{dx} = v + \sqrt{1+v^2} - v$$

i.e., $$x \cdot \frac{dv}{dx} = \sqrt{1+v^2}$$

$$\therefore \quad \frac{dv}{\sqrt{1+v^2}} = \frac{dx}{x}$$

Integrating, we get

$$\int \frac{dv}{\sqrt{1+v^2}} = \int \frac{dx}{x} + \text{const.}$$

i.e., $$\log\left(v + \sqrt{1+v^2}\right) = \log x + \log c$$

$$\therefore \quad v + \sqrt{1+v^2} = cx$$

$$\therefore \quad \frac{y}{x} + \sqrt{\left(1 + \frac{y^2}{x^2}\right)} = cx$$

PART-D

$$y + \sqrt{(x^2 + y^2)} = cx^2$$

is the required solution.

51. Solve: $2xy^2dy - (x^3 + 2y^3)\, dx = 0$

Solution: G.E. is $$\frac{dy}{dx} = \frac{x^3 + 2y^3}{2xy^2},$$

Put $y = vx$; $$\frac{dy}{dx} = v + x \cdot \frac{dv}{dx}$$

∴ G.E. becomes

$$\left[v + x \cdot \frac{dv}{dx}\right] = \frac{x^3 + 2v^3x^3}{2x \cdot v^2x^2} = \frac{1 + 2v^3}{2v^2}$$

i.e., $$x \cdot \frac{dv}{dx} = \frac{1 + 2v^3}{2v^2} - v$$

or $$x \cdot \frac{dv}{dx} = \frac{1}{2v^2}$$

Separating the variables

$$2v^2dv = \frac{dx}{x}; \text{ integrating}$$

$$2\int v^2dv = \int \frac{dx}{x} + \text{constant}$$

i.e., $$2 \cdot \frac{v^3}{3} = \log x + c$$

where $$v = \frac{y}{x}$$

52. Solve $\dfrac{dy}{dx} = \dfrac{x + 2y}{2x + y}$ [B.U. Oct., 1987]

Put $y = vx$; $\dfrac{dy}{dx} = v + x\dfrac{dv}{dx}$

G.E. becomes

$$v + x\frac{dv}{dx} = \frac{x + 2vx}{2x + vx} = \frac{1 + 2v}{2 + v}$$

∴ $$x\frac{dv}{dx} = \frac{1 + 2v}{2 + v} - v$$

$$x\frac{dv}{dx} = \frac{1+2v-2v-v^2}{2+v} = \frac{1-v^2}{2+v}$$

Separating the variables

$$\frac{dx}{x} = \frac{(2+v)}{1-v^2}dv = \frac{2+v}{(1+v)(1-v)}\cdot dv$$

$$\frac{dx}{x} = \left(\frac{1}{2}\cdot\frac{1}{1+v}+\frac{3}{2}\cdot\frac{1}{1-v}\right)\cdot dv$$

(Resolving into partial fractions)

Integrating, we get the solution,

$$\log x = \frac{1}{2}\cdot\log[1+v] - \frac{3}{2}\log(1-v) + c$$

where $$v = \frac{y}{x}$$

53. Solve: $\dfrac{dy}{dx} = \dfrac{y}{x-\sqrt{xy}}$

Solution: This is a Homo. equation

$\therefore$ put $y = vx$ and $\dfrac{dy}{dx} = v + x\dfrac{dv}{dx}$

$$\therefore \quad v + x\frac{dv}{dx} = \frac{vx}{x-\sqrt{xvx}} = \frac{vx}{x-\sqrt{xvx}} = \frac{v}{1-\sqrt{v}}$$

i.e.,

$$x\cdot\frac{dv}{dx} = \frac{v}{1-\sqrt{v}} - v = \frac{v^{3/2}}{1-v^{1/2}}$$

$$\Rightarrow \quad \frac{1-v^{1/2}}{v^{3/2}}\cdot dv = \frac{dx}{x}$$

Integrating, $$\int\left(\frac{1}{v^{3/2}}-\frac{1}{v}\right)dv = \int\frac{dx}{x}+c$$

$$-\frac{2}{\sqrt{v}} - \log v = \log x + c\,; \quad \left(v = \frac{y}{x}\right)$$

or $$2\sqrt{\frac{x}{y}} + \log y + c = 0.$$

PART-D

54. (A similar Example) $\dfrac{dy}{dx} = \dfrac{y}{x + \sqrt{xy}}$

Solution: put $y = Vx;$

Then $$\frac{dy}{dx} = V + x \cdot \frac{dv}{dx}$$

$\therefore$ $$v + x \cdot \frac{dv}{dx} = \frac{vx}{x + \sqrt{xvx}} = \frac{v}{x + x\sqrt{v}} = \frac{v}{\left(1 + \sqrt{v}\right)}$$

$\therefore$ $$x\frac{dv}{dx} = \frac{v}{\left(1 + \sqrt{v}\right)} - v = \frac{v - v - v^{3/2}}{\left(1 + \sqrt{v}\right)}$$

or $$x \cdot \frac{dv}{dx} = \frac{-v^{3/2}}{\left(1 + \sqrt{v}\right)}$$

$\Rightarrow$ $$\frac{1 + v^{1/2}}{v^{3/2}} dv = -\frac{dx}{x}$$

or $$\int \left(\frac{1}{v^{3/2}} + \frac{1}{v}\right) dv = -\int \frac{dx}{x} + c$$

or $$-\frac{2}{\sqrt{v}} + \log v = -\log x + c; \qquad \left(v = \frac{y}{x}\right)$$

or $$-2 \cdot \sqrt{\frac{x}{y}} + \log y + c = 0$$

55. Solve: $\dfrac{dy}{dx} - \dfrac{y}{x} = -\dfrac{y^2}{x^2}$

Solution: Put $y = vx;$

then $$\frac{dy}{dx} = v + x\frac{dv}{dx}$$

$\therefore$ The G.E. becomes

$$v + x\frac{dv}{dx} - v = -v^2$$

or, $$x \cdot \frac{dv}{dx} = -v^2;$$

or $$\frac{dv}{-v^2} = \frac{dx}{x}$$

Integrating, $\frac{1}{v} = \log x + \log c;$

$\therefore$ $\frac{y}{x} = \log (xc)$

$\Rightarrow$ $\boxed{(xc) = e^{y/x}}$ is the solution.

56. Solve: $(x^2 - xy + y^2)\, dx - xydy = 0$ [B.U. Jan. 1993]

Solution:

Given: $(x^2 - xy + y^2)\, dx - xydx = 0$

$$\Rightarrow \quad \frac{dy}{dx} = \frac{(x^2 - xy + y^2)}{xy}$$

(This is a Homogeneous d.c.)

$\therefore$ Put $y = vx \Rightarrow \frac{dy}{dx} = v + x\frac{dv}{dx}$

$\therefore$ The equation becomes

$$v + x\frac{dv}{dx} = \frac{x^2 - vx^2 + v^2x^2}{x^2 \cdot v}$$

$$v + x\frac{dv}{dx} = \frac{1 - v + v^2}{v}$$

$$x\frac{dv}{dx} = \frac{1 - v + v^2}{v} - v$$

$$\Rightarrow \quad x\frac{dv}{dx} = \frac{1 - v}{v}$$

Separating the variables and integrating we get

$$\int \frac{vdv}{1 - v} = \int \frac{dx}{x} + \log c$$

$$\Rightarrow \quad \int \frac{v - 1 + 1}{1 - v}\, dv = \log x + \log c$$

$$\Rightarrow \quad -\int dv + \int \frac{1}{1 - v}\, dv = \log x + \log c$$

$$\Rightarrow \quad -v - \log(1 - v) = \log x + \log c$$

$$\Rightarrow \quad -v = \log(1 - v) + \log x + \log c$$

$$\Rightarrow \quad -v = \log[cx(1 - v)]$$

$$\Rightarrow \quad cx(1 - v) = e^{-v}$$

$$\Rightarrow \quad c(x - y) = e^{-y/x}$$

This is the required solution

PART-D

57. Solve: $\dfrac{dy}{dx} = \dfrac{x^2 - 2xy}{y^2 - 2xy}$

Solution: Put $y = vx$

$$\left[\frac{dy}{dx} = v + x \cdot \frac{dv}{dx}\right]$$

$\therefore$ The G.E. is

$$v + x \cdot \frac{dv}{dx} = \frac{x^2 - 2x \cdot vx}{v^2x^2 - 2x \cdot vx} = \frac{x^2(1-2v)}{x^2(v^2-2v)} = \frac{(1-2v)}{(v^2-2v)}$$

$$\therefore \quad x \cdot \frac{dv}{dx} = \frac{(1-2v)}{(v^2-2v)} - v$$

$$x \cdot \frac{dv}{dx} = \frac{(1-2v) - v^3 + 2v^2}{(v^2-2v)} = \frac{v^3 - v^2 + v - v^2 + v - 1}{(v^2-2v)}$$

or $$\frac{dx}{x} + \frac{(v^2-2v)}{(v-1)(v^2-v+1)}\, dv = 0 \qquad [1]$$

Now, $$\frac{(v^2-2v)}{(v-1)(v^2-v+1)} = \frac{A}{(v-1)} + \frac{Bv+C}{v^2-v+1}$$

$$\therefore \quad (v^2 - 2v) = A(v^2 - v + 1) + (Bv + c)(v - 1)$$

We get $A = -1$; $B = 2$; $C = -1$

$$\int \frac{(v^2-2v)}{(v-1)(v^2-v+1)}\, dv = \int \frac{-1}{(v-1)}\, dv + \int \frac{(2v-1)}{(v^2-v+1)}\, dv$$

$$= -\log(v-1) + \log(v^2 - v + 1)$$

Integrating, (1), we get

Solution: $\log x - \log(v - 1) + \log(v^2 - v + 1) = c$

i.e., $$\frac{x(v^2-v+1)}{(v-1)} = c\,(y^2 - xy + x^2) = c(y - x).$$

58. Solve: $\dfrac{dy}{dx} = \dfrac{3x^2y + y^3}{3xy^2 + x^3}$

Solution: Put $y = vx$; $$\frac{dy}{dx} = \left(v + x \cdot \frac{dv}{dx}\right)$$

$$v + x \cdot \frac{dv}{dx} = \frac{3x^2 \cdot vx + v^3x^3}{3x \cdot v^2x^2 + x^3} = \frac{x^3(3v + v^3)}{x^3(3v^2 + 1)}$$

$$\therefore \qquad x\frac{dv}{dx} = \frac{(3v+v^3)-3v^3-v}{(3v^2+1)} = \frac{(2v-2v^3)}{(3v^2+1)}$$

$$\Rightarrow \qquad \frac{dv}{dx} = \frac{(3v^2+1)}{2v(1-v^2)}\,dv$$

or
$$\frac{dx}{x} + \frac{(3v^2+1)}{2v(v-1)(v+1)}\,dv = 0 \qquad \text{(use P.F)}$$

Integrating
$$\int\frac{dx}{x} + \frac{1}{2}\int\left[-\frac{1}{v}+\frac{2}{v-1}+\frac{2}{v+1}\right]dv = C$$

$$\log x + \frac{1}{2}\{-\log v + 2\log(v-1) + 2\log(v+1)\} = \text{const.}$$

$$\log\left[-\frac{x(v-1)(v+1)}{\sqrt{v}}\right] = \text{const.}$$

$$\therefore \qquad \frac{[(v^2-1)x]}{\sqrt{v}} = c^{\text{const}} = c \quad \text{where} \quad v = \left(\frac{y}{x}\right);$$

Solution:
$$(y^2-x^2) = c\sqrt{xy}$$

59. Solve: $(x^2 - xy)dy - (xy + y^2)dx = 0$

Solution: Put $y = vx$; $\frac{dv}{dx} = \left(v + x\cdot\frac{dv}{dx}\right)$

$\therefore$ The G.E. is

$$\frac{dy}{dx} = \frac{(xy+y^2)}{(x^2-xy)}$$

$$\left[v + x\cdot\frac{dv}{dx}\right] = \left[\frac{x\cdot vx + v^2\cdot x^2}{x^2 - x\cdot vx}\right] = \frac{(v+v^2)}{(1-v)}$$

$$x\cdot\frac{dv}{dx} = \frac{v+v^2}{(1-v)} - v = \frac{v+v^2-v+v^2}{(1-v)} = \frac{2v^2}{(1-v)}$$

$$\frac{dx}{x} = \frac{(1-v)}{2v^2}\,dv = \frac{1}{2}\left[\frac{1}{v^2}-\frac{1}{v}\right]$$

Integrating, we get

$$2\int\frac{dx}{x} + \int\left(\frac{1}{v^2}-\frac{1}{v}\right)dv = C$$

PART-D

$$2 \log x - (1/v) - \log v = C$$

where $v = (y/x)$.

60. Solve: $y^2 + (3xy + y^2)\left(\dfrac{dy}{dx}\right) = 0$

Solution: This is a Homo. equation

Since y is a common factor ($\div y$), we get

$$y + (3x + y)\left(\frac{dy}{dx}\right) = 0$$

$$ydx + (3x + y)dy = 0$$

Put $y = vx$; $\dfrac{dy}{dx} = \left[v + x \cdot \dfrac{dv}{dx}\right]$

$$\therefore \quad vxdx + [3x + vx]\left[v + \frac{dv}{dx}\right] = 0$$

$$x[vdx + (3 + v)\, vdx + (3 + v)\, xdv] = 0$$

$$dx[v + (v^2 + 3v)] + (v + 3)\, xdv = 0$$

$$\frac{dx}{x} + \frac{(v+3)}{(v^2+4v)}dv = 0;$$

$$\frac{dx}{x} + \frac{(v+3)}{(v^2+4v)}dv = 0;$$

$$\frac{dx}{x} + \frac{(v+3)}{v(v+4)}dv = 0$$

$\therefore$ Integrating after resolving into partial fraction

$$\int \frac{dx}{x} + \int \left[\frac{3}{4}\cdot\frac{1}{v} + \frac{1}{4}\cdot\frac{1}{(v+4)}\right]dv = C$$

$$\log x + (3/4) \log v + (1/4) \log (v + 4) = C$$

$$4 \log x + 3 \log v + \log (v + 4) = 4C$$

$$\log x^4 + \log v^3 (v + 4) = 4C$$

$$\log [x^4 v^3 (v + 4)] = 4C = \log k.$$

$$x^4 v^3 (v + 4) = k; \quad \frac{x^4 y^3}{x^3}\left(\frac{y}{x} + 4\right) = k$$

$$\therefore \quad y^3(y + 4x) = k$$

61. Solve: $x^2dy + (2x^2 - 3y^2)\, dx = 0$

Solution: The equation is Homogeneous

Put $y = vx$ and $\dfrac{dy}{dx} = v + x\,\dfrac{dv}{dx}$

$$x^2 \cdot \left[v + x\frac{dv}{dx}\right] + (2x^2 - 3 \cdot v^2x^2) = 0$$

$$\left[v + x \cdot \frac{dv}{dx}\right] = 3v^2 - 2$$

$$\Rightarrow \qquad x \cdot \frac{dv}{dx} = 3v^2 - v - 2 = 3v^2 - 3v + 2v - 2$$

$$= 3v(v - 1) + 2(v - 1)$$

i.e.,
$$x \cdot \frac{dv}{dx} = (v - 1)(3v + 2)$$

$$\therefore \qquad \frac{dx}{x} = \frac{dv}{(v-1)(3v+2)}$$

Integrating, we get

$$\int \frac{dx}{x} - \int \frac{1}{5}\left[\frac{1}{(v-1)} - \frac{3}{3v+2}\right] dv = \text{const.}$$

i.e.,
$$\log x - \left[\frac{1}{5}\right] \log(v - 1) + \left[\frac{1}{5}\right] \log(3v + 2) = \text{const.}$$

i.e.,
$$5 \log x + \log \frac{(3v+2)}{(v-1)} = \text{constant.}$$

$(v = y/x)$

$$\log\left[x^5 \cdot \left\{\frac{3(y/x)+2}{(y/x)+1}\right\}\right] = \text{const.}$$

$$\therefore \qquad x^5 \cdot (3y + 2x) = c(y - x)$$

62. Solve: $(xdy - ydx) + \sqrt{(x^2 + y^2)}\ dx = 0$

Solution: Put $y = vx$; $\frac{dy}{dx} = \left(v + x \cdot \frac{dv}{dx}\right)$

$$[x(vdx + xdv) - vx \cdot dx] + \sqrt{[(x^2 + v^2x^2)]}\ dx = 0$$

$$x\left[vdx + xdv - vdx + \sqrt{(1+v^2)}\ dx\right] = 0$$

$$\therefore \qquad xdv + \sqrt{(1+v^2)}\ dx = 0$$

$$\frac{dx}{x} + \frac{dv}{\sqrt{(v^2+1)}} = 0 \quad \int \frac{dx}{x} + \int \frac{dv}{\sqrt{(v^2+1)}} = C$$

$$\log x + \log\left[v + \sqrt{(v^2+1)}\right] = C = \log k$$

$\therefore$
$$\log\left[x\left(v + \sqrt{v^2+1}\right)\right] = \log k$$

or
$$x\left[v + \sqrt{v^2+1}\right] = k;\ \left[vx + \sqrt{(v^2x^2+x^2)}\right] = k.$$

Solution:
$$\left[y + \sqrt{(y^2+x^2)}\right] = k$$

63. Solve :
$$\frac{dx}{x^3 - 3xy^2} = \frac{dy}{y^3 - 3yx^2}$$

Solution: Put $y = vx;\ dy = vdx + xdv$

$$\frac{dv}{x^3 - 3x\cdot v^2x^2} = \frac{vdx + xdv}{v^3x^3 - 3vx\cdot x^2}$$

$$\frac{dx}{x^3(1-3v^2)} = \frac{(vdx+xdv)}{(v^3-3v)x^3}$$

$\therefore$
$$dx(v^3 - 3v) = vdx(1 - 3v^2) + xdv(1 - 3v^2)$$
$$dx(v^3 - 3v - v + 3v^3 - xdv\ (1 - 3v^2) = 0,$$
$$4(v^3 - v)\ dx + x(3v^2 - 1)dv = 0$$

i.e.,
$$4\cdot\frac{dx}{x} + \frac{3v^2-1}{v^3-v}\,dv = 0$$

$\therefore$
$$4\int\frac{dx}{x} + \int\frac{3v^2-1}{v^3-v}\,dv = \text{const.}$$

$$\log x^4 + \log(v^3 - v) = \log C$$
$$\log[x^4(v^3 - v)] = \log C;\ (v = y/x)$$

$\therefore$
$$x^4(v^3 - v) = C;$$

i.e.,
$$x^4\left\{\left(\frac{y^3}{x^3}\right) - \left(\frac{y}{x}\right)\right\} = c$$

$$x\,(y^3 - x^2y) = c;$$
$$xy\,(y^2 - x^2) = c \text{ is the solution.}$$

Solve:
$$(x^2 - y^2)\,dy = xydx$$

64. Solve
$$x\cos\left(\frac{y}{x}\right)(ydx + xdy) = y\sin\left(\frac{y}{x}\right)(ydx - xdy)$$

Solution: We have, the *G.E.* is

$$\frac{dy}{dx} = \frac{y\left[x\cos\left(\frac{y}{x}\right) + y\sin\left(\frac{y}{x}\right)\right]}{x\left[y\sin\left(\frac{y}{x}\right) - x\cos\left(\frac{y}{x}\right)\right]}$$

$$= \frac{\left(\frac{y}{x}\right)\left[\cos\left(\frac{y}{x}\right) + \left(\frac{y}{x}\right)\sin\left(\frac{y}{x}\right)\right]}{\left(\frac{y}{x}\right)\sin\left(\frac{y}{x}\right) - \cos\left(\frac{y}{x}\right)} \qquad [1]$$

$\therefore$ Put $y = vx$ in (1)

$$v + x\frac{dv}{dx} = \frac{v[\cos v + v\sin v]}{v\sin v - \cos v}$$

i.e.,

$$x\frac{dv}{dx} = \frac{v\cos v + v^2\sin v}{v\sin v - \cos v} - v = \frac{2v\cos v}{v\sin v - \cos v}$$

$$\Rightarrow \quad \frac{v\sin v - \cos v}{v\cos v}\,dv = 2\cdot\frac{dx}{x}$$

Integrating, $\int\left[\frac{\sin v}{\cos v} - \frac{1}{v}\right]dv = 2\log x - \log c$

i.e.,

$$\log c = \log x^2 \cdot v \cdot \cos v$$

or

$$x^2 v\cos v = c$$

where $v = \frac{y}{x}$, we get,

$$xy\cos(y/x) = c$$

is the complete solution.

65. Solve $\left(y - \sqrt{(x^2 + y^2)}\,dx - xdy\right) = 0$

Solution: We can write G.E. as

$$\frac{dy}{dx} = \frac{y - \sqrt{(x^2 + y^2)}}{x}$$

Put $y = vx \cdot v + x\cdot\frac{dv}{dx} = \frac{dy}{dx}$

PART-D

$$\left(v + x \cdot \frac{dv}{dx}\right) = \frac{vx - \sqrt{x^2 + v^2 x^2}}{x}$$

$$x \cdot \frac{dv}{dx} = -\sqrt{(1+v^2)}$$

$$\Rightarrow \qquad \frac{dv}{\sqrt{(1+v^2)}} = -\frac{dx}{x}$$

$$\int \frac{dv}{\sqrt{(1+v^2)}} = -\int \frac{dx}{x} + \text{const.}$$

$$\sinh^{-1}(v) = -\log x + c; \qquad (y = v/x)$$

66. Solve $\quad ydx - xdy = \sqrt{(x^2 - y^2)}\, dx$

Solution: It is a homogeneous equation.

Put $y = vx$;

$$\therefore \qquad dy = Vdx + xdy$$

Hence the given equation becomes

$$Vxdx - Vxdx - x^2 \cdot dv = \sqrt{x^2 - v^2 x^2}\ dx$$

i.e.,
$$-x^2\, dv = x\sqrt{(1-v^2)}\, dx$$

i.e.,
$$\frac{dx}{x} + \frac{dv}{\sqrt{(1-v^2)}} = 0$$

or
$$\int \frac{dx}{x} + \int \frac{dv}{\sqrt{(1-v^2)}} = c$$

or
$$\log x + \sin^{-1}(v) = c$$

Hence
$$\log x + \sin^{-1}\left(\frac{y}{x}\right) = c$$

67. Solve $\quad \dfrac{dy}{dx} = \dfrac{y(x-2y)}{x(x-3y)}$

Solution: It is a homo. equation

Put $y = vx$, $\qquad \dfrac{dv}{dx} = v + x \cdot \dfrac{dv}{dx}$

Hence the given equation becomes

$$V + x \cdot \frac{dv}{dx} = \frac{Vx(x - 2vx)}{x(x - 3vx)}$$

i.e., $$x\frac{dv}{dx} = \frac{v-2v^2}{1-3v} - v = \frac{v-2v^2-v+3v^2}{1-3v} = \frac{v^2}{1-3v}$$

i.e., $$\frac{dx}{x} = \frac{1-3v}{v^2}\,dv = \frac{1}{v^2}\,dv - \frac{3}{v}\,dv$$

Integrating, $$\log k + \log x = -\frac{1}{v} - 3\log v$$

i.e., $$\log(kxv^3) = -x/y$$

Hence $$kx\cdot\frac{y^3}{x^3} = e^{-\frac{x}{y}} \quad \textit{i.e.,} \quad ky^3 = x^2 e^{-\frac{x}{y}}$$

68. Solve $$\frac{dy}{dx} = \frac{3x+2y}{2x-3y}$$

Solution: Put $y = vx$; $$v + x\cdot\frac{dv}{dx} = \frac{3x+2vx}{2x-3vx}$$

or $$v + x\frac{dv}{dx} = \frac{3+2v}{2-3v};$$

or $$\frac{dv}{\frac{3+2v}{2-3v} - v} = \frac{dx}{x}$$

or $$\frac{(2-3v)\,dv}{3+3v^2} = \frac{dx}{x} \quad \text{or} \quad \frac{2}{1+v^2}\,dv - \frac{3v}{1+v^2}\,dv = 3\frac{dx}{x}$$

Integrating, $$2\tan^{-1}(v) - \frac{3}{2}\log(1+v^2) = 3\log x + C$$

or $$2\tan^{-1}\left(\frac{y}{x}\right) - \frac{3}{2}\log\left(1+\frac{y^2}{x^2}\right) = 3\log x + C$$

or $$2\tan^{-1}\left(\frac{y}{x}\right) - \frac{3}{2}\log(x^2+y^2) + 3\log x = 3\log x + C$$

$\therefore$ $$2\tan^{-1}\left(\frac{y}{x}\right) - \frac{3}{2}\log(x^2+y^2) = C$$

PART-D

EXERCISES

Solve the following Homogeneous Differential equation

1. $(1 + e^{x/y})\, dx + e^{x/y}\left(1 - \frac{x}{y}\right) dy = 0$

2. $x^2 \frac{dy}{dx} = \frac{y(x+y)}{2}$

3. $\frac{dy}{dx} = \frac{\sqrt{x^2 - y^2} + y}{x}$

4. $x\left(\cos \frac{y}{x}\right)(ydx + xdy) = y\left(\sin \frac{y}{x}\right)(xdy - ydx)$

5. $\left(x \cos \frac{y}{x} + y \sin \frac{y}{x}\right) y - \left(y \sin \frac{y}{x} - x \cos \frac{y}{x}\right) x \frac{dy}{dx} = 0$

6. $(x^3 - 3xy^2)\, dx = (y^3 - 3x^2y)\, dy$

7. $\left(2\sqrt{xy} - x\right)\frac{dy}{dx} + y = 0$

8. $xdx + ydy = m(xdy - ydx)$

9. $2\frac{dy}{dx} = \frac{x}{y} - 1$

10. $(x^2 + xy)dy = (x^2 + y^2)dx$

11. $(x + y) + (y - x)\left(\frac{dy}{dx}\right) = 0$

12. $xydy - y^2dx - x^2dy = 0$

13. $(2x - 4y + 3)dy + (x - 2y + 1)dx = 0$

14. $(x + y - 1)dy = (x + y + 1)dx$

15. $(2x + y + 1)dx + (4x + 2y - 1)dy = 0$

16. $(2x - 2y + 5)\,(dy/dx) = (x - y + 3)$

NON-HOMOGENEOUS EQUATIONS REDUCIBLE TO HOMOGENEOUS FORM OF FIRST DEGREE IN *X* AND *Y*

Consider the differential equation

$$\frac{dy}{dx} = \frac{ax + by + c}{a^1x + b^1y + c^1}\;;\; \frac{a}{a^1} \neq \frac{b}{b^1} \qquad [1]$$

We find that this equation could have been homogeneous if the constants C and C' are not present. Hence we try to get rid of C and C' from the above equation through the transformation

$$x = X + h\;;\; y = Y + k,$$

where h and k are constants when we put $x = X + h$, $y = Y + k$, then

$$dx = dX,\; dy = dy$$

Hence
$$\frac{dy}{dx} = \frac{dY}{dX}$$

$\therefore$ The given equation reduces to

$$\frac{dY}{dX} = \frac{a(X+h) + b(Y+K) + c}{a'(X+h) + b'(Y+k) + c'}$$

i.e.,
$$\frac{dy}{dx} = \frac{aX + bY + (ah + bk + c)}{a'X + b'Y + (a'h + b'k + c')}$$

We choose h and k

$$ah + bk + c = 0 \text{ and}$$
$$a'h + b'k + c' = 0 \text{ simultaneously}$$

By solving these equations for h and k, we get $h = \dfrac{bc' - b'c}{ab' - a'b}$

$$k = \frac{a'c - a'c}{ab' - a'b}$$

∴ When $\dfrac{a}{a'} \neq \dfrac{b}{b'}$

i.e., when $ab' - a'b \neq 0,$

For these values of h and k, the equation reduces to

$$\frac{dY}{dX} = \frac{aX + bY}{a^1X + b^1Y}$$

Which is Homogeneuos eqn. and can be solved by putting $Y = VX$ and in the solution replace $X = x - h$ and $Y = y - k$; to get the solution of the given equation.

69. Solve: $(x + y + 1)\, dx - (2x + 2y + 3)\, dy = 0$

(OR)

$$\frac{dy}{dx} = \frac{x + y + 1}{2x + 2y + 3}$$ [VTU, August 2001; VTU, August 2002]

Solution: Put $x + y = z$;

$$\therefore \quad 1 + \frac{dy}{dx} = \frac{dz}{dx};$$

$$\Rightarrow \quad \frac{dy}{dx} = \left(\frac{dz}{dx} - 1\right)$$

∴ G.E. becomes
$$\frac{dz}{dx} - 1 = \frac{z + 1}{2(x + y) + 3} = \frac{z + 1}{2z + 3}$$

$$\Rightarrow \quad \frac{dz}{dx} = \frac{z + 1}{2z + 3} + 1 = \frac{3z + 4}{2z + 3}$$

$$\Rightarrow \quad \frac{(2z + 3)}{(3z + 4)}\, dz = dx$$

PART-D

By actual division method for the fraction: $\frac{2z+3}{3z+4}$ is as shown below:

$$3z+4 \overline{\big)\, 2z+3 \,\big(} \frac{2}{3}$$

$$-\left(2z+\frac{8}{3}\right)$$

$$3-\frac{8}{3}=\frac{1}{3}$$

$$\therefore \quad \frac{2z+3}{3z+4} = \left(\frac{2}{3}\right) + \frac{\left(\frac{1}{3}\right)}{(3z+4)} = \frac{2}{3}+\frac{1}{3}\cdot\frac{1}{3z+4}$$

$$\Rightarrow \quad \left[\frac{2}{3}+\frac{1}{3}\cdot\frac{1}{3z+4}\right] dz = dx$$

Integrating, we get

$$\frac{2}{3}z + \frac{1}{9}\log(3z+4) = (x+c)$$

or $$6(x+y) + \log(3x+3y+4) = 9x+k$$

or $$6y - 3x + \log(3x+3y+4) = k \text{ is the required solution}$$

70. Solve: $(x-2y)(dx-dy) = dx+dy$ [VTU, Aug. 2000]

Solution: $(xdx - xdy - 2ydx + 2ydy) = dx + dy$

or G.E. is equivalent to $$\frac{dy}{dx} = \frac{x-2y-1}{x-2y+1} \quad [1]$$

Put $(x-2y) = z$; $1 - 2\cdot\frac{dy}{dx} = \frac{dz}{dx}$

i.e., $$\frac{dy}{dx} = \frac{1}{2}\left[1-\frac{dz}{dx}\right]$$

Hence (1) reduces to

$$\frac{1}{2}\left[1-\frac{dz}{dx}\right] = \frac{z-1}{z+1}$$

i.e., $$\frac{dz}{dx} = 1 - \frac{2(z-1)}{z+1} = \frac{-z+3}{z+1}$$

or $$\left[\frac{z+1}{z-3}\right]\cdot dz = -dx$$

Integrating, $\int\left[1+\frac{4}{z-3}\right]dz = -\int dx + c$

or $z + 4\log(z-3) = -x + c$

i.e., $x - 2y + 4\log(x - 2y - 3) = -x + c$

i.e., $2\cdot(x-y) + 4\log(x-2y-3) = c$

is the required solution

71. Solve $(x + 2y)(dx - dy) = dx + dy$

(similar to previous example) [VTU, March, 2001]

Solution: $(x + 2y)(dx - dy) = dx + dy$

We can write the given equation as:

$$xdx + 2ydx - xdy - 2ydy = dx + dy$$

or $(x + y - 1)\,dx - dy(x + 2y + 1) = 0$

or $$\frac{dy}{dx} = \frac{(x+2y-1)}{(x+2y+1)} \quad [1]$$

Put $x + 2y = z \Rightarrow 1 + 2\cdot\frac{dy}{dx} = \frac{dz}{dx}$

$$\Rightarrow \frac{dy}{dx} = \frac{1}{2}\left[\frac{dz}{dx} - 1\right]$$

$\therefore$ G.E. becomes $\frac{1}{2}\left[\frac{dz}{dx} - 1\right] = \frac{z-1}{z-1}$

$$\frac{dz}{dx} = \frac{2z-2}{z+1} + 1 = \frac{3z-1}{z-1}$$

$$\frac{dz}{dx} = \frac{3z-1}{z-1}$$

or $$\frac{(z+1)}{(3z-1)}\,dz = dx \quad [2]$$

$$3z - 1\,\big)\; z + 1 \;\big(\,\frac{1}{3}$$

$$-\left(z - \frac{1}{3}\right)$$

$$\frac{4}{3}$$

$$\therefore \quad \frac{z+1}{(3z-1)} = \left[\frac{1}{3} + \frac{4}{9}\cdot\frac{3}{(3z-1)}\right]$$

PART-D

$\therefore$ (2) becomes $\left[\frac{1}{3}+\frac{4}{9}\cdot\frac{3}{3z-1}\right] dz = dx$

Integrating, we get

$$\frac{1}{3}z + \frac{4}{9}\log(3z-1) = (x+c)$$

or $\left(\frac{x+2y}{3}\right) + \frac{4}{9}\log(3x+6y-1) = x + c$ is the solution.

72. Solve $\frac{dy}{dx} = \frac{x+2y-3}{2x+y-3}$ [VTU March, 2000]

Solution: Here $\frac{1}{2} \neq 2$ $\left(\because \frac{a}{a'} \neq \frac{b}{b'}\right)$

$\therefore$ Put $\left.\begin{aligned} x &= X+h \\ y &= Y+k \end{aligned}\right\}$ $\left.\begin{aligned} (h+2k-3) &= 0 \\ (2h+k-3) &= 0 \end{aligned}\right\}$ $\therefore \left.\begin{aligned} h &= 1 \\ k &= 1 \end{aligned}\right\}$

$\therefore$ $\frac{dY}{dX} = \frac{X+2Y}{2Y+Y}$ which is a Homogeneous equation.

Put $Y = vX$ $\therefore$ $\frac{dY}{dX} = v + X\frac{dy}{dx}$

or $$v + X\frac{dv}{dy} = \frac{X+2vX}{2X+vX} = \frac{1+2v}{2+v}$$

$\therefore$ $$X\frac{dv}{dX} = \frac{1+2v}{2+v} - v = \frac{1+2v-2v-v^2}{2+v}$$

$\therefore$ $$dv\left(\frac{v+2}{1-v^2}\right) = \frac{dX}{X};$$

$$\left[\frac{v}{v^2-1}+\frac{2}{v^2-1}\right]dv + \frac{dX}{X} = 0$$

$\therefore$ $$\int\frac{v}{v^2-1}dv + \int\frac{2dx}{v^2-1} + \int\frac{dX}{X} = \text{const.}$$

$$\frac{1}{2}\log(v^2-1) + 2\cdot\frac{1}{2}\log\left\{\frac{(v-1)}{v+1}\right\} + \log X = C$$

$$\frac{1}{2}\log(v^2-1) + \log\left\{\frac{(v-1)}{v+1}\right\} + \log X = C$$

$$\log \frac{\left[X\sqrt{(v^2-1)}\,(v-1)\right]}{(v+1)} = \text{const.}$$

$$\sqrt{(v^2-1)}\left[\frac{(v-1)\,X}{(v+1)}\right] = \text{const.}$$

$$\sqrt{(v^2-1)}\left[\frac{(v-1)\,X}{(v+1)}\right] = \text{const.}$$

Squaring, $$\frac{(v^2-1)\,(v-1)^2\,X^2}{(v+1)^2} = C$$

$$(v-1)^3\,X^2 = C\,(v+1);$$

put $v = (Y/X)$

$$(Y-X)^3 = C(Y+X)$$

$$\{(y+1)-(x+1)\}^3 = C\,(y+1+x+1)$$

$$(y-x)^3 = C\,(x+y+2)$$

73. Solve $(3y + 2x + 4)\,dx - (4x + 6y + + 3)\,dy = 0$ [B.U. Aug., 1994]

Solution: The given differential equation can be written as

$$\frac{dy}{dx} = \frac{2x+3y+4}{4x+6y+3}$$

or $$\frac{dy}{dx} = \frac{(2x+3y)+4}{2(2x+3y)+3}\quad \frac{a}{a'} = \frac{b}{b'};\left(\frac{2}{4}=\frac{3}{6}\right)$$

∴ Put $2x + 3y = z;$

$$\Rightarrow \quad 2 + 3\cdot\frac{dy}{dx} = \frac{dz}{dx}$$

$$\Rightarrow \quad \frac{dy}{dx} = \frac{1}{3}\left[\frac{dz}{dx}-2\right]$$

∴ G.E. becomes

$$\frac{1}{3}\left[\frac{dz}{dx}-2\right] = \frac{z+4}{2z+3}$$

$$\Rightarrow \quad \frac{dz}{dx} = \frac{3z+12}{2z+3} + 2$$

$$\Rightarrow \quad \frac{dz}{dx} = \frac{3z+12}{2z+3} + 2 = \frac{7z+18}{2z+3}$$

PART-D

$$\Rightarrow \qquad \int dx = \int \frac{2z+3}{7z+18}\, dz + c \qquad [1]$$

$$7z+18 \,\Big)\, 2z+3 \,\Big(\, \frac{2}{7}$$

$$-\left(2z + \frac{36}{7}\right)$$

$$3 - \frac{36}{7} = -\frac{15}{7}$$

$$\therefore \qquad \frac{2z+3}{7z+18} = \left[\frac{2}{7} - \frac{15}{7}\cdot\frac{1}{7z+18}\right]$$

Now
$$\frac{2z+3}{7z+18} = \frac{1}{7}\left[2 - \frac{15}{7z+18}\right]$$

$$\therefore \qquad \int \frac{2z+3}{7z+18}\, dz = \frac{1}{7}\int\left[2 - \frac{15}{7z+18}\right] dz$$

$$= \frac{1}{7}\left[2z - \frac{15}{7}\log(7z+18)\right]$$

$$= \frac{1}{7}\left[4x + 6y - \frac{15}{7}\log(14x + 21y + 18)\right]$$

$\therefore$ Solution is obtained from (1) as:

$$x = \frac{1}{7}\left[4x + 6y - \frac{15}{7}\log(14x + 21y + 18)\right] + C$$

or
$$3x = 6y - 15\log(14x + 21y + 18) + k \text{ (where } k = 7c)$$

74. Solve $(x - 4y - 9)\, dx + (4x + y - 2)\, dy = 0$ [V.T.U. F/M, 2005]

Solution: $(4x + y - 2)\, dy = -(x - 4y - 9)\, dx$

$$\therefore \qquad \frac{dy}{dx} = \frac{-x + 4y + 9}{4x + y - 2} \qquad ...(1)$$

Put $x = X + h$ and $y = Y + k$, where h and k are constants

$$\therefore \qquad \frac{dy}{dx} = \frac{dy}{dY}\frac{dY}{dX}\frac{dX}{dx} = 1\,.\,\frac{dY}{dX}\,. = \frac{dY}{dX}$$

Equations (1) becomes :

$$\frac{dY}{dX} = \frac{-(X+h) + 4(Y+k) + 9}{4(X+h) + (Y+k) - 2}$$

i.e., $$\frac{dY}{dX} = \frac{(-X+4Y)+(-h+4k+9)}{(4X+Y)+(4h+k-2)} \qquad (2)$$

Let us choose h and k such that $\left\{\begin{array}{r} -h+4k+2=0 \\ 4h+k-2=0 \end{array}\right\}$

Solving these we get, $h = 1$ and $k = -2$

$\therefore$ $$\frac{dY}{dX} = \frac{-X+4Y}{4X+Y}, \text{ (after a minor simplifictions)} \qquad (3)$$

Put $Y = VX \Rightarrow \dfrac{dY}{dX} = V + X\dfrac{dV}{dx}$

$\therefore$ (3) becomes, $$V + X\frac{dV}{dx} = \frac{-X+4VY}{4X+VX}$$

i.e., $$V + X = \frac{X(-1+4V)}{X(4+V)} \quad \text{or} \quad X\frac{dV}{dX} = \frac{-1+4V}{4+V} - V$$

i.e., $$X\frac{dV}{dX} = \frac{-1+4V-4V-V^2}{4+V} \quad \textit{i.e.,} \quad X\frac{dV}{dX} = \frac{-(1+V^2)}{(4+V)}$$

$\therefore$ $$\frac{(4+V)\,dV}{1+V^2} = -\frac{dX}{X} \Rightarrow 4\int\frac{dV}{1+V^2} + \int\frac{V\,dV}{1+V^2} + \int\frac{dX}{X} = c$$

i.e., $$4\tan^{-1} V + \frac{1}{2}\log(1+V^2) + \log X = c$$

$\Rightarrow$ $$8\tan^{-1} V + \log(1+V^2) + 2\log X = 2c$$

i.e., $$8\tan^{-1} V + \log[(1+V^2)X^2] = 2c, \text{ where } V = Y/X.$$

$\therefore$ $$8\tan^{-1}(Y/X) + \log(X^2+Y^2) = 2c = k \text{ (say)}$$

But $X = x - h = x - 1$ and $Y = y - k = y + 2$

$\therefore$ $$8\tan^{-1}\left(\frac{y+2}{x-1}\right) + \log[(x-1)^2 + (y+2)^2] = k.$$

75. Solve $\dfrac{dy}{dx} = \dfrac{6x-4y+3}{3x-2y+1}$

Solution: Put $3x - 2y = V$;

Hence $$3 - 2\frac{dy}{dx} = \frac{dv}{dx}$$

$\therefore$ $$\frac{dv}{dx} = 3 - 2\left(\frac{2v+3}{v+1}\right)$$

PART-D

$$\therefore \quad \frac{dv}{dx} = \frac{-(v+3)}{(v+1)}$$

$$\therefore \quad dx = -\frac{(v+1)}{(v+3)}\,dv = -\left(1 - \frac{2}{v+3}\right)dv$$

$$\therefore \quad x = -v + 2\log(v+3) + C_1$$

$$\therefore \quad x = (2y - 3x) + 2\log(3x - 2y + 3) + C_1$$

$$\therefore \quad 2x - y = \log(3x - 2y + 3) + C$$

76. Solve: $(7y - 3x + 3)dy + (3y - 7x + 7)dx = 0$

or
$$\frac{dy}{dx} = -\frac{3y - 7x + 7}{7y - 3x + 3}$$

Solution: $\left(\because \ \frac{3}{7} \neq \frac{-7}{-3}\right)$

$$\left[\begin{matrix}7k - 3h + 3 = 0 \\ 3k - 7h + 7 = 0\end{matrix}\right\} \therefore \left.\begin{matrix}h = 1 \\ k = 0\end{matrix}\right\} \left.\begin{matrix}x = X + h = X + 1 \\ y = Y = Y\end{matrix}\right\} \left.\begin{matrix}dx = dX \\ dy = dY\end{matrix}\right\}\Bigg]$$

$\therefore$ The G.E. becomes $(7Y - 3X)\,dY + (3Y - 7X)\,dX = 0$

Put $Y = vX;\ dY = (vdX + Xdv)$

$$(7vX - 3X)(vdX + Xdv) + (3vX - 7X)\,dX = 0$$

$$\therefore \quad dX(7v - 3)v + (7v - 3)Xdv + (3v - 7)dX = 0$$

$$dX(7v^2 - 3v + 3v - 7) + (7v - 3)Xdv = 0$$

$$\therefore \quad \frac{dX}{X} + \frac{(7v - 3)}{7(v^2 - 1)}\,dv = 0$$

$$\int\frac{dX}{X} + \int\left\{\frac{v}{(v^2 - 1)}\right\}dv + \frac{-3}{7}\cdot\int\frac{dv}{(v^2 - 1)} = \text{const.}$$

$$\log X + \frac{1}{2}\log(v^2 - 1) - \frac{3}{7}\cdot\frac{1}{2}\log\frac{(v-1)}{(v+1)} = \text{const.}$$

$$14\log X + 7\log(v - 1)(v + 1) - 3\log\left\{\frac{(v-1)}{(v+1)}\right\} = C$$

$$X^{14}(v - 1)^4(v + 1)^{10} = \text{const.}$$

Put $v = \dfrac{Y}{X}$; we get

$$(Y - X)^2(Y + X)^5 = \text{const.}$$

$$\therefore \quad (y - x + 1)^2(y + x - 1)^5 = \text{const.}$$

77. Solve: $(x + y - 3)\, dx - (x + y - 1)\, dy = 0$

Put $x = X + h$; $y = y = Y + k$.

$$\left.\begin{matrix} h + k - 3 = 0 \\ h - k - 1 = 0 \end{matrix}\right\} \quad \left.\begin{matrix} h = 2 \\ k = 1 \end{matrix}\right\} \quad \left.\begin{matrix} x = X + 2 \\ y = Y + 1 \end{matrix}\right\} \quad \begin{matrix} dx = dX \\ dy = dY \end{matrix}$$

$$\frac{h}{\begin{vmatrix} 1 & -3 \\ -1 & -1 \end{vmatrix}} = \frac{k}{\begin{vmatrix} 1 & -3 \\ 1 & -1 \end{vmatrix}} = \frac{1}{\begin{vmatrix} 1 & 1 \\ 1 & -1 \end{vmatrix}}$$

or $$\frac{h}{-1-3} = \frac{k}{-3+1} = \frac{1}{-2}$$

$\Rightarrow$ $$h = 2; \; k = 1$$

$\therefore$ G.E becomes $[(X + 2 + Y + 1) - 3]\, dX - [(X + 2 - Y - 1) - 1]\, dY,$

or $$(X + Y)\, dX - (X - Y)\, dY = 0$$

This is Homo.

Put $Y = vX$; $dy = (vdX + Xdv)$

$\therefore$ $(X + vX)dX - (X - vX)\,(vdX + Xdv) = 0$

After a minor simplification we get,

$$\log X + \left(\frac{1}{2}\right) \log (v^2 + 1) - \tan^{-1} (v) = \text{const.}$$

where $$v = \left(\frac{Y}{X}\right) = \left[\frac{(y-1)}{(x-2)}\right]$$

78. Solve: $(x + y + 2)\, dx - (x - y)\, dy = 0$

Solution:

$$\left.\begin{matrix} (h + k + 2) = 0 \\ (h - k) = 0 \end{matrix}\right\} \quad \left.\begin{matrix} h = -1 \\ k = -1 \end{matrix}\right\} \quad \left.\begin{matrix} x = X - 1 \\ y = Y - 1 \end{matrix}\right\} \quad \begin{matrix} dx = dX \\ dy = dY \end{matrix}$$

$\therefore$ G.E. becomes:

$$[(X - 1) + (Y - 1) + 2]\, dX - [(X - 1) - (Y - 1)]\, dY = 0$$

$$(X + Y)dX - (X - Y)dY = 0$$

This is now Homogeneous.

Put $Y = vX$;

$\therefore$
$$(X + vX)\, dX - (X - vX)\,(vdX + Xdv) = 0$$
$$X\,[(1 + v)\, dX - (1 - v)\,(vdX + Xdv)] = 0$$
$$dX\,(1 + v) - v\,(1 - v)\, dX - (1 - v)\, Xdv = 0$$
$$dX\,[1 + v - v + v^2] + (v - 1)\, Xdv = 0$$
$$(v^2 + 1)\, dX + (v - 1)\, Xdv = 0$$
$$[\div X\,(1 + v^2)]$$
$$\frac{dX}{X} + \frac{(v-1)\, dv}{(v^2 + 1)} = 0$$

PART-D

$$\frac{dX}{X} + \frac{vdv}{v^2+1} - \frac{dv}{v^2+1} = 0$$

$$\therefore \quad \int \frac{dX}{X} + \int \frac{vdv}{v^2+1} - \int \frac{1}{(v^2+1)}\, dv = C$$

$$\log X + \frac{1}{2} \log (v^2+1) - \tan^{-1}(v) = \text{const.}$$

where $v = \left(\frac{Y}{X}\right) = \left[\frac{(y-1)}{(x-1)}\right]$

79. Solve: $(y + x - 2)\, dx - (y - x - 4)\, dy = 0$

Solution:

$$\begin{rcases} h+k-2=0 \\ -h+k-4=0 \end{rcases} \begin{matrix} h=-1 \\ k=3 \end{matrix} \begin{rcases} \\ \end{rcases} \begin{matrix} x = X-1 \\ y = Y+3 \end{matrix} \begin{rcases} \\ \end{rcases} \begin{matrix} dx = dX \\ dy = dY \end{matrix}$$

$\therefore$ G.E. becomes

$$[(Y+3)+(X-1)-2]\, dX - [(Y+3)-(X-1)-4]\, dY = 0$$

$$[Y + X]\, dX - [Y - X]\, dY = 0 \qquad (1)$$

Which is a homo. d.e.]

(1) can be written as $\frac{dY}{dX} = \frac{(X+Y)}{-X+Y}$ (2)

$$\therefore \quad \left(\text{Put } y = vX ; v + X \cdot \frac{dv}{dX} = \frac{dY}{dX}\right)$$

or, $$v + X \cdot \frac{dv}{dX} = \frac{X+vX}{-X+vX}$$

or $$X \cdot \frac{dv}{dX} = \frac{v+1}{v-1} - v = \frac{1+2v-v^2}{v-1}$$

i.e., $$\frac{v-1}{1+2v-v^2} \cdot dv = \frac{dX}{X}$$

Integrating, we get

$$-\int \frac{(1-v)}{1+2v-v^2}\, dv = \log X + C_1$$

i.e., $$\frac{-1}{2} \log (1+2v-v^2) = \log X + C_1$$

i.e., $$\log X^2 (1+2v-v^2) = \log X + C_1$$

i.e., $$\log X^2 (1+2v-v^2) = -2C_1 = \log C_2 \text{ (say)}$$

i.e., $$X^2(1+2v-v^2) = e^{\log c_2} = C \text{ (say)}$$

where $$v = \frac{Y}{X};$$

we get, $$X^2\left(1+2\frac{Y}{X}-\frac{Y^2}{X^2}\right) = C$$

i.e., $$X^2 + 2XY - Y^2 = C$$

i.e., $$(x+1)^2 + 2(x+1)(y-3) - (y-3)^2 = C$$

since $X = x + 1$; $Y = y - 3$

is the C solution

80. Solve $\dfrac{dy}{dx} = \dfrac{x+2y+3}{2x+y+3}$

Solution: $\dfrac{a}{a'} \neq \dfrac{b}{b'}$; therefore we have

$$\left.\begin{matrix}(h+2k+3)=0\\(2h+k+3)=0\end{matrix}\right\}\frac{h}{(6-3)}=\frac{-k}{(3-6)}=\frac{1}{(1-4)} \quad \begin{matrix}\Rightarrow\dfrac{h}{3}=\dfrac{k}{3}=\dfrac{1}{-3};\\ \Rightarrow h=-1;k=-1\end{matrix}\left\}\begin{matrix}x=X-1\\y=Y-1\end{matrix}\right.$$

$$\left.\begin{matrix}dx=dX\\dy=dY\end{matrix}\right\} \text{ G.E. becomes}$$

$$\frac{[X-1+2(Y-1)+3]}{[2(X-1)+(Y-1)+3]} = \frac{dY}{dX}$$

$$\Rightarrow \quad \frac{X+2Y}{2X+Y} = \frac{dY}{dX}. \text{ Which is homogeneous.}$$

or $$(X+2Y)\,dX - (2X+Y)\,dY = 0$$

(Put $Y = vX$)

$$(X+2vX)\,dX - (2X+vX)(vdX+Xdv) = 0$$
$$(X)\,[(1+2v)\,dX - (v+2)(vdX+Xdv)] = 0$$
$$(1+2v)\,dX - v(v+2)\,dX - X(v+2)\,dv = 0$$
$$dX\,[1+2v-v^2-2v] - X(v+2)\,dv = 0$$
$$dX(1-v^2) - X(v+2)\,dv = 0$$
$$-(v^2-1)\,dx - X(v+2)\,dv = 0$$

$$\frac{dX}{X} + \frac{(v+2)\,dv}{(v^2-1)} = 0$$

$$\frac{dX}{X} + \frac{vdv}{v^2-1} + \frac{2}{v^2-1}dv = 0$$

PART-D

$$\int \frac{dX}{X} + \int \frac{v}{(v^2-1)} dv - 2\int \frac{dv}{(v^2-1)} = C$$

$$\log X + \frac{1}{2} \log (v^2 - 1) + \log [(v - 1) \div (v + 1)] = C$$

$$2 \log X + \log (v^2 - 1) + 2 \log [(v - 1) \div (v + 1)] = \text{const.}$$

4.4 EXACT EQUATIONS AND REDUCIBLE TO EXACT FORM

Let us consider the differential equation of the form:

$$M(x, y)dx + N(x, y)dy = 0 \qquad [1]$$

if $\exists$ a function $f(x,y) = f$

such that $\dfrac{\partial f}{\delta x} = M$ and $\dfrac{\partial f}{\partial y} = N$

The d.e. given by (1) takes the form

$$\left(\frac{\partial f}{\partial x}\right)dx + \left(\frac{\partial f}{\partial y}\right)dy = 0 \quad i.e., \quad df(x, y) = 0$$

On integration the complete solution is given by $f(x, y) = C$.

In this case $Mdx + Ndy$ is said to be an exact differential and (1) is called an exact differential equation (Condition for exactness).

The differential equation $Mdx + Ndy = 0$ is exact

if $$\frac{\partial M}{\partial y} = \frac{\partial N}{\partial x}$$

WORKING RULE

1. Verify the condition

$$\frac{\partial M}{\partial y} = \frac{\partial N}{\partial x}; \text{ (Exactness)}$$

2. Integrate M w.r.t. x treating y – as constant
3. Integrate w.r.t. y those terms in N which do not contain x.
4. Equate the sum of the results of (2) and (3) to a constant C, which is the required solution.

In mathematical language, the solution is given by

$$\int_{y-\text{const}} Mdx + \int (\text{Terms in } N \text{ not containing } x)\, dy = c$$

Examples

81. Solve: $[\cos x \tan y + \cos (x + y)]\, dx + [\sin x \sec^2 y + \cos (x + y)]\, dy = 0$

[VTU, March 1999]

Solution: Here,
$$M = \cos x \tan y + \cos (x + y)$$
$$N = \sin x \sec^2 y + \cos (x + y)$$

$$\therefore \qquad \frac{\partial M}{\partial y} = \cos x \cdot \sec^2 y - \sin(x+y)$$

$$\frac{\partial N}{\partial x} = \cos x \cdot \sec^2 y - \sin(x+y)$$

$$\Rightarrow \qquad \frac{\partial M}{\partial y} = \frac{\partial N}{\partial x} \quad \text{(Exactness)}$$

$\therefore$ the given equation is exact.

$$\int_{y-\text{const}} [\cos x \tan y + \cos(x+y)]\, dx + \int o\, dy = C$$

$$\sin x \tan y + \sin(x+y) = C$$

82. Solve $\left\{y\left(1+\frac{1}{x}\right)+\cos y\right\} dx + (x + \log x - \sin y)\, dy = 0$ [VTU, August, 2000]

Solution:

Here
$$M = y\left(1+\frac{1}{x}\right) + \cos y,$$
$$N = x + \log x - \sin y$$
$$\frac{\partial M}{\partial y} = 1 + \frac{1}{x} - \sin y;$$
$$\frac{\partial N}{\partial x} = 1 + \frac{1}{x} - \sin y$$
$$\frac{\partial M}{\partial y} = \frac{\partial N}{\partial x} \quad \text{(Exactness)}$$

$\therefore$ Its G.S. is
$$\int_{y-\text{constant}} \left\{y\left(1+\frac{1}{x}\right)+\cos y\right\} dx + \int o\, dy = c$$
$$y(x + \log x) + x\cos y = C$$

83. Solve $(5x^4 + 3x^2y^2 - 2xy^3)dx + (2x^3y - 3x^2y^2 - 5y^4)\, dy = 0$ [VTU, March, 1999]

Solution:
$$M = 5x^4 + 3x^2y^2 - 2xy^3$$
$$N = 2x^3y - 3x^2y^2 - 5y^4$$
$$\therefore \qquad \frac{\partial M}{\partial y} = 6x^2y - 6xy^2;$$
$$\frac{\partial N}{\partial x} = 6x^2y - 6xy^2$$
$$\Rightarrow \qquad \frac{\partial M}{\partial y} = \frac{\partial N}{\partial x} \quad \text{(Exactness)}$$

$\therefore$ The solution is $\underset{\text{treat} \downarrow y - \text{const}}{\int M dx} \quad \underset{\text{treat} \downarrow y - \text{const}}{\int M dx}$

$$\Rightarrow \quad \underset{(y - \text{const})}{\int (5x^4 + 3x^2y^2 - 2xy^3)\, dx} + \int (-5y^4)\, dy = C$$

$$\Rightarrow \quad x^5 + x^3y^2 - x^2y^3 - y^5 = C$$

84. $xdx + ydy + \dfrac{xdy - ydx}{x^2 + y^2} = 0$

Solution: Rearranging the terms, we get

$$\left[x - \frac{y}{x^2 + y^2}\right] dx + \left[y + \frac{x}{x^2 + y^2}\right] dy = 0$$

$$M = \left(x - \frac{y}{x^2 + y^2}\right), N = \left(y + \frac{x}{x^2 + y^2}\right)$$

$$\frac{\partial M}{\partial y} = -\frac{x^2 - y^2}{(x^2 + y^2)^2}$$

$$\frac{\partial N}{\partial x} = \left[0 + \frac{(x^2 + y^2) \cdot 1 - x \cdot 2x}{(x^2 + y^2)^2}\right] = \frac{y^2 - x^2}{(x^2 + y^2)^2}$$

The equation is exact.

Hence $\underset{y - \text{const}}{\int \left\{\dfrac{(x - y)}{(x^2 + y^2)^2}\right\} dx} + \underset{\text{Terms free from } x}{\int \left\{y + \dfrac{x}{(x^2 + y^2)}\right\} dy} = c$

$$\frac{x^2}{2} - y \cdot \frac{1}{y} \tan^{-1}\left(\frac{x}{y}\right) + \left(\frac{y^2}{2} + 0\right) = C$$

85. Solve $(x^2 - 4xy - 2y^2)\, dx + (y^2 - 4xy - 2x^2)\, dy = 0$

Solution: Here

$$M = x^2 - 4xy - 2y^2;$$
$$N = y^2 - 4xy - 2x^2$$

$$\therefore \quad \frac{\partial M}{\partial y} = -4x - 4y = \frac{\partial N}{\partial x}$$

Hence, (Exact)

Now $\int M dx = \dfrac{1}{3} x^3 - 2x^2y - 2y^2x$

In N, the term free from x is y^2 whose integral w.r.t. y is $\dfrac{1}{3} y^3$

Hence the complete solution is

$$\frac{1}{3}x^3 - 2x^2y - 2y^2x + \frac{1}{3}y^3 = C_1$$

(i.e.) $$x^3 + y^3 - 6xy(x + y) = 3C_1 = C$$

is the solution.

86. Solve $(x^2 + y)\,dx + (y^3 + x)\,dy = 0$ [BU, Oct., 1987]

Solution:

Hence $$M = x^2 + y;$$ $$N = y^2 + x$$

$$\therefore \quad \frac{\partial M}{\partial y} = 1; \quad \frac{\partial N}{\partial x} = 1$$

$$\therefore \quad \frac{\partial M}{\partial y} = \frac{\partial N}{\partial x} \text{ the equation is exact.}$$

$$\therefore \textit{ Solution: } \underset{y\text{-const}}{\int M dx} + \underset{\text{neglect (or omit) (terms containing } x\text{)}}{\int N dy} = \text{constant}$$

i.e., $$\underset{\text{(treat } y - \text{const)}}{\int (x^2 + y^2)\,dx} + \underset{\text{(omit } x \text{ terms)}}{\int (y^3 + x)\,dy} = C$$

$$\left\{\int x^2\,dx + y\int 1\,dx\right\} + \int y^3\,dy = C$$

$$\therefore \quad \frac{x^3}{3} + y \cdot x + \frac{y^4}{4} = C$$

This is the required solution.

87. Solve $(2ax + by + g)\,dx + (2cy + bx + c)\,dy = 0$ [B.U. Aug., 1982]

Solution:

Here $$M = 2ax + by + g;$$ $$N = 2cy + bx + c$$

$$\therefore \quad \frac{\partial M}{\partial y} = b; \quad \frac{\partial N}{\partial x} = b$$

Hence $\dfrac{\partial M}{\partial y} = \dfrac{\partial N}{\partial x}$ and so the equation is exact.

Its solution is given by

$$\underset{y - \text{const}}{\int (2ax + by + g)\,dx} + \underset{\text{omit } x\text{-terms}}{\int (2cy + bx + c)\,dy} = \text{const.}$$

$$\left\{2 \cdot a \cdot \frac{x^2}{2} + byx + gx\right\} + 2 \cdot C \cdot \frac{y^2}{2} + Cy = k$$

$$\therefore \quad ax^2 + byx + gx + Cy^2 + Cy = k$$

PART-D

88. Solve $(2x^2 + 6xy - y^2)dx + (3x^2 - 2xy + y^2)dy = 0$ [B.U. April, 1985]

Solution:

Here
$$M = 2x^2 + 6yx - y^2;$$
$$N = 3x^2 - 2xy + y^2$$

$$\therefore \quad \frac{\partial M}{\partial y} = 6x - 2y$$

$$\frac{\partial N}{\partial x} = 6x - 2y$$

Hence $\frac{\partial M}{\partial y} = \frac{\partial N}{\partial x}$ and so the equation is exact.

$\therefore$ its solution is given by

$$\underset{y-\text{const}}{\int (2x^2 + 6xy - y^2)\,dx} + \underset{\text{(omit x-terms)}}{\int (3x^2 - 2xy + y^2)\,dy} = \text{const.}$$

i.e., $2\int x^2\,dx + 6y\int x dx - y^2\int 1 \cdot dx + \int y^2 \cdot dy$

$$\frac{2x^3}{3} + 6y \cdot \frac{x^2}{2} - y^2 x + \frac{y^3}{3} = C$$

$$\therefore \quad 2x^3 + 9x^2y - 3xy^2 + y^3 = k$$
$$(k = 3C)$$

89. Solve $(3x^2 + 6xy^2)dx + (6x^2y + 4y^3)dy = 0$

Solution:

Here,
$$M = 3x^2 + 6xy^2;$$
$$N = 6x^2y + 4y^3$$

$$\frac{\partial M}{\partial y} = 12xy; \quad \frac{\partial N}{\partial x} = 12xy$$

Hence $\frac{\partial M}{\partial y} = \frac{\partial N}{\partial x}$ and the equation is exact.

$\therefore$ Its solution is given by

$$\underset{(y\text{-const})}{\int M dx} + \underset{(x\text{-terms are neglected})}{\int N dy} = \text{const.}$$

$$\therefore \quad \int (3x^2 + 6xy^2)dx + \int (6x^2y + 4y^3)\,dy = C$$

i.e.,
$$\left\{3\int x^2\,dx + 6y^2\int x dx\right\} + \int 4y^3\,dy = C$$

$$\therefore \quad \frac{3x^3}{3} + 6y^2 \cdot \frac{x^2}{2} + 4 \cdot \frac{y^2}{4} = C$$

The required solution is

$$(x^3 + 3x^2y^2 + y^4 = C)$$

90. Solve $(x + y \cos x)\, dx + \sin x\, dy = 0$

Solution:

$$M = x + y \cos x;$$
$$N = \sin x$$

$$\frac{\partial M}{\partial y} = \cos x; \quad \frac{\partial N}{\partial x} = \cos x$$

$\therefore$ $$\frac{\partial M}{\partial y} = \frac{\partial N}{\partial x} \text{ and so the equation is exact.}$$

$\therefore$ Its solution is given by

$$\int_{y-\text{const}} (x + y \cos x)\, dx + \int_{\text{terms free from } x} \sin x \cdot dy = \text{const}$$

i.e., $$\left\{\int x dx + y \int \cos x\, dx\right\} + 0 = C$$

$\therefore$ the solution is $$\frac{x^2}{2} + y \sin x = C$$

91. Solve $\dfrac{dy}{dx} + \dfrac{y \cos x + \sin y + y}{\sin x + \cos y + x} = 0$ [B.U. Aug., 1996]

Solution: Here we shall first write the equation in the form $Mdx + Ndy = 0$

$$(y \cos x + \sin y + y)dx + (\sin x + x \cos y + x)dy = 0$$

Here $$M = y \cos x + \sin y + y$$
$$N = \sin x + x \cos y + x$$

$$\frac{\partial M}{\partial y} = \cos x + \cos y + 1$$

$$\frac{\partial N}{\partial x} = \cos x + \cos y + 1$$

We see that $$\frac{\partial M}{\partial y} = \frac{\partial N}{\partial x}$$

$\therefore$ equation is exact

Solution $$\int_{y-\text{const}} Mdx + \int_{\text{term free from } x} Ndy = C$$

or $$\int (y \cos x + \sin y + y)\, dx + \int 0 dy = C$$

or $$y \int \cos x\, dx + (\sin y + y) \int 1 dx = C$$

$$y \sin x + (\sin y)\, x = C$$

is the required solution.

PART-D

92. Solve $(3x^2y^2 + x^2)\,dx + (2x^3y + y^2)\,dy = 0$ [BU Feb., 1996]

Solution: Here,
$$M = 3x^2y^2 + x^2,$$
$$N = 2x^3y + y^2$$

$$\therefore \quad \frac{\partial M}{\partial y} = 6x^2y$$

$$\frac{\partial N}{\partial x} = 6x^2y$$

$$\therefore \quad \frac{\partial M}{\partial y} = \frac{\partial N}{\partial x}$$

$\therefore$ the equation is exact,

Its solution is given by

$$\underset{y-\text{const}}{\int M dx} + \underset{\substack{(\text{Neglect } x-\text{terms} \\ \text{in } N)}}{\int N dy} = C_1$$

or
$$\underset{y-\text{const}}{\int (3x^2\,y^2 + x^2)\,dx} + \int (y^2)\,dy = C_1$$

$$= y^2\left[3\cdot\frac{x^3}{3}\right] + \left(\frac{x^3}{3}\right) + \frac{y^3}{3} = C_1$$

or
$$3x^3y^2 + x^3 + y^3 = 3C_1 = C$$

93. Solve $(ax + hy + g)dx + (hx + by + f)\,dy = 0$

Solution: Here
$$M = (ax + hy + g)$$
$$N = (hx + + by + f)$$

$$\therefore \quad \frac{\partial M}{\partial y} = h; \quad \frac{\partial N}{\partial x} = h;$$

Hence the equation is exact and its solution is:

$$\underset{y-\text{const}}{\int (ax + hy + g)\,dx} + \underset{\substack{(\text{terms in } N \\ \text{not containing } x)}}{\int (by + f)\,dy} = C$$

$$\frac{ax^2}{2} + hxy + gx + \frac{by^2}{2} + fy = k$$

$$ax^2 + 2hxy + by^2 + 2gx + 2fy = 2k = C$$

94. Solve $(x^3 + 3xy^2)dx + (3x^2y + y^3)dy = 0$

Solution: Here
$$M = x^3 + 3xy^2;$$
$$N = 3x^2y + y^3;$$

$$\frac{\partial M}{\partial y} = 6xy; \quad \frac{\partial N}{\partial x} = 6xy;$$

$\therefore$ It solution is: $\int_{y-\text{const}}(x^3 + 3xy^2)\,dx + \int y^3\,dy = C$

$$x^4 + y^4 + 6x^2y^2 = k$$

95. Solve $(x^2 - ay)dx - (ax - y^2)dy = 0$

$$M = x^2 - ay;$$
$$N = -(ax - y^2) = -ax + y^2$$

$$\frac{\partial M}{\partial y} = -a;$$

$$\frac{\partial N}{\partial x} = -a;$$

$$\frac{\partial M}{\partial y} = \frac{\partial N}{\partial x}$$

$\therefore$ The equation is exact

Its solution is $\int_{y-\text{const}} Mdx + \int Ndy = C$ (those term is N not containing x)

$$\int(x^2 - ay)\,dx + \int(y^2)\,dy = C$$

$$\frac{x^3}{3} - ayx + \frac{y^3}{3} = C$$

$\therefore$ $$x^3 + y^3 - 3axy = 3C \text{ (or } k)$$

96. Solve $(1 + e^{x/y})\,dx + e^{x/y}\,(1 - x/y)\,dy = 0$

Solution:

$$M = 1 + e^{x/y}$$
$$N = e^{x/y}\,(1 - x/y)$$

$\therefore$ $$\frac{\partial M}{\partial y} = e^{x/y}\left(\frac{-x}{y^2}\right)$$

$$\frac{\partial N}{\partial x} = e^{x/y}\left(-\frac{1}{y}\right) + \left(1 - \frac{x}{y}\right)e^{x/y}\left(\frac{1}{y}\right)$$

i.e., $$\frac{\partial M}{\partial y} = \frac{-x}{y^2}\,e^{x/y};$$

$$\frac{\partial N}{\partial x} = -\frac{1}{y}e^{x/y} + \frac{1}{y}e^{x/y} - \frac{x}{y^2}e^{x/y} = -\frac{x}{y^2}e^{x/y}$$

$\therefore$ $$\frac{\partial M}{\partial y} = \frac{\partial N}{\partial x}$$

$\therefore$ Its is an exact equation

The solution is $\displaystyle\int_{y\text{-const}} (1+e^{x/y})dx + 0 = \text{const}$

(Since N has no term containing x)

$$\therefore \quad x + \frac{e^{x/y}}{(1/y)} = C$$

$$x + ye^{x/y} = C$$

97. Solve $(y^2 e^{xy^2} + 4x^3)dx + (2xy\, e^{xy^2} - 3y^2)dy = 0$

Solution:

$$M = y^2\, e^{xy^2} + 4x^3$$
$$N = 2xy\, e^{xy^2} - 3y^2$$

$$\frac{\partial M}{\partial y} = y^2\, e^{xy^2} \cdot 2xy + e^{xy^2} \cdot 2y$$

$$\frac{\partial N}{\partial x} = 2xy\, e^{xy^2} \cdot y^2 + e^{xy^2} \cdot 2y$$

$$\therefore \quad \frac{\partial M}{\partial y} = \frac{\partial N}{\partial x} \text{ and hence equation is exact.}$$

$\therefore$ The solution is

$$\int_{y-\text{const}} (y^2 e^{xy^2} + 4x^3)\, dx + \int(-3y^2)\, dy = \text{const}$$

i.e.,
$$y^2 \cdot \frac{e^{xy^2}}{y^2} + \frac{4x^4}{4} - \frac{3 \cdot y^3}{3} = \text{const}$$

i.e.,
$$e^{xy^2} + x^4 - y^3 = \text{const}$$

98. *Solve* $(e^y + 1)\cos x\, dx + e^y \sin x\, dy = 0$

Solution:

$$M = (e^y + 1)\cos x;$$
$$N = e^y \sin x$$

$$\therefore \quad \frac{\partial M}{\partial y} = e^y \cos x;$$

$$\frac{\partial N}{\partial x} = e^y \cos x;$$

$$\therefore \quad \frac{\partial M}{\partial y} = \frac{\partial N}{\partial x} \text{ and hence equation is exact.}$$

$\therefore$ The solution is given by

$$\int_{y-\text{const}} Mdx + \int(\text{those terms in } N \text{ not containing } x)\, dx = \text{const}$$

$$\int(e^y + 1)\cos x + 0 = \text{const}$$

(Since there is no term in N not containing x)

$\therefore \qquad (e^y + 1)\sin x = C$

99. Solve: $(x + y)(dx - dy) = dx + dy$

Solution: We have

$$dx - dy = \frac{dx + dy}{(x + y)} = \frac{dv}{v}$$

Put $\left(\begin{array}{l} x + y = v \\ \therefore dx + dy = dv \end{array}\right)$

Integrating, we get $\quad (x - y) = \log v + k$

$$(x - y) = \log(x + y) + k$$

100. Solve $3x(xy - 2)dx + (x^3 + 2y)dy = 0$

Method (I) [B.U. Jan., 1993]

Solution:

Given:

$$M = 3x^2y - 6x$$

$$N = x^3 + 2y;$$

$$\Rightarrow \qquad \frac{\partial M}{\partial y} = 3x^2; \quad \frac{\partial N}{\partial x} = 3x^2$$

$$\therefore \qquad \frac{\partial M}{\partial y} = \frac{\partial N}{\partial x} \quad \Rightarrow \text{ the equation is exact.}$$

$\therefore$ Integrate w.r.t x treating y as constant.

$$\int_{y-\text{const}}(3x^2y - 6x)\,dx = x^3y - 3x^2$$

Integrate the terms in N, which are independent of x, w.r.t. y, we get

$$\int 2y\,dy = y^2$$

$\therefore$ The solution is $x^3y - 3x^2 + y^2 = k$

Method (II):

The equation can be written as

$$3x^2ydx - 6xdx + x^3dx + 2ydy = 0$$

$$\Rightarrow \qquad (y \cdot 3x^2dx + x^3 \cdot dy) - 6xdx + 2ydy = 0$$

$$\Rightarrow \qquad d[yx^3] - d(3x^2) + d(y^2)$$

On integration, we get $\quad yx^3 - 3x^2 + y^2 = k$

EQUATIONS REDUCIBLE TO EXACT FORM

Integrating Factor:

If the equation $Mdx + Ndy$ is not exact, it can often be made exact by multiplying it by a suitable function of x and y, called an integrating factor. The number of such factors is unlimited,.

Sometimes we can find the integrating factors by inspection, with some experience.

The following integrable combinations will be found useful in guessing the integrating factors.

(1) $xdy + ydx = d(xy)$

(2) $\dfrac{xdy + ydx}{xy} = d \log (x/y)$

(3) $\dfrac{xdy - ydx}{x^2} = d(y/x)$

(4) $\dfrac{xdy - ydx}{y^2} = - d\,(x/y)$

(5) $\dfrac{xdy - ydx}{xy} = d \log (x/y)$

(6) $\dfrac{xdy - ydx}{x^2 + y^2} = \dfrac{\dfrac{(xdy - ydx)}{x^2}}{[1 + (y/x)2]} = d \tan - 1\,(y/x)$

(7) $\dfrac{xdy - ydx}{(x^2 - y^2)} = \dfrac{\dfrac{(xdy - ydx)}{x^2}}{[1 - (y/x)^2]} = d\left[\dfrac{1}{2} \log \dfrac{(x + y)}{(x - y)}\right]$

(8) $x\,dx + y\,dy = \dfrac{1}{2} d\,[(x^2 + y^2)]$

(9) $\dfrac{xdx + ydy}{x^2 + y^2} = \dfrac{1}{2}\, d\,[\log (x^2 + y^2)]$

(10) $\dfrac{xdx + ydy}{\sqrt{x^2 + y^2}} = d\left(\sqrt{x^2 + y^2}\right)$

(11) $\dfrac{2xydy - y^2\,dx}{x^2} = d\left(\dfrac{y^2}{x}\right)$

(12) $\dfrac{y^2 dx + 2xydy}{x^2 y^4} = d\left(-\dfrac{1}{xy^2}\right)$

(13) $\dfrac{xdy + ydx}{\sqrt{(1 - x^2y^2)}} = d\,[\sin^{-1}(xy)]$

Solutions with Integrating factors by Inspection

Worked Examples

101. Solve $(y^2 + y + x)dx - ydy = 0$ [B.U. Oct., 1986]

Given: $(y^2 + y)\,dy + (xdy - ydx) = 0$

$\{(\div\, y^2)$; which does not affect the integrability of first group,$\}$

We have $\left(1 + \dfrac{1}{y}\right)dy + \dfrac{xdy - ydx}{y^2} = 0$

$$d(y + \log y) - d(x/y) = 0$$

Integrating, we get $y + \log y - \dfrac{x}{y} = C$

is the required solution.

102. Solve: $xdx = y(x^2 + y^2 - 1)\,dy$

Solution:

Given: $$xdx + ydy = y(x^2 + y^2)dy$$

i.e., $$\frac{1}{2}\,d\,(x^2 + y^2) = y(x^2 + y^2)$$

$$[\div (x^2 + y^2)]$$

or $$\frac{\frac{1}{2}d(x^2 + y^2)}{(x^2 + y^2)} = ydy$$

Integrating, $\frac{1}{2}\log(x^2 + y^2) = \frac{y^2}{2} + C$ **is the solution.**

$$\left[\text{since} \int \frac{f'(x)}{f(x)}\,dx = \log f(x)\right]$$

103. Solve $y(2xy + 1)dx - xdy = 0$

Solution:

Given: $$ydx - xdy + 2xy^2\,dx = 0$$

$$\Rightarrow \quad \frac{ydx - xdy}{y^2} + 2xdx = 0$$

$$\Rightarrow \quad d\left(\frac{x}{y}\right) dx + \int 2xdx = C$$

$$\Rightarrow \quad \left[\frac{x}{y} + x^2 = C\right] \quad \text{or} \quad x(1 + xy) = Cy$$

is the required solution.

104. Solve $ydx - xdy + 3x^2y^2\,e^{x^3}\,dx$

Solution: **Rearranging the terms in the given equation; we get**

$$\frac{ydx - xdy}{y^2} + 3x^2\,e^{x^3}\,dx = 0$$

$$\Rightarrow \quad d\left(\frac{x}{y}\right) + e^{x^3}\,(3x^2)\,dx = 0$$

$$\Rightarrow \quad d\left(\frac{x}{y}\right) + d(e^{x^3}) = 0$$

on integrating, we get

PART-D

$$\frac{x}{y} + e^{x^3} = C \text{ is the required equation.}$$

105. Solve $ye^{x/y} \cdot dx = (xe^{x/y} + y^2)\, dy$

Solution: The given equation can be written as

$$ye^{x/y} \cdot dx = xe^{x/y} \cdot dy + y^2 \cdot dy$$

i.e.
$$e^{x/y}\,(ydx - xdy) = y^2 dy$$

$(\div y^2)$, we get

$$e^{x/y}\left[\frac{ydx - xdy}{y^2}\right] = dy$$

i.e.,
$$e^{x/y} \cdot d(x/y) = dy \quad \text{or} \quad d(e^{x/y}) = dy$$

Integrating $\quad e^{x/y} = y + C$

This is the required solution.

106. Solve $\quad (1 + xy)\, y \cdot dx + (1 - xy)\, x\, dy = 0$

Solution:

Given: $\quad ydx + xdy + xy^2dx - yx^2dy = 0$

$$d(xy) + xy^2dx - yx^2dy = 0$$

$$(\div x^2y^2), \quad \frac{d(xy)}{x^2y^2} + \frac{dx}{x} - \frac{dy}{y} = 0$$

Integrating we get

$$-\frac{1}{xy} + \log x - \log y = C$$

The required solution is

$$-\frac{1}{xy} + \log (x/y) = C$$

107. Solve $\quad (x + 2y^3)\dfrac{dy}{dx} = y$

Solution: $\quad xdy + 2y^3dy = ydx$

or $\quad ydy - xdy = 2y^3dy;$

or, $\quad \dfrac{ydx - xdy}{y^2} = 2ydy \quad$ or $\quad d\left(\dfrac{x}{y}\right) = 2ydy$

Integrating, $\quad \dfrac{x}{y} = 2 \cdot \dfrac{y^2}{2} + C,$

$$\frac{x}{y} = y^2 + C$$

108. Solve $(a^2 - 2xy - y^2)dx - (x + y)^2 dy = 0$

Solution: We have

$$a^2dx - 2xydx - y^2dx - x^2dy - y^2dy - 2xydy = 0$$

or, $$a^2dx - (2xydx + x^2dy) - (y^2dx + 2xydy) - y^2dy = 0$$

or $$a^2dx - d(x^2y) - d\,(y^2x) - y^2dy = 0$$

Integrating $$a^2x - x^2y - y^2x - \frac{y^3}{3} = C$$

109. Solve $\frac{dy}{dx} = \frac{x - y}{x + y}$

Solution: We have

$$(x + y)dy = (x - y)\,dx \quad \text{or,} \quad (xdy + ydx) + ydy = xdx$$

Integrating $$\int d(xy) + \int ydy = \int xdx + C$$

$$xy + \frac{y^2}{2} = \frac{x^2}{2} + C$$

or $$2xy + (y^2 - x^2) = 2C = C',$$

is the required solution.

110. Solve $\frac{dy}{dx} + \frac{y}{x} = x^2$

when $y = 1,\ x = 1$

Solution: We have

$$xdy + ydx = x^3dx$$

$$d(xy) = x^3dx$$

Integrating, we get $$xy = \frac{x^4}{4} + k$$

When $x = 1,\ y = 1$;

$$\therefore \quad 1 = \frac{1}{4} + k \quad \text{or} \quad k = \frac{3}{4}$$

Hence, $$\left(xy = \frac{x^4}{4} + \frac{3}{4}\right)$$

110. (*a*) Solve : $xdx + ydy - \left(\frac{xdy - ydx}{x^2 + y^2}\right) = 0$ [V.T.U. J/A, 2004]

Solution. We know that

$$xdx + ydy = \frac{1}{2}\,d\,(x^2 + y^2) \quad \text{...(1)}$$

PART-D

and $$\frac{xdy - ydx}{x^2 + y^2} = \frac{xdy - ydx}{x^{22}} \cdot \frac{1}{1 + (y/x)^2}$$

$$= d\left(\frac{y}{x}\right) \cdot \frac{1}{1 + (y/x)^2} = d\,\{\tan^{-1}(y/x)\} \qquad \text{...(2)}$$

From (1) and (2), the G.E may be rewnttew as:

$$d\left\{\frac{1}{2}(x^2 + y^2) - \tan^{-1}(y/x)\right\} = 0$$

$$\Rightarrow \qquad \frac{1}{2}(x^2 + y^2) - \tan^{-1}(y/x) = A,$$

111. $(xy \cos xy + \sin xy)\, dx + x^2 \cos xy\, dy = 0$

Solution: We have:

$$\sin (xy)\, dx + x \cos (xy)\,(ydx + xdy) = 0$$

i.e. $$\frac{dx}{x} + \frac{\cos (xy)}{\sin (xy)}(ydx + xdy) = 0$$

i.e. $$\frac{dx}{x} + \frac{\cos z \cdot dz}{\sin z} = 0$$

where, $xy = z$ and $ydx + xdy = dz$

Integrating $\log x + \log \sin z = \log k$

i.e. $\log \{x \sin (xy)\} = \log k$

Hence $x \sin (xy) = k$

112. Solve $e^y dx + (xe^y + 2y)\, dy = 0$

Solution: Here, $M = e^y;\ N = xe^y + 2y$

$$\therefore \qquad \frac{\partial M}{\partial y} = e^y = \frac{\partial N}{\partial x}$$

Hence the equation is exact.

The given equation can be grouped as

$$(e^y dx + xe^y\, dy) + 2ydy = 0$$

i.e. $d(xe^y) + d(y^2) = 0$

on integration, $xe^y + y^2 = C$

is the required solution.

113. Solve $(x + y)^2 \left\{x\dfrac{dy}{dx} + y\right\} = xy\left\{1 + \dfrac{dy}{dx}\right\}$ [VTU, August, 1999]

Solution: **Given:** $(x + y)^2\,[xdy + ydx] = xy[dx + dy]$

i.e., $(x + y)^2\, d(xy) = xyd(x + y)$ *i.e.,* $\dfrac{d(xy)}{xy} = \dfrac{d(x + y)}{(x + y)^2}$

on integrating, we get $\log(xy) = \dfrac{-1}{x+y} + C,$

is the required complete solution

Some Rules for finding the integrating factors of *Mdx* + *Ndy* = 0 to make it exact

Rule (I): If the equation $Mdx + Ndy = 0$ is homogeneous and $Mx + My \neq 0$, then $\dfrac{1}{Mx + Ny}$ is an integrating factor.

Worked Examples (114 – 133) [Relating to Rule (I)]

114. $x^2ydx - (x^3 + y^3)dy = 0$

Solution: The equation is homogeneous, but not exact, here

$$M = x^2y;\ N = -(x^3 + y^3)$$

$$\therefore \quad \text{I.F.} = \frac{1}{Mx + Ny} = \frac{1}{x^3 y - x^3 y - y^4} = -\frac{1}{y^4}$$

$$\therefore \quad -\frac{x^2 y}{y^4} dx + \frac{(x^3 + y^3)}{y^4} dy = 0 \qquad [1]$$

In [1], $\quad M = -\dfrac{x^2 y}{y^4};\ N = \dfrac{(x^3 + y^3)}{y^4}$

$$\therefore \quad \frac{\partial M}{\partial y} = -\frac{(-x^2)(3y^2)}{(y^3)^2} = \frac{3x^2 y^2}{y^6} = \frac{3x^2}{y^4};$$

and $\quad \dfrac{\partial N}{\partial x} = \dfrac{(3x^2 + 0)}{y^4} = \dfrac{3x^2}{x^4}$

$$\therefore \quad \frac{dM}{dy} = \frac{\partial N}{\partial x}\text{, Eqn (1) is exact.}$$

i.e., $\quad -\dfrac{x^2}{y^3} dx + \left[\dfrac{x^3}{y^4} + \dfrac{1}{y}\right] dy = 0$

Its solution is: $\displaystyle\int_{y-\text{const}} + \int[\text{terms in } N \text{ which do not involve } x\,]dy = C$

i.e., $\quad \dfrac{-1}{y^3}\displaystyle\int x^2 dx + \int\left(\frac{1}{y}\right)dy = C$

or $\quad -\dfrac{x^3}{3y^3} + \log y = \text{const}$ is the required solution.

115. Solve $(x^2y - 2xy^2)dx - (x^3 - 3x^2y)dy = 0$

Solution: (Method I)

Given: $(x^2y - 2xy^2) + (3x^2y - x^3)dy = 0$ the equation is homogeneous but not exact; here

$$\therefore \quad M = x^2y - 2xy^2;\ N = (3x^2y - x^3)$$

(Integrating Factor)

$$\therefore \quad \text{I.F.} = \frac{1}{Mx + Ny} = \frac{1}{x^3y - 2x^2y^2 + 3x^2y^2 - x^3y} = \frac{1}{x^2y^2}$$

$$\therefore \quad \frac{(x^2y - 2xy^2)}{x^2y^2}dx - \frac{(x^3 - 3x^3y)}{x^2y^2}dy = 0$$

is exact

or

$$\left[\frac{1}{y} - \frac{2}{x}\right]dx - \left[\frac{x}{y^2} - \frac{3}{y}\right]dy = 0$$

is exact

$\therefore$ complete solution is given by

$$\int_{y-\text{const}} Mdx + \int [\text{terms is } N \text{ not containing } x]\ dx = 0$$

i.e.,

$$\frac{1}{y}\int dx - 2\int \frac{1}{x}\,dx + 3\int \frac{dy}{y} = C$$

i.e.,

$$\frac{x}{y} - 2\log x + 3\log y = C$$

Method (II)

The given equation can be written as.

$$x^2(ydx - xdy) - x^2y^2\left[\frac{2}{x}dx - \frac{3}{y}dy\right] = 0\ (\div x^2y^2),\ \text{we get}$$

$$\frac{ydx - xdy}{y^2} - \left[\frac{2}{x}dx - \frac{3}{y}dy\right] = 0$$

or

$$\int d\left(\frac{x}{y}\right) - \int\left(\frac{2}{x}dx - \frac{3}{y}dy\right) = C$$

(Integrating), we get $\dfrac{x}{y} - 2\log x + 3\log y = C = \log k$

or

$$\frac{x}{y} = \log k\,\frac{x^2}{y^3} \quad \text{or} \quad kx^2 = y^3e^{x/y}$$

116. Solve: $(x^3 - 3xy^2)\,dx - (y^3 - 3\,x^2y)\,dy = 0$

Solution:

Hence, $$M = x^3 - 3xy^2;$$ $$N = -(y^3 - 3x^2y).$$

The equation is homogeneous but not exact; here

$$\therefore \quad \text{I.F.} = \frac{1}{Mx + Ny} = \frac{1}{x^4 + y^4}.$$

$\therefore$ Multiply by $\dfrac{1}{x^4 + y^4}$ we get the exact d.e. in which

$$\therefore \quad \frac{(x^3 - 3xy^2)}{(x^4 - y^4)}dx - \frac{(y^3 - 3x^2y)}{(x^4 - y^4)}dy = 0$$

$\therefore$ Complete solution is given by:

$$\int_{y\text{-constant}} Mdx + \int(\text{those terms in } N \text{ not containing } x)\,dy = \log C_1$$

or $$\frac{1}{4}\log(x^4 - y^4) - \frac{3}{4}\log\left[\frac{x^2 - y^2}{x^2 + y^2}\right] = \log C_1 \qquad [1]$$

$$\therefore \quad (x^4 - y^4)(x^2 + y^2)^3 = C(x^2 - y^2)^3$$

$$\therefore \quad (x^2 - y^2)(x^2 + y^2)(x^2 + y^2)^3 = C(x^2 - y^2)^3$$

$$\therefore \quad (x^2 + y^2)^4 = C(x^2 - y^2)^2$$

Rule (II): If the equation $Mdx + Ndy = 0$, can be written as

$$yf_1(xy)\,dx + xf_2(xy)\,dy = 0$$

i.e. $$M = yf_1(xy);\ N = xf_2(xy)$$

then $\dfrac{1}{Mx - Ny}$ is an integrating factor

Note: $f_1(xy)$, $f_2(xy)$ are functions of (xy).

Worked Examples (117-133) [Relating to Rule (II)]

117. Solve: $(xy^2 + 2x^2y^3)dx + (x^2y - x^3y^2)dx = 0$

Solution: The equation can be written as:

$$(xy + 2x^2y^2)ydx + (xy - x^2y^2)xdy = 0$$

This is of the form

$$f_1(xy)ydx + f_2(xy)\,xdy = 0 \qquad \text{[By Rule II]}$$

$$\therefore \quad \text{I.F.} = \frac{1}{Mx - Ny} = \frac{1}{3x^3y^3}$$

Multiply the given equation by $\dfrac{1}{3x^3y^3}$.

PART-D

$$\therefore \quad \left(\frac{1}{3x^2y}+\frac{2}{3x}\right)dx+\left(\frac{1}{3xy^2}-\frac{1}{3y}\right)dy=0$$

$$\therefore \quad \frac{\partial M}{\partial y}=\frac{-1}{3x^2y^2},\ \frac{\partial N}{\partial x}=\frac{-1}{3x^2y^2}$$

∴ The equation is exact.

∴ The solution is $\int\left(\frac{1}{3x^2y}+\frac{2}{3x}\right)dx-\int\frac{1}{3y}dy=C$

i.e.,
$$\frac{1}{3y}\left(\frac{-1}{x}\right)+\frac{2}{3}\log x-\frac{1}{3}\log y=C$$

118. Solve $(x^2y^2+xy+1)\,ydx+(x^2y^2-xy+1)\,xdy=0.$

Solution: This is in the form: $yf_1(xy)+nf_1(xy)\,dy=0$; where

$$M=(x^2y^2+xy+1)y=x^2y^3+xy^2+y$$
$$N=x^3y^2-x^2y+x$$

$$\therefore \quad \text{I.F.}=\frac{1}{Mx-Ny}$$

$$=\frac{1}{x^3y^3+x^2y^2+xy-x^3y^3+x^2y^2-xy}=\frac{1}{2x^2y^2}.$$

Multiply the given equation by $\frac{1}{2x^2y^2}$.

$$\therefore \quad \left(\frac{y}{2}+\frac{1}{2x}+\frac{1}{2x^2y}\right)dx+\frac{1}{2}\left(x-\frac{1}{y}+\frac{1}{xy^2}\right)dy=0$$

This eqn is exact. ∴ the solution is,

$$xy-\frac{1}{xy}+\log\frac{x}{y}=2C_1=C$$

119. Solve $[xy\sin(xy)+\cos(xy)]\,ydx+[xy\sin(xy)-\cos(xy)]\,xdy=0$ [V.T.U. F/M, 2005]

Solution: This is in the form

$$yf_1(xy)dx+xf_1(xy)\,dy=0;$$

$$\therefore \quad \text{I.F.}=\frac{1}{Mx-Ny}=\frac{1}{2xy\cos(xy)}$$

Multiply the given equation by $\frac{1}{2xy\cos(xy)}$

$$\therefore \quad \frac{1}{2}\left[\tan(xy)+\frac{1}{xy}\right]ydx+\frac{1}{2}\left[\tan(xy)-\frac{1}{xy}\right]xdy=0$$

This equation is exact whose solution is

$$\log \sec (xy) + \log x - \log y = 2C = \log k \quad \text{or} \quad x = ky \cos xy$$

120. Solve $(x^3y^3 + x^2y^2 + xy + 1)\, ydx + (x^3y^3 - x^2y^2 - xy + 1)\, xdy = 0$

Solution: Hence

$$Mx = xy(x^3y^3 + x^2y^2 + xy + 1),$$
$$Ny = xy(x^3y^3 - x^2y^2 - xy + 1),$$

$$\therefore \quad \text{I.F. } \frac{1}{Mx - Ny} = \frac{1}{2x^2y^2(xy+1)}$$

Multiply the given equation by $\frac{1}{2x^2y^2(xy+1)}$.

$$\therefore \quad \frac{(xy+1)(x^2y^2+1)}{2x^2y^2(xy+1)} \cdot ydx + \frac{(xy+1)(x^2y^2-2xy+1)}{2x^2y^2(xy+1)} \cdot xdy = 0$$

i.e.,
$$\frac{1}{2}\left(\frac{x^2y^2+1}{x^2y^2}\right)ydx + \frac{1}{2}\left(\frac{x^2y^2-2xy+1}{x^2y^2}\right)xdy = 0$$

or,
$$\frac{1}{2}\left(y + \frac{1}{x^2y}\right)dx + \frac{1}{2}\left(x - \frac{2}{y} + \frac{1}{xy^2}\right)xdy = 0$$

This equation is exact whose solution is

$$\underset{y-\text{const}}{\int \frac{1}{2}\left(y + \frac{1}{x^2y}\right)dx} + \frac{1}{2} \underset{\text{(Only those free from } x)}{\int \left(-\frac{2}{y}\right)dy} = C$$

$$\frac{1}{2}\left(xy - \frac{1}{xy} - 2\log y\right) = C \quad \left(xy - \frac{1}{xy} - 2\log y\right) = C.$$

Rule (III) Let $Mdx + Ndy = 0$ by the given differential equation.

If $\dfrac{\frac{\partial M}{\partial y} - \frac{\partial N}{\partial x}}{N}$ = a function of x alone (as constant) say $f(x)$.

Then $e^{\int f(x)\,dx}$ is an integrating factor.

121. Solve $(2x \log x - xy)dy + 2ydx = 0$

Solution: Here, $M = 2y; \quad N = 2x \log x - xy$

$$\therefore \quad \frac{dM}{dy} = 2; \quad \frac{\partial N}{\partial x} = 2(1 + \log x) - y.$$

Here
$$\frac{\frac{\partial M}{\partial y} - \frac{\partial N}{\partial x}}{N} = \frac{2 - 2 - 2\log x + y}{2x \log x - xy} = -\frac{1}{x} = f(x)$$

$$\therefore \qquad \text{I.F.} = e^{\int f(x)\,dx} = e^{\int (-1/x)\,dx} = e^{-\log x} = e^{\log x^{-1}} = x^{-1} = \frac{1}{x}.$$

Multiply the given equation by $\frac{1}{x}$ we get

$$\frac{2y}{x}\,dx + (2\log x - y)\,dy = 0$$

So $$\int \frac{2y}{x}\,dx + \int -y\,dy = C \quad \text{or} \quad 2y\log x - \frac{1}{2}y^2 = C. \textit{ Ans.}$$

122. Solve $(x^4e^x - 2mxy^2)dx + 2mx^2y\,dy = 0$

Solution: Here,
$$M = x^4e^x - 2mxy^2,$$
$$N = 2mx^2y;$$

$$\therefore \qquad \frac{dM}{dy} = -4mxy; \qquad \frac{\partial N}{\partial x} = 4mxy$$

$$\therefore \qquad \frac{\left(\frac{\partial M}{\partial y} - \frac{\partial N}{\partial x}\right)}{N} = \frac{-8mxy}{2mx^2y} = -\frac{4}{x} = \text{function of } x \text{ alone.}$$

$$\therefore \qquad \text{I.F.} = e^{-4\int \frac{1}{4}dx} = \frac{1}{x^4}$$

$$\therefore \qquad \frac{x^4e^x - 2mxy^2}{x^4}\,dx + \frac{2mx^2y}{x^4}\,dy = 0 \text{ is exact.}$$

$\therefore$ General solution is:

$$\int e^x\,dx - 2my^2\int \frac{1}{x^3}\,dx = C$$

$$\therefore \qquad e^x + \frac{my^2}{x^2} = C \text{ is the solution.}$$

123. Solve $(4x^2 + y^2 + x)dx + xy\,dy = 0$ [VTU, March, 2000]

Solution:

Given:
$$M = (x^2 + y^2 + x);\ N = xy$$

$$\frac{dM}{dy} = 2y; \quad \frac{\partial N}{\partial x} = y;$$

$$\therefore \qquad \left(\frac{\partial M}{\partial y} - \frac{\partial N}{\partial x}\right) = y$$

$$\therefore \qquad \frac{1}{N}\left(\frac{\partial M}{\partial y} - \frac{\partial N}{\partial x}\right) = \frac{1}{xy}\cdot y = \frac{1}{x}.$$

$$\therefore \qquad e^{\int f(x)dx} = e^{\int\left(\frac{1}{x}\right)dx} = e^{\log x} = x.$$

is an integrating factor.

$$\Rightarrow \qquad (x^3 + y^2x + x^2)\,dx + x^2y\,dy = 0$$

which is an exact.

∴ The solution is given by

$$\underset{y\text{-constant}}{\int (x^3 + y^2x + x^2)\,dx} + \underset{\text{(Neglect } x\text{-terms)}}{\int N dy} = \text{constant}$$

$$\Rightarrow \qquad \frac{x^4}{4} + \frac{y^2x^2}{2} + \frac{x^3}{3} + \int O dy = C$$

$$\Rightarrow \qquad \frac{x^4}{4} + \frac{y^2x^2}{2} + \frac{x^3}{3} = C.$$

124. Solve $\dfrac{dy}{dx} = \dfrac{1}{x^2(1-3y)}$.

Solution: We rewrite the given equation in the differential form:

$$(3x^2y - x^2)\,dx + 1dy = 0, \text{ where}$$

$$M = 3x^2y - x^2;\ N = 1.$$

$$\text{Then} \qquad \text{I.F.} = \frac{\left(\dfrac{\partial M}{\partial y} - \dfrac{\partial N}{\partial x}\right)}{N} = \frac{3x^2 - 0}{1} = 3x^2 = \text{a function of } x\text{-alone}$$

$$\therefore \qquad \text{I.F.} = e^{\int 3x^2} = e^{x^3}.$$

Multiply the given equation by e^{x^3}.

$$\therefore \qquad e^{x^3}(3x^2y - x^2)\,dx + e^{x^3}dy = 0.$$

Which is exact.

∴ Its solution is:

$$\underset{y\text{-const}}{\int M dx} + \int (\text{those free form is } N) dy = C$$

or
$$y = ke^{-x^3} + \frac{1}{3}.$$

(after simplification)

125. Solve $\dfrac{dy}{dx} = 2xy - x$

Solution: We can rewrite it as:

$$(-\,2xy + x)dx + dy = 0.$$

$$\therefore \qquad \text{I.F.} = \frac{1}{N}\left(\frac{\partial M}{\partial y} - \frac{\partial N}{\partial x}\right)$$

PART-D

$$= \frac{(-2x)-(0)}{1} = -2x = \text{function of } x\text{-alone.}$$

Multiply the given equation by I.F.

where

$$\therefore \quad \text{I.F} = e^{\int(-2x)dx} = e^{-x^2}$$

$$(-2\,xye^{-x^2} + xe^{-x^2})\,dx + e^{-x^2}dy = 0$$

which is exact, The solution is given by

$$\int_{y\text{-const}}(-2xye^{-x^2} + xe^{-x^2})dx + \int(0)dy = C$$

i.e.,
$$ye^{-x^2} - \frac{1}{2}e^{-x^2} = C \quad \text{or} \quad ye^{-x^2} = ke^{-x^2} + \frac{1}{2}$$

126. Solve $(2x^3y^2 + 4xy + 2xy^2 + xy^4 + 2y) \cdot dx + 2(y^3 + x^2y + x)\,dy = 0.$

Solution: Please try yourself.

Hint:
$$\frac{1}{N}\left(\frac{\partial M}{\partial y} - \frac{\partial N}{\partial x}\right) = 2x$$

$$\text{I.F.} = e^{\int 2x dx} = e^{x^2}.$$

$\therefore$ G.E. becomes

$$\int(2x^3y^2 + 4x^2y + 2xy^2 + xy^4 + 2y)e^{x^2}dx.$$

$$= \int[2xy^2 + 2x^3y^2)e^{x^2}dy + \int(2y + 4x^2)e^{x^2}dx + \int xy^4e^{x^2}dx$$

$$= x^2y^2e^{x^2} + 2xye^{x^2} + \frac{1}{2}y^4e^{x^2} + C$$

$\therefore$ Solution is: $(2x^2y^2 + 4xy + y^4)\,e^{x^2} = C$

Rule (IV):

Let $Mdx + Ndy = 0$ be the given differential equation.

If
$$\frac{\dfrac{\partial N}{\partial x} - \dfrac{\partial M}{\partial y}}{M} = \text{a function of } y\text{-alone (or constant) say } \phi(y)$$

Then $e^{\int \phi(y)dy}$ is an I.F.

Worked Examples [Related to the Rule (IV)]

127. Solve: $(3x^2y^4 + 2xy)\,dx + (2x^3y^3 - x^3)dy = 0$

Solution: Here, $M = 3x^3y^4 + 2xy;$
$N = 2x^3y^3 - x^2$

$$\therefore \quad \text{I.F.} = \frac{\dfrac{\partial N}{\partial x} - \dfrac{\partial M}{\partial y}}{M} = \frac{6x^2y^3 - 2x - (12x^2y^3 + 2x)}{3x^2y^4 + 2xy}$$

$$= \frac{-2x(3xy^3 + 2)}{xy(3xy^3 + 2)} = -\frac{2}{y}$$

which is a function of y-alone

$$\therefore \qquad \text{I.F.} = e^{\int\left(-\frac{2}{y}\right)dy} = \frac{1}{y^2}$$

Multiply the equation by $\frac{1}{y^2}$; we get

$$\therefore \qquad \left(3x^2y^2 + \frac{2x}{y}\right)dx + \left(2x^3y - \frac{x^2}{y^2}\right)dy = 0 \qquad [1]$$

which is exact and whose solution is,

$$\int_{y\text{-const}} \left[3x^2y^2 + \frac{2x}{y}\right]dx = C$$

$$i.e., \qquad x^3y^2 + \frac{x^2}{y} = C, \quad \text{or} \quad x^3y^3 + x^2 = cy.$$

128. Solve $(y^4 + 2y)\,dx + (xy^3 + 2y^4 - 4x)dy = 0$

Solution:

$$M = y^4 + 2y;$$

$$N = xy^3 + 2y^4 - 4x \quad \frac{dM}{dy} = 4y^3 + 2;$$

$$\frac{\partial N}{\partial x} = y^3 - 4$$

$$\therefore \qquad \text{I.F.} = \frac{1}{M}\left[\frac{\partial M}{\partial y} - \frac{\partial N}{\partial x}\right] = \frac{1 - 3(y^3 + 2)}{y(y^3 + 2)} = \frac{-3}{y}$$

$$\therefore \qquad \text{I.F.} = e^{-\int\frac{3}{y}} = \frac{1}{y^3}$$

Multiply by $\frac{1}{y^3}$, the G.E. becomes

$$\left(y + \frac{2}{y^2}\right)dx + \left(x + 2y - 4\cdot\frac{x}{y^3}\right)dy = 0$$

$$\text{or} \qquad (ydx + xdy) + 2ydy + \left(\frac{2ydx - 4xdy}{y^3}\right) = 0$$

$$i.e. \qquad d(xy) + d(y^2) + d\left(\frac{2x}{y^2}\right) = 0$$

on integrating, we get

$$xy + y^2 + \frac{2x}{y^2} = C$$

or $$xy^3 + y^4 + 2x = Cy^2$$

is the solution.

129. Solve $(xy^2 - x^2)dx + (3x^2y^2 + x^2y) - 2x^3]dy = 0$

Solution:

$$\frac{\partial M}{\partial y} - \frac{\partial N}{\partial x} = -6xy^2 + 6x^2$$

$$\therefore \qquad \frac{\left(\frac{\partial M}{\partial y} - \frac{\partial N}{\partial x}\right)}{M} = \frac{-(6xy^2 - x^2)}{xy^2 - x^2} = -6$$

Hence $$\text{I.F.} = e^{\int(-\alpha)dy} = e^{\int 6dy} = e^{6y}$$

Multiply the G.E. by I.F. = e^{6y}, we get

$$e^{6y}(xy^2 - x^2)\,dx + e^{6y}\,[(3x^2y^2 + x^2y) - 2x^3)]dy = 0$$

is an exact equation.

$\therefore$ Its solution is

$$\left(e^{6y} \cdot y^2 \cdot \frac{x^2}{2} - e^{6y} \cdot \frac{x^3}{3}\right) + 0 = C.$$

Rule (V): If the equation $Mdx + Ndy = 0$ is of the type where,

$$M = (mx^{\alpha}y^{\beta} + m'x^{\alpha'}y^{\beta'})\,y$$

and $$N = (nx^{\alpha}y^{\beta} + n'x^{\alpha'}y^{\beta'})\,x$$

where $\alpha, \beta, \alpha', \beta', m, n, m', n'$ are real constant then I.F. can be found as follows:

(1) First we write the equation as:

$$x^{\alpha}y^{\beta}\,(mydx + nxdy) + x^{\alpha^1}\,y^{\beta^1}\,(m'ydx + n'xdy) = 0.$$

(2) I.F. for expression (1) is $x^{Km-1-\alpha} \cdot y^{Kn-1-\beta}$ and I.F. for expression (II).

$$x^{k^1m^1-1-\alpha^1} \cdot y^{k^1n^1-1-\beta^1}$$

(3) Equate the indices of x and y of both the integrating factors and final k or k^1.

(4) Knowing k or k', we can immediately write the I.F.

130. *Solve* $(y^3 - 2x^2y)dx + (2xy^2 - x^3)dy = 0$

Solution:

$$(y^3 - 2x^2y)dx + (2xy^2 - x^3)dy = 0 \qquad [1]$$

or $$y^2(ydx + 2xdy) + x^2(-2ydx - xdy) = 0$$

For $y^2(ydx + 2xdy)$

We have $\alpha = 0, \beta = 2, m = 1, n = 2$

$$\text{I.F.} = x^{km-1-\alpha}\,y^{kn-1-\beta} = x^{k-1-0} \cdot y^{2k-1-2}$$

$$= x^{k-1} \cdot y^{2k-3} \qquad [2]$$

For $x^2(-2ydx - xdy)$

$\alpha = 2, \beta = 0, m = -2, n = -1.$

$$\text{I.F.} = x^{k^1m^1-1-\alpha^1} \cdot y^{k^1n^1-1-\beta^1} = x^{-2k^1-1-2} \cdot y^{-k^1-1-0}$$
$$= x^{-2k^1-3} \cdot y^{-k^1-1} \qquad [3]$$

The integrating factors (2) and (3) will be same if

$$k - 1 = -2k^1 - 3; \quad \text{and} \quad 2k - 3 = -k^1 - 1$$

on solving $k = 2.$

Substituting in (2), we see that an I.F. of the given differential is

$$x^{2-1}\, y^{4-3} \text{ i.e. } xy$$

∴ on multiplying the given equation (1) by xy we get an exact differential equation,

$$(xy^4 - 2x^3y^2)\, dx + (2x^2y^3 - x^4y)dy = 0$$

or
$$\int (xy^4 - 2x^3y^2)dx = C_1$$

⇒
$$\frac{x^2}{2}\, y^4 - 2 \cdot \frac{x^4}{4}\, y^2 = C_1.$$

or
$$x^2y^2\,(x^2 - y^2) = C. \qquad \textit{Ans.}$$

131. Solve $(3x^2y - y^4)dx + (6x^3 + xy^3)dy = 0$

Solution:

Given:
$$(3x^2y - y^4)dx + (6x^3 + xy^3)\, dy = 0 \qquad [1]$$

i.e.
$$x^2(3ydx + 6xdy) + y^3(-y\, dx + x\, dy) = 0$$

For $x^2\,(3ydx + 6xdy)$

We here $\alpha = 2, \beta = 0, m = 3, n = 6,$

∴
$$\text{I.F.} = x^{km-1-\alpha} \cdot y^{kn-1-\beta} = x^{3k-1-2} \cdot y^{6k-1-0}$$
$$= x^{3k-3} \cdot y^{6k-1} \qquad [2]$$

For $y^3\,(-ydx + xdy)$

$$\alpha = 0, \beta = 3, m = -1, n = 1$$

∴
$$\text{I.F.} = x^{k^1m^1-1-\alpha^1} \cdot y^{k^1n^1-1-\beta^1} = x^{-k^1-1-0} \cdot y^{k^1-1-3}$$
$$= x^{-k^1-1} \cdot y^{k^1-4} \qquad [3]$$

The integrating factors (2) and (3) will be same if

$$3k - 3 = -k^1 - 1;$$

and
$$6k - 1 = k^1 - 4.$$

$$k^1 = (6k + 3)$$

∴
$$3k - 3 = -6k - 3 - 1$$
$$= -6k - 4$$
$$9k = -4 + 3 = -1$$
$$k = -\frac{1}{9}$$

on solving,
$$k = \left(-\frac{1}{9}\right)$$

Substituting is (2) we see that an I.F. of the given differential is

$$x^{-\frac{1}{3}-3} \cdot y^{-\frac{2}{3}-1} - 1 = x^{-\left(\frac{10}{3}\right)} \cdot y^{-\frac{5}{2}}$$

Multiply the given equation (1) by $x^{-\left(\frac{10}{3}\right)} \cdot y^{-\frac{5}{2}}$, we get

$$x^{-\left(\frac{10}{3}\right)} \cdot y^{-\frac{5}{3}} (3x^2y - y^4)dx + x^{-\left(\frac{10}{3}\right)} \cdot y^{-\frac{5}{3}} (6x^3 + xy^3)dy = 0$$

i.e., $$(3 \cdot x^{-4/3} \cdot y^{-2/3} - x^{-10/3} \cdot y^{7/3})dx + [6x^{-1/3} \cdot y^{-5/3} + x^{-7/3} \cdot y^{4/3}]dy = 0$$

This is an exact diff. equation. Its solution is given by

$$\left[3 \cdot \frac{x^{\left(-\frac{4}{3}\right)+1}}{\left(-\frac{4}{3}\right)+1} \cdot y^{-\frac{2}{3}} - \frac{x^{-\left(\frac{10}{3}\right)+1}}{\left(-\frac{10}{3}\right)+1} \cdot y^{7/3}\right] + 0 = C$$

or $$\frac{3 \cdot x^{-\frac{1}{3}} \cdot y^{-\frac{2}{3}}}{\left(-\frac{1}{3}\right)} - \frac{x^{-\left(\frac{1}{3}\right)} \cdot y^{\frac{7}{3}}}{\left(-\frac{7}{3}\right)} = C$$

is the solution.

132. Solve: $(y^2 + 2x^2y)dx + (2x^3 - xy)dy = 0$ [VTU, March 2001]

Solution:

Given: $$(y^2 + 2x^2y)dx + (2x^3 - xy)dy = 0. \qquad [1]$$

or $$y(ydx - xdy) + 2x^2 (ydx + xdy) = 0$$

For $y(ydx - xdy)$,

We have, $\alpha = 0$, $\beta = 1$, $m = 1$, $n = -1$

$$\text{I.F.} = x^{km-1-\alpha} \cdot y^{kn-1-\beta} = x^{k-1-0} \cdot y^{-k-1-1}$$

$$= x^{k-1} \cdot y^{-k-2} \qquad [2]$$

For $$2x^2(ydx + xdy) = x^2(2ydx + 2xdy)$$

$\alpha = 2$; $\beta = 0$; $m = 2$; $n = 2$

$$\text{I.F.} = x^{k^1m^1-1-\alpha^1} \cdot y^{k^1n^1-1-\beta^1} = x^{2k^1-1-2} \cdot y^{2k^1-1-0}$$

$$= x^{2k^1-3} \cdot y^{2k^1-1} \qquad [3]$$

The integrating factors (2) and (3) will be same if

$$\left.\begin{aligned} k - 1 &= 2k^1 - 3; \\ -k - 2 &= 2k^1 - 1; \end{aligned}\right\}$$

Solving, $$k = -3/2$$

On multiplying the G.E. by I.F.

$$= x^{-3/2-1} \cdot y^{+3/2-2} = x^{-5/2} \cdot y^{-1/2}$$

Hence (1) reduces to $[x^{-5/2} \cdot y^{3/2} + 2x^{-1/2} \cdot y^{1/2}]\, dx + [2x^{1/2}y^{1/2} - x^{-3.2} \cdot y^{1/2}]dy = 0$, which is exact.
Hence, the required solution is,

$$\int_{y\text{-const}} (x^{-5/2}\, y^{3/2} + 2x^{-1/2} \cdot y^{1/2})dx + 0 = C$$

i.e., $$-\frac{2}{3}x^{-3/2} \cdot y^{3/2} + 2y^{1/2} \cdot 2 \cdot x^{1/2} = C$$

i.e. $$6x^{1/2}y^{1/2} - x^{-3/2} \cdot y^{3/2} = C$$

133. Solve: $(3y + 2x^2)xdy + 2y(2y + 3x^2)dx = 0$

Solution: The equation can be arranged in the form.

$$3yxdy + 2x^3dy + 4y^2dx + 6x^2ydx = 0 \quad [1]$$

$$(2x^3dy + 6x^2ydx) + (3xydy + 4y^2dx) = 0$$

$$x^2\,[6ydx + 2xdy) + y[4ydx + 3xdy] = 0 \quad [2]$$

For $x^2[6ydx + 2xdy]$

$$\alpha = 2;\ \beta = 0;\ m = 6;\ n = 2$$

$\therefore$
$$\text{I.F.} = x^{km-1-\alpha} \cdot y^{kn-1-\beta} = x^{6k-1-2} \cdot y^{2k-1-0}$$
$$= x^{6k-3} \cdot y^{2k-1} \quad [3]$$

and For $y[4ydx + 3xdy]$

$$\alpha = 0;\ \beta = 1;\ m = 4;\ n = 3$$

$$\text{I.F.} = x^{k'm'-1-\alpha'} \cdot y^{k'n'-1-\beta'} = x^{4k'-1-0} \cdot y^{3k'-1-1}$$
$$= x^{4k'-1} \cdot y^{3k'-2} \quad [4]$$

The integrating factors [3] and [4] will be same if

$$6k - 3 = 4k' - 1$$
$$2k - 1 = 3k' - 2$$

i.e.
$$6k - 4k' - 2 = 0$$
$$2k - 3k' + 1 = 0$$

$\therefore\ \Rightarrow\ (k = 1;\ k' = 1)$

$\therefore$ From (3). $\quad \text{I.F.} = x^3 \cdot y$

$\therefore$ *Solution:*

$$x^5 \cdot y\,[6ydx + 2xdy] + x^3y^2\,[4ydx + 3xdy] = 0$$

which is exact.

Solution is: $\quad 1 \cdot x^6y^2 + 1 \cdot x^4y^3 = C$

or $\quad x^4y^2(x^2 + y) = C$

Exercises on Rule (I) $\dfrac{1}{Mx + Ny}$, $Mn + Ny \neq 0$

1. $(3xy^2 - y^3)dx + (xy^2 - 2x^2y)dy = 0$ $\quad \left(Ans: \dfrac{cy^2}{x^3} = e^{y/x}\right)$

2. $(x^2 - 3xy + 2y^2)dx + (3x - 2y)dy = 0$ $\quad (Ans: x^2 \log x + 3xy = y^2 + (cx^2))$

$$y - xp = x + yp, \left(P = \frac{dy}{dx}\right)$$

$$\left(Ans.\ \frac{1}{2}\log(x^2 + y^2) = C + \tan^{-1}\left(\frac{x}{y}\right) - \frac{6}{\sqrt{7}}\tan^{-1}\left(\frac{x + 2y}{x\sqrt{7}}\right)\right).$$

Exercises on Rule (II)

$$\left(\frac{1}{Mx - Ny} = \text{I.F.}\right)$$

PART-D

Solve the following. D.Es.

1. $(x^2y^2 + xy + 1)ydx + (x^2y^2 - xy + 1)xdy = 0$
2. $(x^2y^2 + 5xy + 2)ydx + (x^2y^2 + 4xy + 2)xdy = 0$
3. $y(xy - 3)dx + x(3xy + 7)dy = 0$
4. $y(xy + 2x^2y^2) + x(xy - x^2y^2)dy = 0$

EXERCISES

1. $(x^2 + y^2 + 2x)dx + 2ydy = 0$
2. $(x^2 + y^2)dx - 2xydy = 0$
3. $\left[\left(y + \frac{y^3}{3}\right) + \frac{x^2}{2}\right] dx + (1 + y)^2 \frac{x}{4} dy = 0$
4. $(20x^2 + 8xy + 4y^2 + 3y^3)ydx - (4)(x^2 + xy + y^2 + y^3)xdy = 0$
5. $(y^4 + 2y)dx + (xy^3 + 2y^4 - 4x)dy = 0$
6. $(xy^3 + y)dx + 2(x^2y^2 + x + y^4)dy = 0$
7. $(xy^2 - x^2)dx + (3x^2y^2 + x^2y - 2x^3 + y^2)dy = 0$
8. $(2x^2y - 3y^2)dx + (2x^3 - 12xy + \log y)dy = 0$
9. $(3x^2y^4 + 12xy)dx + (2x^3y^3 - x^2)dy = 0$
10. $(7x^4y + 2xy^2 - x^3)dx + (x^4 + xy)xdy = 0$

4.5 LINEAR AND BERNOULLI'S EQUATION

The equation

$$Mdx + Ndy = 0$$

i.e. $$M + N \cdot \frac{dy}{dx} = 0 \text{ is linear}$$ (of the first order)

if (1) N does not contain y

and if (2) M involves y to the first degree only.

i.e. if
$$N = f_1(x);$$
$$M = yf_2(x) + f_3(x)$$

Where $f_1(x)$, $f_2(x)$, $f_3(x)$ are functions of x-alone. Thus the general form of a linear equation in dependent variable can be written as:

$$f_1(x)\frac{dy}{dx} + yf_2(x) + f_3(x) = 0$$

i.e. $$\frac{dy}{dx} + y\frac{f_2(x)}{f_1(x)} = -\frac{f_3(x)}{f_1(x)}.$$

i.e. $$\boxed{\frac{dy}{dx} + Py = Q}$$ (linear in y)

Where P and Q are functions of x-alone.

Note: If x is dependent variable, the linear d.e. can be written as

(Linear is x); $\frac{dx}{dy} + P \cdot x = Q$

where P and Q are functions of y.

Working Rule

(1) Make the coeff of $\frac{dy}{dx}\left(\text{or}\frac{dx}{dy}\right)$ as unity

(2) Find P and hence I.F. $= e^{\int Pdx}$ (for the equation which are linear in y)
and I.F. $= e^{\int Pdy}$ (for the equations which are linear in x)

(3) The solution is:

$$y\,(\text{I.F.}) = \int Q\,(\text{I.F.})dx + C \text{ (for linear in } y)$$

and $$x\,(\text{I.F.}) = \int Q\,(\text{I.F.})dy + C \text{ (for linear in } x)$$

A. Worked Examples (Direct applications)

134. Solve $\frac{dy}{dx} + y\cot x = \cos x$ [VTU, August, 2000]

Solution: This is of the standard linear form.

Here, $P = \cot x;\ Q = \cos x;$

$$\int Pdx = \int \cot x dx = \log \sin x$$

$\therefore$ $$e^{\int Pdy} = e^{\log \sin x} = \sin x$$

General solution is;

$\therefore$ $$\int Q e^{\int Pdx} = \int (\sin x)(\cos x)\,dx + C$$

$$= \frac{1}{2}\int \sin 2x dx + C = \frac{1}{4}\cos 2x + C$$

$\therefore$ *Solution:* $$-y e^{\int Pdx} = A + \int Q e^{\int Pdx}\cdot dx$$

i.e. $$y \sin x = -\frac{1}{4}\cos 2x + C$$

135. Solve: $\frac{dy}{dx} + y\cot x = \operatorname{cosec} x$

Solutions: This is of the form $\left(\frac{dy}{dx}\right) + Py = Q$

The solution is $y e^{\int Pdx} = A + \int Q e^{\int Pdx}\,dx$ [1]

Here, $P = \cot x;$

$\therefore$ $$\int Pdx = \int \cot x dx = \log \sin x$$

$\therefore$ $$e^{\int Pdx} = e^{\log \sin x} = \sin x \quad [2]$$

and $$\int Q e^{\int Pdx} dx = \int \operatorname{cosec} x\,(\sin x)\,dx = \int 1\,dx = x \quad [3]$$

PART-D

∴ *Solution:* $y \sin x = (A + x)$

136. Solve: $\frac{dy}{dx} + y \tan x = \sec x.$

Solution: This is a linear differential equation.

$$\frac{dy}{dx} + Py = Q$$

∴ Solution is: $ye^{\int Pdx} = A + \int Qe^{\int Pdx}$

Here $P = \tan x$;

$$\int Pdx = \int \tan x dx = \log \sec x$$

$$\therefore \quad e^{\int Pdx} = e^{\log \sec x} = \sec x.$$

$$\therefore \quad y \sec x = A + \int \sec x \cdot \sec x \, dx$$

$$= A + \int \sec^2 x \, dx = A + \tan x$$

Solution: $y \cdot \sec x = A + \tan x$

137. Solve: $(x^2 + 1) \frac{dy}{dx} + 2xy = 1$

Solution: G.E. can be written as: $\frac{dy}{dx} + \frac{2x}{(x^2+1)} \cdot y = \frac{1}{(x^2+1)}$ [1]

Here, $P = \frac{2x}{(x^2+1)}; \; Q = \frac{1}{(x^2+1)}$

$$\text{I.F.} = e^{\int Pdx} = e^{\int \frac{2x}{x^2+1} dx} = e^{\log(x^2+1)} = (x^2 + 1)$$

The G.S. of (1) is

$$y(x^2 + 1) = \int (x^2 + 1) \frac{1}{(x^2+1)} dx + A$$

i.e., $y(x^2 + 1) = A + x$

138. Solve: $(1 + x^2) \frac{dy}{dx} + xy = \sin h^{-1} x$ [B.U. Feb. 1983]

Solution: Dividing throughout by $(1 + x^2)$,

Given: $\frac{dy}{dx} + \frac{x}{(1+x^2)} \cdot y = \frac{\sin h^{-1}(x)}{(1+x^2)}$

This is of the form:

$$\frac{dy}{dx} + P \cdot y = Q \qquad \text{(Linear in } y\text{)}$$

where $P = \dfrac{x}{1+x^2}$ = functions of x-alone

$Q = \dfrac{\sinh^{-1}(x)}{1+x^2}$ = functions of x-alone

Integrating factor (abbreviated as I.F.) $= e^{\int P dx}$

i.e., $= e^{\int \frac{x}{1+x^2} \cdot dx} = e^{\frac{1}{2}\log(1+x^2)} = (1 + x^2)^{1/2}$

or $= \sqrt{(1+x^2)}$.

Solution is given by

$$y \cdot (\text{I.F.}) = \int Q \cdot (\text{I.F.})\, dx + C$$

$$y \cdot \sqrt{(1+x^2)} = \int \frac{\sinh^{-1}(x)}{(1+x^2)} \sqrt{(1+x^2)}\, dx + C$$

$$= \int \sinh^{-1} x \cdot \frac{1}{\sqrt{1+x^2}}\, dx + C$$

$$= \int t \cdot dt + C$$

$(\sinh^{-1}(x) = t)$

$$\therefore \quad \frac{1}{\sqrt{1+x^2}} \cdot dx = dt$$

i.e. $\quad y\sqrt{(1+x^2)} = \dfrac{t^2}{2} + C.$

where, $\quad t = \sinh^{-1}(x)$

Thus, the solution is

$$y\sqrt{(1+x^2)} = \frac{1}{2}(\sinh^{-1} x)^2 + C.$$

Note: $\quad e^{\frac{1}{2}\log(1+x^2)} = (1 + x^2)^{1/2}$.

In general, we can written

$$e^{n \log X} = X^n \quad \text{and} \quad e^{\log X} = X.$$

139. Solve: $(1 + y^2)dx = (\tan^{-1} y - x)dy$ [B.U. Oct. 1983]

Solution:

Given: $\quad (1 + y^2) \cdot \dfrac{dx}{dy} + x = \tan^{-1}(y)$

or $\quad \dfrac{dx}{dy} + \dfrac{1}{(1+y^2)} \cdot x = \dfrac{\tan^{-1}(y)}{(1+y^2)}$ (Linear in x)

$\therefore$ Compare with $\dfrac{dx}{dy} + P \cdot x = Q$

where $$P = \frac{1}{(1+y^2)};$$

$$Q = \frac{\tan^{-1}(y)}{(1+y^2)} \text{ are functions of } y \text{ alone.}$$

Hence $$\text{I.F.} = e^{\int \frac{1}{1+y^2} \cdot dy} = e^{\tan^{-1}(y)}$$

$\therefore$ The solution is

$$xe^{\tan^{-1}(y)} = \int \frac{\tan^{-1}(y)}{1+y^2} \cdot e^{\tan^{-1}(y)} \cdot dy + C$$

Substituting, $\tan^{-1}(y) = t,$

$$\frac{1}{(1+y^2)} \cdot dy = dt.$$

$$\int \frac{\tan^{-1}(y)}{(1+y^2)} \cdot e^{\tan^{-1}(y)} \cdot dy$$

$$= \int t \cdot e^t dt = t \cdot e^t - \int e^t \cdot dt = t \cdot e^t - e^t$$

$$= \tan^{-1}y \cdot e^{\tan^{-1}(y)} - e^{\tan^{-1}(y)}$$

The required equation is

$$x \cdot e^{\tan^{-1}(y)} = \tan^{-1}(y)\, e^{\tan^{-1}(y)} - e^{\tan^{-1}(y)} + C;$$

$$x = \tan^{-1}(y) - 1 + Ce^{-\tan^{-1}(y)}$$ [1]

140. Solve: (similar to the above example)

$$(1 + x^2)dy = (\tan^{-1} x - y)\, dx$$

Solutions: Dividing by $(x^2 + 1)$ and adjusting, we have,

$$\frac{dy}{dx} = \frac{\tan^{-1}(x)}{(1+x^2)} - \frac{y}{x^2+1};$$

$$\frac{dy}{dx} + \frac{1}{(x^2+1)} \cdot y = \frac{\tan^{-1}(x)}{(1+x^2)}$$ [1]

Here, $$P = \left[\frac{1}{(x^2+1)}\right];$$

$$\int P dx = \int \left[\frac{1}{(x^2+1)}\right] dx = \tan^{-1}(x)$$

$$\therefore \quad e^{\int P\,dx} = e^{\tan^{-1}(x)} \quad [2]$$

$$\therefore \quad \int Q e^{\int P\,dx} = \int \frac{\tan^{-1} x}{(x^2+1)} \cdot e^{\tan^{-1}(x)}\,dx \quad [3]$$

Put $\tan^{-1}(x) = t$;

$$\therefore \quad \left[\frac{1}{(1+x^2)}\right] dx = dt.$$

$$\therefore \quad \int te^t dt = [(e^t t) - \int e^t \times 1 \times dt]$$

$$= [te^t - (-e^t)] = e^t\,[t-1]\ [4]$$

Solution: $ye^{\tan^{-1}(x)} = A + (-1 + \tan^{-1} x)\, e^{\tan^{-1} x}$.

141. Solve: $\dfrac{dy}{dx} + \dfrac{n}{x}\, y = \dfrac{a}{x^n}$;

Solution: This is of the standard liner form. Here $P = \dfrac{n}{x}$;

$$\therefore \quad \int P dx = \int ({}^{n}/_{x}) dx = n \log x;$$

$$\therefore \quad e^{\int P\,dx} = e^{n\log x} = x^n\ [1]$$

$$\int Q e^{\int P\,dx}\, dx = \int \left(\frac{a}{x^n}\right)(x^n) dx = \int a dx = ax \quad [2]$$

$\therefore$ *Solution:* $y e^{\int P\,dx} = A + \int Q e^{\int P\,dx} dx$

i.e. $y(x^n) = (A + ax)$

142. Solve $dr + (2r \cot \theta + \sin 2\theta)\, d\theta = 0$ [V.T.U. J/A. 2004]

Solution: Given : $\dfrac{dr}{d\theta} + (2 \cot \theta)\, r = -\sin 2\theta$

which is linear 'r' [1]

$$\therefore \quad \text{I.F.} = e^{\int (2\cot\theta)\,d\theta} = e^{2\log\sin\theta} = \sin^2\theta$$

$\therefore$ The general solution is given

by $\quad r \sin^2\theta = \int \sin^3\theta \cos\theta\, d\theta + c = -\dfrac{1}{2}\sin^4\theta + c$

or $\quad \boxed{2r \sin^2\theta + \sin^4\theta = 2c.}$

143. Solve $\quad ydx + (3x + 2 - yx)dy = 0$ [Mys. U. Sept. 1984]

Solution:

Given: $\quad y\dfrac{dx}{dy} + 3x + 2 - yx = 0$

PART-D

i.e.
$$y\frac{dx}{dy} + (3-y)\,x = -2$$

$$(\div y);\quad \frac{dx}{dy} + \left(\frac{3-y}{y}\right)x = \frac{-2}{y}$$

This is of the form: $\dfrac{dx}{dy} + P\cdot x = Q.$

where $P = \dfrac{3-y}{y}$; $Q = \dfrac{-2}{y}$ $\Big\}$ are functions of y alone

$\therefore$ (It is linear in x).

Integrating factor (I.F.) $= e^{\int P dy} = e^{\int \frac{3-y}{y}\cdot dy}$

$$= e^{\int\left(\frac{3}{y}-1\right)dy} = e^{3\log y - y} = e^{3\log y}\cdot e^{-y} = y^3\cdot e^{-y}$$

Solution is given by

$$x\cdot(\text{I.F.}) = \int Q\cdot(\text{I.F.})\,dy + C$$

$$x\cdot y^3 e^{-y} = \int \frac{-2}{y}\cdot y^3\cdot e^{-y}dy + C$$

$$xy^3\cdot e^{-y} = -2\int y^2\cdot e^{-y}dy + C$$

$$= -2\left[y^2\left(\frac{e^{-y}}{-1}\right) - 2y\left(\frac{e^{-y}}{1}\right) + 2\left(\frac{e^{-y}}{-1}\right)\right] + C \quad \text{(By parts)}$$

Multiply by e^y. We get the solution as

$$xy^3 = 2y^2 + 4y + 4 + Ce^y.$$

144. Solve $(x + \tan y)\,dy = \sin 2y\,dx$ [B.U. July, 1993]

Solution: It is not fruitful to have y as the dependent variable. We interchange the role of x and y. Dividing by 'dy' we have,

(II) $$(x + \tan y) = \sin 2y\cdot\left(\frac{dx}{dy}\right)$$

or $$\sin 2y\left(\frac{dx}{dy}\right) - x = \tan y$$

Divide by $\sin 2y$,

$$\frac{dx}{dy} - \frac{1}{\sin 2y}\cdot x = \frac{\tan y}{\sin 2y}$$

i.e. $$\frac{dx}{dy} - \frac{1}{2\sin y\cos y}\cdot x = \frac{\sin y}{2\sin y\cos^2 y}$$

or $$\frac{dx}{dy} - \frac{1}{2}x\frac{\sec^2 y}{\tan y} = \frac{1}{2}\sec^2 y \quad [1]$$

(which is linear in x)

$\therefore$ $$\text{I.F.} = e^{\int P dy} = e^{-\frac{1}{2}\int \frac{1}{\tan y}\cdot \sec^2 y dy}$$

$$= e^{-\frac{1}{2}\log(\tan y)} = (\tan y)^{-1/2} = \frac{1}{\sqrt{\tan y}}$$

Solution: $$x\cdot(\text{I.F.}) = \int Q\cdot e^{\int P dx} dy + A$$

i.e. $$\frac{x}{\sqrt{\tan y}} = \frac{1}{2}\int \frac{\sec^2 y}{\sqrt{\tan y}}dy + A \qquad \left(\because \int \frac{f'(x)}{\sqrt{f(x)}}dx = 2\sqrt{f(x)}\right)$$

or $$= \sqrt{\tan y} + A$$

or $$x = \tan y + A\sqrt{\tan y}$$

BERNOULLI'S EQUATIONS

The equation $$\frac{dy}{dx} + Py = Qy^n \quad [1]$$

where P and Q are functions of x alone is called Bernoulli's equations.
(For $n = 0$, $n = 1$, it is already linear).

For any other values of 'n' it can be reduced to a linear equation by the substitution

$$u = y^{1-n} = \frac{1}{y^{n-1}}.$$

This is the same as the solution of the linear equation:

$$\left(\frac{dy}{dx}\right) + Py = Qy^n \quad [1]$$

Divide by y^n; The equation reduces to

$$\left(\frac{1}{y^n}\right)\left(\frac{dy}{dx}\right) + P\left(\frac{1}{y^{n-1}}\right) = Q \quad [2]$$

Put $\frac{1}{y^{n-1}} = y^{-(n-1)} = u$;

$$\therefore \quad -(n-1)\cdot\frac{1}{y^n}\cdot\frac{dy}{dx}=\frac{du}{dx};$$

$$\frac{1}{y^n}\cdot\frac{dy}{dx}=\frac{-1}{(n-1)}\cdot\frac{du}{dx}$$

$$\therefore \quad \left[-\frac{1}{(n-1)}\right]\left(\frac{du}{dx}\right)+Pu=Q$$

i.e.,
$$\left(\frac{du}{dx}\right)+[-(n-1)P]u=[-(n-1)Q] \qquad [3]$$

Let
$$[-(n-1)P]=P';$$
$$[-(n-1)Q]=Q';$$

The equation now reduces to,

$$\left(\frac{du}{dx}\right)+P'=Q' \qquad [4]$$

which is in a Std linear form.

For this linear equation solution is:

$$ue^{\int P'dx}=A+\int Q'e^{\int P'dx}$$

$$\left[\frac{1}{y^{n-1}}\right]e^{-(n-1)\int pdx}=A+(n-1)\int Qe^{-(n-1)\int Pdx}$$

Worked Examples

145. Solve: $(x^2y^3+xy)\dfrac{dy}{dx}=1$

OR

$(x^2y^3+xy)\,dy=dx$ [VTU, Aug., 1999]

Solution: Here, it is not fruitful to have y as the dependent variable. We interchange the role of x and y. Dividing by dy, we have:

$$\left(\frac{dx}{dy}\right)=xy+x^2y^3;$$

$$\left(\frac{dx}{dy}\right)-xy=x^2y^3; \qquad [1]$$

Divide by x^2.

$$\therefore \quad \frac{1}{x^2}\frac{dx}{dy}-\frac{1}{x}\,y=y^3 \qquad [2]$$

Put $\left(-\frac{1}{x}\right) = u;$

$$\therefore \qquad \left(\frac{1}{x^2}\right)\left(\frac{dx}{dy}\right) = \left(\frac{du}{dy}\right)$$

$$\therefore \text{ From (2), } \qquad \left(\frac{du}{dy}\right) + y \cdot u = y^3 \qquad [3]$$

This is of the standard linear form in u and y.

$$\therefore \qquad ue^{\int Pdy} = A + \int Qe^{\int Pdy} dy;$$

Here, $\qquad P = y;$

$$\text{I.F} = e^{\int Pdy} = e^{\int ydy}\, e^{(y^2/2)}$$

and $\qquad \int Qe^{\int Pdy} dy = \int y^3 e^{y^2/2} dy = \int y^2 e^{(y^2/2)} \cdot ydy$

Put $(y^2/2) = t;$

$$\therefore \qquad ydy = dt; \quad y^2 = 2t$$

$$\therefore \qquad \int Qe^{\int Pdy} dy = \int y^2 e^{(y^2/2)} dy$$

$$= 2\int t \cdot e^t\, dt. = 2\,[e^t t - \int e^t \cdot 1dt]$$

$$= 2[te^t - e^t] = 2e^t\,[t - 1]$$

$$= 2e^t\,[t - 1] = 2e^{y^2/2}\,[(y^2/2) - 1]$$

$$= e^{y^2/2}\,(y^2 - 2) \qquad [4]$$

$$\therefore \textit{ Solution: } \qquad \left(-\frac{1}{x}\right) e^{y^2/2} = A + e^{y^2/2}\,(y^2 - 2)$$

146. Solve $(y \log x - 2)\, y\, dx = x\, dy$ [VTU, March, 2001]

Solution: The given equation can be written as:

$$\frac{dy}{dx} = \frac{y^2 \log x - 2y}{x}$$

i.e.

$$\frac{dy}{dx} + \frac{2y}{x} = \frac{y^2 \log x}{x}$$

(Bernoulli's form)

Divide by y^2;

The equation becomes

$$\frac{1}{y^2}\frac{dy}{dx} + \frac{2}{x} \cdot \frac{1}{y} = \frac{\log x}{x} \qquad [1]$$

PART-D

Put $\frac{1}{y} = u;$

$$\therefore \quad -\frac{1}{y^2}\cdot\frac{dy}{dx} = \frac{du}{dx}$$

$$\therefore \quad -\frac{du}{dx} + \frac{2}{x}\cdot u = \frac{(\log x)}{x}$$

i.e.,
$$\frac{du}{dx} - \left(\frac{2}{x}\right)u = -\frac{(\log x)}{x}$$

$$\text{I.F.} = e^{\int\left(-\frac{2}{x}\right)dx} \qquad \text{(Linear is } u\text{)}$$

$$= e^{-2\log x} = \frac{1}{x^2}$$

and
$$\int Q\, e^{\int P dx}\cdot dx + A = \int\left[-\left(\frac{\log x}{x}\right)\left(\frac{1}{x^2}\right)\right] dx + A$$

$$= -\int\frac{\log x}{x^3}\, dx + A = -\left[\log x\cdot\frac{x^{-2}}{-2} - \int\frac{x^{-2}}{-2}\cdot\frac{1}{x}\, dx\right] + A$$

$$= -\left[-\frac{\log x}{2x^2} + \frac{1}{2}\int x^{-3}\, dx\right] + A = -\left[-\frac{\log x}{2x^2} + \frac{1}{2}\cdot\frac{x^{-2}}{-2}\right] + A$$

$$= \frac{\log x}{2x^2} = \frac{x^{-2}}{4} + A$$

$$\therefore \text{ Solution:} \quad u\left(e^{\int P dx}\right) = \int Q e^{\int P dy}\, dx + A$$

Or

$$u\left(\frac{1}{x^2}\right) = A + \frac{\log x}{2x^2} + \frac{1}{4x^2}$$

147. Solve $x^3\frac{dy}{dx} - x^2y + y^4\cos x = 0$ [VTU. Aug., 2001]

Solution:

Given:
$$x^3\frac{dy}{dx} - x^2y = y^4\cos x \qquad [1]$$

Divide by y^4;
$$\frac{x^3}{y^4}\frac{dy}{dx} - \frac{x^2}{y^3} = -\cos x \qquad [2]$$

Put $\frac{1}{y^3} = u; \quad \frac{(-3)}{y^4} \cdot \frac{dy}{dx} = \frac{du}{dx}$

$$\therefore \quad \frac{1}{y^4}\frac{dy}{dx} = \left(-\frac{1}{3}\right)\frac{du}{dx}$$

$\therefore$ [2] becomes

$$-\frac{x^3}{3}\cdot\frac{du}{dx} - x^2 \cdot u = -\cos x$$

$$\therefore \quad \frac{du}{dx} + \frac{3}{x}u = +\frac{3\cos x}{x^3}$$

$\therefore$ *Solution:* $\quad u e^{\int P dx} = A + \int Q e^{\int P dx}\, dx;$

$$P = \left(\frac{3}{x}\right) \qquad \int P dx = \int (3/x) dx = 3 \log x$$

(I.F.) $\quad e^{\int P dx} = e^{3 \log x} = e^{\log x^3} = (x^3)$

and $\quad \int Q \cdot e^{\int P dx} \cdot dx + A = \int \left[+\frac{3\cos x}{x^3}\right] \cdot (x^3)\, dx + A$

$$= +3 \int \cos x dx + A = 3 \sin x + A$$

$$\therefore \quad u(x^3) = A + 3 \sin x$$

or $\quad \frac{x^3}{y^3} = A + 3 \sin x$

or $\quad x^3 = y^3 (A + 3 \sin x)$

148. Solve: $\frac{dy}{dx} - y \tan x = y^4 \sec x$

Solution: Divide by y^4;

$$\frac{1}{y^4}\cdot\frac{dy}{dx} - y^{-3}\tan x = \sec x \qquad [1]$$

Let $\quad \frac{1}{y^3} = u;$

$$\therefore \quad -\frac{3}{y^4}\frac{dy}{dx} = \frac{du}{dx}$$

[1] becomes $\quad -\frac{1}{3}\frac{du}{dx} - u \cdot \tan x = \sec x$

PART-D

$$\Rightarrow \quad \frac{du}{dx} + 3u \tan x = -3 \sec x \quad \text{(Linear is } u\text{)} \qquad [2]$$

$$\therefore \quad P = 3 \tan x; \int P dx = 3 \int \tan x dx$$

$$= 3 \log \sec x = 3 \log \sec^3 x$$

$$\therefore \quad \text{I.F.} = e^{\int P dx} = e^{\log \sec^3 x} = (\sec^3 x)$$

and
$$\int Q e^{\int P dx} dx + A = \int \sec^3 x \cdot (-3 \sec x)\, dx + A$$

$$= -3 \int (1 + \tan^2 x) \cdot \sec^2 x\, dx + A$$

Put $\tan x = t$; $\sec^2 x dx = dt$

$$\Rightarrow \quad u \cdot (\sec^3 x) = -3 \int (1 + t^2) \cdot dt + A$$

$$\therefore \quad u \sec^3 x = -3t - t^3 + A$$

i.e.
$$y^{-3} \sec^3 x = -3t - t^3 + A \qquad \left(\because \ u = \frac{1}{y^3}\right)$$

or
$$\frac{\sec^3 x}{y^3} + 3 \tan x + \tan^3 x = A$$

149. Solve $\dfrac{dy}{dx} + x \sin 2y = x^3 \cos^2 y$ [B.U. July, 93]

Solution: Multiply by $\sec^2 y$, we get

$$\sec^2 y \cdot \frac{dy}{dx} + x \cdot 2 \cdot \sin y \cos y \sec^2 y = x^3$$

i.e.
$$\sec^2 y \cdot \frac{dy}{dx} + 2x \cdot \tan y = x^3 \qquad [1]$$

Put $\tan y = u$;

$$\therefore \quad \sec^2 y \ \frac{dy}{dx} = \frac{du}{dx};$$

$\therefore$ [1] becomes $\dfrac{du}{dx} + 2x \cdot u = x^3$ (linear in u)

$$P = 2x; \quad Q = x^3$$

$$\therefore \quad \text{I.F.} = e^{\int 2x dx}$$

and
$$\int Q \cdot e^{\int P dy} \cdot dx = \int (x^3) \cdot e^{x^2} dx + A = \frac{1}{2} \int x^2 \cdot e^{x^2} \ 2x dx + A$$

$$= \frac{1}{2} \int t\, e^t\, dt + A; = \frac{1}{2} t\, e^t - \frac{1}{2} e^t + A$$

$\therefore$ *Solution:*

$$u(e^{x^2}) = \frac{1}{2} t\, e^t - \frac{1}{2} e^t + A$$

or $$\tan y \cdot e^{x^2} = \frac{1}{2}\, x^2 e^{x^2} + \frac{1}{2}\, e^{x^2} + A$$

or, $$\tan y = \frac{1}{2}\,(x^2 - 1) + A\, e^{-x^2}$$

150. Solve: $y\,(2xy + e^x)\, dx - xe^x\, dy = 0$ [B.U. Mar., 1995]

Solution:

Given: $$xe^x \frac{dy}{dx} - e^x \cdot y = 2\, xy^2$$

i.e. $$\frac{dy}{dx} \cdot y = 2e^{-x} \cdot y^2$$ (Bernoulli's Form)

Divide by y^2; $$\frac{1}{y^2}\frac{dy}{dx} - \frac{1}{x}\cdot\frac{1}{y} = 2e^{-x}$$

Put $\frac{1}{y} = u$;

$$\therefore \quad -\frac{1}{y^2}\frac{dy}{dx} = \frac{du}{dx}\ ; \text{ or } \ \frac{1}{y^2}\frac{dy}{dx} = -\frac{du}{dx}$$

G.E. becomes

$$-\frac{du}{dx} - \frac{1}{x}\cdot u = 2 \cdot e^{-x}$$

or $$\frac{du}{dx} + \left(\frac{1}{x}\right) u = -\, 2e^{-x}$$ (linear is u)

where $$P = \frac{1}{x};\ \int P dx = \log x;$$

$$\therefore \quad \text{I.F.} = e^{\int P dx} = e^{\log x} = x$$

and $$\int Q e^{\int P dx} \cdot dx + \text{const.} = (-2e^{-x}) \cdot x\, dx + A$$

$$= -\, 2 \int xe^{-x}\, dx + A$$

$$= -\, 2\left[x \cdot \left(\frac{e^{-x}}{-1}\right) - 1 \cdot \frac{e^{-x}}{(-1)^2}\right] + A$$

$$= [2x\, e^{-x} + 2e^{-x}] = A$$

PART-D

$\therefore$ *Solution:* $u[x] = \int Q\, e^{\int P dx} \cdot dx + A$

or $\dfrac{x}{y} = [2x + 2]\, e^{-x} + A$

or $y[2x + 2 + Ae^{x}] = xe^{x}$

151. Solve $\dfrac{dy}{dx} - \dfrac{\tan y}{1+x} = (1 + x)\, e^{x} \sec y$ [B.U. Feb., 1996]

Solution: Multiply by cos y; we get

$$\cos y \cdot \frac{dy}{dx} - \frac{1}{1+x} \cdot \sin y = (1 + x)e^{x} \qquad [1]$$

Put $\sin y = u$; $\cos y \cdot \dfrac{dy}{dx} = \dfrac{du}{dx}$

$\therefore$ [1] becomes

$$\frac{du}{dx} - \frac{1}{(1+x)}\, u = (1 + x)\, e^{x} \qquad [2]$$

(linear in u), where

$$P = \frac{-1}{1+x};\; Q = [(1 + x)\, e^{x}]$$

$\therefore$
$$\text{I.F.} = e^{\int P dx} = e^{\int\left(\frac{-dx}{1+x}\right)} = e^{-\log(1+x)}$$

$$= e^{\log(1+x)^{-1}} = \frac{1}{(x+1)}$$

and $\int Q\, e^{\int P dx} \cdot dx + \text{const} = \int (1 + x)\, e^{x} \cdot \dfrac{1}{(1+x)}\, dx + A$

$$= \int e^{x}\, dx + A = e^{x} + A$$

$\therefore$ *Solution:*

$$u\, \frac{1}{(1+x)} = e^{x} + A$$

or; $\dfrac{\sin y}{(1+x)} = e^{x} + A$

or $\sin y = (1 + x)\,(e^{x} + A)$

152. Solve $\dfrac{dy}{dx} + xy = xy^{3}$ [B.V. Mar., 1995, Aug., 1996]

Solution: Divide by y^{3};

$$\frac{1}{y^3} \cdot \frac{dy}{dx} + x \cdot \frac{1}{y^2} = x$$ [1]

Put $\frac{1}{y^2} = u;$

$$\therefore \qquad -\frac{2}{y^3}\frac{dy}{dx} = \frac{du}{dx}$$

(1) becomes

$$\left(-\frac{1}{2}\right) \cdot \frac{du}{dx} + x \cdot u = x$$

i.e. $$\frac{du}{dx} - 2x \cdot u = -2x$$ (Linear in u)

$$\text{I.F.} = e^{\int(-2x)\,dx} = e^{-2\int x dx} = e^{-x^2}$$

and $\int Q\, e^{\int P dx}\, dx + \text{const} = \int (-2x)\, e^{-x^2}\, dx + A = e^{-x^2} + A$

Solution: $u\,(\text{I.F.}) = \int Q\,(\text{I.F.})\, dx + A$

or $$\boxed{\frac{1}{y^2}\, e^{-x^2} = e^{-x^2} + A}$$

or $$\frac{1}{y^2} = 1 + A e^{x^2}$$

153. *Solve* $\frac{dy}{dx} = x^3y^3 - xy$

Solution:

Given: $$\frac{dy}{dx} + xy = x^3y^3$$

Divide by y^3; $\frac{1}{y^3}\frac{dy}{dx} + \frac{1}{y^2}\, x = x^3$

Put $y^{-2} = u$; $-\frac{2}{y^3}\frac{dy}{dx} = \frac{du}{dx}$;

$$\therefore \qquad (-1/2)\,\frac{du}{dx} + u \cdot u = x^3;$$

or, $$\frac{du}{dx} - 2x \cdot u = -2x^3;$$

$$P = -2x; \quad Q = -2x^3$$

PART-D

Solution: $$u\left(e^{\int P dx}\right) = A + \int Q\, e^{\int P dx} \cdot dx$$

i.e. $$u\, e^{\int -2x\, dx} = A + \int -2x^3\, e^{\int -2x\, dx} \cdot dx$$

i.e. $$u e^{-x^2} = A + \int (-2x) \cdot x^2 \cdot e^{-x^2}\, dx = A - \int t e^{-t}\, dt;$$

(Putting $x^2 = t;\ 2xdx = dt$)

$$= A + e^{-t} \cdot t - \int e^{-t} \cdot 1 \cdot dt = A + e^{-t} \cdot t + e^{-t}$$

$$u e^{-x^2} = A + (x^2 + 1)\, e^{-x^2}; \qquad (\because t = x^2)$$

$\therefore$ $$u = \{A e^{x^2} + (x^2 + 1)\};$$

or $$\frac{1}{y^2} = A e^{x^2} + (x^2 + 1)$$

154. Solve $\left(\dfrac{dy}{dx}\right) + x \cdot y = xy^n$

Solution: Divide by y^n; We get

$$\frac{1}{y^n}\left(\frac{dy}{dx}\right) + \left(\frac{1}{y^{n-1}}\right) = x$$

Put $\left(\dfrac{1}{y^{n-1}}\right) = u;$

$\therefore$ $$-\{(n-1)/y^n\}\left(\frac{dy}{dx}\right) = \left(\frac{du}{dx}\right)$$

The equation becomes

$$\left(-\frac{1}{(n-1)}\right) \cdot \left(\frac{du}{dx}\right) + u \cdot x = x$$

$$\left(\frac{du}{dx}\right) - (n-1) \cdot x \cdot u = -(n-1) \cdot x.$$

$\therefore$ Solution is

$$u\, e^{-(n-1)\int x\, dx} = A + \int e^{-(n-1)\int x\, dx} - (n-1)\, x\, dx$$

$$u\, e^{-(n-1)x^2/2} = A + \int d\left[e^{-(n-1)\cdot \frac{x^2}{2}}\right] = A + e^{-(n-1)x^2/2}$$

$\therefore$ $$u = A e^{-(n-1)x^2/2} + 1$$

i.e., $$\left(\frac{1}{y^{n-1}}\right) = A e^{-(n-1)x^2/2} + 1$$

155. Solve $2\left(\frac{dy}{dx}\right) - y \sec x = y^3 \cdot \tan x$

Solution: This is of the Bernoulli's form

Divide by y^3. $\frac{2}{y^3} \cdot \frac{dy}{dx} - \frac{1}{y^2} \sec x = \tan x$ [1]

Put $\frac{1}{y^2} = u;$ $-\frac{2}{y^3}\frac{dy}{dx} = \frac{du}{dx}$

$$\frac{2}{y^3}\frac{dy}{dx} = -\frac{dy}{dx}$$

$$\therefore \quad -\frac{du}{dx} - u \sec x = \tan x;$$

$$\frac{du}{dx} + \sec x \cdot u = -\tan x$$

$$\therefore \quad u e^{\int P dx} = A + \int Q\, e^{\int P dx} dx;$$

$$P = \sec x; \quad \int P dx = \int \sec x \, dx$$

$$= \log(\sec x + \tan x)$$

$$\therefore \quad e^{\int P dx} = e^{\log(\sec x + \tan x)} = (\sec x + \tan x)$$

and $$\int Q\, e^{\int P dx} \cdot dx = \int -\tan x\,(\sec x + \tan x)\, dx$$

$$= -\int [\sec x \tan x + [\sec^2 x - 1]]\, dx$$

$$= -(\sec x + \tan x - x) = (x - \tan x - \sec x)$$

$$u(\sec x + \tan x) = A + (x - \tan x)$$

$$\left(\frac{1}{y^2}\right)(\sec x + \tan x) = A + (x - \tan x - \sec x)$$

156. Solve $\frac{dy}{dx} = y \tan x - y^2 \sec x$

Solution: $\frac{1}{y^2}\frac{dy}{dx} - \frac{1}{y} \tan x = -\sec x$

Put $-\frac{1}{y} = u;$

$$\therefore \quad \frac{1}{y^2}\frac{dy}{dx} = \frac{du}{dx}$$

PART-D

$$\frac{du}{dx} + \tan x \cdot u = -\sec x;$$

$$P = \tan x; \quad \int \tan x\, dx = \log \sec x$$

$\therefore$ *Solution:* $\{u\, e^{\int P dx} = A + \int Q\, e^{\int P dx} \cdot dx\}$

or $$u \cdot e^{\log \sec x} = A - \int \sec x \cdot e^{\log \sec x}\, dx$$

or $$u \cdot \sec x = A - \int \sec x \cdot \sec x dx$$

$$= A - \int \sec^2 x\, dx = A - \tan x$$

$\therefore$ $$-\frac{1}{y^2} \cdot \sec x = A - \tan x$$

i.e. $$\sec x = -Ay + y \tan x$$

157. Solve $1 + [x \tan y - \sec y] \cdot \dfrac{dy}{dx} = 0$

Solution: In this case, the equation in y does not lead to convenient form. So we can try if the dependent and independent variables can be interchanged. Multiplying by $\dfrac{dx}{dy}$, we get

$$\frac{dx}{dy} + x \tan y - \sec y = 0$$

$$\frac{dx}{dy} + \tan y \cdot x = \sec y \qquad \text{(This is linear in } x\text{)}$$

Solution: $x\, e^{\int P dx} = A + \int Q\, e^{\int P dy}\, dy$

Here $P = \tan y$;

$$\int P dy = \int \tan y\, dy = \log \sec y$$

$$e^{\int P dy} = e^{\log \sec y} = \sec y$$

$\therefore$ $$x \cdot \sec y = A + \int \sec y \cdot \sec y\, dy$$

or $$x \cdot \sec y = A + \int \sec^2 y\, dy = A + \tan y$$

$$x \cdot \sec y = A + \tan y$$

158. Solve: $y\left(\dfrac{dy}{dx}\right) + by^2 = a \cos x$

Solution: $$\int e^{ax} \sin bx\, dx = \frac{e^{ax}(a \sin bx - b \cos bx)}{(a^2 + b^2)}$$

$$\int e^{ax} \cos bx dx = \frac{e^{ax}(a \cos bx + b \sin bx)}{(a^2 + b^2)}$$

Put $y^2 = u$; $\quad 2y\left(\frac{dy}{dx}\right) = \left(\frac{du}{dx}\right)$

$$\therefore \quad \frac{1}{2}\left(\frac{du}{dx}\right) + bu = a \cos x$$

$$\left(\frac{du}{dx}\right) + 2bu = 2a \cos x \qquad [1]$$

(This is linear)

$$\therefore \quad \textit{Solution:} \quad u\, e^{\int 2b dx} = A + \int e^{\int 2b dx}\, 2a \cos x\, dx$$

$$ue^{2b \cdot x} = A + 2a \int e^{2bx} \cdot \cos x\, dx$$

$$= A + 2a\, e^{2bx} \frac{(2b \cos x + \sin x)}{(4b^2 + 1)}$$

$$\therefore \quad y^2 e^{2bx} = A + \frac{2ae^{2bx}(2b \cos x + \sin x)}{(4b^2 + 1)}$$

EXERCISES

Solve the following equations: $\left(\text{Form: } \frac{dx}{dy} + Px = Q\right)$

1. $(x + 2y^3)\, y^1 = y$
2. $(1 + y^2)\, dx = (\tan^{-1} y - x)\, dy$
3. $(1 + y^2) + (x - e^{\tan^{-1} y})\, y^1 = 0$
4. $(2x - 10y^3)\, y^1 + y = 0$
5. $e^y dx + (1 + xe^y)\, dy = 0$
6. $\cosh x dy + (y + \cosh x) \sinh x = 0$
7. $\frac{dy}{dx} + y \cos x = 2 \cos x$
8. $\sin x \times \frac{dy}{dx} + 3y = \cos x$
9. $(x + \log y)\, dy + y dx = 0$
10. $(1 + x^2)\, y^1 + y = e^{\tan^{-1}(x)}$

Solve: (Bernoulli's Equation) the following diff. equation

1. $\frac{dy}{dx} + \frac{y}{x} = y^2$
2. $\frac{dy}{dx} = x^3y^3 - xy$
3. $\frac{2dy}{dx} = \frac{y}{x} + \frac{y^2}{x^2}$
4. $\frac{dy}{dx} + y \cos x = y^n \sin 2x$
5. $\frac{dy}{dx} + x \times y = xy^n$
6. $y dx + (ax^2y^2 - 2x)\, dy = 0$
7. $\frac{dy}{dx} - y \tan x = \left[\sin x \cdot \frac{\cos^2 x}{y^2}\right]$
8. $\left(\frac{dy}{dx}\right) - \left(\frac{2y}{x}\right) = (y^3/x^3)$

PART-D

9. $\left(\frac{dy}{dx}\right) + x^3y = e^xy^4$

10. $2(1 + x^2)\, y^1 - y + y^2 = 0$

11. $xy^1 + y = x^3y^6$

12. $(xy\, dx - dy) = y^3\, e^{-x^2}$

13. $\left(\frac{dy}{dx}\right) + 2yx = 2x^3y^3$

14. $\text{Cos } x \left(\frac{dy}{dx}\right) = y\, [\sin(x - y)]$

15. $2xy\, (y^1) = (x^2 + y^2 - 1)$

16. $y\, (2xy + e^x)\, dy = 0$

17. $x^2y - x^3 \times y^1 = y^4 \cos x$

18. $x^2y^1 - xy = y^3 \sin x$

19. $x \times y^1 + y = x^3y^6$

20. $y^1 - \left(\frac{2y}{x}\right) = (y^3/x^3)$

4.6 ORTHOGONAL TRAJECTORIES OF CARTESIAN AND POLAR FORMS: (use of initial condition to be emphasized)

Definition: A family of curves is said to be the orthogonal trajectory of a given family of curves if each member of one family cuts all the curves of the other family at right angles.

Two curves cut at right angles if the tangents to the two curves are at right angles at their point of intersection and the condition for this is that the product of the slopes is equal to (– 1).

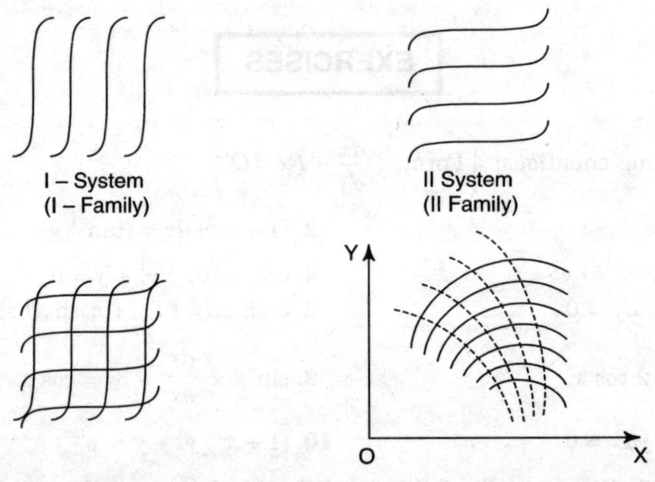

Fig. 4.1

(I) Method of finding the equation of Orthogonal Trajectories (O.T.) (CARTESIAN)

(*i*) Let the cartesian equation of the family of curves F_1 is

$$f(x, y, a) = 0 \qquad [1]$$

(*ii*) Differentiate $f(x, y, a) = 0$ w.r.t. x and eliminate a between their equations and we get an equation

$$F\left(x, y, \frac{dy}{dx}\right) = 0 \qquad [2]$$

(*iii*) Replace $\dfrac{dy}{dx}$ by $\left(-\dfrac{dx}{dy} \quad \text{or} \quad -\dfrac{1}{\left(\dfrac{dy}{dx}\right)} \right)$

i.e. $$F\left[x, y, -\frac{dx}{dy}\right] = 0 \qquad [3]$$

(*iv*) Solve the equation (3) to get the orthogonal trajectory

Note: For the sake brevity, we use the abbreviation (O.T.) for orthogonal Trajectories

(II) Polar Curves:

(*i*) Let f(r, θ, a) = 0 be the given family

(*ii*) Differentiate (i) w. r. to θ and eliminate 'a' to get the equation

$$F\left(r, \theta, \frac{dr}{d\theta}\right) = 0$$

(*iii*) Replace $\dfrac{dr}{d\theta}$ by $-r^2 \dfrac{d\theta}{dr}$ to get

$$F\left[r, \theta, -r^2 \frac{d\theta}{dr}\right] = 0$$

(*iv*) Solve the above eqaution to get the result on O.T.

Worked Examples (Cartesian)

159. Find the orthogonal trajectories of the family of curves

$$\frac{x^2}{a^2} + \frac{y^2}{b^2 + \lambda} = 1, \quad \text{where } \lambda \text{ is a parameter} \qquad \text{[VTU, March, 2000]}$$

Solution: The equation of the family of curves is

$$\frac{x^2}{a^2} + \frac{y^2}{b^2 + \lambda} = 1 \qquad [1]$$

Differentiate w.r.t x,

$$\frac{2x}{a^2} + \frac{2y}{b^2 + \lambda} \cdot \frac{dy}{dx} = 0$$

i.e. $$\frac{x}{a^2} + \frac{y}{b^2 + \lambda} \cdot \frac{dy}{dx} = 0 \qquad [2]$$

To eliminate λ we have from (1)

$$\frac{y^2}{b^2 + \lambda} = 1 - \frac{x^2}{a^2} = \frac{a^2 - x^2}{a^2}$$

$$\therefore \qquad b^2 + l = \frac{a^2 y^2}{a^2 - x^2}$$

Now, from (2) $$b^2 + l = -\frac{a^2 y}{x}\left(\frac{dy}{dx}\right)$$

$$\Rightarrow \qquad \frac{a^2 y^2}{a^2 - x^2} = -\frac{a^2 y}{x}\left(\frac{dy}{dx}\right)$$

or $$\frac{xy}{a^2 - x^2} + \frac{dy}{dx} = 0 \qquad [3]$$

is the d.e. of the given family of curves.

Replacing $\frac{dy}{dx}$ by $-\frac{dx}{dy}$, we get

$$\frac{xy}{(a^2 - x^2)} - \frac{dx}{dy} = 0$$

$$\Rightarrow \qquad ydy - \left(\frac{a^2 - x^2}{x}\right) dx = 0$$

is the differential equation of required orthogonal trajectory, integrating, we get,

$$\int ydy - \int \frac{a^2}{x}\, dx + \int x\, dx = c_1$$

i.e. $$\frac{y^2}{2} - a^2 \log x + \frac{x^2}{2} = c_1$$

i.e. $$y^2 - 2a^2 \log x + x^2 = 2\, c_1 = c$$

or $$x^2 + y^2 - 2a^2 \log x = c$$

160. Find the orthogonal trajectories of $\frac{x^2}{a^2} + \frac{y^2}{a^2 + \lambda} = 1$

where λ is a parameter [VTU, August, 1999]

Solution: Given: $\frac{x^2}{a^2} + \frac{y^2}{a^2 + \lambda} = 1$ [1]

(I) Differentiate (1) w.r.t x,

$$\frac{2x}{a^2} + \frac{2y}{a^2 + \lambda} \cdot \frac{dy}{dx} = 0$$

or $$\frac{xy}{a^2} + \frac{y^2}{a^2 + \lambda} \cdot \frac{dy}{dx} = 0$$

or $$\frac{y^2}{a^2+\lambda} = -\frac{xy}{a^2 p}, \qquad \left(\text{where } p = \frac{dy}{dx}\right)$$

$\therefore$ From (1), $$\frac{x^2}{a^2} - \frac{xy}{a^2 p} = 1$$

[λ is eliminated] [2]

(II) Replace $p\left(=\frac{dy}{dx}\right)$ by $-\frac{1}{p}\left(=\frac{-dx}{dy}\right)$ the differential equation of orthogonal trajectories is

$$\frac{x^2}{a^2} + \frac{xy}{a^2}p = 1.$$

or $$P = \frac{a^2-x^2}{xy} \quad \text{or} \quad \frac{dy}{dx} = \frac{a^2-x^2}{xy}$$

or $$ydy = \left(\frac{a^2}{x} - x\right)dx \qquad [3]$$

(III) Integrating, [3],

$$\frac{y^2}{2} = a^2 \log x - \frac{x^2}{2} + c_1$$

or $$x^2 + y^2 = 2a^2 \log x + 2c_1$$

or $$x^2 + y^2 = 2a^2 \log x + c$$

Which is the required equation

161. Find the differential equation satisfied by the system of parabolas $y^2 = 4a(x + a)$ and show that the orthogonal trajectories of the system belong to the system itself (or that the system is self orthogonal). [B.U. March, 1989]

Solution:

Given: $y^2 = 4a(x + a)$ [1]

Differentiating [1],

$$2y\frac{dy}{dx} = 4a$$

$\Rightarrow$ $$a = \frac{1}{2} \times y \times \frac{dy}{dx}$$

$\therefore$ From [1], $$y^2 = 2y\frac{dy}{dx}\left[x + \frac{1}{2}y \cdot \frac{dy}{dx}\right]$$

or $$y^2 = 2xy\frac{dy}{dx} + y^2\left(\frac{dy}{dx}\right)^2$$

PART-D

or $$y^2 = 2xy\, p + y^2p^2,$$

where $$p = \frac{dy}{dx} \qquad [2]$$

which being free from a (the parameter) is the d.e. of the family [1].

Replacing p by $-\frac{1}{p}$, the d.e. of orthogonal trajectory is

$$y^2 = -\frac{2xy}{p} + \frac{y^2}{p^2}$$

or $$y^2p^2 + 2xyp = y^2 \qquad [3]$$

Which is the same as [2]

Since the d.e. of the family of parabolas [1] and that of the O.T. are same, the family of parabolas [1] is self orthogonal.

The equation of O.T. is again of the form

$$y^2 = 4k\,(x + k)$$

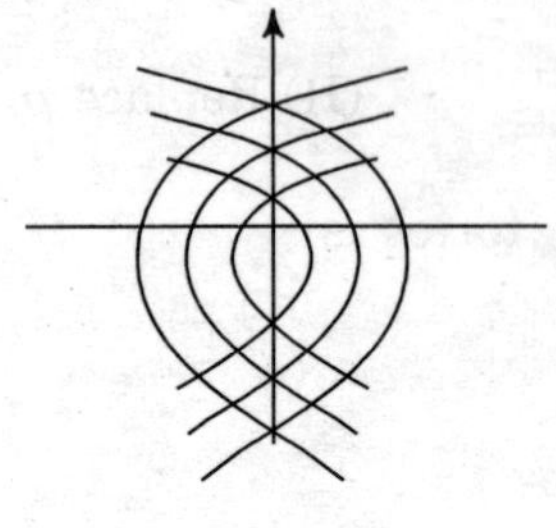

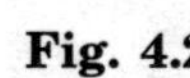

Fig. 4.2

162. Find the orthogonal trajectories of the family of Co-axial Circles

$$x^2 + y^2 + 2\lambda y + c = 0$$

Where l is the parameter and c is constant [B.U. Oct., 1987 B.U. Aug., 1994]

Solution: Given System:

$$x^2 + y^2 + 2\lambda y + c = 0 \qquad [1]$$

(Differentiate once w.r.t. x), we get

$$2x + 2y\left(\frac{dy}{dx}\right) + 2\lambda\left(\frac{dy}{dx}\right) = 0$$

$$\Rightarrow \qquad x + yp + \lambda p = 0; \quad \left[p = \frac{dy}{dx}\right] \qquad [2]$$

$$\therefore \qquad \lambda = -\frac{[x + yp]}{p}$$

To find the d.e. of the system, we have to eliminate λ between [1] and [2], we get

$$x^2 + y^2 + 2y\left\{-\frac{(x + yp)}{p}\right\} + c = 0 \qquad [3]$$

The d.e. of the O.T. is got by replacing p by $-\left(\frac{1}{p}\right)$

or putting $(-1/p)$ in the place of p in (3),

$\therefore$ O.T.:-

$$x^2 + y^2 + 2y\,\frac{\left\{-\left(x - \frac{y}{P}\right)\right\}}{(-1/P)} + c = 0$$

$$\Rightarrow \quad x^2 + y^2 + 2y\,\{p(x - y/p)\} + c = 0$$

$$\Rightarrow \quad x^2 + y^2 + 2xyp - 2\,y^2 + c = 0$$

$$\Rightarrow \quad 2xyp - y^2 + x^2 + c = 0$$

(This is the same as Bernoulli's equation. ∴ we can reduce it to linear form with the substitution $y^2 = u;\ \dfrac{du}{dx} = 2y\,\dfrac{dy}{dx}$)

Hence, we have $x \cdot \dfrac{du}{dx} - u = -(x^2 + c)$

or
$$\frac{du}{dx} - \frac{u}{x} = -\left(u + \frac{c}{x}\right)$$

(This is a linear equation, linear in u), with $P = -\dfrac{1}{x}$

$$Q = -\left(x + \frac{c}{x}\right)$$

$$\therefore \quad \text{I.F.} = e^{\int P\,dx} = e^{\log x^{-1}} = x^{-1} = \frac{1}{x}$$

and
$$\int Q\,e^{\int P\,dx}\,dx + A = -\int\left(x + \frac{c}{x}\right)\frac{1}{x}\,dx + A = -\int\left(1 + \frac{c}{x^2}\right)dx + A$$

$$= -\left[x - \frac{c}{x}\right] + A = \left(\frac{c}{x} - x\right) + A$$

∴ Its solution is

$$\therefore \quad u\,(\text{I.F}) = \left(\frac{c}{x} - x\right) + A$$

or
$$y^2\left(\frac{1}{x}\right) = \frac{(c - x^2)}{x} + A$$

or
$$y^2 = (c - x^2) + Ax$$

or
$$x^2 + y^2 + 2\mu x - c = 0$$

where $(A = 2\mu)$

Which is also a system of Coaxial circles

163. Find the orthogonal trajectories of a family of Coaxial circles $x^2 + y^2 + 2gx + c = 0$, where g is a parameter and c constant. [B.U. August, 1994]

Solution:

Given System:

$$x^2 + y^2 + 2gx + c = 0 \qquad [1]$$

(Differentiate once w.r.t. x), we get

$$2x + 2y\,\frac{dy}{dx} + 2g = 0$$

PART-D

$$\Rightarrow \qquad x + yp + g = 0 \qquad [2]$$

$$\Rightarrow \qquad g = -(x + yP): \left(P = \frac{dy}{dx}\right)$$

To find the d.e. of the system, we have to eliminate 'g' between [1] and [2] or from [1],

$$x^2 + y^2 + 2x\,[-(x + yp)] + c = 0 \qquad [3]$$

The *d.e.* of the O.T. is got by replacing *P* by $\left(-\frac{1}{P}\right)$ or putting $\left(-\frac{1}{P}\right)$ in the place of *P* in [3]

$$\therefore \quad \text{O.T} \;:\; x^2 + y^2 + 2x\left\{-\left(x - \frac{y}{P}\right)\right\} + c = 0$$

$$\Rightarrow \qquad y^2 - x^2 + \frac{2xy}{P} + c = 0$$

$$\Rightarrow \qquad 2xy\left(\frac{dx}{dy}\right) - x^2 = -c - y^2$$

or,
$$\frac{dx}{dy} - \frac{1}{2y}\cdot x = -\frac{y^2 + c}{2xy}$$

or,
$$2x \times \frac{dx}{dy} - \frac{1}{y}\cdot x^2 = -\frac{(y^2 + c)}{y} \qquad [4]$$

put
$$x^2 = u; \quad 2x\,\frac{dx}{dy} = \frac{du}{dy}$$

From (4),
$$\frac{du}{dy} - \frac{1}{y}\cdot u = -\frac{(y^2 + c)}{y}$$

Which is linear in u. $\left(P = -\frac{1}{y}\right)$

$$\therefore \qquad \text{I.F.} = e^{\int\left(-\frac{1}{y}\right)dy} = e^{-\log y}$$

$$= e^{\log y^{-1}} = y^{-1} = \left(\frac{1}{y}\right)$$

$$\therefore \quad \text{Solution is:} \quad u\cdot(\text{I.F.}) = \int Q\cdot e^{\int P dy}\cdot dy + A$$

$$\frac{x^2}{y} = \int -\frac{(y^2 + 1)}{y}\cdot dy + A$$

$$= -\int (1 + cy^{-2})\,dy + A$$

$$\therefore \quad \frac{x^2}{y} = -y - c \times \frac{y-1}{-1} + A$$

or,
$$x^2 = -y^2 + c + Ay$$

$$\Rightarrow \quad x^2 + y^2 - Ay - c = 0$$

$$x^2 + y^2 - 2\mu y - c = 0$$

Put $(A = -2\mu)$

Which is the required O.T. of the given system.

164. Find the O.T. of the system of parabolas

$$y = ax^2; \quad (a = \text{parameter})$$

Solution:

Given System: $y = ax^2$ [1]

(Differentiate w.r.t. x and eliminate a)

i.e.
$$\frac{dy}{dx} = a \times 2x;$$

But $a = (y/x^2)$ from (1)

$$\therefore \quad \frac{dy}{dx} = (y/x^2) \times 2x = \left(\frac{2y}{x}\right)$$

$\therefore$ D.E. of the system:

$$\frac{dy}{dx} = \left(\frac{2y}{x}\right)$$ [2]

So, the D.E. of the system of O.T. is:

$$-\frac{1}{\left(\frac{dy}{dx}\right)} = \frac{2y}{x};$$ [3]

or
$$\boxed{-\frac{1}{P} = \frac{2y}{x}} \qquad \left(\text{Put } P = \frac{dy}{dx}\right)$$

D. Equation of (O.T.) from (3) is

$$= x\,dx + 2y\,dy = 0$$ [3]

Solving, We get

$$-\int x\,dx + \int 2y\,dy = \text{const.}$$

$$\left(\frac{x^2}{2}\right) + y^2 = (\text{const.}):$$

$$x^2 + 2y^2 = (\text{const.}) = 2b^2$$ [5]

The O.T. are the system of ellipses:

$$x^2 + 2y^2 = 2b^2$$

PART-D

165. Find the the O.T. of the parabolas

$$y^2 = 4ax$$
$$(a = \text{parameter})$$

Solution: Given equation of system:

$$y^2 = 4ax \qquad [1]$$

(Differentiate w.r. to x:

$$2y\left(\frac{dy}{dx}\right) = 4a \qquad [2]$$

Eliminate a by division; we get

$$\frac{y^2}{2y\left(\frac{dy}{dx}\right)} = \frac{4ax}{4a}\ ;\ \frac{ydx}{2dy} = x$$

$\therefore$ D.E. of system: $\qquad (y/2) = x\left(\dfrac{dy}{dx}\right) \qquad [3]$

D.E. of O.T. is: $\qquad \left(\dfrac{y}{2}\right) = \dfrac{x\,(-1)}{(dy/dx)} = \dfrac{-x\,dx}{dy}\ ;$

or, $\qquad ydy + 2x\,dx = 0 \qquad [4]$

Solving, $\qquad \int ydy + \int 2x\,dx = (\text{const.})$

or, $\qquad \left(\dfrac{y^2}{2}\right) + x^2 = (\text{const.}) = (b^2/2);$

$$y^2 + 2x^2 = b^2 \qquad [5]$$

$\therefore$ O.T. $\qquad y^2 + 2x^2 = b^2$

166. Find the orthogonal trajectories of the family of circles $x^2 + y^2 = a^2$.

Solution: Given system: $\qquad x^2 + y^2 = a^2. \qquad [1]$

(Differentiating w.r.t. x), we get

$$2x + 2y \cdot \frac{dy}{dx} = 0, \text{ so, D.E. of the system is:}$$

$$\frac{dy}{dx} = -\frac{x}{y} \qquad [2]$$

So, the D.E. of the system of O.T. is

$$-\frac{1}{\left(\frac{dy}{dx}\right)} = -\frac{x}{y} \qquad [3]$$

Hence D.E. of (O.T.) from (3) is

$$\frac{dx}{x} = \frac{dy}{y}$$

Solving, (integrating) we get

$$\log x + \log m = \log y$$

(m = a const. of integration)

$$\therefore \qquad y = mx$$

Thus, O.T. of the given family of circles is the family of straight lines passing through the origin.

167. For the family of curves

$$x^2 + 3y^2 = cy$$

find that member of the orthogonal trajectories which passes through (1, 2)

[B.U., Jan., 1993]

Solution: Given system: $x^2 + 3y^2 = cy$ [1]

(Differentiate w.r.t. x), we get,

$$2x + 6y\frac{dy}{dx} = c \cdot \frac{dy}{dx}$$

$$\Rightarrow \qquad 2x \times \frac{dy}{dx} + 6y = c$$

Substituting in (1), we get

$$\left[2x\frac{dx}{dy} + 6y\right] y = x^2 + 3y^2$$

$$\Rightarrow \qquad 2x \times \frac{dx}{dy} = x^2 - 3y^2$$

$$\Rightarrow \qquad 2xy\frac{dx}{dy} = x^2 - 3y^2 \quad \text{or} \quad \frac{dy}{dx} = \frac{2xy}{(x^2 - 3y^2)} \qquad [2]$$

is the D.E. of the system.

So, D.E. of the system of O.T. is:

$$-\frac{1}{\frac{dy}{dx}} = \frac{2xy}{(x^2 - 3y^2)} \quad \Rightarrow \quad \frac{dy}{dx} = \frac{3y^2 - x^2}{2xy}$$

This is a Homogeneous d.e.

To solve, we put $y = vx$;

$$\Rightarrow \qquad \frac{dy}{dx} = v + x \times \frac{dv}{dx}$$

or

$$v + x \times \frac{dv}{dx} = \frac{3v^2x^2 - x^2}{2x^2v}$$

$$\Rightarrow \qquad v + x \times \frac{dv}{dx} = \frac{3v^2 - 1}{2v};$$

$$\Rightarrow \qquad x\frac{dv}{dx} = \frac{v^2 - 1}{2v}$$

$$\Rightarrow \qquad \frac{dx}{x} = \frac{2v}{v^2 - 1}\, dv$$

on integration we get

$$\log x = \log (v^2 - 1) + \log c$$

$$\Rightarrow \qquad x = c\,(v^2 - 1)$$

$$\Rightarrow \qquad x = c\left(\frac{y^2}{x^2} - 1\right) \Rightarrow x^3 = c\,(y^2 - x^2)$$

This is the required O.T.

To obtain the required member; put $x = 1$ and $y = 2$

We get
$$1 = c\,(4 - 1)$$

$$c = \frac{1}{3}$$

∴ The required member is

$$3x^3 = (y^2 - x^2)$$

168. Find the orthogonal Trajectories of the family of semi-cubical parabolas

$$ay^2 = x^3$$

Solution: Given system: $ay^2 = x^3$ [1]

(Differentiate w.r.t. x), we get

$$2ay\left(\frac{dy}{dx}\right) = 3x^2 \qquad [2]$$

Divide (to eliminate the parameter a), we get

$$\frac{ay^2}{2ay\left(\frac{dy}{dx}\right)} = \frac{x^3}{3x^2}; \quad \frac{ydx}{2dy} = \frac{x}{3}$$

or
$$2x\left(\frac{dy}{dx}\right) = 3y \qquad [3]$$

(D.E. of the system)

The D.E. of O.T. is:

$$2x - \left[-\frac{dx}{dy}\right] = 3y \quad \text{or} \quad -2xdx = 3ydy,$$

$$2xdx + 3ydy = 0 \qquad [4]$$

Solving this (by integration) we get

$$\int 2x dx + \int 3y dy = \text{(const.)}$$

or $$x^2 + 3\left(\frac{y^2}{2}\right) = (c^2/2)$$

or $$2x^2 + 3y^2 = c^2$$ [5]

O.T. $$2x^2 + 3y^2 = c^2$$

169. **Find the orthogonal trajectories of the system of curves:**

$$\left(\frac{dy}{dx}\right)^2 = \frac{a}{x}$$

Solution: Equation of system:

$$\left(\frac{dy}{dx}\right)^2 = \frac{a}{x}$$

$\therefore$ D.E. of O.T. is:

$$\left[-\frac{dx}{dy}\right]^2 = \frac{a}{x} \quad \text{or} \quad \left(\frac{dx}{dy}\right)^2 = \frac{a}{x}$$

$$\therefore \quad \frac{dx}{dy} = \sqrt{\frac{a}{x}}$$

$$\therefore \quad \sqrt{\frac{x}{a}} \times dx = dy$$

or $$\frac{2x^{3/2}}{3\sqrt{a}} = y + c$$

(on Integration)

or $$(y + c)^2 = \frac{4x^3}{9a}$$

$\Rightarrow$ $9a\,(y + c)^2 = 4x^3$ is the O.T.: of the given system

170. **Find the orthogonal trajectories of the system of rectangular hyperbolas $(x^2 - y^2) = a^2$, Where a = parameter**

Solution: Equation of the given system:

$$(x^2 - y^2) = a^2$$ [1]

(Differentiate w.r.t. x), we get

$$2x - 2y\left(\frac{dy}{dx}\right) = 0$$

Hence (a = parameter) is automatically eliminated.

PART-D

So D.E. of the system is:

$$2x - 2y\left(\frac{dy}{dx}\right) = 0;$$

or, $$\left[x - y\left(\frac{dy}{dx}\right)\right] = 0$$ [2]

The D.E. of O.T. is:

$$x - y \cdot \frac{1}{\left(-\frac{dy}{dx}\right)} = 0;$$

or $$x + y\frac{dx}{dy} = 0 \quad \text{or} \quad xdy + ydx = 0$$ [3]

Solving, we get, $xdy + ydx = 0$

or, $$d(xy) = 0; \quad \therefore \quad xy = \text{constant} = k^2.$$

$\therefore$ Orthogonal trajectories:

$$xy = k^2$$ [4]

171. Find the orthogonal trajectories of the system of curves:

$$x^{2/3} + y^{2/3} = a^{2/3};$$

Where a is the parameter

Solution: Equation of system:

$$x^{2/3} + y^{2/3} = a^{2/3}$$ [1]

(Differentiate w.r.t x), we get

$$\frac{2}{3}x^{-1/3} + \frac{2}{3}y^{-1/3} \cdot \frac{dy}{dx} = 0$$ [2]

Thus, 'a' is automatically eliminated, the D.E. of the System is given by [2] that is;

$$\frac{2}{3}\frac{1}{x^{1/3}} + \frac{2}{3} \cdot \frac{1}{y^{1/3}} \cdot \frac{dy}{dx} = 0;$$

or $$\frac{1}{x^{1/3}} + \frac{1}{y^{1/3}} \cdot \frac{dy}{dx} = 0$$ [3]

$\therefore$ The D.E. of O.T. is:

$$\frac{1}{x^{1/3}} + \frac{1}{y^{1/3}}\frac{(-dx)}{dy} = 0;$$

$$x^{1/3}\,dx - y^{1/3}\,dy = 0$$ [4]

Solving (since the variables are seperable)

$$\int x^{1/3}\,dx - \int y^{1/3}\,dy = (\text{const.})$$

$$\frac{x^{4/3}}{(4/3)} - \frac{y^{4/3}}{(4/3)} = \frac{b^{4/3}}{(4/3)} \text{ (say)}$$

$\therefore$ O.T .: $$x^{4/3} - y^{4/3} = b^{4/3}$$

172. **Find the O.T. of the system of curves whose equation is**

$$x^{4/3} + y^{4/3} = a^{4/3}$$

Where a is the parameter

Solution: Equation of the system:

$$x^{4/3} + y^{4/3} = a^{4/3} \quad [1]$$

Differentiate w.r.t. x and eliminate 'a', we get

$$\frac{4}{3}x^{1/3} + \frac{4}{3}y^{1/3} \cdot \frac{dy}{dx} = 0$$

$$x^{1/3} + y^{1/3}\left(\frac{dy}{dx}\right) = 0 \quad [2]$$

$\therefore$ The D.E. of O.T. is: $x^{1/3} + y^{1/3}\left(-\frac{dx}{dy}\right) = 0;$

$$x^{1/3}\,dy - y^{1/3}\,dx = 0$$

or

$$\frac{dx}{x^{1/3}} - \frac{dy}{y^{1/3}} = 0;$$

$$x^{-1/3}\,dx - y^{-1/3}\,dy = 0 \quad [3]$$

The O.T. is obtained by solving [3]; i.e.,

$$\int x^{-1/3}\,dx - \int y^{-1/3}\,dy = \text{(const.)}$$

$$\frac{x^{2/3}}{(2/3)} - \frac{y^{2/3}}{(2/3)} = \text{const.} = \frac{b^{2/3}}{(2/3)}$$

$\therefore$ O.T. $\quad x^{2/3} - y^{2/3} = b^{2/3}$ [4]

173. **Find the orthogonal Trajectory of $xy = c^2$**

Solution: Given system: $xy = c^2$ [1]

(Differentiating w.r.t. x) and eliminate 'c^2', we get

$$x\frac{dy}{dx} + y = 0 \quad [2]$$

$\therefore$ The D.E. of O.T. is:

$$x\left(-\frac{dx}{dy}\right) + y = 0$$

i.e. $\quad -x\,dx + y\,dy = 0$ [3]

Solving (variables are separable) Integrating, we get

$$-\frac{x^2}{2} + \frac{y^2}{2} = \frac{c^2}{2} \quad \text{or} \quad y^2 - x^2 = c^2.$$

174. **Find the orthogonal trajectories of a system of circles, which touch a given straight line at a given point.**

PART-D

Solution: Let us choose the given straight line or the y-axis and the given point (on the given line) as the origin. Since all the circles are to touch the y-axis at the origin, their centres will lie on the x-axis.

Consider a circle of the system. Its centre lies on the x-axis, and so can be written = $(l, 0)$. It touches the y-axis, and so passes through the origin; so, the radius of the circle is l.

Hence, the equation of the circle can be written,

$$(x - \lambda)^2 + (y - 0)^2 = \lambda^2;$$

$$x^2 - 2\lambda x + \lambda^2 + y^2 = \lambda^2$$

$$\therefore \quad x^2 + y^2 - 2lx = 0 \quad \text{or} \quad x^2 + y^2 = 2lx \qquad [1]$$

$\therefore$ The D.E. of the system is obtained by differentiating [1] w.r.t. x, i.e.,

$$2x + 2y\left(\frac{dy}{dx}\right) = 2\lambda$$

$$\Rightarrow \quad x + y\left(\frac{dy}{dx}\right) = l \qquad [2]$$

Divide [2) by (1), to eliminate λ, we get

$$\frac{x + y\left(\frac{dy}{dx}\right)}{x^2 + y^2} = \frac{\lambda}{2\lambda x} = \frac{1}{2x}$$

$$\text{or,} \quad 2x\left[x + y\left(\frac{dy}{dx}\right)\right] = (x^2 + y^2) \qquad [3]$$

$\therefore$ The D.E. of O.T. is given by:

$$2x\left[x + y\left(\frac{dy}{dx}\right)\right] = (x^2 + y^2)$$

$$\text{or,} \quad 2x^2 - 2xy\left(\frac{dy}{dx}\right) = x^2 + y^2$$

$$\text{or} \quad (x^2 - y^2) - 2xy\left(\frac{dy}{dx}\right) = 0;$$

$$(x^2 - y^2)\,dy - 2xydx = 0; \qquad [4]$$

The eqn. [4] is a Homogeneous d.e. we solve it.

Put $y = vx; \ \frac{dy}{dx} = v + x \times \frac{dv}{dx}$ or $dy = (vdx + xdv)$

$$\therefore \quad (x^2 - v^2x^2)(vdx + xdv) - 2xvxdx = 0$$

$$x^2\,[(1 - v^2)(vdx + xdv) - 2vdx] = 0$$

$$(1 - v^2)\,vdx + x\,(1 - v^2)\,dv - 2vdx = 0$$

$$\text{or} \quad dx\,(v - v^3 - 2v) + x\,(1 - v^2)\,dv = 0$$

$$-dx\,(v^3+v)+x\,(1-v^2)\,dv=0$$
$$dx\,v\,(v^2+1)+x\,(v^2-1)\,dv=0$$

$$\frac{dx}{x}+\frac{(v^2-1)}{v(v^2+1)}\,dv=0;\quad \frac{dx}{x}+\left[\frac{-1}{v}+\frac{2v}{v^2+1}\right]dv=0$$

$\therefore$ $$\int\frac{dx}{x}+\int\frac{-1}{v}\,dv+\int\frac{2v}{(v^2+1)}\,dv=\text{(const.)}$$

$$\log x-\log v+\log(v^2+1)=\text{const.}$$

$$\log\left[x\,\frac{(v^2+1)}{v}\right]=\text{const;}$$

$$[x\,(v^2+1)/v]=c\quad\text{or}\quad x\,(v^2+1)=cv;$$

put $$v=(y/x)$$

$\therefore$ $$x\left[\frac{y^2}{x^2}+1\right]=c\times\frac{y}{x}$$

$$(y^2+x^2)=cy;$$

put $$\boxed{c=+2\mu}$$

$\therefore$ $$x^2+y^2-2\mu y=0 \qquad [5]$$

This is clearly, a system of Circles, with the centres on the y-axis, and touching the x-axis at the origin.

[Method 2]

(For solving the D.E. given by equation (4))

$$2xy\left(\frac{dx}{dy}\right)-(x^2)=-y^2$$

Put $x^2=u$; $$2x\,\frac{dx}{dy}=\left(\frac{du}{dy}\right)$$

$\therefore$ $$y\left(\frac{du}{dy}\right)-u=-y^2;\quad\text{or}\quad\left(\frac{du}{dy}\right)-\frac{1}{y}\,u=-y$$

(This is linear)

Its Solution: $$u\,e^{\int Pdy}=A+\int Q\,e^{\int Pdy}\,dy$$

The $$P=\left(-\frac{1}{y}\right);\ \int Pdy=\int(-1/y)\,dy=-\log y$$

$\therefore$ $$\int e^{Pdy}=e^{-\log y}=(1/y)$$

$\therefore$ $$\int Q\,e^{\int Pdy}\times dy=\int(-y)\times(1/y)\,dy=\int -\,dy=-y$$

$\therefore$ *Solution:* $u(1/y) = A - y;$

or $u = Ay - y^2$

or $(x^2 + y^2) = Ay$

or $x^2 + y^2 = 2\mu y; \ (A = 2\mu)$

Other forms of example (16)

(A) Find the O.T. of the system of circles touching the y-axis at the origin.

(B) Find the O.T. of the system of circles $x^2 + y^2 - 2ax = 0$

Where 'a' is a parameter

Solution: The work is the same as in Ex. (16) as above and the O.T. are the system of circle given by

$$x^2 + y^2 - 2\,\mu y = 0$$

175. Find the orthogonal trajectories of the system of curves,

$$\frac{x^2}{(a^2+\lambda)} + \frac{y^2}{a^2} = 1,$$

Where λ is arbitrary

Solution: Given equation: $$\frac{x^2}{(a^2+\lambda)} + \frac{y^2}{a^2} = 1 \qquad [1]$$

(Differentiate w.r.t. x), we get

$$\frac{2x}{(a^2+\lambda)} + \frac{2y}{a^2}\cdot\frac{dy}{dx} = 0;$$

$$\frac{x}{(a^2+\lambda)} + \frac{y}{a^2}\left(\frac{dy}{dx}\right) = 0 \qquad [2]$$

The D.E. of the system is got by eliminating λ between [1] and [2]:

Applying [(1) – x (2)], gives

$$\left[\frac{x^2}{(a^2+\lambda)} + \frac{y^2}{a^2}\right] - \left[\frac{x\cdot x}{(a^2+\lambda)} + \frac{xy}{a^2}\cdot\frac{dy}{dx}\right] = (1-0) = 1;$$

or $$\frac{y^2}{a^2} - \frac{xy}{a^2}\cdot\frac{dy}{dx} = 1;$$

$$\left[y^2 - xy\left(\frac{dy}{dx}\right)\right] = a^2 \qquad [3]$$

This is the D.E. of the system. That of the O.T. is got by replacing $\left(\dfrac{dy}{dx}\right)$ by $\left[-\dfrac{dx}{dy}\right]$

$\therefore$ O.T.: $$y^2 - xy\left[\left(-\frac{dx}{dy}\right)\right] = a^2$$

$$xy\frac{dx}{dy} + y^2 = a^2;\ xy\frac{dx}{dy} + (y^2 - a^2) = 0$$

or, $$xdx + \left[\frac{(y^2 - a^2)}{y}\right] dy = 0$$

or, $$xdx + [y - (a^2/y)]\, dy = 0 \qquad [4]$$

Solution: $$\int xdx + \int \left[y - \left(\frac{a^2}{y}\right)\right] dy = \text{(const.)}$$

i.e. $$\frac{x^2}{2} + \frac{y^2}{2} - a^2 \log y = \text{(const.)}$$

$$(x^2 + y^2) - 2a^2 \log y = \text{(const.)} = k^2$$

or $$(x^2 + y^2) - 2a^2 \log y - k^2 = 0$$

176. Find the orthogonal Trajectories of the system of Co-axial circles $x^2 + y^2 - 2lx + c = 0$ (l = parameter)

Solution: Given system

$$x^2 + y^2 - 2\lambda x + c = 0 \qquad [1]$$

Differentiate w.r.t. x, we get

$$2x + 2y\left(\frac{dy}{dx}\right) - 2\lambda + 0 = 0;$$

$\therefore$ $$l = \left[x + y\left(\frac{dy}{dx}\right)\right] \qquad [2]$$

substitue for λ, from [2] in [1], to eliminate λ (parameter)

$\therefore$ $$x^2 + y^2 - 2x\left[x + y\left(\frac{du}{dx}\right)\right] + c = 0$$

or, $$(y^2 - x^2) - 2xy\left(\frac{dy}{dx}\right) + c = 0$$

This is the D.E. of the given system.

The D.E. of O.T. is

$$(y^2 - x^2) - 2xy\left[-\frac{dx}{dy}\right] + c = 0$$

i.e. $$(y^2 - x^2) + y \times 2x\left(\frac{dx}{dy}\right) + c = 0$$

or, $$2x\left(\frac{dx}{dy}\right) - (1/y)\, x^2 = -(1/y)(y^2 + 1) \qquad [4]$$

PART-D

This is an eqn. which can be reduced to linear form by a suitable substitution

put $x^2 = u;$ $2x\left(\frac{dx}{dy}\right) = \left(\frac{du}{dy}\right)$ [5]

$$\therefore \quad \frac{du}{dy} + \left(\frac{-1}{y}\right) u = -\left(y + \frac{c}{y}\right) \quad [6]$$

This is in the std. form

$$\left(\frac{du}{dy} + Pu\right) = Q$$

The solution is $u\left(e^{\int P dy}\right) = A + \int Q e^{\int P dy}$ [7]

Here $P = (-1/y);$ $Q = -\left(y + \frac{c}{y}\right)$

$$\therefore \quad \text{I.F.} = e^{\int P dy} = e^{-\log y} = \left(\frac{1}{y}\right)$$

and $\int Q e^{\int P dy} \times dy = \int -\left(y + \frac{c}{y}\right) \cdot \frac{1}{y}\, dy$

$$= -\int \left(1 + \frac{c}{y^2}\right) dy = -[y - (c/y)] = [-y + (c/y)]$$

$$\therefore \quad u(1/y) = A - y + (c/y)$$

$$\left(\frac{x^2}{y}\right) = A - y + (c/y)$$

or $x^2 = Ay - y^2 + c;$

$$(x^2 + y^2) - Ay + c = 0$$

(Put $A = 2\mu$)

The eqn. is written

$$(x^2 + y^2) - 2\mu\, y - c = 0 \quad [8]$$

Thus, the O.T. of the given system of Coaxial circles, is the conjugate system of circles:

$$(x^2 + y^2) - 2\mu y - c = 0$$

177. Find the O.T. of the system of circles $(x - 1)^2 + y^2 + 2ax = 0$ where 'a' is the parameter

Solution: Given equation

$$(x - 1)^2 + y^2 + 2ax = 0 \quad [1]$$

(Differentiate w.r.t x), we get

$$2(x-1)+2y\left(\frac{dy}{dx}\right)+2a=0 \quad [2]$$

To eliminate 'a' and get the differential equation, i.e. substitute for 'a' from (2) in (1),

From (2), $2a = [-2(x-1) - 2y\,(dy/dx)]$

$$\therefore \quad (x-1)^2+y^2+x\left[-2(x-1)-2y\left(\frac{dy}{dx}\right)\right]=0$$

$$(x-1)^2+y^2-2x(x-1)-2xy\left(\frac{dy}{dx}\right)=0$$

$$y^2+(x^2-2x+1-2x^2+2x)-2xy\left(\frac{dy}{dx}\right)=0$$

or $$y^2+(1-x^2)-2xy\left(\frac{dy}{dx}\right)=0 \quad [3]$$

The D.E. of O.T. is given by

$$y^2+(1-x^2)-2xy\left[-\left(\frac{dx}{dy}\right)\right]=0$$

$$y\,2x\left(\frac{dx}{dy}\right)-x^2=-(y^2+1)$$

or, $$y\times 2x\left(\frac{dx}{dy}\right)-x^2=-(y^2+1)$$

or $$2x\frac{dx}{dy}+\left(\frac{-1}{y}\right)x^2=-\left(y+\frac{1}{y}\right) \quad [4]$$

put $x^2 = u$; $$2x\left(\frac{dx}{dy}\right)=\frac{du}{dy}$$

$$\therefore \quad \frac{du}{dy}+\left(\frac{-1}{y}\right)u=\left[-\left(y+\frac{1}{y}\right)\right] \quad [6]$$

This is linear and of the form

$$\left(\frac{du}{dy}\right)+pu=Q$$

Solution: $$u\,e^{\int P\,dy}=A+\int Q\,e^{\int P\,dy}\,dy \quad [7]$$

Here, $P = (-1/y)$;

PART-D

$$e^{\int P dy} = e^{-\log y} = \frac{1}{y}$$

$$Q \cdot e^{\int P dy}\, dy = \int -\left(y + \frac{1}{y}\right)\frac{1}{y}\, dy$$

$$= -\int\left(1 + \frac{1}{y^2}\right) dy = -[y - (1/y) = (1/y) - y]$$

Solution: $u \cdot (1/y) = A + (1/y) - y$

$$\frac{x^2}{y} = A + \frac{1}{y} - y;$$

or $x^2 = Ay + 1 - y^2$

i.e. $(x^2 + y^2) - Ay - 1 = 0$

(Put $A = 2\mu$)

O.T.: $(x^2 + y^2 - 2\mu y - 1) = 0$ [8]

178. Find the d.e. of the system of curves

$$x^2 - y^2 - 2\lambda xy = 1, \quad (\lambda = \text{parameter})$$

Show that the d.e. of the O.T. is

$$(x^2 + y^2)\left[x + y\left(\frac{dy}{dx}\right)\right] = \left[x - y\left(\frac{dy}{dx}\right)\right] \text{ and}$$

solve the equation

Solution: Given system

$$x^2 - y^2 - 2\lambda xy = 1 \quad [1]$$

(Differentiate w.r.t. x) we get

$$2x - 2y\left(\frac{dy}{dx}\right) - 2\lambda\left[x\left(\frac{dy}{dx} + y \cdot 1\right)\right] = 0$$

or, $$(x - yp) - l\,(y + xp) = 0; \left[P = \left(\frac{dy}{dx}\right)\right] \quad [2]$$

To find the d.e. of the system, we have to eliminate λ between [1] and [2],

$$\lambda = \frac{(x^2 - y^2 - 1)}{2xy} \text{ from (1)}$$

$$\lambda = \frac{(x - yp)}{(xp + y)} \text{ from (2)}$$

Equating the two to eliminate λ, we get

$$\frac{x^2 - y^2 - 1}{2xy} = \frac{x - yp}{y + xp} \quad [3]$$

The D.E. of O.T. is

$$\frac{(x^2 - y^2 - 1)}{2xy} = \frac{x - y(-1/p)}{y + x(-1/p)} = \frac{xp + y}{(py - x)}$$

$$(x^2 - y^2 - 1)(yp - x) - 2(xy)(y + xp) = 0$$

i.e. $\quad yp\,(x^2 - y^2 - 1) - 2xy\,xp = 2xy^2 + x\,(x^2 - y^2 - 1)$

or $\quad yp\,(x^2 - y^2 - 1 - 2x^2) = (xy^2 + x^3 - 1)$

or, $\quad (x^2 + y^2)\,yp + yp + x\,(x^2 + y^2 - x) = 0$

or $\quad (x^2 + y^2)\left[x + y\dfrac{dy}{dx}\right] = x - yp$

$$(x^2 + y^2)(xdx + ydy) = (xdx - ydy) \qquad [5]$$

To solve this equation

put $(x^2 + y^2) = u$;

$\therefore \quad (2xdx + 2ydy) = du$

$\therefore \quad (x^2 + y^2)(2xdx + 2ydy) = 2xdx - 2ydy$

or $\quad udu = 2xdx - 2ydy$

$\therefore \quad \int udu = \int 2xdx - 2ydy$

$\therefore \quad \int udu = \int 2xdx - \int 2ydy$

$$(u^2/_2) = (x^2 - y^2) + (\text{const.})$$

$$u^2 = 2(x^2 - y^2) + (\text{const.})$$

$$(x^2 + y^2)^2 = 2(x^2 - y^2) + 2a^2$$

or, $\quad (x^2 + y^2)^2 = 2(x^2 - y^2 + a^2)$

Self Orthogonal Curves

A family of curves is said to be self orthogonal if each member of the family cuts every member of the same family at right angles.

i.e. To show that a given system of curves

$f(x, y, a) = 0$ is self orthogonal, it is enough to show that its d.e. $F\left(x, y, \dfrac{dy}{dx}\right) = 0$ and D.E. of O.T. $F\left(x, y, \dfrac{dy}{dx} - \dfrac{dx}{dy}\right)$ are one and the same.

179. Prove that the system of confocal and coaxal parabolas $y^2 = 4a\,(x + a)$ is self-orthogonal.

Solution: Given system: $y^2 = 4a\,(x + a)$ [1]

(Differentiate [1] w.r. to x), we get

$$2y \cdot \frac{dy}{dx} = 4a$$

$$\Rightarrow \quad a = \frac{1}{2}\, y \times \frac{dy}{dx}$$

$\therefore$ From [1], $$y^2 = 2y\frac{dy}{dx}\left[x + \frac{1}{2}y\frac{dy}{dx}\right]$$

or, $$y^2 = 2xy\frac{dy}{dx} + y^2\left(\frac{dy}{dx}\right)^2$$

or $$y^2 = 2xyp + y^2p^2,$$

where $$p = \frac{dy}{dx} \qquad [2]$$

Which being free from a (the parameter), is the d.e. of the family [1]

Replace P by $-\dfrac{1}{P}$,

The D.E. of O.T. is $$y^2 = -\frac{2xy}{P} + \frac{y^2}{P^2} \quad \text{or} \quad y^2P^2 + 2xyP = y^2 \qquad [3]$$

which is the same as [2]. Since the D.E. of the family of parabolas [1] and that of O.T. are same, the family of paraboblas [1] is self orthogonal.

180. Prove that the system of confocal conics

$$\frac{x^2}{a^2 + \lambda} + \frac{y^2}{b^2 + \lambda} = 1$$

is self-orthogonal.

Solution: Given System

$$\frac{x^2}{a^2 + \lambda} + \frac{y^2}{b^2 + \lambda} = 1 \qquad [1]$$

(Differentiate w.r.t. x), we get

$$\frac{2x}{a^2 + \lambda} + \frac{2y}{b^2 + \lambda} \cdot P = 0$$

where $$P = \frac{dy}{dx};$$

or, $$(b^2 + \lambda)\, x + yp\, (a_2 + \lambda) = 0$$

or $$\lambda\, (x + yp) = -\, (a^2yp + b^2\, x)$$

$$\therefore \quad \lambda = -\frac{a^2\, py + b^2 x}{x + yp}$$

$$\therefore \quad a^2 + \lambda = \frac{(a^2 - b^2)\, x}{x + yp},$$

$$b^2 + \lambda = -\frac{(a^2 - b^2)\, yp}{x + yp}$$

∴ From (1)

$$\frac{x^2(x+yp)}{(a^2-b^2)x}-\frac{y^2(x+yp)}{(a^2+b^2)yp}=1$$

or, $$(x+yp)\left(x-\frac{y}{p}\right)=(a^2-b^2) \quad [2]$$

which is free from λ,
is the D.E. of the family [1]

∴ The D.E. of O.T. is obtained by replacing

p by $-\frac{1}{p}$, in [2]

$$\left(x-\frac{y}{p}\right)(x+yp)=(a^2-b^2) \quad [3]$$

This is same as [2]

Since the D.E. of the family of curves [1] and that of O.T. are same, the family of curves [1] is self orthogonal.

181. Find the O.T. of the family of lines with slopes and y-intercept are equal.

Solution: By data, the given system is:

$$y = mx + m \quad [1]$$
$$= m(x+1)$$

Differentiate w.r.t x, we get

$$\frac{dy}{dx}=m$$

∴ [1] becomes, $$y=\frac{dy}{dx}(x+1)$$

∴ The D.E. of O.T. is

$$y=-\frac{dx}{dy}(x+1) \quad \text{or,} \quad ydy+(x+1)\,dx=0$$

Integrating, we get

$$\frac{y^2}{2}+\frac{x^2}{2}+x=k=-\frac{c}{2} \quad \text{or,} \quad x^2+y^2+2x+c=0$$

is the required O.T

$(x+1)^2+y^2=a^1$ is the O.T.

which represents concentric circles with centre (– 1, 0)

182. Determine the 45° trajectories of the family of concentric circles $x^2+y^2=c^2$

Solution: Given $x^2+y^2=c^2$ [1]

Differentiate w.r.t x, we get

$$2x+2y\cdot\frac{dy}{dx}=0$$

PART-D

or $$x + yp = 0$$

where $$P = \frac{dy}{dx}$$ [2]

is the D.E. of given system.

∴ Replacing P by

$$\frac{\frac{dy}{dx} + \tan 45^\circ}{1 - \frac{dy}{dx} \cdot \tan 45 \cdot} = \frac{p+1}{1-p}$$

The D.E. of 45° trajectories is

$$x + y \cdot \frac{1+p}{1-p} = 0 \qquad \text{or} \quad x(1-p) + y(1+p) = 0$$

or, $$(x - y)P = x + y; \quad \text{or} \quad \frac{dy}{dx} = \frac{x+y}{x-y}$$ [3]

This is a Homo eqn. to solve

put $y = vx$; $$\frac{dy}{dx} = v + x \times \frac{dv}{dx}$$

we obtain $$\frac{1-v}{1+v} = \frac{dx}{x}$$

or $$\left(\frac{1}{1+v^2} - \frac{1}{2} \cdot \frac{2v}{1+v^2}\right) dv = \frac{dx}{x}$$

Integrate $\tan^{-1}(v) - \frac{1}{2} \log(1 + v^2)$

$$= \log x + \log c$$

or, $$\tan^{-1}(v) = \log x + \log(1 + v^2)^{1/2} = \log\left(x\sqrt{1+v^2}\right)$$

$$= \log\left(x \cdot \sqrt{1 + \frac{y^2}{x^2}}\right)$$

or, $$\tan^{-1}\left(\frac{y}{x}\right) = \log\left(\sqrt{x^2 + y^2}\right) \text{ is O.T.}$$

183. Find the O.T of a system of concurrent straight lines

Solution: Choose the point of concurrence as the origin. Any straight line thro the origin, has its equation in the form $y = mx$ (1)

Diff. w.r.t x $\qquad \dfrac{dy}{dx} = m$ [2]

to eliminate 'm' divide [2] by [1]

$$\therefore \quad \frac{\left(\dfrac{dy}{dx}\right)}{y} = \frac{m}{mx} = \frac{1}{x};$$

$$\therefore \quad x\left(\frac{dy}{dx}\right) = y \qquad [3]$$

is D.E. of system

$\therefore$ the D.E. of O.T. is

$$x\left(-\frac{dx}{dy}\right) = y$$

$$\Rightarrow \quad -xdx = ydy;$$

$$\text{or} \quad (xdx + ydy) = 0 \qquad [4]$$

(V.S form)

$$\therefore \quad \int xdx + \int ydy = \text{const}$$

$$\text{or,} \quad \left(\frac{x^2}{2}\right) + \left(\frac{y^2}{2}\right) = \frac{a^2}{2}$$

$$\text{or,} \quad x^2 + y^2 = a^2$$

is clearly a system of circles all of which have the same centre (0,0); *i.e.*, a system of concentric circles.

184. Find the O.T. of the curves

$$\left(\frac{x^n}{a^n}\right) + \left(\frac{y^n}{b^n}\right) = C,$$

where C is the parameter

Solution: $\qquad \left(\dfrac{x^n}{a^n}\right) + \left(\dfrac{y^n}{b^n}\right) = C$ [1]

Differentiate w.r.t x, we get

$$\frac{nx^{n-1}}{a^n} + \frac{ny^{n-1}}{b^n} \cdot \frac{dy}{dx} = 0;$$

$$\text{or} \quad \frac{x^{n-1}}{a^n} + \frac{y^{n-1}}{b^n} \cdot \frac{dy}{dx} = 0 \qquad [2]$$

is the D.E of the system

$\therefore$ the D.E. of O.T. is

$$\frac{x^{n-1}}{a^n} + \frac{y^{n-1}}{b^n}\left(-\frac{dx}{dy}\right) = 0;$$

or,
$$\frac{dx}{x^{n-1} \cdot b^n} - \frac{dy}{y^{n-1} \cdot a^n} = 0 \qquad [3]$$

(V.S. form)

Integrating, we get

$$\int \frac{x^{-n+1}}{b^n}\,dx - \int \frac{y^{-n+1}}{a^n}\,dy = \text{const.}$$

$$\Rightarrow \quad \frac{x^{-n+2}}{(-n+2)b^n} - \frac{y^{-n+2}}{(-n+2)a^n} = \text{const}$$

or
$$\frac{1}{b^n\ x^{n-2}} - \frac{1}{a^n\ y^{n-2}} = (2-n)\ (\text{const})$$

or
$$a^n\, x^{2-n} - b^n y^{2-n} = (2-n)\, a^n b^n\ (\text{const}) = C_1$$

Polar Curves (Polar Co-ordinates)

(Orthogonal Trajectories)

(1) Given system :– $f(r, q, a) = 0$

(2) Differentiate eqn:– $F\left[r, \theta, \left(\frac{dr}{d\theta}\right)\right] = 0$

(3) Diff. eqn. of O.T:– $F\left[r, \theta, -r^2\left(\frac{d\theta}{dr}\right)\right] = 0$

(4) Solve (3) this to get the Trajectories

185. Find the orthogonal trajectories of the family of cardioids $r = a\,(1 - \cos\theta)$; where a is the parameter [VTU, March, 2001]

Solution: Given equation of the system:–

$$r = a\,(1 - \cos\theta) \qquad [1]$$

(differentiate [1] w.r.t. θ)

we get,
$$\frac{dr}{d\theta} = a \sin\theta$$

$$\Rightarrow \quad a = \frac{1}{\sin\theta} \cdot \frac{dr}{d\theta}$$

$\therefore$ From (1),
$$r = \frac{1}{\sin\theta} \cdot \frac{dr}{d\theta}\,(1 - \cos\theta)$$

or
$$\frac{dr}{d\theta} = \frac{r\sin\theta}{(1-\cos\theta)} = \frac{2r\sin\left(\frac{\theta}{2}\right)\cos\left(\frac{\theta}{2}\right)}{2\sin^2(\theta/2)}$$

or $$\frac{dr}{d\theta} = r \cot\left(\frac{\theta}{2}\right) \quad [2]$$

is the D.E. of the system.

∴ the D.E. of O.T. is:

$$-r^2 \frac{d\theta}{dr} = r \cot(\theta/2)$$

or $$\frac{dr}{r} = -\tan\left(\frac{\theta}{2}\right) d\theta \quad [3]$$

Integrating [3], we get

$$\log r = -\left[\frac{-\log \cos\left(\frac{\theta}{2}\right)}{\frac{1}{2}}\right] + \log C_1$$

or $$\log r = 2 \log \cos \frac{\theta}{2} + \log C_1$$

or $$\log r = \log C_1 \cos^2 \theta/2$$

or $$r = C_1 \cos^2 \frac{\theta}{2} = C_1 \cdot \frac{1 + \cos\theta}{2}$$

or $$r = C(1 + \cos\theta) \quad \left(C = \frac{1}{2} C_1\right)$$

is the reqd. equation

186. Find the O.T. of the cardioide $r = a(1 + \cos\theta)$; where a is the parameter.

[VTU, August, 2001]

Solution: Given equation of the system:– $r = a(1 + \cos\theta)$ [1]

(Differentiate w.r.t. θ), we get

$$\frac{dr}{d\theta} = -a \sin\theta \quad [2]$$

⇒ $$a = -\frac{1}{\sin\theta} \cdot \frac{dr}{d\theta} \quad [3]$$

∴ From (1) $$a = \frac{r}{1 + \cos\theta} \quad [4]$$

From (3) and (4), we get

$$-\frac{1}{\sin\theta} \frac{dr}{d\theta} = \frac{r}{1 + \cos\theta}$$

or $$\frac{1}{r} \frac{dr}{d\theta} = -\frac{\sin\theta}{1 + \cos\theta} \quad [5]$$

PART-D

To get the D.E. of O.T., we replace $\left(\frac{dr}{d\theta}\right)$ *by* $\left[-r^2\left(\frac{d\theta}{dr}\right)\right]$

$$\therefore \quad \frac{1}{r}\left[-r^2\frac{d\theta}{dr}\right] = -\frac{\sin\theta}{1+\cos\theta}$$

$$\therefore \quad -r\frac{d\theta}{dr} = -\frac{\sin\theta}{1+\cos\theta}$$

or
$$\frac{dr}{r} = \frac{1+\cos\theta}{\sin\theta}\,d\theta = \frac{2\cos^2(\theta/2)}{2\sin(\theta/2)\cos(\theta/2)}\,d\theta$$

$$\therefore \quad \frac{dr}{r} = \cot(\theta/2)\,d\theta$$

Integrating, we get $\log r = 2\log\sin(\theta/2) + \log b$

or
$$\log\frac{r}{b} = \log\sin^2(\theta/2)$$

or
$$\frac{r}{b} = \sin^2(\theta/2)$$

$$\Rightarrow \quad \frac{r}{b} = \frac{1-\cos\theta}{2} \quad \text{or} \quad r = C(1-\cos\theta),$$

where $\frac{b}{2} = C$, which is the requd. O.T.

Evidently, it Represests a Cardioide

187. Find the O.T. of $r\theta = a$

Solution: Given system:– $r\theta = a$ [1]

(Differentiate w.r.t. θ), we get

$$r + \theta\cdot\frac{dr}{d\theta} = 0 \quad \text{or} \quad -\frac{1}{r}\frac{dr}{d\theta} = \frac{1}{\theta}$$

For orthogonal Trajectory, replace $-\frac{1}{r}\frac{dr}{d\theta}$ by $r\frac{d\theta}{dr}$ $\left(\text{or } \frac{dr}{d\theta} \text{ by } -r^2\frac{d\theta}{dy}\right)$, we get

$$\therefore \quad r\frac{d\theta}{dr} = \frac{1}{\theta} \quad \text{or} \quad \frac{dr}{r} = \theta d\theta$$

Integrating, we get

$$\log r = \frac{\theta^2}{2} + \log C$$

or
$$\log\frac{r}{C} = \frac{\theta^2}{2};$$

or $$\frac{r}{C} = e^{\theta^2/2} \quad \text{or} \quad r = C\, e^{\theta^2/2}$$

This is the required equation of O.T.

188. Find the O.T. of $r = a\theta$

Given System: $r = a\theta$ [1]

(Differentiate w.r.t. θ), we get

$$\frac{dr}{d\theta} = a \qquad [2]$$

From [1], $a = \dfrac{r}{\theta}$

Putting this value in (2) to eliminate 'a' we get

$$\frac{dr}{d\theta} = \frac{r}{\theta}$$

For O.T.,

replace $\dfrac{dr}{d\theta}$ by $-r^2 \dfrac{d\theta}{dr}$

$$\therefore \quad -r^2 \frac{d\theta}{dr} = \frac{r}{\theta} \quad \text{or} \quad \frac{dr}{r} = -\theta d\theta$$

Integrating, we get $$\log r = \frac{-\theta^2}{2} + \log C$$

or $$\log \frac{r}{C} = \frac{-\theta^2}{2}$$

$$\Rightarrow \quad \frac{r}{C} = e^{-\theta^2/2} \quad \text{or} \quad r = Ce^{-\theta^2/2}$$

This is the required equation of O.T.

189. A family of parabolas has a common focus and common axis. Find the orthogonal family. (MU 1982)

Or

Find the equation of the system of O.T. of a series of confocal coaxial parabolas $r = \left[\dfrac{2a}{1+\cos\theta}\right]$, where '$a$' is the parameter

Solution: Given system:

$$r = \left[\frac{2a}{1+\cos\theta}\right]; \quad \text{or} \quad r\,(1 + \cos\,\theta) = 2a \qquad [1]$$

Differentiating w.r.t θ, we get

$$r(-\sin\,\theta) + (1 + \cos\,\theta)\left(\frac{dr}{d\theta}\right) = 0 \qquad [2]$$

PART-D

Here, the parameter 'a' is automatically eliminated and so (2) is the D.E. of the system

The D.E. of O.T. is given by putting $\left[-r^2 \cdot \left(\frac{d\theta}{dr}\right)\right]$ in the place of $\left(\frac{dr}{d\theta}\right)$

so, we have, $$-r \sin \theta + (1 + \cos \theta)\left[-r^2 \left(\frac{d\theta}{dr}\right)\right] = 0$$

$$r \sin \theta \, dr + r^2 (1 + \cos \theta) \, d\theta = 0$$

$$\frac{dr}{r} + \frac{(1 + \cos \theta)}{\sin \theta} d\theta = 0;$$

$$\frac{dr}{r} + \frac{2 \cos^2 (\theta/2) \, d\theta}{2 \sin (\theta/2) \cos (\theta/2)} = 0$$

or, $$\frac{dr}{r} + \cot (\theta/2) \, d\theta = 0 \qquad [3]$$

Solution: $$\int \frac{dr}{r} + \int \cot\left(\frac{\theta}{2}\right) d\theta = \text{const.}$$

i.e. $$\log r + \frac{\log \sin (\theta/2)}{(1/2)} = (\text{const}).$$

or, $$\log r + 2 \log \sin (\theta/2) = (\text{const})$$

or, $$\log r + \log \sin^2 (\theta/2) = (\text{const})$$

or, $$\log r + \log \left[\frac{(1 - \cos \theta)}{2}\right] = (\text{const})$$

or, $$\log \left[r \cdot \frac{(1 - \cos \theta)}{2}\right] = (\text{const})$$

or, $$r \, [(1 - \cos \theta)/2] = (\text{const}) = a$$

or, $$r = \left[\frac{2a}{(1 - \cos \theta)}\right] \quad \text{or} \quad \left(\frac{2a}{r}\right) = (1 - \cos \theta)$$

This is a system of parabolas, with the focus of each at the pole and the axis of each along the initial line (the given system is a self orthogonal system of parabolas)

190. Find the O.T. of the class of curves

$$r = (a + \cos n \theta);$$

with parameter a.

Solution: Given System: $$r = a + \cos n\theta \qquad [1]$$

Differentiate w.r.t θ; we get

$$\left(\frac{dr}{d\theta}\right) = 0 + (-\sin n \theta) \cdot n;$$

$$\frac{dr}{d\theta} = -n \sin(n\theta) \qquad [2]$$

Here, 'a' is the parameter is automatically eliminated, and so (2) is the d.e. of the system

So, the D.E. of the O.T. is given by replacing $\left(\frac{dr}{d\theta}\right)$ by $\left(-r^2 \cdot \frac{d\theta}{dr}\right)$,

$\therefore$ O.T.:
$$-r^2 \frac{d\theta}{dr} = -n \sin n\theta$$

$$-n \cdot \frac{dr}{r^2} + \frac{d\theta}{\sin n\theta} = 0 \qquad [3]$$

Solution:
$$n \int \frac{-dr}{r^2} + \int \operatorname{cosec}(n\theta)\, d\theta = (\text{const})$$

or
$$n\left(\frac{1}{r}\right) + \frac{1}{n} \log \tan\left(\frac{n\theta}{2}\right) = (\text{const})$$

or,
$$(n^2/r) = (\text{const}) - \log[\tan(n\theta/2)]$$
$$= \log a + \log[\cot(n\theta/2)] = \log[a \cot(n\theta/2)]$$

$$\boxed{r \log[a \cot(n\theta/2)] = n^2}$$

191. Obtain the O.T. of the curves

$$r = a + \sin 5\theta;$$

('a' variable parameter)

Solution: Given System:
$$r = (a + \sin 5\theta) \qquad [1]$$

(Differentiate w.r.t. θ)

$$\frac{dr}{d\theta} = 0 + \cos 5\theta \cdot 5$$

Here, a is automatically eliminated, and so the d.e. of the system

$$\left(\frac{dr}{d\theta}\right) = 5 \cos 5\theta \qquad [2]$$

$\therefore$ the D.E. of O.T.: $-r^2\left(\frac{d\theta}{dr}\right) = 5 \cos \theta$

or,
$$5\left(\frac{dv}{r^2}\right) + \sec 5\theta \cdot d\theta = 0 \qquad [3]$$

Solution:
$$\int\left(\frac{5}{r^2}\right) dr + \int \sec 5\theta\, d\theta = (\text{const})$$

$$5(-1/r) + (1/5) \log(\sec 5\theta + \tan 5\theta) = (\text{const})$$

or,
$$(-25/r) + \log(\sec 5\theta + \tan 5\theta) = (\text{const})$$

or,
$$(-25/r) + \log(\sec 5\theta + \tan 5\theta) + \log k = 0$$

$$\therefore \qquad \left(\frac{25}{r}\right) = \log [k (\sec 5\theta + \tan 5\theta)]$$

or, $\quad r \log [k (\sec 5\theta + \tan 5\theta)] = 25$ [4]

192. Find the O.T. of the system of curves

$$r^n = a^n \cos n\theta;$$
$$(a = \text{parameter})$$

Solution: Given System: $\quad r^n = a^n \cos n\theta$ [1]

Differentiating w.r.t to θ

$$nr^{n-1}\left(\frac{dr}{d\theta}\right) = a^n (-\sin n\theta)\, n \quad [2]$$

Dividing (1) by (2) (to eliminate a, the paarameter)

$$\frac{r^n}{nr^{n-1}\left(\frac{dr}{d\theta}\right)} = \frac{a^n \cos n\theta}{-na^n \sin n\theta};$$

$$r = -(\cot n\theta)\left(\frac{dr}{d\theta}\right) \quad [3]$$

$\therefore$ D.E. of O.T. is given by substituting

$$\left[-r^2\left(\frac{d\theta}{dr}\right)\right] \text{ for } \left(\frac{dr}{d\theta}\right)$$

that is, $$r = -[\cot n\theta]\cdot\left(-r^2\frac{d\theta}{dr}\right)$$

or, $$\left(\frac{dr}{r}\right) = \cot n\theta\, d\theta \; [4]$$

Solution: $$\int (1/r)\, dr = \int \cot n\theta\, d\theta + (\text{const})$$

i.e. $$\log r = \log (\sin n\theta)\left(\frac{1}{n}\right) + (\text{const})$$

$$n \log r = \log \sin n\theta + (\text{const})$$
$$= \log \sin n\theta + \log b^n \text{ (say)}$$
$$\log r^n = \log b^n (\sin n\theta)$$

$\therefore$ O.T.: $\quad r^n = b^n \sin n\theta$ [5]

193. Obtain the O.T. of the class of curves

$$r^n \sin n\theta = a^n$$

Solution: Equation of the system: $r^n \sin n\theta = a^n$ [1]

Differentiate w.r.t θ and eliminate θ.

It will be more convenient to take logarithms and differentiate

$$\therefore \qquad \log (r^n \sin n\theta) = \log a^n;$$

$$n \log r + \log \sin n\theta = n \log a \quad [2]$$

$\therefore$ $$n \cdot \frac{1}{r} \cdot \frac{dr}{d\theta} + \frac{1}{\sin n\theta} \cos n\theta \cdot n = 0$$

or, $$\left(\frac{dr}{d\theta}\right) + r \cot n\theta = 0 \quad [3]$$

Here 'a' is automatically eliminated, and so (3) is directly the d.e. of the system

The D.E. of O.T. is given by replacing $\left(\frac{dr}{d\theta}\right) by \left[-r^2 \left(\frac{d\theta}{dr}\right)\right]$

$\therefore$ $$-r^2 \left(\frac{d\theta}{dr}\right) + r \cot n\theta = 0$$

or, $$\frac{-d\theta}{\cot n\theta} + \frac{rdr}{r^2} = 0;$$

$$-\tan n\theta + (1/r)\, dr = 0 \quad [4]$$

Solution: $$\int (1/r)\, dr - \int \tan n\theta \, d\theta = \text{(const)}$$

$$\log r + (1/n) \log \cos n\theta = \log b, \text{ (say)}$$

$$n \log r + \log \cos n\theta = n \log b$$

$$\log (r^n \cdot \cos n\theta) = \log b^n$$

$\therefore$ $$r^n \cos n\theta = b^n$$

$\therefore$ O.T.: $$r^n \cos n\theta = b^n$$

194. Find the O.T. of $\quad r = e^{a\theta}$

Solution: $$r = e^{a\theta} \quad [1]$$

or $$1 = ae^{a\theta} \cdot \frac{d\theta}{dr}$$

(Differentiated w.r.t. r)

or $$r \cdot \frac{d\theta}{dr} = \frac{1}{a} = \frac{\theta}{\log r}$$

$\therefore$ Eqn. of O.T. is given by

$$-\frac{1}{r} \cdot \frac{dr}{d\theta} = \frac{\theta}{\log r} \quad \text{or} \quad \frac{\log r}{r}\, dr = -\theta d\theta$$

or integrating, we get

$$\frac{(\log r)^2}{2} = -\frac{\theta^2}{2} + C^2 \quad \text{or,} \quad \theta^2 + (\log r)^2 = C^2$$

195. Find the O.T. of $\quad r^n = a^n \sin n\theta$

$\therefore$ $$nr^{n-1} = a^n \cdot \cos n\theta \cdot (n) \cdot \frac{d\theta}{dr}$$

(on differentiation w.r.t. r)

$$nr^{n-1} = nr^n \cot n\theta \cdot \frac{d\theta}{dr}$$

$$\therefore \quad r\frac{d\theta}{dr} = \tan(n\theta)$$

$\therefore$ Equation of O.T. is $-\frac{1}{r}\frac{dr}{d\theta} = \tan(n\theta)$

or, $$\frac{dr}{r} = -\tan(n\theta)\cdot d\theta$$

$$\therefore \quad \log r = \frac{1}{n}\log\cos n\theta + \log c$$

or, $r^n = c^n \cos n\theta$ is the reqd. O.T.

196. Find the O.T. of

$$r = c\cdot\sin^2\theta$$

Solution: Differentiating w.r.t. θ, and eliminate c, we get

$$r = c\cdot\sin^2\theta \qquad [1]$$

$$\therefore \quad \frac{d\theta}{dr} = 2\,c\sin\theta\cos\theta \qquad [2]$$

Eliminate c from [1] and [2], we get

$$\frac{dr}{d\theta} = 2\cdot\frac{r}{\sin^2\theta}\cdot\sin\theta\cos\theta = 2\,r\cot\theta$$

changing $\frac{dr}{d\theta}$ to $-r^2\frac{d\theta}{dr}$,

the equation to the O.T. is $-r^2\frac{d\theta}{dr} = 2r\cot\theta$ or $2\left(\frac{dr}{r}\right) = -\tan\theta$

Integrating, we get $\log r^2 = \log\cos\theta + \log c = \log c\cos\theta$

$$\therefore \quad \boxed{r^2 = c\cos\theta}$$

is the required O.T.

197. Find the O.T. of the series of logarithmatic spirals $r = a^\theta$; where a varies,

Solution: **Given System:** $r = a^\theta$ [1]

Differentiating [1] wrt θ

$$\frac{dr}{d\theta} = a^\theta\log a = r\log a$$

or $$\frac{dr}{d\theta} = r\cdot\frac{\log r}{\theta} \qquad [2]$$

the D.E. of O.T. is obtained by replacing

$$\frac{dr}{d\theta} \text{ by } -r^2\cdot\frac{d\theta}{dr} \quad -r^2\frac{d\theta}{dr} = \frac{r\log r}{\theta}$$

or $$\frac{\log r}{r}\, dr = -\theta d\theta \qquad [3]$$

Integrating [3] $$\int (\log r)\cdot \frac{1}{r}\, dr = -\frac{\theta^2}{2} + c_1$$

or $$\frac{(\log r)^2}{2} = C_1 - \frac{\theta^2}{2} = \frac{C^2}{2} - \frac{\theta^2}{2}$$

or $$(\log r)^2 = C^2 - \theta^2$$

or $$\log r = \sqrt{C^2 - \theta^2}$$

or $$r = e^{\sqrt{C^2 - \theta^2}}$$

which is the requd. O.T.

198. Find the O.T. of the family of curves

$$r \sin 2\theta = \lambda$$

Solution: **Given System:**

$$r \sin 2\theta = \lambda \qquad [1]$$

Diff. w.r.t. θ, we get

$$\frac{dr}{d\theta}\cdot \sin 2\theta + 2r \cos 2\theta = 0 \quad \text{or} \quad \frac{dr}{d\theta} = -2r \cot 2\theta$$

the D.E. of O.T. is

$$-r^2 \frac{d\theta}{dr} = -2r \cot 2\theta \quad \text{or} \quad \left[4\frac{dr}{r}\right] - 2 \tan 2\theta\, d\theta = 0$$

Integrating, we get

$$4 \log r + \log \cos 2\theta = 4 \log c^4$$

or, $$\log r^4 \cdot \cos 2\theta = c^4 \quad \text{or,} \quad r^4 \cos 2\theta = c^4$$

is the requd. O.T.

199. Find the O.T. of the family of

$$r = \tan(\theta + \alpha) \qquad [1]$$

Diff. w.r.t. θ, we get

$$\frac{dr}{d\theta} = \sec^2(\theta + \alpha) = 1 + r^2$$

$\therefore$ The D.E. of O.T. is

$$-r^2 \cdot \left(\frac{d\theta}{dr}\right) = 1 + r^2 \quad \text{or} \quad d\theta = \frac{1 + r^2}{-r^2}\, dr = \left(-\frac{1}{r^2} - 1\right)$$

Integrating, $$\theta = \frac{1}{r} - r + C \quad \text{or} \quad \theta + r = C + r^{-1}$$

200. Find the orthogonal trajectories of the family of curves $r^2 = a^2 \sin 2\theta$; where a is the parameter

PART-D

Solution: **Given System:** $r^2 = a^2 \sin 2\theta$; take log on both sides, we get

$$2 \log r = 2 \log a + \log \sin 2\theta$$

$$\Rightarrow \quad \frac{2}{r} \cdot \frac{dr}{d\theta} = \frac{2 \cos 2\theta}{\sin 2\theta};$$

i.e. $$\frac{dr}{d\theta} = r \cot 2\theta; \quad [1]$$

This is the D.E. of the given system

The D.E. of O.T. is $-r^2 \cdot \dfrac{d\theta}{dr} = r \cot 2\theta$

i.e. $$-r \cdot \frac{d\theta}{dr} = \cot 2\theta \quad [2]$$

or $$\frac{dr}{r} = -\tan 2\theta$$

on integrating, we get

$$\log r = -\frac{1}{2} \log \sec 2\theta + \log c$$

or $$r^2 = c^2 \cos 2\theta$$
$$r^2 = c^2 \cos 2\theta$$

is the requd. O.T. of the given family of curves

201. Find the O.T. of the family of lemniscates $r^2 = a^2 \cos 2\theta$

Solution: **Given System:** $r^2 = a^2 \cos 2\theta \quad [1]$

(Differentiate w.r.t. θ), we get

$$2r \cdot \frac{dr}{d\theta} = -2a^2 \sin 2\theta$$

$$\Rightarrow \quad a^2 = \frac{-r}{\sin 2\theta} \cdot \frac{dr}{d\theta}$$

Thus, [1] becomes $$r^2 = \frac{-r \cos 2\theta}{\sin 2\theta} \cdot \frac{dr}{d\theta}$$

$$\Rightarrow \quad r \frac{d\theta}{dr} + \cot 2\theta = 0$$

Replace $r \dfrac{d\theta}{dr}$ by $-\left(\dfrac{1}{r} \dfrac{dr}{d\theta}\right)$,

We get $$-\left(\frac{1}{r} \frac{dr}{d\theta}\right) + \cot 2\theta = 0$$

is the D.E. of O.T.

or $$\frac{dr}{r} = \cot 2\theta \, d\theta$$

(V.S form)

Integrating we get $$\int \frac{dr}{r} = \int \cot 2\theta d\theta + \log k$$

$$\Rightarrow \quad \log r = \frac{1}{2} \log (\sin 2\theta) + \log k$$

$$\Rightarrow \quad 2 \log r = \log (\sin 2\theta) + 2 \log k$$

$$\Rightarrow \quad r^2 = k^2 \sin 2\theta \text{ is O.T.}$$

202. Find the O.T. of the system of circles

$$r = C (\cos \theta + 6 \sin \theta)$$

Solution: Given system:- $$r = c (\cos \theta + 6 \sin \theta) \qquad [1]$$

Diff [1] w.r.t. θ, we get $$\frac{dr}{d\theta} = C (- \sin \theta + 6 \cos \theta) \qquad [2]$$

From [1], $$c = \frac{r}{\cos \theta + 6 \sin \theta}$$

To eliminate C, put this value in (2),

$$\therefore \quad \frac{1}{r}\frac{dr}{d\theta} = \frac{6 \cos \theta - \sin \theta}{6 \sin \theta + \cos \theta}$$

To get O.T., replace $\frac{1}{r}\frac{dr}{d\theta}$ by $-r\frac{d\theta}{dr}$

$$-r\frac{d\theta}{dr} = \frac{6 \cos \theta - \sin \theta}{6 \sin \theta + \cos \theta}$$

or $$\frac{dr}{r} = \frac{\cos \theta + 6 \sin \theta}{\sin \theta - 6 \cos \theta}$$

Integrating, we get $$\log r = \log (\sin \theta - 6 \cos \theta) + \log b$$

or $$r = b (\sin \theta - 6 \cos \theta) \text{ is the requd O.T.}$$

203. Find the O.T. of $\left(r + \frac{k^2}{r}\right) \cos \theta = \alpha$, α being the parameter [V.T.U. F/M 2005]

Solution: **Given System:**

$$\left(r + \frac{k^2}{r}\right) \cos \theta = \alpha \qquad [1]$$

Diff [1] w.r.t. θ, we get

$$-\left(1 - \frac{k^2}{r^2}\right)\frac{dr}{d\theta} \cos \theta - \left(r + \frac{k^2}{r}\right) \sin \theta = 0$$

PART-D

or $$\left(r - \frac{k^2}{r}\right)\frac{1}{r}\frac{d\theta}{dr}\cos\theta$$

$$-\left(r + \frac{k^2}{r}\right)\sin\theta = 0$$

For O.T., replace $\frac{1}{r}\frac{dr}{d\theta}$ by $-r\frac{d\theta}{dr}$

$\therefore$ $$-\left(r - \frac{k^2}{r}\right)r\frac{d\theta}{dr}\cos\theta$$

$$-\left(r + \frac{k^2}{r}\right)\sin\theta = 0$$

or, $$\frac{\cos\theta}{\sin\theta}d\theta + \frac{r^2 + k^2}{r(r^2 - k^2)}dr = 0$$

or, $$\frac{\cos\theta}{\sin\theta}d\theta = \frac{(k^2 - r^2) + 2r^2}{r(k^2 - r^2)}dr = \frac{dr}{r} - \frac{-2rdr}{k^1 - r^2}$$

Integrating, we get

$$\log\sin\theta = \log r - \log(r^2 - k^2) + \log C$$

$$= \log\frac{cr}{r^2 - k^2}$$

or $$r^2 - k^2 = Cr\operatorname{cosec}\theta \text{ is the reqd. O.T.}$$

EXERCISES

(I) Find the O.T. of the following equations of curves

(1) $y^2 = cx$ (2) $y = ax^n$

(3) $2x^2 + y^2 = kx$ (4) $3xy = (x^3 - a^3)$

(5) $y^3 = a^2$ (6) $x^3 - 3xy^2 = a$

(7) $x^2 + y^2 = 2ay$ (8) $y = a\sqrt{1 - x^2}$

(9) $y = a + \tan x$ (10) $y = 3x + ce^{-2x}$

(11) $x^2 + y^2 + c^2 = 1 + 2\,cxy$ (12) $\sqrt{x^2 + y^2} = (x + a)$

(II) Find the O.T. passing thro the specific point noted in the following cases

(1) $y = x + ae^{-x}$; (0, 3) (2) $y^2 = c(1 + x^2)$; (– 2,5)

(3) $x^2 + cy^2 = 1$; (2, 1) (4) $x^2 = cy + y^2$; (3 – 1)

(III) Find the O.T. of the curves whose D.E. is $\left(\frac{dy}{dx}\right)^2 = (1/x)$

(IV) Show that the O.T. of the system whose d.e. is $\left(\frac{dr}{d\theta}\right) = F(r, \theta)$, are given by $r^2 + \left(\frac{dr}{d\theta}\right) F(r, \theta) = 0$

(V) Find O.T. of the curves whose equations are in polar coordinates are given below.

(1) $r = e^{a\theta}$ (2) $r = a^{\theta}$

(3) $r = a(1 + \cos\theta)$; (4) $r^2 = a^2 \cos 2\theta$

(5) $r \sin 2\theta = \lambda$; (6) $r = a \cos\theta$

(7) $r = a \sin^2\theta$; (8) $r = \tan(\theta + \alpha)$

(9) $r\theta = a$; (10) $r^n = a^n \cos n\theta$

(VI) Show that the O.T. of the curves $A = r^2 \cos\theta$ are the curves $r = B \sin^2\theta$

4.7 INFINITE SERIES

4.8 CONVERGENCE, DIVERGENCE AND OSCILLATION OF AN INFINITE SERIES

If $\{a_n\} = \{a_1, a_2, a_3 ..... a_n, ...$is an infinite sequence, then the formal expression

$a_1 + a_2 + a_3 ... a_n + ...$ is called the infinite series and is denoted by $\sum_{n=1}^{\infty} a_n$ or $\sum a_n$.

Consider the sequence of partial sums $\{S_n\}$, where

$S_n = a_1 + a_2 + ... + an$, $n = 1, 2, ...$ $(n = 1)$, $S_1 = S_1 = a_1$, $(n = 2)$, $S_2 = a_1 + a_2$,

....

....

$(n = n)$, $S_n = a_1 + a_2 + ... + a_n$

and the sequence $\{S_n\}$ is called the sequence of partial sums

As $n \to \infty$ there are 4 distinct possibilities

(1) $S_n \to$ a finite limit

(2) $S_n \to$ infinity

(3) $S_n \to -$ infinity

(4) S_n may have more than one limit

Now, $\lim_{n \to \infty} S_n = \lim_{n \to \infty} \{a_1 + a_2 + ...+ a_n\}$

$= \lim_{n \to \infty} \Sigma a_n = a_1 + a_2 + ... + a_n + ... \infty$

is called the infinite series Σa_n.

It $a_n > 0$, $\forall$ is then Σa_n in said to be a series of positive terms. A series Σa_n is said to be an alternating series if the terms are alternately positive and negative.

PART-D

Examples

1. $1 + \frac{1}{2} + \frac{1}{4} + \frac{1}{8} + ... \frac{1}{2^{n-1}} + ...$

2. $1 + 3 + 5 + 7 + ... + (2n - 1) + ...$ are infinite series of positive terms

3. $\Sigma \cdot \frac{(-1)^{n-1} \cdot n - 1}{3^{n-1}} = 1 - \frac{1}{3} + \frac{1}{3^2}$ .... is an alternating series

The behaviour of the series Σa_n is defined to be the same as that of the sequence $\{S_n\}$

Definition (I): (convergence):

An infinite series Σa_n is said to be convergent if S_n tends to a finite limit s (as $n \to \infty$) and Σa_n converges to s. [VTU, March, 1999]

Definition (II): (Divergence) [VTU, March, 1999]

If $S_n \to \infty$ or $-\infty$, the series is said to be divergent.

Definition (III): (Oscillatory): Σa_n oscillates if $\lim_{n\to\infty}$ oscillates (finitely or infinitely) between two values or between $+\infty$ and $-\infty$ . . . [VTU, March, 1999]

Examples

1. $\sum_{n=1}^{\infty} \frac{1}{6^{n-1}} = 1 + \frac{1}{6} + \frac{1}{6^2} + \ldots \frac{1}{6^{n-1}} +$ is convergent

since $$\lim_{n\to\infty} S_n = \lim_{n\to\infty} \frac{\left(1 - \frac{1}{6^n}\right)}{1 - \frac{1}{6}} = \frac{(1-0)}{1-\frac{1}{6}}$$

$$= \frac{6}{5} = \text{a unique real number}$$

2. (*i*) $S_n = 1 + 1 + 1 + \ldots.$ is divergent

since $S_n = n;\ \forall n$

and $\lim_{n\to\infty} (S_n) = \lim_{n\to\infty} (n) = \infty$

and (*ii*) $\Sigma n = 1 + 2 + 3 + \ldots. + n + \ldots.$ is divergent

since $\lim_{n\to\infty} (S_n) = \lim_{n\to\infty} \frac{n(n+1)}{2} = \infty$

3. The series

$$\sum_{n=1}^{\infty} (-1)^{n+1} = 1 - 1 + 1 - 1 + 1 - + \ldots. \text{ is an oscillating series}$$

$\therefore$ $S_1 = 1,\ \therefore S_n = 1$ (if n odd)

$S_2 = 1 - 1 = 0\quad S_n = 0$ (if n even)

$\therefore$ $\sum_{1}^{\infty} (-1)^{n+1}$ oscillates finitely;

(between 1 and 0)

4. $\sum_{1}^{\infty} (-1)^{n+1} \times n$

$$S_n = 1 - 2 + 3 - 4 + 5 - \text{..........} \text{ to } n \text{ terms}$$

$$S_n = -\frac{n}{2} \text{ if n is even}$$

$$= \frac{n+1}{2} \text{ if } n \text{ is odd}$$

$\therefore \quad S_n \to -\infty$ (n is even)

$S_n \to +\infty$ (n is odd)

Hence $\Sigma\,(-1)^{n+1} \cdot n$
oscillates infinitely
(between $-\infty$ and $+\infty$)
(an example of infinite oscillation)

Examples (1–5)

Find by definition, whether the following series are convergent or divergent.

1. $1 + \frac{2}{3} + \frac{4}{9} + \frac{8}{27} + \frac{16}{81} + \text{........} \infty$

Solution: It is a G.P. whose common ratio is $= \frac{2}{3}$

$$S_n = 1 + \frac{2}{3} + \frac{4}{9} + \text{....} + \frac{2^n}{3^n}$$

$$= \frac{1\left\{1 - \left(\frac{2}{3}\right)^n\right\}}{\left(1 - \frac{2}{3}\right)} = 3\left[1 - \left(\frac{2}{3}\right)^n\right]$$

$$\therefore \quad \lim_{n \to \infty} s_n = \lim_{n \to \infty} 3\left[1 - \left(\frac{2}{3}\right)^n\right]$$

$$= 3\,[1 - 0] = 3 = \text{a finite quantity,}$$

$\therefore$ Series is convergent

2. $\frac{1}{2 \cdot 3} + \frac{1}{3 \cdot 4} + \text{....} + \frac{1}{(n+1)(n+2)} + \text{.....} + \infty$

Solution:
$$S_n = \frac{1}{2 \cdot 3} + \frac{1}{3 \cdot 4} + \text{....} + \frac{1}{(n+1)(n+2)}$$

Here
$$a_n = \frac{1}{(n+1)(n+2)} = \frac{1}{(n+1)} - \frac{1}{(n+2)}$$

putting $n = 1, 2, 3, \text{.....}n$

PART-D

$$a_1 = \frac{1}{2} - \frac{1}{3}$$

$$a_2 = \frac{1}{3} - \frac{1}{4}$$

$$a_3 = \frac{1}{4} - \frac{1}{5}$$

....

....

$$a_n = \frac{1}{(n+1)} - \frac{1}{(n+2)}$$

Adding, $$S_n = a_1 + a_2 + ... + a_n$$

$$= \left(\frac{1}{2} - \frac{1}{3}\right) + \left(\frac{1}{3} - \frac{1}{4}\right) + ... + \left(\frac{1}{(n+1)}\right) - \left(\frac{1}{(n+2)}\right)$$

$$= \frac{1}{2} - \frac{1}{n+2}$$

(Since all the other terms cancel)

$$\therefore \quad \lim_{n\to\infty} S_n = \lim_{n\to\infty}\left[\frac{1}{2} - \frac{1}{(n+2)}\right] = \frac{1}{2} - 0 = \frac{1}{2}$$

$\Rightarrow$ $\{S_n\}$ converges to $\frac{1}{2}$

$\Rightarrow$ Σa_n converges to $\frac{1}{2}$

3. $1^3 + 2^3 + 3^3 + ...+ n^3 + ... \infty$

Solution: $$S_n = 1^3 + 2^3 + 3^3 + ... + n^3$$

$$= \left[\frac{n(n+1)}{2}\right]^2$$

$$\therefore \quad \lim_{n\to\infty} S_n = \lim_{n\to\infty}\left[\frac{n(n+1)}{2}\right]^2 = \infty$$

$\therefore$ the given series diverges to $+\infty$

4. The geometric series $1 + x + x^2 + ... + \infty$ is

(*i*) convergent if $-1 < x < 1$ *i.e.* $(x) < 1$

(*ii*) divergent if $x \geq 1$

(*iii*) oscillates finitley if $x = -1$

(*iv*) oscillates infinitely if $x < -1$

Solution: (*i*) let $|x| < 1$ *i.e.* $-1 < x < 1$

$$S_n = 1 + x + x^2 + ... \text{ to } n \text{ terms}$$

$$= \frac{(1-x^n)}{(1-x)} = \frac{1}{1-x} - \frac{x^n}{1-n}$$

(using the formula for sum to n term of a G.P)

Since $|x| < 1$, $x^n \to 0$, as $x \to \infty$

$$\therefore \quad \lim_{n\to\infty} S_n = \frac{1}{1-x} - 0 = \frac{1}{1-x}$$

$$= \text{(a unique real number)}$$

$\Rightarrow$ $\{S_n\}$ is convergent

$\therefore$ the given series is convergent

(*ii*) Let $x \geq 1$, suppose, when $x = 1$,

$$S_n = 1 + x + x^2 + ... + x^{n-1} \text{ becomes}$$

$$S_n = 1 + 1 + 1 + ... \text{ to } n \text{ terms} = n$$

$$\therefore \quad \lim_{n\to\infty} S_n = \lim_{n\to\infty} n = \infty$$

$\Rightarrow$ the sequence diverges to ∞

$\therefore$ the given series diverges to ∞.

Suppose $x > 1$, then $x^n \to \infty$ as $n \to \infty$

$$S_n = 1 + x + x^2 + ...+ x^{n-1}$$

$$= \frac{x^n - 1}{x - 1}$$

$$\therefore \quad \lim_{n\to\infty} S_n = \lim_{n\to\infty} \frac{x^n - 1}{x - 1} = \infty; \ (n > 1)$$

$\Rightarrow$ $\sum_{n=0}^{\infty} x^n$ is divergent if $x > 1$

(*iii*) Let $x \leq -1$

If $x = -1$, then the series is $1 - 1 + 1 - 1 + ...$ to ∞

$$\therefore \quad S_n = 1 - 1 + 1 - 1 + ... \text{ to } n \text{ terms}$$

$$= 1 \text{ or } 0 \text{ according as } n \text{ in odd or even.}$$

$$\therefore \quad \lim_{n\to\infty} S_n = 1 \text{ or } 0$$

$\therefore$ $\{S_n\}$ oscillates between two finite values 1 and 0.

$\therefore$ $\sum_{n=0}^{\infty} x^n$ oscillates finitely

(*iv*) If $x < -1$

$$x < -1 \Rightarrow -x > 1$$

PART-D

Let $x = -r$, where $r > 1$

$$S_n = 1 + x + x^2 + \ldots + x^{n-1}$$
$$= 1 + (-r) + (-r)^2 + \ldots (-r)^{n-1}$$

i.e.
$$S_n = \frac{1-(-r)^n}{1-(-r)} = \frac{1-(-1)^n \cdot r^n}{1+r}$$

$$\lim_{n\to\infty} S_n = \lim_{n\to\infty} \frac{1-(-1)^n \cdot r^n}{1+r}$$

Since $r > 1$, $\lim_{n\to\infty} r^n = \infty$

$$\lim_{n\to\infty} S_n = +\infty \text{ or } -\infty$$

According as n is odd or even

$\therefore \sum_{n=0}^{\infty} x_n$ oscillates infinitely if $x < -1$

5. The geometric series

(*i*) $\Sigma \frac{1}{3^{n-1}} = 1 + \frac{1}{3} + \frac{1}{3^2} + \ldots$ converges, since $r = \frac{1}{3} < 1$

(*ii*) $\Sigma\, 2^{n-1} = 1 + 2 + 2^2 + \ldots$ diverges to ∞ since $r = 2 > 1$

(*iii*) $\Sigma\, (-4)^{n-1} = 1 - 4 + 4^2 - 4^3 + \ldots$oscillates between $-\infty$ and $+\infty$, since $r = -4 < -1$

Some General Properties Concerning Infinite Series

The following properties are concerned with the convergence of the infinite series we state them without proof. (The proofs of these properties are beyond the scope of the syllabus)

1. If a series $\Sigma\, a_n$ converges then $\lim_{n\to\infty} a_n = 0$.

Contrapositive: If $\lim_{n\to\infty} a_n \neq 0$, then Σa_n is not convergent. If Σa_n is a series of positive terms, then $\lim_{n\to\infty} a_n \neq 0 \Rightarrow \Sigma a_n$ diverges to ∞.

2. If $a_1 + a_2 + \ldots$ is convergent and has the sum s, then $ka_1 + ka_2 + \ldots$ is convergent and has the sum ks.

3. If $a_1 + a_2 + \ldots$ and $b_1 + b_2 + \ldots$ are both convergent then the series $\Sigma\,(a_n + b_n)$ is convergent and its sum is the sum of the two series.

Series of Positive Terms

An infinite series Σa_n whose terms are all positive is called a series of positive terms (*i.e.* if $a_n > 0$, $\forall\, n = 1, 2, \ldots$)

We consider the following few results for the series of positive terms, which are useful to tests the nature of the series. First we consider the following tests (proofs of these tests are beyond the scope of the syllabus)

4.9 COMPARISON TESTS, P – SERIES TEST, D' ALEMBERTS RATIO TEST, RAABE'S TEST, CAUCHY'S ROOT TEST, CAUCHY'S INTEGRAL TEST.

(All tests without proof) for series of positive terms

1. Comparison Tests (Different Form)

Let Σc_n stand for a convergent series of positive terms and Σd_n for a divergent series. Let Σa_n be another series of positive terms

(*i*) If $a_n \le c_n$, then Σa_n is convergent: (C)

$a_n \ge d_n$, then Σa_n is divergent: (D)

(*ii*) If $\dfrac{a_n}{a_n+1} > \dfrac{c_n}{c_n+1}$, then Σa_n is (C)

$\dfrac{a_n}{a_n+1} > \dfrac{c_n}{c_n+1}$, then Σa_n is (D)

(*iii*) **Comparison test (Limit form)**

If $\lim\limits_{n\to\infty} \dfrac{a_n}{b_n}$ is finite, and not equal to zero, then both the series Σa_n and Σb_n converge or diverge together. i.e. if one series is (C), the other series is also (C) and if one series is (D), the other series is (D)

2. The *P*–series test (or Harmonic series) Def: the series :

$$\sum \frac{1}{n^p} = \frac{1}{1^p} + \frac{1}{2^p} + \frac{1}{3^p} + \ldots + \frac{1}{n^p} + \ldots$$

is called the p – series (or Harmonic series of order p)

Property: The *P*–series $\sum \dfrac{1}{n^p}$ is (i) convergent if $p > 1$ (ii) divergent if $p \le 1$.

Worked Examples (6 to 25)

6. Examine the convergence of the following series.

(*i*) $\dfrac{1}{1^2} + \dfrac{1+2}{1^2+2^2} + \dfrac{1+2+3}{1^2+2^2+3^2} + \ldots$ [VTU, F/M 2005 ; VTU, March, 2000]

Solution: n^{th} term of the given series is: a_n, where

$$a_n = \frac{1+2+3+\ldots+n}{1^2+2^2+3^2+\ldots+n^2} = \frac{\left[\dfrac{n(n+1)}{2}\right]}{\left[\dfrac{n(n+1)(2n+1)}{6}\right]} = \frac{3}{2n+1}$$

Let $$b_n = \frac{1}{n}$$

Now, $$\lim_{n\to\infty} \frac{a_n}{b_n} = \lim_{n\to\infty} \frac{3}{(2n+1)} \times \frac{n}{1} = \frac{3}{2} \ne 0$$

PART-D

$\therefore \sum a_n$ and $\sum b_n$ behave alike.

But $\sum b_n = \sum \frac{1}{n}$ is divergent, (P = 1)

(since $p = 1$). Thus given seires $\sum a_n$ must therefore also be divergent

(*ii*) $1 + \frac{1}{4^{2/3}} + \frac{1}{9^{2/3}} + \frac{1}{16^{2/3}} + \ldots$ [VTU, August, 1999]

Solution: Here, $a_n = \frac{1}{(n^2)^{2/3}} = \frac{1}{n^{4/3}}$

Hence $\sum a_n = \sum \frac{1}{n^{4/3}}$ is conversant $\left(\text{Since } p = \frac{4}{3} > 1\right)$

(*iii*) $\sum\left[\sqrt{n+1} - \sqrt{n}\right]$ [VTU, August, 2000]

Solution: The *n*th term of the series: $a_n = \sqrt{n+1} - \sqrt{n}$.

$$\therefore \qquad a_n = \frac{\left(\sqrt{n+1} - \sqrt{n}\right)\left(\sqrt{n+1} + \sqrt{n}\right)}{\sqrt{n+1} + \sqrt{n}} = \frac{1}{\sqrt{n+1} + \sqrt{n}}$$

Let $$b_n = \frac{1}{\sqrt{n}}$$

Then, $$\lim_{n\to\infty} \frac{a_n}{b_n} = \lim_{n\to\infty} \frac{\sqrt{n}}{\sqrt{n+1} + \sqrt{n}} = \lim_{n\to\infty} \frac{\sqrt{n}}{\sqrt{n}\left(\sqrt{1 + \frac{1}{n}} + 1\right)}$$

But $$\lim_{n\to\infty} \frac{\sqrt{n}}{\sqrt{n}\left(\sqrt{1 + \frac{1}{n}} + 1\right)} = \frac{1}{2}$$

Also, $$\sum b_n = \sum_{n=1}^{\infty} \frac{1}{\sqrt{n}} = \sum_{n=1}^{\infty} \frac{1}{\sqrt{n}}$$

$$= \sum_{n=1}^{\infty} \frac{1}{n^{1/2}} \text{ is divergent}$$

$\left(\text{Since } p = \frac{1}{2} < 1\right)$

Hence by comparison test, $\sum a_n$ is also divergent.

(iv) $\sum \left\{\sqrt[3]{n^3+1}-n\right\}$ [VTU, March, 2001]

Solution:

Let $a_n = \sqrt[3]{n^3+1}-n = (n^3+1)^{1/3}-(n^3)^{1/3}$

$(a^3-b^3) = (a-b)(a^2+ab+b^2)$

$$a_n = \frac{[(n^3+1)^{1/3}-n][(n^3+1)^{2/3}+n(n^3+1)^{1/3}+n^2]}{(n^3+1)^{2/3}+n(n^3+1)^{1/3}+n^2}$$

$$= \frac{n^3+1-n^3}{(n^3+1)^{2/3}+n(n^3+1)^{1/3}+n^2} = \frac{1}{(n^3+1)^{2/3}+n(n^3+1)^{1/3}+n^2}$$

Let $b_n = \frac{1}{n^2}$

$$\therefore \quad \frac{a_n}{b_n} = \frac{n^2}{(n^3+1)^{2/3}+n(n^3+1)^{1/3}+n^2}$$

$$= \frac{n^2}{n^2\left[\left(1+\frac{1}{n^3}\right)^{2/3}+\left(1+\frac{1}{n^3}\right)^{1/3}+1\right]}$$

$$= \frac{1}{\left(1+\frac{1}{n^3}\right)^{2/3}+\left(1+\frac{1}{n^3}\right)^{1/3}+1}$$

$\lim_{n\to\infty} \frac{a_n}{b_n} = \frac{1}{3}$ which is finite and not equal to zero.

$\therefore$ By comparison test, both $\sum a_n$ and $\sum b_n$ converge or diverge together.

But $\sum b_n = \sum \frac{1}{n^2}$ is convergent since $p > 1$.

$\therefore$ $\sum a_n$ is also convergent.

7. Test the convergence of the series

$$\frac{1}{1\cdot 2}+\frac{1}{2\cdot 3}+\frac{1}{3\cdot 4}+\ldots\infty$$

Solution: The nth term of the series is $a_n = \frac{1}{n(n+1)}$

Let $b_n = \frac{1}{n^2};\ \frac{a_n}{b_n} = \frac{1}{n(n+1)}\cdot n^2$

PART-D

i.e.,
$$\frac{a_n}{b_n} = \frac{n}{n+1} \quad \frac{n}{n\left(1+\frac{1}{n}\right)} = \frac{1}{1+\frac{1}{n}}$$

$$\lim_{n\to\infty} \frac{a_n}{b_n} = \lim_{n\to\infty} \frac{1}{1+\frac{1}{n}} = 1$$

(which is finite and not equal to zero)

$\therefore$ By Comparison Test, both the series $\sum a_n$ and $\sum b_n$ converge or diverge together.

But $\sum b_n = \sum \frac{1}{n^2}$ is convergent since $p > 1$.

$\therefore$ $\sum a_n$ is also a convergent series.

8. Test the convergence of the series.

$$\frac{1\cdot 2}{3\cdot 4\cdot 5} + \frac{2\cdot 3}{4\cdot 5\cdot 6} + \frac{3\cdot 4}{5\cdot 6\cdot 7} + \ldots \infty$$

Solution:

Let
$$a_n = \frac{n(n+1)}{(n+2)(n+3)(n+4)}$$

Let
$$b_n = \frac{1}{n}$$

$$\frac{a_n}{b_n} = \frac{n(n+1)}{(n+2)(n+3)(n+4)} \cdot n$$

$$= \frac{n^3\left(1+\frac{1}{n}\right)}{n^3\left(1+\frac{2}{n}\right)\left(1+\frac{3}{n}\right)\left(1+\frac{4}{n}\right)}$$

$$\lim_{n\to\infty} \frac{a_n}{b_n} = \lim_{n\to\infty} \frac{\left(1+\frac{1}{n}\right)}{\left(1+\frac{2}{n}\right)\left(1+\frac{3}{n}\right)\left(1+\frac{4}{n}\right)} = 1$$

(which is finite and not equal to zero)

$\therefore$ By comparison test, both $\sum a_n$ and $\sum b_n$ converge or diverge together.

But $\sum b_n = \sum \frac{1}{n}$ is divergent

since $p = 1$

$\therefore$ $\sum b_n$ is also divergent.

9. Discuss the convergence of

$$\frac{1}{1\cdot 2\cdot 3}+\frac{3}{2\cdot 3\cdot 4}+\frac{5}{3\cdot 4\cdot 5}+\ldots\infty$$

Solution: Here, $$a_n = \frac{2n-1}{n(n+1)(n+2)}$$

Let $$b_n = \frac{n}{n^3}=\frac{1}{n^2}$$

$$\frac{a_n}{b_n} = \frac{2n-1}{n(n+1)(n+2)}\cdot n^2$$

$$= \frac{n\left(2-\frac{1}{n}\right)}{n^3\left(1+\frac{1}{n}\right)\left(1+\frac{2}{n}\right)}=\frac{\left(2-\frac{1}{n}\right)}{\left(1+\frac{1}{n}\right)\left(1+\frac{2}{n}\right)}$$

$\therefore$ $$\lim_{n\to\infty}\frac{a_n}{b_n} = 2 \text{ (which is finite and not equal to zero)}$$

$\therefore$ By comparison test, $\sum a_n$ and $\sum b_n$ converge or diverge together.

But $\sum b_n = \sum \frac{1}{n^2}$ is convergent (since $p > 1$)

$\therefore$ $\sum a_n$ is also convergent.

10. $$\frac{\sqrt{2}-1}{3^3-1}+\frac{\sqrt{3}-1}{4^3-1}+\frac{\sqrt{4}-1}{5^3-1}+\ldots$$

Solution: Here, $$a_n = \frac{\sqrt{n+1}-1}{(n+2)^3-1}$$

Let $$b_n = \frac{\sqrt{n}}{n^3}=\frac{1}{n^{5/2}}$$

Now, $$\lim_{n\to\infty}\frac{a_n}{b_n} = \lim_{n\to\infty}\frac{\sqrt{n+1}-1}{(n+2)^3-1}\cdot\frac{n^3}{\sqrt{n}} = \lim_{n\to\infty}\frac{\sqrt{1+\frac{1}{n}}-\frac{1}{\sqrt{n}}}{\left(1+\frac{2}{n}\right)^3-\frac{1}{n^3}} = 1 \neq 0$$

Hence by limit form of comparison test, $\sum a_n$ and $\sum b_n$ behave alike.

But $$\sum b_n = \sum \frac{1}{n^{5/2}}$$

is convergent $\left(\text{since } p = \frac{5}{2} > 1\right)$

Hence $\sum a_n$ is also convergent.

11. $\sum\left[\sqrt{n^2+1}-\sqrt{n^2-1}\right]$

Solution: Here, $$a_n = \left\{\sqrt{(n^2+1)} - \sqrt{(n^2-1)}\right\}$$

$$= \frac{\left\{\sqrt{(n^2+1)} - \sqrt{(n^2-1)}\right\}\left\{\sqrt{(n^2+1)} + \sqrt{(n^2-1)}\right\}}{\left\{\sqrt{(n^2+1)} + \sqrt{(n^2-1)}\right\}}$$

$$= \frac{(n^2+1)-(n^2-1)}{\left\{\sqrt{(n^2+1)} + \sqrt{(n^2-1)}\right\}} = \frac{2}{\left\{\sqrt{(n^2+1)} + \sqrt{(n^2-1)}\right\}}$$

$$= 0\,(1/n^2) = \left(\frac{1}{n}\right). \text{ Let } b_n = \frac{1}{n}$$

$$\therefore \quad \lim_{n\to\infty}\frac{a_n}{b_n} = \lim_{n\to\infty}\frac{2}{\sqrt{n^2+1}+\sqrt{n^2-1}} \times \frac{n}{1}$$

$$= \lim_{n\to\infty}\frac{2}{\sqrt{1+(1/n^2)}+\sqrt{1+(1/n)^2}} = 1 \neq 0$$

$\therefore$ By the limit form of comparison test $\sum a_n$ and $\sum b_n$ behave alike.

But $\sum b_n = \sum \frac{1}{n}$ is divergent (since $p = 1$).

Hence $\sum a_n$ is also divergent.

12. $\left[\sqrt[3]{n^3+1}-\sqrt[3]{n}\right]$

Solution: Let $$a_n = \left[\sqrt[3]{n^3+1}-\sqrt[3]{n}\right] = [(n^3+1)^{1/3} - (n)^{1/3}]$$

$$(a^3 - b^3) = (a-b)(a^2+ab+b^2)$$

$$\therefore \quad (a-b) = \frac{a^3-b^3}{a^2+ab+b^2}$$

If we take $a = (n^3 + 1)^{1/3}$; $b = (n)^{1/3}$ then a_n assumes the form of $(a - b)$.

$$\therefore \quad a_n = \frac{(n^3+1)-n^3}{(n^3+1)^{2/3}+(n^3+1)^{1/3}\cdot n+n^{2/3}} \text{ by above}$$

i.e.,

$$a_n = \frac{1}{(n^3+1)^{2/3}+\sqrt[3]{n^3+1}\cdot n+n^2}$$

Now, choose $\quad b_n = \dfrac{1}{n^2}$

$$\therefore \quad \lim_{n\to\infty}\frac{a_n}{b_n} = \lim_{n\to\infty}\frac{1}{(n^3+1)^{2/3}+\sqrt[3]{n^3+1}\;n+n^2}\frac{(n^2)}{}$$

$$= \lim_{n\to\infty}\frac{n^2}{\left[n^3\left(1+\frac{1}{n^3}\right)\right]^{2/3}+\sqrt[3]{n^3\left(1+\frac{1}{n^3}\right)}\cdot n+n^2}$$

$$= \lim_{n\to\infty}\frac{n^2}{n^2\left(1+\frac{1}{n^3}\right)^{2/3}+n^2\cdot\sqrt[3]{1+\frac{1}{n^3}}+n^2}$$

$$= \lim_{n\to\infty}\frac{1}{\left[\left(1+\frac{1}{n^3}\right)^{2/3}+\sqrt[3]{1+\frac{1}{n^3}}+1\right]} = \frac{1}{1+1+1} = \frac{1}{3}$$

$$\therefore \quad \lim_{n\to\infty}\frac{a_n}{b_n} = \frac{1}{3} = \text{a finite quantity} \neq 0$$

By comparison test, $\sum a_n$ and $\sum b_n$ behave alike. But $\sum b_n = \sum \dfrac{1}{n^2}$ is convergent (since $p = 2 > 1$).

$\therefore \sum a_n$ is also convergent.

13. Test for convergence

$$\sum_{n=1}^{\infty}\left[\sqrt{n^4+1}-\sqrt{n^4-1}\right]$$

Solution: Let $\quad a_n = \sqrt{n^4+1}-\sqrt{n^4-1}$

Multiply and Divide by $\sqrt{n^4+1}+\sqrt{n^4-1}$

we get, $$a_n = \frac{\left[\sqrt{n^4+1}-\sqrt{n^4-1}\right]\left[\sqrt{n^4+1}+\sqrt{n^4-1}\right]}{\left[\sqrt{n^4+1}+\sqrt{n^4-1}\right]}$$

$$a_n = \frac{(n^4+1)-(n^4-1)}{\sqrt{n^4+1}+\sqrt{n^4-1}} = \frac{2}{\sqrt{n^4+1}+\sqrt{n^4-1}}$$

Let $$b_n = \frac{1}{\sqrt{n^4}} = \frac{1}{n^2}$$

$$\therefore \quad \lim_{n\to\infty}\frac{a_n}{b_n} = \lim_{n\to\infty}\frac{2}{\sqrt{n^4+1}+\sqrt{n^4-1}}\left(\frac{n}{1}\right)^2$$

$$= \lim_{n\to\infty}\frac{2n^2}{\sqrt{n^4\left(1+\frac{1}{n^4}\right)}+\sqrt{n^4\left(1-\frac{1}{n^4}\right)}}$$

$$= \lim_{n\to\infty}\frac{2n^2}{n^2\left[\sqrt{1+\frac{1}{n^4}}+\sqrt{1-\frac{1}{n^4}}\right]}$$

$$= \lim_{n\to\infty}\frac{2}{\left[\sqrt{1+\frac{1}{n^4}}+\sqrt{1-\frac{1}{n^4}}\right]} = \frac{2}{1+1} = 1$$

i.e., $$\lim_{n\to\infty}\frac{a_n}{b_n} = 1 = \text{finite quantity} \neq 0$$

By comparison test $\sum a_n$ and $\sum b_n$ behave alike

But $$\sum a_n = \sum\frac{1}{n^2} \quad (p = 2 > 1) \text{ is convergent.}$$

$\therefore$ a_n is also convergent.

14. $\sum \frac{\sqrt{(n+1)}-\sqrt{n}}{n^p}$

Solution: Let $$a_n = \frac{\sqrt{(n+1)}-\sqrt{n}}{n^p}$$

$$= \frac{\sqrt{(n+1)} - \sqrt{n}}{n^p} \cdot \frac{\sqrt{n+1} + \sqrt{n}}{\sqrt{n+1} + \sqrt{n}}$$

$$= \frac{n+1-n}{n^p\left[\sqrt{(n+1)} + \sqrt{n}\right]} = \frac{1}{n^p\left[\sqrt{(n+1)} + \sqrt{n}\right]} = 0\left(\frac{1}{n^{p+1}}\right)$$

Let $\quad b_n = \dfrac{1}{n^{p+1/2}}$

$$\therefore \quad \lim_{n \to \infty} \frac{a_n}{b_n} = \lim_{n \to \infty} \frac{n^{p+1/2}}{n^p\left[\sqrt{n+1} + \sqrt{n}\right]} = \lim_{n \to \infty} \frac{1}{\left(\sqrt{1+\frac{1}{n}} + 1\right)} = \frac{1}{2}$$

Also, $\sum b_n$ is convergent

if $p + \dfrac{1}{2} > 1$ and divergent

if $p + \dfrac{1}{2} \leq 1$

$\therefore \quad \sum a_n$ is convergent if $p > \dfrac{1}{2}$ and divergent if $p \leq \dfrac{1}{2}$

15. Test the convergence of $\displaystyle\sum_{n=1}^{\infty} \left[\frac{n^p}{\sqrt{n+1} + \sqrt{n}}\right]$

Solution: Let $\quad a_n = \dfrac{n^p}{\sqrt{n+1} + \sqrt{n}}$ and $b_n = \dfrac{1}{n^{\frac{1}{2} - p}}$

$$\text{Then} \quad \lim_{n \to \infty} \frac{a_n}{b_n} = \lim_{n \to \infty} \left(\frac{n^p n^{\frac{1}{2} - p}}{\sqrt{n+1} + \sqrt{n}}\right) = \lim_{n \to \infty} \left(\frac{n^{1/2}}{\sqrt{n}\left[\sqrt{1+\frac{1}{n}} + 1\right]}\right) = \frac{1}{2}$$

Now, we note that

$$\sum b_n = \sum_{n=1}^{\infty} \frac{1}{n^{\frac{1}{2} - p}} \text{ is convergent}$$

if $\quad \dfrac{1}{2} - p > 1$

(*i.e.,*) $\sum b_n$ is convergent if $p < -\dfrac{1}{2}$

PART-D

Hence $\sum a_n$ is convergent if $p < -\frac{1}{2}$.

16. Test for convergence the series

$$\sum_{n=1}^{\infty} \frac{1}{\sqrt{n+1}+\sqrt{n}}$$ [VTU, July/Aug., 2002]

Solution:

$$a_n = \frac{1}{\sqrt{n+1}+\sqrt{n}}$$

Let $$b_n = \frac{1}{\sqrt{n}}$$

$$\therefore \quad \lim_{n\to\infty} \frac{a_n}{b_n} = \lim_{n\to\infty} \frac{\sqrt{n}}{\sqrt{n+1}+\sqrt{n}} = \lim_{n\to\infty} \frac{\sqrt{n}}{\sqrt{n}\left[1+\sqrt{1+\frac{1}{n}}\right]}$$

$$= \lim_{n\to\infty} \frac{1}{\left[1+\sqrt{1+\frac{1}{n}}\right]} = \frac{1}{1+\sqrt{1+0}} = \frac{1}{1} = 1 \neq 0$$

$\therefore$ $\sum a_n$ and $\sum b_n$ behave alike

But $$\sum b_n = \sum \frac{1}{\sqrt{n}} = \sum \frac{1}{n^{1/2}}$$

is divergent

$$\left(\text{since } p = \frac{1}{2} < 1\right)$$

$\therefore$ $\sum a_n$ is also divergent.

17. Discuss the convergence of $\sum_{n=1}^{\infty} \frac{1}{\sqrt{n}} \sin \frac{1}{n}$

Solution: Here $$a_n = \frac{1}{\sqrt{n}} \cdot \sin \frac{1}{n};$$

Let $$b_n = \frac{1}{n^{3/2}}$$

$$\therefore \quad \frac{a_n}{b_n} = \frac{\frac{1}{\sqrt{n}} \sin \frac{1}{n}}{\frac{1}{n^{3/2}}} = \frac{\sin\left(\frac{1}{n}\right)}{\left(\frac{1}{n}\right)}$$

$$\therefore \quad \lim_{n\to\infty} \frac{a_n}{b_n} = \lim_{n\to\infty} \frac{\sin\left(\frac{1}{n}\right)}{\left(\frac{1}{n}\right)} = 1$$

$\therefore$ By comparison test both $\sum a_n$ and $\sum b_n$ converge or diverge together.

But $\sum b_n = \sum \frac{1}{n^{3/2}}$ is convergent. Since p > 1.

$\therefore$ $\sum a_n$ is also convergent.

18. Discuss the convergence of $\sum \frac{1}{n} \sin \frac{1}{n}$.

Solution: Let $a_n = \frac{1}{n} \sin \frac{1}{n}$

Consider $b_n = \frac{1}{n^2}$

$$\therefore \quad \lim_{n\to\infty} \frac{a_n}{b_n} = \lim_{n\to\infty} \left[\frac{\frac{1}{n}\sin\left(\frac{1}{n}\right)}{\frac{1}{n^2}}\right] = \lim_{n\to\infty} \left[\frac{\sin\left(\frac{1}{n}\right)}{\left(\frac{1}{n}\right)}\right]$$

Put $\frac{1}{n} = \theta$

$\therefore$ as $n \to \infty$ $\frac{1}{n} \to 0$ and hence $\theta \to 0$

$$\therefore \quad \lim_{n\to\infty} \frac{a_n}{b_n} = \lim_{n\to\infty} \frac{\sin\theta}{\theta} = 1 \neq 0$$

$\therefore$ By the limit form of comparison cruve, $\sum a_n$ and $\sum b_n$ behave alike.

But $\sum b_n = \sum \frac{1}{n^2}$ is a *p*-series with (p = 2 > 1) and

hence convergent $\therefore$ $\sum a_n$ is also convergent.

19. Discuss the comergence of $\sum_{n=1}^{\infty} \sin\left(\frac{1}{n}\right)$

Solution: Let $a_n = \sin\left(\frac{1}{n}\right)$

PART-D

Let $$b_n = \frac{1}{n}$$

$$\therefore \quad \lim_{n\to\infty} \frac{a_n}{b_n} = \lim_{n\to\infty} \frac{\sin\left(\frac{1}{n}\right)}{\left(\frac{1}{n}\right)}$$

Put $\frac{1}{n} = \theta$;

$\therefore$ We get $$\lim_{n\to\infty} \frac{a_n}{b_n} = \lim_{n\to\infty} \frac{\sin\theta}{\theta} = 1$$

Also, $$\sum b_n = \sum \frac{1}{n} \text{ is } (D)$$

(since $p = 1$)

Hence, by comparison test, $\sum a_n$ is (D)

20. $\sum_{1}^{\infty} \sqrt{n^6+1} - \sqrt{n^6-1}$

Solution: Let
$$a_n = \sqrt{n^6+1} - \sqrt{n^6-1}$$

$$= \frac{\left(\sqrt{n^6+1} - \sqrt{n^6-1}\right)\left(\sqrt{n^6+1} + \sqrt{n^6-1}\right)}{\left(\sqrt{n^6+1} + \sqrt{n^6-1}\right)}$$

$$= \frac{2}{\sqrt{n^6+1} + \sqrt{n^6-1}}$$

Let $$b_n = \frac{1}{n^3}$$

Then, $$\lim_{n\to\infty} \frac{a_n}{b_n} = \lim_{n\to\infty} \frac{2n^3}{\sqrt{n^6+1} + \sqrt{n^6-1}}$$

$$= \lim_{n\to\infty} \frac{2}{\sqrt{\left(1+\frac{1}{n^6}\right)} + \sqrt{\left(1-\frac{1}{n^6}\right)}} = \frac{2}{\sqrt{1}+\sqrt{1}} = \frac{2}{2} = 1 \neq 0$$

Since $\sum b_n = \sum \frac{1}{n^3}$ is (C)

($\because p = 3 > 1$)

$\sum a_n$ is also convergent, by comparison test.

21. $\dfrac{1}{\sqrt{n}} \tan\left(\dfrac{1}{n}\right)$

Let $$a_n = \frac{1}{\sqrt{n}} \tan\left(\frac{1}{n}\right)$$

Let $$b_n = \frac{1}{n^{3/2}}$$

Then $$\lim_{n\to\infty} \frac{a_n}{b_n} = \lim_{n\to\infty} n \tan\left(\frac{1}{n}\right)$$

$$= \lim_{n\to\infty} \frac{\tan\left(\frac{1}{n}\right)}{\left(\frac{1}{n}\right)} = 1 \neq 0$$

Since $\sum b_n = \sum \dfrac{1}{n^{3/2}}$ is (C)

$\sum a_n$ is also (C), by comparison test.

(*i*) $\sum n \sin\left(\dfrac{1}{n}\right)$

Here $$a_n = \left[n \cdot \sin\left(\frac{1}{n}\right)\right] = \frac{\sin\left(\frac{1}{n}\right)}{(1/n)} \to 1, \text{ (as } n \to \infty)$$

Since a_n does not tend to zero. $\sum a_n$ cannot converge. It is a series of positive terms and so is divergent.

22. Examine for convergence of the series $\sum a_n$ where

$$a_n = \frac{1}{n^{\frac{(n+1)}{n}}}$$

Solution: Since $$a_n = \frac{1}{n^{\frac{(n+1)}{n}}}$$

Let $$b_n = \frac{1}{n}$$

Then $$\frac{a_n}{b_n} = \frac{1}{n^{\frac{(n+1)}{n}}} = \frac{1}{n^{1/n}} \to 1; \qquad \text{(as } n \to \infty)$$

PART-D

$\therefore$ $\sum a_n$ and $\sum b_n$ behave alike. But $\sum b_n$ is diveregent, since $\sum\left(\frac{1}{n}\right)$ is (D) ($p = 1$).

Hence $\sum a_n$ is also divergent.

or (II Method), $$a_n = \left(\frac{1}{n^{1+1/n}}\right) = 0\left(\frac{1}{n}\right)$$

and $\left[\because 1+\frac{1}{n} \to 1\right]$

$$\sum\left(\frac{1}{n}\right) \text{ is } (D = \text{divergent})$$

$\therefore$ $\sum a_n$ is divergent.

23. $\sum\left[\sqrt{n^4+1} - n^2\right]$

Solution: Let $$a_n = \sqrt{(n^4+1)} - n^2$$

$$= \frac{\left[\sqrt{(n^4+1)} - n^2\right]\left[\sqrt{n^4+1}+n^2\right]}{\left[\sqrt{n^4+1}+n^2\right]}$$

$$= \frac{\left[n^4+1-n^4\right]}{\sqrt{n^4+1}+n^2} = \frac{1}{\left[\sqrt{n^4+1}+n^2\right]} = 0$$

$$= 0\left[\frac{1}{n^2+n^2}\right] = 0\left(\frac{1}{n^2}\right)$$

$\therefore$ $\sum a_n$ behave like $\left(\sum\frac{1}{n^2}\right)$

$\sum(1/n^2)$ is (C)

(since $p = 2 > 1$)

Hence $\sum a_n$ is also (C)

24. Discuss the convergence of

$$\sum_{n=1}^{\infty} \frac{1^2+2^2 \ldots + n^2}{n^4+1}$$

Solution:

Let $$a_n = \frac{1^2 + 2^2 + \ldots + n^2}{n^4 + 1} = \frac{n(n+1)(2n+1)}{6(n^4+1)}$$

Let $$b_n = \frac{1}{n}$$

$$\therefore \quad \lim_{n\to\infty} \frac{a_n}{b_n} = \lim_{n\to\infty} \frac{n^2(n+1)(2n+1)}{6(n^4+1)} = \lim_{n\to\infty} \frac{\left(1+\frac{1}{n}\right)\left(2+\frac{1}{n}\right)}{6(1+/n^4)} = \frac{1}{3}$$

$\therefore$ $\sum b_n$ is (D)

$\therefore$ $\sum a_n$ is (D) by comparison test.

25. $1 + \frac{1}{2^2} + \frac{2^2}{3^3} + \frac{3^3}{4^4} + \ldots$

Solution: Let $$a_n = \frac{n^n}{(n+1)^{n+1}}; \text{ and } b_n = \frac{1}{n}$$

Then, $$\lim_{n\to\infty} \frac{a_n}{b_n} = \lim_{n\to\infty} \frac{n^{n+1}}{(n+1)^{n+1}}$$

$$= \lim_{n\to\infty} \frac{n}{\left(1+\frac{1}{n}\right)^{n+1}} = \frac{1}{e} > 0$$

Also, $$\sum b_n = \sum \frac{1}{n} \text{ is (D) (since } p = 1)$$

Hence, by comparison test, $\sum a_n$ is (D).

1. Discuss the convergence of the following series whose nth terms are given by

(*i*) $\frac{n+1}{n^3}$ [*Ans.* (C)]

(*ii*) $\frac{1}{1+5n^4}$ [*Ans.* (C)]

(*iii*) $\sin^2\left(\frac{1}{n}\right)$ [*Ans.* (C)]

(*iv*) $\frac{1}{n\sqrt{n^3+1}}$ [*Ans.* (C)]

(v) $\dfrac{n(n+1)}{(n+2)(2n+1)(n^2+1)}$ [*Ans.* (D)]

2. Test the convergence of the series

(*i*) $\dfrac{1}{2^n n^2}$ [*Ans.* (C)]

(*ii*) $\left(\dfrac{1}{n} - \dfrac{1}{n+1}\right)$ [*Ans.* (C)]

(*iii*) $\dfrac{1^2 + 2^2 + \cdots + n^2}{n^6 + 7}$ [*Ans.* (C)]

(*iv*) $\dfrac{(n+a)(n+b)}{n(n+1)(n+2)}$ [*Ans.* (D)]

(*v*) $\dfrac{1+2+\ldots+n}{n^6+7}$ [*Ans.* (D)]

3. Prove that the series $\dfrac{1}{3} + \dfrac{1 \cdot 4}{3 \cdot 6} + \dfrac{1 \cdot 4 \cdot 7}{3 \cdot 6 \cdot 9} + \ldots$ is divergent, but the series

$$\left(\frac{1}{3}\right)^2 + \left(\frac{1 \cdot 4}{3 \cdot 6}\right)^2 + \left(\frac{1 \cdot 4 \cdot 7}{3 \cdot 6 \cdot 9}\right)^2 + \ldots \text{ is convergent.}$$

4. If $\sum a_n$ is a divergent series of positive terms prove that

$\sum \dfrac{a_n}{1 + n^2 a_n}$ is convergent.

D' ALEMBERT'S RATIO TEST

If $\sum u_n = u_1 + u_2 + \ldots + u_n + \ldots$ be a series of positive terms such that

$$\lim_{n \to \infty} \frac{u_{n+1}}{u_n} = l \qquad [1]$$

then the series $\sum u_n$

(*i*) converges if $l < 1$

(*ii*) diverges if $l > 1$

(*iii*) the test fails if $l = 1$.

Note: In case (*iii*) *i.e.*, if $l = 1$, the test fails *i.e.*, the series may converge or diverge but ratio test gives no information in such a situation, we look for some other tests. We can also test the convergence of some infinite series using the following tests.

RAABE'S TEST

Let $\sum u_n$ be a series of positive terms and

$$\lim_{n\to\infty} n\left(\frac{u_n}{u_{n+1}}-1\right) = l \qquad [1]$$

then, if $l > 1$, $\sum u_n$ is convergent and if $l < 1$, $\sum u_n$ is divergent. If $l = 1$, the test fails. (Raabe's test is usually applied when D'Alemberts' test fails)

Cauchy's Root Test (or Radical Test)

Let $\sum u_n$ be a series of positive terms and

$$\lim_{n\to\infty} (u_n)^{1/n} = l \qquad [1]$$

Then if $l < 1$, the series $\sum u_n$ is convergent and if $l > 1$, the series is divergent. (Cauchy's root test fails if $l = 1$)

Examples (26-49)

26. Discuss the convergence of

$$1+\frac{1}{2^2}+\frac{1}{3^3}+\ldots$$

Solution: Here,
$$u_n = \frac{1}{n^n}$$

$$u_{n+1} = \frac{1}{(n+1)^{n+1}}$$

$$\therefore \quad \frac{u_{n+1}}{u_n} = \frac{1}{(n+1)^{n+1}}\cdot\frac{n^n}{1}$$

$$= \frac{n^n}{(n+1)(n+1)^n} = \frac{1}{(n+1)\left(1+\frac{1}{n}\right)^n}$$

Hence
$$\lim_{n\to\infty}\frac{u_{n+1}}{u_n} = \lim_{n\to\infty}\frac{1}{(n+1)}\lim_{n\to\infty}\frac{1}{\left(1+\frac{1}{n}\right)^n} = 0\times\frac{1}{e} = 0 < 1.$$

$\therefore$ By D'Alembert's ratio test; $\sum u_n$ converges.

27. $1+\frac{2!}{2^2}+\frac{3!}{3^3}+\frac{4!}{4^4}+\ldots$ [VTU, March, 1999]

Solution: Here
$$u_n = \frac{n!}{n^n}$$

PART-D

$$u_{n+1} = \frac{(n+1)!}{(n+1)^{n+1}}$$

Now
$$l = \lim_{n\to\infty} \frac{u_{n+1}}{u_n} = \lim_{n\to\infty} \frac{(n+1)!}{(n+1)^{n+1}} \cdot \frac{n^n}{n!}$$

$$= \lim_{n\to\infty} \frac{(n+1)n^n}{(n+1)^{n+1}} = \lim_{n\to\infty} \frac{n^n}{(n+1)^n}$$

$$= \lim_{n\to\infty} \frac{1}{\left(1+\frac{1}{n}\right)^n} = \frac{1}{e}$$

since $(2 < e < 3)$, $\quad l = \frac{1}{e} < 1,$

hence by Ratio, test, $\sum u_n$ is convergent.

28. $1 + \frac{2^2}{2!} + \frac{3^2}{3!} + \frac{4^2}{4!} + \ldots$ [VTU, August, 2001]

Solution: Here
$$u_n = \frac{n^2}{n!}$$

Now
$$\frac{u_{n+1}}{u_n} = \frac{(n+1)^2}{(n+1)!} \times \frac{n!}{n^2} = \frac{n+1}{n^2}$$

$\therefore$
$$\lim_{n\to\infty} \frac{u_{n+1}}{u_n} = \lim_{n\to\infty} \left[\frac{1}{n} + \frac{1}{n^2}\right] = 0 < 1.$$

Hence, by D'Alembert's ratio test, $\sum u_n$ is convergent.

29. Discuss the convergence of

$$\frac{x}{1\cdot 2} + \frac{x^2}{2\cdot 3} + \frac{x^3}{3\cdot 4} + \ldots$$ [VTU, March, 2000]

Solution: Here
$$u_n = \frac{x^n}{n(n+1)}$$

$$u_{n+1} = \frac{x^{n+1}}{(n+1)(n+2)}$$

$$\lim_{n\to\infty} \frac{u_{n+1}}{u_n} = \lim_{n\to\infty} \frac{x^{n+1}}{(n+1)(n+2)} \times \frac{n(n+1)}{x^n}$$

$$= \lim_{n\to\infty} \left(\frac{n}{n+2}\right) x = x$$

$\therefore$ by Ratio test,

$\sum u_n$ is convergent is $x < 1$ and is divergent if $n > 1$,

If $x = 1$, the test fails

Now if $x = 1$, then $u_n = \dfrac{1}{n(n+1)}$

Let $V_n = \dfrac{1}{n^2}$

$$\therefore \quad \lim_{n\to\infty} \frac{u_n}{v_n} = \lim_{n\to\infty} \frac{1}{(n)(n+1)} \times \frac{n^2}{1} = \lim_{n\to\infty} \frac{n}{n+1} = 1 \neq 0$$

Hence, by the limit form of comparison test, $\sum u_n$ and $\sum V_n$ behave alike.

But $\sum V_n = \sum \dfrac{1}{n^2}$ is convergent (since $p = 2 > 1$)

Hence $\sum u_n$ is also convergent

Thus $\sum u_n$ is convergent

if $x \leq 1$

and is divergent if $x > 1$.

30. Discuss the convergence of

$$1 + \frac{x}{2} + \frac{x^2}{5} + \frac{x^3}{10} + \ldots$$ [VTU, August, 2001]

Solution: Here $u_n = \dfrac{x^n}{n^2+1}$;

$$u_{n+1} = \frac{x^{n+1}}{(n+1)^2+1}$$

$$\therefore \quad \lim_{n\to\infty} \frac{u_{n+1}}{u_n} = \lim_{n\to\infty} \frac{x^{n+1}}{(n+1)^2+1} \times \frac{n^2+1}{x^n}$$

$$= \lim_{n\to\infty} \frac{1+1/n^2}{(1+1/n)^2+1/n^2} \,.\, x = x$$

Hence by D'Alembert's ratio test, $\sum u_n$ is convergent if $x < 1$ and divergent if $x > 1$.

The test fails if $x = 1$.

When $x = 1$, $u_n = \dfrac{1}{n^2+1}$

Let $v_n = \dfrac{1}{n^2}$;

PART-D

Here $$\lim_{n\to\infty} \frac{u_n}{v_n} = \lim_{n\to\infty} \frac{n^2}{n^2+1} = 1 \neq 0$$

By comparison test, $\sum u_n$ and $\sum v_n$ behave alike.

But $\sum v_n = \sum \frac{1}{n^2}$ is convergent

Since $p = 2 > 1$.

Hence $\sum u_n$ is also convergent.

Thus $\sum u_n$ is convergent
If $x \leq 1$ and divergent
If $x > 1$.

31. Test for convergence of the series

$$1 + \frac{1+\alpha}{1+\beta} + \frac{(1+\alpha)(1+2\alpha)}{(1+\beta)(1+2\beta)} + \ldots$$ [VTU March, 1999]

Solution: Here $$u_n = \frac{(1+\alpha)(1+2\alpha)\ldots[1+(n-1)\alpha]}{(1+\beta)(1+2\beta)-[1+(n-1)\beta]}$$

$\therefore$ $$u_{n+1} = \frac{(1+\alpha)(1+2\alpha)\ldots[1+(n-1)\alpha][1+n\alpha]}{(1+\beta)(1+2\beta)\ldots[1+(n-1)\beta][1+n\beta]}$$

$\therefore$ $$\lim_{n\to\infty} \frac{u_{n+1}}{u_n} = \lim_{n\to\infty}\left[\frac{1+n\alpha}{1+n\beta}\right] = \lim_{n\to\infty}\left[\frac{\alpha+(1/n)}{\beta+(1/n)}\right] = \frac{\alpha}{\beta}$$

$\therefore$ But ratio test,

$\sum u_n$ is convergent $\frac{\alpha}{\beta} < 1$ *i.e.,* $\alpha < \beta$ or $\beta > \alpha > 0$

$\sum u_n$ diverges if $\frac{\alpha}{\beta} > 1$ and *i.e.,* $\alpha > \beta$ or $\alpha > \beta > 0$

when, $\frac{\alpha}{\beta} = 1$, ratio test fails

when $\frac{\alpha}{\beta} = 1$ *i.e.,* $\alpha = \beta$; $u_n = 1$

Now, $$\lim_{n\to\infty} u_n = 1 \neq 0;$$

$\Rightarrow \sum u_n$ is divergent.

Hence the given series $\sum u_n$ converges if $\beta > \alpha > 0$ and diverges if $\alpha \geq \beta > 0$.

32. Test the convergence $\sum \frac{n^n}{n!}$

Solution: Here $u_n = \frac{n^n}{n!}$ and $u_{n+1} = \frac{(n+1)^{n+1}}{(n+1)!}$

$$\therefore \quad \frac{u_{n+1}}{u_n} = \frac{(n+1)^{n+1}}{(n+1)!} \cdot \frac{n!}{n^n}$$

$$= \frac{(n+1)^n \cdot (n+1) \cdot n!}{(n+1)!\, n^n} = \frac{(n+1)^n}{n^n} = \left[1 + \frac{1}{n}\right]^n$$

$$\therefore \quad \lim_{n\to\infty} \frac{u_{n+1}}{u_n} = \lim_{n\to\infty} \left[1 + \frac{1}{n}\right]^n = e > 1$$

(Since $2 < e < 3$)

$\therefore$ $\sum u_n$ is divergent.

33. Test the convergence of $\sum_{n=1}^{\infty} \frac{2^n \cdot n!}{n^n}$

Solution: Let $u_n = \frac{2^n \cdot n!}{n^n}$; and $u_{n+1} = \frac{2^{n+1}\,(n+1)!}{(n+1)^{n+1}}$

$$\therefore \quad \frac{u_{n+1}}{u_n} = \frac{2^{n+1} \cdot (n+1)!}{(n+1)^{n+1}} \cdot \frac{n^n}{n!\,2^n} = \frac{2 \cdot n^n}{(n+1)^n} = 2 \cdot \left[\frac{n}{n+1}\right]^n$$

$$= 2\left[\frac{1}{n+1}\right]^n = \frac{2}{\left(1 + \frac{1}{n}\right)^n}$$

$$\therefore \quad \lim_{n\to\infty} \frac{u_{n+1}}{u_n} = \lim_{n\to\infty} \frac{2}{\left(1 + \frac{1}{n}\right)^n} = 2 \cdot \lim_{n\to\infty} \frac{1}{\left(1 + \frac{1}{n}\right)^n} = 2 \cdot \frac{1}{e} = \frac{2}{e} < 1$$

($\because$ $2 < e < 3$)

$\therefore$ By ratio test, the series $\sum u_n$ is convergent.

34. $\sum \frac{3^n \cdot n!}{n^n}$

Solution: Here, $u_n = \sum \frac{3^n \cdot n!}{n^n}$;

$$u_{n+1} = \frac{3^{n+1} \cdot (n+1)!}{(n+1)^{n+1}}$$

PART-D

$$\therefore \quad \frac{u_{n+1}}{u_n} = \frac{3^n \cdot 3 \cdot (n+1) \cdot n!}{(n+1) \cdot (n+1)^n} \times \frac{n^n}{3^n \cdot n!}$$

$$= \frac{3n^n}{(n+1)^n} = 3\left[\frac{n}{n+1}\right]^n = 3\left[\frac{1}{\left(1+\frac{1}{n}\right)^n}\right]$$

$$\therefore \quad \lim_{n\to\infty} \frac{u_{n+1}}{u_n} = 3 \lim_{n\to\infty} \frac{1}{\left[1+\frac{1}{n}\right]^n} = 3 \cdot \frac{1}{e} = \frac{3}{e} > 1$$

Hence, by ratio test, $\sum u_n$ is divergent.

35. $\sum \frac{1 \cdot 3 \cdot 5 \ldots (2n-1)}{2 \cdot 4 \cdot 6 \ldots 2n} \cdot x^n$

Solution: Here, $u_n = \frac{1 \cdot 3 \cdot 5 \ldots (2n-1)}{2 \cdot 4 \cdot 6 \ldots 2n} \cdot x^n$

$$u_{n+1} = \sum \frac{1 \cdot 3 \cdot 5 \ldots (2n-1)(2n+1)}{2 \cdot 4 \cdot 6 \ldots 2n\,(2n+2)} \cdot x^{n+1}$$

$$\therefore \quad \frac{u_{n+1}}{u_n} = \frac{1 \cdot 3 \cdot 5 \ldots (2n-1)(2n+1)}{2 \cdot 4 \cdot 6 \ldots 2n\,(2n+2)} \cdot \frac{2 \cdot 4 \cdot 6 \ldots (2n)}{1 \cdot 3 \cdot 5 \ldots (2n-1)} \cdot \frac{x^{n+1}}{x^n}$$

$$= \frac{2n+1}{2n+2} \cdot (x)$$

$$\therefore \quad \lim_{n\to\infty} \frac{u_{n+1}}{u_n} = \lim_{n\to\infty} \left(\frac{2n+1}{2n+2}\right) \cdot x = \lim_{n\to\infty} \left[\frac{1+\frac{1}{2n}}{1+\frac{1}{2n}}\right] \cdot x \to \quad 1 \cdot x = x$$

So, by D'Alembert's Ratio test we have $x < 1$

The series is convergent (C)

$x > 1$; the series is (D)

If $x = 1$; no conclusion

When $x = 1$,

we have $\frac{u_{n+1}}{u_n} = \frac{2n+1}{2n+2} \cdot 1 \to 1$ [1]

Apply Raabe's Test:

We have $$n\left(\frac{u_n}{u_{n+1}}-1\right) = n\left[\left(\frac{2n+2}{2n+1}\right)-1\right]$$

$$= \frac{n[(2n+2-2n-1)]}{(2n+1)} = n/(2n+1)$$

$\therefore$ $$\lim_{n\to\infty} n\left[\left(\frac{u_n}{u_{n+1}}-1\right)\right] = \lim_{n\to\infty}\frac{n}{2n+1} \to \frac{1}{2} < 1$$

$\therefore$ $\sum u_n$ is divergent.

$\therefore$ We get $x < 1$ the series (C) and $x \geq 1$ the series (D)

36. $\frac{1^3}{2^3} + \frac{1^3 \cdot 3^3}{2^3 \cdot 4^3} + \frac{1^3 \cdot 2^3 \cdot 5^3}{2^3 \cdot 4^3 \cdot 6^3} + \ldots$

Solution: Here, $$u_n = \frac{1^3 \cdot 3^3 \cdot 5^3 \ldots (2n-1)^3}{2^3 \cdot 4^3 \cdot 6^3 \ldots (2n)^3}$$

$\therefore$ $$u_{n+1} = \frac{1^3 \cdot 3^3 \cdot 5^3 \ldots (2n-1)^3 \cdot (2n+1)^3}{2^3 \cdot 4^3 \cdot 6^3 \ldots (2n)^3 \cdot (2n+2)^3}$$

$\therefore$ $$\frac{u_{n+1}}{u_n} = \frac{(2n+1)^3}{(2n+2)^3} = \frac{8n^3+12n^2+6n+1}{8n^3+24n^2+24n+8}$$

$\therefore$ $$\lim_{n\to\infty}\frac{u_{n+1}}{u_n} = \lim_{n\to\infty}\frac{(8n^3+12n^2+6n+1)}{(8n^3+24n^2+24n+8)}$$

$\to 1$ (Ratio test fails)

$\Rightarrow$ No conclusion can be drawn from the Ratio test.

Apply Raabe's Test, we have

$$\lim_{n\to\infty} n\left[\left(\frac{u_n}{u_{n+1}}\right)-1\right] = \lim_{n\to\infty}\left[n\left\{\frac{8n^3+24n^2+24n+8}{8n^3+12n^2+6n+1}\right\}-1\right]$$

$$= \lim_{n\to\infty} n\left[\frac{(8n^3+24n^2+24n+8)-(8n^3+12n^2+6n+1)}{(8n^3+12n^2+6n+1)}\right]$$

$$= \lim_{n\to\infty} n\,\frac{(12n^2+18n+7)}{8n^3+12n^2+6n+1}$$

$$= \lim_{n\to\infty}\frac{12n^3+18n^2+7n}{8n^3+12n^2+6n+1}$$

$$\lim_{n\to\infty}\left[\frac{12+(18/n)+(7/n^2)}{8+(12/n)+(6/n^2)+(1/n^3)}\right]$$

$$\to \frac{12+0+0}{8+0+0}=\frac{12}{8}=\frac{3}{2}>1>1$$

Hence by Raabe's Test, the series is convergent.

37. $2x+\dfrac{3x^3}{8}+\dfrac{4x^3}{27}+\ldots+\dfrac{n+1}{n^3}\cdot x^n+\ldots$

Solution: Here, $u_n=\dfrac{n+1}{n^3}x^n$ and $u_{n+1}=\dfrac{(n+2)}{(n+1)^3}\cdot x^{n+1}$

$$\therefore \quad \frac{u_{n+1}}{u_n}=\frac{(n+2)}{(n+1)^3}\times\frac{n^3}{(n+1)}\times\frac{x^{n+1}}{x^n}$$

$$=\frac{n^3(n+2)}{(n+1)^4}\cdot x=\frac{n^4\left[1+\frac{2}{n^4}\right]}{n^4\left[1+\frac{1}{n}\right]^4}\cdot x=\frac{\left[1+\frac{2}{n^4}\right]}{\left[1+\frac{1}{n}\right]^4}\cdot x$$

$$\therefore \quad \lim_{n\to\infty}\frac{u_{n+1}}{u_n}=\lim_{n\to\infty}\frac{\left[1+\frac{2}{n^4}\right]}{\left[1+\frac{1}{n}\right]^4}\cdot x=\frac{(1+0)}{(1+0)^4}\cdot x=x$$

$\therefore$ $x>1$, the series is (D) and if $x<1$, the series is (C)
If $x=1$, (the test fails) no conclusion can be drawn from the Ratio Test.
Apply comparison test,

If $x=1$, $\qquad a_n=\dfrac{n+1}{n^3}=\dfrac{n\left(1+\frac{1}{n}\right)}{n^3}=\dfrac{1}{n^2}\left(1+\dfrac{1}{n}\right)$

Let $\qquad b_n=\dfrac{n}{n^3}=\dfrac{1}{n^2}$,

$$\therefore \quad \lim_{n\to\infty}\frac{a_n}{b_n}=\lim_{n\to\infty}\frac{1}{n^2}\left(1+\frac{1}{n}\right)\cdot\frac{n^2}{1}=(1+0)=1 \text{ (finite)}$$

$\therefore$ by the limit form of comparison test, $\sum a_n$ and $\sum b_n$ behave alike.

$\sum b_n=\sum\dfrac{1}{n^2}$ is a p-series with ($p=2>1$) is convergent.

Hence $\sum a_n$ is also convergent.

38. Examine the convergence the series

$$\sum_{n=1}^{\infty} \frac{1\cdot 3\cdot 5\cdot 7\ldots(2n-1)}{4\cdot 7\cdot 10\cdot 13\ldots(3n+1)}$$

Solution: Here,
$$u_n = \frac{1\cdot 3\cdot 5\cdot 7\ldots[2(2n+1)-1]}{4\cdot 7\cdot 10\cdot 13\ldots[3(3n+1)+1]}$$

$$= \frac{1\cdot 3\cdot 5\cdot 7\ldots(2n+1)}{4\cdot 7\cdot 10\cdot 13\ldots(3n+4)}$$

$$u_{n+1} = \frac{1\cdot 3\cdot 5\cdot 7\ldots(2n-1)(2n+1)}{4\cdot 7\cdot 10\cdot 13\ldots(3n+4)(3n+4)}$$

$$\therefore \quad \frac{u_{n+1}}{u_n} = \frac{1\cdot 3\cdot 5\cdot 7\ldots(2n-1)(2n+1)}{4\cdot 7\cdot 10\cdot 13\ldots(3n+4)(3n+4)}, \frac{4\cdot 7\cdot 10\cdot 13\ldots(3n+1)}{1\cdot 3\cdot 5\cdot 7\ldots(2n-1)}$$

$$= \frac{2n+1}{3n+4}$$

$$\therefore \quad \lim_{n\to\infty} \frac{u_{n+1}}{u_n} = \lim_{n\to\infty} \frac{2n+1}{3n+4} = \lim_{n\to\infty} \frac{\left(2+\frac{1}{n}\right)}{\left(3+\frac{4}{n}\right)} = \frac{2}{3} < 1$$

Hence, by Ratio Test, the given series $\sum u_n$ is (C).

39. Test for convergence

$$\sum_{n=1}^{\infty} \frac{3\cdot 6\cdot 9\ldots 3n}{4\cdot 7\cdot 10\ldots(3n+1)} \frac{5^n}{3n+2}$$

Solution: Here,
$$u_n = \frac{3\cdot 6\cdot 9\ldots 3n}{4\cdot 7\cdot 10\ldots(3n+1)} \cdot \frac{5^n}{3n+2}$$

$$u_{n+1} = \frac{3\cdot 6\cdot 9\ldots 3(n+1)}{4\cdot 7\cdot 10\ldots[3(n+1)+1]} \cdot \frac{5^{n+1}}{3(n+1)+2}$$

$$= \frac{3\cdot 6\cdot 9\ldots 3n(3n+3)}{4\cdot 7\cdot 10\ldots(3n+1)(3n+4)} \cdot \frac{5^{n+1}}{3n+5}$$

$$\frac{u_{n+1}}{u_n} = \frac{(3n+3)(3n+2)}{(3n+4)(3n+5)} (5)$$

$$\therefore \quad \lim_{n\to\infty} \frac{u_{n+1}}{u_n} = \lim_{n\to\infty} \frac{(3n+3)(3n+2)}{(3n+4)(3n+5)} \cdot (5)$$

PART-D

$$= \lim_{n\to\infty} \frac{\left(3+\frac{3}{n}\right)\left(3+\frac{2}{n}\right)}{\left(3+\frac{4}{n}\right)\cdot\left(3+\frac{5}{n}\right)}$$

i.e., $$\lim_{n\to\infty} \frac{u_{n+1}}{u_n} = \frac{(3)(3)}{(3)(3)} \times (5)$$

∴ By Ratio Tests Σu_n. is (D).

40. Find the nature of the series

$$\sum_{n=1}^{\infty} \frac{4\cdot 7\cdot 10\ldots(3n+1)}{1\cdot 2\cdot 3\ldots n}\cdot x^n$$

Solution: Here, $$u_n = \frac{4\cdot 7\cdot 10\ldots(3n+1)}{1\cdot 2\cdot 3\ldots n}\cdot x^n$$

$$\therefore \quad u_{n+1} = \frac{4\cdot 7\cdot 10\ldots[3(n+1)+1]}{1\cdot 2\cdot 3\ldots(n+1)}\cdot x^{n+1}$$

$$= \frac{4\cdot 7\cdot 10\ldots(3n+4)}{1\cdot 2\cdot 3\ldots(n+1)}\cdot x^{n+1}$$

$$= \frac{4\cdot 7\cdot 10\ldots(3n+1)(3n+4)}{1\cdot 2\cdot 3\ldots(n)(n+1)}\cdot x^{n+1}$$

$$\therefore \quad \frac{u_{n+1}}{u_n} = \frac{4\cdot 7\cdot 10\ldots(3n+1)(3n+4)}{1\cdot 2\cdot 3\ldots(n)(n+1)}\cdot x^{n+1} \times \frac{1\cdot 2\cdot 3\ldots n}{4\cdot 7\cdot 10\ldots(3n+1)}\cdot\frac{1}{x^n}$$

i.e., $$\frac{u_{n+1}}{u_n} = \frac{3n+4}{n+1}\cdot x$$

$$\therefore \quad \lim_{n\to\infty}\frac{u_{n+1}}{u_n} = \lim_{n\to\infty}\frac{3n+4}{n+1}\cdot x = \lim_{n\to\infty}\frac{\left(3+\frac{4}{n}\right)}{\left(1+\frac{1}{n}\right)}\cdot x = 3x$$

Hence by Ratio Test,

$\sum u_n$ is (C), if $3x < 1$ or $x < 1/3$

(D) if $3x > 1$ or $x > 1/3$

and the test fails if

$$3x = 1 \quad \text{or} \quad x = \frac{1}{3}$$

If $x = \frac{1}{3}$ we have

$$\frac{u_{n+1}}{u_n} = \frac{(3n+4)}{(n+1)} \cdot \frac{1}{3}$$

$$\therefore \quad \frac{u_n}{u_{n+1}} = \frac{3n+3}{3n+4}$$

$$\text{Now, } \lim_{n\to\infty} n\left(\frac{u_n}{u_{n+1}} - 1\right) = \lim_{n\to\infty} n\left(\frac{u_n}{u_{n+1}} - 1\right)$$

$$= \lim_{n\to\infty} n\left[\frac{-1}{3n+4}\right] = \lim_{n\to\infty} \frac{-1}{\left(3+\frac{4}{n}\right)} = \frac{-1}{3} < 1$$

$\therefore \quad \sum u_n$ is (D)

(when $3x = 1$) by Raabe's Test.

Hence $\sum u_n$ is (C) if $x < 1/3$

and $\sum u_n$ is (D) if $x \geq \frac{1}{3}$

41. Examine for convergence the series

$$\sum_{n=1}^{\infty} \frac{(n!)^2}{2n!} \cdot x^n$$

Solution: Here,
$$u_n = \frac{(n!)^2}{2n!} \cdot x^n$$

$$\therefore \quad u_{n+1} = \frac{[(n+1)!]^2}{(2n+2)!} \cdot x^{n+1}$$

$$\therefore \quad \frac{u_{n+1}}{u_n} = \frac{[(n+1)!]^2 \cdot x^{n+1}}{(2n+2)!} \cdot \frac{2n!}{(n!)^2 \cdot x^n}$$

$$= \left[\frac{(n+1)!}{n!}\right]^2 \cdot \frac{2n!}{(2n+2)!} \cdot x$$

$$= \left[\frac{(n+1)\,n!}{n!}\right]^2 \frac{2n!\,x}{(2n+2)(2n+1)(2n)!}$$

PART-D

i.e., $$\frac{u_{n+1}}{u_n} = \frac{(n+1)^2 \cdot x}{(2n+2)(2n+1)}$$

$$\therefore \quad \lim_{n\to\infty} \frac{u_{n+1}}{u_n} = \lim_{n\to\infty} \frac{n^2\left[1+\frac{1}{n}\right]^2 \cdot x}{n\left(2+\frac{2}{n}\right)\cdot n\left(2+\frac{1}{n}\right)} = \lim_{n\to\infty} \frac{\left[1+\frac{1}{n}\right]^2 x}{\left(2+\frac{2}{n}\right)\left(2+\frac{1}{n}\right)}$$

$$= \frac{(1+0)\,x}{(2+0)(2+0)} = \frac{x}{4}$$

$\therefore$ by D' Alembert's ratio test,

$$\sum u_n \text{ is } \begin{cases} \text{(C), if } \frac{x}{4} < 1, \text{ or } x < 4 \\ \text{(D), if } \frac{x}{4} > 1 \text{ or } x > 4 \end{cases}$$

and test fails if $\frac{x}{4} = 1$

or $x = 4.$

When $x = 4,$

$$\frac{u_{n+1}}{u_n} = \frac{(n+1)^2 \cdot 4}{(2n+2)(2n+1)} = \frac{4n^2 + 8n + 4}{4n^2 + 6n + 4}$$

and we shall apply Raabe's Test.

$$\lim_{n\to\infty} n\left[\frac{u_n}{u_{n+1}} - 1\right] = \lim_{n\to\infty} n\left[\frac{4n^2+6n+2}{4n^2+8n+4} - 1\right]$$

$$= \lim_{n\to\infty} n\left[\frac{4n^2+6n+2-4n^2-8n-4}{4n^2+8n+4}\right]$$

$$= \lim_{n\to\infty} (n)\left[\frac{-2n-2}{4n^2+8n+4}\right] = \lim_{n\to\infty} \frac{-n^2\left(2+\frac{2}{n}\right)}{n^2\left(4+\frac{8}{n}+\frac{4}{n^2}\right)}$$

$$= \lim_{n\to\infty} \frac{-\left(2+\frac{2}{n}\right)}{\left(u+\frac{8}{n}+\frac{4}{n^2}\right)} = + \frac{-(2+0)}{(4+0+0)} = \frac{-1}{2} < 1$$

By Raabe's Test, $\sum u_n$ is (D) when $x = 4$. Hence we conclude that $\sum u_n$ is (C) if $x < 4$ and divergent if $x \geq 4$.

42. Test the convergence of

$$\frac{1}{1!}+\frac{1+2}{2!}+\frac{1+2+2^2}{3!}+\frac{1+2+2^2+2^3}{4!}+\ldots$$

Solution:
$$u_n = \frac{1+2+2^2+\ldots+2^{n-1}}{n!} = \frac{1}{n!}\left[\frac{2^n-1}{2-1}\right] = \frac{1}{n!}(2^n-1)$$

$$\left(\because 1+r+r^2+\ldots+r^{n-1} = \frac{r^n-1}{r-1}\right) \text{ then } r > 1$$

$$\therefore \quad \lim_{n\to\infty}\frac{u_{n+1}}{u_n} = \lim_{n\to\infty}\frac{2^{n+1}-1}{(n+1)!}\cdot\frac{n!}{2^n-1}$$

$$= \lim_{n\to\infty}\frac{2^n\left(2-\frac{1}{2^n}\right)n!}{(n+1)!\cdot n!\cdot 2^n\left(1-\frac{1}{2^n}\right)} = \frac{2-0}{\infty\cdot(1-0)} = 0 < 1$$

Hence theorem is convergent.

43. $\dfrac{2}{3}+\dfrac{2\cdot 4}{3\cdot 5}+\dfrac{2\cdot 4\cdot 6}{3\cdot 5\cdot 7}+\ldots.$

Solution: Here,
$$u_n = \frac{(2\cdot 4\cdot 6\ldots 2n)}{(3\cdot 5\ldots(2n+1))}$$

$$u_{n+1} = \frac{[(2\cdot 4\cdot 6\ldots 2n)(2n+2)]}{[3\cdot 5\cdot 7\ldots(2n+1)(2n+3)]}$$

$$\therefore \quad \frac{u_{n+1}}{u_n} = \frac{2n+2}{2n+3}\ 1. \tag{1}$$

(as $n \to \infty$) $\therefore$ D.A. test fails.

No conclusion can be drawn from the Ratio. Test. Apply the Raabe's Test.

We get
$$\lim_{n\to\infty}\left[n\left(\frac{U_n}{U_{n+1}}-1\right)\right] = \lim_{n\to\infty}\left[n\left\{\left(\frac{2n+3}{2n+2}-1\right)\right\}\right]$$

$$= \lim_{n\to\infty} n\left[\frac{(2n+3-2n-2)}{2n+2}\right]\ \lim_{n\to\infty}\frac{n}{(2n+2)} \to \frac{1}{2} < 1$$

Hence the series is (D).

44. Discuss the convergence of the series

$$1 + \frac{1}{2}x + \frac{1\cdot 2}{2\cdot 5}x^2 + \frac{1\cdot 2\cdot 3}{2\cdot 5\cdot 8}\cdot x^3 + \ldots$$

Solution: Let $$U_n = \frac{1\cdot 2\cdot 3\ldots n}{2\cdot 5\cdot 8\ldots(3n-1)}x^n$$

$$U_{n+1} = \frac{1\cdot 2\cdot 3\ldots n(n+1)}{2\cdot 5\cdot 8\ldots(3n-1)(3n+2)}\cdot x^{n+1}$$

$$\therefore \quad \frac{u_{n+1}}{u_n} = \frac{(n+1)}{(3n+2)}\cdot x$$

$$\therefore \quad \lim_{n\to\infty}\frac{u_{n+1}}{u_n} = \lim_{n\to\infty}\frac{(n+1)}{(3n+2)}\cdot x \to \frac{x}{3}$$

$\therefore$ Series is (C) if $\frac{x}{3} > 1$, or $x > 3$

Series is (D) if $\frac{x}{3} < 1$, or $x < 3$

Series is (?) if $\frac{x}{3} = 1$, $x = 3$

when $x = 3$; (Test fails)

No conclusion can be drawn from Ratio Test, so, we apply next test,

When $x = 3$, we have

(applying Raabe' test)

$$n\left[\frac{U_n}{U_{n+1}} - 1\right] = n\left[\frac{(3n+2)}{n+1}\cdot(1/3) - 1\right] = n\left\{\frac{3n+2-(3n+3)}{3(n+1)}\right\}$$

$$= \frac{n(-1)}{3(n+1)} = \frac{-n}{3x+3} \to (-1/3) < 1$$

$\therefore$ By Raabe's Test, the series is divergent.

Thus the series is

(C) if $x > 3$;

(D) if $x \leq 3$

45. $\frac{1}{2\cdot 3} + \frac{1\cdot 3}{2\cdot 4\cdot 5} + \frac{1\cdot 3\cdot 5}{2\cdot 4\cdot 6\cdot 7} + \ldots$

Solution: (Here, the number of factors in the D^r of any term is one more than in the N^r)

$$u_n = \left\{\frac{1\cdot 3\cdot 5\ldots(2n-1)}{2\cdot 4\cdot 6\ldots 2n(2n+1)}\right\}$$

$$u_{n+1} = \left\{\frac{1 \cdot 3 \cdot 5 \ldots (2n-1)(2n+1)}{2 \cdot 4 \cdot 6 \ldots 2n(2n+2)(2n+3)}\right\}$$

$$\therefore \quad \frac{u_{n+1}}{u_n} = \left\{\frac{(2n+1)(2n+1)}{(2n+2)(2n+1)}\right\} = \left\{\frac{4n^2+10n+6}{4n^2+4n+1}\right\} 1.$$

So, Ratio Test fails

Applying Raabe's Test

We have

$$n\left[\left(\frac{U_n}{Un+1} - 1\right)\right] = \left\{\frac{(4n^2+10n+6)-(4n^2+4n+1)}{(4n^2+4n+1)}\right\}$$

$$= \frac{n(6n+5)}{4n^2+4n+1} = \frac{6n^2+5n}{4n^2+4n+1}$$

$$\rightarrow \quad \frac{6}{4} = (3/2) > 1$$

$\therefore$ By the Raabe's test, the series is convergent.

46. $\dfrac{1^2 \cdot 2^2}{1!} + \dfrac{2^2 \cdot 3^2}{2!} + \dfrac{3^2 \cdot 4^2}{3!} + \ldots + \infty$

Solution:

$$U_n = \frac{n^2(n+1)^2}{n!}$$

$$\therefore \quad U_{n+1} = \frac{(n+1)^2\,(n+2)^2}{(n+1)!}$$

$$\lim_{n\to\infty} \frac{U_{n+1}}{U_n} = \lim_{n\to\infty} = \lim_{n\to\infty} \frac{(n+1)^2\,(n+2)^2}{(n+1)!} \times \frac{n!}{n^2\,(n+1)^2}$$

$$= \lim_{n\to\infty} \frac{n^2\left(1+\dfrac{2}{n}\right)^2}{(n+1)^2\,n^2} = \lim_{n\to\infty} \frac{\left(1+\dfrac{2}{n}\right)^2}{(n+1)^2} = 0\ (< 1)$$

$\therefore$ by Ratio test, the given series is (C).

47. $\Sigma\left(\dfrac{n^3+2}{3^n+2}\right)$

Solution:

$$u_n = \frac{n^2+2}{3^n+2}$$

PART-D

$$\therefore \qquad u_{n+1} = \frac{(n+1)^3+2}{3^{n+1}+2}$$

$$\therefore \qquad \lim_{n\to\infty}\frac{u_{n+1}}{u_n} = \lim_{n\to\infty}\frac{(n+1)^3+2}{3^{n+1}+2}\times\frac{3^4+2}{n^3+2}$$

$$= \lim_{n\to\infty}\frac{n^3\left[\left(1+\frac{1}{n^3}\right)+\frac{2}{n^3}\right]\left(3^n\ \frac{2}{3^n}\right)}{3^n\left[\left(1+\frac{2}{3^n}\right)\cdot n^3\left(1+\frac{2}{n^3}\right)\right]}$$

$$= \lim_{n\to\infty}\frac{\left[\left(1+\frac{1}{n}\right)^3+\frac{2}{n^3}\right]\cdot\left(1+\frac{2}{3^n}\right)}{\left[\left(3+\frac{2}{3^n}\right)\left(1+\frac{2}{3^3}\right)\right]} = \frac{1\cdot 1}{3\cdot 1} = \frac{1}{3}\ (<1)$$

$\therefore$ by ratio test, the given series is (C).

48. Test the convergence of the series

$$\Sigma\sqrt{\frac{1+2^n}{1+3^n}}$$

Solution: Here $\qquad U_n = \sqrt{\dfrac{1+2^n}{1+3^n}}$

Choose $\qquad V_n = \sqrt{(2/3)^n}$

Which is a Geometric series whose common ratio is $\sqrt{2/3} < 1$.

Thus, $\sum V_n$ is (C).

Now, $\qquad \displaystyle\lim_{n\to\infty}\frac{a_n}{b_n} = \lim_{n\to\infty}\sqrt{\frac{1+2^4}{1+3^4}}\cdot\sqrt{\left(\frac{3}{2}\right)^n}$

$$= \lim_{n\to\infty}\frac{\sqrt{1+(1/2)^n}}{\sqrt{1+(1/3)^n}} = \frac{\sqrt{1+0}}{\sqrt{1+0}} = 1$$

$$\left[\because \lim_{a\to 0} a^n = 0, \text{ if } a<1\right]$$

By comparison test $\sum a_n$ and $\sum b_n$ behave alike

But $\sum b_n$ is (C),

Here $\sum a_n$ is also (C)

49. If $a > 1$; and $b > 1$ S.T the infinite series $\sum \frac{\sqrt{(a^n+1)}}{(b^n+1)}$ is (C) when $a < b$, and (D) if $a \geq b$.

Solution: since, x $a > 1$, $a^n \to \infty$;

|||ly $b^n \to \infty$

Let $$a_n = \left[\frac{(a^n+1)^{1/2}}{(b^n+1)^{1/2}} \times \frac{(b^n)^{1/2}}{(a^n)^{1/2}}\right] = \left(\frac{a^n+1}{a^n}\right)^{1/2} \div \left(\frac{b^n+1}{b^n}\right)^{1/2}$$

$$= \left[\frac{1+(1/a^n)}{1+(1/b^n)}\right]^{1/2} \to \left[\frac{1+0}{1+0}\right] = 1$$

$\therefore$ $\sum a_n$; $\sum b_n$ behave alike

But $$\sum b_n = \sum\left[\sqrt{(a/b)}\right]^n = \sum r^n ;$$

$$r = \sqrt{(a/b)}$$

The series is a G. series, which is convergent if the common ratio $r < 1$, and (D) if $r \geq 1$. $\therefore$ $\sum b_n$ is (C); if $r < 1$

$$\sqrt{(a/b)} < 1; \text{ if } (a < b)$$

and is (D) if $r \geq 1$, i.e., if $a \geq b$ $\sqrt{(a/b)} \geq 1$;

EXERCISES

Examine the convergence of the following series:

1. $\frac{2}{3\cdot 4} + \frac{2\cdot 4}{3\cdot 5\cdot 6} + \frac{2\cdot 4\cdot 6}{3\cdot 5\cdot 7\cdot 8} + \ldots$ [*Ans.* (C)]

2. $\frac{1}{3}x + \frac{1\cdot 2}{3\cdot 5}x^2 - \frac{1\cdot 2\cdot 3}{35\cdot 7}x^3 + \ldots$ [*Ans.* (c) when $x < 2$, (D) when $x \geq 2$]

3. $\left(\frac{x^n}{n!}\right)^2$ [*Ans.* (C) $\forall\, x$]

4. $\left[\frac{1}{2}\cdot\frac{x^3}{3} + \frac{1\cdot 3}{2\cdot 4}\cdot\frac{x^5}{5} + \ldots\right]$ [*Ans.* (C) is $(-1 \leq x \leq 1)$ and D outside]

5. $\frac{1}{2\cdot 4}x+\frac{1\cdot 3}{2\cdot 4\cdot 6}x^2+\frac{1\cdot 3\cdot 5}{2\cdot 4\cdot 6\cdot 8}x^3+\ldots$ [*Ans.* for $x \le 1$, (C) for $x > 1$, (D)]

6. $\left(\frac{a}{b}\right)+\frac{a(a+1)}{b(b+1)}x+\frac{a(a+1)(a+2)}{b(b+1)(b+2)}x^2$ [*Ans.* $b-a>1$, (C) $b-a\le 1$, (D)]

7. $1+\left(\frac{\alpha}{\beta}\right)x+\frac{\alpha(\alpha+1)}{\beta(\beta+1)}\cdot x^2+\ldots$ [*Ans.* (C) if $\beta > 1+\alpha$ (D) if $\beta \le 1+\alpha$]

8. $1+\left(\frac{2}{3}\right)x+\left(\frac{2\cdot 3}{3\cdot 5}\right)x^2+\left(\frac{2\cdot 3\cdot 4}{3\cdot 5\cdot 7}\right)x^3+\ldots$ [*Ans.* [C] if $x < 2$ (D) if $x \ge 2$]

9. $1-\left(\frac{1}{2}\right)+\frac{1\cdot 3}{2\cdot 4}-\frac{1\cdot 3\cdot 5}{2\cdot 4\cdot 6}+\ldots$

10. $\frac{1}{\sqrt{1^3+1}}+\frac{1}{\sqrt{2^3+1}}\cdot x+\frac{1}{\sqrt{3^3+1}}\cdot x^2+\ldots$

[*Ans.* The series is *AC* if $|x| < 1$ and also when $x = \pm 1$ i.e., when $|x| \le 1$ or $(-1 \le x \le 1)$]

Worked Example (50 – 65)

Related to Cauchy's root test

If Σu_n be a series of positive terms, then the series is (C) or (D) according as

$$\lim_{n\to 8}(u_n)^{1/n} < 1 \text{ or } > 1$$

50. Discuss the convergence of $\Sigma\left(\frac{n^2}{2^n}\right)$

Solution:
$$u_n = \frac{n^2}{2^n}$$

$\therefore$
$$(u_n)^{1/n} = \frac{(n^2)^{1/n}}{2} = \frac{1}{2}(n^{1/n})^2$$

$$= \frac{1}{2}[e^{1/n \log n}]^2 \qquad (\because A^B = e^{B\log A})$$

$\therefore$
$$\lim_{n\to\infty} u_n^{1/n} = \frac{1}{2}(e^{\circ})^2 = \frac{1}{2} < 1\text{, Hence } \Sigma u_n \text{ is}$$

51. Test for convergence of the series $\left(1-\frac{1}{n}\right)^{n^2}$

Solution: Here
$$u_n = \left(1-\frac{1}{n}\right)^{n^2}$$

$$\therefore \quad (u_n)^{1/n} = \left\{\left(1-\frac{1}{n}\right)^{n^2}\right\}^{1/n} = \left(1-\frac{1}{n}\right)^{n^2}$$

$$\therefore \quad \lim_{n\to\infty} (u_n)^{1/n} = \lim_{n\to\infty}\left(1-\frac{1}{n}\right)^{n} = e^{-1} = \frac{1}{e} = \frac{1}{2.718} \quad i.e < 1$$

Here Σu_n is convergent by Cauchy's Root test or (Radical test).

52. $\dfrac{1}{\left(1+\dfrac{1}{n}\right)^{n^2}}$

Solution: **Here**

$$u_n = \left[\frac{1}{\left(1+\frac{1}{n}\right)^{n^2}}\right]$$

$$\therefore \quad \lim_{n\to 8} (u_n)^{1/n} = \left[\frac{1}{\left(1+\frac{1}{n}\right)^{n^2}}\right]^{1/n} = \frac{1}{\left(1+\frac{1}{n}\right)^{n}} = \frac{1}{e} = \frac{1}{2.718} < 1$$

$\therefore$ **Convergent**

53. $\left(\dfrac{2^2}{1^2}-\dfrac{2}{1}\right)^{-1} + \left(\dfrac{3^3}{2^3}-\dfrac{3}{2}\right)^{-2} + \left(\dfrac{4^4}{3^4}-\dfrac{4}{3}\right)^{-3} + \ldots$

Solution: **Let**

$$u_n = \left\{\frac{(n+1)^{n+1}}{n^{n+1}} - \frac{n+1}{n}\right\}^{-n}$$

$$\therefore \quad (u_n)^{1/n} = \left\{\left(\frac{(n+1)}{n}\right)^{n+1} - \frac{n+1}{n}\right\}^{-1}$$

$$= \left(\frac{n+1}{n}\right)^{-1}\left\{\left(\frac{n+1}{n}\right)^{n} - 1\right\}^{-1} = \left(1+\frac{1}{n}\right)^{-1}\left\{\left(1+\frac{1}{n}\right)^{n} - 1\right\}^{-1}$$

$$\therefore \quad \lim_{n\to\infty} (u_n)^{1/n} = (1+0)^{-1}\,(e-1)^{-1} = (\mathrm{e}-1)^{-1}$$

$$= (2.718 - 1)^{-1} = (1\cdot 718)^{-1}$$

$$= \frac{1}{1.718} < 1$$

Then, by Cauchy's root test, the series is convergent.

54. $1 + \frac{x}{2} + \frac{x^2}{3^2} + \frac{x^3}{4^3} + \ldots$

Solution: Neglecting the first term

Let $$u_n = \frac{x^n}{(n+1)^n}$$

$$\therefore \quad \lim_{n \to 8} (u_n)^{1/n} = \lim_{n \to \infty} \left(\frac{x}{n+1} \right) = 0 < 1$$

For all values of x and hence convergent by Cauchy's root test.

55. $\frac{1}{2} + \frac{2}{3} x + \left(\frac{3}{4} \right)^2 x^2 + \left(\frac{4}{5} \right)^3 x^3 + \ldots$

$1 + \frac{2}{3} x + \left(\frac{3}{4} \right)^2 x^2 + \left(\frac{4}{5} \right)^3 x^3 + \ldots$ [VTU. F/M. 2005]

Here, $$u_n = \left(\frac{n+1}{n+2} \right)^n \cdot x^n$$

$$\lim_{n \to \infty} (u_n)^{1/n} = \lim_{n \to \infty} \left(\frac{n+1}{n+2} \right) \cdot x = x;$$

$\therefore$ If $x < 1$, the series is (C)

If $x > 1$, the series (D)

If $x = 1$;

then $$u_n = \left(\frac{n+1}{n+2} \right)^n = \frac{\left(1 + \frac{1}{n}\right)^n}{\left(1 + \frac{2}{n}\right)^n} \to \frac{e}{e^2} = \frac{1}{e}$$

series of the terms $$= \frac{1}{2.718} > 0, \text{ Hence; (D) as } n \to \infty$$

$$\left(\because \text{ A series of positive terms is divergent } \lim_{n \to \infty} u_n > 0 \right)$$

56. Discus the convergence of $\Sigma \left(\frac{nx}{n+1} \right)^n$

Solution: $$u_n = \left(\frac{nx}{n+1} \right)^n$$

$$(u_n)^{1/n} = \left(\frac{n}{n+1}\right) \cdot x$$

$$\therefore \quad \lim_{n\to\infty} (u_n)^{1/n} = \lim_{n\to\infty} \left(\frac{n}{n+1}\right) \cdot x = \lim_{n\to\infty} \left(\frac{n}{n+1/n}\right) \cdot x = x$$

$\therefore$ **By Cauchy's root test, Σu_n is (C)**

if $x < 1$, and (D), if $x > 1$

When $x = 1$, root test fails

If $x = 1$, $$u_n = \left(\frac{n}{n+1}\right)^n$$

$$\therefore \quad \lim_{n\to\infty} (u_n) = \lim_{n\to\infty} \left(\frac{n}{n+1}\right)^n = \lim_{n\to\infty} \frac{1}{\left(1+\frac{1}{n}\right)^n} = \frac{1}{e} \neq 0$$

$\therefore$ **Σu_n is (D) if $x = 1$**

and Σu_n is (C) if $x < 1$ and (D) if $x \geq 1$.

57. (*a*) $\Sigma \dfrac{x^n}{n^n}$ $(x > 0)$

Solution: $$u_n = \frac{x^n}{n^n};$$

$$\therefore \quad (u_n)^{1/n} = \frac{x}{n}$$

$$\lim_{n\to\infty} (u_n)^{1/n} \to 0 < 1$$

$\therefore$ **By Cauchy's root test, the series is convergent for all values of x.**

57. (*b*) $$\Sigma\left[1+\frac{1}{\sqrt{n}}\right]^{-(n^{3/2})}$$

Solution: Given $$u_n = \left(1+\frac{1}{\sqrt{n}}\right)^{-n^{3/2}}$$

$$\therefore \quad (u_n)^{1/n} = \left[\left(1+\frac{1}{\sqrt{n}}\right)^{-n^{3/2}}\right]^{1/n}$$

$$= \left[1+\frac{1}{\sqrt{n}}\right]^{-n^{3/2}\cdot n^{-1}} = \left[1+\frac{1}{\sqrt{n}}\right]^{-n^{1/2}}$$

PART-D

$$= \left[1+\frac{1}{\sqrt{n}}\right]^{-\sqrt{n}} = \frac{1}{\left[1+\frac{1}{\sqrt{n}}\right]^{\sqrt{n}}}$$

$$\therefore \quad \lim_{n\to\infty} (u_n)^{1/n} = \frac{1}{\lim_{n\to\infty}\left[1+\frac{1}{\sqrt{n}}\right]^{\sqrt{n}}} = \frac{1}{e} < 1$$

$\Rightarrow$ **the given series is convergent by Cauchy's root test.**

58. $\sum_{1}^{\infty} e^{\sqrt{n}}\, x^n$

Solution: Let $\quad u_n = e^{\sqrt{n}}\, x^n$

$$\therefore \quad (u_n)^{1/n} = \left(e^{\sqrt{n}}\, x^n\right)^{1/n} = e^{1/\sqrt{n}} \cdot x$$

$$\therefore \quad \lim_{n\to\infty} (u_n)^{1/n} = x$$

$\therefore$ By Cauchy's root test,

Σu_n coverges if $0 < x < 1$

and diverges if $x > 1$

If $x = 1$, $u_n = e^{\sqrt{n}}$

$$\lim_{n\to\infty} u_n = \lim_{n\to\infty} e^{\sqrt{n}} = \infty \neq 0$$

$\therefore \quad \Sigma\, u_n$ diverges if $x = 1$.

59. $\sum_{1}^{\infty} \left(\frac{n+a}{n+b}\right)^n x^n$

Solution: Here $\quad u_n = \left(\frac{n+a}{n+b}\right)^n \cdot x^n$

$$\therefore \quad (u_n)^{1/n} = \frac{n+a}{n+b} \cdot x$$

$$\therefore \quad \lim_{n\to\infty} (u_n)^{1/n} = \lim_{n\to\infty} \frac{n+a}{n+b} \cdot x = \lim_{n\to\infty}\left[\frac{1+(1/n)}{1+(b/n)}\right] \cdot x = x$$

By root test Σu_n converges

if $0 < x < 1$ and diverges if $x > 1$

If $x = 1$, $u_n = \left(\frac{n+a}{n+b}\right)^n$

$$\lim_{n\to\infty} u_n = \lim_{n\to\infty}\left(\frac{1+a/n}{1+b/n}\right)^n = \frac{e^a}{e^b}\ e^{a-b} \neq 0$$

$\lim\limits_{n\to\infty} u_n$ **does not terd to zero**

$\therefore$ Σu_n **diverges if** $x = 1$

60. $\Sigma \frac{n!}{n^n}$

Solution: Let $u_n = \frac{n!}{n^n}$

$$\therefore \quad \lim_{n\to\infty} (u_n)^{1/n} = \lim_{n\to\infty}\left(\frac{n!}{n^n}\right)^{1/n}$$

Let $l = \lim\limits_{n\to\infty}\left(\frac{n!}{n^n}\right)^{1/n} = \lim\limits_{n\to\infty}\left(\frac{1}{n}\cdot\frac{2}{n}\cdots\frac{n}{n}\right)^{1/n}$

$$\therefore \quad \log l = \lim_{n\to\infty}\frac{1}{n}\left[\sum_{r=1}^{n}\log\left(\frac{r}{n}\right)\right]$$

$$= \int_0^1 \log x\,dx = [x\log x - x]_0^1 = -1$$

$\Rightarrow$ $l = e^{-1} < 1$

$\Rightarrow$ Σu_n is (C)

61. $\Sigma \frac{1}{(\log n)^n}$

Solution: Here $u_n = \frac{1}{(\log n)^n}$

$$(u_n)^{1/n} = \frac{1}{\log n}$$

$$\therefore \quad \lim_{n\to\infty} (u_n)^{1/n} = \frac{1}{\log n} = 0 < 1$$

By Cauchy's root test, Σu_n (C).

62. **Test for convergence** $\sum\limits_{n=1}^{\infty}\left(n\sqrt{n}-1\right)^n$

Solution: Here, $$u_n = \left(n\sqrt{n}-1\right)^n$$

$$\therefore \quad (u_n)^{1/n} = \left\{\left(n\sqrt{n}-1\right)^n\right\}^{1/n} = \left(n\sqrt{n}-1\right)$$

$$\therefore \quad \lim_{n\to\infty} (u_n)^{1/n} = \lim_{n\to\infty} (n^{1/n}-1) = 1-1 = 0 < 1$$

Hence by Cauchy's root test, Σu_n is convergent.

63. Find the nature of the series.

$$\sum_{n=1}^{\infty} a^{n^2}\cdot x^n, a<1$$

Solution: Here $$u_n = a^{n^2}\cdot x^n$$

$$\therefore \quad (u_n)^{1/n} = (a^{n^2})^{1/n}\,(x^n)^{1/n} = a^n \cdot x$$

$$\lim_{n\to\infty}(u_n)^{1/n} = \lim_{n\to\infty} a^n x$$

$$= 0 \qquad (\because a^n \to 0 \text{ as } n\to\infty, \sin a<1)$$

i.e., $$\lim_{n\to\infty}(u_n)^{1/n} = 0 < 1 \text{ so, } \to (C)$$

64. Test for convergence $\sum_{n=1}^{\infty}\left(\frac{n+1}{n}\right)^{n^2}\cdot\frac{1}{3^n}$

Solution: $$u_n = \left(\frac{n+1}{n}\right)^{n^2}\cdot\frac{1}{3^n} = \left(1+\frac{1}{n}\right)^{n^2}\cdot\frac{1}{3^n}$$

$$\therefore \quad (u_n)^{1/n} = \frac{\left\{\left(1+\frac{1}{n}\right)^{n^2}\right\}^{1/n}}{(3^n)^{1/n}} = \frac{(1+1/n)^n}{3}$$

$$\lim_{n\to\infty}(u_n)^{1/n} = \frac{e}{3} = \frac{2.718}{3} < 1,$$

$\therefore$ by Cauchy's root test Σu_n is (C).

EXERCISES

Test the convergence of

1. $\Sigma\frac{n^n}{2^n}$ [*Ans.* (C)]

2. $\sum e^{\sqrt{n}}x^n$; [*Ans.* (C) for $x<1$, (D) for $x>1$]

3. $\Sigma\left(\frac{n+2}{n+3}\right)^n \cdot x^n$; 4. $\Sigma \frac{5^n \cdot n!}{n^n}$ [*Ans.* (D)]

[*Ans.* [(C) for $x < 1$, (D) for $x \geq 1$]

5. $\Sigma\left(1+\frac{1}{\sqrt{n}}\right)^{-n^{3/2}}$ [*Ans.* (C)] 6. $\Sigma \frac{1}{\left(1+\frac{1}{n}\right)^{n^2}}$ [*Ans.* (C)]

7. $\Sigma \frac{1}{n\sqrt{\log n}}$ [*Ans.* (D)] 8. $\left(n\sqrt{n}-1\right)^n$ [*Ans.* (C)]

9. Show that the series $\Sigma \frac{\{(n+1)\,r\}^n}{n^{n+1}}$ is convergent if $r < 1$ and divergent if ($r \geq 1$).

10. Examine the convergence of the series

$$\Sigma \frac{(n+1)(n+2)\ldots(2n)}{n^n}$$ [*Ans.* (D)]

11. Discuss the convergence of $\Sigma(3/2)^n \cdot n^5$ [*Ans.* (D)]

12. Discuss the convergence of $\sum_{n=1}^{\infty}\left[\frac{(n+1)^n\, x^n}{n^{n+1}}\right]$ [*Ans.* (D)]

13. Discuss the convergence

$$\frac{1}{2}x + \left(\frac{2}{3}\right)^2 x^2 + \left(\frac{3}{4}\right)^3 + \ldots$$ [*Ans.* if $x < 1$, (C) if $x \geq 1$, (D)]

14. Discuss the convergence of $\sum_{n=1}^{\infty} \frac{(n+1)^n \cdot x^n}{n^{n+1}}$, $x > 0$ [*Ans.* $x = 1$, (D)]

15. Test the convergence of $\sum_{n=1}^{\infty}\left(1-\frac{3}{n}\right)^{n^2}$ [*Ans.* Σu_n is (C)]

16. $1+\frac{2}{3}x+\left(\frac{3}{4}\right)^2 x^2+\left(\frac{4}{5}\right)^3 x^3+\ldots\ x > 0$ [*Ans.* ($x < 1$) – (C) ($x \geq 1$) – (D)]

Cauchy's Integral test for series of positive terms

Statement: Let f be a non-negative and decreasing, integrable function defined on [1, ∞]. Also let

$$\sigma_n = \sum_{k=1}^{n} f(k) \quad \text{and} \quad I_n = \int_1^n f(x)\,dx$$

then $\lim_{n\to\infty} (\sigma_n - I_n)$ exists lies between 0 and $f(1)$.

Also the series $\Sigma f(n)$ convergence iff the sequence (I_n) converges

PART-D

Moreover in that case the sum of the series lies between

$$I = \lim_{n\to\infty} I_n \quad \text{and} \quad I + f(1).$$

Examples (65 – 70)

65. Show that the series $\sum_{n=1}^{n} \frac{1}{n^p}$, $p > 1$ convergence

and $\sigma = \sum_{n=1}^{n} \frac{1}{n^p}$ lies between $\frac{1}{p-1}$ and $\frac{p}{p-1}$

Solution: Compare $\sum_{n=1}^{n} \frac{1}{n^p}$ with $\sum_{n=1}^{\infty} f(n)$

$$\Rightarrow \qquad f(n) = \frac{1}{n^p} \quad \forall\ n \geq 1$$

Changing n to x; $\qquad f(x) = \frac{1}{x^p};\ \forall\ x \geq 1$ [1]

$f(n) \geq 0$ and is decreasing

$$I_n = \int_1^n f(x)\,dx = \int_1^n \frac{1}{x^p}\,dx = \int_1^n x^{-p}\,dx$$

$$= \left[\frac{x^{-p+1}}{-p+1}\right]_1^n \qquad (\because p > 1)$$

or

$$I_n = \frac{1}{1-p}\left[x^{1-p}\right]_1^n = \frac{1}{1-p}\left[n^{1-p} - 1\right]$$

$\therefore \quad p > 1, \quad \therefore \quad (1 - p) = (\text{–ve})$

$\therefore \qquad \lim n^{(1-p)} = 0$

Hence the sequence $\{I_n\}$ converges to

$$\frac{1}{1-p}[0 - 1] = \frac{1}{p-1} = I$$

$\therefore$ By Integral test, the series is convergent and the sum σ of the series lies between I and $I + f(1)$

i.e., between $\frac{1}{p-1}$ and $\frac{1}{p-1} + 1 = \frac{1}{p-1}$

$[x = 1 \Rightarrow f(1) = 1]$

66. Show that $\lim_{n\to\infty}\left(1 + \frac{1}{2} + \frac{1}{3} + \ldots + \frac{1}{n} - \log n\right)$

exists and lies between 0 and 1

Solution: Let $f(n) = \frac{1}{n}$

Change n to x; $f(x) = \frac{1}{x}$;

$\therefore$ $$\sigma_n = \sum_{k=1}^{n} f(k) = 1 + \frac{1}{2} + \frac{1}{3} + \ldots \frac{1}{n}$$

and $$I_n = \int_1^n f(x)\,dx = \int_1^n \frac{1}{x}\,dx = (\log x)_1^n = \log n - \log 1 = \log n.$$

then, by Integral test,

$$\lim_{n\to\infty} \left(1 + \frac{1}{2} + \frac{1}{3} + \ldots + \frac{1}{n} - \log n\right)$$

$$= \lim_{n\to\infty} (\sigma_n - I_n) \text{ exists}$$

and lies between 0 and $f(1)$; i.e., between 0 and 1.

67. Using the integral test discuss the convergence of the series $\sum \frac{1}{n^2+1}$

Solution: Let $a_n = \frac{1}{n^2+1}$

Consider the function $f(x) = \frac{1}{n^2+1}$

So that $f(x) = a_n$

Clearly $f(x)$ is non-negative and decreasing on $[1, \infty]$

and $$I_n = \int_1^n \frac{dx}{x^2+1} = [\tan^{-1}]_1^n = \tan^{-1}(n) - \tan^{-1}(1)$$

$$\lim_{n\to\infty} I_n = \lim_{n\to\infty} [\tan^{-1}(n) - \tan^{-1}(1)]$$

$$= \frac{\pi}{2} - \frac{\pi}{4} = \frac{\pi}{4}$$

$\therefore$ I_n is convergent

Further $f(1) = 1$; and $I + f(1) = \frac{\pi}{4} + \frac{1}{2} = \left(\frac{\pi+2}{4}\right)$

$\therefore$ By Cauchy's integral test, the given series is convergent and its sum lies between $\frac{\pi}{4}$ and $\left(\frac{\pi}{4} + \frac{1}{2}\right)$

PART-D

68. Discuss the convergence of the series

$$\sum_{n=2}^{\infty} \frac{1}{n(\log n)^{\alpha}}, \text{ where } \alpha \geq 0.$$

Solution: Let
$$a_n = \frac{1}{n(\log n)^{\alpha}}, \text{ where } \alpha \geq 0,\ n \geq 2$$

Consider the function $f(x) = \dfrac{1}{x(\log x)^{\alpha}}$ so that $f(x) = a_n$

Clearly $f(x) \geq 0$ and decreasing on $[2, \infty]$

Case (i) If $\alpha \neq 1$
$$I_n = \int_2^n \frac{dx}{x(\log x)^{\alpha}}$$

put $(\log x = t)$ and Integrate we get

$$= \left[\frac{1}{1-\alpha}(\log x)^{1-\alpha}\right]_2^n = \frac{(\log n)^{1-\alpha}}{1-\alpha} - \frac{(\log 2)^{1-\alpha}}{1-\alpha}$$

$\{I_n\}$ converges if $\alpha > 1$ and diverges if $\alpha < 1$

Hence by cauchy's Integral test, the given series converges if $\alpha > 1$ and diverges if $\alpha < 1$

Case (ii) Let $\alpha = 1$

Hence
$$I_n = \int_2^n \frac{dx}{x \log x}$$

$$\therefore \quad I_n = [\log(\log x)]_2^n = [\log(\log n) - \log(\log 2)] \to \infty, \text{ as } n \to \infty$$

$\therefore$ $\{I_n\}$ diverges and hence the given series diverges.

69. Using the integral test discuss the convergence of

$$\sum \frac{1}{n(n+1)}$$

Solution: Let
$$a_n = \frac{1}{n(n+1)} = f(n)$$

Change n to x; $f(x) = \dfrac{1}{x(x+1)}$

$f(x)$ is positive and decreasing, $\forall\ x \geq 1$

$$\therefore \quad I_n = \int_1^n f(x)\,dx = \int_1^n \frac{dx}{x(x+1)}$$

$$= \int_1^n \frac{(x+1)-x}{x(x+1)}\,dx = \int_0^1 \left[\frac{1}{x} - \frac{1}{(x+1)}\right] dx$$

$$= [\log x - \log (x+1)]_1^n = \left[\log \frac{x}{x+1}\right]_1^n$$

$$= \log \frac{n}{n+1} - \log \frac{1}{2}$$

$$\therefore \qquad I = \lim_{n\to\infty} I_n = \lim_{n\to\infty} \log \frac{n}{n+1} - \log \frac{1}{2}$$

$$= \lim_{n\to\infty} \log \frac{n}{1+\frac{1}{n}} - \log \frac{1}{2} = \log 1 - \log \frac{1}{2} = \log 2 \text{ (finite)}$$

$\therefore$ $\{I_n\}$ is convergent

$\therefore$ By cauchy's integral test the series is also convergent

70. Discuss the convergence of the series $\sum ne^{-n^2}$

Solution: Let $a_n = ne^{-n^2} = f(n)$

$$f(n) = ne^{-n^2}$$

Change n to x; we get

$$\therefore \qquad f(x) = xe^{-x^2}$$

Clearly $f(n)$ is non-negative and monotonic decreasing on $[1, \infty]$

Also $$I_n = \int_1^n xe^{-x^2}\,dx = \frac{1}{2}\left[e^{-1} - e^{-n^2}\right]$$

$$\therefore \qquad I_n \to \frac{1}{2} e^{-1} \text{ as } n \to \infty,$$

$$\Rightarrow \qquad I = \frac{1}{2} e^{-1} \quad \text{and} \quad f(1) = e^{-1};$$

$$\therefore \qquad I + f(1) = \frac{3}{2} e^{-1}$$

$\therefore$ By Cauchy's Integral test, the series Σa_n is (C) and its sum lies between $\frac{1}{2} e^{-1}$ and $\frac{3}{2} e^{-1}$

Test the convergence of the following series using Cauchy's integral test.

1. $\Sigma \dfrac{n}{n^2+4}$
2. $\sum\limits_{n=2}^{\infty} \dfrac{1}{n(\log n)^2}$
3. $\Sigma\, n^2 e^{-n^3}$
4. $\Sigma \dfrac{1}{n^2+1}$

5. $\dfrac{(2/1)}{4}+\dfrac{(3/2)^2}{4^2}+\dfrac{(4/3)^3}{4^3}+\dfrac{(3/4)^4}{4^4}+\ldots$ [convergent (C)

6. $\Sigma\left(1+\dfrac{3}{n}\right)^{-n^2}$ [*Ans.* (C)]

7. $\sum\limits_{3}^{\infty}\dfrac{1}{n\log n\,(\log\log n)^2}$

8. $\sum\limits_{1}^{\infty}\dfrac{n^4}{2n^5+3}$;

9. $\sum\limits_{1}^{\infty}\dfrac{1}{\sqrt{n^2-1}}$

10. $\sum\limits_{1}^{\infty}\dfrac{1}{(n+1)^2}$

4.10 ALTERNATING SERIES, ABSOLUTE AND CONDITIONAL CONVERGENCE, LEIBNITZ'S TEST *(without proof)*

Definition (Alternating series)

A series in which the terms are alternatively positive and negative is called an alternating series.

In symbols $\Sigma\,(-1)^{n-1}\cdot a_n = a_1 - a_2 + a_3 - a_4 + \ldots..$

is an alternating series.

For example, $\sum\limits_{n=1}^{\infty}(-1)^{n-1}\cdot\dfrac{1}{n}=1-\dfrac{1}{2}+\dfrac{1}{3}-\dfrac{1}{4}+\ldots$

is an alternating series.

Definition (2): (Absolute convergence)

A Series Σa_n containing positive and negative terms is said to be absolutely convergent if the series obtained by making all the terms positive, is convergent. That is Σa_n is absolutely convergent if $\Sigma\,|\,a_n\,|$ is convergent.

OR

Let $\Sigma a_n = a_1 + a_2 + \ldots + a_n + \ldots$ [1]

and let $\alpha_n = |\,a_n\,|$

$$\Sigma\,\alpha_n = \alpha_1 + \alpha_2 + \ldots + \alpha_n + \ldots$$
$$= |\,a_1\,| + |\,a_2\,| + \ldots + |\,a_n\,| + \ldots \quad [2]$$

(*i.e*, the series (1) with all its terms taken positively)

If series [2] is convergent, the series [1] is said to be absolutely convergent.

For example,

Consider the series

$$1-\left(\frac{x^2}{2^2}\right)+\left(\frac{x^4}{2^2\cdot 4^2}\right)-\left(\frac{x^6}{2^2\cdot 4^2\cdot 6^2}\right)+\ldots \quad [1]$$

The series has only even powers of x and so is the same for positive or negative values of x

Again, consider the series got by rendering all the terms positive.

$$1+\frac{x^2}{2^2}+\frac{x^4}{2^2\cdot 4^2}+\frac{x^6}{2^2\cdot 4^2\cdot 6^2}+\ldots \qquad [2]$$

$$a_n=\frac{x^{2(n-1)}}{2^2\cdot 4^2\ldots 2^2(n-1)^2}$$

and $$a_{n+1}=\frac{x^{2n}}{2^2\cdot 4^2\ldots 2^2(n-1)^2\cdot(2n)^2}$$

$$\therefore \quad \frac{a_{n+1}}{a_n}=\frac{x^2}{(2n)^2}=\frac{x^2}{4n^2}\to 0, \forall x, \text{ as } n\to\infty$$

$$\therefore \quad \lim_{n\to\infty}\frac{a_{n+1}}{a_n}=0<1$$

Hence by Ratio test,

The series is convergent for all values of x (finite)

Since the "positivised" series itself is convergent, the original series is absolutely convergent. for all values x.

DEFINITION: (3) (CONDITIONAL CONVERGENCE)

The series Σa_n containing positive and negative terms is said to be conditionally convergent or se miconvergent.

if (i) Σa_n is convergent while

(ii) $\Sigma \mid a_n \mid$ is not convergent (Divergent)

For example, consider the series $1-\frac{1}{2}+\frac{1}{3}-\frac{1}{4}+\ldots$ to ∞ is convergent where as

$$\Sigma\left|\frac{(-1)^{n-1}}{n^2}\right| \text{ is divergent}$$

hence $1-\frac{1}{2}+\frac{1}{3}-\ldots$ is conditionally convergent.

Generalised D' Alembert's Ratio test for the Absolute Convergence of Series:

(Statement)

If Σa_n is a series containing both positive and negative terms arbitrarily such that $\lim_{n\to\infty}\left|\frac{a_{n+1}}{a_n}\right|$ $= l$, then Σa_n is absolutely convergent if $l<1$ and is not absolutely convergent if $l>1$.

PART-D

LEIBNITZ'S TEST

The alternating series $\Sigma(-1)^{n-1} \cdot a_n = a_1 - a_2 + a_3 - a_4 + - \ldots \infty$ is convergent if

(*i*) $a_{n+1} < a_n \; \forall_n$

(monotonic deceasing)

and (*ii*) $\lim_{n\to\infty} a_n = 0$

Note: (1) If in the above test $\lim_{n\to\infty} a_n \neq 0$, then the series $\sum_1^{\infty}(-)^{n+1} a_n$ is an oscillating series.

Note: (2) For the convergence of the alternating series $\Sigma(-1)^{n-1} \cdot a_n$, both the conditions (i) and (ii) are necessary, if any one of the above conditions fail, the series is not convergent.

For example, consider the series $\frac{3}{2} - \frac{5}{4} + \frac{9}{8} - \frac{17}{16} + \ldots$

$$a_n = 1 + \frac{1}{2^n} \text{ and } \lim_{n\to\infty} a_n \neq 0$$

$\therefore$ the series is not convergent.

Worked examples

71. Discuss the convergence of the alternating series:

$$\frac{1}{6} - \frac{1}{13} + \frac{1}{20} - \frac{1}{27} + \ldots$$ [VTU, August, 2001]

Solution: Given: $\Sigma(-1)^{n-1} \cdot a_n$;

(This is an alternating series)

where

$$a_n = \frac{1}{6 + (n-1)7} = \frac{1}{7n-1};$$

$$a_{n+1} = \frac{1}{7(n+1)-1} = \frac{1}{7n+6}$$

Now, $7n + 6 > 7n - 1$

So that $\frac{1}{7n+6} < \frac{1}{7n-1}$

$\Rightarrow$ $a_{n+1} < a_n \; \forall_n$. and $\lim_{n\to\infty} a_n = \lim_{n\to\infty} \frac{1}{7n-1} = 0$

$\therefore$ Both the conditions for the convergence of an alternating series are satisfied.

$\therefore$ The given series is convergent.

72. Show that the series

$$1 + \frac{1}{2^2} - \frac{1}{3^2} - \frac{1}{4^2} + \frac{1}{5^2} + \frac{1}{6^2} - \frac{1}{7^2} - \frac{1}{8^2} + \ldots \text{ is convergent}$$

[VTU, March, 2000]

Solution: This positinised series of given series is :

$$1+\frac{1}{2^2}-\frac{1}{3^2}-\frac{1}{4^2}+\frac{1}{5^2}+\frac{1}{6^2}-\frac{1}{7^2}-\frac{1}{8^2}+\ldots$$ whose general term is $u_n=\frac{1}{n^2}$, $S\,u_n = S\,\frac{1}{n^2}$ is a harmonic series ($p = 2 > 1$) and it is convergent.

Thus Le find that the given series is absolutely convergent and so it is convergert.

This ⇒ that given alternating series is convergent.

73. Test the series

$$1-\frac{1}{2}-\frac{1\cdot 3}{2\cdot 4}-\frac{1\cdot 3\cdot 5}{2\cdot 4\cdot 6}+\ldots$$

For (*a*) absolute convergence

(*b*) conditional converogence [VTU, March, 2001]

Let, $$u_n = (-1)^{n-1}\frac{1\cdot 3\cdot 5\ldots(2n-1)}{2\cdot 4\cdot 6\ldots 2n} = (-1)^{n-1}\cdot a_n$$

Where $$a_n = \frac{1\cdot 3\cdot 5\ldots(2n-1)}{2\cdot 4\cdot 6\ldots 2n}$$

Now $$|u_n| = a_n$$

Hence $$a_{n+1} = \frac{1\cdot 3\cdot 5\ldots(2n-1)(2n+1)}{2\cdot 4\cdot 6\ldots 2n\,(2n+2)}$$

$$\therefore \quad \lim_{n\to\infty}\frac{a_{n+1}}{a_n} = 1$$

⇒ Ratio test fails.

Consider $$\lim_{n\to\infty} n\left\{\frac{a_n}{a_{n+1}}-1\right\} = \lim_{n\to\infty} n\left\{\frac{2n+2}{2n+1}-1\right\} = \lim_{n\to\infty}\frac{n}{2n+1}=\frac{1}{2}<1$$

∴ by Raabe's test $\Sigma\,|u_n|$ or Σa_n is divergent.

To test alternating series for convergence.

We have $$\frac{a_{n+1}}{a_n} = \frac{2n+1}{2n+2}<1$$

$$\Rightarrow \quad a_{n+1} < a_n \;\forall\, n$$

(monotonic decreasing), Now, we shall P.T.

$$\lim_{n\to\infty} a_n = 0$$

Since $x^2 > x^2 - 1 \;\forall x$ *i.e.,* $\frac{x}{x+1} > \frac{x-1}{x}$

Let $x = 2, 4. 6, \ldots 2n$, we get

PART-D

$$\frac{2}{3} > \frac{1}{2},$$

$$\frac{4}{5} > \frac{3}{4}$$

$$\frac{6}{7} > \frac{5}{6} \dots \frac{2n}{2n+1} > \frac{2n-1}{2n}$$

Multiplying these inequalities, we get,

$$\frac{2 \cdot 4 \cdot 6 \dots 2n}{3 \cdot 5 \cdot 7 \dots (2n-1)(2n+1)} > \frac{1 \cdot 3 \cdot 5 \dots (2n-1)}{2 \cdot 4 \cdot 6 \dots 2n}$$

$$\Rightarrow \qquad \frac{1}{2n+1} > \frac{1^2 \cdot 3^2 \cdot 5^2 \dots (2n-1)^2}{2^2 \cdot 4^2 \cdot 6^2 \dots (2n)^2} = a_n^2$$

$$\Rightarrow \qquad a_n < \frac{1}{\sqrt{2n+1}}$$

as $n \to \infty$, $a_n \to 0$

$$\therefore \qquad \lim_{n\to\infty} a_n = 0$$

$\therefore$ **By Leibnitz's Test, Σu_n is (C),**

Since it is not absolutely convergent, Σu_n is conditionally convergent.

74. Find the nature of the series

$$x - \frac{x^2}{4} + \frac{x^3}{9} - \frac{x^4}{16} + \dots$$

Solution: Let $\quad u_n = \dfrac{x^n}{n^2}(-1)^{n-1}$

Consider $\quad \left|\dfrac{u_{n+1}}{u_n}\right| = \left|(-1)^n \cdot \dfrac{x^{n+1}}{(n+1)^2} \cdot \dfrac{n^2}{(-1)^{n-1} \cdot x^n}\right| = \dfrac{n^2}{(n+1)^2} \cdot |\, x \,|$

$$\therefore \qquad \lim_{n\to\infty}\left|\frac{u_{n+1}}{u_n}\right| = \lim_{n\to\infty} \frac{n^2}{(n+1)^2} \cdot |x|$$

$$= \lim_{n\to\infty} \frac{1}{\left(1+\frac{1}{n}\right)^2} \cdot |x| = |\, x \,|$$

If $|\, x \,| < 1$, the series is absolutely convergent,

$\Rightarrow \Sigma u_n$ is convergent, if $|\, x \,| < 1$

If $|\, x \,| = 1$, two cases arise.

(*i*) $x = 1$ and

(*ii*) $x = -1$

When $x = 1$, $\Sigma u_n = \Sigma(-1)^{n-1} \cdot \dfrac{1}{n^2}$

$$\Rightarrow \quad \Sigma u_n = 1 - \frac{1}{2^2} + \frac{1}{3^2} - \frac{1}{4^2} + \ldots$$

Consider the "positivised series"

$$\Sigma \mid u_n \mid \; 1 + \frac{1}{2^2} + \frac{1}{3^2} + \frac{1}{4^2} + \ldots$$

Which is convergent $= \Sigma \dfrac{1}{n^2}$

Since ($p = 2 > 1$)

This Σu_n is absolutely convergent and hence convergent.

If $x = -1$, then $\Sigma u_n = \Sigma(-1)^{n-1} \cdot \dfrac{(-1)^n}{n^2}$

$$\Sigma u_n = \Sigma(-1)^{2n-1} \cdot \frac{1}{n^2}$$

or, $$\Sigma u_n = -\left(1 + \frac{1}{2^2} + \frac{1}{3^2} + \frac{1}{4^2} + \ldots\right)$$

which is again convergent.

$\Rightarrow$ the series is convergent for $-1 \le x \le 1$.

75. Test the series

$$1 - \frac{1}{2\sqrt{2}} + \frac{1}{3\sqrt{3}} - \frac{1}{4\sqrt{4}} + \ldots$$

(*i*) absolute convergence

(*ii*) conditional convergence [VTU, July/August, 2002]

Solution: Here $a_n = \dfrac{1}{n\sqrt{n}} = \dfrac{1}{n^{3/2}}$

$$\therefore \quad a_{n+1} = \frac{1}{(n+1)\sqrt{n+1}} = \frac{1}{(n+1)^{3/2}}$$

$$\therefore \quad a_n - a_{n+1} = \frac{1}{n^{3/2}} - \frac{1}{(n+1)^{3/2}}$$

$$= \frac{(n+1)^{3/2} - n^{3/2}}{[n(n+1)]^{3/2}} > 0 \qquad (\because (n+1) > n)$$

i.e., $a_n - a_{n+1} > 0$

$\Rightarrow$ $a_n > a_{n+1}$

PART-D

and $$\lim_{n\to\infty} a_n = \lim_{n\to\infty} \frac{1}{n\sqrt{n}} = 0$$

$\Rightarrow$ The series Σa_n is convergent by Leibnitz's Test.

(*i*) The "positivised" series (absolute series)

is $$1+\frac{1}{2\sqrt{2}}+\frac{1}{3\sqrt{3}}+\frac{1}{4\sqrt{4}}+\ldots$$

$$\therefore \quad a_n = \frac{1}{n\sqrt{n}} = \frac{1}{n^{3/2}}$$

$$\Sigma a_n = \Sigma\frac{1}{n^{2/3}} = \Sigma\frac{1}{n^p}$$

is a *p*-series and hence $\left(P = \frac{3}{2} > 1\right)$ is convergent.

$\Rightarrow$ the alternating series is absolutely convergent.

76. Test the series

$$1-\frac{1}{5}+\frac{1}{9}-\frac{1}{13}+\ldots \text{ for}$$

(*a*) convergence

(*b*) absolute convergence

(*c*) conditional convergence

Solution: (*a*) Here $$a_n = \frac{1}{(4n-3)}$$

($\because$ 1, 5, 9, . . . gives $1 + (n - 1)\,4 = 4n - 3$). (G.T).

$$\therefore \quad a_{n+1} = \frac{1}{4(n+1)-3} = \frac{1}{4n+1}$$

$$\therefore \quad a_n - a_{n+1} = \frac{1}{4n-3} - \frac{1}{4n+1} = \frac{(4n+1)-(4n-3)}{(4n-3)(4n+1)}$$

i.e., $$a_n - a_{n+1} = \frac{4}{(4n-3)(4n+1)} > 0$$

$$\Rightarrow \quad a_n > a_{n+1}$$

$$\lim_{n\to\infty} a_n = \lim_{n\to\infty}\frac{1}{(4n-3)} = \lim_{n\to\infty}\frac{1}{n}\cdot\frac{1}{\left(4-\frac{3}{n}\right)} = 0$$

$\therefore$ the given series is convergent by Leibnitz's Test.

(*b*) The absolute series (positivised series) is given by $1 + \frac{1}{5}+\frac{1}{9}+\frac{1}{13}+\ldots$

$$\therefore \qquad a_n = \frac{1}{4n-3}$$

$$\text{Let} \qquad b_n = \frac{1}{n}$$

$$\therefore \qquad \lim_{n\to\infty} \frac{a_n}{b_n} = \lim_{n\to\infty} \frac{1}{4n-3} \cdot \frac{n}{1} = \lim_{n\to\infty} \frac{1}{\left(4 - \frac{3}{n}\right)} = \frac{1}{4} = \text{finite} \neq 0$$

But $\Sigma b_n = \Sigma \frac{1}{n}$ ($p = 1$) is divergent and hence by comparison test, Σa_n is also divergent.

$\Rightarrow$ the given series is not absolutely convergent.

(*c*) Since the given alternating series is convergent and the absolute series is divergent.

$\Rightarrow$ the alternating series is conditionally convergent.

77. Show that the series

$$\left\{\left[\frac{1}{\sqrt{2}}\right] - \left[\frac{1}{\sqrt{3}}\right] + \left[\frac{1}{\sqrt{4}}\right] - \left[\frac{1}{\sqrt{5}}\right] + \ldots\right\}$$

is convergent, but not absolutely convergent.

Solution: The series is alternating; and the general term is

$$\left[\frac{1}{\sqrt{n+1}}\right]$$

$[n + 1]$ increases to ∞. $\sqrt{n+1}$ increases to ∞.

$$\therefore \qquad \left[\frac{1}{\sqrt{n+1}}\right] \text{ decreases to zero.}$$

Hence, by the Leibnitz's Test, the series is convergent.

Again consider the series got by rendering all the terms we get

$$\left[\frac{1}{\sqrt{2}}\right] + \left[\frac{1}{\sqrt{3}}\right] + \left[\frac{1}{\sqrt{4}}\right] + \ldots$$

This is $\Sigma\left[\frac{1}{n^{1/2}}\right]$ which is divergent since $p = [1/2] < 1$.

Hence, the given series is convergent, but the positivised series is divergent. So the given series is convergent but not absolutely convergent.

78. Examine for convergence the series $\left[\frac{(-1)^n}{n + \sqrt{n}}\right]$

PART-D

Solution: This is the same as the alternating series.

$$\Sigma[(-1)^{n-1} \cdot a_n],$$

where $$a_n = \left\{\frac{1}{\left[n+\sqrt{n}\right]}\right\}$$

Let $$b_n = (n + \sqrt{n});$$

$$b_{n+1} = \left[(n+1)+\sqrt{(n+1)}\right]$$

$\therefore$ $$(b_{n+1} - b_n) = [(n + 1) - n] + \left[\sqrt{(n+1)} - \sqrt{n}\right] = \text{positive for all } x$$

Hence (b_n) is monotonic increasing $\therefore$ $(n + \sqrt{n})$ inceasing to ∞. Hence $a_n = (1/b_n)$ decrease to zero.

Thus, by Leibnitz's Rule, the series is convergent.

79. $\Sigma\left[\dfrac{(-1)^n}{(1+\log n)}\right]$

This is an alternating series.

$$\Sigma (-1)^n \cdot a_n,$$

where $$a_n = \frac{1}{(1+\log n)}$$

$\therefore$ $$a_{n+1} = \frac{1}{1+\log(n+1)}$$

(since log (n) increases to ∞)

$\therefore$ $\dfrac{1}{\log n}$ decrease to 0. *i.e.*, (a_n) diverges to zero.

So, the given series is convergent, by the Liebnitz's Rule.

80. $\Sigma\left[\dfrac{(-1)^n \cdot n}{(2n+3)}\right]$

Solution: Writing ths given series in a component form (term wise), we have

$$\Sigma\frac{(-1)^n \cdot n}{(2n+3)} = \frac{(-1)\cdot 1}{2+3} + \frac{(-1)^2 \cdot 2}{4+3} + \frac{(-1)^3 \cdot 3}{6+3} + \frac{(-1)^4 \cdot 4}{8+3} +$$

$$= -\left[\frac{1}{5}\right]+\left[\frac{2}{7}\right]-\left[\frac{3}{9}\right]+\left[\frac{4}{11}\right]+\ldots$$

The series is alternating and the general term

$$\frac{n}{2n+3} \to \frac{1}{2};$$

it does not $\to 0$. Hence the series caonnot converge.

The series oscillates.

81. Examine the nature of the series

$$1 - \frac{1}{2^p} + \frac{1}{3^p} - \frac{1}{4^p} + \ldots$$

Solution: We can examine the series

When

(*i*) $p > 0$; (*ii*) $p = 0$; (*iii*) $p < 0$

(*i*) $p > 0$; $\therefore n^p \to \infty$

(monotonically increases to ∞)

So $\left(\frac{1}{n^p}\right) \to 0$ and the series is alternating.

So, the series is convergent by the Liebnitz's Test.

(*ii*) $p = 0$; $n^0 = 1$; so the series is $1 - 1 + 1 - 1 + \ldots$

The series oscillates between 0 and 1.

(*iii*) $p < 0$; $n^p \to 0$; $\left[\frac{1}{n^p}\right] \to \infty$.

So, the series oscillates infinitely. Thus, the series is convergent, when $p > 0$.

82. $\Sigma\left[\frac{(-1)^{n-1} \cdot n}{n^2+1}\right]$

Here, the series is alternating. Also the term

$$a_n = \frac{n}{(n^2+1)} = \frac{1}{\frac{(n+1)}{n}}$$

monotonically decreases for ($n > 1$) and tends to 0, as $n \to \infty$.

Hence the given series is convergent by Leibnitz's Test.

Again, $$\Sigma \mid a_n \mid = \Sigma\left|(-1)^{n-1} \cdot \frac{n}{n^2+1}\right| = \Sigma\frac{n}{n^2+1}$$

$$a_n = \frac{n}{n^2+1} ; \text{ Let } b_n = \frac{n}{n^2} = \frac{1}{n}$$

$$\therefore \quad \left(\frac{a_n}{b_n}\right) = \frac{n}{n^2+1} \cdot \frac{n}{1} = \frac{n^2}{n^2+1} = \left(\frac{1}{1+\frac{1}{n^2}}\right)$$

$$\therefore \quad \lim_{n\to\infty}\frac{a_n}{b_n} = \lim_{n\to\infty}\left(\frac{1}{1+\frac{1}{n^2}}\right) = 1 \neq 0$$

$\therefore$ **by comparison test, both Σa_n and Σb_n converge or diverge together.**

But $\Sigma b_n = \Sigma\frac{1}{n}$ is divergent

Since $(p = 1)$

$\Rightarrow$ **Σa_n is also convergent.**

Thus, $\Sigma(-1)^{n-1}\, a_n$ is (C)

and $\Sigma\,|\,(-1)^{n-1}\cdot a_n\,|$ is (D)

$\Rightarrow$ **The given series is conditionally convergent.**

83. Find the nature of the series:

$$\left[1-\frac{1}{\log 2}\right]-\left[1-\frac{1}{\log 3}\right]+\left[1-\frac{1}{\log 4}\right]-\left[1-\frac{1}{\log 5}\right]+\ldots$$

Solution: Here $\quad a_n = 1-\dfrac{1}{\log(n+1)}$

$$\therefore \quad a_{n+1} = 1-\frac{1}{\log(n+2)}$$

$$a_n - a_{n+1} = \frac{1}{\log(n+2)} - \frac{1}{\log(n+1)}$$

$$= \frac{\log(n+1)-\log(n+2)}{\log(n+2)\cdot\log(n+1)} < 0 \; [\because (n+1) < (n+2)]$$

$$\Rightarrow \quad a_n < a_{n+1}$$

Further $\quad \lim\limits_{n\to\infty} a_n = \lim\limits_{n\to\infty}\left[1-\dfrac{1}{\log(n+1)}\right] = 1 - 0 = 1 \neq 0$

Thus, both the conditions of Leibnitz's Test are not satisfied. We conclude that the series oscillates between $-\infty$ and $+\infty$.

84. Prove that

$$\sum_{1}^{\infty}\frac{(-1)^{n+1}\cdot n}{4n-3} \text{ oscillates}$$

Solution: Let $\quad a_n = \dfrac{n}{(4n+3)};$

$$a_{n+1} = \frac{n+1}{(4n+1)}$$

Clearly $a_n > a_{n+1}\ \forall n.$

Hence $\{a_n\}$ is a monotomic decreasing sequence

Also $$\lim_{n\to\infty} a_n = \lim_{n\to\infty} \frac{n}{4n-3} = \frac{1}{4} \neq 0$$

$\therefore$ the given series oscillates.

85. Discuss the convergence of

$$\sum_{n=0}^{\infty} \frac{(-1)^{n-1}}{\log(n+1)}$$

Solution: Let $$a_n = \frac{1}{\log(n+1)}$$

The given series is an alternating series

$$\log(n+2) > \log(n+1)\ \forall_n > 0$$

$$\Rightarrow \quad \frac{1}{\log(n+2)} < \frac{1}{\log(n+1)}$$

$$\Rightarrow \quad a_{n+1} < a_n\ a_n\ \forall_n > 0$$

$\therefore$ a_n is monotonic decreasing.

$$\lim_{n\to\infty} a_n = \lim_{n\to\infty} \frac{1}{\log(n+1)} = 0$$

Hence, by Leibnitz's Test the given series is convergent.

86. $$\frac{x}{1+x} - \frac{x^2}{1+x^2} + \frac{x^3}{1+x^3} - \frac{x^4}{1+x^4} + \ldots$$

$(0 < x < 1)$

Solution: The given series is an alternating series.

$$a_n = \frac{x^n}{1+x^n};\quad a_{n+1} = \frac{x^{n+1}}{1+x^{n+1}}$$

$$\therefore \quad a_n - a_{n+1} = \frac{x^n}{x^{n+1}} - \frac{x^{n+1}}{1+x^{n+1}}$$

$$= x^n\left[\frac{1}{1+x^n} - \frac{n}{1+x^{n+1}}\right] = \frac{x^n(1-x)}{(1+x^n)(1+x^{n+1})}$$

Since $x < 1$, and is positive

$$a_n - a_{n+1} = \text{a positive}$$

$$a_{n+1} < a_n$$

$$\therefore \quad \lim_{n\to\infty} a_n = \lim_{n\to\infty} \frac{x^n}{1+x^n} = 0\ (x^n \to 0 \text{ as } n \to \infty)$$

PART-D

Since $(x < 1)$,

$\therefore$ The series is (C).

87. $1 - \dfrac{1}{2^p} + \dfrac{1}{3^p} - \dfrac{1}{4^p} + \ldots\ p > 0$

Solution: Since $\qquad \Sigma a_n = \Sigma \dfrac{(-1)^{n-1}}{n^p} = \Sigma(-1)^{n-1} \cdot a_n$

Here $\qquad a_n = \dfrac{1}{n^p}$ and $a_{n+1} = \dfrac{1}{(n+1)^p}$

Since $\qquad (n+1)^p > n^p$,

we have $\qquad \dfrac{1}{(n+1)\,p} < \dfrac{1}{n^p}$

i.e., $\qquad a_{n+1} < a_n \ \forall_n$

and $\qquad \lim\limits_{n\to\infty} a_n = \lim\limits_{n\to\infty} \dfrac{1}{n^p} = 0$

Since $p > 0$

$\therefore$ By Leibnitz's Test the given series is (C). The positivised series is

$$\Sigma \mid a_n \mid = \Sigma \frac{1}{n^p}$$

is convergent

(Since if $p > 1$)

and (D) if $p \leq 1$.

The given series is absolutely convergent if $p > 1$ and conditionaly convergent if $0 < p < 1$.

88. $\dfrac{1}{\sqrt{1+1^2}} - \dfrac{1}{\sqrt{1+2^2}} + \dfrac{1}{\sqrt{1+3^2}} + \ldots$ to ∞

Solution: Here, $\qquad a_n = \dfrac{1}{\sqrt{1+n^2}}$;

$$a_{n+1} = \frac{1}{\sqrt{1+(n+1)^2}}$$

We know that $\qquad (n+1)^2 > n^2 \ \forall, n$

$\therefore \qquad 1 + (n+1)^2 > 1 + n^2$

$\therefore \qquad \sqrt{1+(n+1)^2} > \sqrt{1+n^2}$

$\therefore \qquad \dfrac{1}{\sqrt{1+(n+1)^2}} < \dfrac{1}{\sqrt{1+n^2}} \ \forall_n$

$\therefore \qquad a_{n+1} < a_n \ \forall_n$

$$\lim_{n\to\infty} a_n = \lim_{n\to\infty} \frac{1}{\sqrt{1+n^2}} = 0$$

$\therefore$ **both conditions of Leibnitz's Test are satisfied.**

$\therefore$ **The given alternating series converge.**

89. $\frac{1}{2^3} - \frac{1}{3^3}(1+2) + \frac{1}{4^3}(1+2+3) - \frac{1}{5^3}(1+2+3+4) + \text{to } \infty$

Solution:

$$a_n = \frac{1}{(n+1)^3}(1+2+3+\ldots n)$$

$$a_n = \frac{1}{(n+1)^3}\cdot\frac{n(n+1)}{2} = \frac{n}{2(n+1)^2}$$

$$a_{n+1} = \frac{n+1}{2(n+2)^2}$$

Consider, $$a_{n+1} - a_n = \frac{(n+1)}{2(n+1)^2} - \frac{n}{2(n+1)^2} = \frac{(n+1)(n+1)^2 - n(n+2)^2}{2(n+2)^2(n+1)^2}$$

$$= \frac{-n^2-4n+1}{2(n+2)^2(n+1)^2} = \frac{-(n^2+4n+1)}{2(n+2)^2(n+1)^2} < 0$$

$\therefore \qquad a_{n+1} < a_n \ \forall_n$

$$\lim_{n\to\infty} a_n = \lim_{n\to\infty} \frac{n}{2(2n+1)^2} = \lim_{n\to\infty} \frac{n}{2n^2\left(1+\frac{1}{a}\right)^2}$$

$$= \lim_{n\to\infty} \frac{1}{2n^2\left(1+\frac{1}{n}\right)^2} = 0. \text{ Hence } (C).$$

EXERCISES

I. Test the convergence of the following series

1. $1 - \frac{1}{5} + \frac{1}{9} - \frac{1}{13} + \ldots \text{to } \infty$ [*Ans.* (C)]

2. $1 - \frac{1}{\sqrt[3]{2}} + \frac{1}{\sqrt[3]{3}} - \frac{1}{\sqrt[3]{4}} + \text{to } \infty$ [*Ans.* (C)]

3. $1 - \frac{1}{8} + \frac{1}{27} - \frac{1}{64} + \text{to } \infty$ [*Ans.* (C)]

PART-D

4. $1-\dfrac{1}{2^2\cdot\sqrt{2}}+\dfrac{1}{3^2\cdot\sqrt{3}}-\dfrac{1}{3^2\cdot\sqrt{3}}+\ldots$ to ∞ [*Ans.* (C)]

5. $\dfrac{1}{\sqrt{1+1^3}}-\dfrac{1}{\sqrt{1+2^3}}+\dfrac{1}{\sqrt{1+3^3}}\ldots$ to ∞ [*Ans.* (C)]

6. $\Sigma\,(-1)^n\cdot\left(\sqrt{n+1}-\sqrt{n}\right)$ [*Ans.* (C)]

7. $\Sigma\,(-1)^n\cdot\dfrac{1}{\sqrt{n-1}}$ [*Ans.* (C)]

8. $\Sigma\,(-1)^{n-1}\cdot\dfrac{n}{n^2+1}$ [*Ans.* (C)]

9. $\dfrac{1}{1}-\dfrac{1}{2^2}+\dfrac{1}{2^2\cdot 4^2}-\dfrac{1}{2^2\cdot 4^2\cdot 6^2}+\ldots\infty$ [*Ans.* (C)]

10. $\Sigma\,(-1)^n\left[\dfrac{n^2+1}{n^3+1}\right]$ [*Ans.* (C)]

II. Discuss the convergence (Absolute convergence) of the following series

11. $\dfrac{x}{3}-\dfrac{x^2}{5}+\dfrac{x^3}{7}-\dfrac{x^4}{9}+\ldots\infty\ (|\,x\,|\le 1)$ [*Ans. AC* if $|\,x\,|<1$; *NAC* if $x=1$; DIV. if $x=-1$]

12. $1-\dfrac{1}{2^k}+\dfrac{1}{3^k}-\dfrac{1}{4^k}+\ldots$ [*Ans.* C if $k>0$; *AC* if $k>1$]

13. Find the nature of the series $\dfrac{2}{7}-\dfrac{5}{11}+\dfrac{8}{15}-\dfrac{11}{19}+\ldots$ [*Ans.* Oscillatory]

14. Examine the series $\dfrac{1}{7}-\dfrac{1}{12}+\dfrac{1}{17}-\dfrac{1}{22}+\ldots$

for (*a*) convergence
(*b*) absolute converse
(*c*) conditional convergence [*Ans.* C. *CC*, *NAC*]

15. Find the interval of convergence for the following series

(*a*) $\dfrac{1}{1\sqrt{2}}-\dfrac{x^2}{2\sqrt{3}}+\dfrac{x^4}{3\sqrt{4}}-\dfrac{x^5}{4\sqrt{5}}\ldots$ $[-1\le x\le 1]$

(*b*) $1-2x+3x^2-4x^3+\ldots,\ x>0$ $[0<x<1]$

(*c*) $\dfrac{3x}{1}-\dfrac{(3x)^2}{2}+\dfrac{(3x)^3}{3}\ldots$ $\left[\dfrac{-1}{3}<x\le\dfrac{1}{3}\right]$

APPENDIX 1

Illustrative Examples from Engineering Field (Under Parts B and D)

(A) PROBLEMS ON ELECTRICAL ENGG. FILED.

Electric Circuits

The flow of electricity in the simple circuit is as shown in the adjoining diagram. The basic laws that govern the flow of electric currents through a circuit or network, are known as Kirchoff's Laws

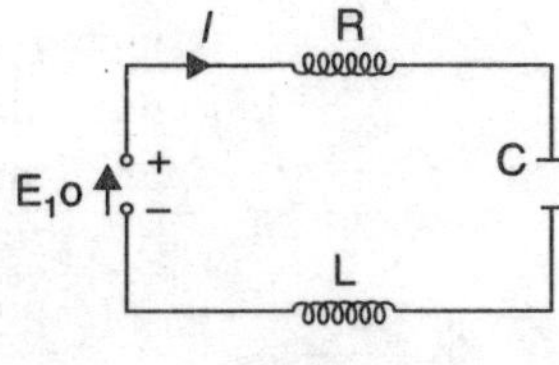

Fig. 1

Notation: t (seconds) = time

Q (Coloumbs) = Quantity of electric charge

I (Amperes) = Current, Rate change of Q

E (volts) = Voltage, (E.M.F).

R (ohms) = Resistance

L (Henrys) = Inductance

C (Farads) = Capacitance

KIRCHOFF'S LAW

(1) The Algebraic sum of all the currents flowing into a junction is equal to zero.

(2) The algebraic sum of all the potential drops in a closed Electric circuit is equal to zero.

(3) Some useful results:

$$L\frac{dI}{dt} + RI + \frac{Q}{c} = E \text{ or } E\ (t) \qquad [1]$$

$$L\frac{d^2Q}{dt^2} + \frac{Q}{c} = E \text{ or } E\ (t) \qquad [2]$$

$$L\frac{d^2I}{dt^2} + R\frac{dI}{dt} + \frac{1}{c}I = \frac{dE}{dt} \qquad [3]$$

1. A capacitance of 4.10^{-3} farads is in series with a 25 ohm resistance and an E.M.F. of (20 cos 6 t). Determine the charge and current at any time, given that the charge is zero at t = 0.

Solution: We have by kirchoff's laws is

$$L\frac{d^2Q}{dt^2}+R\frac{dQ}{dt}+\frac{1}{c}Q=E \quad [1]$$

Here $L = 0, R = 25, E = 20 \cos 6t; c = 4.10^{-3}$

$$\therefore \quad 25\frac{dQ}{dt}+\frac{1000}{4}Q=20\cos 6t;$$

$$25\frac{dQ}{dt}+250\,Q=20\cos 6t$$

$$\frac{dQ}{dt}+10Q=\frac{4}{5}\cos 6t \quad [2]$$

(Linear in Q)

$\therefore$ *Solution:* $Qe^{10t} = A + \int (4/5)\, e^{10t} \cos 6t\, dt$

$$= A+\frac{4}{5}e^{10t}\frac{[10\cos 6t+6\sin 6t]}{(10^2+6^2)}$$

$$\therefore \quad Q=e^{10t}=A+\left(\frac{1}{170}\right)e^{10t}(10\cos 6t+6\sin 6t)$$

or, $$A=Ae^{-10t}+\left(\frac{1}{85}\right)[5\cos 6t+3\sin 6t]$$

when $t = 0$, we have

$Q = 0;$

$\therefore$ $0 = A + (1/85)(5 + 0);$

$A = (-1/17)$

$\therefore$ $Q = (-1/17)\, e^{-10t} + (1/85)\,[5 \cos 6\, t + 3 \sin 6t]$

$$\therefore \quad I=\left(\frac{dQ}{dt}\right)(-1/17)(-10)\,e^{-10t}+(1/85)$$

$$[5t\,(-6\sin 6\,t)+3.6\cos 6t]$$

$$=\frac{10}{17}e^{-10t}+\frac{1}{85}[-30\sin 6t+18\cos 6t]$$

2. A constant E.M.F. E volts is applied to a circuit containing a constant resistance L henries. If the initial current is zero, show that the current builts up to half its theoretical maximum in $(L \log 2)/R$ seconds.

Solution: $$L\frac{dI}{dt}+RI=E$$

$$\Rightarrow \quad \frac{dI}{dt}+\frac{R}{L}\cdot I=\frac{E}{L}$$

(linear D.E. in I)

$$\therefore \qquad I.F. = e^{\int \frac{R}{L} dt} = e^{\frac{Rt}{L}}$$

$$\therefore \quad \textit{Solution:} \qquad Ie^{\frac{Rt}{L}} = \int e^{\frac{Rt}{L}} \cdot \frac{E}{L} dt = \frac{E}{L} e^{\frac{Rt}{L}} \frac{L}{R} + C$$

$$Ie^{\frac{Rt}{L}} = \frac{E}{R} e^{\frac{Rt}{L}} + c$$

Initially, $\qquad t = 0, I = 0$

$$\therefore \qquad 0 = \frac{E}{R} + C \Rightarrow \mathrm{C} = -\frac{E}{R}$$

$$\therefore \qquad Ie^{\frac{Rt}{L}} = \frac{E}{R} e^{\frac{Rt}{L}} - \frac{E}{R}$$

$$I = e^{\frac{E}{R}} \left(1 - e^{\frac{-Rt}{L}}\right)$$

The maximum theoretical current in the circuit is $\frac{E}{R}$. Let after '*t*' seconds the current in the circuit be half its theoretical maximum.

Thus
$$\frac{1}{2}\frac{E}{R} = \frac{E}{R}\left(1 - e^{\frac{-RT}{2}}\right)$$

$$e^{\frac{-RT}{L}} = \frac{1}{2} \Rightarrow T = \frac{(L \log 2)}{R}$$

3. The current in a coil, containing a resistance R, an inductance L and a constant e.m.f. at time t, is given by

$$I = \frac{E}{R}\left(1 - e^{\frac{-Rt}{L}}\right)$$

Obtain a suitable approximation to I when t is very small.

Solution: potential drop across $\qquad R = IR.$

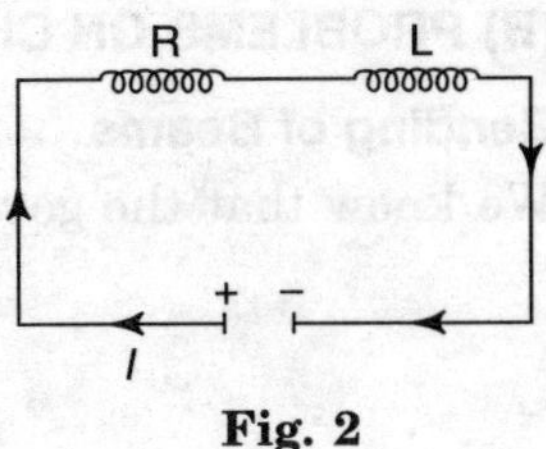

Fig. 2

Voltage produced by inductance $\qquad = L\frac{dI}{dt}$

$$IR + L\frac{dI}{dt} = E$$

$$\frac{dI}{dt} + \frac{RI}{L} = \frac{E}{L}$$

$$I.F = e^{\int \frac{R}{L} dt} = e^{\frac{Rt}{L}}$$

Solution: $$Ie^{\frac{Rt}{L}} = \int \frac{E}{L} e^{\frac{Rt}{L}} dt + c = \frac{E}{L} \cdot \frac{L}{R} e^{\frac{Rt}{L}} + c = \frac{E}{R} e^{\frac{Rt}{L}} + c$$

$\therefore$ $$I = \frac{E}{R} + ce^{\frac{-Rt}{L}} \quad [1]$$

when $t = 0, I = 0,$

$$0 = \frac{E}{R} + c \quad \text{or,} \quad c = -\frac{E}{R}$$

Then [1] becomes, $$I = \frac{E}{R} - \frac{E}{R} e^{\frac{-Rt}{L}} = \frac{E}{R}\left[1 - e^{\frac{Rt}{L}}\right]$$

when t is very small $$I = \frac{E}{R}$$

EXERCISES

1. A resistance of 100 ohms, an inductance of 0.5 henry are connected in series with a battery of 20 volts. Find the current in the circuit as a function of time. $\left[Ans.\ I = \frac{1}{5}\left[1 - e^{-200t}\right]\right]$

2. Solve the equation

$$L \frac{dI}{dt} + RI = E_0 \sin wt$$

where L, R and E_0 are constants and discuss the case when t increases indefinitely

$$\left[Ans.\ I = \frac{E_0 L}{\sqrt{R^2 + L^2 w^2}} \cdot \sin\left[wt - \tan^{-1}\frac{Lw}{R}\right]\right]$$

3. A coil having a resistance of 15 ohms and an inductance of 10 henries is connected to 90 volts supply. Determine the value of current after 2 seconds ($e^{-3} = 0.05$) (*Ans.* 5. 985 amp.)

(B) PROBLEMS ON CIVIL ENGINEERING FIELD

Bending of Beams

We know that the governing differential equation for the beam is given by

$$EI \frac{d^4 y}{dx^4} = q \quad [1]$$

where y is the transverse displacements, E.I. the bending rigidity and q the distributed loading.

It is well known from the mechanics of structure that the forces above and below the neutral surface acting in opposite directions have a tendency to restore the beam to its original position, creating an internal bending moment M given by $\frac{EI}{R}$

where E = modulus of elasticity of the material of the beam.

I = Moment of inertia of cross section of the beam about the neutral axis AB

R = Radius of curvature of elastic curve at $p(x, y)$ [a point on cross sections]

$$R = \frac{\left[1+\left(\frac{dy}{dx}\right)^2\right]^{3/2}}{\frac{d^2y}{dx^2}}$$

If the defection of the beam is small, then $\left(\frac{dy}{dx}\right)^2$ can be taken very small and so neglected.

$$\therefore \qquad R = \frac{1}{\frac{d^2y}{dx^L}}$$

This $M = EI\frac{d^2y}{dx^2}$ is the D.E. of the elastic curve.

The solution of this will have two arb. constants, which can be determined from the bending conditions.

M = moment of all external forces acting on either side of the portions of the beam separated by about cross section about a line.

Shear Force $= \frac{dM}{dx} = EI\frac{d^3y}{dy^3}$

Intencity of Loading $= \frac{d^2M}{dx^2} = EI\frac{d^4y}{dx}$

1. A simply supported beam of span l carrying a uniformly distributed load of w per unit run over the whole span. Find the maximum deflection.

Solution: We know that the moment of the force $\frac{wl}{2}$ about X is $\left[\frac{wl}{2x}\right]$ and moment of the force wx is $-wx \cdot x/2$

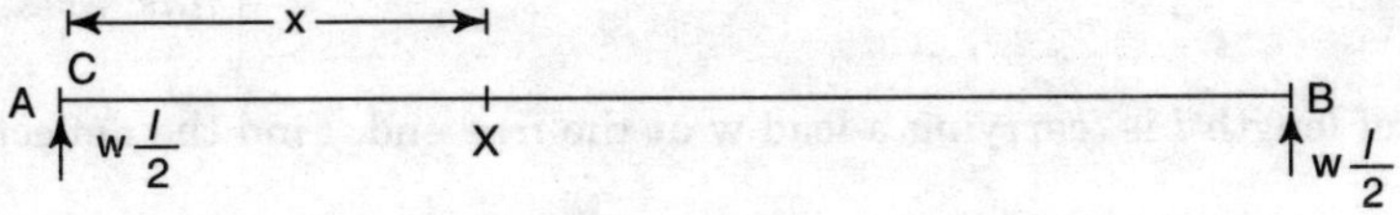

Fig. 3

$$EI\frac{d^2y}{dx^2} = \frac{wl}{2}x - \frac{1}{2}w.x^2 \qquad [1]$$

(Integrating once w.r.t. x), we get

APPENDICES

$$EI\frac{dy}{dx} = \frac{wl}{2}\left(\frac{x^2}{2}\right) - \frac{wx^3}{6} + c_1$$

The maximum deflection occur at middle and hence the slope at that point is zero.

i.e.,
$$\frac{dy}{dx} = 0 \text{ where } x = \frac{e}{L}$$

$$\therefore \quad Q = \frac{wl}{2}\left(\frac{l^2}{8}\right) - \frac{w}{6}\left(\frac{l^3}{8}\right) + c_1 \qquad [2]$$

$$\therefore \quad c_1 = -\frac{wl^3}{24}$$

$$\therefore \text{ [2] becomes:} \quad EI\frac{dy}{dx} = \frac{wlx^2}{4} - \frac{wx^3}{6} - \frac{wl^3}{2y} \qquad [3]$$

Integrating again, w.r.t. x

$$EIy = \frac{wlx^3}{12} - \frac{wx^4}{24} - \frac{wl^3 x}{24} + c_2$$

when $x = 0, y = 0, \Rightarrow c_2 = 0$

Maximum deflection occurs at $x = \frac{l}{2}$

$$EIy = \frac{wl}{12}\left(\frac{l}{2}\right)^3 - \frac{w\left(\frac{l}{2}\right)^4}{24} - \frac{wl^3\left(\frac{l}{2}\right)}{24} = \frac{5}{384}wl^4$$

$$\therefore \quad y = \frac{5\,wl^4}{384\,EI} \text{ is the max. deflection}$$

EXERCISES

1. A horizontal beam of length l is freely supported at both end. Find the equation of its elastic curve and its maximum deflection when the load in w per unit lengths.

$$\left[\textit{Ans}.\text{ Max. deflection}: \frac{5\,wl^4}{24\,EI}\right]$$

2. A cantilever of length l is carrying a load w at the free end. Find the deflection at the free end

$$\left[\textit{Ans}.\ y = -\frac{wl^3}{3\,EI}\right]$$

3. A horizontal beam of length l is freely supported at both ends and is loaded in the middle by a vertical load w. Show that the maximum deflection is $\frac{wl^3}{48\,EI}$.

(C) PROBLEMS ON MECHANICAL ENGG. FIELDS

1. A particle moves in a straight line so that its distance in time '*t*' is given by $s = (1 + 27\,t - t^3)$. Find its velocity and acceleration at the end of 2 seconds. Also find when and where the particle is momentarily at rest. When will it come back to the starting point again.

Solution: $s = 1 + 27t - t^3$,

$$\left(\frac{ds}{dt}\right) = 0 + 27 - 3t^2 = 3\,(9 - t^2)$$

$$\left(\frac{d^2s}{dt^2}\right) = -6t$$

(1) When $t = 2$ secs, velocity = 3 (9 – 2) = 15 cm/sec

Accen = –6.2 = –1.2 cm/sec^2

(2) When the particle is momentarily at rest, velocity = 0

or $(9 - t^2) = 0;\ t^2 = 9;\ t = 3, -3$

$\therefore$ $t = 3$ seconds

at that time, $s = 1 + 27.3 - 33 = 55$

(3) Starting point is the position when

$t = 0;\ s = 1 + 27.0 = 1;$

Again, s will be equal to 1,

at a time when $1 + 27\,t - t^3 = 1$

$$27t - t^3 = 0;$$

$$t\,(t^2 - 27) = 0$$

$$t = 0 \quad \text{and} \quad t^2 = 27;$$

or $$t = 0,\ t = 3\sqrt{3}$$

At the instant when $t = 3\sqrt{3}$,

The particle is at the starting point again.

2. The distance '*x*' descended by a person falling by a parachute satisfies the equation

$$\left(\frac{dx}{dt}\right)^2 = k^2$$

[1 – e – ($2gx/k^2$)], where g and k are constants. Show that

$$x = (k^2/g)\log(\cos h\,(gt/k)],$$

Given $x = 0$ when $t = 0$

Solution:

Let $$u = e^{(gx/k^2)};$$

$$\frac{du}{dt} = e^{(gx/k^2)}\,(g/k^2)\left(\frac{dx}{dt}\right) \quad \text{or} \quad \frac{du}{dt} = \frac{u\cdot g}{k^2}\,\frac{dx}{dt};$$

$$\frac{dx}{dt} = \frac{k^2}{gu}\,\frac{du}{dt} \qquad [1]$$

APPENDICES

$\therefore$ Equation is: $\left(\dfrac{dx}{dt}\right)^2 = k^2\,[1 - e^{-gx}/k^2]$

$$= k^2\,[1 - u^{-2}] = k^2 \left[\frac{1}{u^2}\right] = k^2\,(4^2 - 1)/4^2]$$

$$\therefore \quad \frac{k\sqrt{(u^2-1)}}{u} = \frac{dx}{dt} = \frac{k^2}{gu}\cdot\frac{du}{dt}$$

$$\therefore \quad dt\,\sqrt{u^2-1} = (k/g)\,du;$$

$$\left[\left(\frac{du}{\sqrt{u^2-1}}\right)\right] = (g/k)\,dt$$

$$\int\left[\frac{1}{\sqrt{u^2-1}}\,du\right] = (g/k)\,b + c;$$

$$\cos h^{-1}(u) = (gt/k) + c$$

when $t = 0,\ x = 0;$

$u = e^0 = 1;\ \cos h^{-1} u = \cos h^{-1}(1) = 0$

$\therefore$ $0 = 0 + c;$

or $c = 0$

$$\cos h^{-1}(u) = (gt/k)$$

$$u = \cos h\,(gt/k)$$

$$e^{(gx/k^2)} = \cosh\,(gt/k)$$

$$\left(\frac{gx}{k^2}\right) = \log \cosh\,(gt/k)$$

Hence, $x = (k^2/g)\log[\cosh\,(gt/k)]$

3. The velocity possessed by a body after falling vertically from rest thro' a distance s is found to be $\sqrt{2(gs)}$. Find the height thro' which it has fallen in terms of the time.

Solution: Let the body fall thro' a distance s in time t. Velocity acquired in time

$$t = \sqrt{2(gs)} \text{ (given)}$$

$$\Rightarrow \quad \frac{ds}{dt} = \sqrt{2gs}$$

or $\Rightarrow$ $\dfrac{ds}{\sqrt{s}} = \sqrt{2g}\,dt$

Integrating, $2\sqrt{s} = 2\sqrt{2g}\,.t + c$

when $t = 0, s = 0 \Rightarrow c = 0,$

$\therefore$ $2\sqrt{s} = \sqrt{2g}\cdot t$

squaring $4s = 2gt^2 \Rightarrow s = \frac{1}{2}gt^2$

which gives the height (s) thro' which the body has fallen in terms of time t.

4. A particle moves from a point 0 with an acceleration k times the distance from 0 with a starting velocity $\sqrt{k}$. Find the velocity of the particle at a distance 2 units from 0

Solution: Let f be the acceleration and s, the distance of the particle from 0 at any instant

Given $f = ks$

$\Rightarrow$ $v\frac{dv}{ds} = ks$ or $vdv = ksds$

Integrate $$\frac{v^2}{2} = \frac{ks^2}{2} + c \qquad [1]$$

Initially, when $s = 0,\ v = \sqrt{k}$

$\therefore$ $$\frac{k}{2} = c$$

$\therefore$ From (1), $$\frac{v^2}{2} = \frac{ks^2}{2} + \frac{k}{2}$$

or $v = \sqrt{k(s^2+1)}$

when $s = 2,\ v = \sqrt{k(4+1)} = \sqrt{5k}$.

EXERCISES

1. The acceleration of a moving particle being proportional to the cube of the velocity and negative, find the distance passed over in time t, the initial velocity being v_0, and the distance being measured from the position of the particle at the time $t = 0$

$$\left[Ans.\ s = \frac{\sqrt{2kv_0^2\,t+1}}{kv_0}\right]$$

2. A train starting from rest is accelerated and acceleration at any instant is $\frac{10}{v+1}$ ft/sec, where v is the velocity of the train in *ft*/sec at the instant. Find the distance in which the train will attain a velocity of 30. m.p.w [*Ans.* y = 2936.27 *ft* (approx.)

APPENDICES

APPENDIX 2

Important Results and Formulae to be Remembered

PART A

Analytical Geometry in 3 Dimensions:

1. Distance Formula:

Distance between two points (x_1, y_1, z_1) and (x_2, y_2, z_2) is:

$$d = \sqrt{(x_2 - x_1)^2 + (y_2 - y_1)^2 + (z_2 - z_1)^2}$$

2. Section formula:

(*a*) If (x, y, z) be the point which divides the join of (x_1, y_1, z_1) and (x_2, y_2, z_2) in the ratio $m_1 : m_2$ internally, then

$$x = \frac{m_1 x_2 + m_2 x_1}{m_1 + m_2}, \; y = \frac{m_1 y_2 + m_2 y_1}{m_1 + m_2};$$

$$z = \frac{m_1 x_2 + m_2 x_1}{m_1 + m_2},$$

(*b*) For external division,

$$x = \frac{m_1 x_2 - m_2 x_1}{m_1 - m_2}, \; y = \frac{m_1 y_2 - m_2 y_1}{m_1 - m_2},$$

$$z = \frac{m_1 z_2 - m_2 z_1}{m_1 - m_2}$$

(*c*) Mid-point Formula:

$$\left[\frac{x_1 + x_2}{2}, \frac{y_1 + y_2}{2}, \frac{z_1 + z_2}{2}\right]$$

3. If l, m, n are the d.c s of a line OP and $OP = r$, then coordinates of P are (lr, mr, nr).
4. If l, m, n are the d.c's of a line, then $l^2 + m^2 + n^2 = 1$
5. If a, b, c, are the DR's of a line then the actual D.C's are $\frac{a}{\sqrt{\sum a^2}}, \frac{b}{\sqrt{\sum a^2}}, \frac{c}{\sqrt{\sum a^2}}$,

where $\sum a^2 = a^2 + b^2 + c^2$.

6. $\cos\theta = l_1l_2 + m_1m_2 + n_1n_2$

7. $\sin\theta = \sqrt{(m_1n_2 - m_2n_1)^2 + (n_1l_2 - n_2l_1)^2 + (l_1m_2 - l_2m_1)^2}$

8. For $\perp^r$ line, $l_1l_2 + m_1m_2 + n_1n_2 = 0$

9. For parallel line, $\dfrac{l_1}{l_2} = \dfrac{m_1}{m_2} = \dfrac{n_1}{n_2}$

10. D.R's of PQ: are $[(x_1 - x_2) : (y_1 - y_2) : (z_1 - z_2)]$

11. Projection of PQ in a line whose D.C's are $(l : m : n)$ is

$$(x_2 - x_1)\, l + (y_2 - y_1)\, m + (z_2 - z_1)n.$$

12. Solution of linear equation:

If
$$a_1x + b_1y = c_1$$
$$a_2x + b_2y = c_2$$

are two independent and constant equations; then

$$\frac{x}{-b_1c_2 + b_2c_1} = \frac{y}{-c_1a_2 + a_1c_2} = \frac{1}{a_1b_2 - a_2b_1}$$

(By Cross-multiplication)

or
$$\frac{x}{\begin{vmatrix} c_1 & b_1 \\ c_2 & b_2 \end{vmatrix}} = \frac{y}{\begin{vmatrix} a_1 & c_1 \\ a_2 & c_2 \end{vmatrix}} = \frac{1}{\begin{vmatrix} a_1 & b_1 \\ a_2 & b_2 \end{vmatrix}}$$

or
$$\frac{x}{D_1} = \frac{y}{D_2} = \frac{1}{D}$$

where
$$D = \begin{vmatrix} a_1 & b_1 \\ a_2 & b_2 \end{vmatrix} \neq 0$$

$\therefore$
$$x = \frac{D_1}{D};\; y = \frac{D_2}{D}$$

Again, for the system of equation

$$a_1x + b_1y + c_1z = d_1$$
$$a_2x + b_2y + c_2z = d_2$$
$$a_3x + b_3y + c_3z = d_3$$

where x, y, z are unknown qualities,

Cramer's rule for the solution of the equations is stated as follows:

$$\frac{x}{\begin{vmatrix} d_1 & b_1 & c_1 \\ d_2 & b_2 & c_2 \\ d_3 & b_3 & c_3 \end{vmatrix}} = \frac{y}{\begin{vmatrix} a_1 & d_1 & c_1 \\ a_2 & d_2 & c_2 \\ a_3 & d_3 & c_3 \end{vmatrix}}$$

$$\frac{z}{\begin{vmatrix} a_1 & b_1 & d_1 \\ a_2 & b_2 & d_2 \\ a_3 & b_3 & d_3 \end{vmatrix}} = \frac{1}{\begin{vmatrix} a_1 & b_1 & c_1 \\ a_2 & b_2 & c_2 \\ a_3 & b_3 & c_3 \end{vmatrix}}$$

or
$$\frac{x}{D_1} = \frac{y}{D_2} = \frac{z}{D_3} = \frac{1}{D}$$

where
$$D \neq 0$$

$$\therefore \quad x = \frac{D_1}{D};\ y = \frac{D_2}{D};\ z\ \frac{D_3}{D}$$

13. The General equation of the plane:

$$Ax + By + Cz + D = 0\ [1]$$

14. $(x \cos \alpha + y \cos \beta + z \cos \gamma) = p$. (Normal Form)
where p = length of perpendicular from the origin on the plane: $(\cos \alpha : \cos \beta : \cos \gamma)$ = D.C's of this perpendicular: and

$$\frac{\cos \alpha}{A} = \frac{\cos \beta}{B} = \frac{\cos \gamma}{C} = \frac{-p}{D}$$

15. D.C's of normal to a plane

$$= \cos \alpha : \cos \beta : \cos \gamma$$

$$A : B : C$$

= x–(coefficient) : y–(coefficient)

: z–(coefficient)

16. $p = -\left[\dfrac{D}{\sqrt{A^2 + B^2 + c^2}}\right]$

17. $\left[\left(\dfrac{x}{f}\right) + \left(\dfrac{y}{g}\right) + \left(\dfrac{z}{h}\right)\right] = 1$. (Intercept form).

18. Angle between two planes, is equal to angle between their normals

19. $\cos \theta = \dfrac{AA' + BB' + CC'}{\sqrt{A^2 + B^2 + C^2}\ \sqrt{A'^2 + B'^2 + C'^2}}$

(θ = angle between two planes)

20. Planes parallel: $(A : B : C) = (A' : B' : C')$

21. Any line parallel to a plane, means it is perpendicular to the normal to the plane.
A line of D.C's $(l : m : n)$ is parallel to a plane $(Ax + By + Cz + D) = 0$;
If $(l : m : n)$ is perpendicular to $(A : B : C)$
$\therefore \quad Al + Bm + Cn = 0$

22. Plane parallel to x-axis: x term absent: $(A = 0)$
Plane parallel to y-axis: y terms absent: $(B = 0)$
Plane parallel to z-axis: z term absent: $(C = 0)$

PART B

In the following formulas, *u*, *v* and *y* are differentiable functions, *a*, *c*, and *n* are constants, and 1 n u = log$_e$ *u*.

$$\frac{dc}{dx} = 0.$$

$$\frac{d}{dx}\, x = 1.$$

$$\frac{d}{dx}(cv) = c\,\frac{dv}{dx}$$

$$\frac{d}{dx}\, x^n = nx^{n-1},\ n > 1 \text{ if } x = 0.$$

$$\frac{dy}{dx} = \frac{dy}{du} \cdot \frac{du}{dx}$$

$$\frac{d}{dx}(u + v) = \frac{du}{dx} \cdot \frac{dv}{dx}$$

$$\frac{d}{dx}(uv) = u\,\frac{dv}{dx} + v\,\frac{du}{dx}$$

$$\frac{d}{dx}\left(\frac{u}{v}\right) = \frac{v\,\dfrac{du}{dx} - u\,\dfrac{dv}{dx}}{v^2}$$

$$\frac{du^n}{dx} = nu^{n-1},\ n > 1 \text{ if } u = 0.$$

$$\frac{d}{dx}(\sin u) = \cos u\ .\ \frac{du}{dx}$$

$$\frac{d}{dx}(\cos u) = -\sin u\,\frac{du}{dx}$$

$$\frac{d}{dx}(\tan u) = \sec^2 u\,\frac{du}{dx}$$

$$\frac{d}{dx}(\text{arc}\sin u) = \frac{\dfrac{du}{dx}}{\sqrt{1-u^2}},\ |\,u\,| < 1.$$

$$\frac{d}{dx}(\text{arc}\cos u) = \frac{\dfrac{du}{dx}}{\sqrt{1-u^2}},\ |\,u\,| < 1.$$

$$\frac{d}{dx}(\cos(u) = -\csc u \cot u\,\frac{du}{dx}$$

$$\frac{d}{dx}(\sinh x) = \cosh x.$$

$$\frac{d}{dx}(\cosh x) = \sinh x.$$

$$\frac{d}{dx}(\tanh x) = \text{sech}^2\, x.$$

$$\frac{d}{dx}(\text{ctnh}\, x) = -\text{csch}^2\, x.$$

$$\frac{d}{dx}(\text{sech}\, x) = -\text{sech}\, x \tanh x.$$

$$\frac{d}{dx}(\text{csch}\, x) = -\text{csch}\, x\ \text{ctnh}\, x.$$

$$\frac{d}{dx}(\ln u) = \frac{\dfrac{du}{dx}}{u}.$$

$$\frac{d}{dx}(\log_a u) = \log_a e\ .\ \frac{\dfrac{du}{dx}}{u},\ a > 0 \text{ and } a \neq 1.$$

$$\frac{d}{dx}(e^u) = e^u\ .\ \frac{du}{dx}$$

$$\frac{d}{dx}(a^u) = \ln a\ .\ a^u\ .\ \frac{du}{dx}\quad a > 0.$$

$$\frac{d}{dx}(\cot u) = -\csc^2 u\,\frac{du}{dx}$$

$$\frac{d}{dx}(\sec u) = \sec u \tan u\,\frac{du}{dx}$$

$$\frac{d}{dx}(\text{arc tan } u) = \frac{\frac{du}{dx}}{1+u^2}. \qquad \frac{d}{dx}(\text{arc cot } u) = -\frac{\frac{du}{dx}}{1+u^2}.$$

$$\frac{d}{dx}(\text{arc sec } u) = \frac{\frac{du}{dx}}{u\sqrt{u^2-1}}, \qquad -\pi < \sec^{-1} u < -\frac{\pi}{2}, \ 0 < \sec^{-1} u < \frac{\pi}{2},$$

$$\frac{d}{dx}(\text{arc csc } u) = \frac{\frac{du}{dx}}{u\sqrt{u^2-1}}, \qquad -\pi < \csc^{-1} u < \frac{\pi}{2}, \ 0 < \csc^{-1} u < \frac{\pi}{2}$$

$$\frac{d}{dx}(\sinh^{-1} x) = \frac{1}{\sqrt{x^2+1}}. \qquad \frac{d}{dx}(\cosh^{-1} x) = \frac{1}{\sqrt{x^2-1}}, \ x > 1.$$

$$\frac{d}{dx}(\tanh^{-1} x) = \frac{1}{1-x^2}, \ |x| < 1. \qquad \frac{d}{dx}(\text{ctnh}^{-1} x) = \frac{1}{1-x^2}, \ |x| > 1.$$

$$\frac{d}{dx}(\text{sech}^{-1} x) = \frac{-1}{x\sqrt{1-x^2}}, \ 0 < x < 1. \qquad \frac{d}{dx}(\text{csch}^{-1} x) = \frac{-1}{|x|\sqrt{1+x^2}}, \ x \neq 0.$$

PART C

Integral Calculus

1. $\int df(x) = f(z).$
2. $d \int f(x)\, dz = f(x)\, dz.$
3. $\int 0 \,.\, dx = C.$
4. $\int af(x)\, dx = a \int f(x)\, dx.$
5. $\int (u = v)\, dx = \int udz = \int vdx.$
6. $\int u\, dv = uv - \int v\, du.$
7. $\int \frac{u\, dv}{dx}\, dx = uv - \int v \frac{du}{dx}$
8. $\int f(y)\, dx = \int \frac{f(y)\, dy}{\frac{dy}{dx}}.$
9. $\int u^n\, du = \frac{u^{n+1}}{n+1}, \ n \neq -1.$
10. $\int \frac{du}{u} = \log u.$
11. $\int e^x\, du = e^n$
12. $\int b^n\, du = \frac{b^n}{\log b}.$
13. $\int \sin u\, du = -\cos u.$
14. $\int \cos u\, du = \sin u.$
15. $\int \tan u\, du = \log \sec u = -\log \cos u.$
16. $\int \text{ctn } u\, du = \log \sin u = -\log \csc u.$

17. $\int \sec u\, du = \log(\sec u + \tan u) = \log \tan\left(\dfrac{u}{2} + \dfrac{\pi}{4}\right)$.

18. $\int \csc u\, du = \log(\csc u - \text{etn}\ u) = \log \tan \dfrac{u}{2}$.

19. $\int \sin^n u\, du = \dfrac{1}{2}\, u - \dfrac{1}{2} \sin u \cos u$.

20. $\int \cos^n u\, du = \dfrac{1}{2}\, u + \dfrac{1}{2} \sin u \cos u$.

21. $\int \sec^2 u\, du = \tan u$.

22. $\int \csc^2 u\, du = -\text{ctn}\ u$.

23. $\int \tan^2 u\, du = \tan u - u$.

24. $\int \text{ctn}^2\, u\, du = -\text{ctn}\ u - u$.

25. $\int \dfrac{du}{u^2 + a^2} = \dfrac{1}{a} \tan^{-1} \dfrac{u}{a}$.

26. $\int \dfrac{du}{u^2 - a^2} = \dfrac{1}{2a} \log\left(\dfrac{u-c}{u+a}\right) = -\dfrac{1}{a}\, \text{ctnh}^{-1}\left(\dfrac{u}{a}\right)$, if $u^2 > a^2$.

$$= \frac{1}{2a} \log\left(\frac{a-u}{a+u}\right) = -\frac{1}{a} \tanh^{-1}\left(\frac{u}{a}\right), \text{ if } u^2 > a^2.$$

27. $\int \dfrac{du}{\sqrt{a^2 - u^2}} = \sin^{-1}\left(\dfrac{u}{a}\right)$

28. $\int \dfrac{du}{\sqrt{u^2 - a^2}} = \log(u + \sqrt{u^2 - a^2})$

29. $\int \dfrac{du}{\sqrt{2au - u^2}} = \cos^{-1}\left(\dfrac{a-u}{a}\right)$

30. $\int \dfrac{du}{u\sqrt{u^2 - a^2}} = \dfrac{1}{a} \sec^{-1}\left(\dfrac{u}{a}\right)$

31. $\int \dfrac{du}{u\sqrt{a^2 - u^2}} = -\dfrac{1}{a} \log\left(\dfrac{a + \sqrt{a^2 - u^2}}{u}\right)$

32. $\int \sqrt{a^2 - u^2}\,.\, du = \dfrac{1}{2}\left(u\sqrt{a^2 - u^2} + a^2 \sin^{-1} \dfrac{u}{a}\right)$

33. $\int \sqrt{u^2 \pm a^2}\, du = \dfrac{1}{2}[u\sqrt{u^2 \pm a^2} = a^2 \log(u + \sqrt{u^2 \pm a^2})]$.

34. $\int \sinh u\, du = \cosh u$.

35. $\int \cosh u\, du = \sinh u$.

36. $\int \tanh u\, du = \log(\cosh u)$.

37. $\int \text{ctnh}\ u\, du = \log(\sinh u)$.

38. $\int \text{sech}\ u\, du = \sin^{-1}(\tanh u)$.

39. $\int \text{csch}\ u\, du = \log\left(\tanh \dfrac{u}{2}\right)$.

40. $\int \text{sech}\ u\,.\, \tanh u\,.\, du = -\text{sech}\ u$.

41. $\int \text{csch}\ u\,.\, \text{ctnh}\ u\,.\, du = -\text{csch}\ u$.

42. $\int \frac{xdx}{(x^2+a^2)} = \frac{1}{2}\log(x^2+a^2)$

43. $\int \frac{xdx}{(x^2-a^2)} = \frac{1}{2}\log(x^2-a^2)$

44. $\int \frac{x\,.\,dx}{(a^2-x^2)} = -\frac{1}{2}\log(a^2-x^2)$

45. $\int (ax+b)^n\,.\,dx = \frac{1}{a(n+1)}(ax+b)^{n+1},\ n\neq -1.$

46. $\int \frac{dx}{ax+b} = \frac{1}{a}\log_e(ax+b)$

47. $\int \frac{dx}{(ax+b)^2} = -\frac{1}{a(ax+b)}.$

48. $\int \frac{dx}{(ax+b)^3} = -\frac{1}{2a(ax+b)^2}$

49. $\int x(ax+b)^n\,dx = \frac{1}{a^2(n+2)}(ax+b)^{n+2}$

$$-\frac{b}{a^2(n+1)}(ax+b)^{n+1},\ n\neq -1,-2$$

$$\log\left(\frac{u+\sqrt{u^2+a^2}}{a}\right) = \sin^{-1}\left(\frac{u}{a}\right);\ \log\left(\frac{a+\sqrt{a^2-u^2}}{u}\right) = \text{sech}^{-1}\left(\frac{u}{a}\right);$$

$$\log\left(\frac{u+\sqrt{u^2+a^2}}{a}\right) = \cosh^{-1}\left(\frac{u}{a}\right);\ \log\left(\frac{a+\sqrt{a^2-u^2}}{u}\right) = \text{csch}^{-1}\left(\frac{u}{a}\right)$$

50. $\int \frac{xdx}{ax+b} = \frac{x}{a} - \frac{b}{a^2}\log(ax+b)$.

51. $\int \frac{xdx}{(ax+b)^2} = \frac{b}{a^2(ax+b)} + \frac{1}{a^2}\log(ax+b)$.

52. $\int \frac{xdx}{(ax+b)^2} = \frac{b}{2a^2(ax+b)^2} - \frac{1}{a^2(ax+b)}$

53. $\int x^2(ax+b)^n\,dx = \frac{1}{a^2}\left[\frac{(ax+b)^{n+2}}{n+3}\right.$

$$-2b\frac{(ax+b)^{n+2}}{n+2}+b^2\frac{(ax+b)^{n+2}}{n+1}\Bigg],n\neq-1,-2,-3.$$

54. $$\int\frac{x^2dx}{ax+b}=\frac{1}{a^2}\left[\frac{1}{2}(ax+b)^2-2b(ax+b)+b^2\log(ax+b)\right].$$

55. $$\int\frac{x^2dx}{(ax+b)^2}=\frac{1}{a^2}\left[(ax+b)-2b\log(ax+b)-\frac{b^2}{ax+b}\right].$$

56. $$\int\frac{x^2dx}{(ax+b)^2}=\frac{1}{a^2}\left[\log(ax+b)+\frac{2b}{ax+b}-\frac{b^2}{2(ax+b)^2}\right]$$

57. $\int x^n\,(ax+b)^n\,dx$

$$\frac{1}{a(m+n+1)}\left[x^m\,(ax+b)^{n+1}-mb\int x^{m-1}\,(ax+b)^n\,dx\right],$$

$$=\frac{1}{m+n+1}\left[x^{m+1}\,(ax+b)^n+nb\int x^m\,(ax+b)^{n-1}\,dx\right]$$

$m>0,\ m+n+1\neq0.$

58. $$\int\frac{dx}{x(ax+b)}=\frac{1}{b}\log\frac{x}{ax+b}$$

PROPERTIES OF DEFINITE INTEGRALS

1. $\int_a^b f(x)\,dx$ is said to be a definite integral

$$\int_a^b f(x)\,dx=[\varphi(b)-\varphi(a)];$$

where $\phi'(x)=f(x)$

$$\int_a^b f(x)\,dx=[\phi(x)]_a^b=\phi(b)-\phi(a)$$

2. (i) $$\int_a^b f(x)\,dx+\int_b^c f(x)\,dx=\int_a^c f(x)\,dx$$

(ii) $$\int_a^a f(x)\,dx=[\varphi(a)-\varphi(a)]=0$$

(iii) $$\int_b^a f(x)\,dx=-\int_a^b f(x)\,dx$$

(iv) $\int_a^b k\, f(x)\, dx = k\,[\varphi(b) - \varphi(a)] = k \int_a^b f(x)\, dx$

(v) $\int_a^b k\, dx = k\,(b - a)\,;\ \int_a^b o\, dx = 0$

3. $\int_a^b f(x)\, dx = j(x)]_a^b = [j(b) - j(a)]$

$\int_a^b f(t)\, dt = \varphi(t)]_a^b = [\varphi(b) - \varphi(a)]$

$\therefore \quad \int_a^b f(x)\, dx = \int_a^b f(t)\, dt$

The value of a D.I. depends only on the function, and the limits. It is independent of the variable used.

4. $\int_o^a f(x)\, dx = \int_o^a f(a - x)\, dx$

5. $\int_{-a}^a f(x)\, dx = 0$, when $f(x)$ is an odd function.

$= 2\int_o^a f(x)\, dx,$

when $f(x)$ is an even function.

6. Integration by parts:

$\int uv'\, dx = uv - \int vu'\, dx.$

Rule to evaluate: *(If we choose u as the first function and v′ as the second:)*
Integral of the product of two fucntions: first function multiplied by the integral of the second – Integral of $\frac{d}{dx}$ of the first muliplied by the intergal of the second.

7. (a) $\int_o^{2a} f(x)\, dx = 2\int_o^a f(x)\, dx$, if $f(2a - x) = f(x) = 0$, if $f(2a - x) = -f(x)$.

(b) $\int_o^{\pi} \sin^m x \cos^n x dx = 0$ or, $2\int_o^{\pi/2} \sin^m x \cos^n x\, dx$

according as the power of cos x i.e., n is odd or even.

(c) $\int_o^{\pi/2} \log \sin x\, dx = \int_o^{\pi/2} \log \cos x\, dx = \frac{\pi}{2} \log \frac{1}{2}$

8. $\int_o^{\pi/2} \sin^n x\, dx = \int_o^{\pi/2} \cos^n x\, dx$

$$= \frac{(n-1)}{n} \cdot \frac{(n-3)}{(n-2)} \cdot \frac{(n-5)}{(n-4)} \cdots \frac{4 \cdot 1}{4 \cdot 2}\left(\frac{\pi}{2}\right) \text{ (}n\text{-even)}$$

$$= \frac{(n-1)}{n} \cdot \frac{(n-3)}{(n-2)} \cdot \frac{(n-5)}{(n-4)} \cdots \frac{4}{5} \cdot \frac{2}{3} \text{ [}n\text{-odd]}$$

N^r = start with $(n - 1)$ and go on diminishing it by 2 till you get 1 or 2.

D^r = start with n and go on diminishing it by 2 till you get 2 or 1 (Multiply by $\frac{\pi}{2}$ only when n is even.

VALUES OF SOME CONSTANTS

$\pi = 3.1416$ or 3.142;

$\pi^2 = 9.870$;

$e = 2.71828$

$\log_{10} e = 0.4343$;

$\log \pi = 0.4972$;

$\log \pi^2 = 0.9943$

$\frac{1}{\pi} = 0.3183$; $\log_e 10 = 2.303$;

$\log_e x = 2.303 \log_{10} x$

$\sqrt{2} = 1.414213$;

$\sqrt{3} = 3.162277$.

APPENDIX 3

Additional Examples

PART A

1. Find d.c. of the lines *AB* and *CD* where *A* (1, 2, – 4), *B* (2, 1, – 3), *C* (4, 6, – 1) and *D* (5, 7, 0).

Hence find the acute angle between them.

Solution: We know that

DR's of *AB* are [(1 – 2) : (2 – 1) : (– 4 + 3)] or, (–1 : 1 : – 1)

$\therefore$ d.c. of *AB* are $\dfrac{-1}{\sqrt{3}}, \dfrac{1}{\sqrt{3}}, \dfrac{-1}{\sqrt{3}}$.

and DR's of *CD* are [(4 – 5) : (6 – 7) : (– 1 – 0)] or, (– 1 : – 1 : – 1)

$\therefore$ d.c. of *CD* are $\dfrac{-1}{\sqrt{3}}, \dfrac{1}{\sqrt{3}}, \dfrac{-1}{\sqrt{3}}$.

Let θ be the angle between *AB* and *CD*

$$\therefore \qquad \cos\theta = l_1 l_2 + m_1 m_2 + n_1 n_2$$

$$= \frac{1}{3} - \frac{1}{3} + \frac{1}{3} = \frac{1}{3}$$

$$\therefore \qquad \theta = \cos^{-1}\left(\frac{1}{3}\right)$$

2. If *A* = (1, 3, 5), *B* = (6, 4, 3), *C* = (2, – 1, 4) and *D* = (0, 1, 5) find the projection of *AB* on *CD*. [B.U. Feb., 1994]

Solution: The DR's of *CD* are (– 2 : 2 : 1)

[using $(x_2 - x_1) : (y_2 - y_1) : (z_2 - z_1)$].

Now, the DC's of CD are

$$\left[\frac{-2}{\sqrt{4+4+1}}, \frac{2}{\sqrt{4+4+1}}, \frac{1}{\sqrt{4+4+1}}\right]$$

$$= \left[\frac{-2}{3}, \frac{2}{3}, \frac{1}{3}\right]$$

∴ the projection of AB on CD

$$= -\frac{2}{3}(6-1) + \frac{2}{3}(4-3) + \frac{1}{3}(3-5)$$

$$= -\frac{10}{3} + \frac{2}{3} - \frac{2}{3} = -\frac{10}{3}$$

[projection of AB on CD]

$$= l(x_2 - x_1) + m(y_2 - y_1) + n(z_2 - z_1)$$

3. Find the perpendicular distance from the point (– 1, 3, 9) to the line

$$\frac{x-13}{5} = \frac{y+8}{-8} = \frac{z-31}{1}$$ [VTU, July/August, 2002]

Solution: Here, $p(-1, 3, 9)$

and the given line: $$\frac{x-13}{5} = \frac{y+8}{-8} = \frac{z-31}{1} = r \quad [1]$$

Any point on [1] is

$$M = [5r + 13, -8r - 8, r + 31]$$

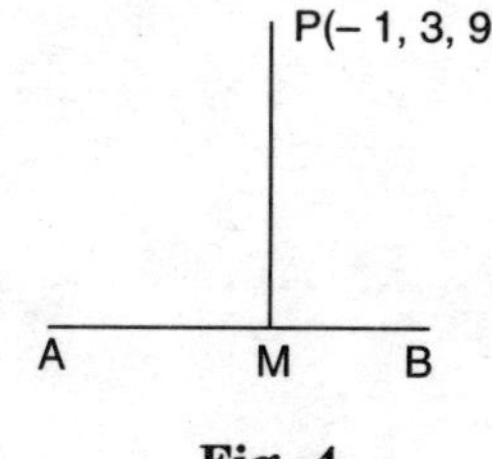

Fig. 4

The DR's of PM are:

$$[5r + 13 + 1, -8r - 8 - 3, r + 31 - 9]$$
$$= [5r + 14, -8r - 11, r + 22]$$

The DR's of given line AB are $(5 : -8 : 1)$

If PM perpendicular to AB, then,

$$5[5r + 14] - 8(-8r + 1) + 1(r + 22) = 0$$

$$\Rightarrow \quad 90r + 180 = 0 \quad \text{or} \quad r = -2,$$

For $r = -2$, PM becomes perpendicular to the lines.

$$\therefore \quad M = (-10 + 13, +16 - 8, -2 + 31)$$
$$= (3, +8, 29)$$

∴ The required perpendicular distance is.

$$PM = \sqrt{(3+1)^2 + (8-3)^2 + (29-9)^2}$$

$$= \sqrt{16 + 25 + 400} = \sqrt{441} = 21.$$

4. Find the distance between the parallel planes

$$2x - 3y + 6z + 12 = 0$$

and $$2x - 3y + 6z - 2 = 0$$

Solution:

Let $$P_1: 2x - 3y + 6z + 12 = 0$$
$$P_2: 2x - 3y + 6z - 2 = 0$$

Since P_1 meets x is:

$$(y = 0, z = 0) \text{ at } 2x + 12 = 0$$

$\Rightarrow$ $(-6, 0, 0)$ is a point on P_1.

The perpendicular distance from (–6, 0, 0) to the plane P_2.

APPENDICES

$2x - 3y + 6z - 2 = 0$ is the required distance and is

$$d = \pm \frac{Ax_1 + By_1 + Cz_1 + D}{\sqrt{A^2 + B^2 + C^2}} = \frac{\pm(-12-2)}{\sqrt{2^2 + (-3)^2 + 6^2}}$$

$$= \pm\left(\frac{-14}{7}\right) = 2.$$

5. Find the angle between the planes

$$P_1 : x - 2y + 3z - 1 = 0$$

and $\quad P_2 : 2x + 4y + 6z + 7 = 0$

Solution: We know that the angle between the planes

$$Ax + By + Cz + D = 0 \text{ and } A'x + B'y + C'z + D' = 0$$

is $\quad \cos\theta;$

where, $$\cos\theta = \pm\left(\frac{AA' + BB' + CC'}{\sqrt{\Sigma A^2}\sqrt{\Sigma A'^2}}\right)$$

$$\therefore \quad \cos\theta = \pm\left(\frac{2-8+18}{\sqrt{14}\sqrt{56}}\right) = \frac{12}{\sqrt{14}\sqrt{4\times 14}} = \frac{3}{7}$$

$$\therefore \quad \theta = \cos^{-1}\left(\frac{3}{7}\right)$$

6. Find the equation of the plane passing through the intersection of the planes

$$3x - 4y + 5z = 7,$$
$$4x + 3y - 7z = 9$$

and passing thro' the point (1, 2, 3).

Solution: The equation of the plane is

$$(3x - 4y + 5z - 7) + k(4x + 3y - 7z - 9) = 0$$

Since (1, 2, 3) lies on it

$$\Rightarrow \quad (3 - 8 + 15 - 7) + k(4 + 6 - 21 - 9) = 0$$

$$\Rightarrow \quad 3 - 20k = 0$$

$$\Rightarrow \quad k = \frac{3}{20}$$

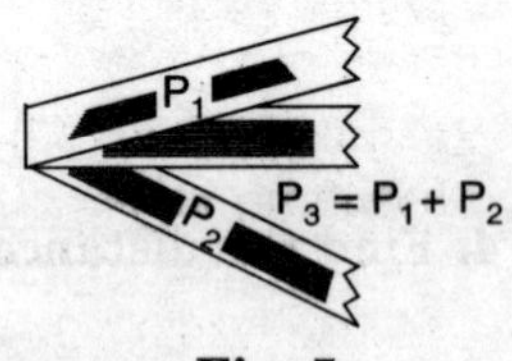

Fig. 5

$\therefore$ The eqn of the plane is

$$(3x - 4y + 5z - 7) + \frac{3}{20}(4x + 3y - 7z - 9) = 0$$

$$\Rightarrow \quad 72x - 71y + 79z - 167 = 0$$

7. Find the equation of the plane thro (1, 0, 2) and parallel to the plane

$$2x + 3y - 4z + 12 = 0$$

Solution:

Given plane:

$$2x + 3y - 4z + 12 = 0 \quad [1]$$

$$2x + 3y - 4z + c = 0 \text{ is parallel to plane [1]}$$

Also (1, 0, 2) lies on it

$$\therefore \quad 2 - 8 + c = 0$$

$$c - 6 = 0$$

$$c = 6$$

$\therefore$ The required planes

$$2x + 3y - 4z + 6 = 0$$

8. Find the equation of the plane which contains the line

$$\frac{x-1}{2} = \frac{y+1}{-1} = \frac{z-3}{4} \text{ and is } \perp^r \text{ to the plane} \quad \text{[B.U. Jan., 1993]}$$

$$x + 2y + z = 12.$$

Solution:

Given: $$\frac{x-1}{2} = \frac{y+1}{-1} = \frac{z-3}{4} \quad [1]$$

Any plane thro the given line is of the form

$$A(x-1) + B(y+1) + C(z-3) = 0 \quad [1]$$

where $$2A - B + 4C = 0 \quad [2]$$

By data (1) is $\perp^r$ to $x + 2y + z = 12$

$$\therefore \quad A \cdot 1 + B \cdot 2 + C \cdot 1 = 0$$

$$\Rightarrow \quad A + 2B + C = 0 \quad [3]$$

Solving [2] and [3] by the method of cross multiplication, we have

$$\frac{A}{-1-8} = \frac{B}{4-2} = \frac{C}{4+1}$$

$$\Rightarrow \quad \frac{A}{-9} = \frac{B}{2} = \frac{C}{5}$$

putting these value in [1], we get

$$-9(x-1) + 2(y+1) + 5(z-3) = 0$$

$$\Rightarrow \quad 9x - 2y - 5z + 4 = 0$$

is the required plane

9. Find the image of the line

$$\frac{x-1}{3} = \frac{y-3}{5} = \frac{z-4}{2}$$

in the plane $$2x - y + z + 3 = 0.$$

Solution: Since $P(1, 3, 4)$ is a point on the line. Let P^1 be the image of P in the plane. $2x - y + z + 3 = 0$

DR's of Normal to the plane are : $(2 : -1 : 1)$

APPENDICES

The DR's PP^1 $(2 : -1 : 1)$.

Hence, the equations of PP^1 is given by

$$\frac{x-1}{2} = \frac{y-3}{-1} = \frac{z-4}{1}$$

$\therefore$ $P = [1 + 2r, 3 - r, r + 4]$ (parametric point)

Mid pt. of PP': $\left[\frac{1+2r+1}{2}, \frac{3+3-r}{2}, \frac{4+r+4}{2}\right] = Q$

i.e., $Q = \left[r+1, 3-\frac{r}{2}, 4+\frac{r}{2}\right]$ and lies on the plane $2x - y + z + 3 = 0$

$\Rightarrow$ $2(r+1) - \left(3-\frac{r}{2}\right)+\left(4+\frac{r}{2}\right) + 3 = 0$

$\therefore$ $3r = -6$ $\therefore$ $r = -2$.

$\therefore$ $P' = [1 - 4, 3 + 2, 4 - 2] = [-3, 5, 2]$

Let us find the point where the line:

$$\frac{x-1}{3} = \frac{y-3}{5} = \frac{z-4}{2}$$

meets the plane [1]

$$2x - y + z + 3 = 0 \quad - [1]$$

Any point on this line is:

$$R = [3r + 1, 5r + 3, 2r + 4]$$

It lies on [1].

$\therefore$ $2(3r + 1) - (5r + 3) + (2r + 4) + 3 = 0$

$\Rightarrow$ $3r + 6 = 0$

$\Rightarrow$ $r = -2;$

$\therefore$ $R = [-5, -7, 0]$

Since R lies in the plane [1]

The image of R is itself

$\therefore$ The line RP' is the image of the line and its equations are

$$\frac{x+3}{-3+5} = \frac{y-5}{5+7} = \frac{z-2}{2-0}$$

i.e., $$\frac{x+3}{2} = \frac{y-5}{12} = \frac{z-2}{2}$$

i.e., $$\frac{x+3}{1} = \frac{y-5}{0} = \frac{z-2}{1}$$

10. Find the distance of the point (1, 2, 3) from the plane $x + y + z = 11$, measured parallel to the line $\frac{x+1}{1} = \frac{y-12}{-2} = \frac{z-7}{2}$.

Solution: The equations of the line thro (1, 2, 3) parallel to the given line are

$$\frac{x-1}{1} = \frac{y-2}{-2} = \frac{z-3}{2} = r \text{ (say)} \quad [1]$$

$$\therefore \quad x = r + 1,\ y = 2 - 2r,\ z = 2r + 3$$

Any point on this line is $(r + 1, z - 2r, 2r + 3)$

If this point lies on the plane $x + y + z = 11$ [2]

then $$r + 1 + 2 - 2r + 2r + 3 = 11 \Rightarrow r = 5.$$

The point of intersection of (1) and (2) is (6, – 8, 13).

Hence, the required distance

$$= \sqrt{(6-1)^2 + (-8-2)^2 + (13-3)^2}$$

$$= \sqrt{25 + 100 + 100} = \sqrt{225} = 15.$$

11. Find the equation of a plane thro the points (2, 2, 1) and (9, 3, 6) and perpendicular to the plane $2x + 6y + 6z = 9$ [VTU, July/August, 2002]

Solution:

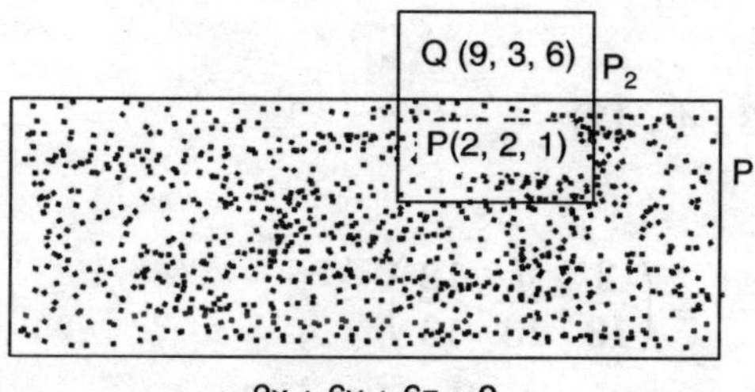

Fig. 5.6

Given: P_1: $2x + 6y + 6z = 9$ [1]

Let P_2: $Ax + By + Cz + D = 0$ [2]

be the eqn. of P_2 which is perpendicular to [1]

Then $2A + 6B + 6C = 0$ [3]

(using: $aa_1 + bb_1 + cc_1 = 0$)

By data, [2] passes thro' points (2, 2, 1) and (9, 3, 6).

$\therefore$ $2A + 2B + C + D = 0$ [4]

and $9A + 3B + 6C + D = 0$ [5]

From [5] – [4], $7A + B + 5C = 0$ [6]

From [3] and [6], we get by cross-multiplication,

$$\frac{A}{30-6} = \frac{B}{42-10} = \frac{C}{2-42}$$

i.e., $$\frac{A}{24} = \frac{B}{32} = \frac{C}{-40}$$

APPENDICES

i.e. $$\frac{A}{3} = \frac{B}{4} = \frac{C}{-5} = k \text{ (say)}$$

$\therefore$ $$A = 3k, B = 4k; C = -5k.$$

From [4], $6k + 8k - 5k + D = 0$

$\therefore$ $$D = -9k$$

putting there values in [2],

$$3kx + 4ky - 5kz = 9k$$

$\Rightarrow$ $$3x + 4y - 5z = 9,$$

is the eqn. off the plane.

12. Determine the D.C's of the normal to the plane $3x + 4y + 12z = 5z$ and hence or otherwise find the length of the $\perp^r$ from the origin on to the plane. [VTU, March, 1999]

Solution:

Given: $3x + 4y + 12z = 52$ [1]

D.R's of the normal to the plane are:

$$= (\text{coeff } x : \text{coeff } y : \text{coeff } z)$$
$$= (3 : 4 : 12).$$

$\therefore$ D.C's are

$$\frac{1}{\sqrt{3^2 + 4^2 + 12^2}} (3 : 4 : 12)$$

$$= \left(\frac{3}{13} : \frac{4}{13} : \frac{12}{13}\right)$$

and the length of $\perp^r$ from the orgin onto the given plane is

$$p = \frac{|-D|}{\sqrt{A^2 + B^2 + C^2}} = \frac{52}{13} = 4.$$

13. Find the distance of the point (1, – 2, 3) from the plane $x - y + z = 0$,

measured parallel to the line $\frac{x}{2} = \frac{y}{3} = \frac{z}{-6}$ the axis being rectangular

Solution: The eqn. of G. line: $\frac{x}{2} = \frac{y}{3} = \frac{z}{-6}$ [1]

Its D.R' are proportional to

$$(2 : 3 : -6)$$

($\div$ each by $\sqrt{4 + 9 + 36} = 7$)

The actual D.C's of line (1) are

$$\left(\frac{2}{7} : \frac{3}{7} : -\frac{6}{7}\right)$$

Now equation of the lines through P (1, –2, 3) and parallel to [1] are

$$\frac{(x-1)}{\left(\frac{2}{7}\right)} = \frac{y+2}{\left(\frac{3}{7}\right)} = \frac{z-3}{\left(-\frac{6}{7}\right)} = r, \text{ (say)} \qquad [2]$$

Since, we have taken actual D.C's in this case r will be the actual distance of any point from the pt. P (1, – 2, 3).

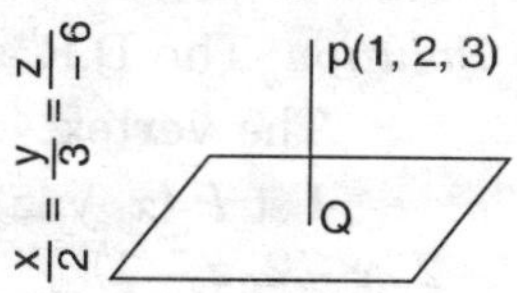

Any pt. as [2] is

$$Q\left[\frac{2}{7}r+1, \frac{3}{7}r-2, -\frac{6}{7}r+3\right]$$

If this lies on plane

$$x - y + z = 0$$

Then, $$\frac{2}{7}r+1-\left(\frac{3}{7}r-2\right)+\left(-\frac{6}{7}r+3\right) = 0$$

or $$-r + 6 = 0$$

$\Rightarrow r = 6$ is the required distance

p(1, 2, 3)

$\frac{x}{2} = \frac{y}{3} = \frac{z}{-6}$

Q

Fig. 7

14. Find the equation to the right circular cone whose axis is

$$\frac{x-1}{2} = \frac{y-2}{3} = \frac{z-3}{4}$$

and one of the generators is

$$\frac{x-1}{-1} = \frac{y-2}{3} = \frac{z-3}{3}$$

[B.U. July, 1993]

Solution: The D.R's of the axis are (2 : 3 : 4)

The D.R's of the generators is (– 1 : 3 : 3)

Angle between the axis and the generators is the semi-vertical angle. This is given by

$$\cos\theta = \frac{2(-1)+3\cdot 3+4\cdot 3}{\sqrt{4+9+16}\cdot\sqrt{1+9+9}} = \sqrt{\frac{19}{29}} \qquad [1]$$

The axis passes thro' (1, 2, 3) and the generators also passes thro' (1, 2, 3), which is the vertex of the cone.

Let $p(x, y, z)$ be any point on the cone. Given AP is a generator and its direction ratios are $(x - 1, y - 2, z - 3)$

$\therefore$ The angle between AP and the axis is also the semi-vertical angle

$$\therefore \quad \cos\theta = \frac{2(x-1)+3(y-2)+4(z-3)}{\sqrt{4+9+16}\sqrt{(x-1)^2+(y-2)^2+(z-3)^2}}$$

$$= \frac{2x+3y+4z-20}{\sqrt{29\left[(x-1)^2+(y-2)^2+(z-3)^2\right]}} \qquad [2]$$

APPENDICES

[1] and [2] $\Rightarrow$

$$(2x + 3y + 4z - 20)^2 = 19\,[(x-1)^2 + (y-2)^2 + (z-3)^2]$$

15. Find the equation of right circular cone whose vertex is the point (2, 2, 1) and the axis with D.R' s (2, 3, 6) with semivertical angle 60°.

Solution: The D.R' s of the axis are (2, 3, 6)

The vertex $V = (2, 2, 1)$.

Let $P(x, y, z)$ be any point on the cone. Then the D.R' s of a generator thro' P are $(x - 2, y - 2, z - 1)$.

Given that the semivertical angle $\theta = 60°$

$$\Rightarrow \quad \cos 60° = \frac{(x-2)\,2 + (y-2)\cdot 2 + (z-1)}{\sqrt{(x-2)^2 + (y-2)^2 + (z-1)^2}} \cdot \frac{1}{\sqrt{x+y+z}}$$

$$\frac{1}{2} = \frac{2x + 2y + z - 9}{3\sqrt{(x-2)^2 + (y-2)^2 + (z-1)^2}}$$

$$\Rightarrow \quad 9\,[(x-2)^2 + (y-2)^2 + (z-1)^2]$$

$$= 4\,(2x + 2 + z - 9)^2$$ is the eqn. of the required cone

16. Find the length of the shortest distance between the lines

$$\frac{x-1}{-1} = \frac{y+2}{-3} = \frac{z+1}{2} \qquad [1]$$

and $$3x + 2y + z = 0 = x + 4y + 3z \qquad [2]$$

Solution: Line [1] = symmetrical form

Line [2] = non- symmetrical form

The S.D. in this case is the length of the $\perp^r$ from any point on the first line [1] onto the plane thro' the 2nd line [2], parallel to the first.

The equation of any plane containing the line is:

$$3x + 2y + z = 0 = x = 4y + 3z$$

is of the form $P_3: P_1 + k\,P_2 = 0$

$$\Rightarrow \quad 3x + 2y + z + k\,(x + 4y + 3z) = 0 \qquad [1]$$

$$\Rightarrow \quad (3 + k)\,x + (2 + 4k)\,y + (1 + 3k)z = 0$$

The D.R's of the line (1) are:

$(-1 : -3 : 2)$, If plane [1] is parallel to the line [1], then

$$(3 + k)(-1) + (2 + 4k)(-3) + (1 + 3k)\,[2] = 0$$

$$\Rightarrow \quad -7k - 7 = 0 \;\; k = -1$$

Substitutely the value of k in [1], we get

$$2x - 2y - 2z = 0$$

$$\Rightarrow \quad x - y - z = 0 \qquad [2]$$

point $A\,(1, -2, -1)$ lies on line [1]

$\therefore$ length of S.D = the length of perpendicular from A onto the plane [2]

$$= \frac{1 + 2 + 1}{\sqrt{1 + 1 + 1}} = \frac{4}{\sqrt{3}}.$$

PART B

17. Find the nth derivative of

$$\cos^{-1}\left[\frac{x - x^{-1}}{x + x^{-1}}\right]$$

Solution: Let $$y = \cos^{-1}\left[\frac{x - x^{-1}}{x + x^{-1}}\right]$$

$$y = \cos^{-1}\left(\frac{x^2 - 1}{x^2 + 1}\right)$$

Let $x = \tan\theta; \Rightarrow \theta = \tan^{-1}(x)$

$\therefore$
$$\begin{aligned} y &= \cos^{-1}(-\cos 2\theta) \\ &= \cos^{-1}[\cos(\pi + 2\theta)] = \pi + 2\theta \\ &= \pi + 2\tan^{-1}(x). \end{aligned}$$

$\Rightarrow$ $$y_n = 2\,\frac{(-1)^{n-1}\cdot(n-1)!}{r^n}\cdot \sin n\,\alpha$$

Line $$r = \sqrt{1 + x^2},\ \alpha = \tan^{-1}\left(\frac{1}{x}\right)$$

Note: If $$y = \tan^{-1}(x)$$

Then $$y_n = \frac{(-1)^{n-1}\cdot(x-1)!}{\dfrac{\sin(n\theta)}{r^n}}$$

18. If $f(x) = \tan x$, then
Show that

$$f^n(0) - {}^n c_2 f^{n-2}(0) + {}^n c_4 f^{n-4}(0) + \ldots$$
$$= \sin\left(\frac{n\pi}{2}\right)$$

Solution:

Given: $\cos x \cdot f(x) = \sin x$

Take nth derivative on both sides, to the left side apply L.T. and to the right side, use standard formula for nth derivative of ($\sin x$), we get

$$\cos x \cdot f^n(x) + {}^n c_1 - (\sin x)\ f^{n-1}(x) + {}^n C_2(-\cos x)\ f^{n-2}(x) + \ldots$$
$$= \sin\left(x + \frac{n\pi}{2}\right)$$

put $x = 0$, on both sides, we get

$$f^n(0) - {}^n c_2 f^{n-2}(0) + {}^n c_n f^{n-4}(0)$$
$$= \sin\left(\frac{n\pi}{2}\right)$$

APPENDICES

19. If $x + y = 1$, then show that

$$\frac{d^n}{dx^n}(x^n y^n) = n!\,[y^n - ({}^xc_1)^2\, y^{n-1} \cdot x + [({}^nc_2)^3 \cdot y^{n-2} \cdot x^2 - + \ldots (-1)^n\, x^n]$$

Solution: To find n^{th} derivative of $x^n y^n$ we use L.T.

For convenience, let $u = y^n,\ v = x^n$

As $$(uv)_n = uv_n + {}^nc_1 u_1 v_{n-1} + {}^nc_2 u_2 v_{n-2} \ldots + u_0 v$$

and note $$u = y^n = (1 - x)^n$$

$\therefore$ $$u_1 = n\,(1 - x)^{n-1}(-1) = -ny^{n-1}$$

and $$u_2 = (-1)^2 \cdot n\,(n-1)\,(1-x)^{n-2}$$

$$u_n = (-1)^n \cdot n!$$

Also, $$V_n = n!$$

$$V_{n-1} = n! \cdot x, \text{ etc}$$

$\therefore$ $$\frac{d^n}{dx^n}(x^n y^n) = \frac{d^n}{dx^n}(y^n x^n)$$

$$= y^n \cdot n! + {}^nc_1\,[-ny^{n-1}]\, n!\, x + {}^nc_2\,[(-1)^2\, n\,(n-1)\, y^{n-2}]\, n! \cdot \frac{x^2}{z} + \ldots [(-1)^n\, n!\, x^n]$$

$$= n!\,[y^n - ({}^nc_1)^2\, y^{n-1} \cdot x + (nc_2)^2\, y^{n-2} \cdot x^2 + \ldots + (-1)^n \cdot x^n]$$

$\therefore$ $$n = {}^nc_1$$

and $$\frac{n\,(n-1)}{2!} = {}^nc_2.$$

20. If $\dfrac{x^2}{a^2+u} + \dfrac{y^2}{b^2+u} + \dfrac{z^2}{c^2+u} = 1$, prove that

$$u_x^2 + u_y^2 + u_z^2 = 2\,(xu_x + yu_y + zu_z)$$

Solution:

Given: $$\frac{x^2}{a^2+u} + \frac{y^2}{b^2+u} + \frac{z^2}{c^2+u} = 1 \qquad [1]$$

$\Rightarrow$ $$x^2\,(a^2+u)^{-1} + y^2\,(b^2+u)^{-1} + z^2\,(c^2+u)^{-1} = 1$$

Diff. [1] w.r.t. 'x' partially, we get

$$2x\,(a^2+u)^{-1} - x^2\,(a^2+u)^{-2} \cdot \frac{\partial u}{\partial x} - y^2\,(b^2+u)^{-2} \cdot \frac{\partial u}{\partial x}$$

$$- z^2\,(c^2+u)^{-2} \cdot \frac{\partial u}{\partial x} = 0$$

or $$\frac{2x}{a^2+u} = \left[\frac{x^2}{(a^2+u)^2} + \frac{y^2}{(b^2+u)^2} + \frac{z^2}{(c^2+u)^2}\right] u_x$$

or $$\frac{2x}{a^2+u} = Vu_x,$$

where $$V = \frac{x^2}{(a^2+u)^2} + \frac{y^2}{(b^2+u)^2} + \frac{z^2}{(c^2+u)^2}$$

or $$u_x = \frac{2x}{V(a^2+u)}$$

$|||^{ly}$, $$u_y = \frac{2y}{V(b^2+u)},$$

$$u_z = \frac{2z}{V(c^2+u)}$$

$$u_x^2 + u_y^2 + u_z^2 = \frac{4}{V^2}\left[\frac{x^2}{(a^2+u)^2} + \frac{y^2}{(b^2+u)^2} + \frac{z^2}{(c^2+u)^2}\right]$$

$$= \frac{4}{V^2}\cdot V = \frac{4}{V} \qquad [2]$$

$$2(xu_x + yu_y + zu_z) = 2\left[\frac{2x^2}{V(a^2+u)} + \frac{2y^2}{V(b^2+u)} + \frac{2z^2}{V(c^2+u)}\right]$$

$$= \frac{4}{V}(1) = \frac{4}{V}$$

From [using [i]] [2] and [3], we have

$$u_x^2 + u_y^2 + u_z^2 = 2(xu_x + yu_y + zu_z).$$

21. If $$x^2 + y^2 + u^2 - v^2 = 0 \text{ and}$$
$$uv + xy = 0,$$

prove that $$\frac{\partial(u, y)}{\partial(x, y)} = \frac{x^2 - y^2}{u^2 + v^2}$$

Solution: Let $$f_1 = x^2 + y^2 + u^2 - v^2; \qquad [1]$$
and $$f_2 = uv + xy \qquad [2]$$

then $$\text{II} = \frac{\partial(f_1, f_2)}{\partial(x, y)} = \begin{vmatrix} \frac{\partial f_1}{\partial u} & \frac{\partial f_1}{\partial v} \\ \frac{\partial f_2}{\partial u} & \frac{\partial f_2}{\partial v} \end{vmatrix}$$

$$= \begin{vmatrix} 2u & -2v \\ v & u \end{vmatrix} = 2(u^2 + v^2)$$

APPENDICES

$$\Rightarrow \quad \frac{\partial(u,v)}{\partial(x,y)} = (-1)^2 \frac{\dfrac{\partial(f_1,f_2)}{\partial(x,u)}}{\dfrac{\partial(f_1,f_2)}{\partial(x,v)}} = \frac{2(x^2-y^2)}{2(u^2+v^2)} = \frac{u^2-y^2}{u^2+v^2}.$$

22. If
$$u = x(1-r^2)^{1/2}$$
$$v = y(1-r^2)^{-1/2}$$
$$w = z(1-r^2)^{-1/2}$$
where
$$r^2 = x^2 + y^2 + z^2$$
prove that
$$\frac{\partial(u,v,w)}{\partial(x,y,z)} = (1-r^2)^{-5/2}$$

Solution:
$$\frac{\partial u}{\partial x} = 1\cdot(1-r^2)^{-1/2} + x\left(-\frac{1}{2}\right)(1-r^2)^{-3/2}(-2r)\frac{\partial r}{\partial x}$$
$$= (1-r^2)^{-1/2} + x\cdot(1-r^2)^{-3/2}\cdot r.\ x/r$$
$$= (1-r^2)^{-1/2} + (1-r^2)^{-3/2}\cdot x^2$$
$$= (1-r^2)^{-1/2} + x^2(1-v^2)^{-3/2}$$
$$= (1-r^2)^{-3/2}[1-r^2+x^2]$$

$|||^{ly}$ we can find:

$$\frac{\partial u}{\partial y} = xy(1-r^2)^{-3/2}; \quad \frac{\partial u}{\partial z} = xz(1-r^2)^{-3/2}$$

and
$$\frac{\partial v}{\partial x} = xy(1-r^2)^{-3/2}$$
$$\frac{\partial v}{\partial y} = (1-r^2)^{-3/2}[1-r^2+y^2];$$
$$\frac{\partial v}{\partial z} = yz(1-r^2)^{-3/2}$$
$$\frac{\partial w}{\partial x} = xz(1-r^2)^{-3/2}$$
$$\frac{\partial w}{\partial y} = yz(1-r^2)^{-3/2}$$
$$\frac{\partial w}{\partial z} = yz(1-r^2)^{-3/2}[1-r^2+z^2]$$

we have,
$$\frac{\partial(u,v,w)}{\partial(x,y,z)} = \begin{vmatrix} (1-r^2)^{-3/2}(1-r^2+x^2), xy(1-r^2)^{-3/2} & xz(1-r)^{-3/2} \\ xy(1-r^2)^{-3/2}, (1-r^2)^{-3/2}(1-r^2+y^2), yz(1-r^2)^{-3/2} \\ xz(1-r^2)^{-3/2} & yz(1-r^2)^{-3/2}, (1-r^2)^{-3/2}(1-r^2+z^2) \end{vmatrix}$$

$$= (1-r^2)^{-9/2} \begin{vmatrix} 1-r^2+x^2 & xy & xz \\ xy & (1-r^2+y^2) & yz \\ xz & yz & (1-r^2+z^2) \end{vmatrix}$$

$$= (1-r^2)^{-9/2}\,(1-r^2)^2 = (1-r^2)^{-5/2}$$

23. If $\quad x + y = 2e^{\theta} \cos \phi$

and $\quad x - y = 2ie^{\theta} \sin \phi$

Verify that

$$\frac{\partial(x,y)}{\partial(\theta,\phi)} \times \frac{\partial(\theta,\phi)}{\partial(x,y)} = 1.$$

Solution: $\quad x + y = 2e^{\theta} \cos \phi;$

$$x - y = 2ie^{\theta} \sin \phi;$$

Solving, we get

$$x = e^{\theta + i\phi};\ y = e^{\theta - i\phi} \qquad [1]$$

$$\therefore \quad \frac{\partial x}{\partial \theta} = e^{\theta + i\phi},\ \frac{\partial y}{\partial \theta} = e^{\theta - i\phi}$$

$$\frac{\partial x}{\partial \phi} = e^{\theta + i\phi};\ \frac{\partial y}{\partial \phi} = -ie^{\theta - i\phi}$$

$$\therefore \quad \frac{\partial(x,y)}{\partial(\theta,\phi)} = \begin{vmatrix} \dfrac{\partial x}{\partial \theta} & \dfrac{\partial x}{\partial \phi} \\ \dfrac{\partial y}{\partial \theta} & \dfrac{\partial y}{\partial \phi} \end{vmatrix} = \begin{vmatrix} e^{\theta+i\phi} & ie^{\theta+i\phi} \\ e^{\theta-i\phi} & -ie^{\theta-i\phi} \end{vmatrix}$$

$$= ie^{2\theta} - ie^{2\theta} = -2ie^{2\phi} \qquad [2]$$

Again solving [1] for θ and ϕ in terms of x and y, we get

$$\theta = \frac{1}{2}\ (\log x + \log y)$$

$$\phi = \frac{1}{2i}\ (\log x - \log y) \qquad [3]$$

$$\frac{\partial \theta}{\partial x} = \frac{1}{2x};$$

$$\frac{\partial \phi}{\partial x} = \frac{1}{2ix};$$

$$\frac{\partial \theta}{\partial y} = \frac{1}{2y},\ \frac{\partial \phi}{\partial y} = -\frac{1}{2iy}$$

APPENDICES

$$\frac{\partial(\theta,\phi)}{\partial(x,y)} = \begin{vmatrix} \frac{\partial\theta}{\partial x} & \frac{\partial\theta}{\partial y} \\ \frac{\partial\phi}{\partial x} & \frac{\partial\phi}{\partial y} \end{vmatrix}$$

$$= \begin{vmatrix} \frac{1}{2x} & \frac{1}{2y} \\ \frac{1}{2ix} & \frac{-1}{2iy} \end{vmatrix} = -\frac{1}{4ixy} - \frac{1}{4ixy} = \frac{-1}{2ixy}$$

Using [1], $\frac{\partial(x,y)}{\partial(\theta,\phi)} = -\frac{1}{2ie^{2\theta}}$ [4]

from [2] and [4], we get

$$\frac{\partial(x,y)}{\partial(\theta,\phi)} \times \frac{\partial(\theta,\phi)}{\partial(x,y)} = -2ie^{2\theta} \cdot \frac{-1}{2ie^{2\theta}} = 1$$

PART C

Examples on Tracing of curves with equations in parametric forms:

Trace the curve: $x = a \cos t + \frac{a}{2} \log \tan^2 \left(\frac{t}{2}\right)$

$y = a \sin t$

1. *Symmetry:* Replace t by $(-t)$ in x, we get

since $x\ (t) = x$

$$x = a \cos (-t) + \frac{a}{2} \log \left(\tan \left[\frac{-t}{2}\right]\right)^2$$

$$= a \cos t + \frac{a}{2} \log \tan^2 \left[\frac{t}{2}\right]$$

$$= x \quad \Rightarrow \quad x\ (-t) = x\ (t)$$

$\Rightarrow$ x is an even function of t [i]

and $y = y\ (t) = a \sin (t)$

or $y(-t) = a \sin (-t) = -a \sin (t) = -y\ (t)$

$\Rightarrow$ $y(-t) = -y(t)$

$\Rightarrow$ y is an odd function of t [ii]

$\Rightarrow$ [i] and [ii] $\Rightarrow$ curve is symmetrical about the x–axis

2. *Origin:* (i) Put $y = 0$

$\Rightarrow$ $a \sin t = 0$

$\Rightarrow$ $t = 0$ (But if, $t = 0$, $x \neq 0$)

$[x \neq 0, y = 0] \Rightarrow$ curve does not pass through origin: at $t = 0$

$\because$ $|\sin t| \leq 1$; $|y| \leq a \Rightarrow$ curve lies bet^n: $y = \pm\, a$.

3. *Asymptote:*

If $\quad y = 0, t = 0,$

then
$$x = a\cos(0) + \frac{a}{2}\log(0)$$
$$= a + \frac{a}{2}[-\infty] \to +\infty$$

i.e., if $x \to \infty, y \to 0$

$\Rightarrow \quad y = 0$ is an asymptote

or $\quad x$ – axis is an asymptote

4. *Region of Existence*

since
$$\frac{dx}{dt} = a(-\sin t) + \frac{a}{2}\cdot\frac{1}{\tan^2\left(\frac{t}{2}\right)}\cdot 2\tan\frac{t}{2}\sec^2\frac{t}{2}\cdot\frac{1}{2}$$
$$= a\left[-\sin t + \frac{1}{2\sin(t/2)\cos(t/2)}\right]$$
$$= a\left[-\sin t + \frac{1}{\sin t}\right] = a\left[\frac{1-\sin^2 t}{\sin t}\right] = a\,\frac{\cos^2 t}{\sin t}$$

and
$$\frac{dy}{dt} = a\cos t; \quad \therefore \quad \frac{dy}{dx} = \frac{\left(\frac{dy}{dt}\right)}{\left(\frac{dx}{dt}\right)}$$
$$= a\cos t\cdot\frac{\sin t}{a\cos^2 t} = \tan t$$

We tabulate corresponding values of x, y and t.

$t = -\pi$	$-\frac{\pi}{2}$	0	$\frac{\pi}{2}$	π
$x = \infty$	0	$-\infty$	0	∞
$y = 0$	$-a$	0	a	0
$\frac{dy}{dx}$ 0	$-\infty$	0	∞	0

If $t = 0$, then the corresponding point is $(-\infty, 0)$

If $t = \pi$, then the corresponding point is $(\infty, 0)$

If $t = \pm\frac{\pi}{2}$; $(0 \pm a)$ are the points

APPENDICES

since $\frac{dy}{dx} = \pm\infty$

$\Rightarrow$ The tangents are | |[el] to y-axis

The shape of the curve is as sketched roughly in the adjusting diagram.

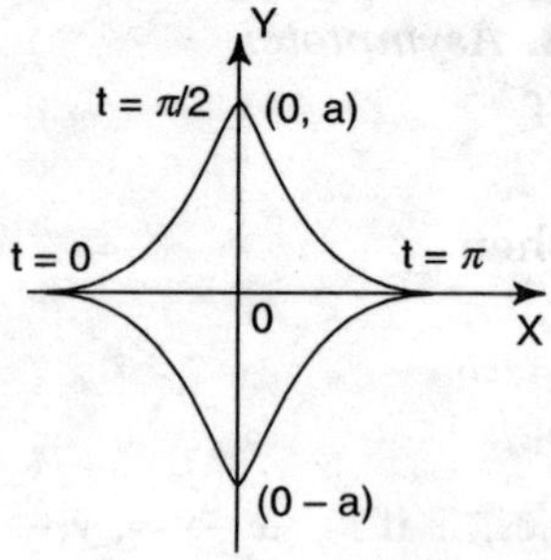

Fig. 8

Trace the curve:

$$x = a\,(1 - t^2);\; y = at\,(1 - t^2)$$

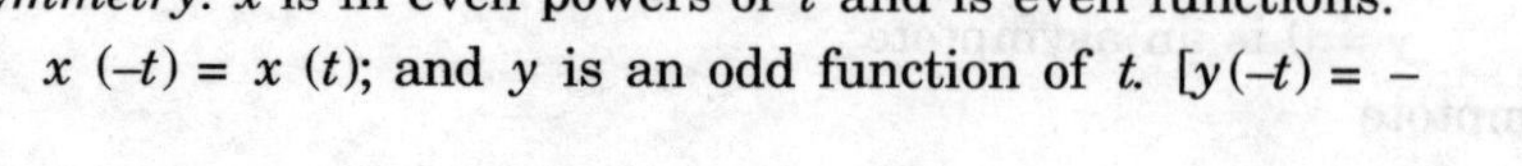

Symmetry: x is in even powers of t and is even functions.

$\because$ $x(-t) = x(t)$; and y is an odd function of t. $[y(-t) = -y(t)]$

$\Rightarrow$ the curve is symmetrical about x–axis

Origin: put $x = 0$

$\Rightarrow$ $t = \pm 1$

and $y = 0$

$\Rightarrow$ the curve passes thro' the origin for real and different values of t

$\Rightarrow$ Origin is a Node.

Again put $y = 0$,

$\Rightarrow$ $at\,(1 - t^2) = 0$

$\Rightarrow$ $t = 0$ or $1, -1$.

when $t = 0, x = a$; when $t = \pm 1, x = 0$.

$\therefore$ The curve meets x–axis at $(0, 0)$ and $(a, 0)$.

Asymptotoe: The curve has on asymptote (because for no value of t, x (or y) $\rightarrow \infty$ and y (or x) $\rightarrow$ finite quantity).

Region of Existence

Since $$\frac{dx}{dt} = -2\,at;\; \frac{dy}{dt} = a\,(1 - 3t^2)$$

$$\therefore \quad \frac{dy}{dt} = \frac{\left(\frac{dy}{dt}\right)}{\left(\frac{dx}{dt}\right)} = \frac{1 - 3t^2}{-2t} = \frac{3t^2 - 1}{2t}$$

If $$\frac{dy}{dt} = 0;\; 3t^2 - 1 = 0 \;\Rightarrow\; t = \pm\frac{1}{\sqrt{3}}.$$

when $$t = \pm\frac{1}{\sqrt{3}};\; x = \frac{2}{3}a,\; y = \pm\frac{2}{3\sqrt{3}}a$$

Thus at the points $\left[\frac{2}{3}a, \pm\frac{2}{3\sqrt{3}}a\right]$ the tangents is | |[el] to x-axis

Again, $\frac{dy}{dx} = \infty$, when $t = 0$

Now when $t = 0, x = a, y = 0$

Thus at the points $(a, 0)$, the tangents is $\perp^r$ to x-axis.

$x \le a$ always. No portion of the curve lies beyond the line $x = a$.

As t increases from 0 to 1, x-decreases from a to 0 and y first increases from 0 and then decreases to zero.

As $t \to \infty, x \to -\infty, y \to -\infty$.

The shape of the curve is as shown in the figure.

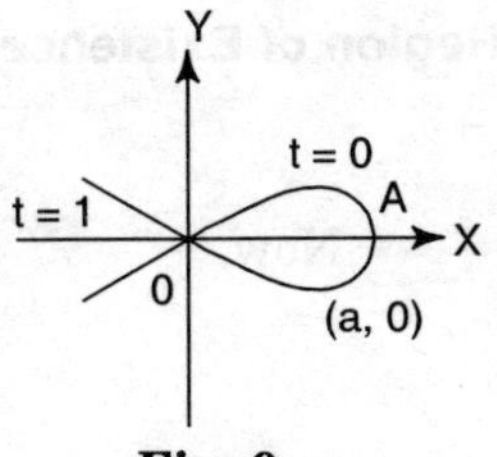

Fig. 9

Trace the Curve

$$x = a(t + \sin t);$$
$$y = a(1 - \cos t)$$

Solution: Given curve is parametric and we note that 't' cannot be easily eliminated from the given equations.

So, the following procedure is adopted to find the slope of the tangent to the curve.

1. *Symmetry:* The curve is symmetric about the y–axis:

$$\because \quad x(-t) = a(-t + \sin(-t)) = -a(t + \sin t) = -x(t)$$

$\Rightarrow$ $x = x(t)$ is an odd function of t

Also $y(-t) = a(1 - \cos(-t)) = a(1 - \cos t) = y(t)$

$\Rightarrow$ $y = y(t)$ is an even function t

2. *Origin:* put $t = 0$, $x = 0$, $y = 0$, curve passes thro (0, 0) $\Rightarrow$ at (0, 0), the slope of the tangent is zero.

$\Rightarrow$ x-axis will be the tangent.

Here, $\dfrac{dx}{dt} = a[1 + \cos t]; \; \dfrac{dy}{dt} = a \sin t$

$$\therefore \quad \frac{dy}{dx} = \frac{a \sin t}{a(1 + \cos t)} = \frac{2 \sin(t/2) \cos(t/2)}{2 \cos^2(t/2)}$$

$$\therefore \quad \frac{dy}{dx} = \tan(t/2)$$

We tabulate corresponding values of x, y and t and $\dfrac{dy}{dx}$.

t	0	$\frac{\pi}{2}$	π	$-\frac{\pi}{2}$	$-\pi$
x	0	$a\left(\frac{\pi}{2}+1\right)$	$a\pi$	$-a\left[\frac{\pi}{2}+1\right]$	$-a\pi$
y	0	a	$2a$	a	$2a$
$\frac{dy}{dx}$	0	1	∞	–1	$-\infty$

Region of Existence

Now, $$\frac{d^2y}{dx^2} = \frac{\frac{1}{2}\sec^2(t/2)}{a(1+\cos t)} = \frac{1}{4a}\sec^4\left(\frac{t}{2}\right)$$

When $t = 0, \dfrac{d^2y}{dx^2} = \dfrac{1}{4a} > 0; \quad (\because a > 0)$

$\Rightarrow$ at the origin, the curve reaches the minimum.

When $0 < t < \pi, y' > 0$

$\Rightarrow$ the function is increasing in $(0, a\pi)$ since the function is periodic with period $\left(\frac{\pi}{2}\right)$ the region of existence is $(-\pi < t < \pi)$

when $t = \pi, x = a\pi, y = 2a$

when $t = -\pi, x = -a\pi, y = 2a$

At the point $(a\pi, 2a)$ the tangent is parallel to the y-axis.

The curve is traced below, $0 \equiv$ vertex:

x–axis is the tangent.

CAB is called base.

Trace the Curve

$$x = a(t + \sin t);$$
$$y = a(1 + \cos t):$$

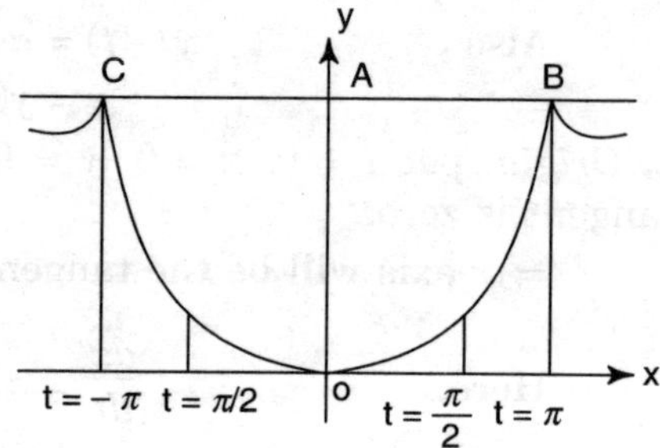

Fig. 10

Solution:

1. *Symmetry:*

Since $x = x(t)$

$\therefore \quad x(-t) = a[-t + \sin(-t)]$
$= -a(t + \sin t)$
$= -x(t)$

$\Rightarrow \quad x(-t)$ is an odd function of t and $y = y(t) = y(t)$

$\therefore \quad y(-t) = a(1 + \cos(-t)) = a(1 + \cos t) = y(t)$

$\Rightarrow \quad y(-t) = y(t)$ is an even function of t

So, the curve is symmetrical about the y-axis.

2. *Origin:* Put $x = 0$;

gives $a(t + \sin t) = 0$

$\Rightarrow \quad t = 0$

$\Rightarrow \quad y = a(1 + \cos 0) = a(1 + 1) = 2a$

$y \neq 0$

$\Rightarrow$ The curve does not pass through the origin.

3. *Region of Existence:*

The base is x-axis, the y-axis passes through the vertex. Tangents will be parallel to y-axis at $(a\pi, 2a)$. At the vertex, the value of θ is 0, and at the cusp (tangents are real and coincident), $\theta = \pi$

The curve is traced below.

The limits for θ are:

$\theta = -\pi$ at B^1

$\theta = 0$ at A

$\theta = \pi$ at B.

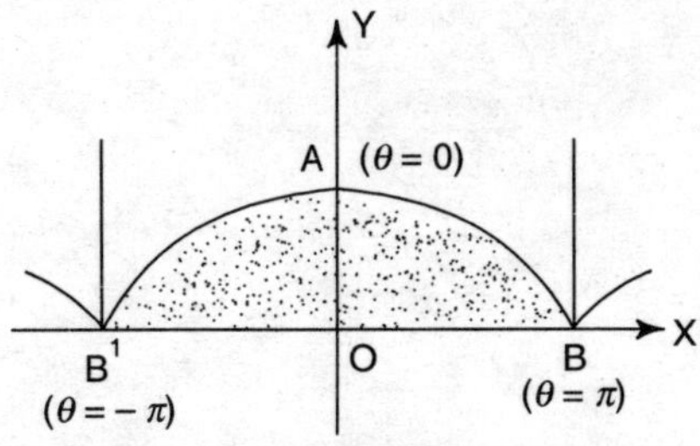

Fig. 11

Trace the Curve

$$x = a\sin^2 t,\ y = a\,\frac{\sin^3 t}{\cos t}$$

Solution:

1. *Symmetry:*

Since $x(-t) = x(t)$ even function of t

and $y(-t) = -y(t)$, an odd function of t

$\Rightarrow$ curve is symmetrical about x-axis.

2. *Origin:* The curve passes thro the origin.

$(\because x = 0 \Rightarrow t = 0, \Rightarrow y = 0)$

3. *Asymptote:*
$$\frac{dy}{dx} = \frac{\frac{dy}{dt}}{\left(\frac{dx}{dt}\right)} = \frac{(1 + 2\cos^2 t)\sin t)}{2\cos^2 t}$$

$$\frac{dy}{dx} = 0;\ \text{when } \sin t = 0, \text{ when } t = 0, \Rightarrow (x = 0, y = 0)$$

$\Rightarrow$ tangent is $||^{el}$ to x–axis at the origin:

$\Rightarrow$ x-axis is the tangent.

Also, $\frac{dy}{dx} = \infty$, if $\cos t = 0$ or $t = \frac{\pi}{2}$, when $t = \frac{\pi}{2}$,

$x = a, y = \infty$. $\therefore$ $x = a$ is an asymptote.

(as $y \to \infty, x \to a$)

4. *Region of Existence:*

$x = a\sin^2 t$, $x \neq$ negative and $x \not> a$ $(\because \sin^2 t \le 1)$

$\Rightarrow$ no position of the curve lies to the left of y-axis and beyond the line $x = a$.

The curve is traced as show in the adjusting diagram.

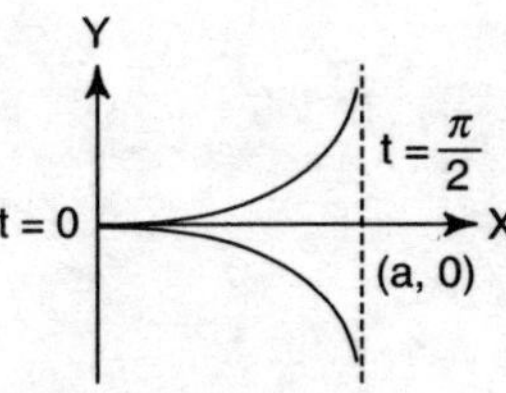

Fig. 12